Communications
in Computer and Information Science 225

Dehuai Zeng (Ed.)

Applied Informatics and Communication

International Conference, ICAIC 2011
Xi'an, China, August 20-21, 2011
Proceedings, Part II

Springer

Volume Editor

Dehuai Zeng
Shenzhen University
Nanhai Ave 3688, Shenzhen, Guangdong, 518060, China
E-mail: zengdh2008@gmail.com

ISSN 1865-0929 e-ISSN 1865-0937
ISBN 978-3-642-23219-0 ISBN 978-3-642-23220-6 (eBook)
DOI 10.1007/978-3-642-23220-6
Springer Heidelberg Dordrecht London New York

Library of Congress Control Number: 2011934039

CR Subject Classification (1998): C, D, F, H, I, J.1, J.2

Typesetting: Camera-ready by author, data conversion by Scientific Publishing Services, Chennai, India

Printed on acid-free paper

Springer is part of Springer Science+Business Media (www.springer.com)

Preface

Computers are firmly rooted in nearly all areas of life today. In company life, documents and products are produced by using computers, communication is by e-mail, companies present their commodities and services on the Internet and copious amounts of data are stored in databases. Different application programs are required for all of these processes.

The 2011 International Conference on Applied Informatics and Communication (ICAIC 2011) held during August 20–21, 2011, in Xi'an, China, provided an excellent platform for scientists and engineers of different countries working in various fields to exchange the latest developments and foster world-wide cooperation in the field of applied informatics and communication. Hundreds of experts and scholars from different countries attended ICAIC 2011.

Being crucial for the development of applied informatics and communication, our conference encompasses numerous research topics and applications: from the core fields of information systems and software engineering, manufacturing and automation engineering, computer-based signal processing and image processing to Communications Technology and other related topics. All of the papers were peer-reviewed by selected experts and 451 were selected for the proceedings of ICAIC 2011. We believe that this book provides a valuable reference for researchers in related fields. The papers describe how efficient, user-friendly, secure and upgradeable information technology systems can be designed, built and meaningfully used and operated in practice.

ICAIC 2011 was organized by the Information Engineering Research Institute, USA, and the proceeding are published by Springer. The conference was held in Xi'an. Xi'an, located in central-northwest China, records the great changes of the country just like a living history book. Called Chang'an (meaning the eternal city) in ancient times, Xi'an is one of the birthplaces of the ancient Chinese civilization in the Yellow River Basin area. It is the eastern terminal of the Silk Road and the site of the famous Terracotta Warriors of the Qin Dynasty. More than 3,000 years of history, including over 1,100 years as the capital city of ancient dynasties, have endowed Xi'an with an amazing historical heritage.

We sincerely thank all the members of the ICAIC Organizing Committee, Academic Committee and Program Committee. Moreover, we also express our sincere appreciation to all the referees, staff and volunteers for their tremendous efforts. Finally, we would like to thank all the authors who have contributed to this volume.

We hope that readers will find lots of interesting and useful information in the proceedings.

June 2011

Dehuai Zeng
Jianwei Zhang
Jun Zhang

Table of Contents – Part II

Intelligent Control and System

Information Processing and Data Mining

Research of the SPIHT Compression Based on Wavelet Interpolation Matching Image

Zhang Xiao-hong and Liu Gang

School of Information Engineering
Jiangxi University of Science and Technology Ganzhou, 341000, China
xiaohongzhang1977@126.com

Abstract. Image compression is a compression coding technology used in digital image. It is applied to reduce redundant information in image data, thus making unit storage and ring data transfer more efficient. Interpolation algorithm is commonly used in image scaling for pixel processing, as to reduce the image distortion and blur. Considering the wavelet transform and wavelet transform coding ideas, a kind of wavelet transform and bilinear interpolation for image matching is presented. Combined with the SPIHT (Set Partitioning In Hierarchical Tree) coding scheme. Large number of experiments show that comparision with the original SPIHT algorithm, the new scheme has more prominent advantages in the peak signal to noise ratio, mean-square deviation and histogram. So this method can improve the effect of recovery image compression efficiently. With generality and flexibility, it is also suitable to process various formats and size of image.

Keywords: image compression, interpolation algorithm, wavelet transform, mixing ratio, lifting wavelet.

1 Introduction

Image compression technology uses fewer bits to express the original image pixels information. Image interpolation technology does not affect the sensory effects in the premise of processing the image to zoom. In recent years, domestic and overseas scholars do a lot of research on image compression and image interpolation, in which image compression technology is widely used in image transmission, image storage and so on; image interpolation technology is widely used in medical imaging systems, infrared thermal imaging, night vision, aerospace and many other fields[1-2]. Due to image compression develops quite rapidly, there have been many classic algorithms, such as EZW (Embedded Zerotree Wavelet Coding) algorithm, SPIHT algorithm [3-4].

Based on existing research, this paper analyses three commonly interpolation algorithms used for image processing and comparison in the performance. Attempts to propose a wavelet transform and bilinear interpolation of the image matching

D. Zeng (Ed.): ICAIC 2011, Part II, CCIS 225, pp. 1–8, 2011.

method, using the low frequency components of bilinear interpolation to compensate for the lack of the wavelet transform combined with interpolation algorithm. For the wavelet interpolation matching image, it uses two SPIHT compression coding algorithms based on different wavelet. The simulation shows that this scheme is simple and fast, improves the quality of image compression, and applies in various image processing.

2 Image Interpolation Theory

In image processing, the interpolation does not produce new image pixel in the situation to change the image pixel value. To facilitate the image processing, supposing the input original image is defined as $f(x,y)$, where x,y for horizontal and vertical coordinate of the original pixel; by interpolation operation, the output image is defined as $f'(x',y')$, where x',y' for horizontal and vertical coordinate of the interpolated image pixel. Interpolated geometric computing is a mapping from f to f'. For the two-dimensional image, the target floating-point pixel coordinates is expressed $f(i+u,j+v)$. This point is decided by its nearby four points $f(i,j)$, $f(i+1,j)$, $f(i,j+1)$, $f(i+1,j+1)$ ((i,j) is the location of the image pixel, and the non-negative integer, $u,v(0\le u,v<1)$ is the floating-point number interval).

The mathematical expression of bilinear interpolation:

$$f(i+u,j+v)=(1-u)(1-v)f(i,j)+(1-u)vf(i,j+1)+u(1-v)f(i+1,j)+uvf(i+1,j+1) \tag{1}$$

According to the existing study results has been known[5], compared with the nearest neighbor interpolation, bilinear interpolation is higher accuracy, better image processing, but also appears jagged and fuzzy phenomena[6-7]. Although the bicubic interpolation is better than the first two, it's interpolation time is several times than other ways[8]. Considering a compromise, this paper studies the bilinear interpolation.

In order to compensate for shortcomings, it uses multi-scale analysis features of wavelet transform and combines its thought with interpolation algorithm. It uses dilation and translation functions to carry on the multi-scale refinement analysis of signal, eventually achieves the high-frequency time to subdivide and low-frequency frequency to subdivide. From the perspective of image processing, wavelet transform has covered the entire frequency domain, fast and so on.

3 SPIHT Compression Coding Theory

The wavelet transformation provides a kind of effective multi-resolution signal processing tool. It is also to achieve a variety of hierarchical image coding algorithm as the foundation. Wavelet image compression generally selects a fixed wavelet basis for wavelet image decomposition, in the wavelet domain researches reasonable quantification programs, scanning and entropy encoding. In the still image wavelet compressed encoding domain, one of the most effective way is Embedded Zerotree Wavelet code (EZW).

SPIHT algorithm is the improvement version of EZW algorithm. Its coding principle produces a series of important charts through the decreasing progressively threshold value sequence to approach each wavelet coefficient. By the EZW algorithm's zero tree structure as foundation, taking a root node and all its successor nodes be classified as a same set and defining the spatial direction tree structure. This structure can well describe the wavelet coefficients between different sub-band spatial relevance. It may carry on a kind fixedly and orderly sorting to the wavelet coefficient. For most images, the compression gain in given case, it can attain the highest PSNR value. This algorithm uses widely quite in the wavelet image compression[9].

The encoding algorithm requires three auxiliary tables that are LSP(list of important nodes), LIP(list of insignificant nodes), LIS(list of insignificant set of nodes). The step description is as follows:

(1) Initialization process: determine initial quantization threshold T ($T=2^n$), initialize $n=0$, ($n=\left\lfloor \log_2\left(\max_{(i,j)\in c}\{|c(i,j)|\}\right)\right\rfloor$), initialize LSP, LIP and LIS, n is the highest bit-plane, $c(i,j)$ is the coefficients of wavelet transform domain.

(2) The n -level coding classification process: to each element in LIP list, if it contains the four coefficient values that are less than T , outputs "0", otherwise "1"; codes each coefficient symbols and highest significant bit, and shifts it to the LSP chain list. To each element in LIS list, if the magnitude value of all leaf nodes are less than T , outputs "0", otherwise "1"; splits the spatial orientation tree, and updates LIP and LSP.

(3) The n -level coding refinement process: to each element of the LSP(excluding joins the new elements in the LSP), outputting its coding plane value in this level.

(4) $n=n-1$, $T=T/2$ to (2), and carry on the encoding of the next level.

4 Lifting Wavelet

Traditional wavelet transforms are all transformations based on real number field. Even the signals to be analyzed are a sequence of pairs of integers, corresponding coefficient of wavelet transforms is real number. Digital image is supplied with 8-bits integers, a new image will be leaded to large distortion easily after the traditional wavelet transform.

In order to reduce the distortion of Reconstruction image, W. Sweldens proposed integer wavelet transform [10] in 1995, it is the Second Generation Wavelet Transform based on lifting model. The process of lifting wavelet transform is divided into three steps: split, predict and update. As shown in Figure 1:

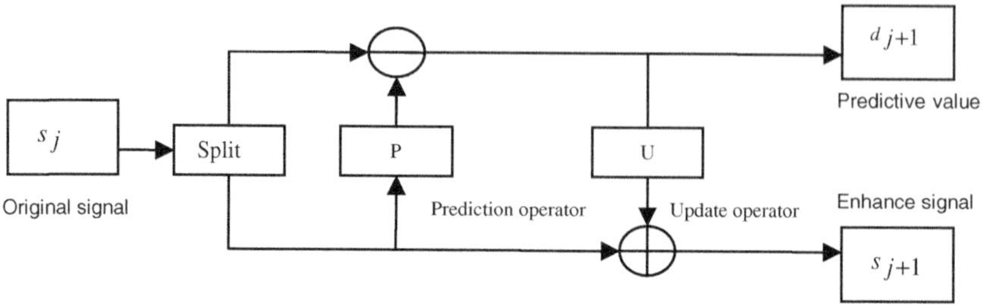

Fig. 1. Lifting process

(1) Split: the original signal is split into two disjoint subsets: even sequence $s_j^{(0)}$ and odd sequence $d_j^{(0)}$.

(2) Predict: Utilization of the correlation between data, the prediction operator P which is nothing to do with the data structure is acted on the samples of even sequence $s_j^{(i)}$ to get predictive value of the samples of odd sequence $d_j^{(i)}$. Original samples of odd sequence subtracted the predictive value to get prediction error of odd sequence which is expressed predictive $d_j^{(i+1)}$, namely: $d_j^{(i+1)} = d_j^{(i)} - P(s_j^{(i)})$.

(3) Update: in order to maintain the overall nature of the raw data, update operator U is acted to $d_j^{(i+1)}$ generate an update collection $s_j^{(i+1)}$ of the samples of even sequence $s_j^{(i)}$. Namely, it is the approximation of the original signal in the low resolution. The process is that: $s_j^{(i+1)} = s_j^{(i)} + U(d_j^{(i+1)})$.

The simplest method to achieve integer wavelet transform is based on wavelet transform which is based on lifting. It involves shift, add and subtract, and it is fast speed and no loss of accuracy.

5 SPIHT Compression Coding Based on Wavelet Interpolation Matching Image

This paper uses multi-scale analysis of wavelet transform features and layers of sub-band decomposition inherent similarities, proposes a kind of image matching fusion method which is based on wavelet transform and the bilinear interpolation. This method uses the low-frequency components of bilinear interpolation to compensate the shortage of the wavelet transform and interpolation. For the wavelet interpolation matching image, we use two SPIHT compression coding algorithms based on different wavelet.

The program flow shown in Figure 2, steps as follows:

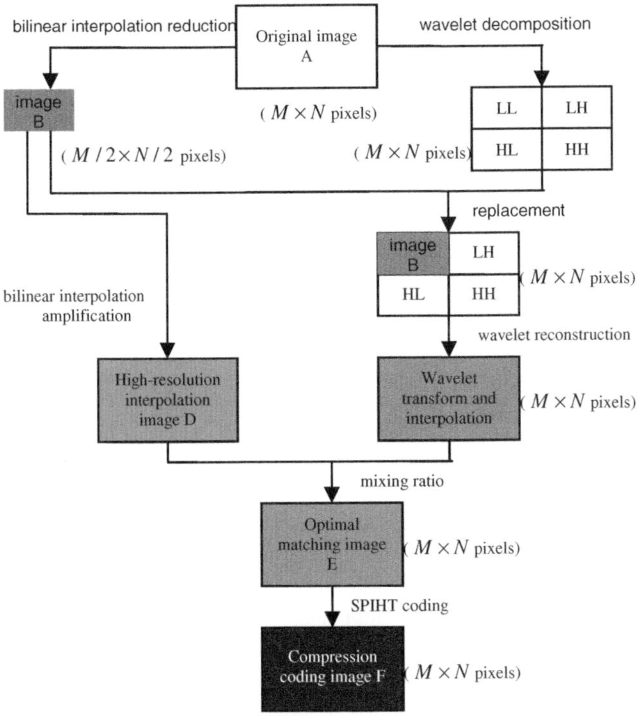

Fig. 2. Flow process of new program

(1) Get the low resolution image B by means of bilinear interpolation of the original image A.
(2) Wavelet transforming the original image A in wavelet transform domain, it will be decomposed into a low frequency component (LL) and three high frequency component (HL, LH, HH), in pixels.
(3) Take three high frequency components(HL, LH, HH) generated in steps(2) and take low-resolution map B in step(1) as a low-frequency component(LL) to get wavelet transform and interpolation of recovery image C by means of anti-transform wavelet.
(4) Get the high-resolution interpolated image D by means of bilinear interpolation of the low-resolution map of B.
(5) Get the optimal results matching image E by means of mixing pairing of image C and image D.
(6) Get compression code image F by means of SPIHT compression code of image E.

In the matching ratio process, it uses the variance matching threshold value α to represent the matching ratio value, easy to control parameter $\alpha(0<\alpha<1)$. Weighing a picture quality standard generally to use mean square error(MSE), peak signal to noise ratio($PSNR$) and image pixel differential value ratio(σ). The original image is set as $\{f(i,j), 1 \le i \le M, 1 \le j \le N\}$, the new image as $\{f'(i,j), 1 \le i \le M, 1 \le j \le N\}$, therefore expressions as follows:

$$MSE = \frac{1}{M \times N} \sum_{i=1}^{M} \sum_{j=1}^{N} \left[f(i,j) - f'(i,j) \right]^2 \tag{2}$$

$$PSNR = 10 * \lg((255)^2 / MSE) \tag{3}$$

$$\sigma = \frac{1}{M \times N} \sum_{i=1}^{M} \sum_{j=1}^{N} \left| f(i,j) - f'(i,j) \right| / f(i,j) \tag{4}$$

6 Experimental Result

6.1 Experimental Simulation Result

This method uses many images to carry on the experiment, here takes the classical image as a example (180*256). For general purposes, setting $M \neq N$ to illustrate the effectiveness of the method, the simulation results show in Figure 3.

It can be seen from the experimental results that, the results of the new program is more closer to the original images significantly.

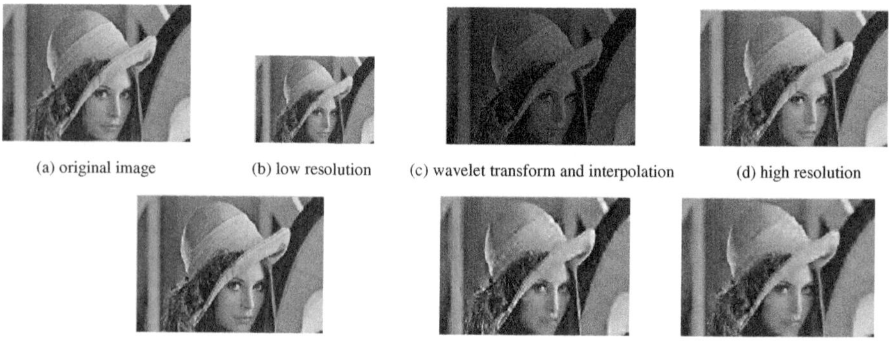

(a) original image (b) low resolution (c) wavelet transform and interpolation (d) high resolution

(e) matching of wavelet transform and bilinear interpolation of lifting wavelet (f) SPIHT Compression encoding (g) SPIHT compression encoding

Fig. 3. Experimental results figure

6.2 Measurement Criteria Comparison

After the multiple repeated experiments, the variance matching threshold value α from 0.1 to 0.9, which produces the peak signal to noise ratio has been in the trend of escalation. Until $\alpha = 0.9$, matching image E achieves a most superior peak signal to noise ratio value. Matching image E carries on SPIHT compression encoding of the traditional wavelet, and SPIHT compression encoding of the lifting wavelet. When wavelet decomposition is at 3-6 level, PSNR achieves optimal result. The Program uses MATLAB platform to establish, running the environment as follows: Pentium IV 3.0GHz, 512MB memory.

Table 1. SPIHT Coding Data Comparison of Wavelet Interpolation Matching Image

Compression ratio	MSE(Mean Square Error)			PSNR(Peak Signal to Noise Ratio)			pixel differential value ratio			Time (ms)
	R	*G*	*B*	*R*	*G*	*B*	*R*	*G*	*B*	
8:1	55.01	46.77	53.85	30.73	31.43	30.82	1.1e-6	2.0e-6	1.9e-6	63.38
16:1	117.2	102.2	113.5	27.44	28.04	27.58	2.2e-6	3.4e-6	3.8e-6	14.16
32:1	182.5	142.8	137.6	25.52	26.58	26.74	4.8e-6	6.0e-6	6.8e-6	6.47
64:1	249.5	193.4	190.9	24.16	25.27	25.32	1.2e-5	1.1e-5	1.3e-5	2.97
128:1	252.9	243.9	248.6	24.10	24.26	24.18	1.7e-5	1.9e-5	2.0e-5	1.47
256:1	253.2	243.6	248.3	24.07	24.26	24.19	2.1e-5	2.0e-5	2.0e-5	1.19

Table 2. SPIHT Coding Data Comparison Based on the Lifting Wavelet Of Wavelet Interpolation Matching Image

Compression ratio	MSE(Mean Square Error)			PSNR(Peak Signal to Noise Ratio)			pixel differential value ratio			Time (ms)
	R	*G*	*B*	*R*	*G*	*B*	*R*	*G*	*B*	
8:1	32.44	37.75	34.45	33.02	32.36	32.76	9.3e-7	1.8e-6	1.6e-6	49.70
16:1	87.55	90.63	86.64	28.71	28.56	28.75	2.0e-6	3.4e-6	3.4e-6	10.52
32:1	153.2	126.8	124.7	26.28	27.10	27.17	4.1e-6	5.6e-6	6.4e-6	5.67
64:1	244.1	180.3	180.6	24.26	25.57	25.56	1.0e-5	1.1e-5	1.3e-5	2.77
128:1	250.5	240.1	244.5	24.14	24.33	24.25	1.6e-5	1.9e-5	2.0e-5	1.02
256:1	250.8	239.6	244.1	24.14	24.34	24.26	2.1e-5	1.9e-5	2.0e-5	0.86

From Table 1, 2, under the similar condition, compared with the traditional compressed encoding, lifting wavelet compression coding mean square error and pixel differential value ratio corresponding reduction, *PSNR* corresponding increasing. The consumption time also drops, shows that this kind of plan improves efficiency, meets users demand.

7 Conclusion

This program carries on the compressed encoding to the wavelet interpolation matching image. Compared to the direct restructuring image, the wavelet interpolation matching image is more suitable for human eye observation, the detail part is richer and more clear, the edge block and denticle are less.

In the wavelet interpolation matching image, we compared SPIHT compression coding algorithms based on different wavelet. Experiments indicates that relative to original algorithm, the new one could improve the PSNR of the reconstructed image in the case of the same compression ratio. PSNR will rise with the increased compression ratio either. Compared to the PSNR of the original algorithm, the scheme can increase PSNR 0.5db above. The new algorithm could make better efficiency in Image compression.

Acknowledgment. This work is jointly supported by the National Natural Science Foundation of China (Grant No. 11062002, 10947117).

References

1. Cha, Y., Kim, S.: Edge-forming Methods for Image Zooming. Journal of Mathematical Imaging and Vision 25, 353–364 (2006)
2. Asamwar, R.S., Bhurchandi, K.M., Gandhi, A.S.: Interpolation of Image Using Discrete Wavelet Transform to Simulate Image Resizing as in Human Vision. International Journal of Automation and Computing 7, 9–16 (2010)
3. Cho, S., Kim, D., Pearlman, W.A.: Lossless Compression of Volumetric Medical Images with Improved Three-Dimensional SPIHT Algorithm. Journal of Digital Imaging 17, 57–63 (2004)
4. Choong, M.K., Logeswaran, R., Bister, M.: Improving Diagnostic Quality of MR Images Through Controlled Lossy Compression Using SPIHT. Journal of Medical Systems 30, 139–143 (2006)
5. Zhao, X.-y., Su, Y., Dong, Y.-q., et al.: Kind of super-resolution method of CCD image based on wavelet and bicubic interpolation. Application Research of Computers 26, 2365–2367 (2009)
6. Liu, X.-y., Zhang, X.-c., Zhou, J.: Image magnification algorithm based on parabolic interpolation and wavelet transformation. Computer Engineering and Applications 44, 48–50 (2008)
7. Dumic, E., Grgic, S., Grgic, M.: The use of wavelets in image interpolation: Possibilities and limitations. Radioengineering 16, 101–109 (2009)
8. Hu, m.+., Tan, J.-q.: Adaptive osculatory rational interpolation for image processing. Journal of Computational and Applied Mathematics 195, 46–53 (2006)
9. Kassim, A.A., Tan, E.H., Lee, W.S.: 3D Color Set Partitioning in Hierarchical Trees. Circuits, Systems, and Signal Processing 28, 41–53 (2009)
10. Wahed, M.E.: Image enhancement using second generation wavelet super resolution. International Journal of Physical Sciences 2, 149–158 (2007)

Research of Dynamic Reconfiguration of Feeder Based the Dynamic Swarm Optimization Algorithm*

Sun Guo-qiang, Wei Zhi-nong, Han Lian-shan, and Ye Fang

College of Energy and Electrical Engineering
University of hohai
Nanjing, Jiangsu Province, China
sunguoqiang2011@live.cn

Abstract. On the basis of reconfiguration algorithms of distribution system in possession, a new reconfiguration algorithms based on dynamic particle swarm optimization (DPSO) is proposed. It adopts the voltage stability of distribution system as the object function, and uses the improved particle swarm optimization algorithm which can trace dynamically the environment's change to make the real-time dynamically reconfiguration of distribution system, and proposes a new coded form which divides the particle module according to the topological structure of distribution system, and then avoids the convergence precociously via the community characters of DPSO. And the method proposed is compared with the one which adopts the power loss of distribution system as the object function, then make a conclusion that it must be considered to find the balance between security and economics under the different conditions simultaneously.

Keywords: Distribution systems, Dynamic reconfiguration, Voltage stability, Dynamic particle swarm optimization algorithm.

1 Introduction

System dispatchers at the control center make the static reconfiguration of distribution system by changing the switch state based on the operation situation of load in normal condition and fault situation. A number of algorithms [1-5]including mathematical programming, method of optimization stream pattern, method of spur track permutation and artificial intelligent methods such as genetic algorithm (GA), simulated annealing (SA), etc, have been proposed to reconfigure distribution system. Actually, load is a dynamic variable in distribution system. The operation scheme of the switch needed by network programming is given by the static reconfiguration technique in off-line state. Although to method of operation with high load curve fluctuation, the reconfiguration schemes are greatly difference in different time-interval. Thus, the scheme only to

* This work is supported in part by National Natural Science Foundation of China(50877024), Natural Science Foundation of Hohai University (2009421411) and Fundamental Research Funds for the Central Universities(2010B05914).

D. Zeng (Ed.): ICAIC 2011, Part II, CCIS 225, pp. 9–18, 2011.

specific load without considering it's relation in different time-interval is lack of practical use.. At present the major problems of distribution network reconfiguration in time varying system are to minimizing real power losses [6-7].

This paper adopts the voltage stability of distribution system as the object function, and introduces the improved dynamic particle swarm optimization algorithm (DPSO) on the basis of reconfiguration algorithms of distribution system in possession. The reconfiguration scheme based on fuzzy set theory and binary particle swarm optimization algorithms to improve power quality and satisfy consumers' demand is proposed [8]. A new coded form which divides the particle module according to the topological structure of distribution system and avoids the convergence precociously via the community characters of DPSO is proposed.

2 Dynamic Reconfiguration Model

The voltage stability phenomenon and insufficient local reactive power supply have been well recognized with the increasing of household appliance facilities in distribution system. Thus shunt capacitor compensation devices are widely utilized. These compensating units supply reactive power which decreased with the square of the voltage amplitude. Accordingly, it is potential possibility that voltage instability and voltage collapse occur in a power system. After general adoption of the market principle, load tap changing (LTC) transformers will be slathered and load throw-off is effectively limited. So the voltage stability problem is greatly prominent in load side and it is very important to make sure voltage is stability in distribution system [9]. The authors in [10-11] have studied the voltage stability index and found it may effectively reduce losses and improve voltage stability.

2.1 The Voltage Stability as the Object Function

The foremost difference between dynamic reconfiguration (DR) and static reconfiguration is that the dynamic reconfiguration not only needs to satisfy constraint condition in fixed time slice but also to consider the relation in several time-intervals and the whole time-interval constraint condition. Thus the optimization voltage stability in several time-intervals as object function is proposed as shown in formulation 1. The voltage stability index has been derived in [10] as shown in 2.

$$\max F = \max \sum_{i=1}^{n} \left[L_0(t_i) - L(t_i) \right] \tag{1}$$

Where n is total number of reconfiguration time-interval; t_i is time-interval i, it's length is Δt_i; $L(t_i)$ is voltage stability index in reconfigured network at time-interval i; $L_0(t_i)$ is voltage stability index in original network at time-interval i.

$$L(t_i) = 4 \begin{bmatrix} \left(x_{eq}(t_i) P_{Leq}(t_i) - r_{eq}(t_i) Q_{Leq}(t_i) \right)^2 + \\ x_{eq}(t_i) Q_{Leq}(t_i) + r_{eq}(t_i) P_{Leq}(t_i) \end{bmatrix} \tag{2}$$

Where $P_{leq}(t_i)$ is the total active power flow in the time-interval i; $Q_{leq}(t_i)$ is the total reactive power flow in the time-interval i; $r_{eq}(t_i)$ is the equivalent resistance; $x_{eq}(t_i)$ is the equivalent reactive resistance.

In (2), it is shown that the value of L is in [0, 1]. If the value approaches 0, it shows the network is comparative stability. Otherwise, it approaches at the edge of the voltage collapse.

The calculation formulations of $r_{eq}(t_i)$, $x_{eq}(t_i)$ are shown in 3.

$$\begin{cases} r_{eq}(t_i)=\sum P_{loss}(t_i)/\begin{bmatrix}\left(P_{Leq}(t_i)+\sum P_{loss}(t_i)\right)^2+\\ \left(Q_{Leq}(t_i)+\sum Q_{loss}(t_i)\right)^2\end{bmatrix} \\ x_{eq}(t_i)=\sum Q_{loss}(t_i)/\begin{bmatrix}\left(P_{Leq}(t_i)+\sum P_{loss}(t_i)\right)^2+\\ \left(Q_{Leq}(t_i)+\sum Q_{loss}(t_i)\right)^2\end{bmatrix} \end{cases} \tag{3}$$

Where $\sum P_{loss}(t_i)$ is the total active power losses, $\sum Q_{loss}(t_i)$ is the total reactive power losses.

2.2 Constraint Condition

The constraint conditions of reconfiguration in distribution system include power flower equations, stability constraint, connectivity constraint and etc [12].

(1) Power flow equations of distribution network

Network configuration must satisfy power flow equations and in this paper it adopts forward and backward substitution method to calculate the power flow of the distribution system.

$$\begin{cases} \Delta P_i = P_i - e_i\sum_{j\in i}\left(G_{ij}e_j - B_{ij}f_j\right) + f_i\sum_{j\in i}\left(G_{ij}f_j + B_{ij}e_j\right) \\ \Delta Q_i = Q_i - f_i\sum_{j\in i}\left(G_{ij}e_j - B_{ij}f_j\right) - e_i\sum_{j\in i}\left(G_{ij}f_j + B_{ij}e_j\right) \end{cases} i=1,2,\ldots\ldots,N \tag{4}$$

where N is the total number of nodes; e and f are the real part and the imaginary part of voltage; G and B are electric conductance and susceptance; P_i ,Q_i, ΔP_i, ΔO_i are the active power, reactive power , increment active power and increment reactive power of the node i.

(2) Constraint of power-up

Distribution network must satisfy the request of power load. Moreover, it cannot have isolated nodes (namely “isolated island”).

(3) Constraint of network topology

Normally, distribution system is designed in closed loop and operated in opened loop, which request distribution network is radicalized after reconfiguration.

(4) Limitation of switch motion

Considering the longevity of switchgear, we should reduce the switch motion times in the term of reconfiguration. According to the operation criterion, the switch motion times must be limited in point as shown in 5.

$$\begin{cases} \sum_{i=1}^{n} \left| x_{kk}^{i+1} - x_{kk}^{i} \right| \le N_{kk}^{i} \\ \sum_{kk=1}^{m} \sum_{i=1}^{n} \left| x_{kk}^{i+1} - x_{kk}^{i} \right| \le N_{s} \end{cases} \tag{5}$$

Where x_{kk}^{i} , x_{kk}^{i+1} are the state variables of the switch kk in the time-interval i and i+1; m is the total number of the switch.

3 Dynamic Reconfiguration Based on Dpso

3.1 Binary PSO Algorithm

In the binary PSO algorithm, the colony and particles may find the best position and then modify their seeking direction of the position and convergence velocity by simulating the social colony behavior and iterative computation according to the following formulas. Ultimately, the optimization position can be found.

$$\begin{aligned} v_{id}^{k+1} = w v_{id}^{k} + c_1 \cdot rand_1^k \left(p_{best,id}^{k} - x_{id}^{k} \right) + \\ c_2 \cdot rand_2^k \left(g_{best,id}^{k} - x_{id}^{k} \right) \end{aligned} \tag{6}$$

Where p_{id} is the best previous position of particle i on dimension d; $g_{best,id}$ is the position of the best particle on dimension d; rand is a quasi-random number selected in [0, 1]; c_1, c_2 are the positive constants; w is the inertia weight.

$$sig\left(v_{id}^{k+1}\right) = 1/\left(1 + \exp\left(-v_{id}^{k+1}\right)\right) \tag{7}$$

Where $\begin{cases} x_{id}^{k+1} = 1 & \rho_{id}^{k+1} < sig\left(v_{id}^{k+1}\right) \\ x_{id}^{k+1} = 0 & \rho_{id}^{k+1} \ge sig\left(v_{id}^{k+1}\right) \end{cases}$, d=1,2…,D, D is the dimension size; $X_j = \left[x_{j1}, x_{j2} .... x_{jD}\right]$ is the state variable of switch; $V_j = \left[v_{j1}, v_{j2}, ... v_{jD}\right]$ is the rate of position change.

In the binary version, PSO virtually seeks hunting direction in the probability. Whereas in DPSO algorithm it adopts dynamic inertia weight to control the convergence speed [14].The inertia weight plays the role of balancing the global search and local search. When it is large, the DPSO is more like a global search method. When it is small, the DPSO is more like a local search method. In this paper, inertia weight is described as this formula. $w = 0.5 + r(t)/2.0$, $r(t)$ is the random number selected in [0, 1]; constant c_1, c_2 both take 1.494 according to the method of Clerc compression factor.

3.2 DPSO

Distribution network reconfiguration is considered as a complex, large-scale and NP-hard combinational optimization problem. Thus it has more complexity in time and space. In order to get the global optimization result, it cost much time and it is also difficult to make the dispatching scheme in time. A strategy, similarly a heuristics algorithm to seek feasible operational schedule is proposed. It can satisfy several dynamic constraint conditions based on PSO instead of pursuing the global optimization in a pure dynamic programming meaning.

The DPSO algorithm can be described as follows:

Step1: Set up the system parameters. According to load curve time-intervals can be divided. For example if the deviation of load in adjacency time step exceeds the defined range, the time-interval can be considered to be end.

Step 2: Initialization: Initialize the positive constants c_1, c_2, and inertia weight w. Initialize the particle in the time-interval i. In this paper it adopts the method of the root node interfusion to ensure the connectivity and radicalization of the distribution network. In the term of reconfiguration based DPSO, the times of the switch motion should be accumulated. If the clamp-on switch motions in the particle state, a new one should be generated.

Step3: Inspect the constraint conditions. If the times of the total switch motion cannot satisfy the formula 5, turn to *step2*, or else go to *step 4*.

Step 4: Inspect the single switch constraint condition as shown in 5. If the times of the single switch motion exceed the defined limitation, this switch is considered as a clamp-on switch and the times are considered to be the maximal value.

Step 5: If i is less than the total number of the time-intervals, the reconfiguration result in this interval can be made as the initial condition of the next reconfiguration, i is accumulated and turn to *step 2*, or else go to stop.

3.3 Encoding Format

According to the similar topological structure of distribution network, the swarm of particles is divided in several sub groups. The purpose is that it adequately makes use of the zonal character of PSO and seeks the optimization result in the sub group to avoid premature convergence. Considering to the multiple branching feeder, the equivalent decoupled method based on branch node is proposed [15].

(1) Node numbering: In this paper it adopts node stratified method to unified deal with the branch.

(2) In distribution network, a large number of nodes make the dimension of adjacent matrix enormous. In order to solve this problem, the technology of adjacent table and structural array is adopted to store the topological information.

It can improve the computation rate and reduce the storage space.

(3) Partition IEEE 13-bus system [16] with the equivalent decoupled method as shown in Figure 1. The array element is defined corresponding switch number in the sub group as follow: $S_0 = [(0)]$, $S_1 = [(1)(4)]$, $S_2 = [(2)]$, $S_3 = [(5)]$, $S_4 = [(10)]$, $S_5 = [(11)]$, $S_6 = [(6)]$, $S_7 = [(7)(12)]$, $S_8 = [(3)(8)]$. The branch where the interconnection switch is divided in independent sub group as follow: $T_0 = [(9)]$, $T_1 = [(13)]$.

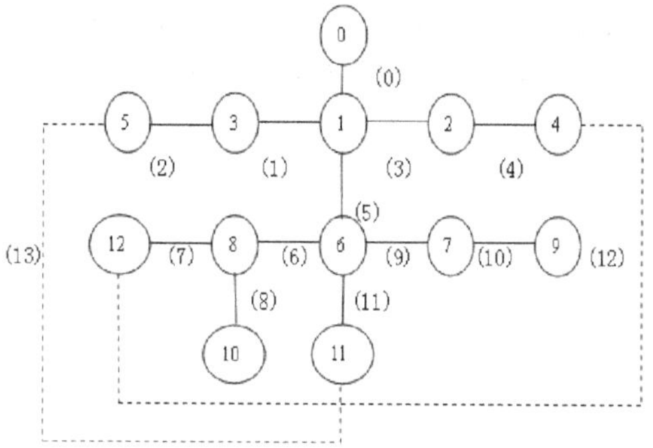

(a) IEEE 13-bus distribution system

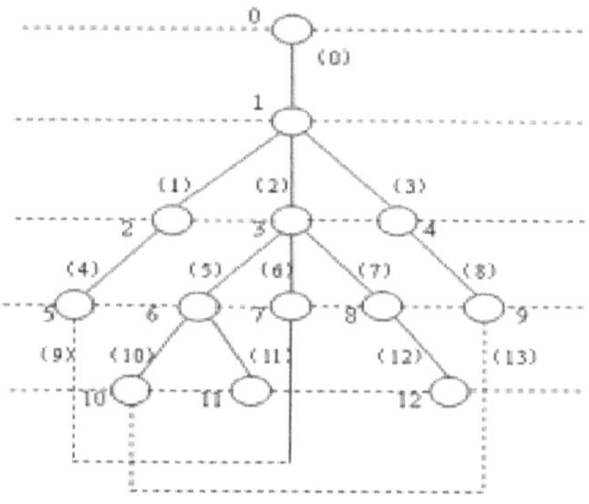

(b) Numbering based on the equivalent decoupled method

Fig. 1. The connecting diagram and numbering diagram of IEEE-13 bus test feeder

According to this encoding manner, two encoding formats can be got as shown in Figure 2. The switch state of particle is open only in the districts DS_1 and DS_2, so the two encodings are considered to have similar topological structure based on discriminant criterion. The discriminant criterion is set down that if the switch state of particles is open in the same encoding district, the particles are considered to have similar topological structure. With the increasing of the network nodes, the advantage of this zonal approach is greatly remarkable.

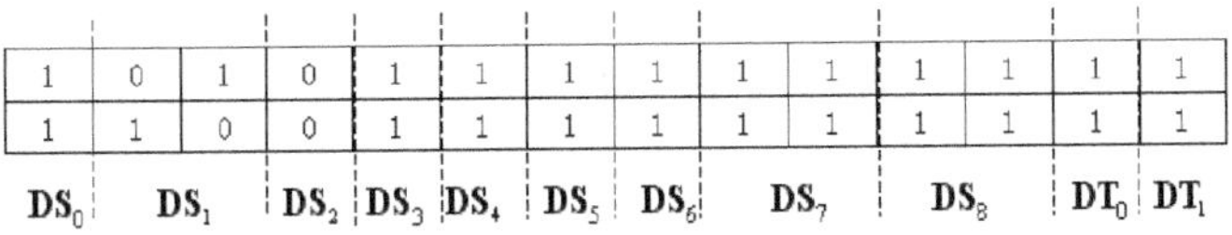

Fig. 2. Encoding format

4 Test Results

The test systems are a 4.16kV system with 13 buses and a 12.66kV system with 69 buses [18]. The system data given in [17] are used to simulate load curve shown in Fig.3. The detailed results are given in Table1. The PE&G69 system is shown in Fig.4.

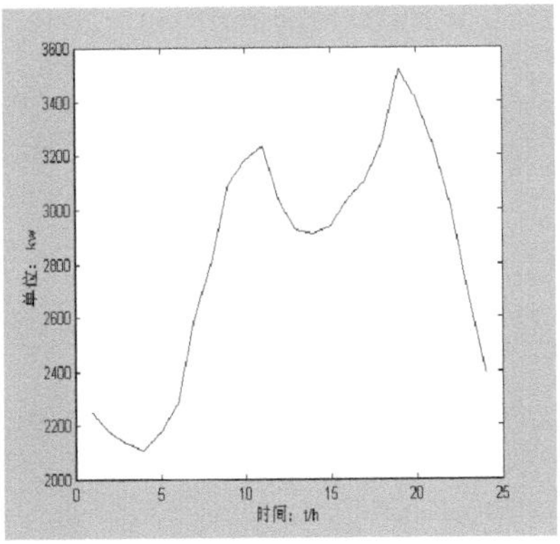

Fig. 3. The simulated curve of daily load

The results in Table 1 show that stability and economy are mutual incompatible. It is unreasonable that unilateral pursuing any side. In the time-interval of balanced load and low-load that the requirement of stability is not prominent, it can use the economy as the object function. The other way round, when the system is in the conditions of irrational parameters and load in the peak, the stability should firstly be considered.

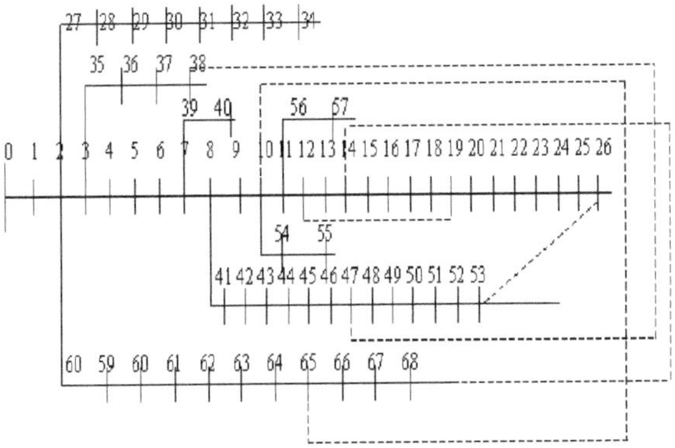

Fig. 4. The PE&G 69 bus system

In the electric market, it is not feasible that breaking the current supply of part customers to protect the system stability. So, dynamic reconfiguration of feeder is a appropriate method in new condition. Considering the factors of network structure and the levels of the load, it is an important task that establishment of appropriate switch on-off table in different time-intervals with different object functions from now on.

Table 1. Test systems optimization results

Object Function	Parameter	IEEE-13 bus	PE&69 bus
Voltage Stability	switch-status	In first time-interval 5/11,8/12 unclosed; in second time-interval 6/8,5/11unclosed; in sixth time-interval 5/11,8/12 unclosed	In first time-interval 15/16,24/25,43/44,10/65,14/68 unclosed; in second time-interval 15/16,24/25,9/10,12/13,38/47 unclosed; in fifth time-interval 13/14,17/18,43/44,10/65,24/25 unclosed
	stability index	the max value is 0.180751 the min value is 0.123452	the max value is 0.04901 the min value is 0.033200
	total loss (kWh)	1553.89	1989.459
Total Loss	switch-status	In first time-interval 6/11,8/12 unclosed; in second time-interval 6/11,6/7unclosed; in forth time-interval 6/11,8/12 unclosed	In first time-interval 13/14,16/17,22/23,46/47,10/65 unclosed; in second time-interval 11/12,18/19,45/46,52/53,10/65 unclosed; in fifth time-interval 11/12,18/19,45/46,24/25,10/65 unclosed; in seventh time-interval 11/12,17/18,44/45,24/25,10/65 unclosed
	stability index	the max value is 0.184475 the min value is 0.127763	the max value is 0.049848 the min value is 0.034589
	total loss (kWh)	1492.797	1962.542

5 Conclusion

A new reconfiguration algorithm based on dynamic particle swarm optimization is proposed in this paper. It adopts the voltage stability of distribution system as the object function, and proposes a new coded form which divides the particle module according to the topological structure of distribution system, and then avoids the convergence precociously via the community characters of DPSO. It shows that stability and economy should be a uniform problem from the results. Based on the change of load and network parameters, different problems should be disposed in different ways. It should change a single optimization problem to a non-linear and multiple objective problem and seek the balance point between them.

References

1. Sarma, N.D.R., Prakasa Rao, K.S.: A New 0-1 Integer Programming Method of Feeder Reconfiguration for Loss Minimization in Distribution Systems. Electric Power System Research 33(2), 125–131 (1995)
2. Bernardon, D.P., Garcia, V.J., Ferreira, A.S.Q., Canha, L.N.: Multicriteria Distribution Network Reconfiguration Considering Subtransmission Analysis. IEEE Transactions on Power Delivery 25(4), 2684–2691 (2010)
3. Yang, H.-T., Tzeng, Y.-T., Tsai, M.-S.: Loss-Minimized Distribution System Reconfiguration by Using Improved Multi-agent Based Particle Swarm Optimization. In: Power and Energy Engineering Conference (APPEEC), 2010 Asia-Pacific Transactions on Power Delivery, vol. 3(3), pp. 1217–1223 (2010)
4. Zhang, C., Zhang, J., Gu, X.: The Application of Hybrid Genetic Particle Swarm Optimization Algorithm in the Distribution Network Reconfigurations Multi-Objective Optimization. In: Third International Conference on Natural Computation, ICNC 2007, pp. 455–549 (2007)
5. Nara, K., Shiose, A., Kitagawa, M.: Implementation of Genetic Algorithm for Distribution Systems Loss Minimum Reconfiguration. IEEE Transactions on Power Systems 7(3), 1044–1051 (1992)
6. Ababei, C., Kavasseri, R.: Efficient Network Reconfiguration Using Minimum Cost Maximum Flow-Based Branch Exchanges and Random Walks-Based Loss Estimations. IEEE Transactions on Power Systems 99, 1 (2010)
7. Carreno, E.M., Romero, R., Padilha-Feltrin, A.: An Efficient Codification to Solve Distribution Network Reconfiguration for Loss Reduction Problem. IEEE Transactions on Power Systems 23(3), 1542–1551 (2008)
8. Chang, R.F., Lu, C.N.: Feeder Reconfiguration for Load Factor Improvement. In: Proceeding of the IEEE Power Engineering Society Transmission and Distribution Conference, vol. 2, pp. 980–984 (2002)
9. Shen, C.-C., Lu, C.-N.: Feeder Reconfiguration for Power Quality Requirement and Feeder Service Quality Matching. In: Proceeding of the IEEE Power Engineering Society Transmission and Distribution Conference, Asia Pacific, vol. 1, pp. 226–231 (2002)
10. Brownwell, G., Clark, H.: Analysis and solution for bulk system voltage instability [J]. IEEE Computer Application in Power 2(3), 31–35 (1989)

11. Kashem, M.A., Ganapathy, V., Jasmon, G.B.: Network Reconfiguration for Enhancement of Voltage Stability in Distribution Networks. IEE Proceedings Gener. Transm. & Distrib. 147(3), 171–175 (2000)
12. Zhang, Y., Rajagopalan, S., Conto, J.: Practical Voltage Stability Analysis. In: Power and Energy Society General Meeting, pp. 1–7. IEEE, Los Alamitos (2010)
13. Yang, L., Guo, Z.: Reconfiguration of electric distribution networks for energy losses reduction. In: Third International Conference on Electric Utility Deregulation and Restructuring and Power Technologies, DRPT 2008, pp. 663–667 (2008)
14. Li, Z., Chen, X., Yu, K., Sun, Y., Liu, H.: A hybrid particle swarm optimization approach for distribution network reconfiguration problem. Power and Energy Society General Meeting (2008)
15. Eberhart, R.C., Shi, Y.: Particle Swarm Optimization:Developments, Applications and Resources. In: Proceedings of the IEEE Congress on Evolutionary Computation, Piscataway, pp. 81–86 (2001)
16. Bommareddy, C., Makram, E.: Power Quality Studies in the Presence of DG. In: Power Systems Conference: Advanced Metering, Protection, Control, Communication, and Distributed Resources, PS 2006, pp. 224–229 (2006)
17. IEEE distribution Planning working Group Report. Radial distribution Test Feeders. IEEE Transactions on Power Systems 6(3), 975–985 (1991)
18. Eajal, A.A., El-Hawary, M.E.: Optimal Capacitor Placement and Sizing in Unbalanced Distribution Systems With Harmonics Consideration Using Particle Swarm Optimization. IEEE Transactions on Power Delivery 25(3), 1734–1741 (2010)

What Competitive Intelligence Problems Exist in Offshore Service Outsourcing Enterprises in Beijing When They Develop International Markets and How to Resolve Them*

Chen Feng

Institute of Scientific and Technical Information of China
No.15 Fuxing Road, Beijing 100038, P.R. China
chenfeng465@126.com

Abstract. Oriented to the needs in that government administration of offshore service outsourcing industry of Beijing City make strategic decisions for getting international competitive advantage against foreign competitors, focusing on competitive intelligence (CI) problems when offshore service outsourcing enterprises in Beijing develop international markets, four kinds of methods are conducted. The first is interviewing with the government administration of offshore service outsourcing industry of Beijing City- Beijing Municipal Bureau of Commerce, the second is interviewing with association of offshore service outsourcing industry of Beijing-Beijing Association of Sourcing Service (BASS), the third is interviewing with a typical leading enterprise in offshore service outsourcing industry of Beijing, and the fourth is desk research. After investigation and research, four major generic CI problems are found out, and four kinds of countermeasure are given.

Keywords: competitive intelligence, offshore services outsourcing, international marketing, countermeasure, case.

1 Introduction

In December25, 2006, Beijing was given the title " Service Outsourcing Base City of China" by Ministry of Commerce, Ministry of Information Industry and the Ministry of Science and Technology of the People's Republic of China. Several days later, the municipal government of Beijing city signed the agreement with the three national Ministry named " the agreement about building Beijing as a leading service outsourcing base City of China". From that time, the municipal government of Beijing city began officially to promote service outsourcing industry as a strategic decision.

---

* This paper is supported by a grant to Dr. CHEN Feng that title is "Studies on Essentials of Theory & Methodology of Industry Competitive Intelligence" from the National Natural Science Foundation of China (No.70973119).

D. Zeng (Ed.): ICAIC 2011, Part II, CCIS 225, pp. 19–24, 2011.

When two topics "what can a competitive intelligence professional do for software SMEs of offshore services outsourcing industry in China" [1] and "on government policy adoption when Chinese SMEs improve competitive intelligence ability and squeeze in global software outsourcing market" [2] were studied, findings showed that competitive intelligence must be depended on highly for developing sccessfully in international market be full of ruthless competitions and risks. The findings are more significant to Beijing than to other base cities of China, because Beijing is a special top base city in China. Beijing occupied the lion's share of service outsourcing industry in China, and ten enterprises of top 15 large service outsourcing enterprises that occupied more than 95% share of China locate in Beijing city. If Beijing cannot became a successful model for developing service outsourcing industry in international market depending on CI, the other base cities are more difficult to became successful models.

In this case, study the topics about "what competitive intelligence problems exist in offshore service outsourcing enterprises in Beijing when they develop international markets and how to resolve them" are significant.

With regard to the user groups of two topic's study findings, decision makers of municipal government and of service outsourcing enterprises are two major groups. For providing a pertinent findings and conclusion, this study chose the decision makers of municipal government as major user group. In other word, the findings and conclusion will focus to provide reference for "what competitive intelligence problems exist in offshore service outsourcing enterprises in Beijing when they develop international markets and how to resolve them" in Beijing municipal government decision makers' shoes.

2 Methodology

In this case, getting reliable raw data and information of study is crucial. As for research environment in China today, interviewing is the top one best method for getting reliable raw data and information of study. Three interviews were conducted as major methods in this study. In addition, desk research was used also.

2.1 Interviewing with the Government Administration of Offshore Service Outsourcing Industry of Beijing City- Beijing Municipal Bureau of Commerce

Beijing Municipal Bureau of Commerce is the official government administration of offshore service outsourcing industry of Beijing City. It is feasible to get reliable raw data and information about whole offshore service outsourcing industry of Beijing from this institution. For working efficiently, outline of interview was designed carefully. This outline of interview included twelve topics and nine relevant questions, example for, "would you please introduce the development of whole offshore service outsourcing industry of Beijing in recent years?", "would you please introduce the competition situation facing to offshore service outsourcing enterprises in Beijing?", "what are CI needs of offshore service outsourcing enterprises in Beijing when they develop international market?" etc.,

With the help of Beijing municipal government officials, the interview was performed. The outline of interview was faxed to interviewees three days ago before the appointment.

2.2 Iinterviewing with the Association of Offshore Service Outsourcing Industry of Beijing-Beijing Association of Sourcing Service (BASS)

Beijing Association of Sourcing Service (BASS) founded in November,2007. BASS enrolls about 100 enterprise members untill to the end of 2008, including all ten enterprises of top 15 large service outsourcing enterprises in China, example for Beyondsoft, Isoftstone, VanceInfo, Founder International Inc. etc.,

It is feasible to get reliable raw data and information about whole offshore service outsourcing industry of Beijing from BASS also. Just like doing in interviewing with Beijing Municipal Bureau of Commerce, outline of interview was designed carefully and faxed to interviewees three days ago before the appointment.

The outline of interview included nine topics and fifteen relevant questions, example for, "would you please introduce the competition situation facing to Beijing's enterprises in international offshore service outsourcing markets?","can Beijing's leading offshore service outsourcing enterprises deal with skillfully international competition with CI ?", "what is the urgent needs of these enterprises when they develop international market?" etc.,

2.3 Interviewing with a Marketing Manager of a Typical Leading Enterprise in Offshore Service Outsourcing Industry of Beijing

Among the ten of top 15 large service outsourcing enterprises of China located in Beijing, company X was choose as a typical enterprise sample to be studied. Interviewing method was used also. The major international market of company X was in USA. The interviewee was a marketing manager of this company. The outline of interview for company A include ten topics and eight relevant questions, example for, "would you please introduce the development of your company in USA market?", "can you get the timely, reliable and actionable competitive intelligence about your competitors and users?", " what helps do your company need from government in CI When you development foreign market", etc.,

2.4 Desk Research

Before launching three interviews, a lot of relevant basic information of offshore service outsourcing industry about Beijing and China were collected and analyzed. The major information resources included webs, reports, books, papers and articles. Example for www.basscom.cn, www.ChinaSourcing.mofcon.gov.cn, "Report on development of China's outsourcing" [3], "Report on China outsourcing base cities and representatives areas" [4] etc.,

3 Findings

Almost all Beijing enterprises in offshore service outsourcing industry have been launching CI activities in to some extent, but study proved that some certain prevalent CI problems existed in these enterprises when they develop international market from the CI professional point of view. Four kinds of CI problems were obvious.

3.1 Enterprises Did Not Put Great Emphasis on CI Generally

For almost all Beijing enterprises in offshore service outsourcing industry, although they pay great attention to information, they can not distinguish CI from information generally, however. In some cases, they thought that CI was equal to information completely, and in some other cases, they thought CI was equal to espionage completely. They can not understand the basic concept and typical characteristics (timely, reliable, actionable) of CI, not to mention the theory and methods of CI systematically, to say nothing of mastering CI skillfully as powerful weapon against competitors in developing international market.

From the CI professional point of view, in any CI best practice company, there must exist a CI Champion in the company[5], [6]. Because the essentiality of CI was underestimated generally, CI Champions hardly existed in enterprises in offshore service outsourcing industry of Beijing. Few enterprises established their formal CI department accordingly.

3.2 Competitive Strategies That Develop International Market Were Scarce in These Enterprises Commonly

Study proved that few enterprises developed their competitive strategies in international market. Although in these enterprises, it was not uncommon that their top leaders were very capable, and they maintained good momentum of development in international market. The business model of these enterprises stayed in traditional phase that oriented to performance indicators in short time, example for growth rate of amount of sales. These enterprises could not think and answer the strategic questions clearly and systemically about long time, example for, "who was I?", "where did I come from, and where I would to go?", " where were my target markets? Who were my target users?, who were my competitors?",etc.,

Because of these problems, lots of enterprise, many enterprises bark up the wrong tree always. They should put the foreign market and users in the first place when they launched promoting activities, in fact, they launched too many promoting activities in home other than abroad unfortunately. The performance turned out to be that brand cognitive of these enterprises was low in international market, although their brand cognitive was high in domestic market.

3.3 CI Professionals for Developing International Market Were Scarce in These Enterprises Prevalently

Study proved that few enterprises owned CI professionals for developing both domestic and international market. Although some staffs especially in marketing department did some CI activities to some extent, but they did the CI oriented to serve their tasks other than to the needs of whole company in strategic level. For these enterprises, no CI professionals existed, not to mention the institutional CI team, to say nothing of more advanced CI professionals, example for CI collectors, CI analysts, CI managers, etc., They did CI activities only limited in a informal stage.

3.4 Free Public CI Products Outside the Enterprises That Can Be Available Were Scarce Prevalently

Enterprises of offshore service outsourcing industry in Beijing longed for CI products in whole CI product line from raw data and information to customized reports when they developed international market, even they did CI activities only limited in a informal stage. They need CI products in whole CI product line including raw data and information, news letters, briefings, competitor files, monthly reports, special reports, consulting reports etc.,

In the view in content of CI, the CI products that enterprises needed were kaleidoscopic, including developing situation of service outsourcing industry, statistics, performance data and indicators of peer enterprises etc.,

Study found that almost all enterprises could not meet the needs of CI products by themselves. They pressed for getting lots of CI products with the help of other actors in Beijing. Major other actors in Beijing are like the office departments of national government, the office departments of local government, industry associations, public institutes of information etc.,

Example for, some authoritative reports of offshore service outsourcing industry published by consulting company as a kind of CI product were of great value to enterprises, but maybe the price of only one copy of this report run to several hundred thousand US$. It was too expensive to be bore by a single enterprise of Beijing. If a public institute of information bought the report and provided it to enterprises as a kind of free public CI product, both the institutes of information and enterprises of Beijing will have a ball. Many public institute of information of Beijing today own superfluous money but do not know how to spend the money out.

4 Conclusions

After finding out what competitive intelligence problems exist in offshore service outsourcing enterprises in Beijing, the more significant thing is to know that how to resolve them. Four kinds of countermeasures were given as conclusions in the end of study.

(1). Give "competitive intelligence" instead of "information" a unambiguous law position in policy laws and regulations related the public social information service system issues about service outsourcing industry of Beijing.

(2). Aim to the whole service outsourcing industry rather than single corporations get international competitive advantage, study the competitive strategy problems in whole service outsourcing industry level, and develop competitive strategy of service outsourcing industry of Beijing oriented to international market that promote interplaying between the competitive strategy of service outsourcing of Beijing in whole industry level and in single enterprise level.

(3). Base on the promoting service outsourcing industry programs both of China and Beijing, launch a special program for educate and train competitive intelligence professionals for offshore service outsourcing enterprises of Beijing, and make the most of the budgeted special funds that use to educate and train general professionals in service outsourcing industry.

(4). Oriented to the need of the whole service outsourcing industry getting international competitive advantage, focusing on the function that provide free public competitive intelligence products and services for enterprises of Beijing, under the direction of competitive intelligence theory and methodology, launching all leading competitive intelligence actors of the whole service outsourcing industry of Beijing that include official department, enterprises, society of service outsourcing industry, competitive intelligence organizations etc., to join up, set up and operate a competitive intelligence system in offshore service outsourcing industry of Beijing.

References

1. Chen, F.: What Can a Competitive Intelligence Professional Do for Software SMEs of Offshore Services Outsourcing Industry in China. In: Zhang, S., Guo, S., Zhang, H. (eds.) Advances in Management of Technology: Proceedings of The International Conference on Management of Technology Taiyuan 2007, pp. 153–157. Aussino Academic Publishing House, Marrickville (2007)
2. Chen, F.: On Government Policy Adoption When Chinese SMEs Improve Competitive Intelligence Ability and Squeeze in Global Software Outsourcing Market. In: Zhang, S., Zhang, H. (eds.) Advances in Management of Technology: Proceedings of The International Conference on Management of Technology Taiyuan 2008, pp. 380–384. Aussino Academic Publishing House, Marrickville (2008)
3. China Council for International Investment Promotion (CCIIP). Report on development of China's outsourcing. Shanghai Jiao Tong University Press, Shanghai,China, pp 1–101 (2007)
4. China Council for International Investment Promotion (CCIIP). Report on China outsourcing base cities and representatives areas. China Council for International Investment Promotion (CCIIP), China, pp 1–81 (2008)
5. Matteo, M.R., Drkman, I.K.: Building Credibility, Champions, and a Mandate for Competitive Assessment. In: Competitive Intelligence Review 10th Anniversary Retrospective Edition, pp. S19–S23. Society of Competitive Intelligence professionals, Alexandria (1996)
6. Prescott, J.E., Herring, J., Panfely, P.: Leveraging Information for Action: A Look into the Competitive and Business Intelligence Consortium Benchmarking Study. Competitive Intelligence Review 1, 5–9 (1998)

NN Assessment on the Government Responsibility Performance of Reverse Logistics Management

Zhang Fengrong[1] and Xiao Xinwei[2]

[1] Department of Politics and Law Northeast Normal University
Changchun, China
[2] Department of Chemistry, Changchun University of Science and Technology
Changchun, China
fengrong65@yeah.net

Abstract. The spillover of Quasi-public goods attributes make reverse logistics become a part of the responsibilities of the Government at present. Therefore, the evaluation of reverse logistics management efficiency has become an important indicator to evaluate the responsibility of local governments. In order to provide a more scientific and effective decision making of resources distribution, this paper establishes a comprehensive evaluation model of thinking closer to human being by introduces RBF neural network model which has powerful vector classification and fast computation to the qualitative and quantitative analysis based on the TOPSIS evaluation results of reverse logistics management efficiency. The example shows that the model is more generalized to promote the use of.

Keywords: Neural Network(NN), Assessment, Reverse Logistics, Government Responsibility.

1 Introduction

The development our country's logistics industry is much later. At present, ability for third-party logistics enterprise developing to forward logistics market is still limited, so in a short period, to achieve something in the field of reverse logistics is obvious difficult. As an economic activity, reverse logistics also has external effect. External effect of reverse logistics is originated from its attribute of quasi-public goods, such attribute can let reverse logistics bring benefit for the company and also add the profit or effectiveness (welfare) of other enterprises, consumers and society, that is, externalization of partial effect or overflow of partial benefit. Due to external effect of reverse logistics may cause the standard profit of reverse logistics enterprise is less than social profit, in this case, it will weaken the starting power of reverse logistics enterprise so that cause insufficient supply and bring loss of social welfare. As for the enterprise, requiring batch of reverse logistics is far less than forward logistics, and dispersed, little batch and relative cheap waste materials cause reverse logistics activity difficult to obtain scale merit; as for the society, the growing waste materials not only produce environmental external effect, but also threaten human beings' health and safety. While the special external effect attribute of reverse logistics can lead to mismatch between enterprise target and social objective, so that market mechanism is difficult to play a role in the allocation of resources of reverse logistics and appears

D. Zeng (Ed.): ICAIC 2011, Part II, CCIS 225, pp. 25–31, 2011.

so-called *market failure*, so as to recovery processing task of most enterprises can only be undertaken by the government and become a part of the government liability. Therefore, assessment on the performance of reverse logistics management has been an important target for measuring local government liability. [1][2][3]

Assessment on the performance of reverse logistics management is belonging to multi-target appraisal and decision-making. A common decision method is proposed on the basis of principle of optimality and famous American scientist in management H. A. Simon's principle of Satisfying. In those methods, some of them only consider the advantages or disadvantages of scheme or assessment's object in assessing, thus the results of assessment maybe not the best, such as method of Trade-off, Goals and Objectives Achievement Method, Expectancy method, entropy method and analytic hierarchy process (AHP); some methods are over simple in the procedure of assessment, or subjective weighting or constraint condition is over harsh, that make these methods are hard to widely use in practice, for instance, comprehensive target method, principal component analysis, efficiency coefficient method, data envelopment analysis, etc. TOPSIS neural network assessment method applied in this paper is calculated to overcome the defects above-mentioned assessment methods, establish a comprehensive assessment model combining qualitative with quantification closing to human thinking model, and improving validity and reliability of assessment results.

2 Target System of Assessment and Decision-Making on the Performance of Reverse Logistics Management under the View of Government Liability

Assessment on the government responsibility performance of reverse logistics management pays much attention on investment cost, income profit and service level, so this paper select fixed assets investment (ten thousand Yuan), annual disposal cost (ten thousand Yuan) and number of invested personnel (year·people), annual income of renewable materials for sale (ten thousand Yuan), recovery rate of the waste materials (%)and number of people in household for sharing such service (year·ten thousand households) as assessment index, shown as Fig.1, calculation and quantification of all assessment indexes attribute value can refer to reference [4] and [5].

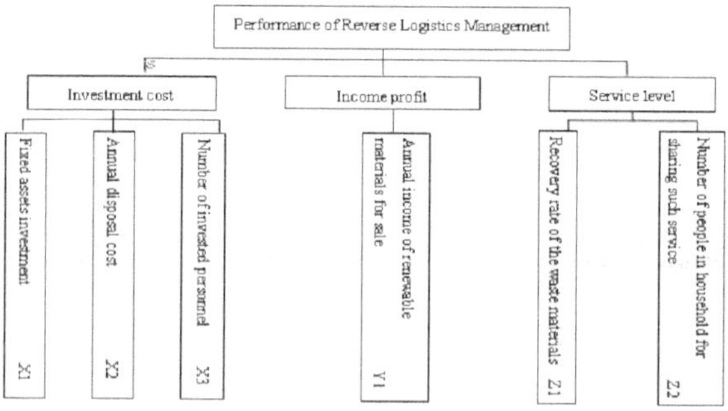

Fig. 1. Target System of Assessment and Decision-making on the Performance of Reverse Logistics Management

3 TOPSIS Neural Network Assessment Model of the Performance of Reverse Logistics Management

The basic thought of TOPSIS neural network assessment on the performance of reverse logistics management under the view of the government liability is that: calculate approach degree of all assessment schemes and ideal solution by TOPSIS method, select RBF (Radial Basis Function) neural network model, and taking basic target attribute value describing reverse logistics management activities as input vector, taking approach degree standing for comprehensive assessment target as the output value of neural network, and then using enough vector samples to train the network, so that different input vectors get different output value. Weighing coefficient belonging to the well-trained network is a correct internal relation obtained by application study of network itself. At this moment, the network can make comprehensively assessment of the performance of reverse logistics management of different local governments as an effective method combining qualitative with quantitative analysis. Modeling theory of TOPSIS neural network assessment is shown as following in detailed:

3.1 TOPSIS Method

TOPSIS is *the ordering method closing to ideal value*, is a statistical method applying for comparing assessment according to multi-target and multi-object proposed by Hwang and Yoon in 1981. Through constructing ideal solution and negative ideal solution of assessment, that is, optimum solution and the worst solution of several targets, this method will compare the actual solution of all schemes with optimum solution and the worst solution. If the solution of some scheme is most closing to the optimum solution and keeping far away from the worst solution, then the solution is the concentrated ideal solution of this scheme, the corresponding scheme is the best scheme. [6][7]

Given decision matrix of multi-target decision-making issues is $A = (y_{ij})_{n\times m}$, in this equation, y_{ij} is No. j targeted value in No. I scheme, i=1,2...n; j=1,2...m. Generally speaking, the targeted value in A is divided into benefit high yields target and cost low yield target, the former is the bigger the ideal, and the latter is the smaller the ideal. The target that the bigger the ideal for membership grade of A is:

$$s_{ij} = \frac{y_{ij} - y_{i\min}}{y_{i\max} - y_{i\min}} \tag{1}$$

The target that the smaller the ideal for membership grade of A is:

$$s_{ij}^{'} = 1 - y_{ij} = \frac{y_{i\max} - y_{ij}}{y_{i\max} - y_{i\min}} \tag{2}$$

According to equation (1), (2) and matrix A, we can get targeted membership grade matrix $S = (s_{ij})_{n\times m}$,

Based on the optimal relativity to make sure ideal solution and negative ideal solution of the target, and then construct ideal solution vector X+ and negative ideal vector X-:

$$Z^{+}=\{\max s_{ij}|i=1,2,...,n,\ j=1,2,...,m\}=\{z_1^{+},z_2^{+},...,z_m^{+}\} \tag{3}$$

$$Z^{+}=\{\min s_{ij}|i=1,2,...,n,\ j=1,2,...,m\}=\{z_1^{-},z_2^{-},...,z_m^{-}\} \tag{4}$$

The distance from each solution to ideal solution is:

$$D^{+}=\sqrt{\sum_{j=1}^{m}\left[\omega_j(s_{ij}-z_j^{+})\right]^2} \qquad (i=1,2,...,n) \tag{5}$$

The distance from each solution to negative ideal solution is:

$$D^{-}=\sqrt{\sum_{j=1}^{m}\left[\omega_j(s_{ij}-z_j^{-})\right]^2} \qquad (i=1,2,...,n) \tag{6}$$

In this equation, ω_j is the weighing of No. j target.
The relative approach degree of some assessment scheme for ideal solution is:

$$D_i=D_i^{-}/(D_i^{-}+D_i^{+}) \qquad (i=1,2,...,n) \tag{7}$$

Di is called approach degree. If assessment scheme approaches to ideal solution, the approach degree closes to 1; on the contrary, if the assessment scheme approaches to the negative ideal solution, the approach degree closes to 0. Taking approach degree as judgment basis, makes order for the scheme, and makes assessment and decision-making on all assessment schemes.

3.2 Principle of RBF Network Model

RBF network is proposed by Moody J, Darken C in 1980s, a one type of forward neural network, having the function of strong vector classifier and the ability of rapid calculation; and it can approach to any nonlinear functions by any precision. RBF network is consisted of three layers, the first layer is input layer, consisting of signal source node; the second layer is hidden layer (radial basic layer), number of implicit units is determined by the issues described; the third layer is output layer, response for the input model. The signal is transferred from input layer to radial basic layer, produce partial response by kernel function, and linear output at output layer. The output of No. I node in the hidden layer of RBF network is:

$$r_i(x)=R_i(\|(x-c_i)\|/\sigma_i) \qquad (i=1,2,...,k) \tag{8}$$

In equation (8), x is n dimension input variable; C_i is the center of No. I primary function, having the same number of dimension with x; a is No. i perceptual variable, determining the width and the size of C_i that this function surrounding the center point; k is the number of perceptual unit, $\|\bullet\|$ is vector norm, and is Euclid Norm generally.

Kernel function in hidden layer is radial symmetry, and there are many types. But the most common used is Gaussian function, shown as the following equation:

$$R_i(x) = \exp\left[-\|x - c_i\|^2 /(2\sigma^2)\right] \qquad (i = 1,2,...,k) \tag{9}$$

This paper takes Gaussian function as basic function, and then the output of RBF network model is:

$$y_q = \sum_{i=1}^{k} \omega_{iq} \exp\left[-\|x - c_i\|^2 /(2\sigma^2)\right] \qquad (q = 1,2,...,m) \tag{10}$$

In equation (10), q is the number of output node; ω is the weight connecting the hidden layer and the output layer. RBF network adjusts centrality parameter c_i and weight ω through input error and output error, so that reach the adjustment on network internal coefficient. [8][9]

4 Empirical Analysis

4.1 TOPSIS Assessment Results

This paper will add three samples from H city to L city based on the empirical data of A~G cities in reference [4], adopting TOPSIS neural network assessment method to assess the performance of reverse logistics management. In TOPSIS assessment, fixed assets investment (X1), annual disposal cost (X2) and number of invested personnel (x3) are cost low yields target, annual income of renewable materials for sale (Y1), recovery rate of the waste materials (z1) and number of people in household for sharing such service (Z2) are benefit high yields target, targeted value after the transformation of matrix is shown as table 1.

Table 1. Targeted Data of the Performance Of Reverse Logistics Management for Ten Cities

Cities	X1	X2	X3	Y1	Z1	Z2
A	0.3425	0.3477	0.3574	0.5529	0.5109	0.5446
B	0.3995	0.3299	0.3647	0.4810	0.3623	0.4684
C	0.3138	0.3249	0.3971	0.3262	0.3251	0.3540
D	0.3396	0.5360	0.3723	0.2488	0.3994	0.2996
E	0.4719	0.3574	0.3403	0.2598	0.2787	0.3595
F	0.3677	0.3676	0.3574	0.2930	0.4412	0.2723
G	0.3892	0.3385	0.4467	0.3759	0.2647	0.2560
H *	0.2806	0.3151	0.3426	0.3388	0.2447	0.3601
I *	0.3004	0.3139	0.3339	0.3131	0.2643	0.3159
J *	0.2906	0.3148	0.3392	0.3257	0.2540	0.3355

Using above-mentioned TOPSIS calculating method, calculating each targeted value and the approach degree Di of optimum value respectively, and ordering the assessment results of A~G cities. The calculating results are shown in table 2.

Table 2. Calculation and Ordering of Approaching Degree

Cities	D+	D-	Di	Order
A	0.2454	0.4879	0.6647	1
B	0.2959	0.3414	0.5291	2
C	0.4408	0.1499	0.2505	7
D	0.4336	0.2575	0.3796	3
E	0.4662	0.1926	0.2937	5
F	0.4402	0.1959	0.3237	4
G	0.4702	0.1826	0.2756	6
H *	0.0593	0.0896	0.407	
I *	0.0896	0.0593	0.3485	
J *	0.0571	0.0506	0.3763	

According to the ordering of assessment results on the approaching degree of A~G cities, we can see that C city, E city, F city and G city have lower approach degree of optimum value, and it is indicated that the performance of reverse logistics management is much worse. A city has the highest assessment value, the follows are B city and D city, the performance of reverse logistics management of those three cities is much better. Above analysis is mostly in line with the assessment results in reference [4].

4.2 RBF Neural Network Simulation Evaluation

Taking the samples of A~G cities as training samples, 6 targeted values as network input, and assessment as network output; the number of nodes in hidden layer is confirmed 6; establishing network structure 6-6-1; after centralizing the data, given training speed is 1.7, number of weighing seed is 4, Sigma parameter is 0.1, the weight is 0.7, the biggest frequency of iteration is 10000; and the error becomes $3.67e^{-16}$ through studying neural network; output result meets the requirement of training precision, the error is in the available arrangement. In order to verify the reliability for selecting RBF network as the method of assessment on the performance of reverse logistics management, selecting H~J cities as the tested samples, and input the parameter and weighting of train-welled RBF network primary function to RBF network.The assessment results are shown in table 3.

Table 3. Simulation Assessment Result

Cities	A	B	C	D	E
Expected value	0.665363	0.308024	0.29233	0.535683	0.37261
Result	0.66536	0.30802	0.29233	0.53568	0.37261
relative error	-0.000005	-0.000014	0.000001	-0.000005	0.00001
Cities	**F**	**G**	**H ***	**I ***	**J ***
Expected value	0.253753	0.279753	0.60165	0.398351	0.46985
Result	0.25375	0.27975	0.60171	0.39868	0.47007
relative error	-0.000012	-0.00001	0.000101	0.000826	0.000464

Through simulation operation, the three output results of the test samples is much satisfied, it is indicated that this network has much stronger generalization ability, meanwhile, it can used for the assessment on the performance of reverse logistics management of other districts.

5 Conclusion

Reverse logistics is a developing industry and there are many unconfirmed factors in China, so the management of reverse logistics is the government's current responsibility, which constitutes one of the government performance evaluations. This paper establishes a TOPSIS neural network model for evaluating the performance of the local government's reverse logistics management. From the view of evaluating result, this method has a certain theoretical significance and reference value in actual application. In addition, we can see that using TOPSIS neural network to make decision-making has many advantages: (1) RBF network model not only has the function of fast convergence but also effectively resolves the divergent issue of primary simulated value; (2) It absorbs advantages of RBF neural network and TOPSIS analysis utilizing the sensitive of differences indicators of TOPSIS analysis improves the precision of input data of RBF network so as to reduce the error efficiently; (3)This method has broad practical application and direction in government decision-making, which can be used for policy formulation and performance monitoring.

Acknowledgement. It is a project supported by Science Development Program of Science Department of Jilin Province and NSFC of Jilin Province.

References

[1] Wang, Y., Liu, Z.: Research on Reverse Logistics External Effect and Its Government Interfere. Journal of Transportation Engineering and Information 12, 76–81 (2006)
[2] Zhou, Q.: Research on Necessity and the Barrier Mechanism of Reverse Logistics Development in Our Country. Ecological Economics 4, 66–68 (2007)
[3] Zhou, Q.: Research on the Government Strategy of Reverse Logistics Development in Our Country. Economic Review 1, 31–33 (2007)
[4] Pan, D.: Assessment on the Performance of the Third-party Reverse Logistics Based on Circular Economy. Master thesis of Nanchang University 6, 22–30 (2007)
[5] Tang, Q., Feng, M.: Practical Statistical Analysis and Its DPS Data Processing System. Scientific and Technical Publishers (2002)
[6] Zhang, B., Da, Q.: Research on the Performance of Reverse Logistics Management for Local Government Based on DEA. Soft Science 1(21), 66–69 (2007)
[7] Zhang, F.R., Fet, A.: Main Factor TOPSIS Analysis in Decision-making on the Efficiency of Third-party Reverse Logistics Enterprise. Australia: M&D Forum 11, 492–496 (2008)
[8] Zhang, F., Annik, M.F., Wang, J.: Fuzzy RBF assessment on productive efficiency of environmental impacted enterprise. In: IEEE International Conference on Mechatronics and Automation, ICMA 2009, vol. 1, pp. 465–470. IEEE press, Los Alamitos (2009)
[9] Shao, J., Liu, X., Li, D.: Comprehensive Assessment on Sustainable Utilization Level of Regional Water Resource Based on RBF Network. Water Resources and Power 12(6), 9–11 (2006)

Credit Scoring Based on Eigencredits and SVDD*

Haizhou Wei and Jianwu Li

Beijing Key Lab of Intelligent Information Technology, School of computer
Beijing Institute of Technology
Beijing, 100081, China
haizhou_wei@sina.cn

Abstract. Credit risk evaluation is an active research topic in financial risk management, and credit scoring is an important analytical technique in credit risk evaluation. In this paper, a new two-stage method is introduced to perform credit scoring. Eigencredits are firstly constructed based on creditworthy examples through principal component analysis to extract the principal features of creditworthy data. Then, support vector domain description (SVDD) is further used to describe creditworthy examples. Preliminary experiments based on two real data sets from UCI repository show the effectiveness of the proposed method.

Keywords: Credit scoring, Eigencredits, Support vector domain description (SVDD).

1 Introduction

Credit risk evaluation is crucial for financial institutions because high risks associated with inappropriate credit decisions may result in great losses. Thus, with the rapid growth and increased competition in credit industry, it is necessary to perform fast automatic decisions on credit risk evaluations for financial institutions, especially when facing millions of credit card or consumer loan applications simultaneously.

In fact, automatic Credit scoring has gained more and more attention in the recent decade. The objective of credit scoring models is to assign credit applicants to either a 'good credit' group that is certain to repay financial obligation or a 'bad credit' group with high possibility of defaulting on the financial obligation. Thus, the credit scoring model is developed on the basis of observed historical data including good and bad examples. A well designed model should have a high classification accuracy to evaluate the new applicants as good or bad.

Initially, most statistical and optimization techniques had been widely employed to build the credit scoring models such as linear discriminant analysis [1] and logistic regression [2]. Although these methods are relatively simple and explainable, the ability to discriminate good customers from bad ones is still an argumentative problem. In recent years, some new methods from artificial intelligence have also

* This work was supported by the foundation of Beijing Key Lab of Intelligent Information Technology.

D. Zeng (Ed.): ICAIC 2011, Part II, CCIS 225, pp. 32–40, 2011.

been used for developing the credit scoring systems, such as decision tree [3], the artificial neural networks [4], k-nearest neighbor [5], genetic algorithm [6], Artificial Immune System Algorithm [7] and support vector machines [8]. Among these artificial intelligence methods, the decision tree, artificial neural networks and support vector machine are generally regarded as the most efficient credit scoring models. Additionally, some two-stage scoring models [9] are presented to overcome the shortcoming of the single scoring model. These models have also shown promising prediction results.

All these methods mentioned above are based on binary-class classifiers. However, there exists a special extension of the classification problem, the data domain description, also called one-class classification. In domain description the task is not to distinguish between classes of objects like in binary classification problems, but to give a description of a set of objects. This description should cover the class of objects represented by the training set, and ideally should reject all other possible objects in the object space. The data domain description performs better than two class classifiers when negative samples are too few or too many. This approach is usually used for outlier detection or novelty detection.

Credit scoring can also be considered as an outlier detection problem. If the new applicant is an outlier object, he or she is more likely to be with bad credit. In this work we will introduce one-class classifier — support vector data description (SVDD) to model credit scoring.

In order to describe the creditworthy examples better, we preprocess the training data using a method called "Eigencredits". This idea is inspired by "Eigenfaces" [10], which is proposed for face recognition. Just like the Eigenfaces approach, in this work we map new examples into a feature space spanned by the Eigencredits which are the principal components of the initial training creditworthy vectors. In feature space, the creditworthy vectors gather more tightly. This seems to be more suitable to be described by SVDD.

The rest of this paper is organized as follows. Section II gives an introduction on Eigencredits. Section III describes the SVDD briefly. Experimental results are presented in Section IV. Section V discusses the advantages of the proposed model and concludes this paper.

2 Eigencredits

2.1 The Eigencredits Approach

Much of the previous work on credit scoring has ignored the issue of just what aspects of the applicants' information are important for identification. This suggested us to find the significant local and global features. In mathematical terms, we wish to find the principal components of the distribution of creditworthy vectors.

Good credit samples, being similar in overall configuration, should not be randomly distributed in the huge samples space and thus can be described by a relatively low dimensional subspace. The main idea of the principal component analysis is to find the principal directions which best account for the distribution of good credit samples within the entire samples space. These principal directions define the subspace of creditworthy samples, which is called as "credit space". Therefore, we also call the principal components as "Eigencredits".

Each individual example can be represented exactly in terms of a linear combination of all Eigencredits. Each example can also be approximated using only the "best" Eigencredits-those which have the largest Eigenvalues, and which therefore account for the most variance within the set of creditworthy vectors. Any example can be approximately reconstructed by storing a small collection of weights for Eigencredits. The weights are found by projecting the example vector onto each Eigencredit. Each individual, therefore, can be characterized by the Eigencredits to form an extremely compact representation.

2.2 Calculating Eigencredits

Let the training set of good credit samples be $\{\Gamma_1, \Gamma_2, \Gamma_3, \dots, \Gamma_N\}$. Each sample vector has M variables. The average vector of the set is $\Psi = \frac{1}{N}\sum_{i=1}^{N}\Gamma_i$. The vectors are centered by $\Phi_i = \Gamma_i - \Psi$, and then are subject to principle component analysis.

To obtain the principle components, we need to calculate the M eigenvectors u_k and eigenvalues λ_k, respectively, of the covariance matrix

$$C = \frac{1}{N}\sum_{i=1}^{N}\Phi_i\Phi_i^T = \frac{1}{N}AA^T, \tag{1}$$

where the $A = [\Phi_1\Phi_2 \dots \Phi_N]$.

The associated Eigenvalues allow us to rank the eigenvectors according to their importance in characterizing the variation among the credit vectors from large to small. Then we define

$$\alpha = \sum_{i=1}^{p}\lambda_i / \sum_{i=1}^{M}\lambda_i \cdot \tag{2}$$

Through setting the value of α, we can determine the best p eigenvectors.

A new example Γ is transformed into its Eigencredit components (projected into "credit space") by a simple operation,

$$\omega_k = u_k^T(\Gamma - \Psi), \tag{3}$$

for $k = 1, \dots, p$.

The weights form a vector $\Omega^T = [\omega_1, \omega_2, \dots, \omega_p]$ which describes the contribution of each Eigencredit in representing the input example vector Γ. Then the vector is further provided for SVDD to evaluate whether or not it is creditworthy, as addressed in the following.

3 SVDD

3.1 Normal SVDD

SVDD was first presented in 1999 by Tax and Duin [11], inspired by the support vector machines by Vapnik [12]. SVDD defines a model which gives a closed hypersphere around the data. The sphere is characterized by center $\boldsymbol{a}$ and radius $R > 0$. The volume of the sphere is minimized by minimizing R^2.

To allow the possibility of outliers in the training set, the distance from x_i to the center $\boldsymbol{a}$ should not be strictly smaller than R, but larger distances should be penalized. Therefore, we introduce slack variables $\xi_i \geq 0$ and the minimization problem changes to:

$$F(R, a, \xi_i) = R^2 + C\sum_i \xi_i \,. \tag{4}$$

This has to be minimized under the constraints

$$\xi_i \geq 0,\ \|x_i - a\|^2 \leq R^2 + \xi_i \ \forall i \,. \tag{5}$$

In (4), the variable C gives the trade-off between the volume of the sphere and the number of outliers.

They form a constrained optimization problem with the training data x_i and we can construct the Lagrange multiplier method on the problem with Lagrange multiplier $\alpha_i \geq 0$ and $\gamma_i \geq 0$. Finally, we obtain:

$$a = \sum_i \alpha_i x_i \,, \tag{6}$$

$$\max_{\alpha_i} L = \sum_i \alpha_i (x_i \cdot x_i) - \sum_{i,j} \alpha_i \alpha_j (x_i \cdot x_j) \,, \tag{7}$$

with constraints $0 \leq \alpha_i \leq C, \sum_i \alpha_i = 1$.

Eq. (6) states that the center of the sphere is a linear combination of data objects, with weight factors α_i which are obtained by optimizing Eq. (7). Only for a small set of objects the equality in Eq. (5) is satisfied: these are the objects which are on the boundary of the sphere itself. For those objects the coefficients α_i will be non-zero and are called the support objects. Only these objects are needed in the description of the sphere. The radius R of the sphere can be obtained by calculating the distance from the center of the sphere to a support vector.

A test object z is accepted when the distance to the center is smaller than the radius. Expressing the center of the sphere in terms of the support vectors, we accept objects when

$$(z \cdot z) - 2\sum_i \alpha_i (z \cdot x_i) + \sum_{i,j} \alpha_i \alpha_j (x_i \cdot x_j) \leq R^2 \tag{8}$$

3.2 SVDD with Negative Samples

Negative examples (objects which should be rejected) if available, can be incorporated into the training procedure to improve the description. In contrast with the target examples which should be within the sphere, the negative examples should be outside the sphere.

In the following the target objects are enumerated by indices *i*, *j* and the negative examples by *l*, *m*. For further convenience, assume that target objects are labeled $y_i = 1$ and outlier objects are labeled $y_l = -1$. Again we allow for errors in both the target and the outlier set and introduce slack variables ξ_i and ξ_l:

$$\min_{R,\xi_i,\xi_j} F(R,a,\xi_i,\xi_l) = R^2 + C_1\sum_i \xi_i + C_2\sum_l \xi_l, \tag{9}$$

With the constraints:

$$\|x_i - a\|^2 \le R^2 + \xi_i, \|x_l - \alpha\|^2 \le R^2 - \xi_l,\ \xi_i \ge 0, \xi_l \ge 0\ \forall i,l. \tag{10}$$

Then we use Lagrange multiplier method and we obtain

$$\begin{aligned} \max_{\alpha_i,\alpha_j,\alpha_l,\alpha_m} L = & \sum \alpha_i (x_i \cdot x_i) - \sum_l \alpha_l (x_l \cdot x_l) \\ & - \sum_{i,j} \alpha_i \alpha_j (x_i \cdot x_j) + 2\sum_{l,j} \alpha_l \alpha_j (x_l \cdot x_j) \\ & - \sum_{l,m} \alpha_l \alpha_m (x_l \cdot x_m) \end{aligned} \tag{11}$$

With the constraints:

$$\sum_i \alpha_i - \sum_l \alpha_l = 1, 0 \le \alpha_i \le C_1, 0 \le \alpha_l \le C_2\ \forall i,l. \tag{12}$$

When we finally define new variables $\alpha_i' = y_i \alpha_i$ (index i now enumerates both target and outlier objects), the SVDD with negative examples is identical to the normal SVDD. The constraint given in Eq. (12) changes to $\sum_i \alpha_i' = 1$ again the testing function Eq. (8) can be used.

The method just presented only computes a sphere around the data in the input space. However, normally, data are not spherically distributed. Since the problem is stated completely in terms of inner products between vectors, the method can be made more flexible. Inner products of objects $(x_i \cdot x_j)$ can be replaced by a Kernel function $K(x_i \cdot x_j)$ which satisfies Mercer's theorem. This implicitly maps the objects x_i into a new feature space which a better, tighter description can be obtained. Therefore, we can replace all inner products $(x_i \cdot x_j)$ by $K(x_i \cdot x_j)$.

4 Experiments

4.1 Real World Credit Data Sets

Two real world data sets, Australian and German credit dataset, which are from UCI Machine Learning repository [13], are used to test the performance of the Eigencredits and SVDD model. The Australian credit data set consists of 307 good credit samples and 383 bad credit samples. The German credit data set consists of 700 good credit samples and 300 bad credit samples.

4.2 Performance of Eigencredits and SVDD Model on the Whole Dataset

In all experiments, the program ran under matlab 6.5 with prtools [14] and dd_tools [15] toolbox. We set $\alpha = 0.95$ in Eq. (2) and adopt Gaussian kernel for SVDD.

The classification results of the proposed model are compared with two other standard binary-class classifiers: support vector machine (SVM) and backpropagation artificial neural network (BPANN).

We randomly divide each dataset into two parts, training dataset taking 3/4 of the total dataset and testing dataset which taking 1/4 of the total dataset.

Let the number of creditworthy cases classified as good be GG and classified as bad with GB, denote the number of default cases classified as good with BG and as bad with BB. Five commonly used evaluation criteria measuring the efficiency of the classification are as follows:

$$\mathrm{Pr}\,ecision(\mathrm{Pr}) = \frac{GG}{GG+BG} \times 100\%$$

$$\mathrm{Re}\,call(\mathrm{Re}) = \frac{GG}{GG+GB} \times 100\%$$

$$F1 = \frac{2 \cdot Precision \cdot Recall}{Precision + Recall}$$

$$Specificity(Sp) = \frac{BB}{BG+BB} \times 100\%$$

$$Percentage\,Correctly\,Classified(PCC) = \frac{GG+BB}{GG+GB+BG+BB} \times 100\%$$

The three methods used identical training and testing sets. For each method, we used 5-fold cross validation in the training set to search for the best parameters. Finally, the experimental results are shown in Tables 1 and 2. For the convenience, we use SVDD to express the Eigencredits and SVDD model briefly.

Table 1. Results of SVDD, SVM and BPANN on Australian dataset

Method	Pr%	Re%	Sp%	F1%	PCC%
SVDD	86.7	85.6	89.5	86.1	87.7
SVM	83.3	92.1	85.3	87.5	88.3
BPANN	88.9	84.2	91.6	86.5	88.3

Table 2. Results of SVDD, SVM and BPANN on German dataset

Method	Pr%	Re%	Sp%	F1%	PCC%
SVDD	77.0	84.0	41.4	80.3	71.2
SVM	80.2	87.5	49.5	83.6	76.1
BPANN	77.4	88.7	39.4	82.7	73.9

From the two tables above, we can conclude that SVM and BPANN outperform SVDD model on the whole dataset. This is not surprising, because SVM and BPANN are constructed to give the best separation between the given two classes. The dataset we use in the experiment is not fit for the Eigencredits and SVDD model because bad credit samples take nearly 50% of the total samples. This is more suitable for binary-class classifiers.

4.3 Performance of Eigencredits and SVDD Model with SVM and ANN on the Datasests Which Contain Few Bad Credit Samples

In this section, we reduce the bad credit samples in such a way that the bad credit samples take about 10% of the total number of samples. Then we perform the same process as mentioned in section B and the results are shown in Table 3 and Table 4.

Table 3. Results of SVDD, SVM and BPANN on reduced Australian dataset

Method	Pr%	Re%	Sp%	F1%	PCC%
SVDD	94.7	93.4	60.0	94.0	89.6
SVM	89.4	100	10.0	94.1	89.5
BPANN	93.7	97.4	50.0	95.5	90.7

Table 4. Results of SVDD, SVM and BPANN on reduced German dataset

Method	Pr%	Re%	Sp%	F1%	PCC%
SVDD	93.1	85.1	42.1	88.7	80.9
SVM	90.2	100	00.0	94.9	90.2
BPANN	90.6	93.7	10.5	92.1	85.6

From table 3 and table 4 we can find that when the bad credit samples are much less than the good credit samples, Eigencredits and SVDD model has lower Recall. But, Eigencredits and SVDD model has much higher Specificity which means Eigencredits and SVDD model can well detect the small amount of bad credit samples. This is very crucial for financial institutes as bad credit applicants will probably bring great losses. In many practical circumstances, bad credit samples are much less encountered than good credit samples. According to external press released from several major banks in America, bad credit samples take less than 10% of the total samples. This is suitable for Eigencredit and SVDD model.

5 Discussion and Conclusions

We further analyze the Eigencredits and SVDD model in detail, and at least three advantages for this method can be found. Firstly, positive examples (creditworthy data) are available easily for any financial institution, but negative examples are generally complex. It is not realistic for a financial institution to encounter all kinds of negative instances before the construction of classifiers. Therefore, binary-classifiers can not adapt to the negative instances which are not described by the negative training samples. Thus, SVDD should be a better choice to this situation. Secondly, when using SVDD model, banks can control the risk of credit card by adjusting the radius of the hypersphere of SVDD model. This is very easy to obtain different risks model. Finally, SVDD model can easily implement the incremental learning. When new data arrived, the combination of support vector and new data is used for retraining, which is reasonable since the model by support vector is identical to the model obtained by all data.

The Eigencredits and SVDD model is more suitable for the case that the good credit samples have similar characteristics. For a particular group of people, for example, the students or college teacher (people who have the same job), the good credit samples among them have more similarity. So Eigencredits and SVDD model is more appropriate in this case. The banks can set different models for different groups of people to get better prediction results.

In this paper we propose to use Eigencredits and SVDD model to solve the credit scoring problem. As the experiment results indicated, Eigencredits and SVDD model can well detect the bad credit samples when bad credit samples are much less than good credit samples. Meanwhile, Eigencredits and SVDD model has three additional advantages. Future research will aim at discussing the performance of Eigencredits and SVDD model in certain group of samples.

References

1. Myers, J.-H., Forgy, E.-W.: The development of numerical credit evaluation systems. Journal of the American Statistical Association 58(303), 799–806 (1963)
2. Wiginton, J.-C.: A note on the comparison of logit and discriminantmodels of consumer credit behavior. J. Financ. Quant. Anal. 15(3), 757–770 (1980)
3. Zhou, X.-Y., Zhang, D.-F., Jiang, Y.: A new credit scoring method based on rough sets and decision tree. In: Washio, T., Suzuki, E., Ting, K.M., Inokuchi, A. (eds.) PAKDD 2008. LNCS (LNAI), vol. 5012, pp. 1081–1089. Springer, Heidelberg (2008)
4. West, D., Dellana, S., Qian, J.-X.: Neural network ensemble strategies for financial decision applications. Computers and Operations Research 32(10), 2543–2559 (2005)
5. Henley, W.-E., Hand, D.-J.: A k-nearest-neighbor classifier for assessing consumer credit risk. The Statistician 45(1), 77–95 (1996)
6. Zhang, D.-F., Huang, H.-Y., Chen, Q.-S., Jiang, Y.: A comparison study of credit scoring models. Natural Computation 1, 15–18, 24–27 (2007)

7. Chuang, C.-L., Lin, R.-H.: Constructing a reassigning credit scoring model. Part 1. Expert Systems with Applications 36(2), 1685–1694 (2009)
8. Huang, C.-L., Chen, M.-C., Wang, C.-J.: Credit scoring with a data mining approach based on support vector machines. Expert Systems with Applications 33(4), 847–856 (2007)
9. Lin, S.-L.: A new two-stage hybrid approach of credit risk in banking industry. Expert Systems with Applications 36(4), 8333–8341 (2009)
10. Turk, M., Pentland, A.: Eigenfaces for Recognition. Journal of Cognitive Neuroscience 3(1), 71–86 (1991)
11. Tax, D., Duin, R.: Support vector domain description. Pattern Recognition Letters 20, 11-13, 1191–1199 (1999)
12. Vapnik, V.: Statistical Learning Theory. Wiley, Chichester (1998)
13. Blake, C., Keogh, E., Merz, C.-J.: UCI Repository of Machine Learning Databases., `http://www.ics.uci.edu/~mlearn/MLRepository.html`
14. Duin, R.-P.-W., et al.: PRTools4.1, A Matlab Toolbox for Pattern Recognition (2007), `http://www.prtools.org/`
15. Tax, D.-M.-J.: dd_tools_1.5.0 (2006), `http://homepage.tudelft.nl/n9d04/dd_tools.html`

A New Morning Exercise Weather Forecast Model

Jingming Xia*, Ping Wang, and Miao Jiang

Department of Information Engineering
Nanjing University of Information Science and Technology
Nanjing, Jiangsu Province, China
jingming468@sohu.com

Abstract. According to the defects of current forecast model, the essay puts forward a new real-time broadcast system based on wireless sensor network. The system consists of wireless sensor network node, SINK node, terminal analysis and evaluation system, community display platform. Sensor network nodes collect the real-time environmental and meteorological data from morning exercise sites, and then through RF and GPRS communication model, it transmits data to terminal analysis and evaluation system, and broadcasts to nearby communities. The system realizes the fine forecast of morning exercise index, and offers a new idea for large cities' functional meteorology fine forecast model.

Keywords: wireless sensor network, morning exercise index, GPRS.

1 Introduction

With the development of modern technology and the improvement of the living standards, people are increasingly concerned about the impact of living and surrounding environment on the life of human beings while they pursue the material life quality, which puts forward higher requirement for the weather service. In view of this situation, the urban environment and weather forecast technology as a new edge science rapidly develops throughout the country. Morning exercise weather index forecast technology is one kind of urban environment weather forecast technology. The current morning exercise weather index forecast is that meteorological department comprehensively analyses the sky conditions, wind, ground conditions, air cleanliness, temperature changes and other weather conditions in the next day morning. Then it gives the corresponding values, and reports the weather conditions through television, Internet, radio or other public media.

This prediction model exits the following problems:

1) Morning exercise time is mainly focused within three hours in the next morning, the current weather forecast model is 24-hours, this model of weather forecast for the next morning is always not accurate and precise on complex weather conditions.

* This work is partially supported by Jiangsu Product Research Fund Grant # BY2009105 to Jingming Xia.

D. Zeng (Ed.): ICAIC 2011, Part II, CCIS 225, pp. 41–49, 2011.

2) Air quality is the important indicator to determine the suitability of doing morning exercises, but now it is lack of real-time air quality monitoring devices in the city's main morning exercise sites.

3) It is not accurate to characterize the morning exercise suitability of a city with a single index, because the city scale of large cities and central cities is too large, the air quality and terrain vegetation have big differences in different morning exercise sites. For example, it is not suitable to do morning exercises in the woods in cloudy day due to overmuch carbon dioxide, but less affected in the public square.

As a new peer-to-peer morning exercise index forecast model, the system uses wireless sensor network to collect meteorological and air quality data in morning exercise sites, then transfers the real-time assessment information to community display platform through GPRS network.

2 The Structure of Morning Exercise Index Forcast System

This system consists of wireless sensor network node, SINK node, transmission network and terminal evaluation display platform. The overall structure of the system is shown in Fig. 1.

Considering the local meteorological factors of city are distributed averagely, the meteorological elements of morning exercise site can directly collect the data of automatic weather stations nearby the morning exercise site to save the cost. In each morning exercise site of city (public squares, urban forest parks, the surrounding mountains) to deploy a dense wireless sensor network node, which is responsible for collecting air quality data of morning exercise site, then it transmits the data to the SINK node from sensor nodes by way of multi-hop type, and SINK node as the convergence node within the network should have some integration of data storage and

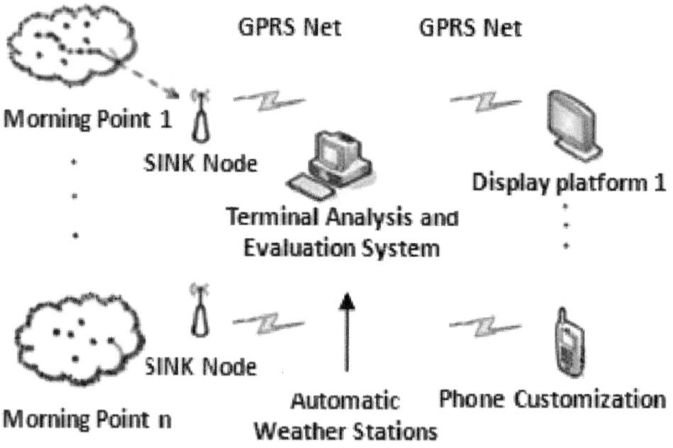

Fig. 1. Structure of morning exercise index real-time broadcast system

data computing, in addition SINK node also connect sensor network with end-user equipment. In this system the data of convergence node will be transmitted to the analysis terminal and evaluation platform by the GPRS wireless transmission technique. This platform will analyse the uploaded data and then transmit all the index information of morning exercise sites around the community within three kilometers to the community, then it uses the LCD liquid crystal to broadcast, the community that can't use broadcast system can customize information through mobile phone.

The system designs the SINK node as the center and arranges the network nodes as tree structure around the morning exercise site, and transport protocol uses ZigBee network transport protocol in order to enhance the robustness and reliability of the network and avoid the network suffering man-made damage.

3 The Design of Hardware Node and Module

3.1 The Design of Sensor Node Module

Sensor node is the basic unit of wireless sensor networks. Sensor node module consists of the main atmospheric pollutants sensors, data acquisition and processing module, wireless transceiver module and power module. The hardware structure of the sensor node is shown in Fig. 2.

DSM501 dust sensor and electrochemical gas sensor of British city technology are selected in this system, the use of conductivity sensors to detect gas concentrations, it has high accuracy and low price, and it can be used continuously for more than three years in the air, its products measure the main pollutants O_3, SO_2, CO_2, NO_x and other polluting gases concentration, data is converted by the A/D converter to digital signal then it is sent to processing unit.

The battery-powered is used to supply wireless sensor network energy, data-processing unit with ultra-low power 16-bit Texas Instruments microcontroller MSP430F149, which has the 8 12-bit ADCs, 60K FLASH and 2KRAM, and it can work normally in -40 ~85 degrees outside with low power consumption. The microcontroller also has a working model and five sleeping models, it can turn off unused modules of MSP430F149 through the application and switch work model and sleep model according to the request. The system only needs to be opened between 4:00 to 8:00 am and the rest is dormant period. The sensor signal is converted to digital signal by the operational amplifier exit through A/D converter. MSP430F149 builds in temperature sensor to detect the ambient temperature and can correct the number of data of air according to the need.

Wireless transceiver module realizes close quarters communication by Chipcon's CC2430 chip which contains 2.4GHZ high-performance RF transceivers and has solidified a Zigbee network stack to simplify system development, plusing only a small amount of peripheral circuits can constitute transceiver module. The host controller

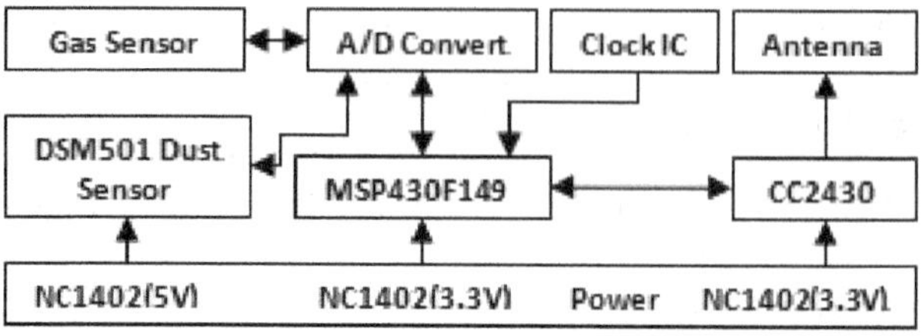

Fig. 2. Structure chart of sensor node function

MSP430F149 controls CC2430 working station and achieve reading and writing cache data through the SPI interface(SI, SO, SCLK, Csn), FIFOP pin and FIFO pin set sending and receiving FIFO's buffer, SFD pin control clock and timer's input, CCA pin control Clear Channel estimates. CC2430 and MSP430F149 interface is shown in Fig. 3.

The system uses two AA batteries power for dual supply, and uses step-up DC/DC chip NCP1402 to output voltage 5V and 3.3V respectively. 5V supplies for the data acquisition unit and 3.3V is used for the CC2430 and the MSP430F149. Because the data acquisition unit and the wireless transceiver unit consume large power, so the system uses analog switch ISL43111 to control sensors' power supply. When there is no task acquisition, close sensors' power and set the transceiver to sleep.

3.2 The Design of SINK Node Module

SINK node is a bridge between sensor network and the terminal processor, which needs certain data processing capabilities and completes communication with sensor network and terminal. SINK node primarily consists of embedded microprocessor module, wireless communication module and external expansion memory, shown in Fig. 4.

SINK node is responsible for processing and transmitting data of wireless sensor network and GPRS network and also responsible for building the network, node addition and deletion and other tasks, which needs to have strong processing capacity. The processor chip uses Samsung's S3C2410 processor. S3C2410 is based on ARM9 core, which supports the Windows CE operating system and can be configured on the PPP dial-up. With the three-channel UART S3C2410 is very easy to communicate with GPRS and wireless sensor network.

Wireless communication module contains two parts. The communication between SINK nodes and sensor network still uses CC2430, and the system consisted of SINK nodes and evaluation and dissemination of terminal uses GPRS wireless communication model, using Siemens' MC55 tri-band GPRS communication module. MC55 embeds TCP/IP protocol, which can be easily connected with the Internet to realize the communication between SINK nodes and terminal.

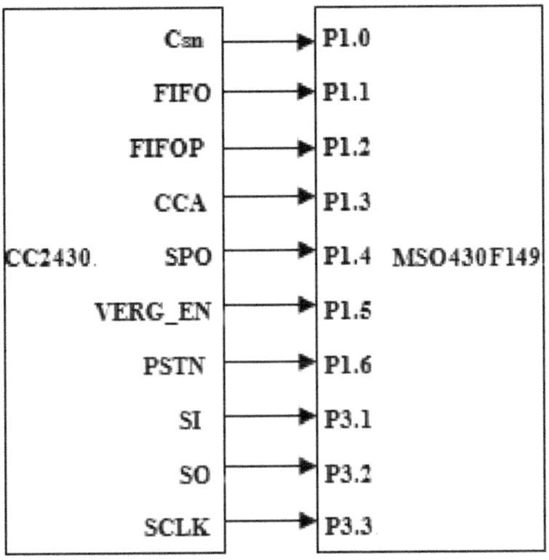

Fig. 3. CC2430 connected with MSP430F149

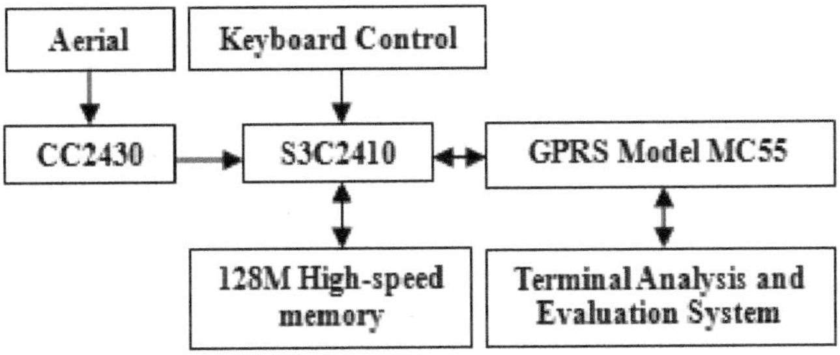

Fig. 4. SINK node structure and function

The module includes SMS, voice communication, wireless Internet access and data transfer function, providing the serial port, which can be directly connected with S3C2410, and the communication function can be operated by calling the AT command. Windows CE system embeds TCP/IP stack. GPRS initialization just uses AT command to establish the channel, set the baud rate, and can make the upper layer protocol open GPRS link which can use PPP protocol for data link layer's connectivity and data transmission. After the establishment of GPRS data link layer, the socket communication can be completed. Because the system requires relatively high real-time, it chooses UDP protocol on the top of IP layer.

CC2430 uses 2.4G frequency band, and the maximum transfer rate is 250Kbit/s, while the MC55 maximum transfer rate is only 171.2Kbit/s. In real-time data transmission, the system should establish memory buffer which is in the charge of the system. The system adds 128Mexternal data memory as the system expanded memory in the SINK node.

3.3 Terminal and Community Display Module

Terminal mainly includes high-performance computer, GPRS communication module, data receiving management analysis and evaluation system, router, database. Terminal computer receives the IP packet air quality data from SINK node every half an hour through the GPRS module.

According to the existing condition of communities, community display module can be separately designed. This system provides the design of display platform based on S3C2410 and VGA LCD TV for the community which has LCD TV bulletin boards. S3C2410 LCD has its own scanning interface, using high-performance video D/A to converter ADV7120 chip which can be converted to VGA interfaces. ADV7102 includes 3 independent single-chip 8-bit high-speed D/A converters to process RGB data, in conversion circuit 3 groups correspond 8-bit data input to accept RGB video data, using standard TTL level, video control signal line includes composite video sync signal SYNC, blanking signal BLANK, white-level reference signal REFWHITE and pixel clock signal CLOCK, external 1.23V D/A reference voltage and output satisfactory adjustment, in the output the analog RGB signals with high impedance adopts output current source model. Synchronous reference current output signal Isync encoded in the green video signal synchronization information, the connection diagram shown in Fig. 5. Setting the display to work in 16-bit color model displays the 60HZ 640X480 resolution image, display screen information, timing control and data formats controlled by the register from LCDCON1 to LCDCON5. Residents can also customize the morning exercise SMS by their needs, the terminal computer will send the SMS to users through GPRS module and charge monthly fee.

4 The Design of System Software

4.1 Software Design of Sensor Node

Sensor node is mainly responsible for collecting and sending data, transmitting other nodes' data and communicating with the SINK node. In the software design, besides considering the function realization, energy conservation should be also considered. Sensors collect environmental data every 30 minutes, close the sensor power and initialize the CC2430 after finishing collecting data. At first, the data of the node should be sent and then test whether there are the transmitted data, after finishing the work the node goes to sleep to wait for the next round of collection. The modularization programming includes data collection subroutine, wireless transceiver subroutine and clock control subroutine. Flow chart shows in Fig. 6.

4.2 Software Design of SINK Node

The SINK node of each morning exercise site receives data through CC2430 and sends data to the terminal through GPRS. Data transmission via GPRS should go through the Internet, thus transferring data package must conform to TCP/IP protocol, the transmitted data are in turn progressing with UDP protocol header encapsulation, IP protocol and PPP protocol encapsulation header package, packaged MC55 data format is shown in Table 1. Software adopts the bottom-up thought to progressively achieve every function functions, all functions use methods of modular design and are ultimately managed by Windows CE operating system. PPP software package works on the Windows CE operating system, which controls the MC55's connection to network. The program structure diagram is shown in Fig. 7.

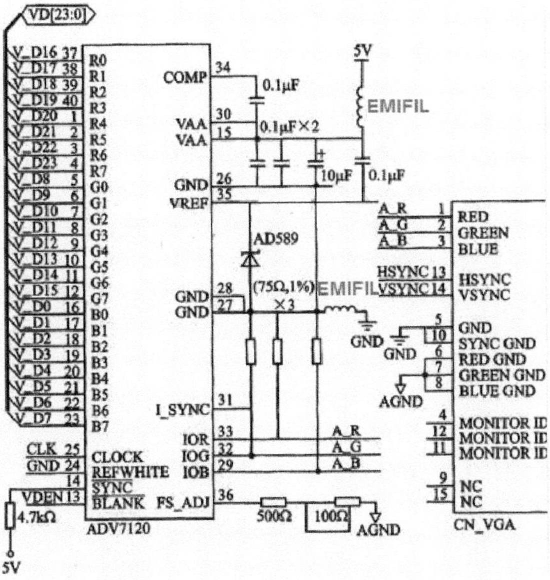

Fig. 5. ADV7102 driver VGA interface circuit diagram

Table 1.

PPP header	IP header	UDP header	Data	Check code

4.3 Software Design of Terminal and Display System

The core program of terminal is actually the management information system. Database uses SQL, and the users' interface uses VC++ to program. The database is responsible for storing the data received from SINK node and calculating the air cleanliness index (ww) according to the concentration of various pollutants, while via the Internet or GPRS, the database automatically receives the real-time weather data from local weather stations. The received data include sky conditions (tk), ground conditions (dm), average wind (v), the condition of temperature changes (bt), data can be classified by terminal analysis and evaluation system, in accordance with formula (1) the system calculates the morning exercise index.

$$Ycl=0.7tk+1.6dm+0.6v+1.0ww+0.8bt+1.1 \quad (1)$$

Every half hour, the system sends the latest morning exercise index of the main city's morning exercise sites and all morning exercise sites within three kilometers around to the corresponding community through the external GPRS network of computer. Community display system is based on EVC and Windows CE to develop, releasing the index of morning exercise sites through VGA LCD TV. According to the index community residents can decide whether to do morning exercises and where to do morning exercises in time.

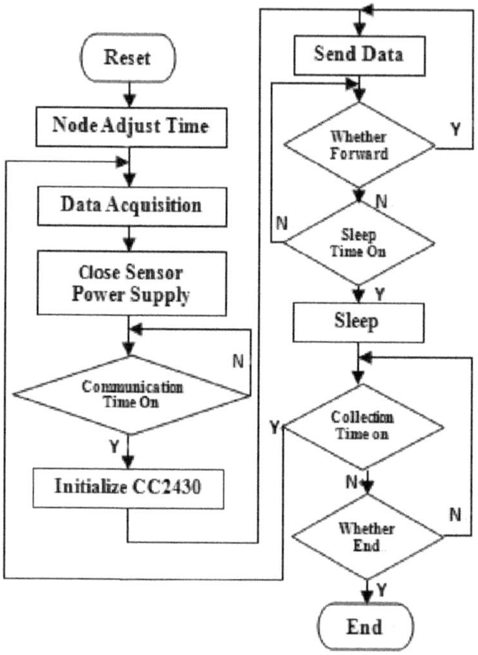

Fig. 6. Flow chart of sensor node applications

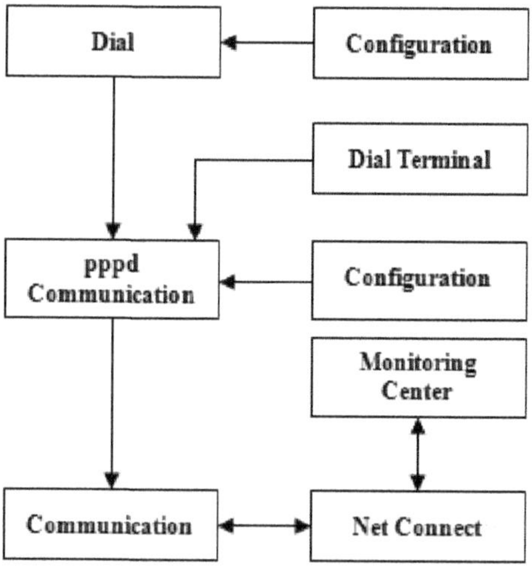

Fig. 7. Structure diagram of GPRS dial-up program

5 Conclusions

With the expansion of the city, the exact real-time broadcast of functional meteorology is a new model of weather broadcast and forecast in the future. In this article, the design pattern of morning exercise index real-time broadcast system based on wireless sensor networks has broken through the traditional weather forecast, which achieves the goal of real-time broadcast morning exercise index. The system has been tested in part of morning exercise sites and communities in Nanjing. This system has the feature of working stability, simple maintenance, lower energy consumption, higher precision measurement of gas concentration, wide usable temperature ranges, lower BER of transmission, immediate broadcast and so on. Changing the sensors and part of software and hardware, the system also can be used in accurate weather report like the urban atmospheric local pollution, the traffic conditions and so on.

References

1. Li, Y., Li, L., Wang, C.: The research of wireless sensor networks multi-path routing algorithm. Wuhan University of Technology (Transportation Science & Engineering) 1, 369–371 (2009)
2. Moninger, W.R., Mamrosh, R.D., Pauley, P.M.: Automated meteorological reports from commercial aircraft. Bulletin of AMS 84(2), 203–216 (2003)
3. Zhong, Y., Tang, B., Tan, C., et al.: The research of wireless smart sensor arrays used in atmospheric environmental monitoring. Communication 6, 137–140 (2008)
4. Dun, F., Wang, L., Yin, H.: Air pollution gases integrated monitoring system based on embedded system and GPRS transmission. Measurement and Control Technology 25, 66–69 (2006)
5. Bu, T., Lv, L., Wang, W.: Agricultural environment monitoring system based on TinyOS wireless sensor network. Agricultural Network Information 2, 23–26 (2009)
6. Liu, X., Ha, L.: Morning exercise index forecast. Chifeng College 4, 60–62 (2008)

Research on Yangtze River Dangerous Articles Logistics Port Hierarchical Layout Evaluation of Jiangsu Province Based on Disaster Theory

Zhang Peilin, Mao Jian, and Yuan Ziwen

School of Transportation
WUHAN University of Technology
WUHAN P.R. China
peilin.zhang@yahoo.cn

Abstract. The distinct along the Yangtze River of Jiangsu Province is China's key development areas of petrochemical industry. Reasonable hierarchical layout of dangerous articles ports plays an important role in use coastal resources effectively and meeting hinterland chemical industry. Breaking traditional port layout theory, dangerous articles logistics port hierarchical layout evaluation model has been made based on disaster economic theory and gray clustering method. Followed with case study of 8 ports in Jiangsu province, evaluation results and development suggestions have also been given.

Keywords: disaster economic, dangerous article logistics, layout evaluation.

1 Introduction

In recent years, related researches on hierarchical layout of container port groups, oil and ore terminals location are increasing. Feng Qingli(2000) made a regional port-level model and algorithm based on fuzzy clustering analysis with objective function, which got a conclusion that Shangdong province should take Qindao port as main port, Rizhao and Yantai as subsidiary ports in construction plan of coastal ports. Jiang Haihong(2003) developed a evaluation model of large oil terminal construction site with AHP and fuzzy comprehensive assessment. Besides, an example related to the oil terminal site ranking of east and south China ports has been made. Wang Chong(2005) has made a location and task division analysis to ports on circum-Bohai region with principal components-cluster, resulting in a division of three levels of the ports. At present, dangerous articles transport of Yangtze River in Jiangsu has the following main issues: first, the layout of oil, liquid chemicals terminal is scattered; ports utilization differs severely; structural contradictions are obvious; unreasonable use of port coastline resources exists. Second, since dangerous articles' characters, a number

D. Zeng (Ed.): ICAIC 2011, Part II, CCIS 225, pp. 50–58, 2011.

of flammable, explosive and virulent articles may cause huge losses if transport accident or disaster happened; reasonable ports layout based on disaster reduction and prevention has not formed yet, safety supervision and emergency measures are also in deficiency. Consequently, it is necessary to optimize chemical transport system ports layout along the Yangtze River in Jiangsu, in order to form a reasonable transportation system, promote the port restructuring and integration of resources, and enhance the safety and environmental protection capacity. This essay breaks traditional transport economic theory and applies disaster economic theory points. Besides, it aims on port economic and security optimization taken with actual situation of Jiangsu ports dangerous articles logistics. Finally, dangerous articles logistics ports layout along the Yangtze River in Jiangsu is evaluated with AHP and grey clustering evaluation model.

2 Grey Clustering Method Hierarchical Layout Evaluation Model of Dangerous Articles Port

2.1 Establishment of Evaluation Indicator System

For regional port hierarchical layout, influence factors to ports hierarchical layout are intricate, complicated, various, and interrelated constraints, all these factors determine regional port hierarchical layout together. Meanwhile, dangerous articles transportation has its specialty: bulk liquid dangerous articles are flammable, corrosive, toxic. What's more, they have large quantities, varieties and characters. Once any problem happens in transport, it can lead to serious results such as ships sinking, people dying and pollution to terrestrial, atmospheric and water caused by bulk noxious liquid substances.

Based on the evaluation criteria and indicators choosing principles such as accuracy, comparability, direction, independence, viability and hierarchy, and the above influence factors analysis of dangerous articles logistics port hierarchical layout, considering relevant literature and consulting on expert s who have engaged in port research for many years, hierarchical layout evaluation index system of dangerous articles logistics port has been divided into two levels, which is showed in table1.

2.2 Hierarchical Layout Evaluation Model

Grey clustering is a multi-index evaluation method in gray system theory, which can follow different indicators to make comprehensive evaluation so as to determine evaluation object category. In this paper, whitenization weight functions gray clustering has been applied. Based on whitenization weight functions, whitening value of clustering objects have been treated overall by this method. Besides, whitening values are concluded by a number of gray types to determine which category the cluster objects belong to. Grey whitening weight function clustering steps are as follows:

Table 1. Hierarchical Layout Evaluation Index System

First class indicators		**Second class indicators**	
Natural conditions	A1	Geographical position	A11
		Shoreline condition	A12
Port technical conditions	A2	Berth number	A21
		Throughput	A22
		Dangerous articles storage ability	A23
		Collection and transmission ability	A24
Environment safety	A3	Distance among dangerous source	A31
		Disaster Prevention and Safety Management System	A32
		Climate and geology factor	A33
Demand factor	A4	Existing industrial scale	A41
		Trade environment	A42
Hinterland condition	A5	Hinterland GDP	A51
		Hinterland city population	A52

a) Construct the coefficient matrix D:

In coefficient matrix D, d_{ij} is a whitening value of No. i clustering object to No. j clustering indicator (i =1,2,3…, s ; j =1,2,3…, m), D is a matrix comprised of d, which is also called coefficient matrix:

$$D = \begin{pmatrix} d_{11} & d_{12} & \dots & d_{1m} \\ d_{21} & d_{22} & \dots & d_{2m} \\ \dots & \dots & \dots & \dots \\ d_{s1} & d_{s2} & \dots & d_{sm} \end{pmatrix}$$

b) Determine the type and range of gray category:

According to the demands of dangerous articles logistics ports hierarchical assessment, gray category will be divided into three types which are pivotal ports, major ports, and general port. Each value range of gray category is:

$$A \in \left[a_i^1, a_i^2\right]$$
$$B \in \left[a_i^2, a_i^3\right]$$
$$C \in \left[a_i^3, a_i^4\right]$$

The value of $a_i^l\left(l = 1,2,\ldots\ldots4\right)$ is determined according to the practical problems demand of each specific indicators or qualitative research results.

c) Determine gray whitening weight function of each index

General whitening weight function is divided into three basic types that are upper limit measure, middle measure, and lower limit measure three types of measure. Images and specific expression of whitening weight function are as follows:

$f^1 : [a_1, b_1, \infty]$ is upper limit measure whitening weight function, its expression is:

$$f^1(x) = \begin{cases} 0, & x\langle a \\ \dfrac{x - a_1}{b_1 - a_1}, & x \in \left[a_1, b_1\right] \\ 1, & a \geq b_1 \end{cases} \tag{1}$$

$f^2 : [a_1, b_1, c_2]$ is middle measure whitening weight function, its expression is :

$$f^2(x) = \begin{cases} 0, & x \notin \left[a_1, c_2\right] \\ \dfrac{x - a_1}{b_1 - a_1}, & x \in \left[a_1, b_1\right] \\ \dfrac{c_2 - x}{c_2 - b_1}, & x \in \left[b_1, c_2\right] \end{cases} \tag{2}$$

$f^3 : [0, b_2, c_2]$ is middle measure whitening weight function, its expression is :

$$f^3(x) = \begin{cases} 0, & x \notin [0, c_2] \\ 1, & x \in [0, b_2] \\ \dfrac{c_2 - x}{c_2 - b_2}, & x \in [b_1, c_2] \end{cases} \tag{3}$$

d) Calculate the clustering weight matrix ϖ_{jk}

$\omega_{jk} = \dfrac{\lambda_{jk}}{\sum_{j=1}^{m} \lambda_{jk}}$, ω_{jk} is clustering weight of number k grey category to j indicator. λ_{jk} is marginal value of number k grey category to j indicator.

e) Calculate gray clustering coefficient σ_{ik}**, and determine gray category fo each cluster object**

$\sigma_{ik} = \sum_{j=1}^{m} f_{jk}(d_{ij})\omega_{jk}$, σ_{ik} is clustering coefficient of number k grey category to i indicator.

f) Cluster Analysis

If σ_{ik}^* meets $\sigma_{ik}^* = \max_k \sigma_{ik} = \max\{\sigma_{i1}, \sigma_{i2}, \ldots, \sigma_{in}\}$, clustering object i belongs to grey class k^*. Then judge which level the number i evaluation unit belongs to. Choose the maximum element σ_j^k in $\sigma(i)$, that is $\sigma_j^k = \{\max \sigma_i^k\}$. According to proximity principle, comment level corresponded to σ_j^k be the final results of the terminal.

3 Example Application on Ports ALONG Yangtze River in Jiangsu

Located in the north of Shanghai International Shipping Centre, ports along Yangtze River in Jiangsu are spread over both sides of Yangtze River from Nanjing port to the mouth of Yangtze Rive about 360 kilometers. These ports have one hundred and more opening ten-thousand-ton level berths. According to statistics, at present ports in Jiangsu take in 90% foreign trade ships which pass in and out Yangtze River. Furthermore, dangerous articles transportation is increasing continuously. This paper will take 8 dangerous articles ports as example.

3.1 Calculate Coefficient Matrix D

Based on survey and statistics analysis, optimized coefficient matrix is as follows:

$$D=\begin{vmatrix} 0 & 100 & 100 & 65 & 100 & 86 & 83 & 100 & 62 & 100 & 70 & 51 & 78 \\ 67 & 63 & 23 & 49 & 39 & 43 & 33 & 54 & 51 & 24 & 60 & 13 & 0 \\ 100 & 91 & 58 & 100 & 26 & 100 & 100 & 92 & 34 & 11 & 100 & 100 & 79 \\ 22 & 44 & 28 & 12 & 9 & 29 & 0 & 69 & 26 & 20 & 10 & 11 & 23 \\ 56 & 12 & 4 & 9 & 17 & 14 & 83 & 62 & 57 & 13 & 20 & 24 & 37 \\ 89 & 54 & 25 & 29 & 47 & 57 & 67 & 77 & 74 & 6 & 80 & 30 & 100 \\ 44 & 19 & 8 & 7 & 10 & 57 & 33 & 38 & 100 & 9 & 0 & 11 & 59 \\ 22 & 0 & 8 & 0 & 0 & 0 & 100 & 54 & 83 & 22 & 40 & 14 & 53 \end{vmatrix}$$

3.2 Determine Grey Whitening Weight Function

Formations (1) to (3) have given the detail steps to whitening weight function. Results are listed below through calculation:

Whitening weight function of geographical position f_{1k} is:

$f^1(0, 30, \infty)$, $f^2(0, 30, 100)$, $f^3(0, 75, 100)$

Whitening weight function of shoreline condition f_{2k} is:

$f^1(0, 20, \infty)$, $f^2(0, 20, 100)$, $f^3(0, 60, 100)$

Whitening weight function of berth number f_{3k} is:

$f^1(0, 10, \infty)$, $f^2(0, 10, 100)$, $f^3(0, 50, 100)$

Whitening weight function of throughput f_{4k} is:

$f^1(0, 15, \infty)$, $f^2(0, 15, 100)$, $f^3(0, 45, 100)$

Whitening weight function of dangerous articles storage ability f_{5k} is:

$f^1(0, 15, \infty)$, $f^2(0, 15, 100)$, $f^3(0, 40, 100)$

Whitening weight function of collection and transmission ability storage ability f_{6k} is:

$f^1(0, 30, \infty)$, $f^2(0, 30, 100)$, $f^3(0, 50, 100)$

Whitening weight function of distance among dangerous source f_{7k} is:

$f^1(0, 50, \infty)$, $f^2(0, 50, 100)$, $f^3(0, 90, 100)$

Whitening weight function of disaster prevention and safety management system f_{8k} is:

$f^1(0, 20, \infty)$, $f^2(0, 20, 100)$, $f^3(0, 50, 100)$

Whitening weight function of climate and geology factor f_{9k} is:

$f^1(0, 10, \infty)$, $f^2(0, 10, 100)$, $f^3(0, 30, 100)$

Whitening weight function of existing industrial scale f_{10k} is:

$f^1(0, 10, \infty)$, $f^2(0, 10, 100)$, $f^3(0, 20, 100)$

Whitening weight function of trade environment f_{11k} is:

$f^1(0, 20, \infty)$, $f^2(0, 20, 100)$, $f^3(0, 60, 100)$

Whitening weight function of hinterland GDP f_{12k} is:

f^1 (0, 15, ∞), f^2(0, 15, 100), f^3 (0, 50, 100)
Whitening weight function of hinterland city population f_{13k} is:
f^1 (0, 25, ∞), f^2(0, 25, 100), f^3 (0, 70, 100).

3.3 Calculate Clustering Weight ϖ_{jk}

Based on $\omega_{jk} = \frac{\lambda_{jk}}{\sum_{j=1}^{m} \lambda_{jk}}$, the calculating result of clustering weight ϖ_{jk} is showed in the following table 2:

Table 2. Clustering weight Calculating

k / j	1	2	3
1	0.111	0.109	0.067
2	0.074	0.087	0.058
3	0.037	0.072	0.078
4	0.056	0.065	0.087
5	0.056	0.058	0.063
6	0.111	0.072	0.070
7	0.185	0.130	0.112
8	0.074	0.072	0.074
9	0.037	0.043	0.054
10	0.037	0.029	0.032
11	0.074	0.087	0.088
12	0.056	0.072	0.065
13	0.093	0.101	0.045

3.4 Judgment of Grey Types

According to the model, dangerous articles ports along Yangtze River in Jiangsu hierarchical judgment have been divided into three grey types: pivotal ports, major ports, and general ports. Take Nanjing port as example, the calculation process of grey type's judgment is as follows:

First, calculate the clustering coefficient σ_{1k}(k=1, 2, 3)

σ_{11}=f_{11}(50)×0.111+ f_{21}(4. 7) ×0.074+ f_{31}(55)×0.037+ f_{41}(585) ×0.056+ f_{51}(230)×0.056+ f_{61}(85)×0.111+ f_{71}(65)×0.185 +f_{81}(95)×0.074+ f_{91}(878)×0.037+ f_{101}(584)×0.037+ f_{111}(80)×0.074+ f_{121}(3775)×0.056+ f_{131}(624.46)×0.093 =0.705

Referring to the same calculate process,

$$\sigma_{12} = 0.717$$

$$\sigma_{13} = 0.712$$

Based on preferred principle, $\max\sigma_{1k} = 0.717$, consequently, the hierarchical attribute of Nanjing port belongs to number 2 level, namely major port. The last calculate result is showed in table 3.

Table 3. Calculate result

Ports name	Pivotal ports	Major ports	General ports	Result
Nanjing	0.705	0.717	0.712	Major port
Jiangyin	0.681	0.667	0.666	Pivotal port
Suzhou	0.724	0.713	0.687	Pivotal port
Zhenjiang	0.622	0.626	0.635	General port
Changzhou	0.645	0.651	0.660	General port
Nantong	0.613	0.703	0.654	Major port
Taizhou	0.603	0.676	0.692	General port
Yangzhou	0.685	0.643	0.717	General port

4 Conclusion

This essay breaks traditional transport economic theory and applies disaster economic theory points. With AHP and grey clustering evaluation model, dangerous articles logistics ports layout along the Yangtze River in Jiangsu is evaluated. Finally, the analysis result of layout plan is as follows: Suzhou port and Jiangyin port are pivotal ports; Nanjing port and Nantong port are major ports; Taizhou port, Changzhou port, Yangzhou port and Zhenjiang port are general ports. The research result of this essay will play an instructive role in Jiangsu ports layout and strategy plan.

References

1. Qing, F.: A Study on Regional Port Hierarehieal Layout Baseon Fuzzy Clustering Algorithm. Shipping and Port Management 28 (November 2006)
2. Aversa, R., Botter, R.C., Haralambides, H.E.: A Mixed Integer Programming Model on the Location of a Hub Port in the East Coast of South Ameriea. Maritime Eeonomies & Logisties 7 (2005)
3. Yang, H.-l.: Researches of hierarchy layout on the coastal container terminals in China. Journal of Dalian Maritime University 27(2) (May 2001)

4. TcLirn, Thanopoulou, H., Beynon, M., Beresford, A.: An APPlieation of AHP on TranshiPment Port Seleetion: A Global PersPeetive. In: 2004 Maritime Eeonomies & Logisties, Cardiff University, UK, pp. 70–79 (2004)
5. Orlovsly, S.A.: Decision-Making with a Fuzzy Perference Relation. Fussy Sets and Systems, 155–167 (1978)
6. Song, D.-W., Yeo, K.-T.: A Competitive Analysis of Chinaese Container Ports Using the AHP. In: Maritime Eeonomies & Logisties, University of Hong Kong, Hong Kong, pp. 34–52 (2004)

Decision Rule Acquisition Based on Variable Rough Set Model in IFDISs*

Da-kuan Wei[1], Lu-jin Tang[1], and Bing Huang[2]

[1] Department of {Computer and Communication Engineering & Mathematics and Computing Science}
Hunan University of Science and Engineering
Yongzhou Hunan 425100, China
[2] School of Information Science Nanjing Audit University Nanjing 211815, China
Dakuan_Wei@163.com

Abstract. In this paper, a sort of complicated information system-the incomplete and fuzzy decision information system (IFDIS) will be studied; the focal point researching on IFDIS falls into decision rule acquisition which has important reality significance and application value. On the base of the gamma-tolerance relation and the philosophy of variable precision rough set model in incomplete information system, the gamma-tolerance relation-based variable precision rough set model in incomplete information system is presented. Subsequently, the variable rough set model in IFDIS is proposed and its basic properties are discussed. The method of precision reduct in IFDIS based on the variable rough set model is provided. The notion and approach of decision rule acquisition focused on optimal and pessimal decision rules are presented. Finally the algorithm for rule acquisition will be given and illustrated it valid by an example.

Keywords: Rough set, IFDIS, gamma-tolerance relation, decision rule acquisition, Variable rough set model.

1 Introduction

Rough set theory, an excellent tool to deal with imprecision, vagueness and uncertainty, proposed by Poland mathematician Professor Z. Pawlak in 1982 [1], has been extensively applied in many fields, such as knowledge discovery [2, 3], intelligent decision [4, 5], machine learning [6, 7], data mining [8, 9], approximation reasoning [10, 11], etc.. In fact, most application based rough set theory may centers on information system, and according to whether or not there are null values, information system can be divided into two categories –complete and incomplete. The classical rough set theory may only process complete information system because of weak power of data acquisition, but there are rather more incomplete or fuzzy information system in real world, therefore the Pawlak's rough set theory must be extended in order to overcome the problem.

* This paper is partially supported by Hunan Nature Science Foundation (09JJ6099), the Scientific Research Projects of Hunan Provincial Education Department (10A041, 09C443) and Jiangsu Nature Science Foundation (BK2009395).

D. Zeng (Ed.): ICAIC 2011, Part II, CCIS 225, pp. 59–72, 2011.

At the present, there are mainly two approaches to extend the classical rough set theory from complete information system to incomplete information system. The first is indirect method that transforms an incomplete information system to complete information system by adding attribute values to missing values [12]. The second is direct method that extends the related concepts in classical rough set theory to those in incomplete information system, which has been studied with great concentrations by many researchers all over the world. For examples, M.Kryszkiewicz proposed tolerance relation in incomplete information system to extend the rough set model [13, 14], R.Slowinski et al generalized definition of rough approximations based on non-symmetric similarity relation [15], G.Y. Wang proposed extension rough set model based on limited tolerance relation [16, 17].

The above several models are obtained due to remove one of conditions in equivalent relation. The other extension models in incomplete information system are also acquired by generalizing the equivalent relation to general binary relation. For instance, the following models can be extended in compete information system such as the rough set model based on neighborhood relation [18-20], the covering generalized rough set model [21-23], the multi-granulation-based rough set model [24], and so on.

In incomplete information system, there exists such a system which is called an incomplete and fuzzy decision information system (IFDIS in short) whose conditional attribute values may be null and decision attribute values are fuzzy. As for this information system, we have ever deeply studied its rough set models and knowledge reductions [25-27], but so far the decision rules acquisition in IFDIS has not been discussed by researcher and practitioners in both home and abroad. This paper, based on the previous researching on IFDIS, introduces a new approach by virtue of variable precision rough set model [28] and $\gamma-$ tolerance relation [29] in incomplete information system to construct a new kind of rough set model which will be called variable rough set model based on $\gamma-$tolerance relation in IFDIS. It is a generalization of rough set model in incomplete decision information system and is also a generalization of rough set mode in complete fuzzy decision information system. By use of notion of precision reduct, the reducts in IFDIS be obtained, furthermore the optimal and pessimal decision rules can be acquired.

The paper organization is as follows: Section 2 introduces the concepts related to incomplete information system and $\gamma-$tolerance relation. Section 3 gives the variable precision rough set model based on $\gamma-$tolerance relation and its properties. Section 4 introduces concept of incomplete and fuzzy decision information system on the base of incomplete information system and fuzzy decision information system. Section 5 proposes the variable rough set model in IFDIS and its basic properties. Section 6 provides the method of precision reduct in IFDIS based on the variable rough set model. Section 7 presents the notion and method of decision rule acquisition focused on optimal and pessimal decision rules. Section 8 shows a sample to illustrate the algorithm for rule acquisition. Finally, section 9 concludes this paper.

2 Incomplete Information System and $\gamma-$ Tolerance Relation

In this section, the concept of incomplete information system and M.Kryszkiewcz's tolerance relation are firstly introduced.

Definition 1. Let $S=(U,A,V,f)$ be an information system, where U is a nonempty finite set of objects called universe of discourse, A is a nonempty finite set of conditional attributes; and for every $a\in A$, such that $f:U\to V_a$, $V=\bigcup_{a\in A} V_a$, where V_a is the value set of the attribute a. If there exists at least an attribute $a\in A$ such that V_a contains null values (a null value is usually denoted by $*$). Then the system S is called an incomplete information system.

Definition 2. Let $S=(U,A,V,f)$ be an incomplete information system, $B\subseteq A$. Then

$$T(B)=\{(x,y)\in U\times U \mid \forall b\in B, b(x)=b(y)\vee b(x)=*\vee b(y)=*\}$$

is called a M.Kryszkiewcz's tolerance relation or a tolerance relation in short [13, 14].

Obviously, $T(B)=\bigcap_{b\in B} T(\{b\})$ holds, T is reflexive and symmetric, but not transferable.

Let $T_B(x)=\{y\in U \mid (x,y)\in T(B)\}$, and then $T_B(x)$ is called the tolerance class of the object $x\in U$ with respect to the set $B\subseteq A$ of conditional attributes. $T_B(x)$ is constituted by all objects in universe U which is possibly indiscernible with x.

From the definition of tolerance relation, we can see that if the attribute values of two objects with respect to every attribute in attribute set are equivalent or the attribute values of at least one object include null values, then the two objects must belong to the same tolerance class, in other words, the null may take arbitrary values. Therefore the tolerance relation is too loose to suiting for the reality. With regards to this, we are going to introduced $\gamma-$ tolerance relation.

Let $S=(U,A,V,f)$ be an incomplete information system, $b\in B\subseteq A$, $V_b=\{b(x)\mid x\in U\}$, $V_b'=\{b(x)\mid b(x)\neq *, x\in U\}$, and it be possible that V_b contains the same elements, $|V_b|$ represents the number of non-null elements in V_b, and the same elements is calculated by the number of these same elements, $|b(x)|$ expresses the number of non-null attribute value $b(x)$.

Definition 3 [29]. Let $S=(U,A,V,f)$ be an incomplete information system, $x,y\in U$, $b\in B\subseteq A$, and there be at least a non-null value in V_b, Then the probability that $b(x)$ equals $b(y)$, denoted by $p_b(x,y)$, is defined as follows:

$$p_b(x,y)=\begin{cases} 1 & b(x)\neq *\wedge b(y)\neq *\wedge b(x)=b(y)\\ \dfrac{|b(y)|}{|V_b|} & b(x)=*\wedge b(y)\neq *\\ \dfrac{|b(x)|}{|V_b|} & b(y)=*\wedge b(x)\neq *\\ 1 & b(x)=*\wedge b(y)=*\\ 0 & b(x)\neq *\wedge b(y)\neq *\wedge b(x)\neq b(y)\end{cases}$$

Definition 4 [29]. Let $S=(U,A,V,f)$ be an incomplete information system, $x,y\in U$, $B\subseteq A$, the probability that x and y have the same attribute values on B, denoted by $p_B(x,y)$, is defined as follows:

$$p_B(x,y)=\prod_{b\in B} p_b(x,y).$$

$p_B(x,y)$ is called the probability that x equals y on B as well.

Evidently, if $C\subseteq B$, then $p_B(x,y)\le p_C(x,y)$.

Definition 5 [22]. Let $S=(U,A,V,f)$ be an incomplete information system, $B\subseteq A$, and $\gamma\,(0\le\gamma\le 1)$ be a constant. Then $T_\gamma(B)=\{(x,y)\mid x,y\in U\wedge p_B(x,y)\ge\gamma\}$ is said to be $\gamma-$ tolerance relation on B.

$T_\gamma^B(x)=\{y\in U\mid p_B(x,y)>\gamma\}=\{y\in U\mid (x,y)\in T_\gamma(B)\}$ is called $\gamma-$ tolerance class of $x\in U$.

3 Variable Precision Rough Set Model Based on $\gamma-$ Tolerance Relation

The paper [28] proposes the variable precision rough set model based on M.Kryszkiewcz's tolerance relation, and the $\gamma-$ tolerance relation-based variable precision rough set model is presented in this section.

Definition 6 [30]. Let $X,Y\subset U$, $0.5<\delta\le 1$. Then X is said to be majority inclusion in Y [28, 30] if and only if $D(Y/X)\ge\delta$, where

$$D(Y/X)=\begin{cases}|X\cap Y|/|X| & |X|>0\\ 1 & |X|=0\end{cases}.$$

Where $|\cdot|$ is the cardinal number of a set.

Definition 7. Let $S=(U,A,V,f)$ be an incomplete information system, $T_\gamma^B(x)$ be $\gamma-$ tolerance class of $x\in U$, $X\subseteq U$. Then the $\delta-$ upper and the $\delta-$ lower approximation of X, denoted by $\overline{T}^B_{\gamma\delta}(X)$ and $\underline{T}^B_{\gamma\delta}(X)$ respectively, are defined as follows:

$$\overline{T}^B_{\gamma\delta}(X)=\{x\mid x\in U\wedge D(X/T_\gamma^B(x))>1-\delta\},$$

$$\underline{T}^B_{\gamma\delta}(X)=\{x\mid x\in U\wedge D(X/T_\gamma^B(x))\ge\delta\}.$$

$\underline{T}^B_{\gamma\delta}(X)$ is also called the $\delta-$positive region of X based on $\gamma-$ tolerance relation, denoted by $POS^B_{\gamma\delta}(X)$.

The $\delta-$ boundary region and $\delta-$ negative region of X based on $\gamma-$ tolerance relation, represented by $BNR^{B}_{\gamma\delta}(X)$ and $NEG^{B}_{\gamma\delta}(X)$ separately, are defined as follows:

$$BNR^{B}_{\gamma\delta}(X)=\{x \mid x\in U \wedge 1-\delta < D(X/T^{B}_{\gamma}(x)) < \delta\},$$

$$NEG^{B}_{\gamma\delta}(X)=\{x \mid x\in U \wedge D(X/T^{B}_{\gamma}(x)) \leq 1-\delta\}.$$

As $\delta=1$, the variable precision rough set model is the common rough set model for incomplete information system.

From definition 5 and definition 7, we can obtain the following conclusions.

Let $X,Y \subset U$, $0.5<\delta\leq 1$, $0\leq\gamma\leq 1$, then the following properties hold:

(1) $\underline{T}^{B}_{\gamma\delta}(X)\subseteq\overline{T}^{B}_{\gamma\delta}(X)$;

(2) $\underline{T}^{B}_{\gamma\delta}(\phi)=\overline{T}^{B}_{\gamma\delta}(\phi)=\phi$, $\underline{T}^{B}_{\gamma\delta}(U)=\overline{T}^{B}_{\gamma\delta}(U)=U$;

(3) $\overline{T}^{B}_{\gamma\delta}(X)\bigcup\overline{T}^{B}_{\gamma\delta}(Y)\subseteq\overline{T}^{B}_{\gamma\delta}(X\bigcup Y)$,

$\overline{T}^{B}_{\gamma\delta}(X)\bigcap\overline{T}^{B}_{\gamma\delta}(Y)\supseteq\overline{T}^{B}_{\gamma\delta}(X\bigcap Y)$;

(4) $\underline{T}^{B}_{\gamma\delta}(X)\bigcup\underline{T}^{B}_{\gamma\delta}(Y)\subseteq\underline{T}^{B}_{\gamma\delta}(X\bigcup Y)$,

$\underline{T}^{B}_{\gamma\delta}(X\bigcap Y)\subseteq\underline{T}^{B}_{\gamma\delta}(X)\bigcap\underline{T}^{B}_{\gamma\delta}(Y)$.

4 Incomplete and Fuzzy Decision Information System (Ifdis)

On the base of the concept of incomplete information system and notion of complete fuzzy decision information system, we shall introduce the definition of incomplete and fuzzy decision information system as follows.

Definition 8 [25-27]. Incomplete and fuzzy decision information system (IFDIS in short), denoted still by S, is defined as:

$$S=(U,A,V,F;D,W,G).$$

Where: (1) U is a nonempty finite set of objects, i.e., $U=\{x_1,x_2,\ldots,x_n\}$;

(2) A is a nonempty finite set of conditional attributes, i.e., $A=\{a_1,a_2,\ldots,a_m\}$;

(3) $F=\{f_1,f_2,\ldots,f_m\}$ is a set of conditional attribute mappings, $f_j:U\to V_j$, $j\leq m$, and there is at least a V_j containing null values $*$, $V=\bigcup_{1\leq j\leq m} V_j$;

(4) D is a nonempty finite set of decision attributes, i.e., $D=\{d_1,d_2,\ldots,d_p\}$;

(5) $G=\{g_1,g_2,\ldots,g_p\}$ is a set of fuzzy decision mappings, $g_k:U\to W_k$, $W_k\in F(U)$, $W=\bigcup_{1\leq k\leq p} W_k$ and $W_k(x_i)\in[0,1]$, $i\leq n$, $k\leq p$, $F(U)$ is the whole of all fuzzy sets of U.

If $p=1$, i.e., let $D=\{d\}$, $G=\{g\}$, $W\in F(U)$. Then $S=(U,A,V,F;\{d\},W,\{g\})$ is called an incomplete and fuzzy (single) decision information system which is researched in this paper. While $p>1$, S is called an incomplete and fuzzy multi-decision information system.

Especially, S is incomplete information system as D, W and G are empty sets; furthermore S is complete fuzzy decision information system as (U,A,V,F) is a complete information system.

5 The Variable Rough Set Model in Ifdis

Definition 9. Let $S=\{U,A,V,F;\{d\},W,\{g\}\}$ be an IFDIS, $B\subseteq A$, $0<\delta\le 1$, $T_\gamma^B(x)$ be $\gamma-$ tolerance class of $x\in U$ with respect to B in incomplete space (U,A,V,F). For every $W\in F(U)$, let

$$\overline{T}^B_{\gamma\delta}(W)(x)=\max\{\delta\cdot W(y)\mid y\in T_\gamma^B(x)\},\ \underline{T}^B_{\gamma\delta}(W)(x)=\min\{\delta\cdot W(y)\mid y\in T_\gamma^B(x)\}.$$

Where $W(y)$ represents membership function value of y. $\overline{T}^B_{\gamma\delta}(W)$, $\underline{T}^B_{\gamma\delta}(W)$ are called $\delta-$upper approximation set and $\delta-$lower approximation set of fuzzy decision set W based on $\gamma-$ tolerance relation in incomplete space (U,A,V,F) separately.

With the parameter δ change, $\delta-$ upper approximation set and $\delta-$ lower approximation set are various, for this reason, the rough set model is called the variable rough set model in IFDIS. Obviously, $\overline{T}^B_{\gamma\delta}(W)$ and $\underline{T}^B_{\gamma\delta}(W)$ are a pair of fuzzy sets.

From definition 9, as $\gamma=1$, $\delta=1$, $\overline{T}^B_{\gamma\delta}(W)$ and $\underline{T}^B_{\gamma\delta}(W)$ are the usual upper approximation set and lower approximation set IFDIS respectively [27].

The properties of the variable rough set model in IFDIS are as follows:

Property 1. Let $S=\{U,A,V,F;\{d\},W,\{g\}\}$ be an IFDIS, $B\subseteq A$, $0\le\gamma\le 1$, $0<\delta\le 1$. Then the following conclusions hold:

(1) $\underline{T}^B_{\gamma\delta}(W)\subseteq\overline{T}^B_{\gamma\delta}(W)$;

(2) If $W_1\subseteq W_2$, then $\overline{T}^B_{\gamma\delta}(W_1)\subseteq\overline{T}^B_{\gamma\delta}(W_2)$, $\underline{T}^B_{\gamma\delta}(W_1)\subseteq\underline{T}^B_{\gamma\delta}(W_2)$.

Property 2. Let $S=\{U,A,V,F;\{d\},W,\{g\}\}$ be an IFDIS, $B\subseteq A$, $0\le\gamma\le 1$, $0<\delta\le 1$. Then

$\overline{T}^B_{\gamma\delta}(\overline{T}^B_{\gamma\delta}(W))\supseteq\underline{T}^B_{\gamma\delta}(\overline{T}^B_{\gamma\delta}(W))\supseteq\overline{T}^B_{\gamma\delta}(\underline{T}^B_{\gamma\delta}(W))\supseteq\underline{T}^B_{\lambda\delta}(\underline{T}^B_{\gamma\delta}(W))$ Proof According to Property 1, we can easily prove that $\overline{T}^B_{\gamma\delta}(\overline{T}^B_{\gamma\delta}(W))\supseteq\underline{T}^B_{\gamma\delta}(\overline{T}^B_{\gamma\delta}(W))$, $\overline{T}^B_{\gamma\delta}(\underline{T}^B_{\gamma\delta}(W))\supseteq\underline{T}^B_{\gamma\delta}(\underline{T}^B_{\gamma\delta}(W))$ are valid.

Now we only prove $\underline{T}^{B}_{\gamma\delta}(\overline{T}^{B}_{\gamma\delta}(W)) \supseteq \overline{T}^{B}_{\gamma\delta}(\underline{T}^{B}_{\gamma\delta}(W))$. In fact, for every $x \in U$, then

$$\overline{T}^{B}_{\gamma\delta}(\underline{T}^{B}_{\gamma\delta}(W))(x) = \max_{y\in T^B_\gamma(x)} \{\delta \cdot \underline{T}^{B}_{\gamma\delta}(W)(y)\}$$

$$= \max_{y\in T^B_\gamma(x)} \{\delta \cdot \min_{z\in T^B_\gamma(y)} \{\delta \cdot W(z)\}\} \le \min_{y\in T^B_\gamma(x)} \{\delta \cdot \max_{z\in T^B_\gamma(y)} \{\delta \cdot W(z)\}\}$$

$$= \min_{y\in T^B_\gamma(x)} \{\delta \cdot \overline{T}^{B}_{\gamma\delta}(W)(y)\} = \underline{T}^{B}_{\gamma\delta}(\overline{T}^{B}_{\gamma\delta}(W))(x).$$

Hence, $\underline{T}^{B}_{\gamma\delta}(\overline{T}^{B}_{\gamma\delta}(W)) \supseteq \overline{T}^{B}_{\gamma\delta}(\underline{T}^{B}_{\gamma\delta}(W))$.

Property 3. Let $S = \{U, A, V, F; \{d\}, W, \{g\}\}$ be an IFDIS, $B \subseteq A$, $0 \le \gamma \le 1$, $0 < \delta \le 1$. For every $x \in U$, then:

$$\overline{T}^{B}_{\gamma\delta}(W^C)(x) + \underline{T}^{B}_{\gamma\delta}(W)(x) = \delta, \quad \underline{T}^{B}_{\gamma\delta}(W^C)(x) + \overline{T}^{B}_{\gamma\delta}(W)(x) = \delta.$$

Where W^C is the complementary set of the fuzzy set W .

Proof. $\overline{T}^{B}_{\gamma\delta}(W^C)(x) = \overline{T}^{B}_{\gamma\delta}(W^C(x))$

$$= \max\{\delta \cdot (1 - W(y)) \mid y \in T^B_\gamma(x)\}$$

$$= \delta \cdot \max\{1 - W(y) \mid y \in T^B_\gamma(x)\}$$

$$= \delta \cdot (1 - \min\{W(y) \mid y \in T^B_\gamma(x)\})$$

$$= \delta - \underline{T}^{B}_{\gamma\delta}(W)(x),$$

$$\Rightarrow \overline{T}^{B}_{\gamma\delta}(W^C)(x) + \underline{T}^{B}_{\gamma\delta}(W)(x) = \delta.$$

Similarly, the other formula can be proved.

According to definition 4, we immediately obtain:

Property 4. Let $S = \{U, A, V, F; \{d\}, W, \{g\}\}$ be an IFDIS, $B \subseteq A$, $0 \le \gamma \le 1$, $0 < \delta_1 \le \delta_2 \le 1$. Then

(1) $\underline{T}^{B}_{\gamma\delta_1}(W) \subseteq \underline{T}^{B}_{\gamma\delta_2}(W) \subseteq \underline{T}^{B}_{\gamma 1}(W)$;

(2) $\overline{T}^{B}_{\gamma\delta_1}(W) \subseteq \overline{T}^{B}_{\gamma\delta_2}(W) \subseteq \overline{T}^{B}_{\gamma 1}(W)$.

6 Knowledge Reduction

In this section, we shall firstly make use of cut set in fuzzy set theory to introduce the concept of precision, then present the notion of precision reduct in IFDIS.

Definition 10. Let W be a fuzzy set defined on universe set U , for every $\alpha \in [0,1]$, then

$$W_\alpha = \{x \mid x \in U \wedge W(x) \ge \alpha\}$$

is called a $\alpha -$ cut set of W [30].

Property 5. Let $S=\{U,A,V,F;\{d\},W,\{g\}\}$ be an IFDIS, $B\subseteq A$, $0\le\gamma\le 1$, $0<\delta\le 1, 0\le\alpha, \beta\le 1$, and assume that:

$$\overline{T}^{B}_{\gamma\delta}(W)_{\beta}=\{x\in U\mid \overline{T}^{B}_{\gamma\delta}(W)(x)\ge\beta\},\ \underline{T}^{B}_{\gamma\delta}(W)_{\alpha}=\{x\in U\mid \underline{T}^{B}_{\gamma\delta}(W)(x)\ge\alpha\}.$$

Then (1) $\overline{T}^{B}_{\gamma\delta}(W)_{\beta}$ and $\underline{T}^{B}_{\gamma\delta}(W)_{\alpha}$ monotonously decrease with regard to β and α respectively under the notion of set inclusion; (2) $\underline{T}^{B}_{\gamma\delta}(W)_{\alpha}\subseteq\overline{T}^{B}_{\gamma\delta}(W)_{\alpha}$; (3) $\alpha\ge\beta\Rightarrow \underline{T}^{B}_{\gamma\delta}(W)_{\alpha}\subseteq\overline{T}^{B}_{\gamma\delta}(W)_{\beta}$.

Proof. From the definition 10, the conclusions (1), (2) and (3) can be easily proved.

Especially, as $\alpha\ge\beta$, (1) $\overline{T}^{B}_{\gamma\delta}(W)_{\beta}=\phi\Rightarrow\underline{T}^{B}_{\gamma\delta}(W)_{\alpha}=\phi$; (2) $\underline{T}^{B}_{\gamma\delta}(W)_{\alpha}=U\Rightarrow\overline{T}^{B}_{\gamma\delta}(W)_{\beta}=U$.

Property 6. Let $S=\{U,A,V,F;\{d\},W,\{g\}\}$ be an IFDIS, $B\subseteq A$, $0\le\gamma\le 1$, $0<\delta\le 1, 0\le\alpha, \beta\le 1$. Then:

(1) $\overline{T}^{A}_{\gamma\delta}(W)_{\beta}\subseteq\overline{T}^{B}_{\gamma\delta}(W)_{\beta}$; (2) $\underline{T}^{B}_{\gamma\delta}(W)_{\alpha}\subseteq\underline{T}^{A}_{\gamma\delta}(W)_{\alpha}$.

Proof. (1) $B\subseteq A\Rightarrow T^{A}_{k}(x)\subseteq T^{B}_{k}(x)$ for $\forall x\in U\Rightarrow$ $\max\{\delta\cdot W(y)\mid y\in T^{A}_{\gamma}(x)\}\le\max\{\delta\cdot W(y)\mid y\in T^{B}_{\gamma}(x)\}\Rightarrow$ $\overline{T}^{A}_{\gamma\delta}(W)(x)\le\overline{T}^{B}_{\gamma\delta}(W)(x)$ $\Rightarrow\overline{T}^{A}_{\gamma\delta}(W)\subseteq\overline{T}^{B}_{\gamma\delta}(W)\Rightarrow\overline{T}^{A}_{\gamma\delta}(W)_{\beta}\subseteq\overline{T}^{B}_{\gamma\delta}(W)_{\beta}$.

(2) The formula can be analogously proved.

Definition 11. Let $S=\{U,A,V,F;\{d\},W,\{g\}\}$ be an IFDIS, $B\subseteq A$, $0\le\gamma\le 1$, $0<\delta\le 1, 0\le\beta\le\alpha\le 1$. Then (α,β) precision degree of W is defined as:

$$\alpha^{B}_{\gamma\delta}(\alpha,\beta)=\left|\underline{T}^{B}_{\gamma\delta}(W)_{\alpha}\right|/\left|\overline{T}^{B}_{\gamma\delta}(W)_{\beta}\right|.$$

And ruled: as $\overline{T}^{B}_{\gamma\delta}(W)_{\beta}=\phi$, then $\alpha^{B}_{\gamma\delta}(\alpha,\beta)=1$.

Obviously, $0\le\alpha^{B}_{\gamma\delta}(\alpha,\beta)\le 1$, and $\alpha^{B}_{\gamma\delta}(\alpha,\beta)$ is monotonously non-increasing w. r. t. α and is monotonously non-decreasing w. r. t. β.

From property 6 and definition 11, we instantly have:

Property 7. Let $S=\{U,A,V,F;\{d\},W,\{g\}\}$ be an IFDIS, $B\subseteq A$, $0\le\gamma\le 1$, $0<\delta\le 1$, $0\le\beta\le\alpha\le 1$. Then $\alpha^{A}_{\gamma\delta}(\alpha,\beta)\ge\alpha^{B}_{\gamma\delta}(\alpha,\beta)$.

Definition 12. Let $S=\{U,A,V,F;\{d\},W,\{g\}\}$ be an IFDIS, $B\subseteq A$, $0\le\gamma\le 1$, $0<\delta\le 1$, $0\le\beta\le\alpha\le 1$. If $\alpha^{A}_{\gamma\delta}(\alpha,\beta)=\alpha^{B}_{\gamma\delta}(\alpha,\beta)$ and $\alpha^{B}_{\gamma\delta}(\alpha,\beta)\ne\alpha^{C}_{\gamma\delta}(\alpha,\beta)$ for every $C\subset B$. Then B is said to be a $(\alpha,\beta)-$precision reduct under variable rough set model in IFDIS.

From definition 12 known, precision reduction ensures the ratio of the number of the certain objects that variable membership values are not less than α to that of the possible objects that variable membership values are not less than β is invariable.

From Property 7, definition 11 and 12, we can easily associate that the following results hold.

Property 8. Let $S=(U,A,V,F;\{d\},W,\{g\})$ be an IFDIS, $B \subseteq A$, $0 \le \gamma \le 1$, $0 < \delta \le 1, 0 \le \beta \le \alpha \le 1$. Then B is a $(\alpha,\beta)-$precision reduct under variable rough set model in IFDIS if and only if B is the minimal proper subset of conditional attributes which simultaneously satisfies the two equations $\overline{T}^{B}_{\gamma\delta}(W)_{\beta} = \overline{T}^{A}_{\gamma\delta}(W)_{\beta}$ and $\underline{T}^{B}_{\gamma\delta}(W)_{\alpha} = \underline{T}^{A}_{\gamma\delta}(W)_{\alpha}$.

7 Decision Rule Acquisition in Ifdis

It is significant researches to acquire decision rules in IFDIS in decision analysis, intelligent decision, data mining, and so on. In this section, we mainly discuss two types of decision rule acquisitions after knowledge reducing: optimal decision rules and pessimal decision rules.

Definition 13. Let $S=\{U,A,V,F;\{d\},W,\{g\}\}$ is an IFDIS. Then the rule

$$R_I(x): \wedge_{a \in A}(a,a(x)) \rightarrow (d,W(x)) \text{ //supported by object } x$$

is called a initial fuzzy decision rule (with respect to object x) of S, and also denoted by R_I in short.

Definition 14. Let $S=\{U,A,V,F;\{d\},W,\{g\}\}$ is an IFDIS, B is a $(\alpha,\beta)-$precision reduct in IFDIS. Then the rules

$$R_O(x): \wedge_{b \in B}(b,b(x)) \rightarrow (d,\overline{T}^{B}_{\gamma\delta}(W)(x)) \text{ //supported by object } x$$

$$R_P(x): \wedge_{b \in B}(b,b(x)) \rightarrow (d,\underline{T}^{B}_{\gamma\delta}(W)(x)) \text{ //supported by object } x$$

are called optimal decision rule and pessimal decision rule (with respect to object x) of S, and also denoted by R_O and R_P in short respectively.

Both optimal decision rules and pessimal decision rules are derived from initial fuzzy decision rules; therefore they are collectively called initialized decision rules.

The optimal decision rules can provide deciders with excellent decision rules to make a strategic decision. For this reason, researches on them have important value in intelligent decision analysis theory and application.

The algorithm and example to find the optimal (pessimal) decisions in IFDIS are given as follows.

ALGORITHM

Input: IFDIS $S=\{U,A,V,F;\{d\},W,\{g\}\}$, the valid values: $\gamma(0\le\gamma\le 1)$, $\delta(0<\delta\le 1)$ and $\alpha,\beta(0\le\beta\le\alpha\le 1)$.

Output: all of optimal decision rules $R_O(x_i)$ and pessimal decision rules $R_P(x_i)$ of S, $x_i\in U$.

Step 1: Initialize data $U,A,V,W,B=\phi$.

Step 2: Compute the subsets of conditional attributes A : $B_1,\ldots,B_q$, where $|B_r|=|A|-1$ ($r=1,\ldots,q$).

Step 3: Calculate all $\gamma-$ tolerance classes $T_\gamma^A(x_i)$ and $T_\gamma^{B_r}(x_i)$ ($1\le i\le n, 1\le r\le q$), in virtue of the formula $T_\gamma^A(x_i)=\{y\in U \mid p_A(x_i,y)>\gamma\}=\{y\in U \mid (x_i,y)\in T_\gamma(A)\}$.

Step 4: Compute upper-approximation $\overline{T}^A_{\gamma\delta}(W)$, $\overline{T}^{B_r}_{\gamma\delta}(W)$ and lower-approximation $\underline{T}^A_{\gamma\delta}(W)$, $\underline{T}^{B_r}_{\gamma\delta}(W)$ ($r=1,\ldots,q$) of the fuzzy set W according to the formulae

$$\overline{T}^A_{\gamma\delta}(W)(x_i)=\max\{\delta\cdot W(y)\mid y\in T_\gamma^A(x_i)\},\ \underline{T}^A_{\gamma\delta}(W)(x_i)=\min\{\delta\cdot W(y)\mid y\in T_\gamma^A(x_i)\}.$$

Step 5: Compute cut sets $\overline{T}^A_{\gamma\delta}(W)_\beta, \underline{T}^A_{\gamma\delta}(W)_\alpha, \overline{T}^{B_r}_{\gamma\delta}(W)_\beta$, and $\underline{T}^{B_r}_{\gamma\delta}(W)_\alpha$ ($r=1,\ldots,q$).

Step 6: If $\overline{T}^A_{\gamma\delta}(W)_\beta\subset\overline{T}^{B_r}_{\gamma\delta}(W)_\beta$ or $\underline{T}^A_{\gamma\delta}(W)_\alpha\supset\underline{T}^{B_r}_{\gamma\delta}(W)_\alpha$ ($r=1,\ldots,q$), then go to step 7 (A itself is a reduct of the system S); Otherwise, $\overline{T}^A_{\gamma\delta}(W)_\beta=\overline{T}^{B_r}_{\gamma\delta}(W)_\beta$ and $\underline{T}^A_{\gamma\delta}(W)_\alpha=\underline{T}^{B_r}_{\gamma\delta}(W)_\alpha$, go to step 2 (consider continuously the corresponding the subsets of B_r until finding a $B\subseteq B_r\subseteq A$ satisfies that $\overline{T}^B_{\gamma\delta}(W)_\beta=\overline{T}^{B_r}_{\gamma\delta}(W)_\beta=\overline{T}^A_{\gamma\delta}(W)_\beta$ and $\underline{T}^B_{\gamma\delta}(W)_\alpha=\underline{T}^{B_r}_{\gamma\delta}(W)_\alpha=\underline{T}^A_{\gamma\delta}(W)_\alpha$, but $\overline{T}^B_{\gamma\delta}(W)_\beta\subset\overline{T}^C_{\gamma\delta}(W)_\beta$ or $\underline{T}^B_{\gamma\delta}(W)_\alpha\subset\underline{T}^C_{\gamma\delta}(W)_\alpha$ for every $C\subset B_r\subseteq A$).

Step 7: Write out the precision reduct B.

Step 8: Output the optimal decision rules and pessimal decision rules of S:

$R_O(x_i)$: $\wedge_{b\in B}(b,b(x_i))\rightarrow(d,\overline{T}^B_{\gamma\delta}(W)(x_i))$ //supported by object x_i,

$R_P(x_i)$: $\wedge_{b\in B}(b,b(x_i))\rightarrow(d,\underline{T}^B_{\gamma\delta}(W)(x_i))$ //supported by object x_i.

8 Example

The following data table gives an example of IFDIS [27].

Table 1. Incomplete and Fuzzy Decision Informaaion System

	x_1	x_2	x_3	x_4	x_5	x_6	x_7	x_8	x_9
a_1	1	1	3	1	*	3	3	3	2
a_2	2	*	2	2	2	1	2	1	3
a_3	1	3	3	*	1	*	*	2	*
d	0.8	0.6	0.9	0.7	0.8	0.5	0.7	0.6	0.5

Where $U=\{x_1,x_2,\ldots,x_9\}$, $A=\{a_1,a_2,a_3\}$,

$W=0.8/x_1+0.6/x_2+0.9/x_3+0.7/x_4+0.8/x_5+0.5/x_6$
$+0.7/x_7+0.6/x_8+0.5/x_9$.

If $\gamma=0.2,\delta=0.9,\alpha=\beta=0.5$. Then $R_\gamma^A(x_1)=\{x_1,x_4,x_5\}$, $R_\gamma^A(x_2)=\{x_2,x_4\}$, $R_\gamma^A(x_3)=\{x_3,x_7\}$, $R_\gamma^A(x_4)=\{x_1,x_2,x_4\}$, $R_\gamma^A(x_5)=\{x_1,x_5,x_7\}$, $R_\gamma^A(x_6)=\{x_6,x_8\}$, $R_\gamma^A(x_7)=\{x_3,x_7\}$, $R_\gamma^A(x_8)=\{x_6,x_8\}$, $R_\gamma^A(x_9)=\{x_9\}$;

$\overline{T}_{\gamma\delta}^A(W)=0.72/x_1+0.63/x_2+0.81/x_3+0.72/x_4+0.72/x_5$
$+0.54/x_6+0.81/x_7+0.54/x_8+0.45/x_9$,

$\underline{T}_{\gamma\delta}^A(W)=0.63/x_1+0.54/x_2+0.63/x_3+0.54/x_4+0.63/x_5$
$+0.45/x_6+0.63/x_7+0.45/x_8+0.45/x_9$;

$\overline{T}_{\gamma\delta}^A(W)_\beta=\{x_1,x_2,x_3,x_4,x_5,x_6,x_7,x_8\}$,

$\underline{T}_{\gamma\delta}^A(W)_\alpha=\{x_1,x_2,x_3,x_4,x_5,x_7\}$.

Let $B_1=\{a_1,a_2\}$, we can calculate

$\overline{T}_{\gamma\delta}^{B_1}(W)_\beta=\{x_1,x_2,x_3,x_4,x_5,x_6,x_7,x_8\}$, $\underline{T}_{\gamma\delta}^{B_1}(W)_\alpha=\{x_1,x_2,x_3,x_4,x_5,x_7\}$,

Then $\overline{T}_{\gamma\delta}^A(W)_\beta=\overline{T}_{\gamma\delta}^{B_1}(W)_\beta, \underline{T}_{\gamma\delta}^A(W)_\alpha=\underline{T}_{\gamma\delta}^{B_1}(W)_\alpha$.

Furthermore we can still compute

$\overline{T}_{\gamma\delta}^{B_1}(W)_\beta\subset\overline{T}_{\gamma\delta}^{\{a_1\}}(W)_\beta, \overline{T}_{\gamma\delta}^{B_1}(W)_\beta\subset\overline{T}_{\gamma\delta}^{\{a_2\}}(W)_\beta$;

$\underline{T}_{\gamma\delta}^{B_1}(W)_\alpha\supset\underline{T}_{\gamma\delta}^{\{a_1\}}(W)_\alpha$ and $\underline{T}_{\gamma\delta}^{B_1}(W)_\alpha\supset\underline{T}_{\gamma\delta}^{\{a_2\}}(W)_\alpha$.

Therefore $B_1=\{a_1,a_2\}$ is a (0.5, 0.5) reduct of the IFDIS.

Let $B_2=\{a_2,a_3\}$, $B_3=\{a_1,a_3\}$. By similarly calculating, we can conclude: B_2 and B_3 are not the reductions of S.

Further as $\gamma=0.2,\delta=0.9,\alpha=\beta=0.5$, the optimal decision rules of S is obtained as follows:

$(a_1,1)\wedge(a_2,2)\rightarrow(d,0.72)$ // supported by object x_1,

$(a_1,1)\wedge(a_2,*)\rightarrow(d,0.63)$ // supported by object x_2,

$(a_1,3)\wedge(a_2,2)\rightarrow(d,0.81)$ // supported by object x_3,

$(a_1,1)\wedge(a_2,2)\to(d,0.72)$ // supported by object x_4 ,

$(a_1,*)\wedge(a_2,2)\to(d,0.72)$ // supported by object x_5 ,

$(a_1,3)\wedge(a_2,1)\to(d,0.54)$ // supported by object x_6 ,

$(a_1,3)\wedge(a_2,2)\to(d,0.81)$ // supported by object x_7 ,

$(a_1,3)\wedge(a_2,1)\to(d,0.54)$ // supported by object x_8 ,

$(a_1,2)\wedge(a_2,3)\to(d,0.45)$ // supported by object x_9 .

Analogously, the pessimal decision rules of S can be gained:

$(a_1,1)\wedge(a_2,2)\to(d,0.63)$ // supported by object x_1 ,

$(a_1,1)\wedge(a_2,*)\to(d,0.54)$ // supported by object x_2 ,

$(a_1,3)\wedge(a_2,2)\to(d,0.63)$ // supported by object x_3 ,

$(a_1,1)\wedge(a_2,2)\to(d,0.54)$ // supported by object x_4 ,

$(a_1,*)\wedge(a_2,2)\to(d,0.63)$ // supported by object x_5 ,

$(a_1,3)\wedge(a_2,1)\to(d,0.45)$ // supported by object x_6 ,

$(a_1,3)\wedge(a_2,2)\to(d,0.63)$ // supported by object x_7 ,

$(a_1,3)\wedge(a_2,1)\to(d,0.45)$ // supported by object x_8 ,

$(a_1,2)\wedge(a_2,3)\to(d,0.45)$ // supported by object x_9 .

9 Conclusions

The incomplete and fuzzy decision information system (IFDIS) is a sort of complicated information system, and researching on decision rule acquisition in IFDIS has important reality significance and application value. In this paper, on the base of the previous discussing on $\gamma-$ tolerance relation and philosophy of variable precision rough set model in incomplete information system, the variable precision rough set model based on $\gamma-$ tolerance relation is presented. After introducing concept of incomplete and fuzzy decision information system under the incomplete information system and fuzzy decision information system, the variable rough set model in IFDIS is proposed and its basic properties are discussed. The method of precision reduct in IFDIS based on the variable rough set model is provided. The notion and approach of decision rule acquisition focused on optimal and pessimal decision rules are presented. Finally the algorithm for rule acquisition is given and illustrated it valid by an example.

References

[1] Pawlak, Z.: Rough sets. International Journal of Computer and Information Sciences 11, 341–356 (1982)

[2] Zhong, N., Skowron, A.: A rough set set-based knowledge discovery process. Int. J. App. Math. Sci. 11(3), 603–619 (2001)

[3] Wu, W.Z., Mi, J.S., Zhang, W.X.: A New Rough Set Approach to Knowledge Discovery in Incomplete Information System. In: Proceedings of the Second International Conference on Machine Learning and Cybernetics, Xi'an, November 2-5, pp. 1713–1718 (2003)

[4] Pawlak, Z.: Rough set approach to knowledge-based decision support. European Journal of Operational Research 99, 48–57 (1997)

[5] Slowinski, R.: Intelligent Decision Support-handbook of Applications and Advances of the Rough Sets Theory. Kluwer Academic Publishers, Boston (1992)

[6] Shan, N., Ziarko, W.: An incremental learning algorithm for constructing decision rules. In: Rough Set, Fuzzy Sets and Knowledge Discovery, pp. 326–334. Springer, Heidelberg (1994)

[7] Grzymala-Busse, J.W.: LERS-A system learning from examples based on rough sets. In: Intelligent Decision Support: Handbook of Applications and Advance of the Rough Sets Theory, pp. 3–18. Kluwer Academic Publishers, Dordrecht (1992)

[8] Pawlak, Z.: Drawing conclusions from data-rough set way. International Journal of Intelligent Systems 16, 3–11 (2001)

[9] Chan, C.C.: A rough set approach to attribute generalization in data mining. Journal of Information Science 107, 169–176 (1998)

[10] Pawlak, Z.: Rough set: Theoretical Aspects of Reasoning about Data. Kluwer Academic Publishers, Boston (1991)

[11] Lin, T.Y., Liu, Q.: First-order rough set logic I: Approximate reasoning via rough sets. Fundamenta Informaticae 27(2-3), 137–154 (1996)

[12] Hong, T.-P., Tseng, L.-H., Wang, S.-L.: Learning rules from incomplete training examples by rough sets. Expert Systems with Applications 22, 193–285 (2002)

[13] Kryszkiewicz, M.: Rough set approach to incomplete information systems. Information Sciences 112, 39–49 (1998)

[14] Kryszkiewicz, M.: Rules in incomplete information systems. Information Sciences 113, 271–292 (1999)

[15] Slowinski, R., Vanderpooten, D.: A Generalized Definition of Rough Approximation Based on Similarity. IEEE Transactions on Knowledge and Data Engiering 12, 331–336 (2000)

[16] Wang, G.Y.: Rough Set Theory and Knowledge Acquisition. Xi'an JiaoTong University Press, Xi'an (2001)

[17] Wang, G.Y.: Extension of Rough Set under Incomplete Information systems. Journal of Computer Research and Development 39(10), 1238–1243 (2002)

[18] Yao, Y.Y.: Relational interpretation of neighborhood operators and rough set approximation operators. Information Sciences 111, 239–259 (1998)

[19] Hu, Q.H., Yu, D.R., Liu, J.F., Wu, C.X.: Neighborhood rough set based heterogeneous feature subset selection. Information Sciences 178, 3577–3594 (2008)

[20] Hu, Q.H., Yu, D.R., Xie, Z.X.: Numerical Attribute Reduction Based on Neighborhood Granulation and Rough Approximation. Journal of Software 19(3), 640–649 (2008)

[21] Zhu, W., Wang, F.Y.: Reduction and axiomization of covering generalized rough sets. Information Sciences 152, 217–230 (2003)

[22] Zhu, W.: Relationship between generalized rough sets based on binary relation and covering. Information Sciences 179, 210–225 (2009)

[23] Zhu, W.: Relationship among basic concepts in covering-based rough sets. Information Sciences 179, 2478–2486 (2009)

[24] Qian, Y.H., Liang, J.Y., Dang, C.Y.: Incomplete Multigranulation Rough Set. IEEE Transactions on Systems, Man, and Cybernetics—Part A: Systems and Humans 40(2), 420–431 (2010)

[25] Wei, D.-k., Zhou, X.-z.: Rough set model in incomplete and fuzzy decision information system on improved-tolerance relation. In: 2005 IEEE the Proceedings of the International Conference on Granular Computing, Beijing, China, pp. 278–283 (July 2005)
[26] Wei, D., Tang, L.: Attribute reduction based on inclusion degree for incomplete and fuzzy decision information system. Journal of Communication and Computer 3(5), 22–28 (2006)
[27] Wei, D., Zhao, y., Zhou, X.: A rough set approximation to incomplete and fuzzy decision information system. In: 2006 IEEE Proceedings of the 6th World Congress on Control and Automation, Dalian, China, June 21-23, pp. 4219–4223 (2006)
[28] Zhang, H.Y., Liang, J.Y.: Variable Precision Rough Set Model and a Knowledge Reduction Algorithm for Incomplete Information System. Computer Science 30(4), 153–155 (2003)
[29] Wei, D.-k.: Knowledge reduction in incomplete systems based on γ–tolerance relation. In: Lang, J., Lin, F., Wang, J. (eds.) KSEM 2006. LNCS (LNAI), vol. 4092, pp. 504–513. Springer, Heidelberg (2006)
[30] Zhang, W.X., Leung, Y., Wu, W.Z.: Information System and Knowledge Discovery. Science Press, Beijing (2003)

Synchronization of a Complex Network for Neighboring Agents with Time-Varying Delays

Jie Zhang, Man Li, and Bo Liu

College of Science, North China University of Technology, Beijing 100144, P. R. China
J.Zhang25@ovi.com

Abstract. This paper gives a complex dynamical network model. In this network, the agent interacts with its neighbors with time-varying delays. Using the Lyapunov method and the matrix inequality, we discuss the synchronization of the proposed network based on the synchronous target with and without time-varying delays, respectively. Furthermore, we verify our theoretical results by some simulations.

Keywords: Complex network, time-varying delays, synchronization.

1 Introduction

In recent years, the synchronization problem of complex dynamical networks has attracted many researchers. The synchronization studies [1]-[9] of the complex network can be applied many fields, such as robotics, information collection, wireless networks, and so on. In [11], the authors analysed that the synchronization of a scale-free dynamical network is robust against random removal of nodes.

More recently, Wang and Chen discussed a dynamical complex network model and also investigated its synchronization in [12]. In [13], the authors proposed a new complex network model for reputation computation in Virtual Organizations and investigated its convergence dynamics. However, in many networks, time-varying delays may affect the synchronization, as a result of the moving of the nodes, the congestion of the communication channels, the asymmetry of interactions, and the finite transmission speed. In [10], the authors studied average consensus problems in undirected networks with a common constant communication delay and fixed topology. In this paper, we consider the synchronization of a complex dynamical network with local-interaction and time-varying delays for the neighbours. In such a network, each node communicates with its neighbouring nodes to exchange information, and time-varying delays exist in the neighbouring nodes. We consider the synchronization of the network with the synchronous target with time-varying delays. Specially, we also consider the synchronization of the network with the synchronous target without time-varying delays.

The rest of this paper is designed as follows. In Section 2, we give a model of the complex dynamical network with local-interaction and time-varying delays for the neighbours. In Section 3, we consider the synchronization of the network with the synchronous target with time-varying delays. In Section 4, we consider the

D. Zeng (Ed.): ICAIC 2011, Part II, CCIS 225, pp. 73–81, 2011.

synchronization of the network with the synchronous target without time-varying delays. In Section 5, the simulations verify our theoretical results. Conclusion is finally summarized in Section 6.

2 Preliminaries and Model Formulation

We consider a complex dynamical network with time-varying delays as follows:

$$\dot{x}_i(t) = f(x_i(t)) + \sum_{j\in N_i} c_{ij} a_{ij} \Gamma(x_j(t-\tau(t)) - x_i(t)) + u_i(t), \tag{1}$$

where $x_i(t) = (x_{i1}(t), x_{i2}(t), \cdots, x_{in}(t))^T \in R^n (i = 1,2,\cdots,N)$ represent the state vector of the node i at time t ; N_i is the set of all those coupling with the node i ; $f(\bullet) \in R^n$ is continuously differentiable; $c_{ij} > 0$ is the feedback gain between nodes i and j; $A = (a_{ij}) \in R^{n\times n}$ describes the outer-coupling structure of the network, if $i \neq j$, $a_{ij} > 0$, otherwise $a_{ij} = 0$. $\Gamma = diag(\gamma_1, \gamma_2, \cdots \gamma_n) \in R^{n\times n}$ is a inner-coupling matrix, $\gamma_i \geq 0$ represents that the node links through its $i-th$ state with their neighbours. $u_i(t)$ is the control input with a simple linear feedback law. In this paper, we consider two cases as follows:

Case 1. The synchronous target with time-varying delays

$$u_i(t) = -c_{ii} d_i \Gamma(x_i(t) - s(t-\tau(t))), \tag{2}$$

Case 2. The synchronous target without time-varying delays

$$u_i(t) = -c_{ii} d_i \Gamma(x_i(t) - s(t)), \tag{3}$$

where $d_i \geq 0$ is the feedback gain between nodes i and the synchronous target $x_1(t) = x_2(t) = \cdots = x_N(t) = s(t) \in R^n$ is the synchronous state, and satisfies $\dot{s}(t) = f(s(t))$; $\tau(t)$ is the time-varying delay, and $\tau(t)$ satisfies the following Assumption [10]:

***Assumption 1*:** $0 \leq \tau(t) \leq h,\ \dot{\tau}(t) \leq d \leq 1,\ t \geq 0.$

For the next discuss, we give the following Lemma:

***Lemma 1*:** [10] For any real vector function $x(t) \in R^n$, there exists a $n\times n$ symmetrical matrix $W > 0$ such that $[x(t) - x(t-\tau(t))]^T W [x(t) - x(t-\tau(t))] \leq h \int_{-\tau(t)} \dot{x}^T(s) W \dot{x}(s) ds$, $t \geq 0$, where $\tau(t)$ satisfy $0 \leq \tau(t) \leq h$.

3 The Synchronous Target with Time-Varying Delays

In this section, we consider the synchronization of the complex network for the synchronous target with time-varying delays. In order to discuss the synchronization of the network, we give the matrix $L=(l_{ij})_{N\times N}$ and $D=(d_{ij})_{N\times N}$, where $l_{ij}=\begin{cases}-c_{ij}a_{ij}, & i\neq j\\ 0, & i=j\end{cases}$, and $d_{ij}=\begin{cases}0, & i\neq j\\ c_{ii}d_i, & i=j\end{cases}$.

In the following discussion, $A>0(\geq,<,\leq)$ means A is a positive (or semi positive, negative, semi-negative) definite matrix. We analyze the synchronization condition of the proposed network by means of the following theorem.

***Theorem 1**:* The complex network (1) of N nodes (n-dimension) is synchronized for the control input (2), under Assumption 1, if h satisfies

$$\Phi=\begin{bmatrix} M+M^T+dP & N+(1-d)P & R\\ N^T+(1-d)P & -(1-d)P-\frac{1}{h}Q & 0\\ R^T & 0 & 0\end{bmatrix}+h\begin{bmatrix} M^TQM & M^TQN & R^TQM\\ N^TQM & N^TQN & R^TQN\\ M^TQR & N^TQR & R^TQR\end{bmatrix}\leq 0, \quad (4)$$

where $M=D_f(s)\otimes I_n-D\otimes\Gamma$, $N=L\otimes\Gamma$, $R=(L-D)\otimes\Gamma$, and $D_f(s)$ is the *Jacobian* of f evaluated at $s(t)$, and $P>0$, $P=P^T$, $Q>0$, $Q=Q^T$.

Proof: Let $e_i(t)=x_i(t)-s(t)$, we have

$$\begin{aligned}\dot{e}_i(t)&=f(x_i(t))-f(s(t))-c_{ii}d_{ii}\Gamma[e_i(t)+(s(t)-s(t-\tau(t)))]\\&\quad+\sum_{j\in N_i}c_{ij}a_{ij}\Gamma[e_j(t-\tau(t))-(s(t)-s(t-\tau(t)))-e_i(t)]\cdot\end{aligned}\quad (5)$$

Linearizing the network (5) on synchronous solution $s(t)$, we have the following equation:

$$\dot{\eta}_i(t)=D_f(s)\eta_i(t)-c_{ii}d_{ii}\Gamma[\eta_i(t)+(s(t)-s(t-\tau(t)))]$$
$$+\sum_{j\in N_i}c_{ij}a_{ij}\Gamma[\eta_j(t-\tau(t))-(s(t)-s(t-\tau(t)))-\eta_i(t)].$$

Noting that $\eta(t)=(\eta_1^T(t),\eta_2^T(t),\cdots,\eta_N^T(t))^T$, we have

$$\dot{\eta}(t)=(D_f(s)\otimes I_n)\eta(t)-(L\otimes\Gamma)\eta(t-\tau(t))+((L-D)\otimes\Gamma)(\xi(t)+\eta(t)),$$

where $\xi(t)=(s(t)-s(t-\tau(t)))\otimes(\underbrace{1,1,\cdots,1}_{N})^T$.

Let $\varepsilon(t)=\eta(t)-\eta(t-\tau(t))$. The above equation can be rewritten as follows:

$$\dot{\eta}(t)=[(D_f(s)\otimes I_n)-D\otimes\Gamma]\eta(t)+(L\otimes\Gamma)\varepsilon(t)+((L-D)\otimes\Gamma)\xi(t), \quad (6)$$

For simplicity, we note that $L \otimes \Gamma = N$, $(D_f(s) \otimes I_n) - D \otimes \Gamma = M$, $(L - D) \otimes \Gamma = R$. So the equation (6) is $\dot{\eta}(t) = M\eta(t) + N\varepsilon(t) + R\xi(t)$.

Select a Lyapunov function [10]:

$$V(t) = \eta^T(t)\eta(t) + \int_{-\tau(t)} \eta^T(s)P\eta(s)ds + \int_{-h}^{0}\int_{+\theta} \dot{\eta}^T(s)Q\dot{\eta}(s)dsd\theta,$$

where $P > 0$, $P = P^T$, $Q > 0$, $Q = Q^T$. Then, we have

$$\begin{aligned}\dot{V}(t) &= (M\eta(t) + N\varepsilon(t) + R\xi(t))^T \eta(t) - \int_{-h} \dot{\eta}^T(s)Q\dot{\eta}(s)ds \\ &\quad + \eta^T(t)(M\eta(t) + N\varepsilon(t) + R\xi(t)) + \eta^T(t)P\eta(t) \\ &\quad - (1-\dot{\tau}(t))(\eta(t) - \varepsilon(t))^T Q(M\eta(t) + N\varepsilon(t) + R\xi(t)) \\ &\le \eta^T(t)(M^T + M + P - (1-d)P)\eta(t) + \varepsilon^T(t)(-(1-d)P - \frac{1}{h}Q)\varepsilon(t) \\ &\quad + \eta^T(t)(2N + 2(1-d)P)\varepsilon(t) + \eta^T(t)(P + P^T)\xi(t) \\ &\quad + hy^T(t)\begin{bmatrix} M^TQM & M^TQN & R^TQM \\ N^TQM & N^TQN & R^TQN \\ M^TQR & N^TQR & R^TQR \end{bmatrix} y(t) = y^T(t)\Phi y(t)\end{aligned}$$

where $y(t) = (\eta^T(t), \varepsilon^T(t), \xi^T(t))^T$. According to (4) $\Phi \le 0$, we can find a positive constant α such that $\dot{V}(t) \le y^T(t)\Phi y(t) \le -\alpha\|y(t)\|^2 \le -\alpha\|\eta(t)\|^2 \le 0$, and $\dot{V}(t) = 0$ if and only if $\eta(t) = 0$. Furthermore, $V(t)$ is a decreasing function, and $V(t) \le V(0)$. We can have $\eta(t) \to 0$ as $t \to \infty$. From above discussion, we have $\lim_{t\to\infty}\|x_i(t) - s(t)\| = 0$, i.e. $x_i(t) \to s(t)$. Then, the complex network (1) with time-varying delays is synchronized. The proof is completed.

The special case: The network (1) satisfies the assumption as follows.

Assumption 2: $D_f^T(s) = D_f(s)$, $c_{ij} = c$, $a_{ij} = a_{ji}$ and $P = Q = I_n$.

We give a corollary based on Assumption 2 and Theorem 1.

Corollary 1: The complex network (1) of N nodes (n-dimension) is synchronized for the control input (2), under Assumptions 1 and 2, if h satisfies

$$\Phi^* = \begin{bmatrix} 2M + dI_n + hMM & N + (1-d)I_n + hMN & R + hRM \\ N + (1-d)I_n + hNM & -(1-d+\frac{1}{h})I_n + hNN & hRN \\ R + hMR & hNR & hRR \end{bmatrix} \le 0,$$

where $M = D_f(s) \otimes I_n - D \otimes \Gamma$, $N = L \otimes \Gamma$, $R = (L - D) \otimes \Gamma$, and $D_f(s)$ is the *Jacobian* of f evaluated at $s(t)$.

4 The Synchronous Target without Time-Varying Delays

In this section, we consider the synchronization of the complex network for the synchronous target without time-varying delays about the network (1). Under this condition, the network (1) with the control input (3) can be rewritten

$$\dot{x}_i(t)=f(x_i(t))+\sum_{j\in N_i}c_{ij}a_{ij}\Gamma(x_j(t-\tau(t))-x_i(t))$$
$$-c_{ii}d_i\Gamma(x_i(t)-s(t)). \tag{7}$$

The same as Theorem 1, we give the following theorem.

Theorem 2: The complex network (1) of N nodes (n -dimension) with the control input (3) is synchronized, under Assumption 1, if h satisfies

$$\Phi=\begin{bmatrix} M+M^T+dP & N+(1-d)P & N \\ N^T+(1-d)P & -(1-d)P-\frac{1}{h}Q & 0 \\ N^T & 0 & 0 \end{bmatrix}+h\begin{bmatrix} M^TQM & M^TQN & N^TQM \\ N^TQM & N^TQN & N^TQN \\ M^TQN & N^TQN & N^TQN \end{bmatrix}\le 0, \tag{8}$$

where $M=D_f(s)\otimes I_n-D\otimes\Gamma$, $N=L\otimes\Gamma$, and $D_f(s)$ is the *Jacobian* of f evaluated at $s(t)$, and $P>0$, $P=P^T$, $Q>0$, $Q=Q^T$.

Proof: Let $e_i(t)=x_i(t)-s(t)$, we have

$$\dot{e}_i(t)=f(x_i(t))-f(s(t))-c_{ii}d_{ii}\Gamma e_i(t)$$
$$+\sum_{j\in N_i}c_{ij}a_{ij}\Gamma[e_j(t-\tau(t))-(s(t)-s(t-\tau(t)))-e_i(t)].$$

Linearizing the above network on synchronous solution $s(t)$, we have the following equation.

$$\dot{\eta}_i(t)=D_f(s)\eta_i(t)-c_{ii}d_{ii}\Gamma\eta_i(t)$$
$$+\sum_{j\in N_i}c_{ij}a_{ij}\Gamma[\eta_j(t-\tau(t))-(s(t)-s(t-\tau(t)))-\eta_i(t)].$$

Noting that $\eta(t)=(\eta_1^T(t),\eta_2^T(t),\cdots,\eta_N^T(t))^T$, and $\xi(t)=(s(t)-s(t-\tau(t)))\otimes(\underbrace{1,1,\cdots,1}_{N})^T$, we have

$$\dot{\eta}(t)=(D_f(s)\otimes I_n)\eta(t)-(L\otimes\Gamma)\eta(t-\tau(t))$$
$$+((L-D)\otimes\Gamma)\eta(t)+(L\otimes\Gamma)\xi(t).$$

Let $\varepsilon(t) = \eta(t) - \eta(t - \tau(t))$, then

$$\dot{\eta}(t) = M\eta(t) + N\varepsilon(t) + N\xi(t),$$

where $M = D_f(s) \otimes I_n - D \otimes \Gamma, N = L \otimes \Gamma$.

Select the Lyapunov function as Theorem 1, and calculate derivative for the both sides of the Lyapunov function. Then, we have

$$\dot{V}(t) \le y^T(t)\Phi' y(t) ,$$

where $y(t) = (\eta^T(t), \varepsilon^T(t), \xi^T(t))^T$, According to (8), we can find a positive constant α' such that $\dot{V}(t) \le 0$ and $\dot{V}(t) = 0$ if and only if $\eta(t) = 0$. The same as Theorem 1, we have that the complex network (7) with time-varying delays is synchronized.

Corresponding to Corollary 1, we give the second special case.

Corollary 2: The complex network (1) of N nodes (n-dimension) is synchronized for the control input (3), under Assumptions 1 and 2, if h satisfies

$$\Phi'^{*} = \begin{bmatrix} 2M + dI_n + hMM & N + (1-d)I_n + hMN & N \\ N + (1-d)I_n + hNM & -(1 - d + \frac{1}{h})I_n + hNN & 0 \\ N & 0 & 0 \end{bmatrix} \le 0 ,$$

where $M = D_f(s) \otimes I_n - D \otimes \Gamma, N = L \otimes \Gamma$, and $D_f(s)$ is the *Jacobian* of f evaluated at $s(t)$.

5 Simulations

We give some simulation results about the above discussions in this section. In these figures, the curves are the trajectories of all the nodes of the network and the synchronous target. For the convenience, we let $c_{ij} = c = 0.6$ for all i, j, $\tau(t) = \tau$, and $N = 10$.

In Fig. 1, we consider the synchronization of the complex network for the synchronous target with time-varying delays. In subfigures (a1) (b1) (c1), the coupling structure of the network is $a_{ij} = a_{ji}$. These subfigures show the trajectories of the nodes with different time delays, and $\tau_{a1} < \tau_{b1} < \tau_{c1}$. From the figures, if we have the bigger delay, the synchronization of the network needs the longer time. The needed time of the synchronization equals the lag-time. In subfigures (d1) (e1) (f1), the coupling structure of the network is $a_{ij} \ne a_{ji}$. These subfigures show the trajectories of the nodes with different time delays, and $\tau_{d1} < \tau_{e1} < \tau_{f1}$. From these figures, we can find the following phenomenon. After several times (the size of one time is the size of the lag-time), the network arrives at synchronization, and the bigger delay needs the longer time.

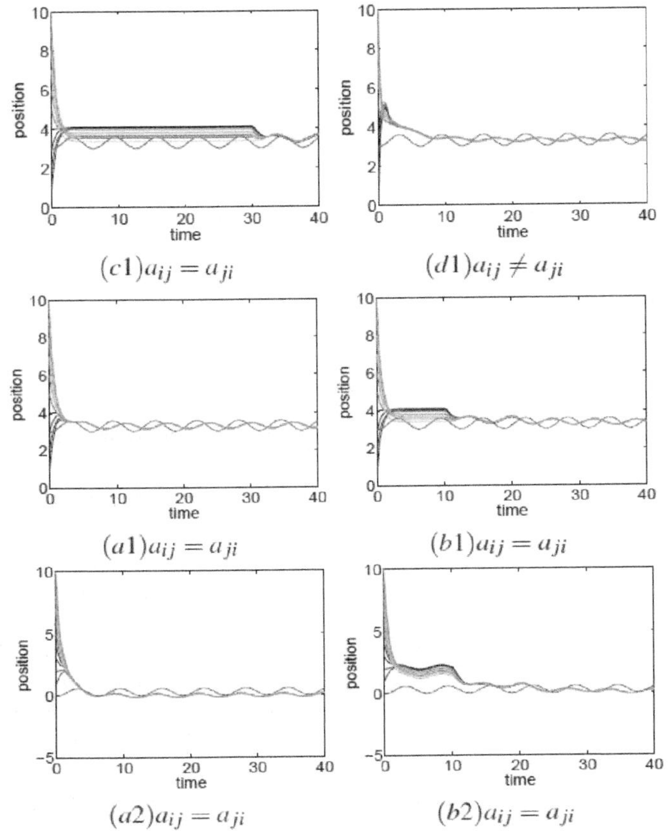

Fig. 1. The trajectories of the nodes of the network (1) with time delays.

In Fig. 2, we consider the synchronization of the complex network for the synchronous target without time-varying delays. In subfigures (a2) (b2) (c2), the coupling structure of the network is $a_{ij} = a_{ji}$. These subfigures show the trajectories of the nodes with different time delays, and $\tau_{a2} < \tau_{b2} < \tau_{c2}$. In subfigures (d2) (e2) (f2), the coupling structure of the network is $a_{ij} \neq a_{ji}$. These subfigures show the trajectories of the nodes with different time delays, and $\tau_{d2} < \tau_{e2} < \tau_{f2}$. We can get the same results as Fig. 1.

From these figures, we find that the bigger time delay need the longer time for the synchronization of the complex network and the network can arrive at the synchronization finally.

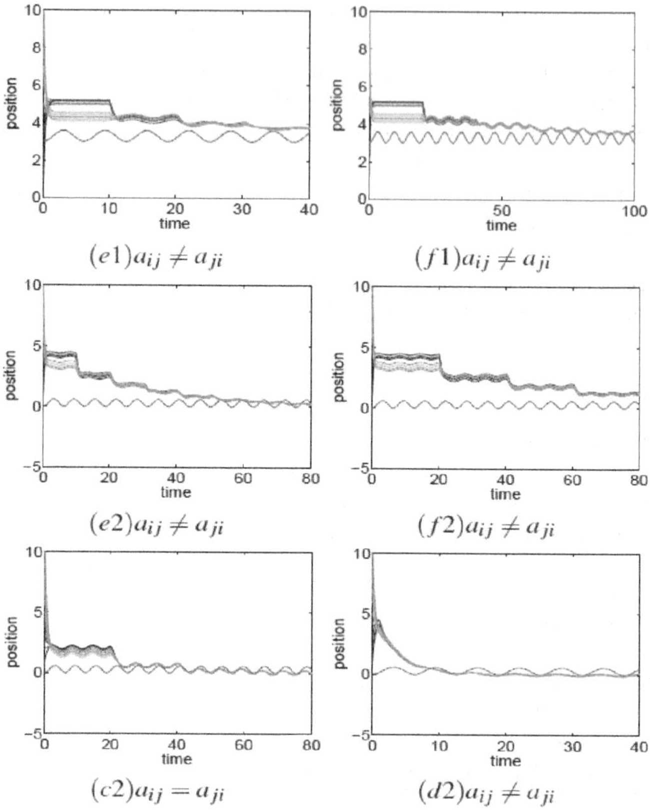

Fig. 2. The trajectories of the nodes of the network (1) with time delays, and the synchronous target without delay.

6 Conclusion

In this paper, we consider the synchronization of a complex dynamical network with time-varying delays for the neighbors. We discuss the synchronization of the network for the synchronous target with time-varying delays. We also discuss the synchronization of the network for the synchronous target without time-varying delays. Specially, we give some simulations to verify our theoretical results.

Acknowledgment. This work was supported by the Beijing Natural Science Foundation Program (1102016), the Beijing Municipal Training Programme for the Talents (2009D005002000006), Funding Project for Academic Human Resources Development in Institutions of Higher Learning Under the Jurisdiction of Beijing Municipality (PHR20110855) , the Beijing College Scientific Research and entrepreneurship Plan Project, and the College Scientific and technological Activities of North China University of Technology.

References

1. Wang, X.F., Chen, G.: Complex networks: Small-world, Scale-free, and beyond. IEEE Circuits Syst. Mag. 3(1), 6–20 (2003)
2. Barahona, M., Pecora, L.M.: Synchronization in small-world networks. Phys. Rev. Lett. 89(5), 54101 (2002)
3. Li, C., Chen, G.: Synchronization in general complex dynamical networks with coupling delays. Phys. A 343, 263–278 (2004)
4. Li, Z., Feng, G., Hill, D.: Controlling complex dynamical networks with coupling delays to a desired orbit. Phys. Lett. A 359, 652–656 (2006)
5. Lu, W., Chen, T.: Synchronization of coupled connected neural networks with delays. IEEE Trans. Circuits Syst. I, Reg. Papers 51(12), 2491–2503 (2004)
6. Chen, G., Zhou, J., Liu, Z.: Global synchronization of coupled delayed the neural networks and applications to chaotic CNN models. Int. J. Bifurc. Chaos 14(7), 2229–2240 (2004)
7. Cao, J., Chen, G., Li, P.: Global synchronization in an array of delayed neural networks with hybrid coupling. IEEE Trans. Syst., Man, Cybern. B, Cybern. 38(2), 488–498 (2008)
8. Yu, W., Cao, J., Lü, J.: Global synchronization of linearly hybrid coupled networks with time-varying delay. SIAM J. Appl. Dyn. Syst. 7(1), 108–133 (2008)
9. Wang, L., Kong, X., Shi, H., Dai, H., Sun, Y.: LMI-based criteria for synchronization of complex dynamical networks. J. Phys. A: Math. Theor. 41, 1751–8113 (2008)
10. Sun, Y.G., Wang, L.: "Average consensus in networks of dynamic agents with switching topologies and multiple time-varying delays. Syst. Control. Lett. 57, 175–183 (2008)
11. Barabasi, A., Albert, R., Jeong, H.: Mean-field theory for scale-free random networks. Physcia A 272, 173–187 (1999)
12. Wang, X., Chen, G.: Synchronization in small-world dynamics networks. Int. J. Bifur. Chaos Appl. Sci. Eng. 12, 187–192 (2002)
13. Cao, J., Yu, W., Qu, Y.: A new complex network model and convergence dynamics for reputation computation in virtual organizations. Phys. Lett. A 356(6), 414–425 (2006)

An Optimized Method for Transient Electromagnetic Field

Zhou Nannan[1], Xue Guo-qiang[1], and Li Xiu[2]

[1] Institute of Geology and Geophysics, Chinese Academy of Science, Beijing 100029
[2] Chang'an University, Shan'xi Xi'an 710049
N.N.Zhou@dr.com

Abstract. During the forward transformation of wave-field, two-step optimized algorithm is used for decreasing the number of integral coefficients and sampling in dispersed digital integral, as while as ensure transient electromagnetic field's calculation precision. This algorithm solves the ill-posed problem produced by large quantity of integral coefficients and depresses the ill character of the first calculation function. During wave-field inverse transformation, normalized algorithm is introduced to the inverse transformation by adopting the deviation theory and the Newton iterative form to make the transformed wave field stable and reliable. It is concluded that this algorithm is effective and useful.

Keywords: Transient electromagnetic field, Wave-field conversion, Optimization, normalization.

1 Introduction

Many kinds of technology have been introduced into transient electromagnetic method for the past few years, such as pseudo-seismic interpretation, electromagnetic migration and comprehensive inverse with migration. These methods extend connotation of transient electromagnetic method and make it possible that drawing more beneficial information for imaging underground target from survey data than conventional TEM (transient electromagnetic method). Especially, wave-field transformation is the focus for its importance and significance in electromagnetic migration and layered imaging.

Diffusion equation of transient electromagnetic field depicts the induction diffusion character in electromagnetic vortex field. However, it is inferior that clarifying the electric boundaries through migration and inversion based on diffusion equation. So, it is the focus that to seek for a mathematics method to converse transient electromagnetic field to wave field. Li Xiu supplies the definition of transformation: extracting propagating character in electromagnetic response and depressing attenuation and loosening character in electromagnetic wave propagation. Through that, it is easy to introduce migration imaging and Born approximate inversion into transient EM-field inversion interpretation. Wave-field transformation is based on the mathematics conversion formula:

D. Zeng (Ed.): ICAIC 2011, Part II, CCIS 225, pp. 82–87, 2011.

$$f(x,y,z,t) = \frac{1}{2\sqrt{\pi t^3}} \int_0^\infty \tau e^{-\frac{\tau^2}{4t}} u(x,y,z,\tau)d\tau \tag{1}$$

This formula is the wave-field conversion style. In this formula, $u(x,y,z,\tau)$ is wave field of $f(x,y,z,t)$; τ is the wave-field time, corresponding to EM-field time; $f(x,y,z,t)$ is the field component of transient electromagnetic field. The importance of this formula is not the forward simulation,but the inverse transformation. However, inverse transformation process is ill-posed and unstable. In order to solve this problem, it is necessary to adopt optimized algorithm and normalized method. This paper supplies an optimized algorithm on wave-field transformation.

2 Optimized Algorithm in Wave-Field Forward Transformation

2.1 Numerical Method

Wave-field transformation formula could be discretized to following style:

$$f(x,y,z,t_i) = \sum_{j=1}^{n} u(x,y,z,\tau_j) a(t_i,\tau_j) h_j \tag{2}$$

$$\text{where } a(t_i,\tau_j) = \frac{1}{2\sqrt{\pi t_i^3}} \tau_j e^{-\frac{\tau^2{}_j}{4t_i}} \tag{3}$$

is named kernel function; h_j integral coefficient;

It is crucial that getting one group of τ_j and h_j makes them meet (2) most, because kernel function decreases quickly as τ increases and wave-field value $u(x,y,z,\tau)$ is usually continuous function, whose variability amplitude is not large and affects calculation results less. Consider a special circumstance: defining $u(x,y,z,\tau)=1$. After that, formula(1) becomes

$$f(x,y,z,t) = \frac{1}{2\sqrt{\pi t^3}} \int_0^\infty \tau e^{-\frac{\tau^2}{4t}} d\tau \tag{4}$$

Using special integral $\int_0^\infty xe^{-ax^2}dx = \frac{1}{2a}$, formula(4) can be transformed into:

$$f(x,y,z,t) = \frac{1}{\sqrt{\pi t}} = \frac{1}{2\sqrt{\pi t^3}} \int_0^\infty \tau e^{-\frac{\tau^2}{4t}} d\tau \tag{5}$$

Then, go for the discretized formula of (5) through numerical integral:

$$f(x,y,z,t_i) = \frac{1}{\sqrt{\pi t_i}} = \sum_{j=1}^{n} a(t_i,\tau_j) h_j \tag{6}$$

Choose one group of τ_j and h_j (j=1,2,…,n)by optimized method and take them into formula (2), then work out one set of transient electromagnetic values $f(x,y,z,t_i)$ (i=1,2,…,m).

2.2 Optimized Method

Select m and n as sampling time tracks and integral coefficients respectively. M must be larger than n for the need of full-rank matrix in wave-field transformation formula. For the purpose of getting optimum value to (6), choose the best group of τ_j and h_j (j=1,2,…,n) through optimized method.

For one set of t_i and τ_j (i=1,2,…,m; j=1,2,…,n), $a_{ij}h_j = f_i$ (7)

$$\text{where } a_{ij} = \frac{1}{2\sqrt{\pi t^3_i}}\tau_j e^{\frac{\tau_j^2}{4t_i}}; f_i = \frac{1}{\sqrt{\pi t_i}} \tag{8}$$

Last formula's matrix form is: **A·H=F** (9)

where $A=[a_{ij}]_{m\times n}$ is coefficient matrix; $F=\{f_i\}_m$ is known transient electromagnetic vector; $H=\{h_j\}_n$ is integral coefficient vector.

Choose τ_j and h_j to make objective function

$$\varphi = \|F - A \bullet H\|^2 \tag{10}$$

minimum.

There are two steps to acquire suitable τ_j and h_j. The first step is to give one set of τ_j (j=1,2,…,n) and get the value of a_{ij}. Then, work out h_j through minimizing φ. The necessary and sufficient condition for minimizing formula(10) is that H satisfies

$$A^T A \bullet H = A^T \bullet F \tag{11}$$

Solve last equation and get one set of h_j (j=1,2,…,n), which makes φ minimum in given τ_j. The second step is to research how to gain one group of τ_j (j=1,2,…,n) for objective function's minimum under the condition of h_j (j=1,2,…,n) known. Provided that $V_i = a_{ij}h_j$ (12) and its matrix form $V = A \bullet H$ (13) are feasible. Develop V into taylor series and choose the first order of $\delta\tau_j$, then, get the equation:

$$V = V_0 + \frac{\partial V}{\partial \tau}\delta\tau \tag{14}$$

where $V_0 = (v_i^{(0)})_m, \frac{\partial V}{\partial \tau} = \left(\frac{\partial v_i}{\partial \tau_j} \right)_{m \times n}, \delta\tau = (\delta\tau_j)_n$ where $v_i^{(0)}$ is one set of V_i under the value of ${\tau^{(0)}}_j$; $\delta\tau_j$ is tiny increment of τ_j. Take last equation into objective function:

$$\varphi = \left\| F - V^{(0)} - \frac{\partial V}{\partial \tau} \delta\tau \right\|^2 \tag{15}$$

where φ is $\delta\tau$'s function. Then , $\delta\tau$ is worked out by minimizing φ. After that, the equation is inferred:

$$(B + \alpha I) \bullet \delta\tau = C \tag{16}$$

where

$B = \left(\frac{\partial V}{\partial \tau} \right)^T \left(\frac{\partial V}{\partial \tau} \right), C = \left(\frac{\partial V}{\partial \tau} \right)^T (F - V^{(0)})$,;**I** is identity matrix; α is damping coefficient. Solve function (16) and acquire new value of $\tau^{(1)}$:

$$\tau^{(1)} = \tau^{(0)} + \delta\tau . \tag{17}$$

Take $\tau^{(1)}$ into equation(14), repeat above-mentioned process and obtain secondary estimate value. Ultimately, the most optimized τ is worked out.

3 Regularizing Algorithm in Inverse Transformation of Wave-Field

In practice, it's the goal to get wave-field through transient electromagnetic field. At first, ill-posed problem (2) is transformed into regularizing equation(18) by Tikhonov regularizing theory.

$$A^T A + \alpha(\delta) I) \bullet U = A^T F \tag{18}$$

Through equation (18), approximation solution of wave-field comes out. But there is a key parameter $\alpha(\delta)$ must be worked out by deviation principle. $\alpha(\delta)$ makes bias match error vector and if $F^h \in R(A_h), \left\| F^h - F_\delta^h \right\| \le \delta, and \left\| F_\delta^h \right\|^2_{L^h} > \delta^2$, there must be one $\alpha(\delta)$ satisfies equation (19):

$$\Delta_\eta(\alpha) = \varphi(\alpha\ (\eta)) - \left(\delta + \hbar \left\| U_\alpha^h \right\|_{\omega^h} \right)^2 = 0 \tag{19}$$

Take equation (19) into Taylor series at α_k :

$$\Delta_\eta(\alpha) = \Delta_\eta(\alpha_k) + \Delta'_\eta(\alpha_k)\Delta\alpha_k = 0 \text{ , where } \Delta\alpha_k = -\frac{\Delta_\eta(\alpha_k)}{\Delta'_\eta(\alpha_k)}. \tag{20}$$

Then change regularizing parameter by Newton iterative computation:

$$\alpha_{k+1} = \alpha_k + \Delta\alpha_k = \alpha_k - -\frac{\Delta_\eta(\alpha_k)}{\Delta'_\eta(\alpha_k)} \quad \text{k=0,1,2,...,n,} \tag{21}$$

$$\text{where } \Delta'\eta(\alpha) = -(\frac{dU_\alpha^h}{d\alpha})U_\alpha^h\left[\alpha + \hbar^2 + \hbar\delta / \left\|U_\alpha^h\right\|_{\omega^h}\right]. \tag{22}$$

After foregoing steps, discretizing regularizing algorithm comes out as the final result:

(1) Give δ and $\hbar$ to $A_{h,} F_\delta^h$ as input error and start computing with initial regularizing parameter α_0 and controlled precision ε ;

(2) Discretize formula(1) into following equation: $A_h \bullet U^h = F_\delta^h$, where k=0;

(3) Solve regularizing equation $(A_h^T A_h + \alpha_k I)U_h = A_h^T F_\delta^h$ and get the value of $U_{\alpha_k}^h$;

(4) Get $\Delta_\eta(\alpha_k)$ through equation(19) and skip to step(8) if $|\ \Delta_\eta(\alpha_k)\ | < \varepsilon$;

(5) Solve equation $(A_h^T A^h + \alpha_k I)(U_{\alpha_k}^h)' = -U_{\alpha_k}^h$; get solution of $(U_{\alpha_k}^h)'$;

(6) Compute $\Delta'_\eta(\alpha_k)$ and α_{k+1} according to formula(22) and (21);

(7) Definite K=K+1, then turn to step(3);

(8) Definite $U_{\alpha^*}^h = U_{\alpha_k}^h$ and output result. Put the algorithm into end.

4 Application

On application, the aim of previously described algorithm is to give the transient electromagnetic field. Take wave-field value as one and get the transient electromagnetic field $f(x, y, z, t) = \frac{1}{t\sqrt{\pi}}$, where m=n=20. Separate the actual measurement time of TEM into seven segments. Then, estimate the regularizing parameter by deviation principal and Newton iterative method. The mean square relative errors of seven time sections are 7.3%,8.7%,4.2%,5.6%,8.0%,4.7%,2.2% respectively. By these results, the degree of sensitivity in some time section is not bad. So, compress the time range of time section and add the number of time sections. After that, calculation precision will be more accurate.

5 Conclusion

This paper brings optimized algorithm method and regularizing method into wavefield transformation. These methods control and decrease the number of dispersed integral coefficients and abscissa in the forward transformation under the condition of keeping calculation precision. After that, the ill-posed problem is improved in inversion matter.

References

1. Wang, J.Y., Oldenburg, D., Levy, S.: Magnetotelluric pseudo-seismic interpretation. Oil Geophysical Prospecting 20(1), 66–79 (1995)
2. Levy, S., Oldenburg, D., Wang, J.: Subsurface image using magnetotelluric data. Geophysics 53, 104–117 (1988)
3. Wang, J.Y., Fang, S.: Concerning magnetotelluric average velocity problem. Prospectingfor Petroleum 25(1), 84–89 (1986)
4. Wang, J.Y.: Magnetotellufic Pseudo-Seismic Interpretation. Petroleum Industry Press, Beijing (1990)
5. Zhdanov, M.S., Matusevich, V.Y., Frenkel, M.A.: Seismic and electromagnetic migration, Nauka (1988) (in Russian)
6. Zhdanov, M.S., Booker, J.R.: Underground imaging by electromagnetic migration. In: 63rd Ann. Internat Mtg. Expl. Geophys, Expanded Abstracts, pp. 355–357 (1993)
7. Zhdanov, M.S., Li, W.D.: 2D finite-diference time domain electromagnetic migration. In: 67th SEG EM2, vol. 1, pp. 370–373 (1997)
8. Lee, S., Memechan, G.A.: Phase-field imaging: The electromagnetic equivalent of seismic migration. Geophysics 52(5), 678–693 (1987)
9. Chen, B.C., Li, J.M., Zhou, F.T.: Wave-field conversion method for transient electromagnetic field. Oil Geophysical Prospecting 34(5), 546–554 (1999) (in Chinese)
10. Xiao, T.Y., Yu, S.G., Wang, Y.F.: Numerical Solution of Inversion Problem, pp. 143–153. Science Press, Beijing (2003) (in Chinese)

A Novel Matrix Model of 802.11 Wireless LANs and Coverage Rate Optimization Using Improved GA

Maoheng Sun[1,*], Qian Li[1], Xuan Zhou[2], Jian Wang[3,4], Yu Zheng[1], and Yanwei He[1]

[1] College of Electronics and Information Engineering, Tongji University, Shanghai 201804, China
[2] Key Laboratory of Advanced Control and Optimization for Chemical Processes, Ministry of Education, East China University of Science and Technology, Shanghai 200237, China
[3] State key laboratory of Mechanical System and Vibration, Shanghai Jiao Tong University, Shanghai, 200240, China
[4] Cyber Physical System R&D Center, The Third Research Institute of Ministry of Public Security, Shanghai, 201204, China
sunmaoheng@sina.cn

Abstract. AP placement is a vital design issue for deploying an 802.11 Wireless Local Area Networks system. In this paper, a comprehensive solution including modeling of WLAN and optimization of coverage rate is presented to enhance economy, practicability and adaptability of WLAN system. A novel matrix model of WLAN system (MMWS) is established by mapping a power matrix to a corresponding effective coverage element (ECE) matrix in a target area. Providing that transmitting power and user demands are given, a optimization algorithm comprising some adapted GA operators is proposed to deploy the logical Access Points (APs) for obtaining maximized coverage rate (CR). Meanwhile, a new fitness function is provided to assess the network coverage and improve the crossover/mutation operations. By experiments on synthetical data, performance of our method is evaluated and analyzed in detail. The results indicate that the effective CR of a target area obtained by the proposed method has exceeded 95%, and the co-channel interference has been avoided to the utmost extent. It is evident that the proposed solution is practical to optimize the CR of WLANs.[1]

Keywords: IEEE 802.11 WLANs, Matrix model of WLAN system, MMWS, Coverage rate optimization, Improved GA.

[*] Corresponding author.

[1] Sponsed by following projects:

1. Major program of Science and Technology Commission of Shanghai Municipality (No. 09DZ1122202);
2. National Natural Science Foundation projects (No. 60804029);
3. National Natural Science Foundation projects (No. 50805094);
4. National Basic Research Program of China (No. 2006B705400);
5. National Science and Technology Support Projects of China (No. 2006BAH02A17).

D. Zeng (Ed.): ICAIC 2011, Part II, CCIS 225, pp. 88–96, 2011.

1 Introduction

In 1997, IEEE published IEEE 802.11 standard[1], the first Wireless Local Area Network (WLAN) specification, which has been developed and perfected rapidly[2, 3]. The aim of WLAN design is to satisfy the traffic demands from the mobile terminals while remaining the cost as low as possible[4].A genetic design procedure for deploying Access Points (APs) in a multi-cell WLAN system often consists of four steps[5]:

1) Partitioning the entire target area into many grids.
2) Selecting candidate positions for each Access Point.
3) Measuring signal strength of each candidate[6, 7].
4) Determining positions of APs and assigning the channels[8].

Within step 4) of the design procedure above, there are two basic design principles. Firstly, APs should be placed so that there are no coverage gaps in the target area with minimized coverage overlaps between and among APs. Another requirement is that the cost of equipment and installation is preferable to be as low as possible, i.e. number of APs should be minimized.A crucial problem of equilibrium between performance and economy has come into being that how to deploy and position APs, and how to obtain best coverage rate (CR) once the transmitting power and user demands are given. A systematic and synthetical approach should be proposed to solve this problem rather than utilize traditional simulation tools only. Although many methods of base station (BS) deploying in the cellular networks based on CDMA or TDM have been put into practice[2], they cannot be directly transplanted into the CSMA/CA WLANs. As one of most popular heuristic methods, genetic algorithm (GA) is a promising approach to optimize the CR of WLANs.In this paper, a novel matrix model is proposed to enhance the adaptability of WLANs, and a CR optimization method using improved GA is implemented.

2 Matrix Model of WLAN System

For obtaining a maximum CR[3], we propose a matrix model of WLAN system (MMWS) that is propitious to optimize AP locations. As **Fig. 1** illustrated, a target area is plotted as a square. Each grid is located by its central coordinate. The asterisks symbolize AP candidate positions initialized by a random generation process.

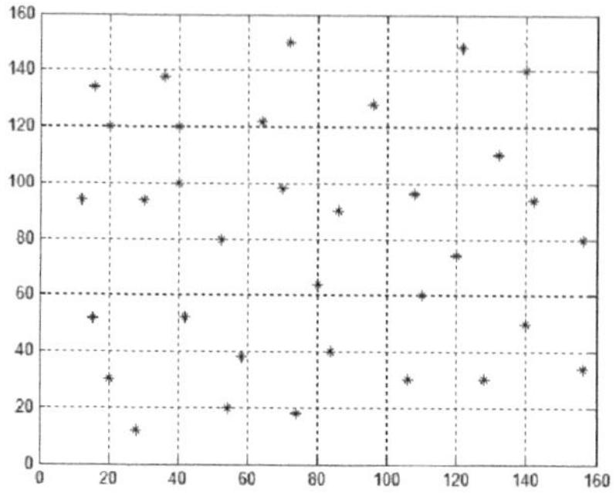

Fig. 1. A target area and candidate AP locations

The mathematical model of MMWS can be established based on some parameters. The count of grids is m, number of candidate locations n, and the required number of APs k. The effective covered grids in the target area has a form of union, shown as Eq. (1). Where, D_i denotes the number of covered grids of each AP.

$$D = \bigcup_k D_i, \quad i = 1, 2, \cdots, n \tag{1}$$

Eq.(2) gives the total effective CR, R_{cov}, which is the ratio of the sum of ECE and the count of grids.

$$R_{cov} = \frac{D}{m} = \frac{\bigcup_k D_i}{m}, i = 1, 2, \cdots, n \tag{2}$$

To simulate the strength of received signal, the two-ray-ground model is adopted. The power of received signal P_r can be calculated by Eq. (3).

$$P_r = \frac{P_t G_t G_r h_t^2 h_r^2}{d^4 L} \tag{3}$$

Where,

P_t — power of sending signal;
G_t — transmitting antenna gain
G_r — receiving antenna gain;
h_t — transmitting antenna height;
h_r — receiving antenna height;
d — transmission distance of the radio signal;
L — channel disturbance.

A target area can be represented by an m×n matrix S, shown as Eq.(4). Where m is the count of grids in the target area, n is the count of the candidate locations of APs. The value of element s_{ij} is the signal power of ith grid received from the jth AP position.

$$S = \left[s_{ij} \right]_{m \times n} \tag{4}$$

If the power that a grid receives from a given AP is higher than -60 dBm, that grid will be regarded as an effective coverage element (ECE) subordinated to that AP position. Correspondingly, matrix S can be transformed to a matrix A by binarization with power thresholding, shown as Eq. (5). The value of element a_{ij} is 1 if the ith grid is an ECE of the jth AP position, otherwise 0.

$$A = \left[a_{ij} \right]_{m \times n}, \quad a_{ij} = \{0,1\} \tag{5}$$

Eq.(6) exemplifies a matrix of power and its corresponding matrix of ECE.

$$S = \begin{pmatrix} -63.5 & -58.3 & -46.1 & \cdots & -46.8 \\ -79.6 & -65.4 & -55.1 & & -48.5 \\ & \vdots & & \ddots & \vdots \\ -50.8 & -100.3 & -64.2 & \cdots & -46.7 \end{pmatrix}_{m \times n}$$
$$\Rightarrow A = \begin{pmatrix} 0 & 1 & 1 & \cdots & 1 \\ 0 & 0 & 1 & & 1 \\ & \vdots & & \ddots & \vdots \\ 1 & 0 & 0 & \cdots & 1 \end{pmatrix}_{m \times n} \tag{6}$$

3 Key Algorithms of MMWS System

3.1 Improvement of Genetic Operators

Three basic operators in traditional GA are employed in our method, reproduction, crossover and mutation. In a reproduction operation, the individual strings are copied from the parents to their offspring according to their fitness value. The probability of reproduction operation is linear with the fitness value of an individual, i.e. higher fitness results in more offspring in the next generation. Crossover operator exchanges genes of two individuals from a certain point that is often selected randomly. Mutation operator changes the bit of the individual with a comparatively low probability. Crossover and mutation improve the searching capability and adaptability of GA algorithm by simulating diversity of the population resembling natural species.

Here crossover is regarded as a principal operator since it has strong capability on global searching, meanwhile mutation as the auxiliary operator due to its local searching capability. Fixed crossover and mutation probabilities are used in traditional GA, which may bring many troubles to the algorithm. For instance, a low probability of mutation and crossover often significantly slows down the convergence speed, or leads to premature solution. On the contrary, a high mutation probability could result in divergence of the procedure. In order to avoid these problems, adaptive crossover and mutation operators are proposed in our improved GA. Crossover and mutation rates of a pair of parents are adjusted automatically depending on the similarity between them. A high similarity will reduce the crossover rate of parents and vice versa. Adaptive alternation of crossover rate improve the effectiveness of operation and global search. Similarly, mutation rate against the fitness of an individual have a same varying relationship between individual similarity and crossover rate. Adaptive mutation operation is conducive to maintain the diversity of population, accelerates the convergence speed and prevents the algorithm from falling into local optimum. The proposed crossover and mutation rates are calculated as Eq. (7) and Eq. (8).

$$c_{i,n+1} = \frac{c_{i,n}}{\ln\left(e + s_{i,j}\right)}, \quad j = 1, 2, \ldots, N \tag{7}$$

$$m_{i,n+1} = \left(1 - \ln f_{i,n}\right) m_{i,n} \tag{8}$$

Where,

N — the sum of individuals in the population;

$c_{i,n}$ — the crossover rate between the ith and jth individuals in the nth generation;

$s_{i,j}$ — the similarity between the ith and jth individuals. Hamming distance is regarded as the similarity;

$m_{i,n}$ — the mutation rate of the ith individual in the nth iteration;

$f_{i,n}$ — the fitness of the ith individual in the nth generation.

3.2 Fitness Function Optimization

Instead of real values, we adopt a new encoding mechanism, in which each individual is treated as a union of k independent parts, where k is the required number of APs. Each part represented by a string corresponds to the index of a selected AP. Assuming the total number of AP is n, the length of the string, l, satisfy the condition given in Eq.(9). Hence, the length of an individual will be $k \times l$ bits.

$$2^{l-1} < n < 2^{l} \tag{9}$$

The decoded value of each part is calculated by Eq. (9) and sometimes it can be larger than the total number of APs n, which will cause overflow of the part number data. A linear mapping method is proposed to handle this problem described schematically in **Table 1**.

Table 1. Coding of Individuals

Decimal	Encoded	Crossover, mutation	Offspring	Decoded
0	00···00	⇒	00···01	$0 < k < n$
1	00···01	⇒	01···00	
⋮	⋮	⇒	⋮	⋮
n	10···11	⇒	11···11	Overflow

Since the decoding value of each part should be bounded to the range of [1, n], a linear scaling is introduced into the decoding mechanism to calculate the index of an AP. The mapping function is given as Eq.(10). Where function *int* maps its parameter to a nearest integer, I_D', which refers to the decimal value of a part and satisfy $0 \le I_D' \le n$.

$$\frac{I_D'}{I_D} = \frac{C_n^k}{2^l - 1} \Rightarrow I_D = \mathrm{int}\left(I_D \cdot \frac{C_n^k}{2^l - 1} \right) \tag{10}$$

The fitness function is calculated according to the expression of genes. Fitness function f_1(I) has the same definition with CR.

$$f_1(I) = CR = \frac{Cov}{m} \tag{11}$$

Cov refers to the number of effective coverage grids, while *m* is the sum of grids in the entire area. This fitness function performs poorly in GA due to its tendency to converge prematurely in the early phases, besides, it may lead to slow convergence speed. The following linear scaling function, f_2(I), is proposed to handle these problems.

$$f_2(I) = a \cdot f_1(I) + b \tag{12}$$

The linear scaling coefficient *a*, *b* make f_2(I) to satisfy:

$$\begin{cases} f_2^{avg} = f_1^{avg} \\ f_2^{\max} = c \cdot f_1^{avg} \end{cases} \tag{13}$$

Where, f_1^{avg} and f_2^{avg} are average of f_1(I) and f_2(I), respectively. f_1^{max} is maximum of f_1(I). *c* is a constant to control converge speed. From Eq. (12) and Eq. (13), we can get:

$$\begin{cases} f_{avg2} = a \cdot f_{avg1} + b \\ c \cdot f_{avg1} = a \cdot f_{\max 1} + b \end{cases} \tag{14}$$

Here, the value of *a*, *b* are obtained, and the value of f_2(I) can be evaluated by Eq. (12). On the basis of f_1(I), f_2(I), here a new fitness function, f_3(I), is proposed as Eq. (15). The value of f_3(I) accords with exponential growth within a specific range.

$$f_3(I) = \exp\left(\frac{Cov}{m} \right) \tag{15}$$

4 Performance Evaluation and Result Analysis

In the simulation environment, the entire area is a square of 160m×160m, which is partitioned into 6400 symmetrical grids with the identical size 2m×2m, plotted in

Fig. 1. There are 40 location candidates from which 3 sites will be selected to place the APs. Accordingly, the current optimization problem is equivalent to searching suitable locations from location candidates in order to maximize the overall radio coverage. Therefore, the sum of the feasible solutions is C_{40}^{3}=9880.

Here, a two-ray-ground model is utilized to discriminate ECEs. The parameters of the proposed algorithm are provided as follows. Initial crossover rate and initial mutation rate are 0.1 and 0.6, respectively. Number of iteration is limited to 200 rounds. Parameter *c* in Eq. (13) is 3.

Each fitness function is calculated for 20 times, and the average value is used. Results of different fitness functions are plotted in **Fig. 2**, **Fig. 3** and **Fig. 4**, respectively. **Fig. 2** indicates the algorithm with f_1(I) converges in 40 generations, and is insensitive to the variance of population size (PS). It is obviously observed from **Fig. 3** that the convergence rate of f_2(I) is slower than that of f_1(I). Compared to PS=5, the algorithm performs better when PS=50. It means that population size has some effects on the algorithm with f_2(I), but they are not monotone linear relationship. An excessive size of population could easily lead to the degeneration of the algorithm efficiency. **Fig. 4** depicts that convergence rate of f_3(I) is faster than the previous two fitness functions, and the effect of population size on the performance of algorithm is weaker as well. The algorithm with a large number of population performs just a slightly better than with a small population size.

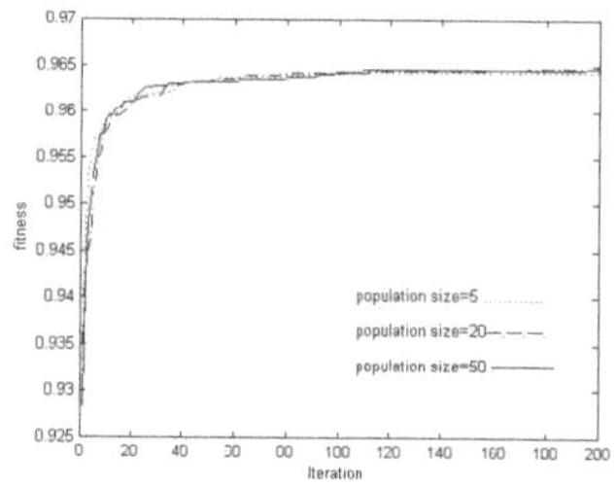

Fig. 2. Value of f_1(I) with different population size

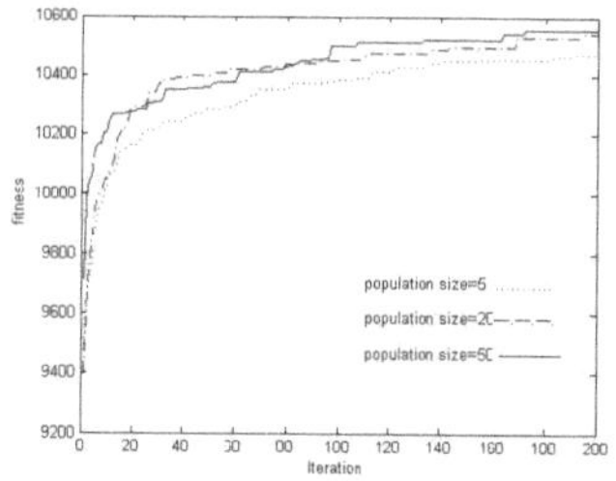

Fig. 3. Value of f_2(I) with different population size

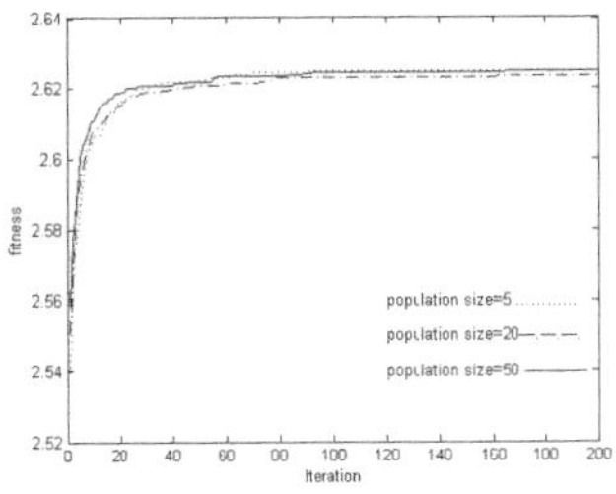

Fig. 4. Value of $f_3(\mathrm{I})$ with different population size

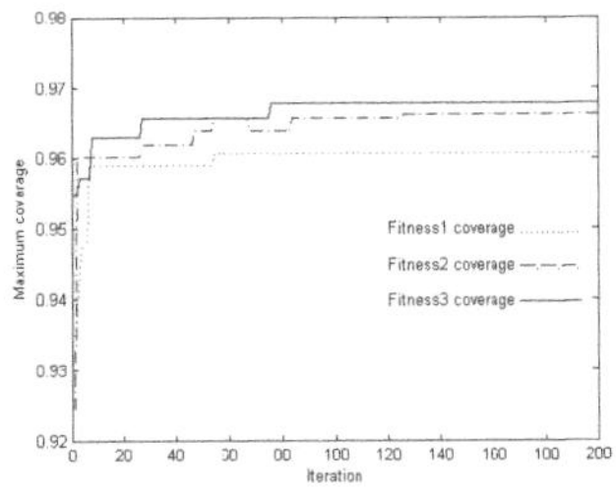

Fig. 5. Max CRs of the three fitness functions

The maximum coverage curves of the three fitness functions are shown in **Fig. 5**, which indicates that the CR of $f_3(\mathrm{I})$ is the largest one, and $f_2(\mathrm{I})$ follows. A noise appears in the 60th iteration. Fitness function just reflects the CR and guides the direction of convergence, but is not completely positive correlated with the CR. Therefore, the difference between $f_2(\mathrm{I})$ and the direction of convergence leads to a noise.

The simulation results of the three fitness functions are listed in Table II. It indicates that the maximum CR of $f_3(\mathrm{I})$ is larger than $f_1(\mathrm{I})$ and $f_2(\mathrm{I})$, and the convergence speed of $f_3(\mathrm{I})$ is faster than $f_2(\mathrm{I})$, and weigh against $f_1(\mathrm{I})$. In conclusion, algorithm with $f_3(\mathrm{I})$ performed better than the others do.

Table 2. Comparison among 3 Fitness Functions

	f_1(I)	f_2(I)	f_3(I)
Fitness	0.9643	1.5690e+004	2.6318
CR	96.43%	96.56%	96.80%
Covered grids	6172	6180	6195
Final iteration	47	124	55

5 Conclusions

The design of an 802.11 WLAN system should ensure complete coverage of the target area. AP locations have to be selected so that there are no coverage gap throughout the target area and with minimized co-channel coverage overlap. By mapping a power matrix to a corresponding ECE matrix, we firstly established a matrix model of WLAN system, which is easy to be analyzed and optimized by heuristic methods. Conventional GA operators are improved with embedding some heuristic criteria and employing a proposed fitness function. Given transmitting power and user demands, an adaptive GA is implemented to place AP locations with maximized effective coverage rate in a synthetical target area. Evaluation results and comparison among different fitness functions have evidently proved the practicability, adaptability and efficiency of proposed solution on optimization of coverage rate and avoidance of the co-channel interference.

References

[1] Crow, B.P., Widjaja, I., Kim, J.G., et al.: IEEE 802.11 wireless local area networks. IEEE Communications Magazine 35(9), 116–126 (1997)

[2] Lieska, K., Laitinen, E., Lahteenmaki, J.: Radio coverage optimization with genetic algorithms. In: The Ninth IEEE International Symposium on Personal, Indoor and Mobile Radio Communications, pp. 318–322. IEEE, Boston (1998)

[3] Lee, Y., Kim, K., Choi, Y.: Optimization of AP placement and channel assignment in wireless LANs. In: 27th Annual IEEE Conference on Local Computer Networks, pp. 831–836 (November 2002)

[4] Prommak, C., Kabara, J., Tipper, D., et al.: Next generation wireless LAN system design. Proceedings of MILCOM 2002, Anahem, California, USA, pp. 473–477 (October 2002)

[5] Hills, A.: Large-scale wireless LAN design. IEEE Communications Magazine 39(11), 98–107 (2001)

[6] Ling, X., Yeung, K.L.: Joint access point placement and channel assignment for 802.11 wireless LANs. IEEE Transaction on Wireless Communications 5(10), 1536–1576 (2006)

[7] Armour, S., Doufexi, A., Lee, B.-S., et al.: The impact of power limitations and adjacent residence interference on the performance of WLANs for home networking applications. IEEE Transction on Consumer Electronics 47(3), 502–511 (2001)

[8] Rodrigues, R.C., Mateus, G.R., Loureiro, A.A.F.: On the design and capacity planning of a wireless local area network. In: 2000 IEEE/IFIP Network Operations and Management Symposium, pp. 335–348. IEEE, Honolulu (2000)

The Matchup Study among the Reservoir's Input Volume, Reservoir Capacity and the Deployed Data Based on SWAT Model

A Case Study of Zhelin Reservoir

Chongliang Sun, Juanle Wang, and Jia Du

Institute of the Geographic Sciences and Natural Resources Research CAS
Beijing, China
C.L.Sun@usa.com

Abstract. The capacity is an important parameter of the reservoir, the variety of which plays a key role in the circulation of reservoir. The two most important factors, which affect the reservoir's capacity, are the input volume and the output volume. This paper simulates the input volume with the help of SWAT model which take the basin's underlay into account, and get the real output volume data from the practical deployed management. Under such basis, the paper analyzes the month's input volume, the month's output volume, and the change of the reservoir's capacity of Zhelin reservoir. The study shows the matchup is pretty good among the reservoir's input volume, reservoir capacity and the deployed data in Zhelin reservoir. This study could help the decision-makers to forecast the right output volume under a special weather condition to deploy the reservoir' capacity to ensure it's safety and make the greatest profit at the same time.

Keywords: Reservoir's capacity, Input volume, Deployment, Hydrological model, Matchup.

The capacity is an important parameter of the reservoir, the variety of which plays a key role in the circulation of reservoir. And it has influence on many aspects, such as the electricity generation, the normal drainage to the lower reaches of a river, supplying water to the public, adjust the runoff of the lower reaches of the river, pisciculture in the reservoir, and the controlling flood in many areas. With the rapid development of modern information technology, especially the 3S technology, it is possible for us to simulate the basin's runoff with hydrological model and on such basis to matchup the capacity change of the reservoir to make a automate adjustment of the reservoir's capacity. And the SWAT model, as a distributed hydrological model, has been studied adequately both on theory and practice[1~3]. This paper will make a comprehensive analysis among the reservoir's input volume based on SWAT model, the output volume, and the capacity variation to expect to forecast the output volume under the given weather conditions.

D. Zeng (Ed.): ICAIC 2011, Part II, CCIS 225, pp. 97–104, 2011.

1 The Study Method

This paper uses the water-balance method to analyze the relationship among the reservoir's monthly input volume, the reservoir's monthly output volume, and the reservoir's capacity variation. The reservoir's monthly input volume would be acquired from the SWAT distributed hydrological model taking account of the underlying conditions of the basin. The reservoir's output volume would be acquired from the practical observed data on the spot, as well as the reservoir's volume variation. At the same time, the water-balance equation of a reservoir can be described as the following:

$$V_{\mathrm{var}} = V_{in} - V_{out} \tag{1.1}$$

where,

V_{var}: Net monthly variation of the reservoir's capacity, million m^3;
V_{in}: monthly input volume modeled by SWAT, million m^3;
V_{out}: monthly output volume acquired from the observed data, million m^3.

Form the above equation of (1-1), the reservoir's monthly input volume can be calculated by the following equation:

$$V_{in} = V_{\mathrm{var}} + V_{out} \tag{1.2}$$

Based on the above equation, the paper would use a complete basin with a reservoir at the end of the basin and the reservoir should be suitable to model several hydro-process.

2 Introduction of the Study Area

This paper select Zhelin reservoir basin to perform the model and its validation. The study area is located in the north-east of Jiangxi province, China, that is the Xiuhe river basin which is a sub-branch of the Yangtze river, with a reservoir at it's low reach. And the reservoir just locates in the east longitude 115.5° and the north latitude 29.2°. This basin is like a strip with 176 km in west-east direction, longer than 84 km which is the average width in south-north direction. And the altitude is between 10-1200m. The basin is with an area of 9340 km^2, and the main river length of 353 km. In this basin, the multi-year average observed flow amount is 10.85 billion m^3. And the basin is composed of 60 percent of mountain, 30 percent of hill, 7 percent of hillock and the other 3 percent of valley plain. The following map is the boundary of this study area. The fig 1 shows the study area location and the shape information, etc.

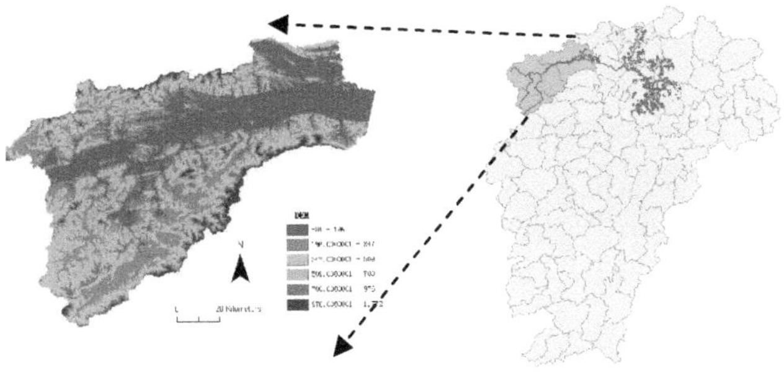

Fig. 1. The shape and location of the research area

In this area, the average temperature yearly is 16-17 degrees, and the maximum appears in July, about 29 degrees. The monthly minimum average temperature is almost in January, about 6-7 degrees. The yearly average humidity is 80%, disperse at the whole area evenly. And the yearly average wind velocity is 2.1 grade, and the upper is low when the lower part is stronger.

3 The SWAT Model and Its Parameter Calibration and Validation

SWAT(Soil and Water Assessment Tool)was developed by USDA-ARS, Doctor Jeff Amonld in 1994. And it is a long-term distributed hydrological model to help to manage the water resource, etc. The model have many input parameters, but for this study we just to analyze the runoff, so we just use several relative parameters such as the precipitation, temperature, evapotranspiration, land-use, soil, relative humidity, etc. and almost all of such data come from the data-sharing network[4], and the soil property data come from the literature[5], the spatial data of the soil property would be interpolated by the the cubic spline interpolation[6,7]. The model takes the inlet area threshold 4500ha, and produce 119 sub-basins totally.

Because the SWAT model has a different performance in different region, so the parameters should be calibrated according to the practical territory. With the J.E.Nash[8] theory, we select the following parameters to validate, the model efficiency parameter E_{ns}, the evenly error R_e, and the relative parameter R^2. And the R^2 would be calculated in excel, the other two calculating formula as the following:

$$R_e = \frac{Q_{sim,i} - Q_{obs,i}}{Q_{obs,i}} \times 100\%$$

$$Ens = 1 - \frac{\sum_{i=1}^{n} \left(Q_{obs,i} - Q_{sim,i}\right)^2}{\sum_{i=1}^{n} \left(Q_{obs,i} - \overline{Q}_{obs}\right)^2}$$

Where, $Q_{obs,i}$ is the practical runoff process

$Q_{sim,i}$ is the simulating runoff process

$\overline{Q}_{obs}$ is the even practical observed data

This study take the zhelin basin data from 2000 to 2004 to calibrate and validate the parameters. Thereinto, the data from 2001-2004 would be used to calibration and the data of 2000 will be used to the validation. And the calibrated parameters is as the following.

Table 1. Calibrated parameters of the model

parameter	The value
ESCO	0.1
SCS curve number CN_2	-7
a parameter ALPHA_BF	0.041
SOL_AWC	0.04

After the calibration, the R_e ,E_{ns} , R^2 take the parameters respectively, during the validation period 0.101, 0.808, and 0.925. And the simulation result is pretty well in zhelin basin. The runoff contrast from SWAT and the observed data can be described by fig 2.

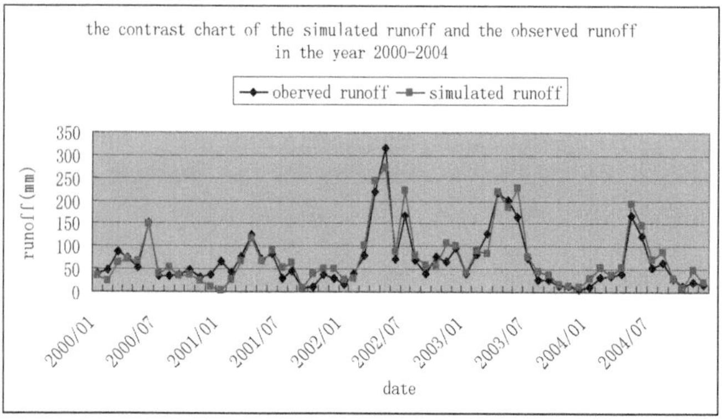

Fig. 2. The sketch map of the relationship between monthly observed runoff and the modeled runoff of Zhelin basin of 2000 to 2004

4 The Model Result

In this paper, we used the SWAT model to simulate the monthly input volume of the reservoir taking account into the underlying conditions. At the same time, we take the output volume data and the reservoir's capacity variation data from the observed data. And the matchup study could be performed now. The monthly input volume, the output volume, and the reservoir's variation data could be described by the tab2, as well as the fig 3.

Table 2. The data of monthly reservoir volume variety and the modeled runoff as well as the reservoir outlet amount between the year of 2000 to 2004 in zhelin basin (Unit: million m^3)

date	Reservoir volume change	Simulated input	Outlet volume	Water increment
2000/01	-94.6	329.702	469.1726	-139.471
2000/02	55	234.434	438.8799	-204.446
2000/03	31.24	597.8534	884.7868	-286.933
2000/04	75.01	670.7988	613.6985	57.1003
2000/05	169.85	621.6704	317.1201	304.5503
2000/06	920.16	1392.22	494.1109	898.1095
2000/07	115.04	376.5888	212.47	164.1188
2000/08	-101.984	483.0648	433.7136	49.3512
2000/09	-728.248	320.0818	1080.821	-760.74
2000/10	-18.848	341.0034	470.1465	-129.143
2000/11	-248.52	218.2758	546.915	-328.639
2000/12	-533.92	100.1248	881.699	-781.574
2001/01	-29.4	23.1632	648.5707	-625.408
2001/02	-768.264	251.5262	1160.166	-908.64
2001/03	150.792	621.0166	568.7864	52.2302
2001/04	862.152	1088.577	291.4696	797.1074
2001/05	406.48	614.7588	254.6272	360.1316
2001/06	400.73	832.4742	362.7076	469.7666
2001/07	-15.66	492.3114	263.872	228.4394
2001/08	69.93	604.765	201.5243	403.2407
2001/09	-22.95	76.8682	85.3571	-8.4889
2001/10	22.68	368.6498	72.7922	295.8576
2001/11	-162.114	472.5106	393.3538	79.1568
2001/12	-449.726	475.7796	730.6804	-254.901
2002/01	-147.29	248.0704	290.7769	-42.7065
2002/02	-46.42	284.7766	187.9029	96.8737
2002/03	94.16	950.2516	658.2931	291.9585
2002/04	1191.66	2281.015	860.6438	1420.371

Table 2. (*continued*)

2002/05	-88.968	2552.715	3047.516	-494.8
2002/06	35.088	826.7768	448.1287	378.6481
2002/07	747.968	2084.595	588.1744	1496.42
2002/08	-123.728	750.936	755.6436	-4.7076
2002/09	-87.64	554.329	452.0791	102.2499
2002/10	-114.24	545.456	392.4285	153.0275
2002/11	210.168	980.7934	417.6735	563.1199
2002/12	-250.488	938.4832	1151.741	-213.258
2003/01	-1045.44	400.4058	1427.829	-1027.42
2003/02	-168.28	840.0396	928.4269	-88.3873
2003/03	-238.52	800.8116	1436.448	-635.636
2003/04	466.816	2053.866	1572.319	481.5472
2003/05	360.594	1736.026	1532.671	203.3551
2003/06	738.79	2126.438	793.614	1332.824
2003/07	-85.68	727.8662	773.843	-45.9768
2003/08	-484.848	432.5354	727.736	-295.201
2003/09	-256.552	349.1292	499.0735	-149.944
2003/10	-550.52	146.0776	684.8182	-538.741
2003/11	-607.63	121.3266	735.1252	-613.799
2003/12	-471.8	106.6628	525.4702	-418.807
2004/01	-137.758	269.926	235.1934	34.7326
2004/02	45.57	492.5916	85.0875	407.5041
2004/03	-225.15	358.8428	544.0729	-185.23
2004/04	237.054	507.5356	53.0585	454.4771
2004/05	1167.634	1795.708	180.5733	1615.135
2004/06	937.92	1356.168	210.1363	1146.032
2004/07	282.8	669.4912	110.2536	559.2376
2004/08	-82.6	824.6286	556.5411	268.0875
2004/09	-73.92	279.5462	338.2703	-58.7241
2004/10	-142.24	81.1646	256.4773	-175.313
2004/11	-465.76	452.056	661.8645	-209.809
2004/12	-1369.4	193.0578	1522.947	-1329.89

Form the above datum and the fig3, we can see the matchup among the three factors is pretty good. In the year of 2002, 2003, and 2004, when the precipitation is relatively more than the other year, the reservoir's variation almost appears after the simulated input volume increasing time, which means that the simulation effect bring forward the practical runoff by the precipitation. And the fig 3 also shows the water volume ascending value which is the difference between the input volume and the output volume has the same changing trend with the reservoir's volume variation.

Totally, except for the few periods when the runoff lag or influenced by some other factors, the three parameters such as the simulated input volume by SWAT,

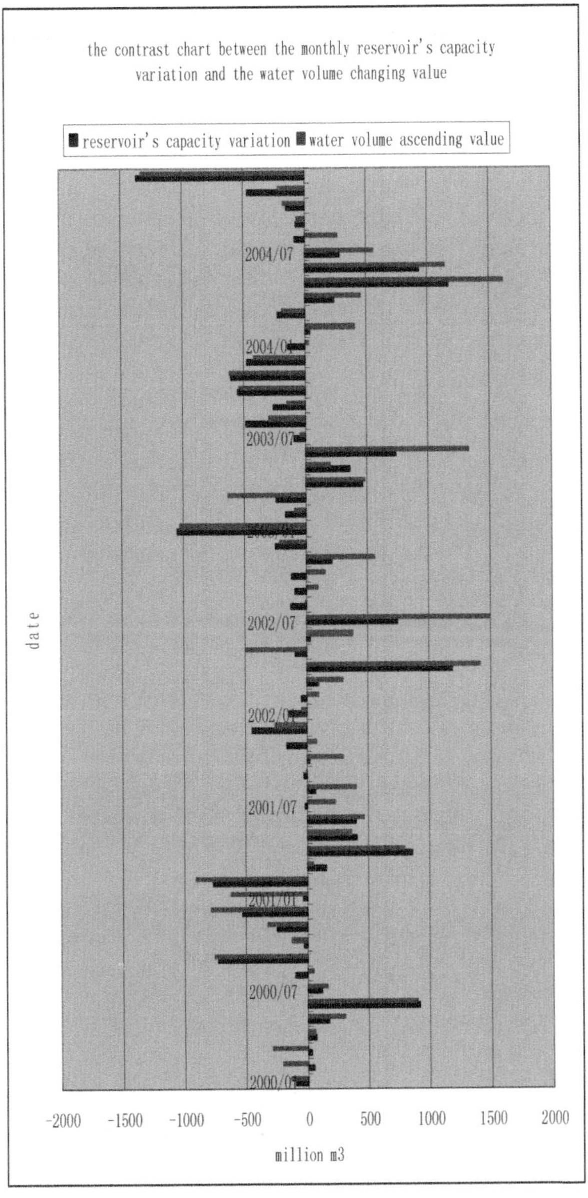

Fig. 3. The chart of monthly reservoir volume variety and the modeled runoff as well as the reservoir outlet amount between the year of 2000 to 2004 in zhelin reservoir

The observed reservoir's volume variation, and the observed output volume, could be matched up well in zhelin reservoir. And the result shows the SWAT model can simulate the runoff of the basin well. On such basis, we can forecast the output volume based on the simulated runoff, which is the input volume of reservoir, with the change of the weather conditions, and to realize the automation of the reservoir capacity adjustment. It is significant for the production on the spot, etc.

5 Conclusion and Discussion

This paper studied the matchup among the simulated input volume by SWAT, the observed reservoir's volume variation, and the observed reservoir's output volume. From the above result, we can see the study method of this paper is efficient and good for the simulation of the runoff. But there also still exist some point to be improved. One of these is the improvement of the simulation precision. And the other future work would be focused on the automatic retrieval of the reservoir's capacity information, especially based on the RS data and the DEM, etc. And that should be a main task for the researchers to pay their effort to.

Acknowledgment. The paper was sponsored by earth system data sharing program(2005DKA32300);the National Natural Science Foundation of China(40801180□ 40771146) .I would thanks to Professor Zhu Xingming, Ph.D Huang Yaohuan, the institute of water and hydrology research(iwhr), China. Prof Zhu gave us some useful advices and Ph.D Huang helps me to debug the program to run the data. At the same time, I would thank the model programmers who had devoted that much effort to the model.

References

1. Badas, M.G., Sulis, M., Deidda, R., et al.: Evaluation of SWAT Stream flow Components for the Araxisi Catchment (Sardinia, Italy) [EB/OL]. In: The Second International SWAT Conference, pp. 27–32
2. Heuvelmans, G., Muys, B., Feyen, J.: Analysis of the spatial variation in the parameters of the SWAT model with application in Flanders, Northern Belgium. Hydrology and Earth System Sciences 8(5), 931–939 (2004)
3. Van Liew, M.W., Veith, T.L., Bosch, D.D., et al.: Suitability of SWAT for the conservation effects assessment project: comparison on USDA agricultural research service watersheds. Journal of Hydrology Engineering 12(2), 173–189 (2007)
4. `http://www.geodata.cn`
5. The land-use administration bureau of Jiangxi province, the soil category. The Agricultural Scientific Press of China, Beijing (1991) (in chinese)
6. Wei, H., Zhang, Z., Yang, J.: The establishment of soil database of SWAT. The Hydrological Technology 38(6), 15–18 (2007) (in chinese)
7. Cai, Y., Zhang, K., Li, S.: The transferring method study among the different granule size of soil. Journal of Soil 40(4), 514–517 (2003) (in chinese)
8. Nash, J.E., Sutcliffe, J.V.: River flow forecasting through conceptual models. Journal of Hydrology 10, 282–290 (1970)

OMMP: A Novel Approach to Multihomed Multicasting

Xiao Chen and Nianzu Liu

School of Mathematics and Information
Shanghai Lixin University of Commerce
No.2800, Wenxiang Road, Shanghai, China
liunianzu88@163.com

Abstract. We proposed a multihomed proxy model to analyze the overlay multicast routing problem in a fine granularity so that we can perform the traffic balancing on a link basis rather than on a node basis. We also compared three path selection algorithms, which do the necessary preparation before the routing process can start. Through simulation results, we show that the alternative path algorithm results in a much better performance in both tree diameter and traffic balancing. In the end we give the OMMP architecture to combine all above notions into a practical framework.

Keywords: Routing algorithm, Multicast routing, Multihomed routing.

1 Introduction

The overlay multicast routing problem is usually considered under a singlehomed proxy model in most literatures [1][6]. When an MSN (Multicast Service Node) has multiple access links, this model is no longer applicable. The objective of our work is to propose not only a multihomed proxy model to describe the overlay multicast routing problem accurately but also a systematic approach to solve this problem under our proposed model. First we choose a proper physical path for each overlay link based on the current network status and the routing policy. Then the routing process calculates the multicast tree according to the output of the first process. In our model the capacity of each MSN is translated to the access link's capacity rather than the node's degree bound so that our routing algorithm can achieve a high utilization among the multiple access links. We also design an OMMP(Overlay Multicast over Multihomed Proxies) architecture to assign different tasks to different entities in the physical network. By this way we form a realistic framework to put all above notions into practice.

The rest of this paper is organized as follows. Section 2 proposes the multihomed proxy model and draws a comparison between our model and the original one. In section 3 we enumerate three different path selection algorithms and analyze their performance from different perspectives. We illustrate the combination of path selection and multicast routing by a modified CT(compact tree) algorithm in section 4. The OMMP architecture is also described in this section. Section 5 shows our simulation results.

D. Zeng (Ed.): ICAIC 2011, Part II, CCIS 225, pp. 105–112, 2011.

2 Overlay Network Model

While multiple access links enhance the availability and utilization of the MSNs, they introduce several new problems. As we have seen in the above example, since the access bandwidth for each MSN in each multicast session is allocated on the specific access links, the degree constraint model for the access resource limitation in [3] is not applicable any more. The access bandwidth consumption is not represented by the degree of the node in the multicast tree but by the actual link load. Therefore the resource allocation balancing is performed on a link basis which makes great difference in the generated tree topology. In addition, since two MSNs can be connected through different pairs of access links, the network distance between the two nodes could vary due to different choices. In order to make a tradeoff between the network latency and the resource allocation balancing, it is important to choose the proper unicast path as the overlay link for multicast tree construction. This effort also adds to the overhead of the routing process.

In order to solve all above problems, we first need another model for the overlay network. Suppose the overlay network in our research is defined as a ternary structure $OG = (V,A,E)$. V represents the set of MSNs while A stands for the set of access links attached to the MSNs. Let *ve* denote the function that maps a certain access link to its attached MSN such that for every $l \in A$ there is one and only one MSN $v \in V$ and $ve(l)=v$. We also use le(v) to denote the set of access links of MSN $v \in V$. E is a set of overlay links, each of which corresponds to a shortest unicast path between two MSNs through a pair of access links $(l1, l2)$. Note that in our model there may exist multiple overlay links between two nodes due to different choices of the access links. Function $c(l1, l2)$ calculates the cumulative cost along the shortest unicast path through access links of $l1$ and $l2$ between the nodes $ve(l1)$ and $ve(l2)$. For each access link l, we use $s_{max}(l)$ and $s_T(l)$ to represent the maximal available bandwidth resource and the consumed bandwidth for tree T over link l respectively. To simplify our model, when two nodes is connected by one link directly, this shared access link is counted twice in A. Based on these definitions, we will introduce a revised version of the MDDL and LDRB[1] problems as formulations of our specific overlay multicast routing problem.

Definition 1: Minimum diameter, link stress limited spanning tree problem (MDLSL):

Given an overlay network model $OG=(V,A,E)$, a link capacity bound $s_{max}(l) \in N$ for each access link $l \in A$ and a cost $c(l1, l2)$ for each overlay link $(l1, l2) \in E$, find a spanning tree T of OG of minimum diameter, subject to the constraint that $s_T(l) \leq s_{max}(l)$ for all $l \in A$

Suppose that each node in V only has one attached access link in A. Then the MDLSL problem will amount to the MDDL problem which has been proven to be NP-complete[3] for a fixed degree bound. Consequently our MDLSL problem is also NP-complete under an assumption of a fixed link stress bound.

The second formulation of the problem aims to find the most balanced allocation of the link stress over all related access links. Similar to the definition of residual degree in [1], we define the residual capacity at access link l with respect to a tree T to be $rest(l)= s_{max}(l)–s_T\ (l)$. Then the most balanced allocation indicates a maximal smallest residual capacity among all access links. Such an allocation promises a least likelihood of the session block.

Definition 2: Limited Diameter, residual capacity balanced spanning tree problem (LDRCB):

Given an overlay network model $OG=(V,A,E)$, a link capacity bound $s_{max}(l) \in N$ for each access link $l \in A$, a cost $c(l1, l2)$ for each overlay link $(l1, l2) \in E$ and a diameter bound of B, finding a spanning tree T of OG with diameter $\leq$ B that maximizes $min_{l \in A}\ rest(l)$, subject to the constraint that $s_T\ (l)\ \leq\ s_{max}(l)$ for all $l \in A$

LDRCB is also an NP-complete problem because its special case is the LDRB problem, which has been proven NP-complete [1].

3 Path Selection Algorithm

3.1 Shortest Path First

Intuitively the shortest possible path algorithm is to select the shortest eligible path as the overlay link for the concerned two nodes without considering whether the residual capacity for each access link is balanced. One straightforward way is to remove all ineligible access links from the underlying network, which has no enough bandwidth for the request, and calculate the shortest path between these two nodes. The corresponding pair of access links for this path will stand for the overlay link for the two nodes in our model. This approach will surely try to assign a minimum cost for each overlay link. It is intended as a greedy heuristic for the MDLSL problem, which tries to find the minimum diameter tree as long as there is enough resource.

3.2 Broadest Path in k-Shortest Ones

Although the shortest path first algorithm provides a good output in the beginning, its performance will deteriorate soon as some hot lines are exhausted in terms of capacity. In most cases, a longer tree diameter doesn't matter too much as long as it is within a specific bound. Based on this notion we give a heuristic here for the LDRCB problem, which tries to optimize the workload allocation as well as the tree diameter. Suppose nodes v1 and v2 have n1 and n2 eligible access links for a session request respectively. Then there are potentially n1*n2 candidate paths for the overlay link between these two nodes. Sort all these eligible paths in ascending order according to their distance. The path with the largest capacity in the first k paths then will be our choice for the overlay link.

Input:
  $G = (V, E)$
  $v1, v2 \in V$
  $VL1 = le(v1)$
  $VL2 = le(v2)$
  B {Bandwidth Requirement}
Output:
  Primary and secondary path
  for the overlay link between v1 and v2
Procedure:
for all $l \in VL1$ such that $cap(l) < B$ **do**
  $VL1 \leftarrow VL1 - \{l\}$
end for
for all $l \in VL2$ such that $cap(l) < B$ **do**
  $VL2 \leftarrow VL2 - \{l\}$
end for
if $|VL1| * |VL2| > 0$ **then**
  P $\leftarrow$ the set of all paths between v1 and v2
  Sort P in ascending order according to the distances
  Pk $\leftarrow$ the set of the first k paths
  Sort Pk in descending order according to the capacity
  (pi is the ith path in Pk)
  $i \leftarrow 2$
  while $i \leq k$ and $p1.distance \leq pi.distance$ **do**
    $i \leftarrow i + 1$
  end while
  if $i > k$ **then**
    $i \leftarrow argmin_{i \in \{2..k\}}$ {pi.distance | pi shares no access links with p1}
    OR
    $i \leftarrow argmin_{i \in \{2..k\}}$ {pi.distance}
  end if
end if
Primary Path $\leftarrow p1 \in Pk$
Secondary Path $\leftarrow pi \in Pk$

Fig. 1. Alternative Path Algorithm

3.3 Alternative Path Algorithm

Both of the above two algorithms select only one underlying path as the overlay link. This could be insufficient in some cases. For example, one overlay link may run out of bandwidth before it is used for tree construction because it shares the same access link with some other overlay link, which has already consumed the access bandwidth in the tree. On the other hand, even though all selected underlying paths have enough bandwidth, it might be difficult for them to build a tree within a stringent diameter bound. Our solution to this issue is to have two underlying paths prepared for each overlay link. The primary path has a larger capacity while the secondary one has a shorter distance. Therefore the upper layer overlay routing algorithm may have an alternative when some parameter of a certain overlay link is unsatisfactory. Although this approach cannot avoid all those situations stated above completely, it can drastically reduce their occurrence according to our simulation results. The detailed algorithm is shown in Fig. 1.

4 Routing Algorithm and Ommp Architecture

4.1 Integrated Routing Algorithm

The path selection algorithm and the multicast routing algorithm work in two different layers. The former selects one path among multiple paths between two MSNs to be the overlay link. The latter selects a collection of overlay links from the full-meshed overlay network to build the tree. For simplicity we choose the CT(compact tree) heuristic introduced in[1] for this purpose.

The CT algorithm is similar to Prim's minimum spanning tree. It first chooses a node as the root and proceeds to add a node to the tree, which leads to a minimum diameter. In detail, it uses $d(v)$ to denote the distance of the longest overlay path from node $v \notin T$ to any other node $u \in T$. The node v with the minimum value of $d(v)$ is chosen each time to be connected to the tree T through node $n(v)$ if the degree constraint is not violated. This process goes on until all nodes are connected to the tree. Since a randomly chosen root may not lead to a minimum diameter tree, this procedure is iterated n times for all n nodes respectively. A tree with the minimum diameter in all n trees is chosen as the final multicast tree.

Because the original CT only considers the degree constraint, we need to transform it to recognize link level bandwidth limitation. We order that the path selection algorithm supplies an m-element capacity vector Cap for an n-node overlay network, (for first and second algorithm $n \leq m \leq 2(n \quad 1)$; for alternative path, $n+1 \leq m \leq 4(n \quad 1)$). This vector corresponds to the capacity of all access links used in an overlay network. *Link(v,u)* denotes the index of the access link from node v to u on the overlay link (v, u). Then each time an overlay link (v, u) is used to build the tree, the value of *Cap[Link(v, u)]* and *Cap[Link(u, v)]* is updated to *Cap[Link(v, u)]-req* and *Cap[Link(u, v)]-req* respectively where *req* stands for the bandwidth requirement of the session. By this way an overlay link (v, u) could be used if and only if its related two access links have enough capacity.

Therefore a basic path selection algorithm should at least supplies the value of *dist(v, u)* and *Link(v, u)* (also *Link(u, v)*) for each overlay link (v, u) as well as a capacity vector *Cap* for all involved access links. All these information informs the overlay network layer of the distance of each overlay link, the access links of each overlay link and the capacity of each access link. According to these knowledge a routing algorithm can build a multicast tree without knowing the details of the underlying network. Since the alternative path algorithm calculates a secondary path for each overlay link, it needs another two functions of *dist'(v, u)* and *Link'(v, u)* to denote the distance and access links of the secondary path.

In the above algorithm, some secondary paths are used to substitute their primary paths for a shorter tree diameter when the diameter limitation is violated during the tree building process. We call it a shrinking process. In the shrinking process, the value of $d(u)$ for some $u \in W$ may be changed. In order to reduce the time complexity, we can record a list of nodes on each overlay link such that only those nodes need to update their longest path distance when this overlay link is shrieked. In addition each step in above algorithm is executed only if the overlay link has enough capacity in vector *Cap*.

4.2 OMMP Architecture

Although we have separated the solution of MDLSL and LDRCB problem into two tightly coupled processes, namely path selection and multicast routing, we still need architecture to allocate different tasks to different entities in our physical network. Before we go ahead, we should at first analyze the characteristics of these two different tasks. The path selection is strongly associated with the status of the underlying network. In order to select a most suitable path, the algorithm has to measure concerned parameters of each candidate, including bandwidth capacity and network latency. This process demands several network transmissions and can't be completed immediately. Furthermore the path selection is done for each pair of MSNs in the overlay network. For n-node overlay network, there will be n(n-1)/2 path selections. Consequently this process can't be executed on a session basis. On the other hand, when parameters are determined for all overlay links, the routing algorithm can be executed immediately by any single node. Based on these considerations, we design our OMMP architecture like Fig. 2.

In OMMP we refer to the backbone network connecting all MSNs as a core network. We also assume that there is a manager node who has a knowledge of the topology details of the core network. This assumption is realistic when all MSNs belong to a single autonomous system(AS), such as a Diffserv domain [2]. Even if the MSNs belong to different domains, we can use a hierarchical structure in [5] to abstract the core network. As we stated above, the bottleneck in overlay multicast is access links of each MSN. Then each MSN only needs to monitor the status of its own access links and report it to the manager periodically. When the manager has collected the information of the access links from all MSNs, it is able to generate a full-meshed overlay network spanning all MSNs over the core network for a certain bandwidth requirement. It means the manager selects paths for overlay links in a centralized way before any session request arrives. As soon as a session request comes, the manager can immediately give a partial overlay network of the group members to the initiator MSN, who in turn carries out the routing task and calculates the distribution tree.

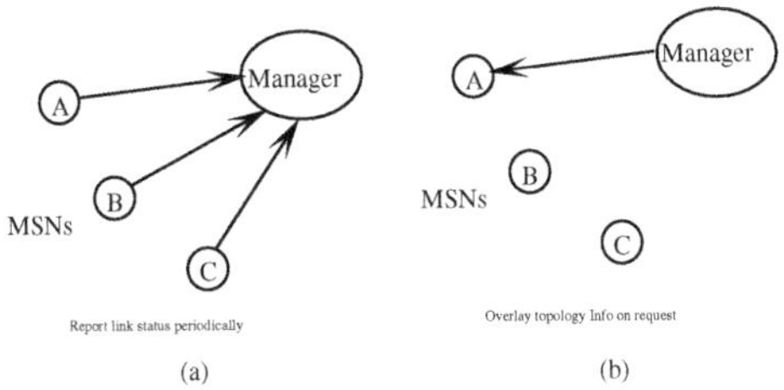

Fig. 2. OMMP Architecture

5 Simulations

5.1 Performance on Rejection Rate

The first experiment measures the rejection rate as the tree diameter limitation changes. From Fig. 3 we can see that the S algorithm always has a least rejection rate. This is because 100 sessions is not a big burden for our designed overlay network. When there is enough capacity, the S algorithm is surely the first choice especially for a strict diameter bound. We also find that the curve for the A algorithm is very close to that of the B algorithm. One more thing to mention is that the curves drop sharply when the diameter bound is above 5. This indicates that the relative delay penalty is less than 3 in our overlay multicast.

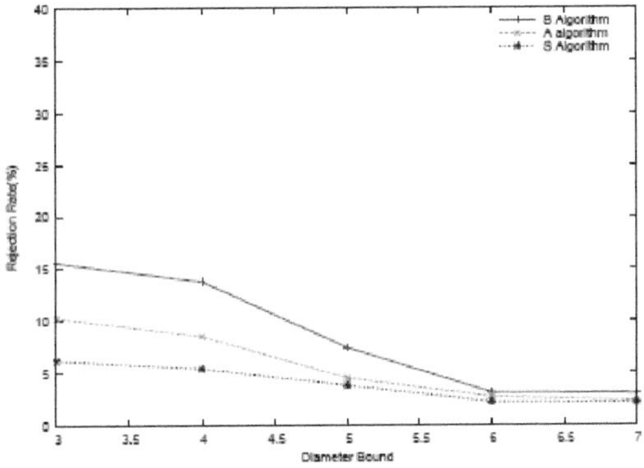

Fig. 3. Rejection VS Diameter Bound

5.2 Performance on Traffic Balancing

Another performance criterion for these path selection algorithms is how much they can balance their traffic load among their access links. This is important for keeping a low possibility of traffic congestion. In order to measure it accurately, we invent a parameter named balance degree. It equals to the difference of the residual capacity between the least loaded access link and the most loaded access link of a single MSN. The balance degree for an overlay network is obtained by averaging the value among all MSNs. When this value becomes bigger, the overlay network has a less balanced traffic over all its access links. We measure this value as the session number grows gradually. The simulation result is shown in Fig. 4. According to this figure we can infer that the S algorithm is less efficient in traffic balancing.

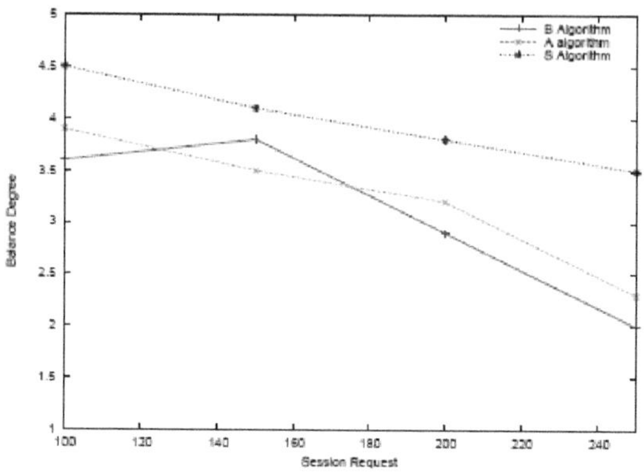

Fig. 4. Balance Degree VS Session Requests

References

1. Shi, S.Y., Turner, J.S.: Routing in overlay multicast networks. In: Proc. IEEE INFOCOM (June 2002)
2. Blake, S., et al.: An Architecture for Differentiated Services. RFC2475 (December 1998)
3. Shi, S., Turner, J., Waldvogel, M.: Dimensioning server access bandwidth and multicast routing in overlay networks. In: Proceedings of NOSSDAV (June 2001)
4. Deering, S.: Multicast Routing in Internetworks and Extended LANs. In: SIGCOMM Summer 1988 Proceedings (August 1988)
5. Li, P.: Mohapatra: QRON: QoS-aware routing in overlay networks. IEEE JSAC (2003)
6. Banerjee, S., Kommareddy, C., Kar, K., Bhattacharjee, B., Khuller, S.: Construction of an efficient overlay multicast infrastructure for real-time applications. In: IEEE INFOCOM, San Francisco, CA (April 2003)
7. Estrin, et. al.: Protocol Independent Multicast-Sparse Mode (PIM-SM): Protocol Specification, RFC 2362 (June 1998)
8. Waxman, B.M.: Routing of multipoint connections. IEEE J. Select. Areas Commun. 6(9) (December 1988)
9. Chen, X., Shao, H., Wang, W.: DIRA: Distributed Insertion & Relocation Routing Algorithm for Overlay Multicast in Diffserv Domain (submitted)

Measuring Governance Efficiency of Corporate Using Extension Mathematical Theory

Xie An and Fan Zeming

Management School
Northwestern Polytechnical University
Xi'an, 710072, P. R. China
xiean9563886@sina.com

Abstract. High-precision measurement of corporate governance efficiency provided valid benchmark and decision-making foundation for improving governance efficiency. Key factors were extracted from a variety of factors of influence governance efficiency, and they were unified dimensional processing using extension element transformation theory. Measurement model of corporate governance efficiency was constructed by extension set theory and the extension matrix theory. The method is effective and feasible through application demonstrates, also it resolves the uniform dimensional measurement and consistency of judgment matrix of the governance efficiency effectively, and can be widely applied in governance efficiency measurement of the same corporate or different corporate.

Keywords: corporate development, governance efficiency, measurement model, extension mathematical theory.

1 Introduction

Governance and technology play a very important role in the development of corporate. As for the measurement of technology efficiency, scholars have done lots of researches and got fruitful findings (Ruggiero, 2000; He, et al., 2004; Li, et al., 2002; Fan, et al., 2003). Due to the measurement of corporate governance efficiency is a complex and important system involving many mutual-coupling and unknown or uncertain factors, its study is relatively less. In other fields, the study of the multi-index comprehensive measurement (Wang, et al., 2001; Samanta, et al., 2005) can be roughly divided into the subjective and objective methods. Both have their own characteristics with strengths and weaknesses and apply to different occasions. Of which, the subjective approach reflects the subjective judgments or intuitive judgments by the experts and decision makers on the factors for evaluation of decision-making, for example, expert advice, the Delphi method (Lycurgus, 1999), the AHP (John, et al., 1993; Prasanta, et al., 2001; Peng, et al., 2004; Chang, et al., 2007), the fuzzy AHP (FAHP)(Lu, et al, 2006; Qureshi, 2009), etc. The objective method is based on the statistical data analysis of objective information to determine the weights, such as the principal component analysis (Thierry, 1999; Philipp, et al, 2000), Information Entropy (Labudda, 1994; Shuiabi, et al., 2005; Zhang, et al., 2007),

D. Zeng (Ed.): ICAIC 2011, Part II, CCIS 225, pp. 113–123, 2011.

multi-objective optimization methods (Pentti, 1987, Huo, et al., 2005). Among them, the AHP Method is becoming more scientific by conducting further mathematical processing for the subjective judgments of experts. But the AHP cannot decouple the factors and deal with uncertain reasoning. The Fuzzy Clustering Method (Nbrphy, 2000) is applicable for Classification of fuzzy factors and its weight in the same level. Its weak point is unable to determine the weight of single factor and deal with uncertain reasoning too. Aiming at this, the research presented a measurement method of corporate governance efficiency based upon extension mathematical theory. The theoretical pillar is the matter element theory and the extension set theory (Hou, et al., 2005), and its logic cells are the matter element. The extension set theory and the conversions theory of matter elements is two important branches of the extension mathematical theory. Extension Set Theory can judge the degree of a certain set by the measurement value of one's feature, also it can evaluate fine and quantitative by use the correlation functions value in the interval (-∞,+∞). So we would make use of extension set theory in measurement of governance efficiency, not only can solve the two extreme cases of human judgment, but also make up the lack that the fuzzy AHP (Lu, et al, 2006; Qureshi, 2009) that cannot normally considered judgment matrix consistency. So a new way to determine index weight of corporate governance efficiency was provided. Using matter-element conversion theory, we can transform not only the objects, characteristics of the objects and the value of objects, but also the qualitative and essence.

2 The Measurement of Governance Efficiency

2.1 The Extension Section Number and Its Judgment Matrix

Definition 1. E(U) is the whole extension sets in the fixed domain U. Suppose $a = \langle a^-, a^+ \rangle \in E(U)$, then the simple function between u and a $K_a(u)$ can be expressed as:

$$K_a(u) = \begin{cases} \dfrac{2(u - a^-)}{a^+ - a^-}, & u \le \dfrac{a^- + a^+}{2} \\ \dfrac{2(a^+ - u)}{a^+ - a^-}, & u \ge \dfrac{a^- + a^+}{2} \end{cases}$$

and $a = \langle a^-, a^+ \rangle = \{x | 0 < a^- < x < a^+\}$ is called extension section number. The sign $\langle a^-, a^+ \rangle$ can include a^+, a^-, and can also exclude them.

Especially, when $a^+ = a^-$, the extension number a is a common positive number.

Between the two extension section numbers $a = \langle a^-, a^+ \rangle$ and $b = \langle b^-, b^+ \rangle$, when and only when $a^- = b^-$,

$a^+ = b^+$, then $a = b$.

The rules of the calculation of the extension are as follows:

Theorem 1. Suppose $a=\langle a^-,a^+\rangle, b=\langle b^-,b^+\rangle$ are two extension numbers, then

(1) $a\oplus b=\langle a^- +b^-,a^+ +b^+\rangle$;

(2) $a\otimes b=\langle a^-b^-,a^+b^+\rangle$;

(3) $\forall\lambda\in R^+,\lambda a=\lambda\langle a^-,a^+\rangle=\langle\lambda a^-,\lambda a^+\rangle$;

Definition 2. Suppose $a=\langle a^-,a^+\rangle$, $b=\langle b^-,b^+\rangle$ are two extension numbers the degree of the possibility of $a\geq b$ is defined as:

$$V(a\geq b)=\sup_{u\geq v}(K_a(u)\Lambda K_b(v))$$

Theorem 2. Suppose $a=\langle a^-,a^+\rangle$, $b=\langle b^-,b^+\rangle$ are two extension numbers, then $V(a\geq b)$ is calculated as follows:

$$V(a\geq b)=\frac{2(a^+ -b^-)}{(b^+ -b^-)(a^+ -a^-)}$$

Especially, when $b^- < a^+$, $V(a\geq b)$ is positive number, standing for the degree of the possibility of $a\geq b$; But when $b^- > a^+$, $V(a\geq b)$ is negative, standing for the degree of the impossibility, $a\geq b$. When $b^- = a^+$, then $V(a\geq b)=0$.

Using the extension section number as the elements, the vector and the matrix are respectively called the extension section number vector and the extension section number matrix, which adhere to the rules of the common number matrix or vector. Suppose $A=(a_{ij})_{n\times n}$ is the extension section number matrix, that is, $a_{ij}=\langle a_{ij}^-,a_{ij}^+\rangle$, expressed as $A^-=(a_{ij}^-)_{n\times n}$, $A^+=(a_{ij}^+)_{n\times n}$ and expressed as $A=\langle A^-,A^+\rangle$, so do with section vector $x=(x_1,x_2,\cdots,x_n)^T$, that is, $x_i=(x_i^-,x_i^+)$.

Definition 3. Suppose A is an extension section number matrix, and λ is an extension section number. If there is an extension section number vector x which justifies the equation $Ax=\lambda x$, then call λ as a feature value of A, and call x a feature vector of A corresponding to λ.

Theorem 3. If $Ax=\lambda x$, then $A^-x^-=\lambda^-x^-$, $A^+x^+=\lambda^+x^+$.

Theorem 4. Suppose $A=\langle A^-,A^+\rangle$, if λ^-,λ^+ are respectively A^-,A^+ the largest feature values, then

(1) $\lambda=\langle\lambda^-,\lambda^+\rangle$ is the feature value of A;

(2) $X=\langle kx^-,mx^+\rangle$ is the whole feature values of A corresponding to λ.Among them,

x^-, x^+ are respectively any positive feature values of A^- and A^+ corresponding to λ^-, λ^+, and k, m are the whole positive numbers $0 < kx^- \leq mx^+$.

Definition 4. Call $A = (a_{ij})_{n\times n}$ an extension section number judgment matrix, if $i, j = 1,2,\cdots,n$, then get:

$$a_{ij} = \langle a_{ij}^-, a_{ij}^+ \rangle, \text{ and } {}^1\!/_9 \leq a_{ij}^- \leq a_{ij}^+ \leq 9$$

Suppose $A = (a_{ij})_{n\times n}$ is an extension section number judgment matrix. $w = (w_1, w_1, \cdots, w_n)^T$ is the extension section number power vector corresponding to A. If a_{ij} objectively reflects the proportion between w_i and w_j, not a approximate value, then $a_{ij} = \dfrac{w_i}{w_j}$ $(i, j = 1,2,\cdots,n)$. Now it's easy to know that $a_{ij} a_{jk} = a_{jj} a_{ik}$ is justified by any $i, j = 1,2,\cdots,n$.

Definition 5. Suppose $A = (a_{ij})_{n\times n}$ is an extension section number matrix, if with any $i, j, k = 1,2,\cdots,n$, all get

$$a_{ij} = {}^1\!/_{a_{ji}}, \quad a_{ij} a_{jk} = a_{jj} a_{ik}$$

Then call A an unanimous extension section number matrix, and call the above equation the unanimous condition.

Obviously, when $a_{ij}^- = a_{ij}^+$, the unanimous extension section number matrix is the common unanimous matrix.

2.2 The Principles of the Measurement Mode

1) To construct the judgment matrix, when assign the integers 1-9 and their derivatives, take the vagueness of people's judgment into consideration. Specifically, when compare two formulas' important, the traditional hierarchical analytic methods only judge two extreme situations. But in reality, people's judgment is always within a certain area. For example, when compare A and B, people always think that the importance of A and that of B is between 4.5 and 5.5. And this is more approximate to the reality.

2) Using the values of the related functions expanded to (-∞, +∞) can make the assessment more thorough and easier to express in quantity. Thus it offers a new approach to assess the program from a changing angle.

3) Based on the theory of the extension set and methods, the approach how to construct the judgment matrix when people study the indetermination of the degree of the relative importance makes up for the lack of the unanimous of the judgment matrix using vague AHP method.

2.3 The Construction of the Measuring System

Owing to the complexity, un-linearity and the complexity of environment of corporate system, the exterior and inside of corporate organization exists a great deal of factors that influence the corporate governance efficiency. In order to evaluate the condition of corporate governance efficiency precisely, the writer synthesized both the qualitative indexes and quantitative indexes of the corporate governance efficiency and both the process indexes and result indexes that reflect the governance activity and established the measuring index system of corporate governance efficiency. Thus, the corporate governance activity was reflected completely from different angle.

On this basis, this example combines the traditional governance efficiency evaluation selection standard, chooses the corporate financial capacity, corporate marketing, corporate organizational capital, corporate human resource governance, corporate informationization, corporate sustainability to consider the corporate governance efficiency influential factor as the primary index of comprehensive evaluation.

1 Corporate informationization (B_1). The main evaluation index on the level of corporate informationization includes: the degree of importance that the corporate executives attach on the establishment of corporate information system (C_1), the degree of health of corporate internal information system (C_2), the degree of health of corporate operation information system (C_3), corporate informationization benefit (C_4).

2 Corporate marketing (B_2). The main evaluation index on the level of corporate marketing capacity includes: market share capacity of corporate product (C_5), degree of corporate service satisfaction (C_6).

3 Corporate organizational capital (B_3). The main evaluation index on the level of corporate organizational capital capacity includes: corporate basic governance level (C_7), corporate culture (C_8), corporate image and brand (C_9).

4 human resource governance (B_4).The main evaluation index on the level of corporate human resource governance includes: basic quality of corporate leadership (C_{10}), corporate staff quality (C_{11}), corporate human resource policy (C_{12}).

5 Corporate sustainability (B_5). The main evaluation index on the level of corporate sustainable growth includes: corporate strategy and policy evaluation (C_{13}), corporate originality evaluation (C_{14}), corporate technical equipment upgrading level evaluation (C_{15}), corporate service hard environment evaluation (C_{16}) and corporate development capacity (C_{17}).

6 The corporate Financial Analysis (B_6). The main indexes to be assessed within the corporate financial analysis include corporate repaying debt ability (C_{18}), the corporate profit ability (C_{19}), corporate assets carrying ability (C_{20}).

The index system of measurement plan is in a tree structure, the measurement is bottom up from the deepest measurement index, till the total measurement objective, therefore, the measurement of corporate governance efficiency is a process of multiple index and multi-level comprehensive measurement. According to the research technical route, confirm the governance efficiency after field investigation, data collection, literature consulting and expert consulting.

2.4 Construct the Extension Judgment Matrix

$a_{ij} = \langle a_{ij}^-, a_{ij}^+ \rangle$, the element of $A = (a_{ij})_{n \times n}$, is an extension section number. In order to express in quantity every element among the extension judgment matrix, $(a_{ij}^- + a_{ij}^+)/2$, the middle value of the extension judgment matrix is the integer put forward by T. J. Saaty using the AHP method to compare and judge. Because of the different units of the quantities among the system of the indexes that are hard to compare, a united assessment standards should be established. This system uses the theory of the matter-element transformation [9-10] to covert all the indexes into the united units without specific quantities.

2.5 To Calculate the Weight Vector

In the process of constructing the extension section number judgment matrix, we employ the method of proportion standards. Suppose $a_{ij}^t = \langle a_{ij}^{-t}, a_{ij}^{+t} \rangle \ (i, j = 1,2,\cdots,n_k; t = 1,2,\cdots,T)$ is an extension section number given by t experts. According to the equation:

$$A_{ij}^k = \frac{1}{T} \otimes (a_{ij}^1 + a_{ij}^2 + \cdots + a_{ij}^T) \tag{1}$$

Get the total extension section number in the k level, and then get the total extension judgment matrix from the total factors of c in k level corresponding to the h factor in k-1.

For the above total extension section number judgment matrix $A = \langle A^-, A^+ \rangle$, in k level, the steps to get the power vector with the unanimous condition are as follows:

Step 1: To get the positively unanimous vectors x^-, x^+ corresponded with the maximized value A^+, A^-.

Step 2: Use $A^- = (a_{ij}^-)_{n_k \times n_k}$, $A^+ = (a_{ij}^+)_{n_k \times n_k}$ to calculate

$$k = \sum_{j=1}^{n} \sqrt{1 \Big/ \sum_{i=1}^{n} a_{ij}^+} \qquad m = \sum_{j=1}^{n} \sqrt{1 \Big/ \sum_{i=1}^{n} a_{ij}^-} \tag{2}$$

Step 3: To get the weight vector

$$S^k = (S_1^k, S_2^k, \cdots, S_{n_k}^k)^T = \langle kx^-, mx^+ \rangle \tag{3}$$

Step 4: The single ranking of the levels

According to the theorem 2, calculate

$$V(S_i^k \geq S_j^k)(i = 1,2,\cdots,n_k; i \neq j), \text{if } \forall\, i = 1,2,\cdots,n_k; i \neq j,$$

$$V(S_i^k \geq S_j^k) \geq 0, \text{ then}$$

$$P_{jh}^k = 1, P_{ih}^k = V(S_i^k \geq S_j^k), \quad i, j = 1,2,\cdots,n_k \tag{4}$$

Among them, P_{jh}^{k} refers to the single ranking of the i factor in the k level corresponding to the h factor in the level k-1. After converted get $P_h^k = (P_{1h}^k, P_{2h}^k, \cdots P_{n_k h}^k)^T$, which means the single power vector of the i factor in k level to the h factor in the k-1 level.

Step 5: The total ranking of the levels

After get all the $P_h^k = (P_{1h}^k, P_{2h}^k, \cdots P_{n_k h}^k)^T$, when $h = 1, 2, \cdots, n_{k-1}$, we get the rank matrix $n_k \times n_{k-1}$.

If the ranking weight vector of the k-1level to the total object is $W^{k-1} = (W_1^{k-1}, W_2^{k-1}, \cdots, W_{n_{k-1}}^{k-1})^T$, then W^k, the ranking weight vector of the k level to the total object can be worked out by the following formula:

$$W^k = (W_1^k, W_2^k, \cdots, W_{n_k}^k)^T = P^k W^{k-1} \tag{5}$$

Also have:

$$W^k = P^k P^{k-1} \cdots P^3 W^2 \tag{6}$$

3 The Calculation Governance Efficiency

3.1 Determining of the Indexes Weights

The indexes of corporate governance efficiency are from the Section 3.2, and the main steps are as follows.

Step 1: the experts marked the comparison score value among all the indexes and, the extension interval judgment matrix A was obtained. Using transformation of Eq. (1) gets:

$$A^- = \begin{bmatrix} 1 & 2.85 & 5 & 2.05 & 1 & 2.1 \\ 0.205 & 1 & 1.43 & 0.725 & 0.255 & 0.655 \\ 0.14 & 0.57 & 1 & 0.41 & 0.17 & 0.42 \\ 0.27 & 1.2 & 1.71 & 1 & 0.305 & 0.76 \\ 0.835 & 2.85 & 4.775 & 2.04 & 1 & 2.1 \\ 0.31 & 1.36 & 2.185 & 0.975 & 0.385 & 1 \end{bmatrix},$$

$$A^+ = \begin{bmatrix} 1 & 4.9 & 7 & 4.1 & 1.25 & 3.2 \\ 0.335 & 1 & 3.52 & 0.84 & 0.35 & 0.74 \\ 0.2 & 0.695 & 1 & 0.585 & 0.21 & 0.46 \\ 0.49 & 1.385 & 2.44 & 1 & 0.49 & 1.025 \\ 1 & 4.115 & 6.07 & 3.39 & 1 & 2.635 \\ 0.475 & 1.54 & 2.385 & 1.28 & 0.48 & 1 \end{bmatrix}$$

To get the positively unanimous vectors x^-, x^+ corresponded with the maximized value A^+, A^-.

$$x^- = (0.3114, 0.0899, 0.0571, 0.1113, 0.2990, 0.1313),$$
$$x^+ = (0.3352, 0.0954, 0.0519, 0.1130, 0.2868, 0.1177)$$

Using transformation of Eq. (2) gets: k=2.1538, m=2.4676;

According to Eq. (3) and Eq. (4), To get the single ranking of the levels of the criterion level each index to the total goal by unanimous.

$$W^2 = (0.1343, 0.0738, 0.0154, 0.1086, 0.2618, 0.4061)$$

Step 2: By the score value of all indexes the experts gave the score value of object layer C relatives to index layer B, the judgment matrixes B_1□ B_2□ B_3□ B_4□ B_5□ B_6 were obtained.

$$P_{B_1}^3 = (0.1075, 0.0082, 0.6646, 0.2197),\ P_{B_2}^2 = (0.4653, 0.5347),$$
$$P_{B_3}^3 = (0.0340, 0.4829, 0.4831),\ P_{B_4}^3 = (0.0182, 0.4192, 0.5626),$$
$$P_{B_5}^3 = (0.0915, 0.0296, 0.3692, 0.2837, 0.2262)$$
$$P_{B_6}^3 = (0.0540, 0.6767, 0.2693)$$

Step 3: To get the total ranking weight vector of the criterion level on each index to the total goal.

Using transformation of Eq. (5) and (6) gets:

$$w_3^1 = (0.0144, 0.0011, 0.0893, 0.0295, 0.0343, 0.0395, 0.0005, 0.0074, 0.0075, 0.0020, 0.0455, 0.0611, 0.0240, 0.0077, 0.0967, 0.0743, 0.0592, 0.0219, 0.2748, 0.1094)$$

B. Calculation of Corporate Governance Efficiency

Suppose the determining measurement sets of the goal layer is v_i, the relations among the vector index level, the weights and measurement level of governance efficiency measurement can be expressed as:

$$v_i = \sum_{j=1}^{n} w_3^1 \cdot x_{ij} \tag{7}$$

Where v_i is the measurement value of governance efficiency in the ith object, and the greater v_i is, the higher governance efficiency is. x_{ij} is the sample data of the jth indexes in the ith object.

4 The Experimental Validation

The experiment data comes from the information system of the corporate.

(1) The sample data processing of the measurement index of corporate governance efficiency.

To collect the indexes of the corporate governance efficiency from 2003 to 2005, due to dimension of the sample data of each index from corporate internal is different, it cannot compare directly. Therefore, the matter unit conversion theory [5] is used to unify dimension. Index data of transformation as shown in Table 1.

Table 1. The index sample data of the corporate

x	x_1	x_2	x_3	x_4	x_5	x_6	x_7	x_8	x_9	x_{10}
2003	1.08	0.83	0.84	0.81	1.25	1.23	1.17	1.04	1.06	2.07
2004	1.07	0.92	0.93	0.89	1.21	1.24	1.17	1.03	1.23	3.24
2005	1.09	0.80	0.85	0.90	1.00	1.26	1.22	1.07	1.15	1.02
x	x_{11}	x_{12}	x_{13}	x_{14}	x_{15}	x_{16}	x_{17}	x_{18}	x_{19}	x_{20}
2003	1.09	1.22	1.13	0.96	1.17	0.23	1.11	0.80	0.63	0.86
2004	1.09	1.23	2.19	1.56	1.16	3.88	1.11	0.84	1.26	0.98
2005	0.99	1.15	1.23	0.99	1.20	1.02	1.09	0.89	0.93	0.85

(2) The calculation of the governance efficiency

According to the determinate weight and the index sample data of the corporate, the measurement value of governance efficiency of the i th year can be calculated using Eq. (7). The value of v_i is shown in the following table 2.

Table 2. The integrated measurement value the corporate governance efficiency

Year	2003	2004	2005
v_i	0.8630	1.3644	0.9960

From table 2 we can know the governance efficiency of the corporate is the best in 2004, and it is the worst in 2003.

(3) The reliability analysis of the experimental results

Firstly, due to the same corporate in different years, the change of the corporate scale and the various elements were little in consecutive three years, and the result of the experiment is aroused by the control, organize, plan, coordinate and encourage of corporate, and that is the function of governance. Therefore main experiment result is reliable.

Secondly, because the experiment compared was different year in same corporate, the influence from the external environment is removed. Therefore, the experiment results were reliability too.

Finally, the sample data of the index used in the experiments comes from the information system database of the corporate. So the experimental data source is reliability too.

5 Conclusion

Aiming at the key problem that corporate governance efficiency cannot be measured validly at present, extension mathematical theory was used to measure and evaluate of corporate governance efficiency, and an effective judging system of corporate governance efficiency is constructed so as to realize more science and standardization for the measurement of the corporate governance efficiency. In practical applications, establishing of the correlation function will need to fusion into experts' expertise. Comparing the identified governance efficiency with the actual operating conditions of corporate governance, the satisfied result showed the measuring method of the governance efficiency closing to the objective reality. This method can be applied not only to calculate governance efficiency of different years of one corporate, but also to calculate governance efficiency of different corporate. Therefore it has certain theoretical significance and application value.

Acknowledgment. We are grateful to the editor and the anonymous reviewers for their constructive and helpful comments and suggestions.

References

1. Chang, J.E., Jiang, T.L.: Research on the Weight of Coefficient through Analytic Hierarchy Process. Journal of Wuhan University of Technology 29(1), 153–155 (2007)
2. Fan, C.Y., Han, X.M., Tang, W.H.: The Extract of Expert Judging Information and the Integral Method of Target Scaling in AHP. Journal of Air Force Engineering University (Natural Science Edition) 4(1), 49–56 (2003)
3. He, F., Chen, R., He, L.C.: The Measurement of Chinese Technical Efficiency: The Application of Stochastic Frontier Production Function. Systems Engineering-theory & Practice 14(5), 46–50 (2004)
4. John, C.S., Tang, I.N.: Sector priority and technology choice in the Korean machinery industry. International Journal of Technology Governance 8(3/4/5), 333–341 (1993)
5. Hou, Q., Shen, Y.Z.: The Choosing of Supplier Based on Extension Judgment Theory. Sci/tech Information Development & Economy 15(4), 156–158 (2005)
6. Huo, Y.B., Han, Z.J.: Research on Giving Weight for Multi-indicator Based on GME Principle And GA. Application of Statistics and governance 24(3), 39–50 (2005)
7. Labudda, K.D.: Fuzzy Multi-objective Approach to Power System State Estimation. International Journal of Engineering Intelligent Systems for Electrical Engineering and Communications 2(2), 83–91 (1994)
8. Li, G.J., Yan, H.: DEA Model for Measuring Input-and-output-based Technical Efficiency. Systems Engineering Theory & Practice 1(1), 26–32 (2002)
9. Lu, L.P., Wang, S.F.: Application of Fuzzy Hierarchy Method in Urban Intersection Evaluation. Journal of Wuhan University of Science and Technology 29(3), 286–288 (2006)
10. Lycurgus, L.L.: The Future Impact of Technology in the Greek Health Sector. International Journal of Healthcare Technology and governance 1(3/4), 409–414 (1999)
11. Nbrphy, C.K.: Combining Belief Functions When Evidence Conflicts. Decision Support Systems 29(1), 19–24 (2000)

12. Qureshi, M.N.: Framework for Benchmarking Logistics Performance Using Fuzzy AHP. International Journal of Business Performance and Supply Chain Modeling 1(1), 82–98 (2009)
13. Pang, F.H., et al.: Evaluation of Water Quality by Standard Deviation Weight Fuzzy Assessment on Water Source Area in Middle Line Project of Transferring Water from South to North. Journal of Northwest A & F University(Natural Science Edition) 36(2), 229–234 (2008)
14. Peng, G.F., Li, S.H., Sheng, M.K.: AHP in Evaluating Government Performance: Determining Indicator Weight. China Soft Science (6), 136–139 (2004)
15. Pentti, M.: Environmental Problems of Modern Societies. International Journal of Technology governance 2(2), 263–278 (1987)
16. Philipp, A., Peter, R.: Acceptance of Modern Biotechnology in Developing Countries: a case study of the Philippines. International Journal of Biotechnology 2(1/2/3), 115–131 (2000)
17. Prasanta, K.D., et al.: Multiple Attribute Decision Making Approach to Petroleum Pipeline Route Selection. International Journal of Services Technology and governance 2(3/4), 347–362 (2001)
18. Ruggiero, J.: Measuring technical efficiency. European Journal of Operational Research (121), 138–150 (2000)
19. Samanta, B., Roy, T.K.: Multi-objective entropy transportation model with trapezoidal fuzzy number penalties, sources, and destinations. Journal of Transportation Engineering 131(6), 419–420 (2005)
20. Shuiabi, E., Thomoson, V., Bhuiyan, N.: Entropy as a Measure of Operational Flexibility. European Journal of Operational Research 165(3), 696–707 (2005)
21. Thierry, G.: Public Research Industry relationships: Efficiency Conditions in Current Innovation. International Journal of Technology governance 17(3), 334–350 (1999)
22. Wang, G.H., Liang, L.: Deriving Information from Judgment Matrix and Regulating It According as Consistent Rule. Systems Enging-theory Methodology Application 10(1), 68–71 (2001)
23. Zhang, C., Zhu, W.D., Yang, S.L.: The Chinese Commercial Bank's Operational Risk Measurement Model on Entropy. Forecasting 26(5), 55–58 (2007)

Moving Target Tracking System Based on DSP*

Guo Jichang, Chen Minjun, and Wu Xiaojia

School of Electronic Information Engineering
Tianjin University
Tianjin, 300072, China
guojichang66@126.com

Abstract. The detecting and auto-tracking of Moving target is an important topic in the area of visual applications. A moving target auto detecting and tracking system based on DSP is proposed in this paper. The hardware block diagram and algorithm realization were introduced, and the system functions and several key techniques were analyzed. Test and measurement result show that the proposed system can track the target under normal illumination. It can track the moving object with 5cm/s screen speed. The realized system meets the real-time and reliability requirements, and it can be used in fields like supervision system, robot visual and etc.

Keywords: moving target, auto-tracking, difference, matching.

1 Introduction

Moving target auto-tracking system is an important topic in the area of visual applications. In real life, there are mass of information included in the motion of moving objects. We often more concerned about moving objects on many occasions, such as road traffic monitoring, security of important occasions, car's autopilot, the product line monitoring of factory and so on. As a result, moving target detecting and tracking problems are of great practical significance and value.

This paper realized a moving target auto-tracking system based on DSP.

2 Hardware Design

The hardware block diagram of an automatic target tracking system, with High-speed digital signal processor DM642 as the core, is showed in Figure 1.

---

* This work is supported by Natural Science Foundation of Tianjin #10ZCKFGX00700 to Guo Jichang.

D. Zeng (Ed.): ICAIC 2011, Part II, CCIS 225, pp. 124–130, 2011.

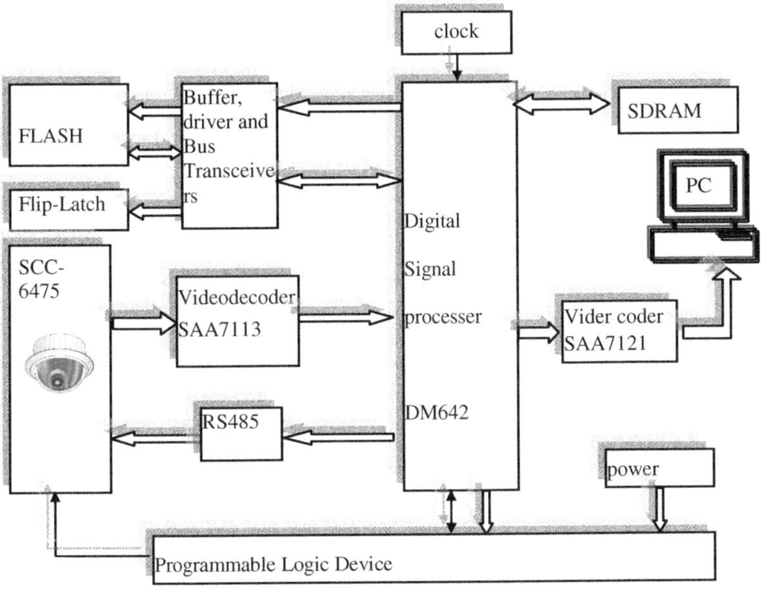

Fig. 1. Hardware block diagram of Moving target tracing system

2.1 Circuits Fundamentals of Moving Target Tracking System

Fundamentals and the main working process of the system are as follows: In the first place, we should initialize the system after power-on (or reset).Then, DM642 begins to gather image information and store them, Select a frame in every few frames, and determine wether there is a moving target or not according to the difference of two neighbourhood frame selected. If there is no moving target, the system continues to make recurrence of collecting. On the contrary, if there is a moving target in the image, we should get the characteristic template of the differential image after some necessary process such as binary, projection, extracting the template, and so on. Then a frame image can be collect after some frames. The characteristic template will be used to complete target matching in the selected frames, so that we can acquire motion vector of the moving target. And then combined with the camera parameters calculate horizontal and vertical angle of rotation, and output it to the camera Holder controller through the RS485 interface. After received control instructions, Holder controller will adjust the camera's focal length and direction so that it can realize the target of tracking.

2.2 Function Module

According to their function, moving target tracking system can be divided into 5 modules: video decoder module, DSP data processing and storing module, scheduling control module, Holder control module and video coding output module.

The most important part of video decoder module is the video decoder circuits which can convert the camera's analog video signals into digital video signals. The camera output a standard signal of PAL video format, and video decoder output digital video signals with 720×576 pixels, 25 frames/second. DM642 has a special interface for these signals.

DSP data processing and storing modules mainly include DSP chip, DSP specific power supply module, SDRAM, FLASH, EMIFA interface and some peripheral circuits such as system reset, and the main function is to complete the image difference, binarization, projection, template matching, tracking, etc.

The main task of scheduling control module is to distribute external device address, and latch some address, so as to ensure normal work of the system.

Video encoding module use SAA7121 module. This module converts the standard digital video signals from DM642 into analog signals, and directly input them to the monitor for display.

After processing, the main task of Holder control module is to transform the output of parallel instructions from DM642 into serial RS485 signals which are transmitted to the Holder controller of camera. Then it controls the horizontal rotation, vertical rotation and the focal length of the holder to complete the target tracking.

2.3 Essential Components and Functions

1) FLASH ROM curing program

In order to run the system off-line, ROM curing program must be used. This system uses a piece of AMD's AM29LV320B FLASH for storing boot and application program. The system need to boot-up from FLASH, so FLASH must be configured in the CE1 space. Because the EMIFA of DM642 has only 20 address lines and addressable space is 1M byte. If a FLASH of 4M bytes is needed to expand, a paging process is needed. In this system, FLASH is divided into 512Kb×8 pages. Page address is given by the CPLD which offer three paging address lines. The scope of each page address is: 0x90000000-0x90080000, that is to say, the address of FLASH overlap to each other. So when reading, writing or burning a FLASH program, paging is necessary.

Process of BOOT is as follows: After resetting, the EDMA controller of DM642 copy FLASH programs into the RAM which is started from address zero. For DM642, 1Kb program can be copied from CE1 space. After transfer completed, The CPU begin to execute programs from address zero. BOOT select pins BOOTMOOD [1:0] set to 11, at this time, the system will copy from the FLASH and execute the programs. System software is far greater than 1Kb, so only a few initializations and jumping instructions can be put into the space of 1Kb. The 1Kb program is Bootloader. Bootloader's main function is to initialize the various EMIFA space of DM642, copy the whole system software into SDRAM, jump to SDRAM and execute the system software.

2) Video Decoder and Video Encoder

In this system the video decoder is SAA7113. The main function of SAA7113 is to decode the analog video signal into standard digital video signal. Video encoder chip, SAA7121, can encode standard digital video signal (CCIR601, BT656) into composite TV signal in NTSC or PAL system, and it supports S-Video video output at the same time.

3) Logic Control Module chip

In the present system, CPLD use Xilinx's XC9500 series chip and the external circuits which mainly completed five functional modules: baud rate selection interface module, DM642 interface module, FLASH interface module, UART interface module and the Extended User Interface modules.

4) UART (Universal Asynchronous Receiver and Transmitter)

For high-speed processing of DSP, if waiting inquiry mechanism is used to communicate with the external serial ports, it will greatly reduce the efficiency of the system. The Solution is to add a parallel-series conversion chip: TL16C752. DM642 send the parallel sequences of control instructions to TL16C752, then TL16C752 transfer these instructions into UART format and send them to the Holder Controller.

The main features of TL16C752 are as follows: Two sets of built-in UART system can work independently; when working, baud rate is up to 1.5Mbps (24MHz crystal) or 3Mbps (48MHz crystal); with 64 bytes sent / received FIFO in-house, it can decrease of the number CPU interruptions; The start and stop level of Receiving FIFO can be realized through software programming; with two control modes, the software and hardware flow, etc.

3 Software Components and Functions

The software includes two parts: the target identification and tracking. Algorithm flow diagram is shown in Figure 2. In the follow, the function of these two parts was described.

The process of moving target tracking is to analyse the image sequence taken by camera and identify which part of data is the moving target information. Then calculate the target's location in the image and the turning angle which camera required or other values (such as holder stepping value) so as to aim at the target.

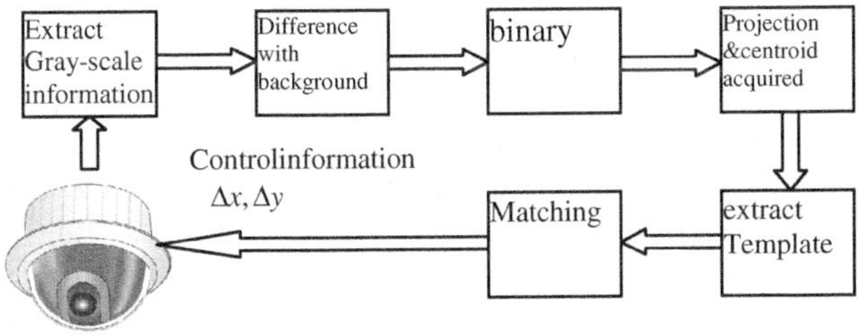

Fig. 2. Software flow chart

Moving targets show different moving features. According to these features, some algorithms can be used to orient and forecast target. In the end, target tracking was realized. Precision, speed, reliability are the three important measurements of tracking algorithms. Each existing method of moving target tracking has its own advantages and has different applications too [2][3][4]. In Practice, the selection of the tracking algorithm should be based on the characteristics of applications.

After extracting the greyscale image information, the following processing steps are: differential with background, binary, projection to obtain the geometric centre, template extraction as shown in the above flow chart. The aim of background difference is to distinguish image data of the moving objects from the background information. The aim of binary is to reduce the quantity of the image data which need to dispose and get the extremum of image information. In order to obtain the centroid of moving target and improve the accuracy of the following process, projection method is used in this system. Based on the accuracy of geometric centre, the system will extract a template with stable image information and access to the centroid of the moving target as far as possible. Then through gliding in the new collected greyscale image (with the moving target), the motion vector can be acquired according to the algorithm formula. Based on the motion vector, the instructions can be generated to drive the camera to rotate so that the moving target tracking was realized. This is an integrated process.

Notice that not every image process is complete like that, it may return to execute once or several times when it accords with some conditions.

4 The Key Issue in the Design

4.1 Speed of Data-Processing

The output of decoder is the image sequences with resolution 720x576. Each pixel has 256 greyscales, that is, each grey value takes up 8bit (1 Byte). In other words, the amount of data required to deal with per second is: 25x720x576x1 Byte, that is, 9.887Mbyte/s. For this system, it is difficult to deal with such amount of data in real time and it is unnecessary. So it is necessary to reduce the amount of data. Two methods can be used to do in this system:

1) Less Frames

The rate of input images is 25frames/s. whether it is necessary to process every frame? In general, the monitored targets are often people walking in normal speed which are more than 2 meters away from the camera. Under such circumstances, if he is walking in constant speed 1m/s, when we get all of the 25 frames in 1 second, we find it difficult to distinguish the subtle differences between consecutive frames. Our naked eyes can't see a greater difference until the two frames are 7 or 8 fames away. The system processes a frame every 6 frames, so that the image data need to be processed reduced significantly, and the result is satisfactory.

2) Lower Resolution Ratio of image

In present system, the original image size is 720x576 points. It is not necessary to adopt such a high resolution ratio, so the system gets a point every two points that makes the size decline to 360x228 points, that is, the data of each image is reduced by 3/4. More importantly, in the following application of algorithm, it greatly reduces the amount of operation and improve the speed of operation which is good for real-time.

4.2 Threshold Issue of Binary

In the differential image, only the region which includes moving target is bright. The rest region is gloomy. Dividing line in differential images can be seen as a noise subjected to Gaussian distribution [5]. As the background is much bigger than the noise region in differential images and it has constant distribution but moving targets haven't, so we can use the Mean and standard deviation of differential image to estimate noises. Specific formula:

$$\left|I(x,y,t)-\mu(t)\right|>3\sigma(t) \tag{1}$$

$\mu(t)$ And $\sigma(t)$ in formula above is as:

$$\mu(t)=\frac{1}{m\times n}\sum_{x=0}^{m-1}\sum_{y=0}^{n-1}I(x,y,t) \tag{2}$$

$$\sigma(t)=\sqrt{\frac{1}{m\times n}\sum_{x=0}^{m-1}\sum_{y=0}^{n-1}[I(x,y,t)-\mu(t)]^2} \tag{3}$$

According to this threshold formula, select a value for the threshold:

$$threshold=\mu+3\sigma(t) \tag{4}$$

Set it to be 255 when the greyscale value is bigger than the threshold or 0 when it is smaller.

$$I(x,y,t)=\begin{Bmatrix}0, if I(x,y,t)<threshold\\ 255, otherwise\end{Bmatrix} \tag{5}$$

In this formula $I(x,y,t)$ is binary pixel value of differential image, $\mu(t)$ is the mean of the image, and $\sigma(t)$ is the standard deviation.

If fully according to formula (5), the threshold based on greyscale value will hardly change during the entire process. However, in the actual environment, the light will change (especially outdoor scenes), and this change will cause changes of pixels' greyscale value, and result in decreased accuracy of detection. So it is necessary to take detection and compensation with changes of light into consideration.

A simple compensation is used in this system. It allows threshold change by plus or minus 10% when light changed. The specific way is: when it has bright light and greater greyscale value, threshold multiplies by 1.1; when it has gloomy bright and smaller greyscale value, threshold multiply by 0.9. Another threshold is needed, when the sum of all pixels' greyscale value is bigger than this threshold, the system adopts plus compensation, otherwise minus compensation is adopted.

4.3 Size and Shape of Template

In theory, the shape of the template should be similar to the moving targets, so that the algorithm can be more accurate [7]. But in practice it is impossible. Firstly, moving targets are multiple. Secondly, the shape of moving target is usually irregular. If the algorithm must depict the irregular image, it is not only complicated and time-consuming, but more importantly, it is not necessary. The range of error is acceptable, as long as the system can extract digital image features of the moving target and following algorithms are effective.

It is unsuitable when the size of template is too big or too small. If the template is very big, the amount of calculation will greatly increase. But if it is too small, effective image information will be greatly reduced, so does the success rate of match, and it easily led to wrong tracking. Through experimenting again and again, rectangular template was selected which size is 40×64.

5 Conclusion

On condition that the illuminance of the environment and the speed of moving target are all normal, the moving target tracking system based on this design scheme can realize the real-time tracking of moving target, and basically solve the problem of target error recognition. In the system, the selections of the image frames, the binarization of the threshold, the size and shape of template was considered. All these factors and parameters are of great importance to the real-time, identification and tracking performance of the system. The system can be used in varieties of intelligent video supervision and it has broad prospects for applications.

References

1. Texas Instruments. TMS320DM642 Video/Imaging Fixed-Point Digital Signal Processor (2006)
2. Wang, H.-S., Mersereau, R.M.: Fast Algorithms for the Estimation of Motion Vectors. IEEE Trans. Image Processing 3(8), 43–438 (1999)
3. Meiyuan, Sunhuaijiang, Xiadeshen: An Image fast refining algorithm based on modified template. Transition of China Images 9(11), 1306–1311 (2006)
4. Cai, Q., Mitiche, A., Aggarwal, J.K.: Tracking Human Motion in an Indoor Environment. In: The 2nd International Conference on Image Processing, Washington D.C, vol. 1(1), pp. 215–218 (1995)
5. Zhuhui, Lizaiming: Moving target detection technique in video sequence. Signal Process 18(5), 449–451 (2002)
6. Songkeou, Huangfenggang, Lanxiaoting: Binary image target centriod fast search and tracking algorithm. Pattern Recognition and Artificial Intelligence 2(11), 161–168 (1998)
7. Lijun, Zhouqibo, Gejun, et al.: A research for automatic tracking of moving target in dynamic template matching algorithm. Infrared Technology 27(4), 328–332 (2005)
8. Zhanghui, Wangqiang, Zhuzhigang: Fast detection, traction and discrimination of moving target. Transaction of Tsinghua University 42(10), 1401–1405 (2002)
9. Chenminjun: The research and realization of moving target tracking algorithm based on DSP[Thesis of master's degree]. Tianjin University, Tianjin (2006)

Robot Arm Safety Improvement by Position/Torque Switching Control

Wu Haibin[1], Zheng Hongqing[1], Yang Jianming[2], and Li Feng[2]

[1] School of Mechanical Engineering and Automation, Fuzhou University
Fuzhou, Fujian Province, China
[2] Department of Mechanical System, Meijo University
Nagoya, Japan
156483885@qq.com

Abstract. With the development of robot technology, it is an inevitable trend in the future that Robots will come into the normal life. Robots must be safe enough and ensure no hazard to person in order to coexist with people harmoniously. This paper provides a method to promote the robot safety by taking position-torque-position switching control. Namely, robot will work in normal position control mode for fast and accurate respond, until it impacts to humans. Once robot strikes to humans, it will change to torque control mode from position mode immediately. Robot will contact with humans very softly in torque control mode, so it is very safe. When humans leave from robot, it resumes to position control mode, and executes remainder tasks. Signal of striking to humans can be detected by torque sensor fixed in joint axis. Withdrawing of humans can be detected by measuring robot speed variety. The merit of above control method is that, at the condition of ensuring normal task performance, the robot safety has been enhanced remarkably. This method has been experimented at a 1DOF robot arm, and gain the anticipate results.

Keywords: robot safety, position/torque switching control, performance, sliding mode control.

1 Introduction

For normal industrial robot arms, the stiffness is designed to be very high in order to precisely pursuit trajectories planned. But high stiffness implies that the collisions will be very serous, when robots strike to humans. This is the reason why humans are forbidden to go into the robot workspace. Normal robot arms don't have enough safety to coexist with humans. With the development of robotics and severe aging population problem, humans hope robots can coexist friendly with everyone in houses and offices, helping to convey, sweep, and take care of older and handicapped. This kind of robots can be called human-friendly robot(HFR). In the last few years, there have been a lot of interests in this area[1,2,3].

D. Zeng (Ed.): ICAIC 2011, Part II, CCIS 225, pp. 131–139, 2011.

Improving the robot safety of physical interaction with humans may reduce the performance, such as position pursuit precision, frequency band of dynamic response[4]. Researchers have taken many methods to compete the trade-off between safety and performance, and try to resolve this problem. Ref.[5] proposed that some non-contact sensors were equipped on the robot arm to detect whether there are humans to be close to robot, and then decide how to control the robot motion. Since then, this idea has been developed continuously. Dana Kulic[6] presented a new implementing way, in which the real-time danger degree must be calculated during the time that humans are close to robot based on nonlinear danger factor, and then trajectories must be adjusted on line. This method has still been developing. The second method to improve the robot safety is to increase the joint compliance, by which the stiffness or impedance have to be reduced. There are many different approaches to realize it. For example, Dongjun Hyun[7] proposed to design a kind of special changeable stiffness machine that is installed in the robot joint. Ref.[8] proposed to employ Macro-Micro actuations in parallel for every robot joint. Ref.[3] proposed two distributed actuations design. All of them are the detail design methods to improve joint compliance, and play an important role in the area of robot safety[9,10,11]. But every one of them is imperfect. For instance, some of them may cause the frequency band narrowed, or machine structure is too complicated to use, or there is a significant impact to humans that may be hazard.

This paper presents one intelligent method to improve the robot safety completely depending on control, which don't need to increase new machine or sensors. One new control strategy, namely position-torque-position switching control method has been proposed. In this method, position control is a normal task mode, and torque control is safe mode in which there is no hazard to humans. This paper take 1DOF robot arm as the experiment platform and the test has reached the prospective results.

2 Joint Mathematic Model and Control Algorithm of Robot Arm

Series type industrial robot arms normally include several degrees, some of which are used to position, and some other of which are used to adjust gesture. The safety of robot arms can be guaranteed if the joints used for positioning have the compliance to a certain extent. In this sense, this paper take 1DOF robot arm as the research objective whose one end is fixed and can rotate around the axis, and the other end is simplified to an equivalent load. Without the loss of generality, we design the 1DOF robot arm to rotate in vertical plane, so the torque caused by equivalent load is changeable with the angle. The arm is driven by a DC motor, and a harmonic gear and a torque sensor are equipped between motor and arm link, as shown in Fig.1.

This paper proposed position-torque-position switching control strategy. Robot arms keep the position control mode like normal industrial robots to ensure the performance, before humans collide to robots and after humans leave the robots. Only during the collision between humans and robots, robots keep torque control mode, in which the force acting on humans is limited in a very low value, so it is very safe. Robot arms will finish the whole position pursuit task, and not be suspended by the collision, as shown in Fig.2.

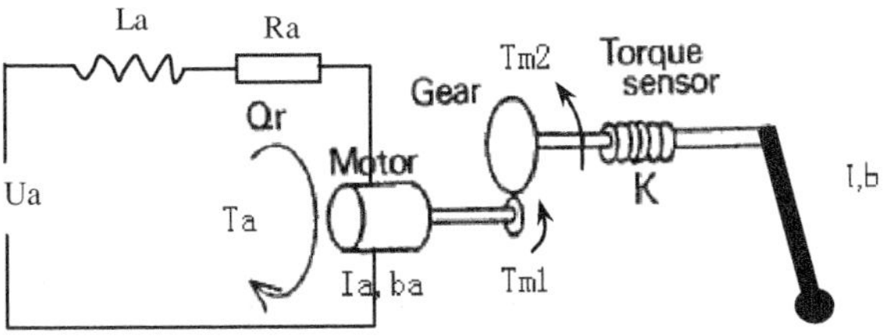

Fig. 1. 1DOF robot arm model

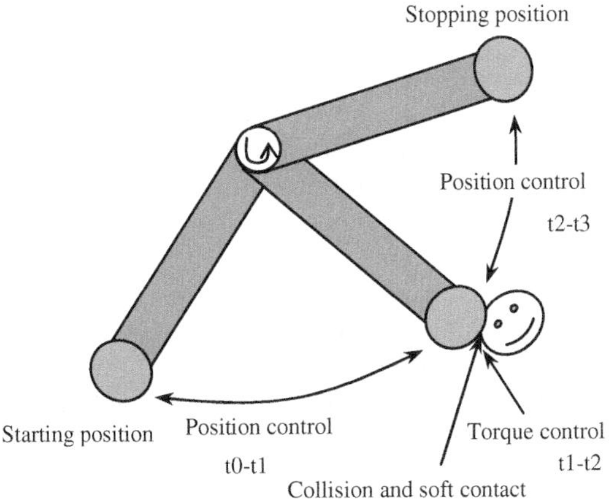

Fig. 2. Position-torque-position switching control diagram

A torque sensor was equipped on the joint axis to detect the actual torque outputting to the load. The mathematic model is given as followed

$$I_a \cdot \ddot{\theta}_a + b_a \cdot \dot{\theta}_a = T_a - T_{m1} \tag{1}$$

$$I \cdot \ddot{\theta} + b \cdot \dot{\theta} = T_{m2} - T_l \tag{2}$$

$$T_{m2} = K(\theta - \frac{\theta_a}{N}) \tag{3}$$

$$U_a = L_a \frac{di_a}{dt} + R_a \cdot i_a + V_e \tag{4}$$

Where $V_e = K_e \cdot \omega$ is the motor induced electromotive force, $T_a = K_t \cdot i_a$ is motor torque, $T_{m2} = N \cdot T_{m1}$ is the input torque to joint axis, $\theta = \theta_a / N$ is the angle of joint axis, $T_l = mgR \cdot \sin\theta$ is the load torque produced by equivalent weight, I_a is the motor inertia, b_a is the motor damping coefficient, T_{m1} is the load torque acting on the motor axis, N is the gear ratio, L_a is the motor inductor, R_a is the motor resister, U_a is the motor input voltage, i_a is the motor current, θ_a is the angle of motor axis, θ is the angle of link, I is the load inertia, b is the load damping coefficient. K_e is induced electromotive force constant. K_t is motor torque constant. R is arm length.

The sliding mode control(SMC) method has been used, in order to reduce the friction influence and speed fluctuation during position control. It is supposed that there is a ramp reference position input as followed

$$r = r(t) = Vt \tag{5}$$

So speed reference $\dot{r} = V$ is a constant, and position error and speed error are respectively

$$e = r - \theta,\ \dot{e} = \dot{r} - \dot{\theta} \tag{6}$$

Sliding mode surface is defined as

$$S = Ce + \dot{e} \tag{7}$$

Proportional switching control input has been used as

$$u(t) = (\alpha|e| + \beta|\dot{e}|)\,\mathrm{sgn}(S) \tag{8}$$

Where C , α , β are the positive constants, θ can be detected by encoder, and $\dot{\theta}$ is derived from differentiating θ.

3 Switching from Position to Torque Control

The key to implement the position-torque-position switching control is how to switch from position to torque control, and from torque to position control appropriately. In position pursuit mode, the feedback control loop need only position sensor and don't need torque senor. But in torque control mode, in order to achieve safety, the torque/force impacting on humans must be limited strictly, so torque sensor is inevitable. In this scheme, a torque observer, whose input is real time position, velocity, acceleration, and load torque, whose output is torque observation value, is constructed based on known mathematic model in position control mode. The torque observation value can be compared with the detected torque by torque sensor. If they are different obviously, the observer model can be revised until they are much closed to each other. When robot arm collides to humans, the motor torque output will increase instantly in the position control mode, but at the same time, torque observation value may change very little. Just for this reason, we can take the difference between torque senor and torque observation value as the signal of colliding to humans. When the difference reaches a threshold, robot arm will switch from position to torque control immediately. The threshold value determines the switching sensitivity. The larger the threshold, the better resisting the little disturbance or unsmoothed rotation, but the heavier impacting to humans. Fig.3 shows the switching process.

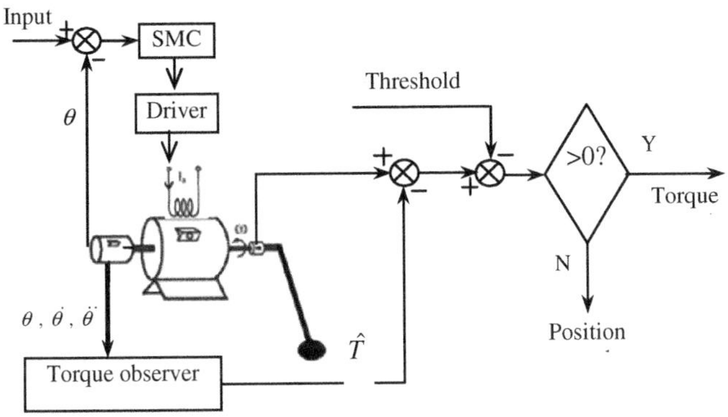

Fig. 3. Switching from position to torque

In the torque control mode, system has lost the position pursuit capability thoroughly. In order for robot arms to keep the press force to humans equal to the threshold approximately, the setting torque must be correspond to the actual position of robot arm because the load torque is different in different position. In static state, the threshold value is equal to the torque acting on humans and the static friction torque.

4 Torque Control Algorithm

The purpose of torque control is to impose a constant pressure on humans. During torque control mode, the necessary driving torque is variable depending on different arm position. To keep the press to humans non-fluctuated, the setting torque in torque control loop must be adjusted according to current position, which can be achieved by real time detecting arm position and then calculated as

$$T_s = \hat{T} + T_b \tag{9}$$

$$\hat{T} = I \cdot \ddot{\theta} + b \cdot \dot{\theta} + mgR \cdot \sin\theta \tag{10}$$

Where T_b is the torque pressed on humans, $\hat{T}$ is torque observation value. Torque control theory and the setting torque calculation are shown in Fig.4.

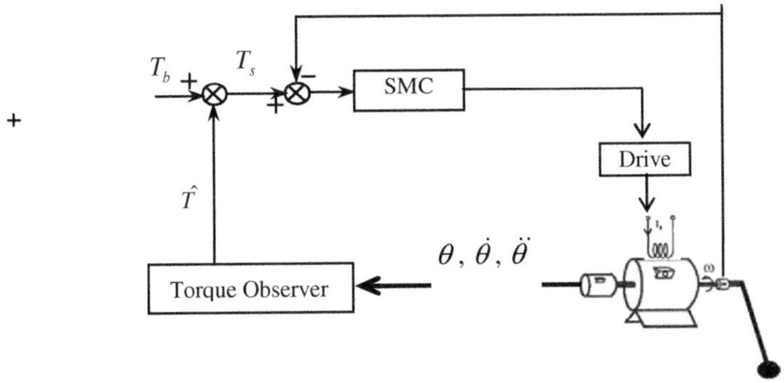

Fig. 4. Torque control theory

During torque control, sliding mode control method has also been taken to resolve friction problem, especially in lower speed. E_T is torque error input of sliding mode controller, but it is hard to achieve good results just depending on E_T. We propose an integration of E_T so that the sliding mode surface is defined as followed

$$S = C \cdot E_T + E_I \tag{11}$$

Where

$$E_I = \int E_T dt \tag{12}$$

Proportional switching control input is also used as

$$u(t) = (\alpha|E_T| + \beta|E_I|)\operatorname{sgn}(S) \tag{13}$$

Where C, α, β are the positive constants.

5 Switching from Torque to Position Control

The reason why robot arm provides a constant press to humans is to achieve a switching signal from torque to position control when humans are detached from arm. As you know, if there is no mutual force between arm and humans, the arm will be unable to perceive the withdrawing of humans, and can't resume the rest position pursuit task. When humans are detached from arm, arm has to follow the setting torque and lead to accelerative rotating suddenly and quickly. This scheme just extract the velocity variation as the switching signal from torque to position control. Namely, it is transformed to position control when the velocity reached a threshold. The time during torque control is uncertain, i.e. the time between t1 and t2 in Fig.5. But in torque control mode, the setting position is still variable with time. For example, at the beginning of torque control, the setting position is at point A, and the setting position has reached point p' at the end of torque control, whereas arm actual position is at point p. the error between arm setting position and actual position may be very large. In such case, the current setting position of resuming to position control from torque control should not be the above setting position point p', but the current actual position point p. The objective position point is invariable, and arm trajectory must be searched renewably.

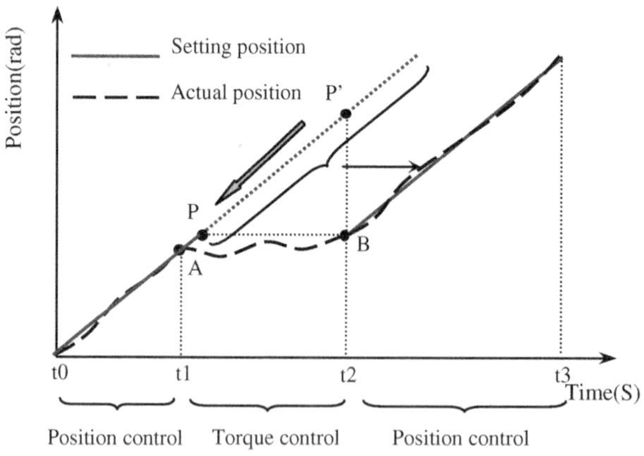

Fig. 5. Switching from torque to position control

6 Experiment and Results

A 1DOF robot arm has been established in order to verify the position-torque-position switching control strategy and torque control algorithm proposed in this paper. The experiment system consists of DC motor(Maxon Motor) equipping with 256P encoder, harmonic gear(HI-T Drive, RS-14-50-(003)C-SP type) of ratio 50, torque sensor(KYOWA, TP-2KMCB type) of stiffness K 690 N · m/rad, 0.5m length arm, 0.5kg equivalent weight load, PC control computer, AD and DA transformation card(Interface, PEX-361316, 16-bit resolution) used for sampling torque sensor signal and outputting motor analogue voltage control signal whose sample period is 5ms. In position sliding mode control, $C=20$, $\alpha=20$, $\beta=0.5$. In torque sliding mode control, $C=1$, $\alpha=0.15$, $\beta=0.1$.

The experiment is designed to pursuit 0-5π trajectory task, any point of which may collide to humans, and then at any time humans may be detached from arm, but position pursuit task can be resumed and reach to the objective. The position pursuit of whole course is shown in Fig.6, where A is the switching point from position to torque control, and B is the switching point from torque to position control. From Fig.6 it can be seen that the first and second position track segments are both very closed to settings, and torque control does not influence the positioning task. But the setting position of the latter position trajectory must begin from arm current actual position, and the rest unfinished trajectory must be conveyed as the setting position.

During the whole switching control process, the variety of torque is shown in Fig.7, where A is correspond to A in Fig.6, and ΔT is the torque threshold switching from position to torque control (0.4Nm in program). The torque keeps changeless approximately in torque control mode. Point B in Fig.7 correspond to point B in Fig.6, is the time that humans are detached from arm. The rotating velocity of robot arm is shown in Fig.8 during the whole motion, where the velocity has the fluctuation of about $\pm$0.05Rad/S in position control mode, because of variable load torque in different position. There are oscillations when switching from position to torque control, and the velocity oscillation is more severe at the time of switching from torque to position control.

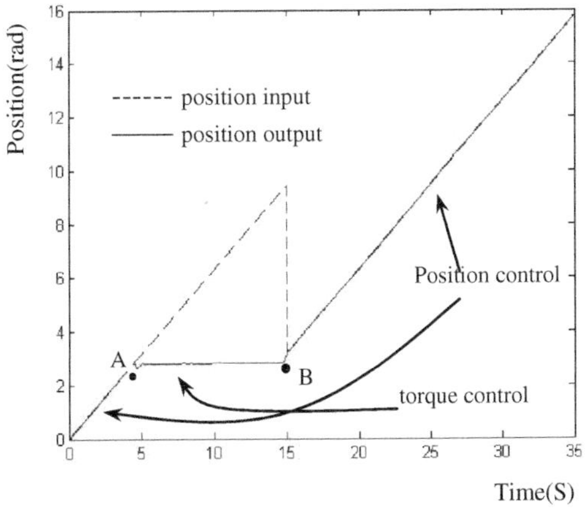

Fig. 6. Position pursuit of switching control process

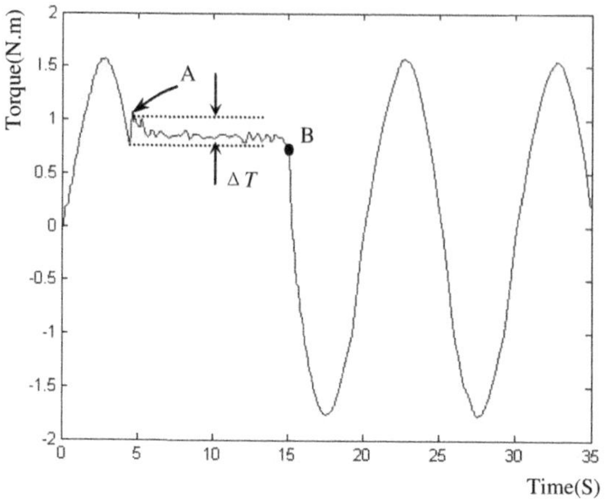

Fig. 7. Load torque variation during switching control

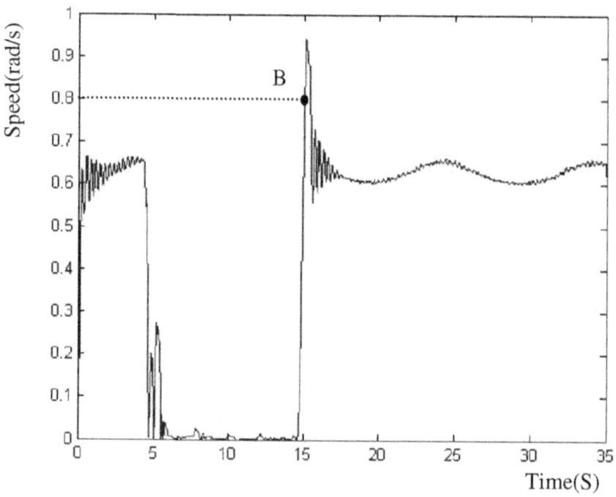

Fig. 8. Angular velocity variation during switching control

7 Conclusion

When robots and humans coexist in same workspace, robots must have enough safety to ensure harmless to humans. A new intelligent control strategy, position-torque-position switching control is proposed in this paper. The advantage of this method is that the robot arm safety has been improved by controlling strategy, at the condition of no additional equipments. Furthermore, robot arm can resume to finish the rest task from the collision when humans are detached from arm. The effectiveness has been verified by the mathematic modelling and experiments in 1DOF robot arm.

References

[1] Yamada, Y., Suita, K., Imai, K., Ikeda, H., Sugimoto, N.: A failure-to-safty robot system for human-robot coexistence. Robotics and Autonomous Systems 18, 283–291 (1996)

[2] Lim, H.-O., Sunagawa, M., Takeuchi, N.: Development of human-friendly robot with collision force suppression mechanism. In: ICROS-SICE International Joint Conference, pp. 5712–5716 (2009)

[3] Zinn, M., Khatib, O., Roth, B.: A new actuation approach for human friendly robot design. In: Proceedings of the IEEE International Conference on Robotics & Automation, pp. 249–254 (2004)

[4] Schiavi, R., Grioli, G., Sen, S., Bicchi, A.: VSA-II: a novel prototype of variable stiffness actuator for safe and performing robots interacting with humans. In: IEEE International Conference on Robotics and Automation, pp. 2171–2176 (2008)

[5] Espiau, B., Boulic, R.: Collision avoidance for redundant robots with proximity sensors. Robotics Research 3, 243–252 (1986)

[6] Kulic, D., Croft, E.: Pre-collision safety strategies for human-robot interaction. Auton Robot 22, 149–164 (2007)

[7] Hyun, D., Yang, H.S., Park, J., Shim, Y.: variable stiffness mechanism for human-friendly robots. Mechanism and Mechine Theory 45, 880–897 (2010)

[8] Morrel, J.B.: Parallel coupled micro-macro actuators, PhD Thesis, Massachusetts Institute of Technology, Cambrigde, MA (1996)

[9] Pratt, G., Williamson, M.: Series elastics actuators. In: Proc. of IEEE/RSJ International Conference on Intelligent Robots and Systems, vol. 1, pp. 399–405 (1995)

[10] Kilicaslan, S., Kemal Ozgoren, M., Kemal Ider, S.: Hybrid force and motion control of robots with flexible links. Mechanism and Machine Theory 45, 91–105 (2010)

[11] Pun, J., Semercigil, S.E.: Joint stiffness control of a two-link flexible arm. Nonlinear Dynamics 21, 173–192 (2000)

The Design of Voice Alarm Terminal Based on MSP430

Wu Qiyan, Miao Changyun, and Wu Zhigang

School Of Information and Communication Engineering
Tianjin Polytechnic University
Tianjin, China
qiyanwu@fastmail.fm

Abstract. This paper proposes a design of voice alarm terminal which is based on MSP430.It design the hardware circuit with MSP430F149 microcontroller as its core and uses the program structure and functional modular methods to design its software. The Voice Alarm Terminal can integrate with the enterprise VoIP system, but also can send the appropriate alarm sound depending on different events and broadcast through the VoIP terminal, so that personnel-related can hand danger timely. It has extensive application value.

Keywords: MSP430F149, MI22420, Alarm terminal, VoIP.

1 Introduction

With the social development, a variety of alarm terminal come into being and is widely used in industry, agriculture, transportation and medical. When the danger occurs, it can remind us of dealing with danger timely to avoid big losses. Many companies have established an alarm system. In order to facilitate scheduling, they also establish the VoIP system which is independent of the alarm system. The dangerous situations occurring, the relevant staff is not conducive to the timely processing of danger. Meanwhile, it also makes the construction costs increase greatly.

Based on this situation, a voice alarm terminal is designed. It can connect with the VoIP system. Once the danger occurs, the alarm terminal sends a request to media gateway controller(MGC) through alarm gateway, then MGC sends PA (Power Amplifier) call command to all or specified parts of VoIP terminal. Each VoIP terminal broadcasts the alarm sound generated by the alarm terminal to make relevant staff deal with danger timely. Because the alarm terminal combines with the VoIP system, the enterprise is not necessary to establish additional alarm system, significantly saving in the construction costs. The alarm terminal is designed in this paper for the VoIP system which has been established in enterprises interior and has a wide range of application values.

D. Zeng (Ed.): ICAIC 2011, Part II, CCIS 225, pp. 140–147, 2011.

2 Design of the Alarm System Based on the Voice Alarm Terminal

The alarm system accesses an alarm terminal through the alarm gateway in the VoIP system. It concludes VoIP and alarm launched. The VoIP is achieved using simplified soft-switching network. It can accomplish a single call, group call, all call and PA call and so on and makes up of MGC and VoIP terminal. The part of alarm launched mainly completes alarm sound generation and access. It is decomposed into two parts: alarm gateway and alarm terminal. This paper is going to focus on the alarm terminal design. The topology structure of alarm system is shown as Fig 1.

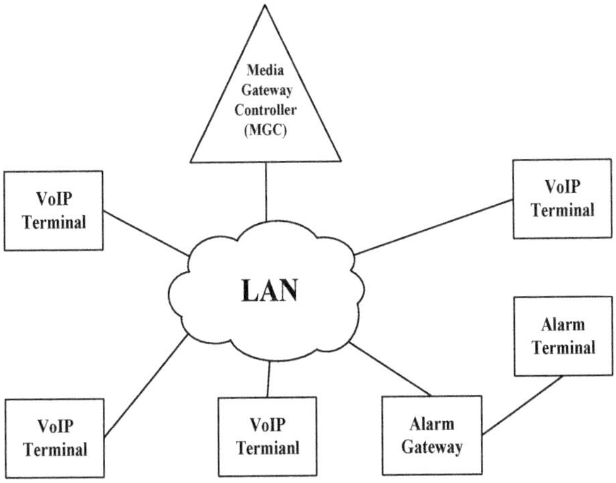

Fig. 1. The topology structure of alarm system

Media gateway controller controls the construction and the release of the VoIP terminal and alarm gateway connection, according the received signal.

The VoIP terminal consists of media gateway (MG) and voice terminal. Voice terminal is responsible for the generation of voice. Media gateway is responsible for media conversion as well as the connection between PSTN and IP network.

Alarm gateway is used to request MGC to send group call or all call command to make VoIP terminal automatically hook and enter PA alarm state.

Alarm terminal is responsible for playing an alarm sounds according to alarm PLC code.

3 Hardware Design for Voice Alarm Terminal

The hardware circuit of this voice alarm terminal constituted by MCU control circuit, voice playback circuit, voice storage circuit, level conversion circuit, alarm code access circuit, 8-way manual alarm button, key and display circuit, as shown in Fig 2.

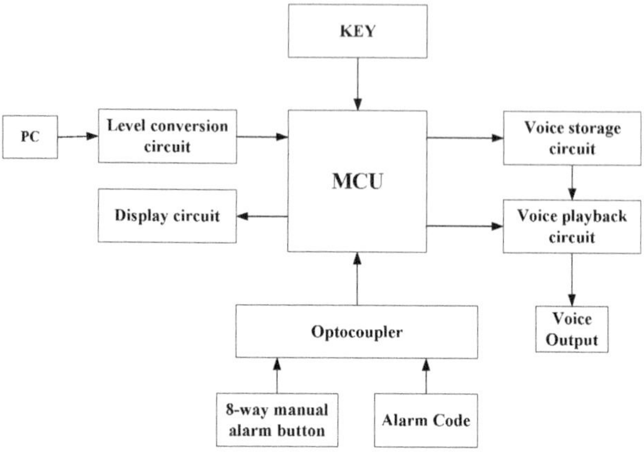

Fig. 2. Hardware circuit diagram for voice alarm terminal

3.1 Voice Playback Circuit

ML22420 voice chip is used to design voice playback circuit. It is the serial ROM interface built-in 4-channel mixed audio playback, and external ROM for storing voice data voice playback LSI. In the alarm terminal, its various operations carried out are controlled by an external MCU. MCU Serial Peripheral Interface (SPI) connects with ML22420 SPI, MCU as the host and ML22420 as a slave. MCU sends command to ML22420 and receives responses from ML22420 to control ML22420 voice chip. In the concrete realization of the time, a series of different command codes is sent to control the voice chip to complete different operations. The alarm sound it plays is from the voice storage chip W25X16. Its schematic is shown in Fig 3.

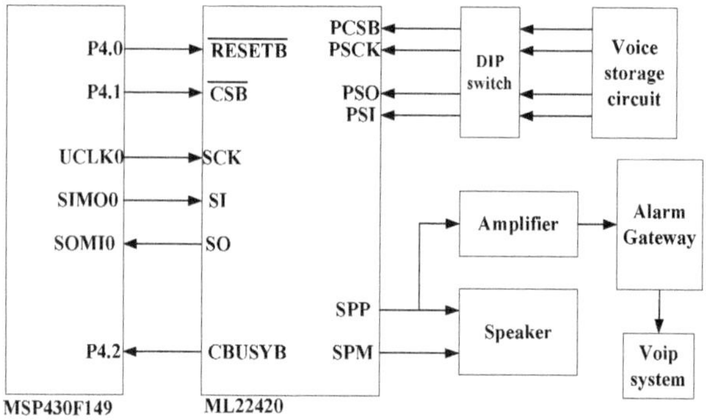

Fig. 3. The schematic of the voice playback circuit

3.2 Voice Storage Circuit

Voice storage circuit mainly uses W25X16 (16M-bit) Serial Flash memories from Winbond Electronics Corporation. The W25X16 series offers flexibility and performance well beyond ordinary Serial Flash devices and supports the standard SPI, and a high performance dual output SPI using four pins: Serial Clock, Chip Select, Serial Data I/O and Serial Data Out.

W25X16 is mainly used to store voice data which is sent over the MCU SPI. ML22420 doesn't have recording function. Therefore, the alarm voice data stored in the W25X16 is pre-recorded in the PC, converted the encoding of a variety of formats (such as PCM, ADPCM) through the conversion software, segmented and entered through RS232. Because the recording and playback share SPI, a DIP switch is added among MCU SPI, ML22420 and W25X16. When recording, the first four DIP switch are dialed to ON position and after four are allocated to the OFF position. When normally playing, the position of DIP switch is opposite to recording.

3.3 Level Conversion Circuit

In the PC, serial RS232 level are ± 10V and the signal voltage of the general MCU application system is TTL level which are 0 and +5 V. Therefore, the MCU serial can not be directly connected to the PC serial. The level must be shift. Level conversion circuit makes use of MAX232 chip from TI Company to achieve the level translation between the MCU and PC.

3.4 MCU Control Circuit

MCU control circuit is made up of MSP430F149 MCU from TI Company and its peripheral interface circuit. The MSP430F149 are microcontroller configurations with two built-in 16-bit timers, a fast 12-bit A/D converter, two universal serial synchronous/asynchronous communication interfaces (USART), and 48 I/O pins. It mainly provides the interface to the external circuit, is the control core in the whole system. The schematic of MCU control circuit is shown in Fig 4.

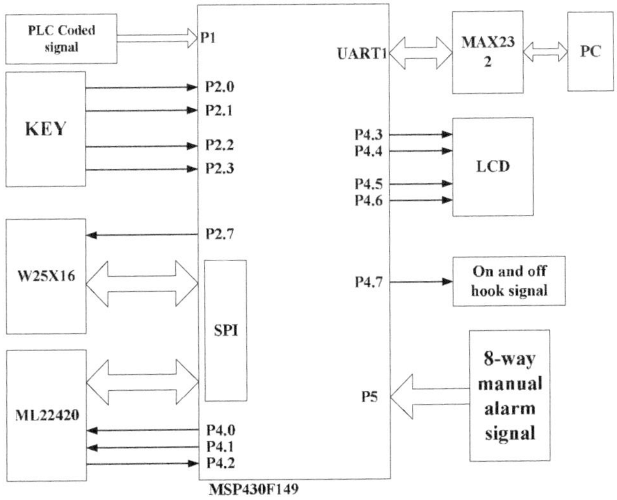

Fig. 4. The schematic of MCU control circuit

In the MCU control circuit, P1 port is used to receive coding signals from the PLC; P2.0 ~ P2.4 port are used to read the function key input and achieve the corresponding functions; MCU SPI connects with W25X16 and ML22420, transmits voice data which is received from MCU UART1 port to voice storage chip and send control commands to voice playback chip ML22420; UART1 port connects with RS232 port in the PC through MAX232 in order to receive the alarm voice data from PC; P4.0 port is used as a reset signal for ML22420 and ML2240 is idle or busy judged by P4.2 port. P4.3~P4.6 port connected with the LCD is used to make LCD display various states. P4.7 port is used to generate on- hook and off-kook signal. The signal from 8-way manual alarm button is received by P5 port.

3.5 Others Circuit

External alarm signal is converted into the corresponding alarm code by PLC, and then connects with P1 port in MCU through the optoelectronic coupled devices TLP521 which can be controlled to achieve the corresponding alarm. The purpose of using TLP521 achieves the front-end of alarm code and alarm circuit isolation.

Keyboard input circuit is used to adjust the alarm volume, speaker / LINE output options, and reset. It is through the MCU P2.0 ~ P2.3 port to achieve these functions.

Manual alarm button connected with P5 port through the TLP521 to play the appropriate alarm sound.

Display circuit mainly uses SG12864 LCD module. In the normal working hours, it shows the state such as the name of enterprise, IP address. On the alarm terminal operations, it shows the current ongoing operations, such as recording. When danger occurs, it can display alarm type, playback mode of the alarm and so on.

4 Software Design for Voice Alarm Terminal

The software for voice alarm terminal is designed with the program structure and function modular design method and prepared using C language. The overall design includes the main procedure, interrupt service routine and task subroutines.

4.1 Main Procedure

The main procedure completes the initialization setting for the clock module port, flash memory, I/O ports, LCD, timers, SPI and voice playback chip ML22420, and sends on-hook signal to VoIP first. Then, it enters the main loop and deals with the corresponding subroutines according to the appropriate flag to achieve the set function. Its flow chart is shown in Fig 5.

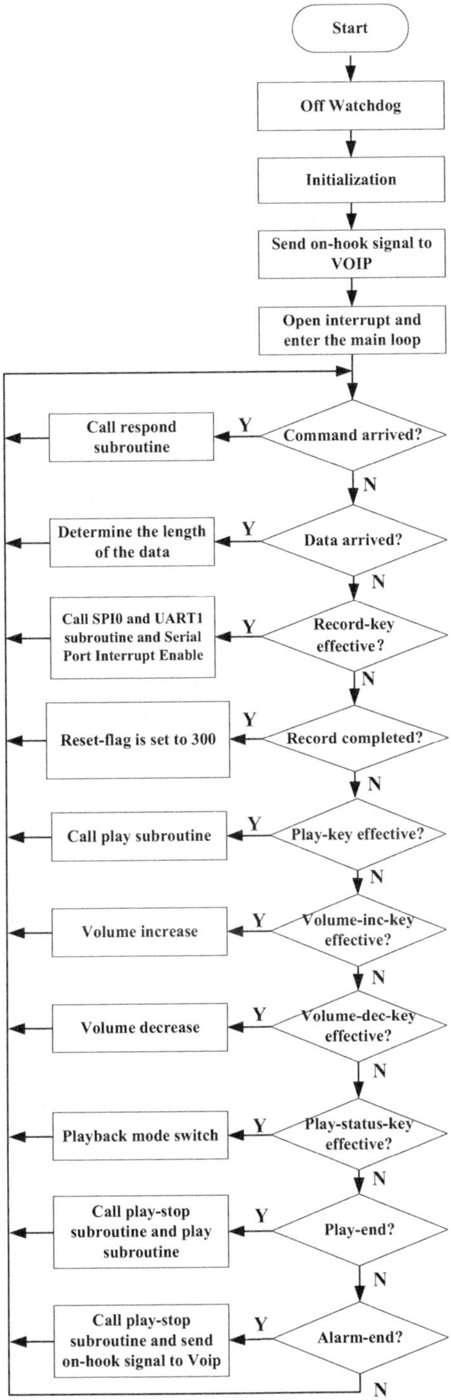

Fig. 5. The flow chart of main procedure

4.2 Interrupt Service Routines (ISR)

Interrupt service routine includes Timer_A interrupt service routine and UART1 interrupt service routine. Timer_A interrupt service routine mainly finishes the setting of function keys and the detection of the manual alarm button, playback code, reset flag, the timer counter flag, the number of LED flash when playing voice and play_start and alarm_end flag for the corresponding treatment. Its flow chart is shown in Fig 6.

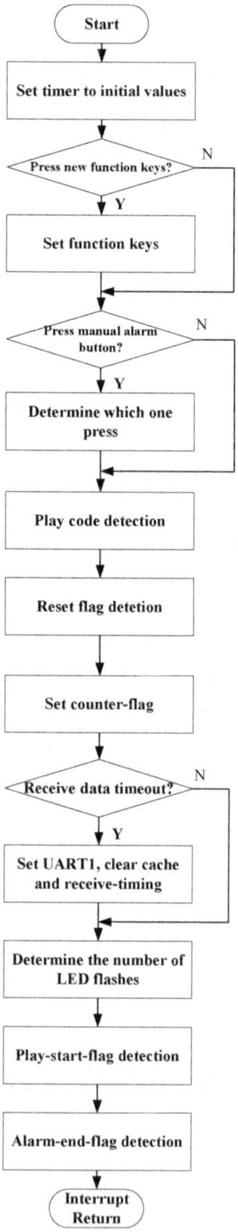

Fig. 6. The flow chart of Timer_A interrupt service routine

UART1 interrupt service routine is used to complete the data communication between PC and MCU. In the main procedure, the alarm sound is recorded through calling the appropriate subroutine.

4.3 Task Subroutines

Task subroutines consist of ML22420 task, FLASH of MCU task, voice storage task, display task, LCD module driving task, SPI task and UART task and so on.

ML22420 task subroutine mainly completes voice data reading and writing, voice loop playback, voice broadcast channel selection and the state reading, and the initialization of ML22420 and so on.

Reading, writing and erasing FLASH are carried out by FLASH of MCU task subroutine.

Voice storage task subroutine is mainly used to finish the voice storage chip reading and writing.

The SPI task subroutine realizes the initialization of SPI port and voice data communications tasks among the MCU, ML22420 and W25X16.

UART task subroutine is made use of initialing the USART port and responding the received data.

5 Conclusion

The voice alarm terminal based on MSP430 involved in this paper is not only to fuse the enterprise VoIP system, but also can achieve a variety of alarm, automatic alarm and manual alarm. Its alarm sound can play through speaker and reach the VoIP terminal through the line to play. The terminal has a low cost, simple operation, easy maintenance, expansion and flexible characteristics. It is not only fit to various types of factories, enterprises and other industrial control sites, but also in the education system, science and technology systems, electrical systems, and the banking system has a wide range of promotion and application prospects.

References

1. Li, B., Miao, C.-y., Li, Y.-f., Wu, Z.-g.: Research of VOIP Transmission Terminal based on DSP. Microcomputer Information 26(6-2), 118–119 (2010)
2. Guifu, C., Jun, S., Long, Y.: Design of alert system terminal based on single chip microcomputer and DTMF. Electronic Test 4, 69–71 (2009)
3. Shen, J., Yang, Y., Zhai, X.: MSP430 series ultra low power 16-bit Microcontroller Principles and Applications. TsingHua University Press, Beijing (2004)
4. Zhang, X.: MSP430 series of utility C Programming Language. Posts & Telecom Press, Beijing (2005)
5. Lu, C.: The design of multi-channels signal detection and voice alarm system based on AT89C51. Journal of Jiamusi University (Natural Science Edition) 27(2), 181–184 (2009)
6. Zhang, X., Tong, J., Wang, Z., Jia, J.: An Intelligent Annunciator Based on Public Telephone Net. Journal of BeiHua University (Natural Science) 5(3), 279–281 (2004)

Multidisciplinary Design Optimization Based on Game Theory

Qingguo Wen and Baowei Song

Room 703 School of Marine Engineering
Northwestern Polytechnical University
Xi' an, Shanxi Province, China
qg-wen@in.com

Abstract. A novel multidisciplinary design optimization based on Game Theory is proposed in this paper. Game Theory studies strategic situations, and Nash equilibrium is one of the best important concepts of it. Nash equilibrium theory is applied to model the relations among the design players, the design variable and object have been regarded as the game decision-space and game side; the response of system has been looked as the payoff. The method allows that each discipline adopts its own design optimization technique, and need not systematic coordination. So it reduces the calculation task and improves operational efficiency. At last an example has been chosen to test the novel method by using iSIGHT: a gear reducer box design problem. The result shows that the new method is easy to be realized on computer and gets a better optimized result than collaborative optimization (CO) method. The new method is feasible.

Keywords: Game Theory, Multidisciplinary Design Optimization, Collaborative Optimization.

1 Introduction

Multidisciplinary Design Optimization (MDO) has been extensively used in complex system design. The complex system is composed of several disciplines or several sub-system, which involves a great deal of feedback processes and iterative processes. The coupling between these processes and troublesome task of disciplinary analysis leads to the high cost and low efficiency. MDO shows great potentialities to be applied in complex system design.

To solve this problem, a novel multidisciplinary design optimization based on game theory was presented in this paper. The new method is based on Nash equilibrium, which is one of the best important concepts of game theory. Comparing with other MDO, the method allows that each discipline adopts its own design optimization technique, and need not systematic coordination; Furthermore, the system analysis is allotted to sub-systems, so the design cycle and the calculation load of design are reduced.

The previous MDO[1]-[3] which we used extensively, for instance, collaborative optimization(CO) method. It needs a whole system analysis when the calculation goes

D. Zeng (Ed.): ICAIC 2011, Part II, CCIS 225, pp. 148–154, 2011.

one step, and the variance between the coupling variables is corresponded by the system restriction. So the cost of design is increased, and the system restriction has some subjectivity, therefore, the optimization model of system becomes worse.[4] It is shown that the multidisciplinary design optimization based on game theory has advantage of the arithmetic theory.

2 Nash Equilibrium

At the beginning of 1950's, the master of game theory John Nash raised the most popular concept for strategic form games: Nash equilibrium which was included two famous papers about non-cooperative games.

Suppose that you have a strategy *b* that is dominated by another strategy, say *a*. We have seen that it is never a good idea to play *b* because no matter what the other player does, you can always do better with *a*. Now suppose you actually have some idea about the other player's intentions. In that case, you would choose a provided it does better than *b* *given what the other player is going to do*. You don't, in other words need to know that *a* performs better than *b* against all strategies of the other player; you simply need to know that it performs better against the specific strategy of your opponent. Indeed, *a* is called a *best response* against the other player's known strategy if it does better than any of your other strategies against this known strategy.

Typically you will not know exactly what the other player intends to do; at best you will have a guess about his strategy choice. The same logic applies, however; what you really care about is how *a* performs vis-à-vis *b* or any other strategy for that matter when played against your *guess* about your opponent's strategy. It only pays to play a best response against that strategy which you believe your opponent is about to play.

Of course, your guess might be wrong. And then you would be unhappy and you would want to change what you did. But suppose you and your opponent guessed correctly, and you each played best responses to your guesses. In that case, you would have no reason to do anything else if you had to do it all over again. In that case, you would be in a Nash equilibrium.

Definition: A strategy s_i^* is a best response to a strategy vector s_{-i}^* ($s_{-i}^* = (s_1^*, s_2^*, \cdots, s_{i-1}^*, s_{i+1}^*, \cdots, s_n^*)$) of the other players if $u_i(s_i^*, s_{-i}^*) \geq u_i(s_i, s_{-i}^*)$, for all s_i [5]. In other words, s_i^* is a "dominant strategy" in the very weak sense that it is a best strategy to play provided the other players do in fact play the strategy combination s_{-i}^*. We need a condition to ensure that player *i* is correct in his conjecture that the other players are going to play s_{-i}^*. And, likewise, the other players are correct in their conjectures. This analysis gives us the following definition:

The strategy vector $s^* = (s_1^*, s_2^*, \cdots, s_n^*)$ is a Nash equilibrium if $u_i(s_i^*, s_{-i}^*) \geq u_i(s_i, s_{-i}^*)$ for all and s_i all *i*.

Game theory studies strategic situations which players are interrelated and the equilibrium of strategy. In other words, when a choice of one player or subsystem is affected by other players or subsystems, also other players or subsystems impact the strategy of one player and the equilibrium. Comparing with MDO, there have design objects which are interrelated, game theory is the very ology which solves and attempers the conflict of design object.

Nash equilibrium theory is applied to model the relations among the design players, the design variable and object have been regarded as the game decision-space and game side, and the response of system has been looked as the payoff upon one strategy.

For illustration, there are n players, in other words, the number of variables is n. The design variables of system could be regard as strategy group $(s_1, s_2, \cdots, s_n)$. Finding system Nash equilibrium points will be proceeded as follows:

1) Assuming $k=0$, choosing the initialization strategy group $s^0 = (s_1^0, s_2^0, \cdots, s_n^0) \in (S_1 \times S_2 \cdots \times S_n)$;

2) Every discipline i figures out the optimum point $u_i(s_1^k, s_2^k, \cdots, \overline{s}_i^k, \cdots, s_n^k) = \min u_i(s_1^k, s_2^k, \cdots, s_i^k, \cdots, s_n^k)$, $s_i \in S_i$. Those disciplines are optimized by their own optimization algorithm depending the information of design points and restriction. Every design variable is invariable except s_i^k, then $\overline{s}_i^k$ is the best strategy of sub-game.

3) Setting the new strategy $s^{k+1} = (\overline{s}_1^k, \overline{s}_2^k, \cdots, \overline{s}_n^k)$, going to step 2.

4) If $\left\| s^{k+1} - s^k \right\| \le \varepsilon$ (ε is convergence precision), or $k > N$ (N is a big integer), it is shown that all sub-games could not optimize their own design object, then s^{k+1} is the game solution of system, ceasing the process. Contrarily, the process goes to step 2.

Now, s^{k+1} is the system equilibrium point, every s_i of s^{k+1} will never change his own strategy for optimizing design object.

3 Example Analysis by iSIGHT Software

The chosen example is one of the ten standard examples which are evaluated arithmetic capability of MDO[6]. There is a Speed Reducer Test Problem. The well-known problem represents the design of a simple gear box such as might be used in a light airplane between the engine and propeller to allow each to rotate at its most efficient speed. The objective is to minimize the speed reducer weight while satisfying a number of constraints imposed by gear and shaft design practices. The math model is specified as follows:

$$\begin{aligned} f(x) &= C_{f1}x_1x_2^2(C_{f2}x_3^2 + C_{f3}x_3 - C_{f4}) - \\ &C_{f5}x_1(x_6^2 + x_7^2) + C_{f6}(x_6^3 + x_7^3) + \\ &C_{f1}(x_4x_6^2 + x_5x_7^2) \\ C_{f1} &= 0.7854 \quad C_{f2} = 3.3333 \\ C_{f3} &= 14.9334 \quad C_{f4} = 43.0934 \\ C_{f5} &= 1.5079 \quad C_{f6} = 7.477 \end{aligned} \tag{1}$$

$g_1 : 27.0/(x_1 x_2^2 x_3) - 1 \le 0$
$g_2 : 397.5/(x_1 x_2^2 x_3^2) - 1 \le 0$
$g_3 : 1.93 x_4^3/(x_2 x_3 x_6^4) - 1 \le 0$
$g_4 : 1.93 x_5^3/(x_2 x_3 x_7^4) - 1 \le 0$
$g_5 : A_1/B_1 - 1100 \le 0$
$g_6 : A_2/B_2 - 850 \le 0$
$g_7 : x_2 x_3 - 40.0 \le 0$
$g_8 : 5.0 \le x_1/x_2 \le 12.0$
$g_9 : (1.5x_6 + 1.9)/x_4 - 1 \le 0$
$g_{10} : (1.1x_7 + 1.9)/x_5 - 1 \le 0$

$$A_1 = \left[\left(\frac{745.0x_4}{x_2 x_3} \right)^2 + 16.9\times 10^6 \right]^{0.5}, B_1 = 0.1x_6^3 \quad A_2 = \left[\left(\frac{745.0x_5}{x_2 x_3} \right)^2 + 157.5\times 10^6 \right]^{0.5}, B_2 = 0.1x_7^3$$

$2.6 \le x_1 \le 3.6 \quad 0.7 \le x_2 \le 0.8$
$17 \le x_3 \le 28 \quad 7.3 \le x_4 \le 8.3$
$7.3 \le x_5 \le 8.3 \quad 2.9 \le x_6 \le 3.9$
$5.0 \le x_7 \le 5.5$

Following literature [6], the best result of this problem is: design variable=(3.5, 0.7, 17, 7.3, 7.71, 3.35, 5.29); object function $f(x)_{\min} = 2994$.

The problem is decomposed into three lower-level subproblems:

$$f_1(x) = C_{f1} x_1 x_2^2 (C_{f2} x_3^2 + C_{f3} x_3 - C_{f4}) \tag{2}$$

$$f_2(x) = C_{f6} x_6^3 + C_{f1} x_4 x_6^2 - C_{f5} x_1 x_6^2 \tag{3}$$

$$f_3(x) = C_{f6} x_7^3 + C_{f1} x_5 x_7^2 - C_{f5} x_1 x_7^2 \tag{4}$$

The frames of Collaborative Optimization (CO) and the novel method are specified as follows:

1) Fig 1 shows CO frame on iSIGHT software. Calculation1, Calculation2 and Calculation3 are subproblem I, II & III, HighLevelCalculation is system level. The system level design variable is z1 z2 and z3, the design variable of subproblem I is x1 x2 and x3, the design variable of subproblem II is x1 x2 x3 x4 and x6, the design variable of subproblem III is x1 x2 x3 x5 and x7.

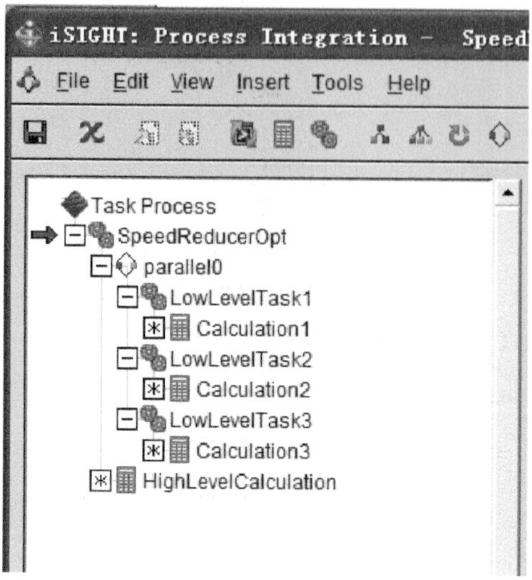

Fig. 1. CO frame on iSIGHT software

The system level model:

$$\min f = f_1 + f_2 + f_3$$

$$f_1(x) = C_{f1}x_1x_2^2(C_{f2}x_3^2 + C_{f3}x_3 - C_{f4}) \quad (5)$$

$$f_2(x) = C_{f6}x_6^3 + C_{f1}x_4x_6^2 - C_{f5}x_1x_6^2 \quad (6)$$

$$f_3(x) = C_{f6}x_7^3 + C_{f1}x_5x_7^2 - C_{f5}x_1x_7^2 \quad (7)$$

$$s.t. \quad J_1^* = (x_1^* - z_1)^2 + (x_2^* - z_2)^2 + (x_3^* - z_3)^2$$

$$J_2^* = (x_1^* - z_1)^2 + (x_2^* - z_2)^2 + (x_3^* - z_3)^2$$

$$J_3^* = (x_1^* - z_1)^2 + (x_2^* - z_2)^2 + (x_3^* - z_3)^2$$

The subproblem level model:

$$F_1 = J_1 + f_1$$

$$F_2 = J_2 + f_2$$

$$F_3 = J_3 + f_3$$

For converging fast with CO, there is a relaxation factor ε =0.01 in system level, the equation restrictions become: $J_1^* \le \varepsilon$, $J_2^* \le \varepsilon$, $J_3^* \le \varepsilon$.

2) The frame of new method is similar as CO's, except the variables of system level are x2,x3 ,x4,x5; the variables of subproblem I is x1; the variables of subproblem II is x6; the variables of subproblem III is x7; and the system restriction is canceled. So the new method has no coupling variables between system level and subproblems or among subproblems by redistributing variables. It is different form CO's.

After finishing this two frames, the optimization arithmetic could be utilized which is provided by iSIGHT software. The system level adopts Modified Method of Feasible

Directions (MMFD), the subproblems adopt Sequential Quadratic Programming (NLPQL). The result is summarized in Table 1. The convergence process of CO is shown in Fig 2, the new method is Fig 3.

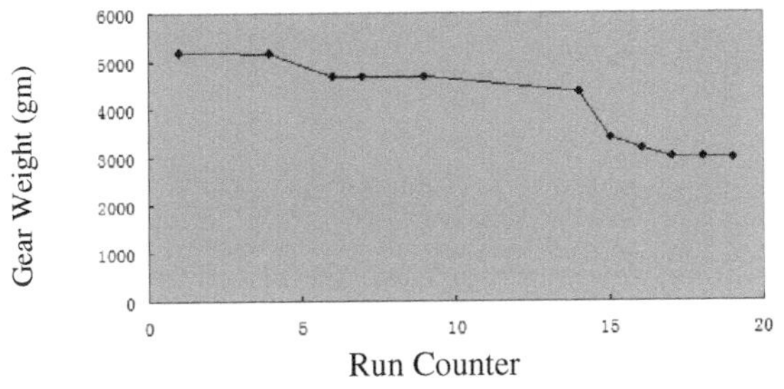

Fig. 2. Convergence process of CO

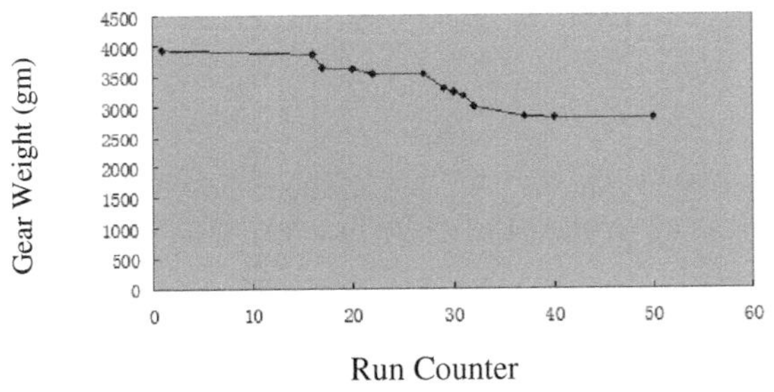

Fig. 3. Convergence process of new method

Table 1. Comparison of result

Type	Parameters								
	x_1	x_2	x_3	x_4	x_5	x_6	x_7	f	Steps
Initialization	3.5	0.71	25.0	7.9	7.6	3.0	5.09999		
Reference	3.5	0.7	17.0	7.3	7.71	3.35	5.29	2994	
CO	3.5	0.7	17.0	7.3	7.3	3.35063	5.28588	2984.85	27
New method	3.5	0.7	17.0	7.3	7.3	3.35021	5.0	2813.75	60

The best point of new method is f=2813.75, it is evident that the new method gets better point than the given literature's and CO's. But the number of steps about the new method is larger than the latter. In other words, the convergence speed of the new method is slower than CO. Because CO adopts a system restriction and the system converges more quickly. For instance, $(y-y')^2 \leq \varepsilon$, it accelerates the process of convergence. Despite few steps, it costs more to calculate on computer for CO. There are the new method and CO running on the same computer, it is shown that every step of CO costs more than the former. Because of the system restriction, it brings big calculation workload to CO, so the whole convergence achieved time of CO is longer than the new method.

4 Conclusion

This paper has shown the novel multidisciplinary design optimization based on game theory. The method allows that each discipline adopts its own design optimization technique, and need not systematic coordination, so the system analysis is allotted to sub-systems, the design cycle and the calculation load of design are reduced. A comparison of result between new method and CO suggests that the new method has some advantages in optimum result and compute time.

References

1. Yukish, M., Bennett, L., Simpson, T.W.: Requirements on MDO imposed by the undersea vehicle conceptual design problem, US: AIAA pp. 2000-4816(2000)
2. Kodiyalam, S., Sobieski, J.S.: Bi-level integrated system synthesis with response surfaces, US: AIAA pp. 99-1306-wip(1999)
3. Hulme, K.F., Bloebaum, C.: A Comparison of solution strategies for simulation based multidisciplinary design optimization, US: AIAA 98-4977 (1998)
4. Wang, Z., Chen, X., et al.: Research on the Theory and Application of Multidisciplinary Design Optimization of Flight Vehicles, pp. 229–280. National Defence Industry Press (2006)
5. Dutta, P.K.: Strategies and Games: Theory and Practice, pp. 63–71. MIT Press, Cambridge (1999)
6. Padula, S.L., Alexandrov, N., Green, L.L.: MDO test suite at NASA langley research center. AIAA 96-4028 (1996)

Theoretical Model Analysis of the Governing Relationship of the Inlaid Fibre Plank – Concrete Bond Structure

Zuquan Yang[1] and Zailin Guo

[1] Urban construction college, Wuhan University of Science and Technology, Wuhan, Hubei, China
z.yang88@yahoo.com.cn

Abstract. The reinforcement technology of the surface layer embedded carbon fibre plank is a kind of new application form used in the carbon fibre plank reinforced concrete structure components. In this page, the theoretical model of the governing relationship of the inlaid fiber plank --concrete bond structure has been illustrated according to the test research with the epoxy resin and epoxy mortar as the bonder, and the processing of the data. Some results are got by the theoretical analysis of the model:(1)the failing load of the briquette using the bonder of cement mortar is much lower than that of the briquette using the bonder of epoxy resin and epoxy mortar;(2)the BPE model can well simulate the bond slippery curve of the failing process of the briquette;(3)the limit bond stress will increase with the increasing of the slot size, to the briquette with the epoxy resin --concrete interface as the failing interface.

Keywords: Surface layer embedded, Carbon fibre plank, Inlaid reinforcement method, Failing load, Bond slippery.

1 Introduction

The bond governing relationship of the carbon fibre plank --concrete bond structure was initially developed from the bond property of the steel bar --concrete structure. For the research of the carbon fibre strengthened composite material, it was mainly focused on the CFRP bars, but little done to the CFRP planks. Some local and foreign scholars have also posed some theoretical model for the CFRP planks, as:1, BPE model;2, modified BPE model;3, Malvar model;4, CMR model;5, Continuous curve model. In these models, the BPE [1] model and modified BPE model are best known to the scholars.

2 The Purpose and Method of the Theoretical Analysis

In this test, the failure forms of the briquette with epoxy resin and epoxy mortar as the bonder are splitting failure and interface failure in the interface of the bonder and concrete; for the briquette with cement mortar, is the interface failure in the interface of

D. Zeng (Ed.): ICAIC 2011, Part II, CCIS 225, pp. 155–162, 2011.

the bonder and CFRP planks. The bond slip governing relationship describes the relationship of the local bond stress and the slip on the nodes in the failure interface.

The theoretical research in this paper is done with respect to the following two fields: the relationship of the local bond shear stress and the slip; researching the influencing factors of the bond property and the influencing regularities, on the theory stage.

Generally, the analysis of the governing relationship of the CFRP--concrete bond structure is based on one bond slip governing relationship model. The BPE model is adopted in this paper to take up the theoretical analysis of the governing relationship of this test, according to the research on the inlaid CFRP plank briquette with the epoxy resin as the bonder done by De Lorenzis, because the test in this mainly discuss the bond slip governing relationship of the epoxy resin and CFRP plank structure, the epoxy mortar and CFRP plank structure, the cement mortar and CFRP plank structure .

The scholar De Lorenzis used the modified BPE model, and got the model curve as shown in Figure 1[2]. The bond property of the CFRP plank and concrete is discussed in this paper using this curve model, and good theoretical result is got.

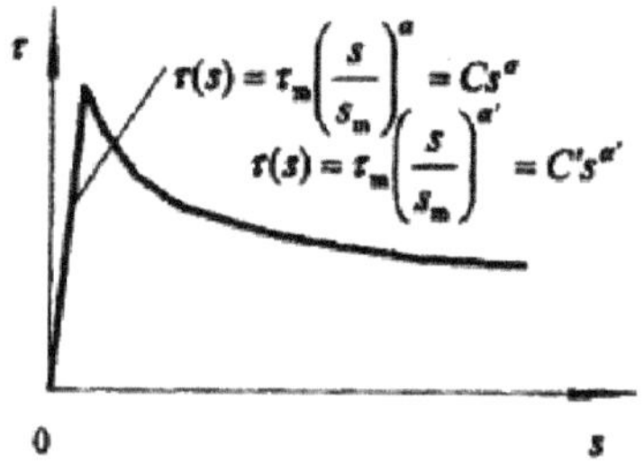

Fig. 1. De Lorenzis governing relationship model

The stress and displacement mentioned are the stress and displacement on the failure interface. The displacement measured in this test is the relative slip of the CFRP plank and the concrete, but the real slip equals to the relative slip of the CFRP plank and the bonder and the relative slip of the bonder and the concrete[3].

Since the relative slip of the bonder and the concrete is not possible to be measured, some assumptions are done in the theoretical analysis of this test:

1) when the interface is failed, the measured slip is from the slip of th interface;

2) the average slippage value is the average value of the value on the free end and the loading end. The De Lorenzis governing relationship in Figure 1 can be expressed as:

(1) if the failure forms is splitting failure, the equations of the bond stress--slip relationship curve is:

$$\tau_s = \tau_u (\frac{s}{s_u})^{\alpha} = cs^{\alpha} \quad (0 \le s \le s_u) \tag{1}$$

In the expression: τ_s is bond stress, S *is the local slip*, τ_u and S_u *are the local bond stress and slip on the peak value point*, α is a parameter valued between 0 and 1.

(2) if the failure form is the interface failure on the interface of the epoxy resin and concrete interface, the equation of the curve is:

$$\tau_s = \tau_u (\frac{s}{s_u})^\alpha = cs^\alpha \ (0 \le s \le s_u) \quad \text{(ascending branch)}$$

$$\text{and } \tau_s = \tau_u (\frac{s}{s_u})^{\alpha'} = c's^{\alpha'} \ (s \ge s_u) \text{ (descending branch)} \qquad (2)$$

In the expression: τ_s is bond stress, S *is the local slip*, τ_u and S_u *are the local bond stress and slip on the peak value point*, α is a parameter valued between 0 and 1, α' is a parameter valued between -1and 0.

3 Discussion of the Interface Bond Slip Governing Relationship

3.1 Characteristic Values of the Governing Relationship

It is shown in the equations of the curve of the governing relationship model of the interface bond slip, two crucial value τ_u and S_u are characteristic values of the CFRP plank and concrete bond structure.

According to the research of the bond property of the steel bars, the limit bond strength of the plain steel bar is:

$$\tau_u = (1 + 2.1d / l_a) \times f_t \qquad (3)$$

In the expression: d is the diameter of the steel bar, l_a is the bond length, f_t is the tensile strength designed value.

According to the research of the bond property of the concrete and steel bar, the limit bond strength of the ribbed steel bars and concrete is:

$$\tau_u = \lfloor -0.11(c/d - 4.5)^2 + 2.25 \rfloor \times f_t^{0.8} \qquad (4)$$

In the expression: d and f_t are the same as in the Expression 3, c is the perimeter of the steel bar.

The failure forms for the briquettes with the epoxy resin and epoxy mortar as the bonder are: the splitting failure of the concrete and the interface failure on the interface of the bonder and the concrete. According to the test data, the bond strength τ_u is dependent on the strength grades of the concrete, the bond length of the CFRP plank, the size of the slot, and the size of the CFRP plank. The related parameters of th briquettes are shown in Table 1, according to the qualitative analysis of the failure form when the CFRP plank and the concrete structure fails, based on the Expression 3 and Expression 4.

Table 1. Parameters of the briquette and the test characteristic values

Briquette Number	Concrete f_t (MPa)	S_1 (mm)	S_2 (mm)	S_1+S_2/l
HY_C20_80_10	1.11	100	62.2	2.03
HY_C20_80_20	1.11	110	61.8	2.15
HY_C20_200_10	1.11	100	62.4	0.81
HY_C20_200_20	1.11	110	62.2	0.86
HY_C40_80_10	1.60	100	61.8	2.02
HY_C40_200_10	1.60	100	61.8	0.81
HY_C40_200_20	1.60	110	62.2	0.86
HS_C20_80_20	1.11	110	62.0	2.15
HS_C20_200_10	1.11	100	61.4	0.81
HS_C20_200_20	1.11	110	61.8	0.86
HS_C40_80_10	1.60	100	62	2.02
HS_C40_200_10	1.60	110	62	0.86
HY_C20_80_10	1.61	9.52	0.38	
HY_C20_80_20	1.78	10.69	0.413	
HY_C20_200_10	1.60	5.85	0.34	
HY_C20_200_20	1.77	6.2	0.43	
Briquette Number	S_1/S_2	τ_u(MPa)	S_u (mm)	
HY_C40_80_10	1.62	6.42	0.25	
HY_C40_200_10	1.62	5.9	0.26	
HY_C40_200_20	1.77	6.15	0.28	
HS_C20_80_20	1.77	10.13	0.48	
HS_C20_200_10	1.63	6.2	0.46	
HS_C20_200_20	1.78	6.4	0.44	
HS_C40_80_10	1.61	6.41	0.21	
HS_C40_200_10	1.77	3.6	0.31	

In the table, S_1 is the bond perimeter of the crossed section of the bond interface of the bonder and the concrete, S_2 is the bond perimeter of the crossed section of the bonder and the CFRP plank, l is the bond length of the CFRP plank, f_t is the measured tensile strength designed value of the concrete.

t is shown in the table that, the limit bond strength is related to the $S1+S2/d$ and $S1/S2$. By analyzing the test data and fitting the data, the following are got:

When the briquettes with the epoxy resin and epoxy mortar as the bonder produce interface failure on the interface of the bonder and the concrete:

$$\tau_u = 4.4\times\sqrt{\frac{S_1+S_2}{l}}\times f_t^{0.15} \tag{5}$$

When produce the splitting failure:

$$\tau_u = \frac{S_1}{S_2}\times 4.4\times\sqrt{\frac{S_1+S_2}{l}}\times f_t^{0.15} \tag{6}$$

In the expressions, S_1 is the bond perimeter of the crossed section of the bond interface of the bonder and the concrete, S_2 is the bond perimeter of the crossed section of the bonder and the CFRP plank, l is the bond length of the CFRP plank, f_t is the measured tensile strength designed value of the concrete.

$$s_u = 0.17\left(\frac{s_1}{s_2}\right)\times\frac{1}{f_t}\times l^{0.1} \tag{7}$$

In the expression, l is the bond length of the CFRP plank, f_t is the measured tensile strength designed value of the concrete.

The comparison of the fitted value and the real value of the governing relationship curve of the interface bond slip is done in Table 2, and it shows that the two are basically coincided.

Table 2. The comparison of the characteristic values of the load-slip curve

Briquette Number	Concrete f_t (MPa)	τ_u (MPa)		
		Test value	Theoretical value	error
HY_C20_80_10	1.11	10.69	10.76	0.6%
HY_C20_80_20	1.11	9.52	9.38	-1.4%
HY_C20_200_10	1.11	5.847	5.93	1.4%
HY_C20_200_20	1.11	6.2	6.71	8.2%
HY_C40_80_10	1.60	5.9	6.05	2.5%
HY_C40_200_10	1.60	6.15	6.9	12.1%
HY_C40_200_20	1.60	6.42	6.44	0.3%
HS_C20_80_20	1.11	10.13	10.52	4.5%
HS_C20_200_10	1.11	6.2	5.92	-4.5%
HS_C20_200_20	1.11	6.4	6.7	4.6%
HS_C40_80_10	1.60	6.41	6.46	0.7%
HS_C40_200_10	1.60	6.6	6.96	5.4%

Briquette Number	Su (MPa)		
	Test value	Theoretical value	error
HY_C20_80_10	0.4125	0.4	-4.0%
HY_C20_80_20	0.38	0.36	-6.2%
HY_C20_200_10	0.34	0.39	14.7%
HY_C20_200_20	0.5	0.43	-13.2%
HY_C40_80_10	0.26	0.27	4.1%
HY_C40_200_10	0.28	0.3	7.4%
HY_C40_200_20	0.25	0.247	-1.1%
HS_C20_80_20	0.48	0.4	-17.5%
HS_C20_200_10	0.46	0.39	-15%
HS_C20_200_20	0.44	0.43	-1.4%
HS_C40_80_10	0.21	0.24	18%
HS_C40_200_10	0.31	0.27	-12.6%

For the briquette with the cement mortar as the bonder, the failure forms are all interface failures on the interface of the CFRP plank and cement mortar, the characteristic values of the governing relationship are also the bond strength τ_u and the related slippage S_u. Since the limit load is relatively small when the failing happens, no cracks are produced on the interface of the concrete and the cement mortar. It can be illustrated that the limit bond strength of the cement mortar is closely related to the surface shape of the plank, the size of the plank, the bond length, the strength grades of the cement mortar, but not to the strength grades of the concrete, the size of the slot. The statistical regression analysis of the characteristic values of the interface bond slip curve of the briquettes with cement mortar as the bonder is not done, because the limit of the number of the briquettes.

3.2 Analysis of the Governing Relationship of the Interface Bond Slip

The local scholar Rong Li and the others have done the pull-off test to the inlaid plank reinforced concrete briquette, and deduced the local bond slip curve on theory stage, according to the measured strain on the nodes on the CFRP plank along the bond length range[4], and got that the interface bond slip curve of th CFRP plank-concrete structure is divided into three branches: the ascending branch, the descending branch, and the level branch.

According to the analysis of the test data, the basic form of the interface bond slip governing relationship curve is the same as the De Lorenzis curve. For the briquettes with the epoxy resin and epoxy mortar as the bonder, if the failure form is the splitting failure, the interface bond slip relationship in the ascending branch is:

$$\tau_s = \tau_u \left(\frac{s}{s_u} \right)^{0.5} \qquad (0 \le s \le s_u) \tag{8}$$

In the expression, τ_u and S_U are respectively the bond strength and the relevant slip, the calculation method is the same as in the previous expressions.

The bond slip governing relationship model curve is divided into two branches, if the failure form is the interface failure on the bonder-concrete interface, as shown in Figure 2(a). The interface bond slip relationship in the ascending branch is:

$$\tau_s = \tau_u \left(\frac{s}{s_u} \right)^{0.5} \qquad (0 \le s \le s_u) \tag{9}$$

In the descending branches:

$$\tau_s = \tau_u \left(\frac{s}{s_u} \right)^{-0.6} \qquad (s \ge s_u) \tag{10}$$

The failure form is the interface failure on the CFRP plank-cement mortar interface for the briquette with the cement mortar as the bonder. It is shown in the test data curve that the bond slip governing model curve is also divided into two branches as the ascending branches and descending branches shown in Figure 2(b), though no analysis of the characteristic values of the governing equations has been done in this paper. The

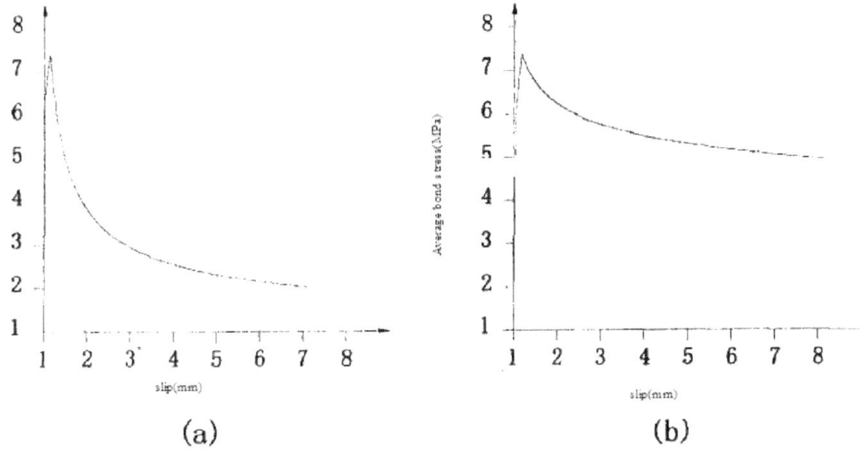

Fig. 2. The governing relationship curve of the briquette with the cement mortar as the bonder

ascending branch of the bond slip relationship curve is the same as that of the briquette with the epoxy resin and epoxy mortar as the bonder:

$$\tau_s = \tau_u \left(\frac{s}{s_u} \right)^{0.5} \qquad (0 \le s \le s_u) \tag{11}$$

The decending branch of the bond slip relationship curve is also the same as that of the briquette with the epoxy resin and epoxy mortar as the bonder, but with different coefficients:

$$\tau_s = \tau_u \left(\frac{s}{s_u} \right)^{-0.15} \qquad (s \ge s_u) \tag{12}$$

In the expression, τ_u and S_u are the bond strength and the relevant slip.

4 Conclusion

Some of the following conclusions are got according to the theoretical analysis in the paper:

(1) The failure load for the briquette with the cement mortar as the bonder is much less than that for the briquette with epoxy resin and epoxy mortar as the bonder.

(2) The BPE model can well simulate the bond slip curve of the fail precess of the briquette. The curve is sectional type: for the splitting failure of the concrete, only the ascending branch appears, since the briquette suddenly fails when limit resin-concrete interface.

load reaches; for the interface failure, the model can describe the whole process, including the ascending branch and the descending branch.

(3) The limit bond stress value will increase with the increasing of the slot size, for the interface failure on the on the epoxy resin-concrete interface.

References

1. Wang, J., Yang, Y., Yue, Q.: The FRP composite material and its application analysis in the civil engineering. Huaqiao University Academic Journal 34(4), 5–10 (2004), 26 (1), 1–6 (2005)
2. Li, R., Teng, J., Yue, Q.: A new field for the application of the FRP material reinforced concrete structure-inlaid strengthening method (NSM). Industrial Architecture 34(4), 5–10 (2004)
3. Guo, Z., Cao, S.: The development of the Bond character of the FRP-concrete structure. Unique Construction 22(2), 70–74 (2005)
4. Li, R., Teng, J., Yue, Q.: Test analysis of the interface bond character of the CFRP plank-concrete structure. Industrial Architecture 35(8), 31–34 (2005)

Analysis of Variability of Degree of Compaction and Air Void Content in Asphalt Pavement of Baosan Expressway

Donglin Shu[1], Shaopeng Wu[1], and Hua Yu[2]

[1] Key Laboratory for Silicate Materials Science and Engineering of Ministry of Education
Wuhan University of Technology
Wuhan, Hubei Province, China
[2] School of Mechanical and Electrical Engineering
Wuhan University of Technology
Wuhan, Hubei Province, China
shu66_dl@sohu.com

Abstract. Degree of compaction (DOC) and air void content (VV) play an important role in the construction of asphalt pavements. The strict control of DOC and VV is a fundamental guarantee of improvement of the pavement performances. The variability reflects the overall management and construction level. In this paper, the variability of the DOC, VV and Marshall density, obtained from various sections and different layers of Baosan expressway, was investigated. The obtained data showed that a certain variability of the indexes as mentioned above existed during the construction of asphalt pavements. The coefficients of variability differed with different sections as well as layers.

Keywords: Degree of compaction, Air void content, Variability, Asphalt pavement.

1 Introduction

The construction quality of asphalt pavements is an important guarantee of high-performance pavements. However, asphalt pavements are designed based on some assumptions. One of these assumptions is that asphalt pavements are homogenous [1]. The design life of asphalt pavements in China is 15 years. However, it is found that most of them occur serious premature distresses in the early years after trafficking. The distress happens locally and the distribution and seriousness is non-uniform. This indicates that the assumption of homogeneity fails to explain the pavement damage. Among other reasons, the direct one can be attributed to the segregation of asphalt mixtures.

The segregation of asphalt mixtures can be divided into three categories: material segregation, temperature segregation and compacted segregation [2]. Material

D. Zeng (Ed.): ICAIC 2011, Part II, CCIS 225, pp. 163–170, 2011.

segregation is primarily concerned with the variation of raw material, which leads to the change of gradation. Temperature segregation happens during mixing, transport, paving and compaction process of asphalt mixtures. The result of temperature segregation is that some local regions appear large temperature change [3]. Compaction segregation is referred to the large change in Degree of compaction (DOC) and air void content (VV) of asphalt pavements due to unreasonable setting of compaction parameters as well as the difference of compaction force at difference regions. In general, the coefficients of DOC and VV are used to evaluate the degree of segregation of asphalt mixtures. Furthermore, DOC is determined on the basis of Marshall specimen density. As a result, the variation of Marshall specimen density may be one of the important reasons of the variability of DOC.

In the current specifications and construction practice, DOC and VV are regarded as two important indexes during construction. Although these two indexes are strictly controlled within the specified ranges, little attention is paid to the effect of variability on the construction quality. The engineering practice has proved that the pavements which have low coefficient of variability (CV) of DOC and VV show less probability for premature distress compared to those with high CV. Therefore, it is significant to investigate the variability of DOC, VV and Marshall specimen density to promote construction quality, pavement performance as well as timely preventive maintenance.

This paper will investigate the variability of DOC, VV and Marshall specimen density based on the data obtained from three different sections that were constructed by different contractors in No.207 National Highway. The effect of different contractors and layers will also be discussed. Hereafter, DOC is determined by comparing the bulk density of the cores obtained in-situ to Marshall specimen density. The Marshall specimens are compacted in laboratory by using asphalt mixture in plant. VV is calculated by comparing bulk density of the core to maximum theoretical density of loose asphalt mixture obtained in the mixing plant.

2 Variability Analysis

2.1 Variability of DOC

The coefficient of variability (CV) is used as the most precise parameter to investigate the variability of some random variables of asphalt pavements. It not only reflects the degree of scatter, but also the instability of random variables. Hereafter, the variation of DOC will be evaluated by means of CV. The test of DOC was carried out at different sections and various layers according to the construction specification of the Ministry of transport[4]. Figs 1 to 3 give the sampling statistics of the variability of DOC that obtained from various sections and layers. As shown in these figures, x denotes the mean DOC; S is standard deviation; C_v stands for CV; LM1 stands for Contact 1 of the of the pavements; LM3 stands for Contact 3 of the pavements; LM4 stands for Contact 4 of the pavements.

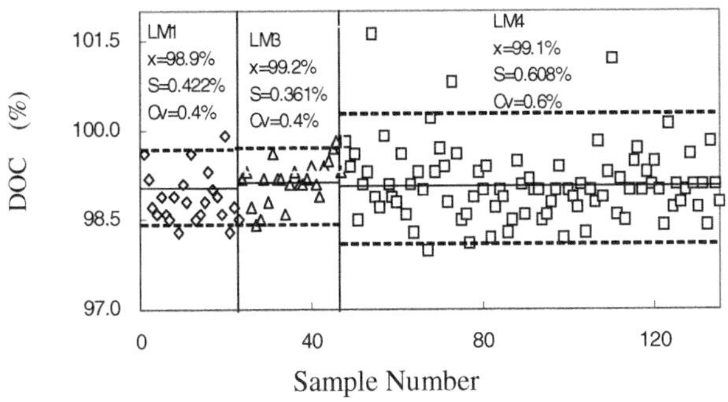

Fig. 1. DOC sampling statistics of the surface layer

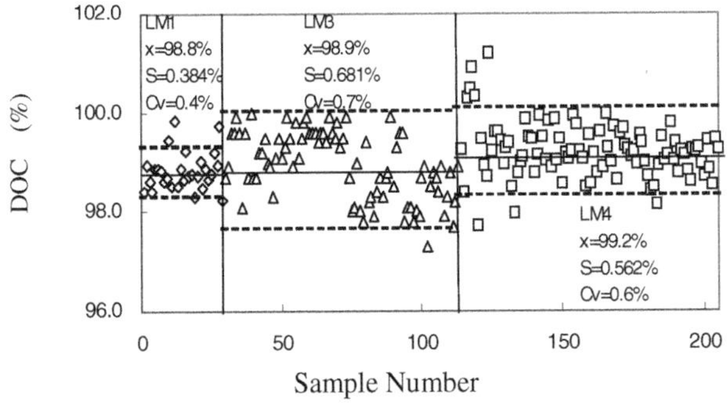

Fig. 2. DOC sampling statistics of the middle layer

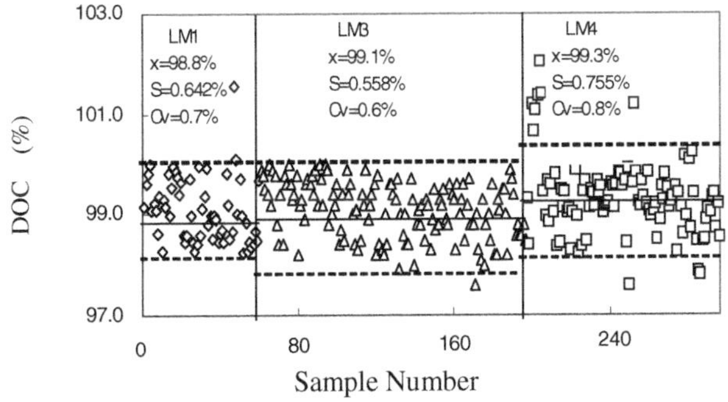

Fig. 3. DOC sampling statistics of the underside layer

As one can see from Figs 1 to 3, DOC can meet the specification requirements. However, certain variability exists in the DOC of asphalt pavements. The difference depends on the construction level, technological level and management mode of the contractors. For example, The CV of LM1 varies from 0.4% to 0.7% and the same for LM3; however, LM4 shows greater CV, from 0.6% to 0.8%. By comparing the CV of these sections, it can be concluded that LM1 and LM3 show better construction level, technological level as well as management mode than LM4.

The CV of DOC also depends on the locations of the layer. For example, the surface layer is varied from 0.4% to 0.6%; the middle layer ranges from 0.4% to 0.7%; the underside layer shows a relatively larger range, from 0.6% to 0.8%. The above analysis indicates that the CV of DOC decreases in the bottom-up direction. The reason for this may be due to the difference in the mixture gradation used in these layers. For example, the surface layer, the middle layer and the underside layer are used AC-16, AC-20 and AC-25, respectively. The number indicates the nominal maximum size aggregate that is used in the mixture. It can be concluded that the coarser aggregates are applied, the more difficult for compaction and thus resulting in higher CV.

2.2 Variability of VV

Figs 4 to 6 give the sampling statistics of the variability of VV that obtained from various sections and layers. It should be noted that in these figures, y stands for the mean VV.

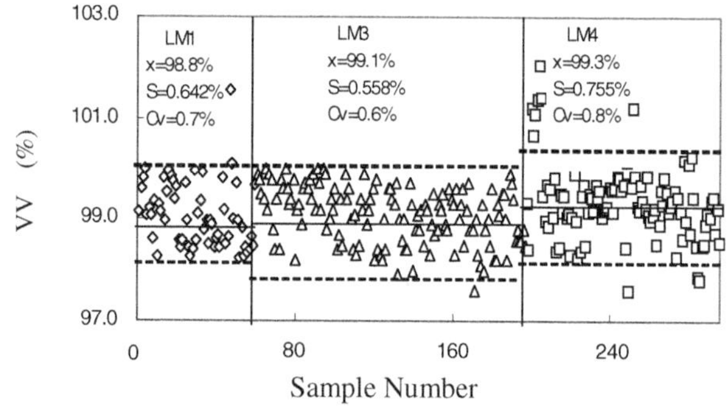

Fig. 4. VV sampling statistics of the surface layer

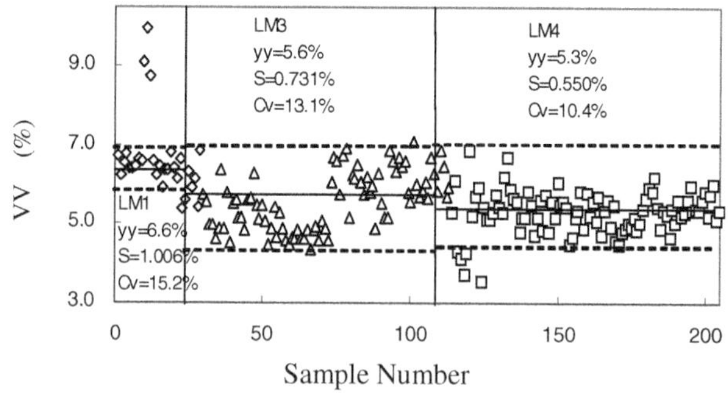

Fig. 5. VV sampling statistics of the middle layer

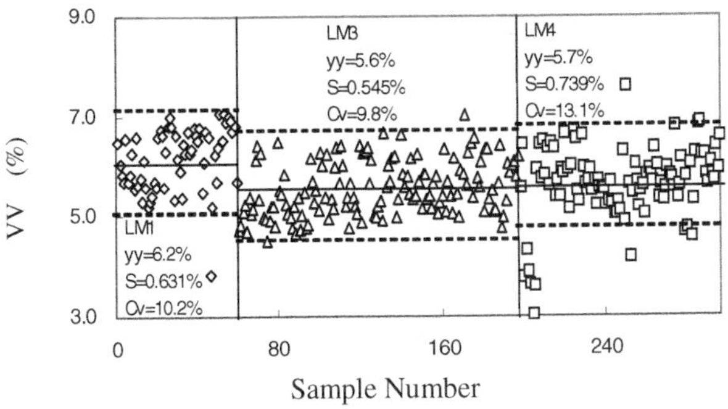

Fig. 6. VV sampling statistics of the underside layer

As illustrated, VV can meet the specification requirements. However, there is certain variability in VV of asphalt pavements. The difference depends on the contractors. For example, the CV of LM1 is varied from 8.6% to 15.2%; the CV of LM3 is varied from 7.1% to 13.1%; LM4 ranges from 10.1% to 13.1%. By comparing the CV of these sections, it can be concluded that LM1 and LM3 also show better construction level, technological level as well as management mode than LM4.

The CV of VV also depends on the locations of the layer. For example, the surface layer is varied from 7.1% to 10.1%; the middle layer ranges from 10.4% to 15.2%; the underside layer ranges from 9.8% to 13.1%. Obviously, CV of the surface layer is minimum. Because the nominal maximum aggregate size of the surface layer is minimum so that it can be compacted relatively easily.

Additionally, as can be indicated from the above discussion, a good correlation exits between DOC and VV. The CV of DOC tends to increase with the increasing of the CV of VV.

2.3 Variability of Marshall Specimen Density

Figs 7 to 9 show the sampling statistics of the variability of Marshall specimen density that obtained from various sections and layers. In these figures, y stands for the mean VV, and ρ stands for the mean density with the unit is g/cm^3.

As illustrated in Figs 7 to 9, VV can meet the specification requirements. However, certain variability exists in Marshall specimen density of asphalt pavements. The difference depends on the contractors. For example, the CV of LM1 varies from 0.1% to 0.2%; the CV of LM3 is 0.3%; LM4 ranges from 0.5% to 1.1%. By comparing the CV of these sections, it can be concluded that the result is in agreement with the previous conclusion.

The CV of Marshall specimen density also depends on the locations of the layer. For example, the surface layer is varied from 0.1% to 1.1%; the middle layer ranges from 0.2% to 0.9%; the underside layer ranges from 0.2% to 0.5%. There is no obvious difference among structure layers. The above indicates that the effects of the sampling methods and operators may play an important role on determination of Marshall specimen density. In order to diminish the variability of compactness, the variability of Marshall standard density should be controlled strictly.

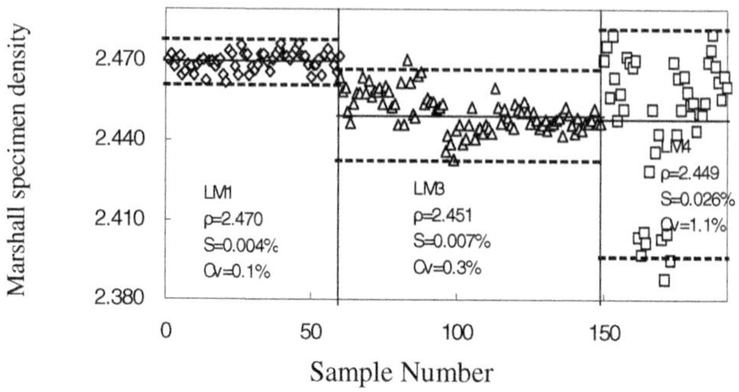

Fig. 7. Marshall specimen density sampling statistics of the surface layer

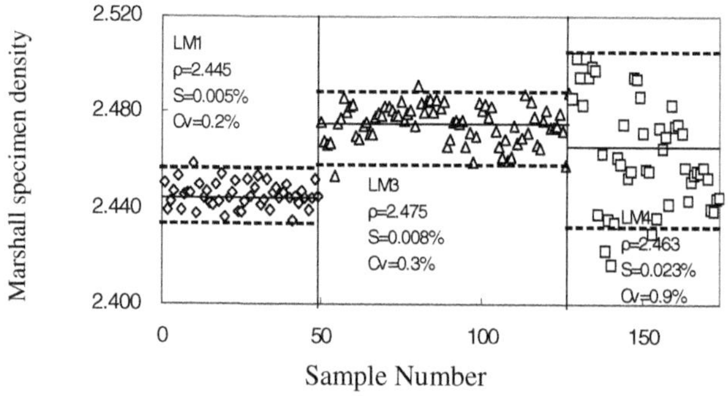

Fig. 8. Marshall specimen density sampling statistics of the middle layer

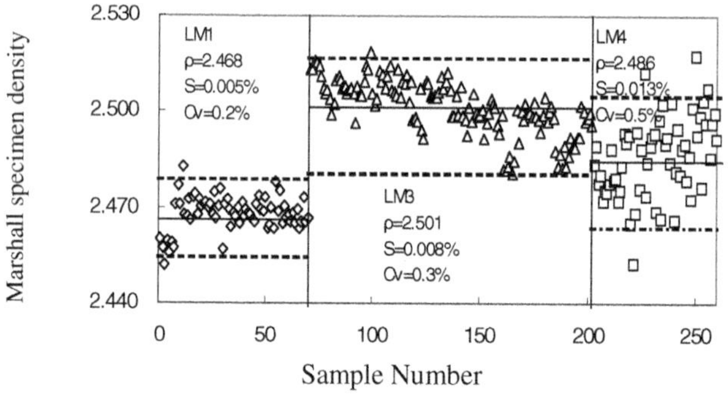

Fig. 9. Marshall specimen density sampling statistics of the underside layer

2.4 Comprehensive Analysis of Pavement Variability

Table 1 lists the CV of DOC, VV and Marshall specimen density. As given in Table 1, the variability of Contact 1 on these 3 parameters is lower than that of Contact 3, and Contact 4 show highest value. The variability of these 3 parameters increases in the bottom-up direction. The CV of Marshall specimen density is less than that of DOC, and is far less than that of VV. By the comparing of these 3 indexes, it is more reasonable to evaluate the overall level of pavement by means of VV.

Table 1. The total variation of DOC, VV and marshall specimen density

Contact Number	Structure Layer	CV (%)		
		DOC	VV	Marshall specimen density
LM1	surface layer	0.4	8.6	0.1
	middle layer	0.4	15.2	0.2
	underside layer	0.7	10.2	0.2
LM3	surface layer	0.4	7.1	0.3
	middle layer	0.7	13.1	0.3
	underside layer	0.6	9.8	0.3
LM4	surface layer	0.6	10.1	1.1
	middle layer	0.6	10.4	0.9
	underside layer	0.8	13.1	0.5

3 Conclusions

1) Certain variability exists in DOC, VV and Marshall specimen density of asphalt pavement during construction.

2) Under the statistics analysis, the CV of DOC is varied from 0.4% to 0.8%; the CV of VV is varied from 8.6% to 15.2%; the CV of Marshall specimen density ranges from 0.1% to 1.1% in Baosan Expressway.

3) The difference of the variability depends on the construction level, technological level and management mode of various contractors.

4) The CV of each index decreases in the bottom-up direction of various layers.

5) The variability of asphalt pavement should be measured by VV, but not only by DOC.

References

1. Fang, F.: Pavement Engineering. China Communications Press, Beijing (2001) (in Chinese)
2. Liu, H., Zhang, J.: Effect of machinery and material to segregation in pavement construction. Journal of Wuhan University of Technology 31(4), 159–162 (2009) (in Chinese)
3. Xie, W., Wang, P., Wang, Y.: Material segregation due to temperature chang in asphalt concrete paving. Construction Machinery and Equipment 4, 45–47 (2003) (in Chinese)
4. JTGF40-2004, Standard specification for construction and acceptance of highway asphalt pavement (in Chinese)

Research of University Course Scheduling System Based on Evolutionary Algorithm

Hongbin Guo[1] and Jingfeng Yan[2]

[1] Dean's Office, Xuchang University
Xuchang, Henan, 461000, China
[2] School of Computer Science and Technology, Xuchang University
Xuchang, Henan, 461000, China
guohongbin56@126.com

Abstract. This paper proposes the method of university course scheduling system based on evolutionary algorithm by constructing an effective data model with courses as the core of scheduling and lesson plans as the basis of scheduling. The features of evolutionary algorithm used in this paper are as follows: 1) the two strategies of hard constraints and soft constraints are taken into account for fitness function design; 2) the selection strategy of stochastic ranking is proposed to improve the convergence rate of population. The simulation experiment proves that this algorithm is, to some extent, universal since it can find automatically the model to solve this problem on the basis of the actual situation of a university.

Keywords: evolutionary algorithm, Course Scheduling, genetic algorithm.

1 Introduction

In the context of enrolment expansion in institutions of higher education, the number of students increases sharply, and the lagged development of hardware and software gives rise to resource shortage and exerts a lot of pressure on educational administration, especially the core of educational administration, course scheduling. And how to schedule courses in a time-saving, effort-saving and efficient manner is a tough problem for each university. Designed on the basis of improved evolutionary algorithm and the actual situation of universities, this system is, to some extent, universal since it finds automatically with algorithm the model to solve the problem. The current course scheduling in many universities is completed manually on the basis of past experience. Due to the heavy workload, it takes considerable time, and sometimes goes by contraries. Some universities have developed simple course scheduling systems with computer platform; however, inefficient algorithms such as exhaustive approach are mainly used. In this case, long time is usually spent and good solution is not necessarily found. Therefore, it's most urgent that a highly efficient course scheduling system should be created. With the application of genetic algorithm of evolutionary algorithms, this system solves the problem of course scheduling properly.

D. Zeng (Ed.): ICAIC 2011, Part II, CCIS 225, pp. 171–177, 2011.

Firstly proposed by Professor Holland of the University of Chicago in 1962, genetic algorithm is a self-adaptive artificial intelligence technology which mimics the process of natural evolution for extreme values. It borrows some ideas from genetics, improves the adaptability of individuals through mechanisms such as natural selection, inheritance and mutation, and embodies the process of evolution of "survival of the fittest in natural selection" in nature. And it is widely used in many fields such as function optimization, combinatorial optimization, production scheduling, machine learning, image processing, and pattern recognition.

Course scheduling aims to work out the proper times and locations of courses according to relative information such as courses, teachers and classes, in which case, the whole process of genetic algorithm is fully presented.

2 Ideas and Framework Flowchart

This system takes courses (a specific course) as the core of scheduling and lesson plans (lesson plan for the specific course) as the basis of scheduling. It connects courses, teachers and classes according to lesson plans to create individual codes, extracts relative constraint information (for instance, whether multimedia facilities are needed for a specific course, the number of students in a specific class), works out randomly the times and locations of courses according to the constraints, creates the initial course scheduling population, converges towards the desirable course scheduling scheme after such operations as fitness testing, selection, crossover and mutation, and finally generates the course scheduling information of the specific course. And the class schedule for the whole school is created with the conduction of the above operations for all lesson plans.

The algorithm framework is shown in Fig. 1.

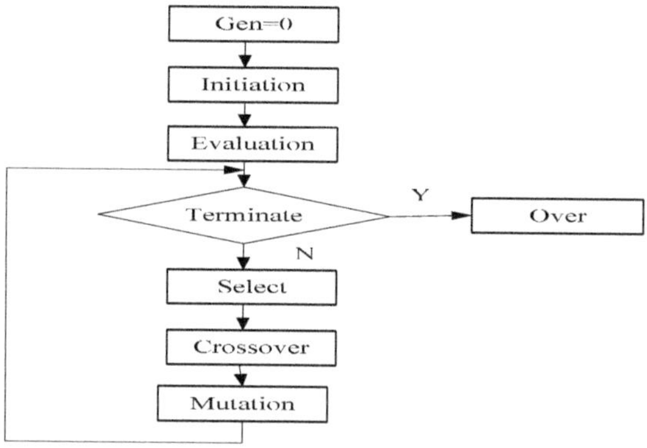

Fig. 1. Flow chart of algorithm framework

3 Introduction to Key Technology

3.1 Encoding

Individual codes shall not only include the required information, but also be simplest in form and arranged properly for the purposes of shortest codes, largest information storage and easiest operation.

Example of original codes:

03	05	08	2001	3001
Department	Major	Grade	Course No.	Teacher No.

Fig. 2. Original individual codes

Fig.2 is the fixed part of individual codes, namely, it is fixed prior to course scheduling. The codes are explained as follows:

Department: the department which runs a specific course.

Major: the major which runs a specific course.

Grade: the grade in which a specific course is run.

Course No.: for obtaining other information of a specific course.

Teacher No.: for obtaining other information of a specific teacher who teaches a specific course.

3.2 Individual Initialization

Initialization of individuals refers to the random generation of the times and locations of courses on the basis of original individual information, and the results of generation shall meet certain conditions of constraints. Then, the results of generation will be added to the original individuals to form complete individual codes for fitness function testing.

Example of complete individual codes:

03	05	08	2001	3001	C405	123153
Department	Major	Grade	Course No.	Teacher No.	Classroom No.	Time of course

Arrangement of course locations (generation of classroom No.):

(1) Determination of classroom locations. Classroom locations will be determined according to the units which run courses, that is, each department has its own teaching block, and classrooms will be allocated within this building.

(2) Matching of constraints on hardware facilities. For courses which require multimedia facilities, the randomly generated classrooms shall be equipped with these facilities.

(3) Matching of constraints on student number. Classroom locations at which classrooms have more seating capacity than the number of course attenders will be checked, and one of them will be selected randomly.

Scheduling of course times:

The strategy of form-filling (in the database, the properties of occupied time are available in terms of teachers, grades and class schedule) is adopted to avoid the three hardware conflicts of course scheduling (namely, a teacher cannot teach two courses at the same time, two courses cannot be offered to the same class, and courses cannot be offered to two classes in the same classroom).

(1) Checking whether there is conflict of time for teachers. During course scheduling for the first time, a teacher's time will be recorded. And the following times generated for courses must be checked first to make out whether there is conflict between the scheduled and the to-be-scheduled.

Before the method of checking is introduced, its storage is briefly explained. The scheduled time is stored with a string (with two digits of numbers as a unit in which the first one indicates the day of the week and the second one, the order of a lesson, for example, 2352 represents the first two periods on the afternoon of Tuesday and the second two periods on the morning of Friday).

The methods are as follows:

During checking, the string of scheduled time is extracted, truncated and segmented to create several individuals of course times, between which and the randomly generated course times a comparison is made to decide whether there is conflict, if any, the randomly generated time will be discarded; if not, selected.

(2) Checking whether there is conflict of time for classrooms. The method is the same as that for teachers.

(3) Checking whether there is conflict of time for classes. The method is the same as that for teachers, too.

3.3 Fitness Function

Fitness function is designed on the basis of constraints of course scheduling which include hard constraints and soft constraints. Hard constraints include the following:

(1) A teacher cannot teach two courses at the same time.
(2) Two courses cannot be offered to the same class.
(3) Courses cannot be offered to two classes in the same classroom.

Since hard constraints have been avoided in course scheduling, fitness function design mainly centers on soft constraints which include the following:

(1) Consideration will be shown for teachers who are the elderly or who have to care for their babies or who live far away from the school, and the first two periods in the morning and the second two periods in the afternoon will not be scheduled for such teachers as much as possible.

(2) For those who have both administrative and teaching duties as well as young teachers, the management's administrative duties and young teachers' scientific research shall be taken into account, and courses will be scheduled together as much as possible for them so that they can devote more effort to the above-mentioned duties.

(3) For general teachers, their occupational health shall be taken into account, and courses shall be scheduled as separately as possible.

(4) In view of the capacity of students, closely related courses shall be separated as much as possible to avoid mutual interference, and unrelated courses will be scheduled together.

(5) In view of instructional efficiency, the fundamental courses (such as advanced mathematics and college physics) and specialized courses will be scheduled for the morning, and elective courses, practical courses and laboratory courses will be scheduled for the afternoon.

(6) In view of the utilization rate of classrooms, the spare capacity of classrooms shall be minimized.

Fitness function design is based on the constraint requirements which must be met. When a certain weight is set, the larger the weigh is, the better the result is.

4 Evolutionary Operator Design

In this paper, a total of N=32 (N is the number of schemes) initial schemes are generated when the class schedule of a certain course is randomly generated, the number of schemes reduces by half each time the selection operator is executed, and all operators recur three times (N=32, N=16, N=4). At last the optimal scheme is selected.

Firstly, N (all course scheduling schemes) is tested by fitness function to select three optimal schemes (copies shall be saved for the three optimal schemes, as there is update each time an operator is executed, in which way the last scheme is obtained), then N/2-3 schemes (namely, 13 schemes for the first time, 5 for the second time and 1 for the third time) are randomly selected. In this way, good schemes are saved, and the diversity of schemes is maintained.

4.1 Selection Strategy

This paper proposes the selection strategy of stochastic ranking to avoid the local optimum of population. And the specifics are as follows:

The algorithm creates population of initialization first, evaluates the individuals of the population, ranks the individuals according to the stochastic ranking strategy, then performs with mutation operator genetic operation for the individuals selected to create sub-individuals, and select the best individual as new individual from sub-individuals, if the new individual is better than the worst individual of the population, the worst individual will be replaced by the new individual to create new population and continue the evolution until the conditions for termination are met. And the algorithm is described as follows:

Step 1. population of initialization
Step 2. calculate fitness of each individual
Step 3. find the best and worst individual
Step 4. while terminate do
Step 5. stochastic ranking of all the individuals
Step 6. select α individuals in order
Step 7. for i := 1 to α do

Step 8. mutation of the above chosen individuals
Step 9. od
Step 10. select the best individual of mutation individuals
Step 11. if new individual is better then
Step 12. new individual replaces the worst individual
Step 13. fi
Step 14. find the best and worst individual
Step 15. od

4.2 Crossover Operator

Crossover operator mainly deals with the times of courses, and the times of crossover are decided by the number of schemes (N/2). The crossover of any two schemes and the crossover of time-individuals in a scheme are decided randomly. With fitness function testing, the three optimal schemes will be updated.

4.3 Mutation Operator

Mutation operator mainly deals with the times of courses. Every scheme shall be mutated, and the number and the content of mutated time-individuals are randomly generated (required to pass the testing by three hard constraints). With fitness function testing, the three optimal schemes will be updated.

With one of the most effective algorithms at the present time-evolutionary algorithm, this system solves the problem of course scheduling. And the class schedule created by this system can meet the requirements of practical teaching. From the outset of system design, universality is taken into account, and many constraints can be set manually, so this system can essentially meet the requirements of course scheduling of regular institutions of higher learning. Since there is little such software in China, and courses are still scheduled manually in a considerable number of schools, the introduction of this software can save human and material resources to the maximum and optimize the use of educational resources. Therefore, its market prospects are quite bright.

5 Conclusion

With one of the most effective algorithms at the present time-evolutionary algorithm, this system solves the problem of course scheduling. And the class schedule created by this system can meet the requirements of practical teaching. From the outset of system design, universality is taken into account, and many constraints can be set manually, so this system can essentially meet the requirements of course scheduling of regular institutions of higher learning. Since there is little such software in China, and courses are still scheduled manually in a considerable number of schools, the introduction of this software can save human and material resources to the maximum and optimize the use of educational resources. Therefore, its market prospects are quite bright.

References

1. Tang, Y., Tang, X.: Class Schedule Based on Genetic Algorithm. Computer Applications 1, 93–94, 97 (2002)
2. Fang, H.L.: Genetic Algorithms in Timetabling and Scheduling, Ph.D. Thesis, Department of Artificial Intelligence, University of Edinburgh, UK (2004)
3. Burke, E.K., Elliman, D.G., Weare, R.F.: A Genetic Algorithm Based University Timetabling System. In: East-West Conference on Computer Technologies in Education, Crimea, Ukraine, pp. 35–40 (2006)
4. Su, Y.: The System of Optimized Class Schedule Based on Genetic Algorithm. Journal of Henan University (natural science) 35(1), 77–79 (2005)
5. Tao, T., Li, H., Xiong, Z.: Application of Multidimensional Collision to the Arranging Courses Algorithm. Journal of East China Geological Institute 4, 256–259 (2001)
6. Wu, Z., Chen, S., Sun, X.: Retrospective Algorithm and Intelligent Curriculum Arrangement. Computer Engineering 3, 792–801 (1999)

Design of 3D Vehicle Intelligent Guidance Simulation System Based on Microsoft Robotic Developer Studio*

Wenqiang Ma, Man Zhang, Zhaoxiang Liu, and Gang Liu

Key Laboratory of Modern Precision Agriculture System Integration of Education Ministry
China Agricultural University
Beijing 100083, China
mawq1234@sohu.com

Abstract. Simulation can make accurate predictions for the dynamic behavior of the vehicle, and provide analytical basis for the design of vehicle navigation systems. This study based on MRDS, a new type of three-dimensional vehicle guidance simulation system was developed and a vehicle model was created. The system could achieve accurate simulation results in real time through simulation experiment. The study results showed that the MRDS based on three-dimensional vehicle navigation simulation system could achieve the simulation of vehicle navigation and control successfully, and provide help to the design of the real navigation systems.

Keywords: Simulation, navigation control, vehicle model.

1 Introduction

Vehicle Intelligent Guidance, as a Multi-disciplinary integration technology, is a complex area of artificial intelligence. The cost of the hardware of vehicle automatic guidance system is quite expensive, while simulation can make accurate predictions for the dynamic behavior of the vehicle and provide analytical basis for the design of vehicle navigation system. So the development of simulation system for vehicle navigation is very meaningful.

According to the different implementations, simulation technology can be divided into three categories [1]. The first category is based on the joint dynamics simulation software and Matlab. For example, Chen Xinrong from Beijing University of Technology, using of ADAMS (Automatic Dynamic Analysis of Mechanical Systems) software and Matlab interface, simulated a plane two liberty degree robot arm successfully and effectively[2]. In the second category, the simulation platform is fully completed by developers. For example, Ling Jialiang, from Central South University, using VC+ + and OpenGL, developed an industrial robot simulation software and achieved functions such as robot modeling, collision detection and dynamic graphics simulation [3]. Ran Xin, form Ocean University of China, developed an outdoor mobile

* This work is partially supported by NSFC Grant #30900869 and "863"Project Grant #2006AA10A304 to Zhang Man.

D. Zeng (Ed.): ICAIC 2011, Part II, CCIS 225, pp. 178–185, 2011.

robot navigation simulation software, and on this basis, using a combination of Direct3D with VC++, further developed 3D version of the software [4]. The third category is integrated simulation platform, such as MRDS. It is a service-oriented visual robot programming kit, developed by Microsoft Corporation. Powered by the AGEIA physX engine, MRDS can provide physics for the simulation environment. And MRDS also includes a .NET-based, services-oriented runtime, which can reduce burden of dealing with multi-threaded tasks for developers, and improve the system stability [5]. In some developed countries, MRDS began to be applied to education and the development of commercial products. Simon.Blackmore from U.K, combined MRDS with Google Earth, developed a agricultural Robot simulation software SAFAR (Software Architecture For Agricultural Robots), in order to provide a platform and method for the design of agricultural Robot effectively [6].

This study aims to develope a 3D Vehicle Intelligent Guidance Simulation System based on MRDS and VS2008 programming environment, and test the system by a variety of navigation and control methods.

2 Overview of MRDS

MRDS is a Windows-based visual robot programming kit that offers a service-oriented runtimes. The structure of MRDS has three lays. The bottom lay is simulation engine, which is based on the AGEIA physX engine, along with Microsoft XNA Framework and DirectX 9 runtime. The middle lay include Concurrency and Coordination Runtime (CCR) and Decentralized Software Services (DSS), which implement multi-threaded asynchronous processing and interact with simulation services in real-time. The top lay include VSE (Visual Simulation Environment) and VPL (Visual Programming Language).

2.1 Concurrency and Coordination Runtime

Robots are typically equipped with multiple sensors and actuators that are coordinated to accomplish a common goal. So the implication of robot must handle a great mumble of multi-threaded tasks. The CCR is a lightweight library that is supplied as a .NET DLL. It is designed to handle asynchronous communication between loosely coupled services that are running concurrently. The CCR eliminates many of the issues related to multi - threading. It uses its own threading mechanism, which is much more efficient than the Windows threading model [7]. With CCR, developers are shielded from all of the work being done under the covers to manage asynchronous operations and deal with parallel execution.

2.2 Decentralized Software Services

DSS is a lightweight application model that combines the traditional network architecture and characteristics of network services[8]. DSS, which is built on top of the CCR, is responsible for controlling the basic functions of robotics applications. The application created by DSS contains several independent, parallel running services. Service is the basic unit of simulation applications, the principle shown in Fig.1.

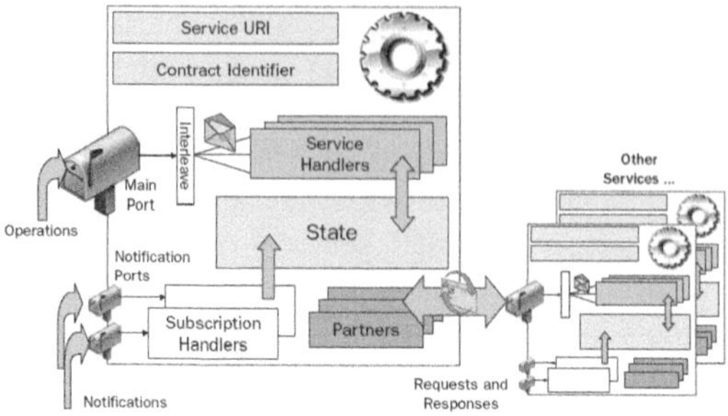

Fig. 1. Principle of Service

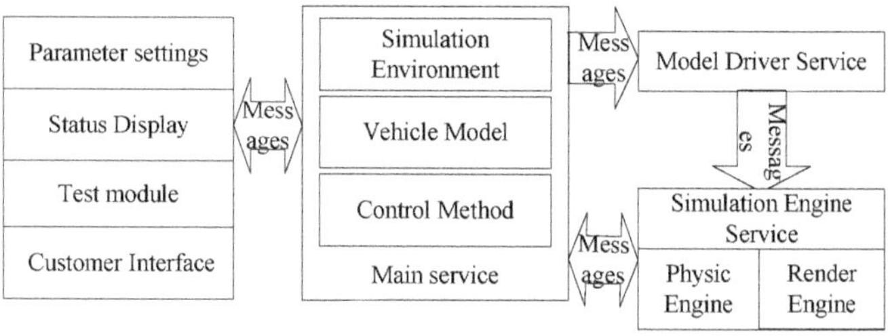

Fig. 2. Flowchart of the System

Each service has a state associated with it and certain types of messages. When a service receives a message, it may change its state and then send additional messages and notifications to other services. In addition, Services also have partner services so that they can send messages to those services and receive responses. DSS uses DSSP (Decentralized Software Services Protocol), which is based on SOAP (Simple Object Access Protocol). DSSP support subscription that is needed by MRDS to request event notifications be sent to a subscribing service.

3 Design and Implementation of System

3.1 General Description and Flow Chart

The system has four basic service components, include the simulation environment and model service, model-driven service, interactive service and the simulation engine service. Service communicate with each other through message. They were all running on one DSS node, which provide loading environment for services. The flow chart shown in Fig.2.

3.2 The Creation of Vehicle Model

1) Structure and Physical Model of Vehicle: The model with complex physical structure requires a large number of kinematical and dynamic calculations in the simulation process, leading to decrease of the real-time capability of the simulation

system. In this system, the simple physical structure (Box, capsule, etc.) included in MRDS kit and three-dimensional vector types has been used to create complex mechanical structure. Then use 3D modeling tools, such as Maya, 3D Studio Max, have been used to generate the geometry of the model. As shown in Fig.3, a hybrid modeling approach has been used in the simulation platform that can display rendering, structure and physical models to satisfy the needs of visualization, body build and collision detection. Moreover, the model also has a series of physical parameters, including mass, friction coefficient and so on, so that the model has a high quality of simulation relative to the real world.

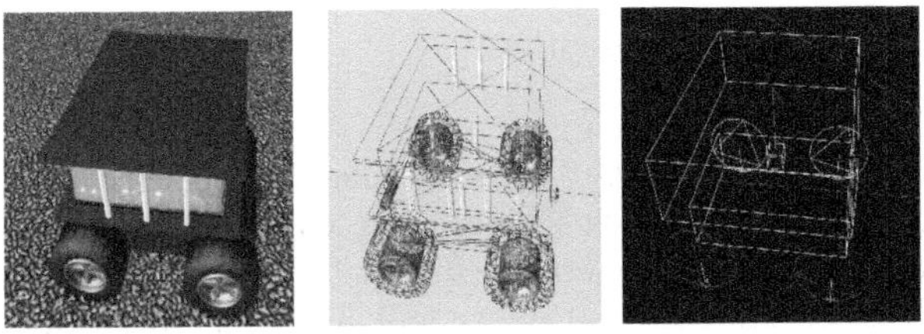

Fig. 3. Display of Hybrid Model

The vehicle model of the system has included body, wheels and several sensors. The body is the main part of the body's appearance, containing most of the mass of the model. The body also include two properties, position and current heading, which could be read in real-time in order to simulate the function of GPS and electronic compass. As main actuator of the vehicle model, the model of wheels has been created separately with the body. Wheels include speed and steer angle property which are used to meet different needs of drive method. The vehicle model also has combined with several sensor services that are provided by MRDS kit, such as web camera and IR sensor, which would be used to achieve more complex control methods.

2) The Model-driven Service: The model-driven service has connected the vehicle model and simulation engine. Its main role is to achieve messages subscription and processing between them. Furthermore, the model-driven service has provided a four-wheel simulation drive method, and steer-turning method of the front wheels. The interleave arbiter that provided by CCR runtime, has been used in he model-driven service, in order to subscribe messages from model to operation handles. Executed as multi-threaded tasks, operation handles used to call specific driver functions, and send notification messages to other services.

3.3 The Design of Simulation Environment and Customer Interface

1) Simulation Environment: In order to set the appropriate programming model, the project of simulation environment service has been created by DSSNewService, which is a command line tool contained a variety of parameters of programming model. Next, the project needed to add several DLLs that defined many basic types and functions used to support simulation engine and adjunction of model. The simulation engine has been claimed as a pattern of simulation environment service for the purpose of

adjunction of visual interface. Then, Add ground, sky, light source and vehicle model to the visual interface by the Start method in the service.

2) Customer Interface: Windows form as custom interface has been used to interact with services in the system. The functions of the custom interface have included selection of navigation method, testing model-driver, display parameters of model, start and stop simulation. Because services are running on DSS node, which can't interact directly with Windows form. So, a Windowsform service has been created in the system, in order to hook up Windows messages. In addition, other services have also needed to add Windowsform serviceport to receive and subscribe operations for Windows messages.

3.4 The Navigation Control Algorithm

Straight-line navigation is a method that based on the current position of vehicle to calculate the lateral error, as input of control method, between the vehicle and pre-set routes. The principle shown in Fig.4. The system has used pre-set straight-line as route, and calculate the lateral error by the equation (1) and (2). Where A, B, C is coefficients of the scheduled line; (x, y), (x_0, y_0), (x_1, y_1) are the coordinates of current location of vehicle, start and end location of route. The u is cross product of current position coordinate and scheduled line, to define polarity (right is positive and left is negative) of vehicle position relative to the route. The control variable was steer angle of front wheels.

$$D = \left| \frac{Ax + By + C}{\sqrt{A^2 + B^2}} \right| ; \tag{1}$$

$$u = (y_0 - y)(x_1 - x) - (x_0 - x)(y_1 - y); \tag{2}$$

The system has used PID and Fuzzy method as navigation algorithm. The formula of PID method is shown in equation (3) and (4).

$$e_{k+1} = e_k + delt \tag{3}$$

$$delt = K_p(D_{k+2} - D_{k+1}) + K_i D_{k+2} + K_d(D_{k+2} - 2D_{k+1} + D_k) \tag{4}$$

Where the *e* is the expected angle, *delt* is increment of control, *D* is lateral error. Parameters of PID method can adjust in customer interface.

In Fuzzy method, the lateral error and heading angle error have been selected as input variables. The universe of fuzzy set of the lateral error is [-30cm,30cm], and quantitative factor of it is 0.2. The universe of fuzzy set of the heading angle error is [-12°, 12°], and quantitative factor of it is 0.2. The control variable is steer angle. The universe of fuzzy set of it is [-12°, 12°], and scale factor of it is 0.2. The rule shown in Fig.5 and Table 1.

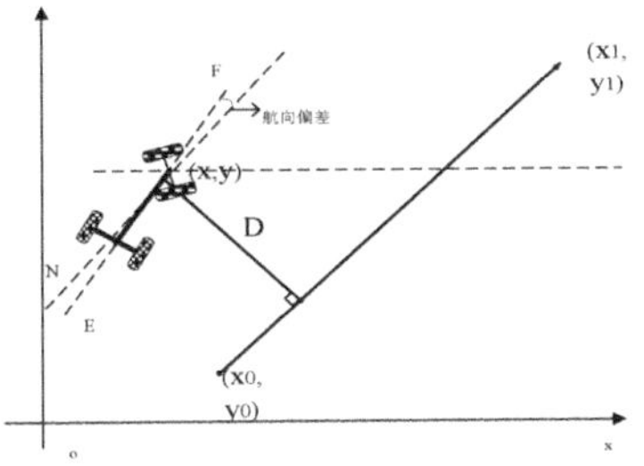

Fig. 4. Principle of Straight-line navigation

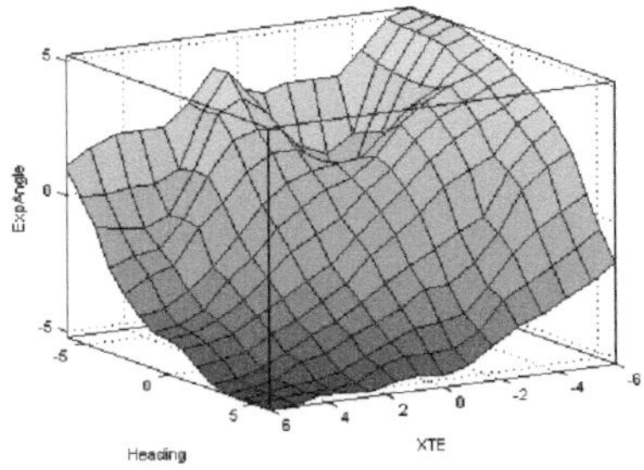

Fig. 5. Surface of Fuzzy rules

Table 1. Fuzzy Rules

Heading angle error ψ	Lateral Error E						
	NB	NM	NS	ZO	PS	PM	PB
NB	PB	PB	PB	PB	PM	PM	PS
NM	PB	PB	NB	PM	PM	ZO	NS
NS	PB	PB	PS	PS	ZO	NM	NM
ZO	PB	PM	PS	ZO	NS	NM	NM
PS	PM	PS	ZO	NS	NM	NM	NB
PM	PS	ZO	NM	NM	NB	NB	NB
PB	NS	NM	NB	NB	NB	NB	NB

4 Simulation and Analysis

In the simulation, the PID and Fuzzy method have been selected as navigation control algorithm. The start position coordinate of route was (1,0) ,and the end position coordinate was (2,-10), marked by cone barricades. The initial position of the vehicle model was (0,0); the initial lateral error was 1 meter and heading angle error was 0; the speed of vehicle model was 0.4m/s. The scene and data of simulation shown in Fig.6 and Fig.7.

When using the PID control method, the vehicle could approach the pre-set route quickly and enter into stable state under the situation of large initial error. In the stable state, the maximum value of lateral error was 3cm, and the average value of it was less than 1cm. Because the condition of simulation environment was better than the condition of reality, the effect of control was also better than the actual navigation experiment. Considering PID method has strong dependence of condition of environment, the simulation also has adopted fuzzy control method which is better in term of adaptation of complex condition. In fuzzy control simulation, the vehicle also could be quickly approaching the route. In the stable state, the maximum lateral error of vehicle model was 6.7cm, and the average of it was 5.3cm. In the further research, a more complex simulation environment will be built to test the effect of fuzzy control.

Fig. 6. Scene of simulation

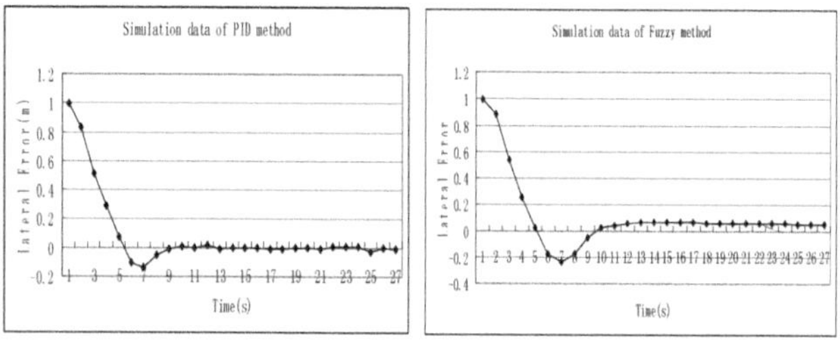

Fig. 7. Data of simulation

5 Conclusions and Perspectives

In this study, a simulation system based on MRDS for vehicle guidance has been developed. Using MRDS service-oriented, flexible and interactive features, a vehicle guidance simulation software platform has been constructed;, The created vehicle model, containing a variety of sensor models, and achieved navigation simulation through PID and fuzzy algorithms. The simulation results showed that the models were accurate and real-time, and could provide the basis for the design of the actual navigation systems. In the further research, the environment models would be created to improve the simulation effect and the navigation control method would be improved to meet the needs of complex conditions of actual environment.

References

1. Huang, L., Yi, Y.-b., Zhang, H.: Motion simulation based on Microsoft Robotics Studio. Journal of Mechanical & Electrical Engineering 27(8), 122–126 (2010)
2. Chen, X., Yu, Y., Jiang, C., Liu, Y.: Robot simulation based on virtual prototype. Journal of Beijing University of Technology 1(30), 38–40 (2004)
3. Ling, J.: The Design and Implementation of Industrial Robot Simulation System based on OpenGL, Central South University (2009)
4. Ran, X.: Outdoor Autonomous Mobile Robot Navigation System and Its Simulation Software Design and Implementation, Ocean University of China (2008)
5. Microsoft Corporation, Microsoft Robotics Dev Studio 2008 R3 SDK (2008)
6. http://mobotsoft.com
7. Johns, K.: Professional Microsoft Robotics Studio (2008)
8. Tong, G., He, H., Shou, Z.: The soccer robot Microsoft 3D simulation platform based on Robotics Studio. Journal of University of Jinan 21, 1–4 (2007)
9. Morgan, S.: Programming Microsoft Robotics Studio (2007)

A Lyrics to Singing Voice Synthesis System with Variable Timbre*

Jinlong Li[1], Hongwu Yang[1], Weizhao Zhang[1], and Lianhong Cai[2]

[1] Department of Physics and Electronic Engineering
Northwest Normal University
lanzhou, Gansu Province, China
[2] Department of Computer Science and Technology
Tsinghua University, Beijing, China
lijinlong_88@sina.com

Abstract. In this paper, we present a singing voice synthesis system, which can convert lyrics to singing voice. As the synthetic song's timbre is too monotonous, a new singing voice morphing algorithm based on GMM (Gaussian Mixture Model) was presented accordingly. The MOS test shows that the average MOS score of synthesized song is above 3.3 before timbre conversion. The professional singer's timbre can be added proportionally by changing the scale factor k in the system. The ABX test demonstrates that the accuracy can be up to 100% in the case of k=0 or k=1, and it can be higher than 64.5% in the case of 0<k<1. The experiments also show the mean of GMM has greater impact on a singer's timbre than weight ratio and covariance.

Keywords: Singing voice synthesis, melody control model, timbre modification, GMM, MIDI.

1 Introduction

With the development of technology in HMI (Human-Machine Interaction) system, people's demand for music is not limited to appreciation level, and a number of amateurs hoping to create singing voice through signal processing. Therefore, the singing voice synthesis (SVS) has been a research hotspot. The aim of the SVS technology is to make a computer sings like people. Several techniques have been proposed in this field, for example, Tokuda [1] presented a HMM-based speech synthesis system which is applied to English. Zhou [2] constructed a mandarin singing voice synthesis system by using concatenation technique. Gu[3] can synthesize the signals of Mandarin Singing Voice based on HNM (Harmonic plus Noise Model) and enhanced scheme. Macon [4] adopted sinusoidal model in the waveform synthesis portion of the system, and they considered vibrato and spectral tilt variation. Saitou [5] investigated the acoustic differences between speaking and

* This work is partially supported by the National Science Foundation of China (NSFC) under grant NO.60875015 and the Key Project of Chinese Ministry of Education under grant MO.208146.

D. Zeng (Ed.): ICAIC 2011, Part II, CCIS 225, pp. 186–193, 2011.

singing voice (English or Japanese), and proposed an automatic speech-to-singing synthesis system, and their experimental results showed that the system can convert speaking voice into singing voice naturally. However, there are many differences between Chinese and other language, and Chinese has its own complexity. Another problem is that the synthetic naturalness is easy to be affected by the environment and recording equipment. Therefore, we propose a lyrics-to-singing synthesis system based on current research achievements. In our work, lyrics are converted into speech by using TTS firstly. Then, we analyze differences between Chinese song and speech, and construct melody control models on the basis of analysis results. Finally, modify acoustic parameters of speech signal by comparing with the corresponding information from the MIDI file. In addition, the system can also achieve timber conversion by modifying spectrum based on GMM [6], and it may be useful in some actual application fields nowadays.

2 System Design

2.1 System Framework

In our system, lyrics are converted into speech under the technology of TTS (text to speech), and the melody parameters of the song are extracted from the MIDI file firstly. Then, the melody control model is adopted to convert the lyrics to song by modifying the acoustic parameters of speech signal. Lastly, spectrum parameters extracted from the synthesized song and the target song are modeled by GMM respectively, and the song's timbre can be converted by mixing their model parameters.

The framework of the proposed lyrics-to-singing synthesis system is shown in Fig. 1. The inputs of the system are lyrics of a song and its corresponding MIDI file, and the output is the synthesized singing voice. The STRAIGHT [7, 8] algorithm was proposed by Japanese scholar in 1997. It can analyze speech signal more precisely, which is based on source / filter model. Because STRAIGHT algorithm has lots of advantages, such as unique fundamental frequency extraction, elimination of periodic interference during spectrum extraction, higher parameter extraction accuracy, and large modification amplitude, we chose STRAIGHT algorithm as our synthesis tool.

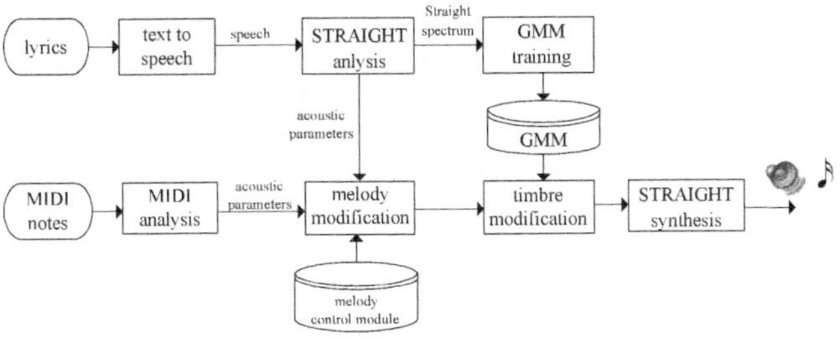

Fig. 1. System framework

The system contains four modules: 1) Text to speech module. During the process of the TTS, lyrics of a song were analyzed firstly; then, the waveform of syllables were selected from the speech database according to the pinyin and intonation information; at last, the syllables of lyrics were concatenated to obtain synthesized speech. 2) MIDI analysis module. A MIDI file is a kind of descriptive data file, which can be used to exchange information between computer and electronic instrument. It contains 16 channels, and each channel defines several performance parameters of an instrument. Therefore, we can extract the thematic characteristics from one of the MIDI channels. 3) Melody control module. 4) Spectrum modification module. The last two modules are the core of this system, and they will be introduced in the following pages.

2.2 Melody Control Module

Speaking and singing are two main communication ways for people, which generated by the same pronunciation organs. However, melody of them is not the same with each other. Music contains four elements: pitch, duration, volume, and timbre. Therefore, the key points of the singing voice synthesis are to solve expression and cooperation among these elements. As pitch and duration can reflect the melody of a song largely, and our melody control module focuses on these two elements.

Firstly, extracting F0 [9] by getting a matrix of n*7 from the corresponding MIDI file. In theory, this F0 contour can reflect the theme of a song. But vibrato is also an important factor in affecting the naturalness of the synthesized song, and the emergence of proper vibrato can make singing more sweet and full. Vocal vibrato is generated from the friction between vocal cords and breath during the process of relaxation and tension. According to existing research [10], the fluctuation frequency of vibrato is 5 to 8 times per second. In order to get synthesized singing voice with higher naturalness, lots of research and analysis study about vibrato have been done in this paper. According to our experiments, it can be found that: 1) Vibrato accounts for about 70 percent of the vowel's duration; 2) The appearance of vibrato has relation with a syllable's duration, and the experiments show that when a syllable's duration exceeds 0.5s, vibrato can be produced basically. The STRAIGHT algorithm is used to modify F0 according to the existing research results and the analysis of our study.

Table 1. Duration modification strategy

Duration Modification	
Before	**After**
dur_c	k*dur_c
dur_b	dur_b
dur_v	dur_n – (k*dur_c) – dur_b

Secondly, there are also differences in syllable duration between speech and singing voice. Extending or shortening the duration is decided by the information extracted from corresponding notes in score. In order to make each syllable of song more natural, we modify duration [11] of speech in our research, such as table 1.

2.3 Spectrum Modification Module

Spectrum of speech signal can reflect important features of the voice, such as formant frequency and formant bandwidth, which are closely related with personality information of each person. Sound track is a resonant cavity, and spectral envelope can reflect the characteristics of its resonance.

Currently, there are several methods for modifying spectral envelope, VQ (vector quantization), HMM (Hidden Markov Model), ANN (artificial neural network), speaker interpolation, DFW (dynamic frequency warping), GMM [12], etc. GMM is based on Bayesian decision theory, and it converts classification problem into the problem of estimating the distribution of data. It can decompose the problem of training and model matching into many sub-problems, such as the choice of expression, the training of model parameters and the calculation of the probability. GMM can approximate the probability density function of arbitrary shape smoothly, and the algorithm complexity is lower relatively. The introduction of GMM can make the modification of spectrum simplified. Therefore, GMM is used to fit and represent spectral envelope in this research.

We choose each frame of synthesized singing voice as a feature vector and extract STRAIGHT spectrum firstly. Then, EM algorithm is adopted to estimate parameters of the Gaussian model, such as average point, variance and proportion weights of each Gaussian distribution. The original signal spectral envelope and the spectrum after GMM approximation are shown in Fig.2. It is easy to find that GMM can represent the speech spectrum relatively accurate.

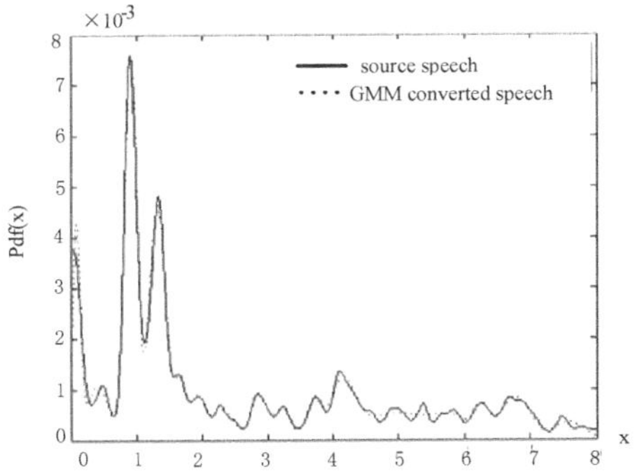

Fig. 2. Comparison of the spectrum envelope

Spectrum of speaker voice carries important information, and adjusting the spectrum is the primary content of the voice conversion [13] technology currently. Similarly, modifying singing voice's spectrum, the timber of a singer can be changed accordingly. Therefore, we try to change the singer's timber by modifying the spectrum of the singing voice.

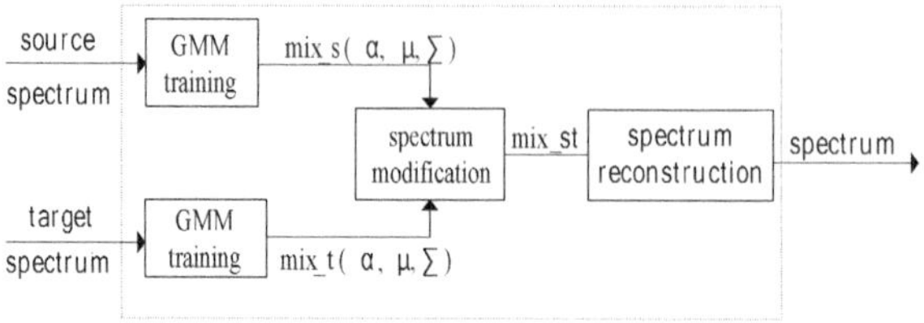

Fig. 3. Spectrum modification process

The Fig.3. shows the spectrum modification process, GMM is used to train STRAIGHT spectrum of the source and target song, model parameters mix_s and mix_t can represent their spectrum respectively. Then, we can adjust the mean μ, variance Σ and weights α of the two models according to actual needs. Specific modification method can be expressed as follow:

$$mix\_st = mix\_s * (1-k) + mix\_t * k\ . \tag{1}$$

Finally, the STRAIGHT algorithm is used again to synthesize the new song according to the new parameters. The synthesized singing voice is saved as PCM (*.wav) file for doing subjective listening tests.

3 Perception Test

In this study, perceptual evaluation was carried out by conducting two subjective tests (MOS and ABX) with 8 students. In the MOS test, we have 7 songs synthesized by using Only_dur, Only_F0, and Dur_F0 respectively. 1 to 3 are female singing voice; 4 to 7 are male singing voice. Subjects are asked to assess naturalness and clarity of the synthesized song on a five scale. The MOS score are as demonstrated in table 2.

Table 2. MOS test results

Method / MOS	Only_dur	Only _F0	Dur_F0
Song1	1.29	2.44	3.25
Song2	1.25	2.56	3.27
Song3	1.13	2.38	3.64
Song4	1.34	1.74	3.35
Song5	1.07	1.69	2.98
Song6	1.05	2.88	3.29
Song7	1.30	2.81	3.32
Mean	1.20	2.34	3.30

Where Only_dur represents the synthetic way only with duration modification; Only_F0 represents the synthetic way only with F0 modification; Dur_F0 represents the synthetic way with melody modification (duration and F0). From the table 2, it is obvious that our melody control model is acceptable, and the synthetic results can achieve an average MOS score of 3.3. The experiment results can also show that F0 has greater effect on the naturalness of the synthesized song relatively, which is consistent with people's perception.

In the ABX test, we have 7 songs synthesized by modifying their timbre. Actually, singing voice have already been synthesized after melody control module, ABX test is just used to evaluate the spectrum modification algorithm. Subjects are asked to assess X (singing voice after timbre conversion) is more like A (singing voice after melody conversion) or B (singing voice of the professional singer). X can be generated from 5 methods, such as: M1, M2, M3, M4 and M5. In table 3, M1: only covariance morphing; M2: only weights morphing; M3: only mean value morphing; M4: mean value and weights morphing; M5: morphing on all GMM Parameters. The ABX test results are demonstrated in table 3.

Table 3. ABX test results

Ratio / Method	k=0	k<0.5	k>0.5	k=1
M1	100%	64.5%	68.5%	100%
M2	100%	68.5%	69.5%	100%
M3	100%	83%	80%	100%
M4	100%	91.5%	88%	100%
M5	100%	94%	92.5%	100%

From Table 3, When the scale factor $k = 0$, no matter in what way, X can be judged as A with probability of 100 percent; when $k<0.5$, timbre of A account for larger proportion, which is higher than 64.5 percent, when $k>0.5$, timbre of professional singer account for larger proportion, which is higher than 68.5 percent, when $k=1$, no matter in what way, X can be judged as A with probability of 100 percent. The results also show that the accuracy can be higher under the way M3 and M4. That is to say, mean of the GMM has greater impact on a singer's timbre relatively.

4 Conclusions and Future Works

To improve the naturalness and the clarity of synthesized mandarin singing voice, this paper presents a melody modification algorithm firstly, and vibrato analysis is an important part. Subject evaluation results show that the naturalness of synthesized singing voice is close to the real song relatively. We also find that F0 has a greater impact on the naturalness of the synthesized singing voice, and this is consistent with the perception of people's ear. Then, a spectrum morphing algorithm based on GMM is proposed to convert timbre of the synthesized song. The results show that GMM can represent spectrum of the signal with higher precision, and we can join a professional singer's timbre proportionally by adjusting the scale factor k. The experiments also show the mean of GMM has greater impact on a singer's timbre than weight ratio and covariance.

However, F0 and spectral are discussed separately. The modification of spectrum may change F0 to some extent, and it may decrease naturalness and clarity of singing voice. We will consider this problem and improve the quality of synthesized singing voice in our future works.

Acknowledgment. Author would like to acknowledge the support of the National Science Foundation of China (NSFC) under grant No. 60875015 and the Key Project of Chinese Ministry of Education under grant No. 208146.

References

1. Tokuda, K., Zen, H., Black, A.W.: An HMM-based speech synthesis system applied to English. In: Proc. 2002 IEEE Workshop on Speech Synthesis, Santa Monica, CA, pp. 41–46 (2002)
2. Zhou, S.-s., Chen, Q.-q., Wang, D.-d., et al.: Acorpus-based concatenative mandarin singing voice synthesis system. In: Proc. Seventh International Conference on Machine Learning and Cybernetics, Kunming, China, pp. 2695–2699 (July 2008)
3. Gu, H.Y., Liau, H.L.: Mandarin Singing Voice Synthesis Using an HNM Based Scheme. In: Proc. International Congress on Image and Signal Processing, Sanya, China, pp. 347–351 (2008)
4. Macon, M.W., Jensen-Link, L., Oliverio, J., et al.: A Singing Voice Synthesis System Based on Sinusoidal Modeling. In: Proc. IEEE International Conference on Acoustics, Speech, and Signal Processing, Munich, Germany, pp. 435–438 (1997)
5. Saitou, T., Goto, M., Unoki, M., et al.: Speech-to-sing synthesis: converting speaking voices to sing voices by controlling acoustic features unique to sing voices. In: Proc. 10th National Conference on Man-Machine Speech Communication, Lanzhou, China, pp. 477–482 (August 2009)
6. Kawanami, H., Iwami, Y., Toda, T., et al.: GMM-based Voice Conversion Applied to Emotional Speech Synthesis. In: Proc. European Conference on Speech Communication and Technology, Geneva, Switzerland, pp. 2401–2404 (2003)
7. Kawahara, H., Estill, J., Fujimura, O.: Aperiodicity extraction and control using mixed mode excitation and group delay manipulation for a high quality speech analysis, modification and synthesis system straight. In: Proc. International Workshop on Models and Analysis of Vocal Emissions for Biomedical Application, Firentze Italy, pp. 13–15 (September 2001)
8. Saitou, T., Goto, M., Unoki, M., et al.: Vocal Conversion from speaking voice to singing voice using STRAIGHT. In: Proc. Interspeech, Antwerp, Belgium, pp. 4005–4006 (2007)
9. Saitou, T., Unokiand, M.M., Akagi: Development of an F0 control model based on F0 dynamic characteristics for singing-voice synthesis. Speech Communication 46, 405–417 (2005)

10. Lai, W.-h.: F0 Control Model for Mandarin Singing Voice Synthesis. In: Proc. 2th International Conference on Digital Telecommunications, San Jose, California, pp. 12–15 (2007)
11. Cai, L.-h., Hou, j., Liu, r., et al.: HMM Parametric Singing Synthesis with Pitch Instruction. In: Proc. 8th National Conference of Multimedia Technology, Xi'an, China, pp. 219–225 (2009)
12. Chen, Y., Chu, M., Chang, E., et al.: Voice conversion with smoothed GMM and map adaptation. In: Proc. Eurospeech, Geneva, Switzerland, pp. 2413–2416 (2003)
13. Cano, P., Loscos, A., Bonada, J., et al.: Voice Morphing System for Impersonating in Karaoke Applications. In: Proc. International Computer Music Conference, Rio de Janeiro, Brazil, pp. 109–112 (2000)

Face Recognition Using Spatially Smooth and Maximum Minimum Value of Manifold Preserving*

Hao Wen[1] and Youkui Wen[2]

[1] School of Communication Engineering University of XiDian
XiAn, Shaanxi Province, China
[2] School of Economic Management University of XiDian
XiAn, Shaanxi Province, China
wenhao112233@sina.com

Abstract. Subspace learning based face recognition methods have attracted considerable interests in recently years. However, the accuracies of previous methods are not so high, because they don't utilize the manifold of face image data sufficiently and neglect some particular characters of the images. Thus a new method to form graph of data is proposed in this paper, and the method is used to develop two face recognition algorithms. At the same time the pixels correlation in images is considered sufficient under the constrain of spatially smooth in the two developed algorithms. So the features of the projected subspace based on our algorithms have better classification ability. Therefore, the right recognition rates are enhanced by the two proposed algorithms. This is further confirmed by experiments.

Keywords: manifold learning, face recognition, spatially smooth, subspace learning, pattern recognition.

1 Introduction

Face recognition has been one of the most important issues in pattern recognition over the last several decades[1]. Subspace analysis approaches received considerable research interests and have been widely investigated for face recognition[2]. Among them, the Principle Component Analysis (PCA)[3] and Linear Discriminant Analysis (LDA)[4] have been the two most popular methods. However, these two methods fail to discover the underlying manifold structure. Recently there has been a considerable interest in geometrically motivated approaches to data analysis in high dimensional spaces. The geometrical and topological properties of the sub-manifold from random points ("scattered data") lying on this unknown sub-manifold are believed to serve as a better statistical estimate. Along this direction, many subspace learning algorithms have been proposed for face recognition. Some popular ones include ISOMAP[5], Locality Preserving Projection (LPP)[6], and neighbourhood Preserving Embedding (NPE)[7]. Yan presented a general formulation to unify different subspace algorithms

* This work is partially supported by the National Foundation of China No. 70373046, and Ministry of Education, Humanities and Social Sciences Planning Fund No.09YJA630127.

D. Zeng (Ed.): ICAIC 2011, Part II, CCIS 225, pp. 194–204, 2011.

within a common framework named as graph embedding framework, and he used the graph embedding framework as a general platform for developing new subspace algorithm and presented the Marginal Fisher Analysis (MFA)[8]. The shortcoming of these algorithms is that the manifolds of between-classes and within the same class data are not utilized sufficiently or simultaneously. Another shortcoming of these methods is that the characters of the specific structure of the images are neglected. An image intrinsic is a matrix or 2-order tensor, and image should not merely be treated as a vector. To solve the second shortcoming, tensor-based approaches have been proposed [9], [10], but they only consider the relationship between pixels in the same row (or column) [11], thus fail to fully explore the spatial information of images. Therefore, penalized subspace analysis has been adopted for spatially smoothing subspace learning in face recognition as outlined, e.g in [11], [12], [13].

Based on the above review, A new method for constructing graph of data is proposed in this paper with using class labels to properly account for between-classes and within-class manifold of data. The maximum and minimum values of manifold of data are preserved with the new graph. At the same time, the discrectized Laplacian operator penalty is used to ensure the spatially smooth of image data. Thus, two new face recognition algorithms, using maximum minimum value of manifold preserving and spatially smooth, are developed in this paper. The features of the projected subspace based on our two algorithms have stronger recognition abilities and the right recognition rates are enhanced with the two proposed algorithms.

The remainder of the paper is organized as follows: The Laplacian penalty is clarified in Section II. Two algorithms are proposed in Section III. Experiments used to evaluate the proposed algorithms are presented in Section IV followed by conclusions and future works in Section V.

2 Laplacaian Smoothing

Let f be a function defined on a region of interest, $\Omega \subset R^d$. The Laplacain operator L is defined as follows:

$$Lf(t) = \sum_{i=1}^{d} \left(\left(\partial^2 f \right) / \left(\partial^2 t_i^2 \right) \right) \tag{1}$$

The Laplacian penalty functional, denoted by J , is defined by:

$$J(f) = \int_{\Omega} [Lf]^2 \, dt \tag{2}$$

Intuitively, $J(f)$ measures the smoothness of the function f over the region Ω . In this paper, the primary interest is in the image which is intrinsically a two-dimensional signal. Therefore, in the following $d = 2$. Let a image be represented as $n_1 \times n_2$ matrix. So It can be represented as a vector in $n = n_1 \times n_2$. Let $a_i \in R_n$ be the basis vectors (projection functions) obtained by any subspace method. Without loss of generality, a_i can also be considered as a function defined on a $n_1 \times n_2$ lattice. For a face image, the region of interest Ω is a two dimensional rectangle, which for notational convenience we take it to be $[0,1]^2$. A lattice is defined on Ω as follows:

Let $h=(h_1,h_2)$, where $h_1=1/h_1$ and $h_2=1/h_2$ then Ω_h consists of the set of two-dimensional vectors $t_i=(t_{i1},t_{i2})$ with $t_{i_j}=(i_j-0.5)h_j$ for $1\le i_j\le n_j$ and $1\le j\le 2$. There are a total of $n=n_1\times n_2$ grid points in this lattice. Let $\boldsymbol{D}_j$ be an $n_j\times n_j$ matrix that yields a discrete approximation to $\partial^2/\partial t_j^2$. Thus if $\boldsymbol{u}=(u(t_1),\cdots,u(t_{n_j}))$ is an n_j dimensional vector, which is a discretized version of a function $u(t)$, then $\boldsymbol{D}_j$ has the property that: $[\boldsymbol{D}_j\boldsymbol{u}]_i\approx((\partial^2u(t_i))/(\partial t^2))$, for $i=1,\cdots,n$. There are many possible choices of $\boldsymbol{D}_j$, the modified Neuman discretization is applied in our paper.

$$\boldsymbol{D}_j=\frac{1}{h_j^2}\begin{pmatrix}-1 & 1 & & & & & 0\\ 1 & -2 & 1 & & & & \\ & 1 & -2 & 1 & & & \\ & & \cdots & \cdots & \cdots & & \\ & & & 1 & -2 & 1 & \\ & & & & 1 & -2 & 1\\ & & & & & 1 & -1\end{pmatrix} \tag{3}$$

Give $\boldsymbol{D}_j$, a discrete approximation for two-dimensional Laplacian L is the $n\times n$ matrix:

$$\boldsymbol{Q}=\boldsymbol{D}_1\otimes\boldsymbol{I}_2+\boldsymbol{D}_2\otimes\boldsymbol{I}_1 \tag{4}$$

where $\boldsymbol{I}_j$ is $n_j\times n_j$ identity matrix for $j=1,2$. "$\otimes$" in (4) is the Kronecker product [14]. For a $n_1\times n_2$ dimensional vector $\boldsymbol{a}$, it is easy to check that $\|\boldsymbol{Qa}\|^2$ is proportional to the sum of the squared differences between nearby grid points of $\boldsymbol{a}$ in its matrix form. It provides a measure of smoothness of $\boldsymbol{a}$ on the $n_1\times n_2$ lattice. Let $\boldsymbol{A}$ is defined as the projection matrix, then J is the discretized Laplacian regularization functional:

$$J(A)=\boldsymbol{A}^T\boldsymbol{Q}^T\boldsymbol{QA} \tag{5}$$

3 Spatially Smooth and Maximum Minimum Value of Manifold Preserving

3.1 The New Within-Class and Between-Class Graphs

In this section the two face recognition algorithms will be described. The two algorithms are based on new method to form graph of data. The new method are introduced first.

We first present our method of forming within-class graph and between-classes graph. Let $\boldsymbol{X}$ be defined as the original vector assembly of data samples, and

X_i, X_j be two vectors of X, U_i and U_j be any two different classes, then the graph of X under supervised condition is first formed as follows:

$$W_w(i,j)=\begin{cases}\left(\langle X_i,X_j\rangle/\left(\|X_i\|\|X_j\|\right)\right) & X_i,X_j\in U_i, i\neq j\\ 0 & others\end{cases} \tag{6}$$

$$W_b(i,j)=\begin{cases}\left(\langle X_i,X_j\rangle/\left(\|X_i\|\|X_j\|\right)\right) & X_i\in U_i, X_j\in U_j, i\neq j\\ 0 & others\end{cases} \tag{7}$$

W_w and W_b are within-class graph matrix and between-class graph matrix respectively. The generic problem of linear dimensionality reduction is the following. Given a set $\{X_1,\cdots X_m\}$ in R^n, find a transformation matrix $A=(a_1,a_2,\cdots,a_d)\in R^{n\times d}$ that maps these m points to a set of points $Y_1,\cdots,Y_m$ in R^d $(d<<n)$, such that Y_i "represents" X_i, where

$$Y_i = A^T X_i \tag{8}$$

Y_i is an any low dimensional vector which is obtained with projecting any X_i to low dimensional vector. According to the idea of classification, the data within the same class are assumed to be close but the data between classes are assumed to lie apart. So the following two objective functions can be defined as:

$$\min \sum_{i,j}\|Y_i - Y_j\|^2 W_w(i,j) \tag{9}$$

$$\max \sum_{i,j}\|Y_i - Y_j\|^2 W_b(i,j) \tag{10}$$

We note that the W_w and W_b defined in (6) and (7) are not suitable for optimal classification. Hence, (9) and (10) can not yield optimal results if the (6) and (7) are utilized. Therefore, new graphs of W_w and W_b, named as W_{nw} and W_{nb} are proposed based on (6) and (7). The procedure of forming W_{nw} and W_{nb} is to first add W_w and W_b to obtain a new function, W_g:

$$W_g = W_w + W_b \tag{11}$$

Then, the maximum and minimum positive values, $W_{gi\max}$ and $W_{gi\min}$, are found for any i row of W_g. Finally W_{nw} and W_{nb} are defined as:

$$W_{nw}(i,j)=\begin{cases}W_{gi\min} & X_i,X_j\in U_i, i\neq j\\ 0 & others\end{cases} \tag{12}$$

$$W_{nb}(i,j)=\begin{cases}W_{gi\max} & X_i\in U_i, X_j\in U_j\\ 0 & others\end{cases} \tag{13}$$

Equations (11) and (12) are the new within-class and between-classes graphs of our two face recognition algorithms.

3.2 Maximum Minimum Value of Manifold Preserving

This subsection describes first the proposed face recognition algorithms based on W_{nw} and W_{nb} without employing a spatially smooth constrain. Equations (9) and (10) can be rewritten as follows:

$$\min \sum_{i,j} \|Y_i - Y_j\|^2 W_{nw}(i,j) \tag{14}$$

$$\max \sum_{i,j} \|Y_i - Y_j\|^2 W_{nb}(i,j) \tag{15}$$

and the following relation can be derived based on:

$$\frac{1}{2}\sum_{ij}\left(\|Y_i - Y_j\|^2 W_{nw}(i,j)\right) = tr\left(A^T X (L_{nw}) X^T A\right) \tag{16}$$

The " tr " in (16) is the trace of a matrix, D_{nw} is an diagonal matrix $D_{nw}(i,i) = \sum_j W_{nw}(i,j)$ and $L_{nw} = D_{nw} - W_{nw}$. A similar relation using $\mathbf{W}_{nb}$ can also be derived as follows:

$$\frac{1}{2}\sum_{ij}\left(\|Y_i - Y_j\|^2 W_{nb}(i,j)\right) = tr\left(A^T X (L_{nb}) X^T A\right) \tag{17}$$

According to (14), (15), (16), (17) the follow objective function can be defined

$$A^* = \arg\max \left(\frac{A^T X (L_{nb}) X^T A}{A^T X (L_{nw}) X^T A} \right) \tag{18}$$

The transformation matrix, $\mathbf{A}$, consists of eigenvectors of the corresponding n max eigenvalue of the follow generalized eigenvector problem.

$$X(L_{nb})X^T A = \lambda X(L_{nW})X^T A \tag{19}$$

The objective function (18) is an important component of our algorithm. Another objective functions will be described. In order to enhance the compactness of the within-class graph[15], a new constrain defined in (20) can be added to (19)

$$\max\left(A^T X W_{nw} X^T A\right) \tag{20}$$

and a new objective function is obtained:

$$A^* = \arg\max \left(\frac{A^T X (\beta L_{nb} + (1-\beta) W_{nw}) X^T A}{A^T X (L_{nw}) X^T A} \right) \tag{21}$$

The transformation matrix, $\mathbf{A}$, in (21) consists of eigenvectors of the corresponding n max eigenvalue of the following generalized eigenvector problem.

$$X\left(\beta L_b - (1-\beta)W_w\right)X^T A = \lambda X L_{nw} X^T A \tag{22}$$

The β in (21) or (22) is a constant and belongs to [0,1]. In the following the two algorithms with the constrain are described.

3.3 Spatially Smoothing and Maximum Minimum Value of Manifold Preserving

The constrain specified in (4) can be added to objective functions (18) or (21) to yield:

$$A^* = \arg\max\left(\frac{A^T X(L_{nb})X^T A}{(1-\alpha)A^T X(L_{nw})X^T A + \alpha\left(A^T Q^T QA\right)}\right) \tag{23}$$

$$A^* = \arg\max\left(\frac{A^T X(\beta L_{nb} + (1-\beta)W_{nw})X^T A}{(1-\alpha)A^T X(L_{nw})X^T A + \alpha\left(A^T Q^T QA\right)}\right) \tag{24}$$

The α in (23) and (24) is a constant and belongs to [0,1]. The $\mathbf{A}$, in (23) or (24) consist of the eigenvectors of the corresponding n max eigenvalue of the follow generalized eigenvector problem respectively.

$$X(L_{nb})X^T A = \lambda\left((1-\alpha)X(L_{nw})X^T + \alpha\left(Q^T Q\right)\right)A \tag{25}$$

$$\begin{aligned} &X(\beta L_{nb} - (1-\beta)W_{nw})X^T A \\ &= \lambda\left((1-\alpha)X(L_{nw})X^T + \alpha\left(Q^T Q\right)\right)A \end{aligned} \tag{26}$$

Equations (25) and (26) describe the two algorithms based on spatially smoothing and new graphs. The specific procedure for using the two algorithms for face recognition is described as follows.

Input: The training vector data assemble, X_R and the testing vector data assemble, X_E . Also input the class label of data in each assemble, α and β , respectively.

Output: Right recognition rate.

Step 1 Solve the transformation, A , based on the input X_R and (25) or (26);

Step 2 Calculate (8) with $\mathbf{A}$ to get the projection assembles Y_R and Y_E of X_R and $\mathbf{X}_E$. The feature extraction is accomplished in this step;

Step 3 classify the Y_E by the nearest neighbour criterion utilizing Y_R .

For convenience, the algorithm using (25) is named as Spatially Smooth and Maximum Minimum Value of Manifold Preserving 1 (SSMMP1), and the algorithm using (26) is named as Spatially Smooth and Maximum Minimum Value of Manifold Preserving 2 (SSMMP2). Experimental verification of the two algorithms is given in the next section.

4 Experiments

To evaluate our algorithms, we compare them with the currently spatially smooth subspace algorithms named as S-NPE, S-LPP, S-LDA, S-MFA[11]; OSSLDA[12]; PF-LOG, PF-DOG, PL-LOG, PL-DOG[13], respectively. In our experiments, we use the ORL and Yale databases for face recognition. In all the experiments, the original images were normalized such that the two eyes were aligned at the same position. Then, the facial areas were cropped into the final image for matching. The size of each cropped image in all the experiments is 32×32 pixels, with 256 gray levels per pixel.

4.1 Face Recognition on ORL Database

The ORL face database consists of 400 face images of 40 indivliduas (10 samples per person). The images have different variations including expressions and with/without glasses. The images have some tilting and rotation up to 20 degrees. For each individual, $l(=3,5)$ images were randomly selected as training samples, and the rest were used for testing. We repeated this process 20 times for all the mentioned algorithms and calculate the average right recognition rate. The best performance obtained by these algorithms as well as the corresponding dimension of the optimal subspace and the corresponding parameters are shown in Table 1. Plots of right recognition rate versus dimension for each algorithm are shown in Fig. 1. The x-coordinate in Fig. 1 denotes the dimension, and the y-coordinate in Fig. 1 denotes the right recognition rate. For clarity, the range of the y-coordinate value in Fig. 1 is constrained as 80%-95% and 75%-100% respectively, and the scope of the x-coordinate value in Fig. 1 is constrained as 0-120 and 0-200 respectively.

Table 1. Right Recognition rate of Different Algorithm on ORL Database

Algorithm	3Train	5Train
S-LPP	92.79% (28), $\alpha=0.65$	97.55% (17), $\alpha=0.65$
S-NPE	92.82% (77), $\alpha=0.6$	97.1% (28), $\alpha=0.3$
S-LDA	93.64% (39), $\alpha=0.35$	98.15% (36), $\alpha=0.3$
S-MFA	93.79% (50), $\alpha=0.25$	98.15% (68), $\alpha=0.25$
OSSLDA	93.53% (35), $\alpha=0.01$	97.95% (38) $\alpha=0.01$
PF-LOG	93.18% (39), $\alpha=0.1$	97.85 (29), $\alpha=0.1$
PF-DOG	93.82% (39), $\alpha=0.1$	98.15% (29), $\alpha=0.1$
PL-LOG	91.29% (16), $\alpha=0.1$	96.8% (14), $\alpha=0.1$
PL-DOG	92.79% (28), $\alpha=0.1$	97.2% (16), $\alpha=0.1$
SSMMP1	92.93% (68), $\alpha=0.4$	98.15% (43), $\alpha=0.4$
SSMMP2	94.04% (51), $\alpha=0.25$, $\beta=0.2$	98.25%(46), $\alpha=0.4$, $\beta=0.2$

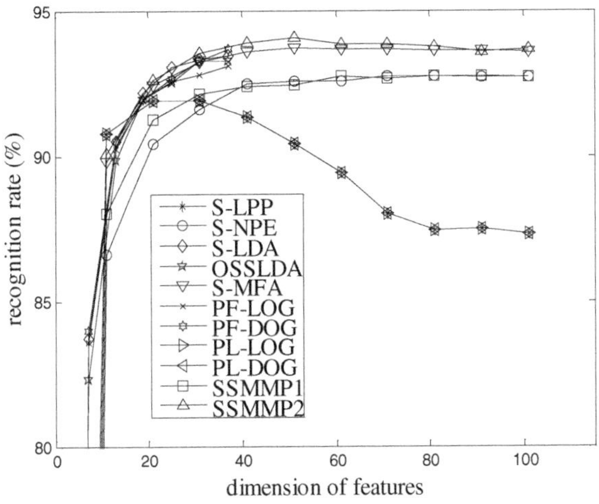

(a) 3train recognition rate on ORL database

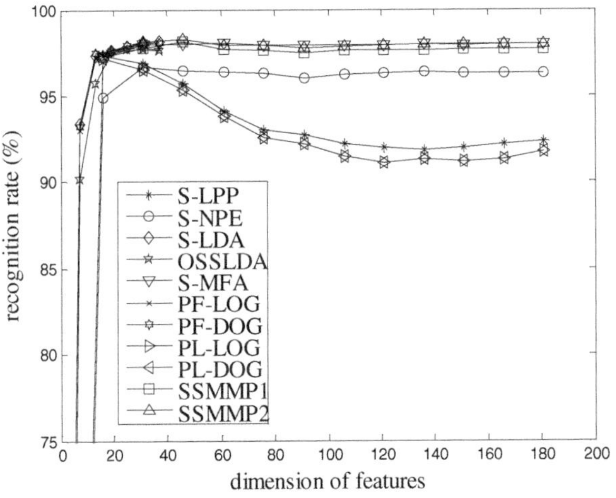

(b) 5train recognition rate on ORL database

Fig. 1. Plots of Right Recognition Rate versus Dimension for Each Algorithm Using ORL Database

As shown in Table 1 and Fig.1, the recognition rate of SSMMP1 is not higher than any other published algorithm's. The recognition rate of SSMMP2 is the highest at any train number. So the formula (20) should be added in this experiment.

4.2 Face Recognition on Yale Database

The Yale face database contains 165 images (11 samples per person). The images demonstrate variations in lighting condition, facial expression, with/without glasses.

For each individual, $l(=3,5)$ images are randomly selected for training and the rest are used for testing. The experiment is designed the same as before. The best performance obtained by these algorithms as well as the corresponding dimension of the optimal subspace and the corresponding parameters are shown in Table 2. Plots of right recognition rate versus dimension for every algorithms are shown in Fig. 2. For clarity, the scope of y-coordinate value in Fig. 2 is constrained as 40%-75% ,40%-85%, and the scope of the x-coordinate value in Figrure2 is constrained as 0-50 , 0-75, respectively.

Table 2. Right Recognition Rate of Different Algorithm on Yale Database

Algorithm	3Train	5Train
S-LPP	71.33% (15), $\alpha=0.7$	79.44% (17), $\alpha=0.75$
S-NPE	71% (33), $\alpha=0.85$	78.89% (28), $\alpha=0.6$
S-LDA	72% (14), $\alpha=0.8$	80.67% (14), $\alpha=0.7$
S-MFA	72.17% (20), $\alpha=0.75$	80.89% (14), $\alpha=0.7$
OSSLDA	71.58% (14), $\alpha=0.01$	80% (14), $\alpha=0.01$
PF-LOG	71.92% (12), $\alpha=0.1$	80.22% (14), $\alpha=0.1$
PF-DOG	72.58% (14), $\alpha=0.1$	80.89% (14), $\alpha=0.1$
PL-LOG	71.5% (15), $\alpha=0.1$	78% (14), $\alpha=0.1$
PL-DOG	72.17% (15), $\alpha=0.1$	79.44% (12), $\alpha=0.1$
SSMMP1	73.67% (21), $\alpha=0.6$	81.56% (13), $\alpha=0.6$
SSMMP2	73% (20), $\alpha=0.8$, $\beta=0.2$	81.44% (13), $\alpha=0.6$, $\beta=0.2$

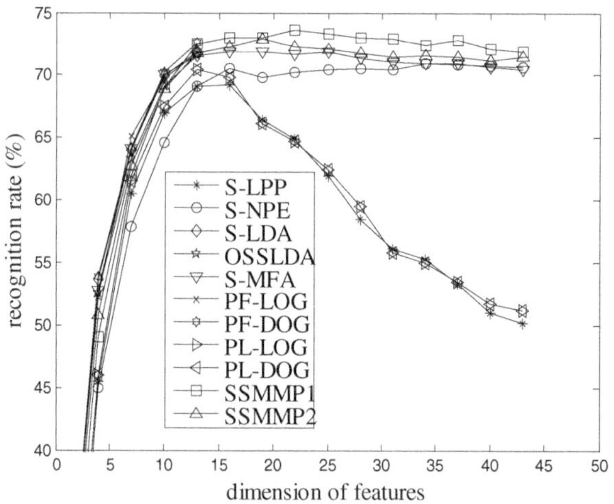

(a) 3train recognition rate on Yale database

Fig. 2. Plots of Right Recognition Rate versus Dimension for Each Algorithm Using Yale Database

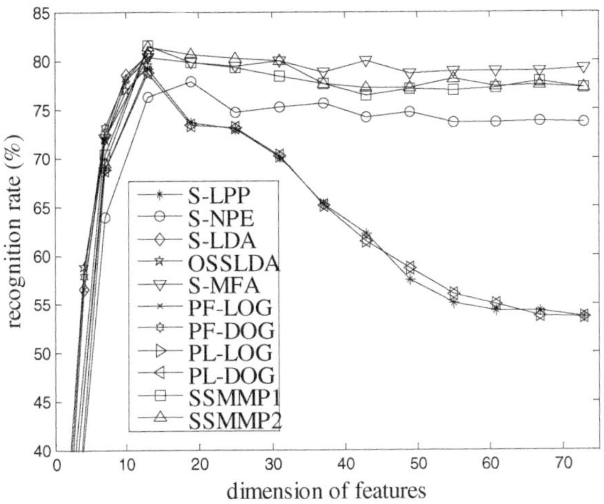

(b) 5train recognition rate on Yale database

Fig. 2. *(continued)*

Our algorithms outperformed all other algorithms for all cases. Results of SSMP1 is higher than the that of SSMP2, so the constrain of formula (20) should not be added to the Yale database.

4.3 Analysis

The recognition rates of the new proposed algorithms are higher than other published relevant algorithms. Because the new graphs are mainly designed for classification. It is shown that the robustness of SSMMP1 is not better than SSMMP2. But SSMMP1 only has one parameter need to be determined.

5 Conclusion

A new graph is proposed that form the basis for two face recognition algorithms. The features extracted by the new algorithms have good intra-class compactness and good interclass separation ability. The right recognition rates are enhanced with the algorithms, which are validated by experiments. But optimal parameters determination is difficult. So how to determinate the parameter is the future work.

References

1. Zhao, W., Chellappa, R., Phillips, P.J., Rosenfeld, A.: Face Recognition: A Literature Survey. ACM Computing Surveys 35(4), 399–458 (2003)
2. Xu, D., et al.: Concurrent Subspaces Analysis. In: Proc. IEEE Int. Conf. Comput. Vis. Pattern Recog., pp. 203–208 (June 2005)

3. Turk, M., Pentland, A.: Face Recognition Using Eigenfaces. In: Proc. IEEE Int. Conf. Comput. Vis. Pattern Recog., Maui, Hi., pp. 586–591 (June 1991)
4. Ye, J., Janardan, R., Park, C., Park, H.: An Optimization Criterion for Generalized Discriminant Analysis on Undersampled Problems. IEEE Trans. Pattern Anal. Mach. Intell. 26(8), 982–994 (2004)
5. Cai, D., He, X., Han, J.: Isometric Projection. In: Proc. 22nd Con. Artif. Intell., pp. 528–533 (2007)
6. He, X., Yan, S., Hu, Y., Niyogi, P., Zhang, H.: Face Recognition Using Laplacianfaces. IEEE Trans. Pattern Anal. Mach. Intell. 27(3), 328–340 (2005)
7. He, X., Cai, D., Yan, S., Zhang, H.: Neighborhood preserving embedding. In: Proc. IEEE Int. Conf. Comput. Vis. Pattern Recog., Beijing, China, pp. 1208–1213 (October 2005)
8. Yan, S., et al.: Graph Embedding and Extensions: A General Framework for Dimensionality Reduction. IEEE Trans. Pattern Anal. Mach. Intell 29(1), 40–51 (2007)
9. Vasilescu, M.A.O., Terzopoulos, D.: Multilinear analysis of image ensembles: TensorFaces. In: Proceedings of the 7th European Conference on Computer Vision, pp. 447–460 (May 2002)
10. He, X., Cai, D., Niyogi, P.: Tensor subspace analysis. In: Advances in Neural Information Processing Systems, vol. 18. MIT Press, Cambridge (2005)
11. Cai, D., He, X., Hu, Y., Han, J., Huang, T.: Learning a Spatially Smooth Subspace for Face Recognition. In: Proc. IEEE Int. Conf. Comput. Vis. Pattern Recog., Minneapolis, MN, pp. 1–7 (June 2007)
12. Hou, C., Nie, F., Zhang, C., Wu, Y.: Learning an Orthogonal and Smooth Subspace for Image Classification. IEEE Signal Process. Lett. 16(4), 303–306 (2009)
13. Zuo, W., Liu, L., Wang, K., Zhang, D.: Spatially smooth subspace face recognition using LOG and DOG penalties. In: Yu, W., He, H., Zhang, N. (eds.) ISNN 2009. LNCS, vol. 5553, pp. 439–448. Springer, Heidelberg (2009)
14. Horn, R., Johnson, C.: Topics in Matrix Analysis. Cambridge Univ. Press, Cambridge (1991)
15. Cai, D., He, X., Zhou, K.: Locality Sensitive Discriminant Analysis. In: Proc. 20th International Joint Conference on Artificial Intelligence, Hyderabad, India, pp. 708–713 (January 2007)

SCADA of VAV/CAV Air-Conditioning System Base on Lonworks Technology

Fei Wang, Zhi-jun Zou, Hao Li, and Xin Wang

School of Enviroment and Architecture
University of Shanghai for Science and Technology
Shanghai, China
f.wang23@yahoo.cn

Abstract. Constant Air Volume (CAV)and Variable Air Volume (VAV) air-conditioning system become more and more widely used because of it's air volume controlling property and it's energy saving property.

In the whole VAV and CAV air-conditioning system, CAV and VAV damper is important component.CAV and VAV damper can control the supply air volume or return/exhaust air volume. And there is lots of parameter can be tested in CAV and VAV damper such as air volume ,damper statues, damper position and temperature of room(VAV only). These parameters are craticals for the whole system.

Since there are too many parameters to feedback , it cost too much to transform the parameters through analog signal, which will take too many cables and too many channel of DDC controller. The problem can be solved by using lonworks technology. Lonwork technology is a kind of open control network technology. Using lonworks technology DDC controller can acquire the parameters with a couple of cable.

This paper introduces how to apply lonworks technology technology to CAV and VAV air-conditioning system, and presents framwork of CAV and VAV control system based on lonworks technology, including the component of whole system such as Variable Frequency Drivers (VFD) , CAV VAV damper and network topology. Whitin the proposed framwork CAV and VAV control system can be built easliy and its performance good.

Keywords: Lonworks, VAV, CAV, SCADA.

1 Vav/Cav Air-Conditioniing System

Constant Air Volume (CAV)and Variable Air Volume (VAV) air-conditioning system is a kind of air conditioning system which can control the supply air volume. Supply volume of VAV AC system can be adjusted according to air-conditioning load of room [1]. When the air-conditioning load reduces, the supply volume will be turned down and vice versa.So VAV AC system can save energy. Becuse of the energy-saving property, VAV AC system is always used in office AC system.

CAV AC system can control supply air volume and return air volume accurately. Fixed supply air and return air is good to keep room pressure stable.So CAV AC system is always used in which the room pressure is demanded accurately. For example clean room, biochemistry laboratory. CAV AC sytem can set max supply air

D. Zeng (Ed.): ICAIC 2011, Part II, CCIS 225, pp. 205–211, 2011.

(occupied mode) when persons work in room for comfort and health. It can also set min supply [1]air(unoccupied mode) when no person work in room for only room pressure.In this situation, CAV system can also save energy.

According to the propertis of CAV/VAV AC system mentoined above, this kind of AC system is more and more widely used.

2 Vav and Cav Damper

VAV and CAV damper are very important componet in VAV AC system and CAV AC system, which can control the supply air volume. It's a base of VAV and CAV AC system.

2.1 CAV Damper

Fig.1 is the schmatic of CAV damper.CAV actuator, which adjust the damper angle according the feedback of flowmeter. When testing air volume is less than setpoint, the actuator can turn up the damper. When testing air vloume is more than the setpoint, the actuator can turn down the damper.

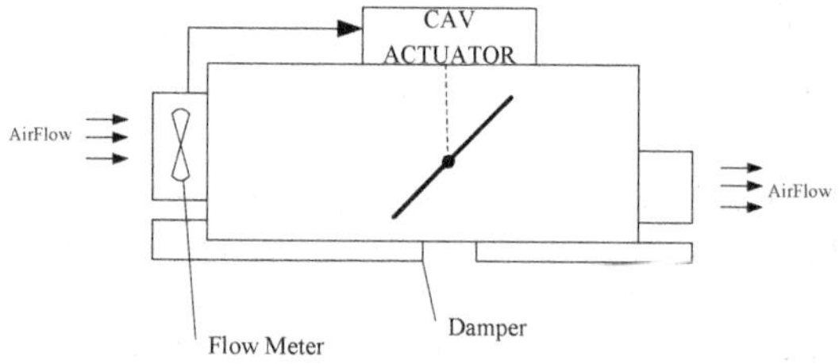

Fig. 1. Consititution of CAV damper

2.2 VAV Damper

There is a lot kinds of VAV damper. Fig 2 is a schmatic of a single-duct VAV damper. VAV actuator can also adjust the damper angle according the feedback of flowmeter .But VAV damper is different with CAV damper. Air volume of CAV damper is constant. Air volume of VAV is adjusted according to error between room temperature and its setpoint[2].

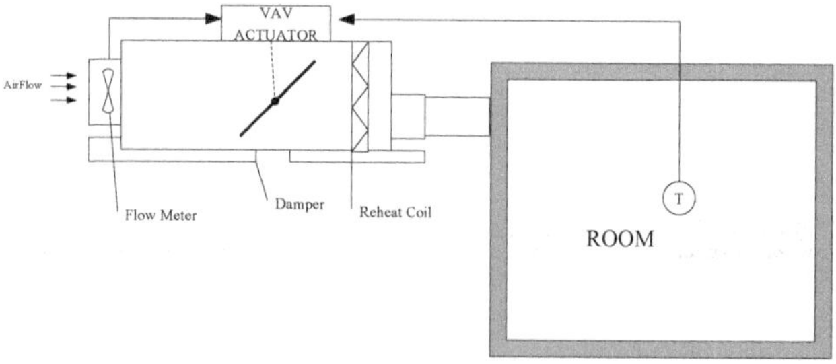

Fig. 2. Consititution of VAV damper

[1] This work is partially supported by key project of Shanghai Education Committee (J50502).

For VAV and CAV AC system, they need to get a lots useful infromation from VAV damper and CAV damper, such as actual air volume of CAV damper, damper angle. CAV damper have some input parameters such as actualstate(occupied or unoccupied)[4]. Fig 3 is a detail of input and output parameters of CAV damper.

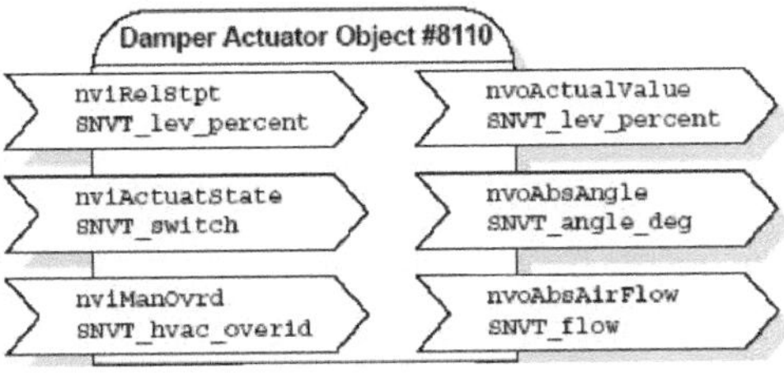

Fig. 3. CAV input and output parameter details

VAV damper has much more input and output parameters, such as damper air volume, damper angle, room temperature, temperature setpoint, air flow control point, damper mode(summer mode or winter mode), reheat coil valve state[4]. Fig 4 is the detail(some of them are optional).

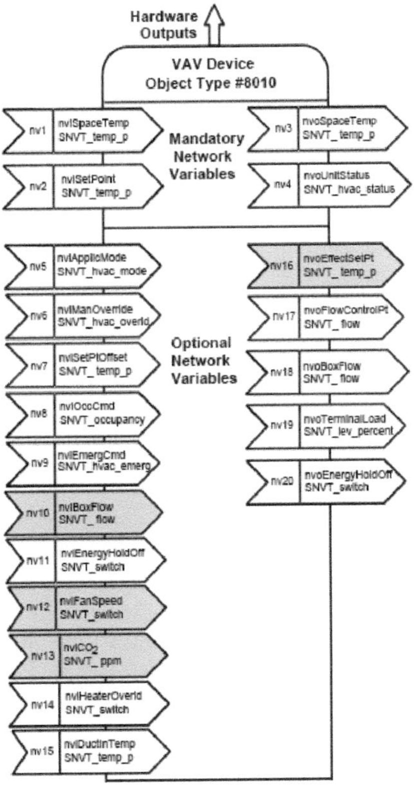

Fig. 4. VAV input and output parameter details

To transform parameters, every parameters need a couple of cable, and every parameters need an IO channel of DDC controller. If the control system of VAV AC system need all of the parameters, it cost too high. So we need a kind of commuication technology to tansform the parameter of VAV and CAV system with in a copule cable.

3 Lonworks Technology

LonWorks technology was developed by Echelonin 1990s and is regarded as an open standard in control network. Each LonWorks device consists of a Neuron Chip, a transceiver and the application electronics. The beauty of this technology is that the communication protocol, called LonTalk protocol, which is incorporated in the chip as a part of firmware[6]. As mentioned before, the protocol follows OSI model [5]. Together with the fact that the transceivers are standard, LonWorks devices of different frequency bands can be connected in a common network and are able to talk to each other. Each Neuron Chip reserves I/O hardware and processing power for an application-specific task is implemented using standard object interfaces. It retrieves raw data from field devices, executes a sequence of control actions and commands an actuator. The concept of an open system is always conserved and thus all LonWorks products are interoperable even from different vendors. To speed up the development time, Echelon provides several Neuron Chip control modules for developers. They are compact and contain all necessary components for LonWorks communication. One of these modules, TP/XF-78F, using Neuron Chip 3150 as its core central processing unit, is used in the prototype for the new monitoring system.

Neuron Chip is a system-on-a-chip with multiple microprocessors, random access memory (RAM), read-only memory (ROM), communication and I/O interface ports. ROM contains an operation system, the LonTalk communication protocol and an I/O function library. The chip has non-volatile RAM for configuration data and for the application program, both of which are downloaded over the communication network. From the control point of view, the Neuron Chip incorporates 11 I/O ports with different I/O firmware programmable functions. This leads to convenient direct interface with external hardware such as relays and A/D converters through the I/O ports. The local operation network technology is constituted by nodes performing sensing and control functions. The exchanged messages in these networks are of small size and they are optimized for the number of exchanged messages per second, thus, allowing real-time control.

The transceiver is an electronic module that provides the physical interface between the communications port of the Neuron Chip and physical medium, called a channel, which transports the digital communication packets to other devices. All devices connected to a specific channel must have compatible transceivers running at the same bit rate. Transceivers are available from Echelon and other manufacturers for a variety of media, including single twisted pair,power line, RF, infrared, fiber optics and coaxialcable. LonWorks systems may have multiple channels of the same or different types of media and channels, which are connected by LonWorks routers.The application electronics is basically the control circuit connected to the

Neuron Chip through the I/Opins such as the sensors, actuators, relays and A/D converters.

4 Ahu Controll of Cav/Vav Ac System

Beside the VAV/CAV damper, controlling system need to controll AHU of VAV/CAV AC system. Fig.5 is the schmatic of a typical AHU controlling system of VAV AC system. It controls supply air tempertrature and return air humidity with cold coil valve and heat coil valve, controls supply air pressure with VFD(Variable Frequency Driver)[7], monitors fan state (run/stop, fault, speed, local/romote), whole supply air volume. Fan state can also transformed use lonworks tecnology.

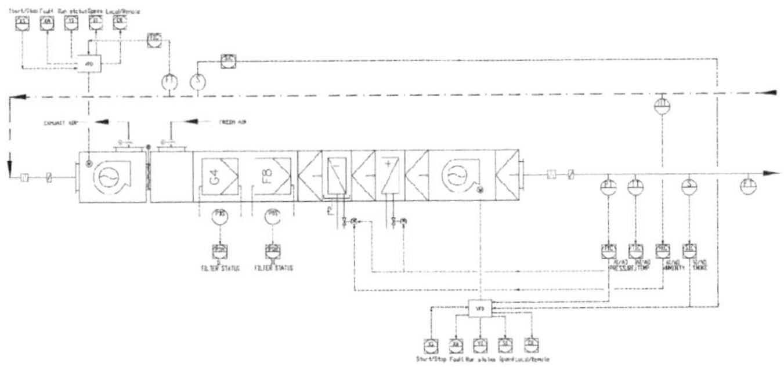

Fig. 5. The schmatic of AHU controlling system of VAV AC system

5 Network Topology of Scada System within Lonworks Technology

Fig.6. is the network topology of VAV/CAV control system with in LonWorks tecnology.Main component of the system is as Table 1.

Table 1. Main Component of The Control System

Component name	Component model
DDC	Siemens MEC-200
VAV ACTUATOR	Honeywell W7751H
CAV ACTUATOR	Belimo LMV-D2-MP
GATEWAY	Belimo UK24Lon
VFD	Danfoss VLT

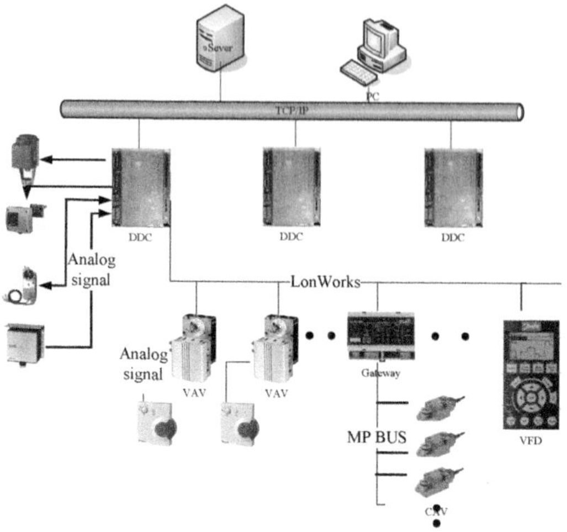

Fig. 6. Network topology of VAV/CAV control system

AHU parameter is transformed by sensor through analog signal. LonWorks net work connect CAV actuator ,VAV actuator,VFD and DDC controller.The parameters of VAV/CAV damper are transformed through LonWorks network.MP BUS is a communication bus from Belimo company,it can connect to Lonworks network with a gateway.DDC controllers and PC are connected with ethernet, they transformer infromation in TCP/IP protocal. The control system is really used in SIP SuZhou China ,and it performance good so far.

6 Conclusion

This paper introduces how to applicate LonWorks tecnology into VAV/CAV control system. What's the advantage to transfrom parameter between DDC controller and VAV/CAV dampers using LonWorks. The paper propose a solution of VAV/CAV control system.

Acknowledgment. The authors would like to thank the key project of Shanghai Education Committee (J50502) for the support of this research work.

References

1. ASHRAE, 1999 ASHRAE Handbook Applications, ch. 40, America Society of Heating, Refrigerating and Air-Conditioning Engineers Inc., Atlanta, GA (1999)
2. Wang, S.: Dynamic simulation of building VAVair-conditioning system and evaluation of EMCS on-linecontrol strategies. Building and Environment 34, 681–705 (1999)
3. Mitchell, B.W.: Distributed LonWorks control and monitoring system for environment and security. Applied Engineering in Agriculture 15(4), 345–350 (1999)

4. Lonworks Protocol. Echelon Corp., CA (1993)
5. Bijendran, N.J., Ashok, K.A.: Open Systems Interconnection. Elsevier, New York (1990)
6. Echelon, Neuron Chip Data Book, Echelon, USA (February 1995)
7. Yee, K.F., Chan Cary, W.H., Yik Francis, W.H., Chan Himson, Y.T.: A study on the energy saving of converting a VAV system from inlet guidevane to variable fan speed control. Transactions Hong Kong Institution of Engineers 9(2), 18–24 (2002)

Research and Development of Anycast Communication Service

Yuanyuan Zhang[1], Zhenwei Yu[1], and Ying Qian[2]

[1] Department of Mechanical Electronic & Information engineering
University of Mining and Technology(Beijing)
Beijing, China
[2] Network and Modern Education Technology Center
Normal University of Science & Technology
Qinhuangdao, Hebei Province, China
yuanyuan_z@sohu.com

Abstract. Anycast is one of three forms of service in the future. The definition, function and type of anycast service are introduced. Based on the analysis of previous research method and their shortcomings about application-layer anycast and network-layer anycast, it's proposed anycast service in a new architecture framework, and given the latest research progress of anycast in the new framework.

Keywords: Anycast, Anycast router, Network architecture, Overlay, Load balance.

1 Introduction

With more and more users sharing and querying information by the WWW, a popular site may occur blocking due to excessive access to the users. In order to reduce network load and simplify some kind of the network application, replicated servers are often deployed in the network with the same content to provide good service to users. A new standard communication mode facing next-generation networks----- anycast appeared. The anycast service has been proposed in RFC 1546.Anycast is defined in the next generation network (IPv6) addressing architecture as a special routing model which allows a sender to access at least one in a group which shares the same anycast address. Ideally, the packet is sent to the nearest one in the group, where "nearest" can be defined according to the routing metric used by the routing protocol. For example, in Figure 1, sender 1 and sender 2 are sending packets to the same anycast group. If hop count is the routing metric, the anycast routing protocol will deliver sender1's packet to server 1 and sender 2's packet to server 2, as they are the nearest servers, respectively.

D. Zeng (Ed.): ICAIC 2011, Part II, CCIS 225, pp. 212–217, 2011.

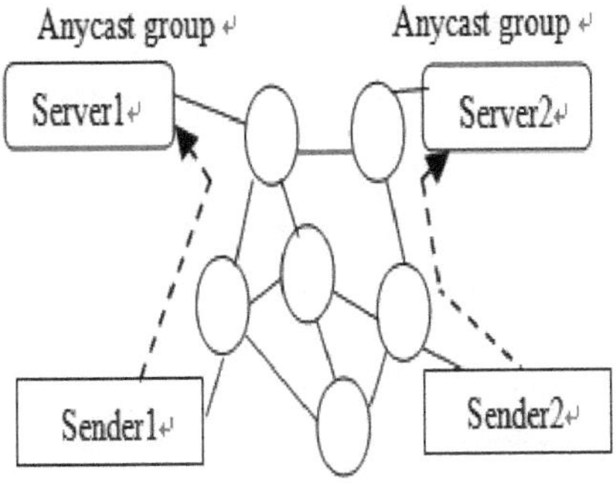

Fig. 1. An example of network-layer anycast

Anycast is a very useful service, and plays an important role in many applications. With the new network applications, new applications emerging, its demand is also growing. However, research on anycast has just started, there are still problems restricting the implementation of this service in many ways.

2 Research Content of Anycast

2.1 Scalability

Unicast routing protocols reduce the size of the routing table by prefix aggregation of the same destination address. Contrary to IP multicast, routes for IP anycast groups cannot be aggregated—the routing infrastructure must support one route per IP anycast group. In other words, route aggregation cannot be done for anycast addresses since different anycast addresses share the same network prefix may have different next hop. For example, if the router supports 10,000 distinct anycast groups, 10,000 additional routing entries will appear in the routing table.

2.2 Routing

Routing problem is the core of anycast service, as past research involved. From the point of Anycast service definition, it can be seen that the main function of this service is for users to make the right choice. As there is no difference between Ipv6 Anycast address space and the unicast address space, it's usually considered that it's not necessary to change to change the unicast routing protocol. However, multiple hosts share one anycast address, so there must be some kind of new protocols for anycast.

2.3 Stateless Connection Service

The stateless connection service means that anycast packets are delivered to the nearest server, but routing process does not depend on previous packets, Due to the stateless nature of the IP protocol, subsequent packets from the same source node targeted at the same anycast group may arrive at different group members.

Consequently, native IP anycast cannot be applied directly to support distributed session-based services.

2.4 Anycast in the New Framework

Previous anycast can not solve these problems above fundamentally. Thus, with the new network system architecture appearing in recent years, many scholars began to research a new architecture to solve the problem in anycast service. I3 (Internet Indirect infrastructure) delivers the packet not directly to the destination node, but its essence is to build a Internet overlay network, it needs the support of peer-to-peer network routing system accordingly.

3 The Realization in the New Framework

3.1 Two Types of Anycast

Currently anycast communication has two main research directions: one on the network layer using anycast router to improve network efficiency and robustness of communications; the other is on application layer by managing means to achieve the anycast service, optimize network performance. Network layer anycast implementation is simple, but need to modify the router configuration and lack of flexibility in the selection criteria; application layer anycast does not understand the network topology changes and load conditions and can not determine access to the server congestion based on the distance, which is fatal for quality of network service. Anycast previous study was according to these two forms of anycast service respectively, but still can not fundamentally solve about two questions: First, session-based applications can not use the advantage of this anycast service. Secondly, anycast caused the poor scalability and IP anycast can not be aggregated.

3.2 Overlay Anycast Network Structure

Overlay Network technology is a method to structure network, with a specific technology, independent of a particular level. It can construct a virtual network in accordance with some kind of need on the basis of the existing physical network, in order to acquire new functions which is not or difficult available in original network, and ensure maximum compatibility with the existing network. Without the physical structure, overlay network only need the logical integration to provide the necessary services, and solve the current network system structure problems more flexibly. Therefore, the article combines this point of view with the anycast address. anycast service structure will be divided into three layers, as shown in Figure 2.

The main function of each layer is defined as follows:

1) Underlying physical network layer: it is the basis of construction of anycast overlay layer, it provides basic physical link and transmission by existing TCP/IP protocol and network communications and the necessary communication node to the upper (Anycast overlay layer) layer covers all the basic physics of the original seven network layer network architecture.

2) Anycast overlay (Overlay Network): It is the core layer of anycast network, a virtual layer built on the underlying (basic physical network layer). It maintains the network topology, routing algorithm to generate a node at the routing table, and provides routing function for its upper (the user application layer) layer.

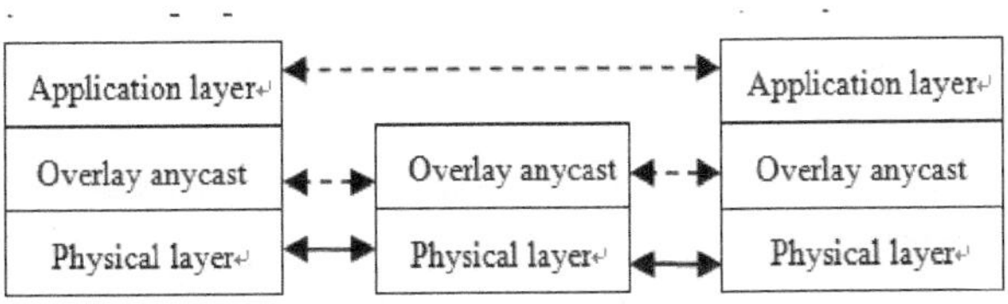

Fig. 2. Abstract model of overlay anycast service

3) The anycast application layer: It is responsible for providing service and applications interface for the end user.

It is the interface between end users and anycast network and realizes anycast service users need.

3.3 Related Research in the Overlay Framework

Reference [1] to overcome the limitations of IP anycast network layer, it combined advantages of application layer anycast and proposed a new architecture PIAS (Proxy IP Anycast Service) based on overlay proxy.The basic idea of PIAS is to solve the scalability problem of network-layer anycast by deploying an overlay of proxies. Specially, a large number of anycast proxies are deployed around the Internet and the anycast address information are logged at the proxies. Routers are not responsible for delivering anycast packets. Rather, anycast packets are delivered to the nearest anycast proxy which in turn delivers the packet directly to an AS using unicast. Although the burden of routers has been alleviated, deploying a large number of proxies may take a while.

In order to solve the scalability problem, Reference [5] proposed the concept "routing group".An anycast address is mapped to a routing group using SHA-1as a based hash function in our routing framework. When a router receives an anycast packet, it first uses the hash function to determine which group it belongs to. Anycast routers of the same routing group will self-organize into an overlay network. By dividing anycast routers and addresses into $\sqrt{N}$ routing groups, the size of routing table for anycast can be bounded by O($\sqrt{N}$), where N denoted the number of anycast groups. But it divided the router group logically and do not fundamentally solve the scalability problem.

Reference [3] proposes a location-aware and load-aware anycast architecture--LALA based on P2P network. It created an anycast forwarding according to the algorithm Scribe, when the client node sends a anycast request, it makes the request forwards to the first high-level tree note, that is, the parent node, if found the parent node overload, it automatically adjusts the child nodes, expands the scope of the child nodes, absorb more anycast servers, and ultimately find the optimal node.

Based on the PIAS architecture, Reference [9] improved the server proxy deployment through new deployment options, which expands the range of anycast service while PIAS is fit for a small range of network anycast service. PIAS address the optimal placement and dimensioning of such an anycast architecture using an integer linear program (ILP). Unfortunately, due to the complexity of the formulation, an exact solution can only be computed for relatively small networks. The article proposes a near-optimal heuristic approach to tackle anycast infrastructure dimensioning and proxy placement in large networks. It is proved that the new method is better than the ILP deployment.

Reference [15] proposed an overlay anycst service model based on direct probing--POA. POA maintained a neighbor ring, which including k members, for each nod. When anycast request arrived, POA probed the best server according to neighbor information. POA' probing was client-oriented and iterative, compared with Meridian, it can cut down the query latency from 5t2log(R) to 3t2log(R), and accelerate the constringency of neighbor selection.

Reference [16] present DOAT (Distributed Overlay Anycast Table), a delay-aware, application layer anycast system designed to return accurate results in short time.

To estimate network delay distance, DOAT nodes use a network coordinate system to find their location on a multi-dimensional (usually two or three dimensional) delay space. This space is then mapped into a single-dimensional DOAT coordinate using a space filling curve, in order to simplify searching operations. This coordinate is used to construct a locality-aware overlay. Scalability is addressed by aggregating group memberships using Bloom filters and query resolution is accelerated using efficient routing.

4 Conclusion

Anycast service is one of the three networks service in the future, it plays an important role in replication server, load balance, multicast communication and video on demand. Although anycast communication for the latest research direction is in the new architecture to achieve, the current study does not solve two problems inherent in anycast fundamentally: scalable and session-based service. The development of research in the future is how to construct anycast architecture combining the flexibility of the application layer anycast and efficiency of the network layer anycast and how to achieve a reasonable deployment of anycast nodes Research is still difficult and hot, and anycast still faces both bright future and great challenges.

References

1. Ballani, H., Francis, P.: Towards a Deployable IP Anycast Service. In: Proceedings of ACM SIGCOMM (2005)
2. Charlie Hu, Y., Zhang, R.: Anycast in Locality-Aware Peer-to-Peer Overlay Networks (2006)
3. Freedman, M., Lakshminarayanan, K., Maziéres, D.: OASIS: Anycast for Any Service. In: Proceedings of the 3rd Symposiumo Networked Systems Design and Implementation (NSDI 2006), SanJose, California, United States (2006)

4. Wu, C.-J., Hwang, R.-H., Ho, J.-M.: A Scalable Overlay Framework for Internet Anycasting Service. In: The 22nd Annual ACM Symposium on Applied Computing, SAC 2007 (2007)
5. Nandi, A.: LALA:Location Aware Load Aware Overlay anycast. Master thesis of Rice university 8, 19–22 (2004)
6. Stevens, T., Vermeir, J., De Leenheer, M., Develder, C., DeTurck, F., Dhoedt, B., Demeester, P.: Distributed Service Provisioning Using Stateful Anycast Communications. In: Proc. 32nd IEEE Conference on Local Computer Networks (LCN), Dublin, Ireland (2007)
7. Lin, C.-Y., Lo, J.-H., Kuo, S.-Y.: Load-Balanced Anycast Routing. In: IEEE ICPADS (2004)
8. Tsui, K.C., Liu, J., Shi, J.: Anycast-Based Cooperative Proxy Caching: Preliminary Results. In: IEEE/WIC/ACM WI (2004)
9. Ballani, H., Francis, P.: Understanding IP Anycast. In Cornell CIS Technical Report TR2006-2028 (2006)
10. Dr. George Riley. BGP-Anycast Simulations. In: CAIDA Project (2007)
11. Stevens, T., De Leenheer, M., Develder, C., De Turck, F., Dhoedt, B., Demeester, P.: ASTAS: Architecture for Scalable and Transparent Anycast Services. Journal of Communications and Networks 9(4), 457–465 (2007)
12. Ma, z., Chen, c.: An overlay anycast service model based on direct probing. Journal of Shanxi University of Science & Technology 27(5) (2009)
13. Mykoniati, E., Latif, L., et al.: Distributed Overlay Anycast Tables using Space Filling Curves. In: Proceedings of the 28th IEEE International Conference on Computer Communications Workshops (2009)

Constellation Design and Performance Simulation of LEO Satellite Communication System

Jun Liang, Nan Xiao, and Jiwei Zhang

Telecommunication Engineering Institute
Air Force Engineering University
Xi'an, China
gzxl135@yahoo.com.cn

Abstract. LEO satellite communication weighs more and more in communication system now and future. Since different types of constellations have rather different performances in coverage ability and inter-satellite link (ISL) building conditions, a polar and an inclined circular low orbit constellation were designed in this work. Coverage ability and ISL performances of constellations were simulated by STK while delay, throughput and jitter of satellite network were simulated by NS2. The results of simulations were compared and analyzed which indicates that, in a constellation with 32 satellites (orbit altitude 1450km), polar orbit constellation is more suitable to build global LEO satellite communication system than inclined circular orbit because of its nicer performance in global coverage, ISL building conditions and network identity.

Keywords: words-Constellation design, Inter-Satellite Link, Coverage analysis, Performance Simulation.

1 Introduction

Satellite communication has earned great attentions in most countries all over the world, especially in military regions, because of its particular advantages such as larger coverage ranges, higher mobility, greater transmission capacity and more flexible network organizing modes [1]. Powerful satellite communication ability is not only needed by national economy and military safety, but also represents the general national power of a country [2]. As a great country, China has crucial effects on international politics, military affairs and global economy, however, the satellite communication ability of China is still weak and needs improvement urgently because developing global LEO satellite communication system has very important and especial meanings for China.

GEO satellite requires highly in EIRP of end users and G/T value of ground receivers because of its long transmission path and high signal attenuation, which result

D. Zeng (Ed.): ICAIC 2011, Part II, CCIS 225, pp. 218–227, 2011.

in difficulties of communication through handsets, real-time transmissions of messages and twi-jump communications between mobile stations. However, LEO satellite cuts down the requirements of EIRP and G/T value for its lower orbit altitude and less transmission loss, which enables it to support small ground stations communicating with satellite directly. At the same time, short transmission delay allows twi-jump communication between mobile stations without data exchange processing on satellites [3]. As a result, LEO satellite communication system has been regarded widely all over the world and develops fleetly, such as Iridium and Teledesic.

A constellation design which could realize global coverage was brought forward but the effects of constellation on ISL building was overlooked in paper [4]; the changing of ISL and satellite reverse movements was analyzed while the network performance was not mentioned and studied in paper [5]. In this work, a polar and an inclined circular orbit constellations were designed to make comparison between different constellations and the ISL and satellite network performance parameters were simulated by STK and NS2 which proves the advantages of polar orbit constellation in LEO global satellite communication system constructing.

2 Constellation Design

Constellation design is one of the basic questions in satellite communication system foundation, which affects the cost of system, the ISL performance, the service range and the communication quality etc. It usually uses retrograde orbit for easily control of satellite orbit [6] and the relation between satellite and Earth rotation period

$$T_s / T_e = k / n \tag{1}$$

where k, n are integer, T_s is period of satellite, T_e is period of Earth rotation and T_e=86164s. The altitude of satellite h can be derived from the Keplerian law

$$T_s = 2\pi\sqrt{\frac{(R_e + h)^3}{\mu}} \tag{2}$$

where the Earth radius R_e=6378.137km, μ is Kepler's constant. When k=2, n=25, T_s and h can be computed from equation (1) that T_s=6893s, h=1450km. These would be the assumed satellite orbit parameters in this work.

2.1 Polar Constellation Design

The design of polar constellation is mainly based on coverage band idea. The coverage performance between contiguous orbits has big difference between direct and converse orbits. (see Fig.1.)

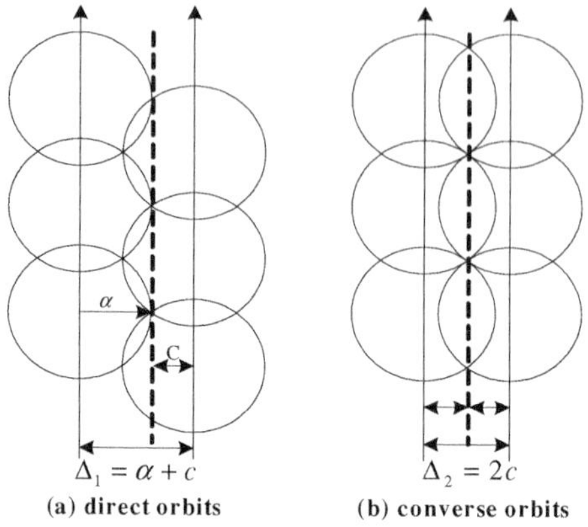

Fig. 1. Geometry Relations of Coverage Between Contiguous Orbits

$$c = \arccos\left[\frac{\cos\alpha}{\cos(\pi / S)}\right] \tag{3}$$

where c is half-width of satellite coverage band, S is number of satellites in one orbit. Δ_1 and Δ_2 are longitude difference of direct and converse orbits.

As satellites in separated orbits has the same phase which leads that satellites may crash in pole, so near-polar orbit is used (orbit obliquity between 80°~100°) in polar constellation design. Near-polar constellation still can be designed based on coverage band idea because the orbit obliquity is still close to 90°, but the longitude difference between direct and converse orbits should be adjusted as

$$\Delta_1' = \arcsin(\sin\Delta_1 / \sin i) \tag{4}$$

$$\Delta_2' = \arccos\left(\frac{\cos\Delta_2 - \cos^2 i}{\sin^2 i}\right) \tag{5}$$

In order to realize global coverage, the constellation should accords with

$$(n-1)\Delta_1' + \Delta_2' = \pi \tag{6}$$

where n is numbers of orbits. It can be computed from (5), (6) and (7) that

$$(n-1)\cdot\arcsin\left(\frac{\sin(a+c)}{\sin i}\right) + \arccos\left(\frac{\cos 2c - \cos^2 i}{\sin^2 i}\right) = \pi \tag{7}$$

The orbit obliquity is set as i=86.4°, the minimum ground elevation E=10°. Numbers of satellites with different orbits in global coverage can be computed from equation (3), (6), and (7) with Matlab (see Table 1).

Table 1. Satellite Numbers With Different Orbits

Scheme	n	Satellite Sums	Δ_1'	Δ_2'	Global Coverage Rate
A	3	63	60.34°	59.21°	100%
B	4	32	48.34°	35.19°	100%
C	5	35	40.36°	19.24°	100%
D	6	36	34.91°	8.34°	100%

Table 1. shows that it needs at least 32 satellites to realize global coverage with the given elevation and orbit altitude. Theoretically speaking, the numbers of satellites in one orbit needed in scheme B is $S=\pi/c=7.0264$, but actually, $S=8$, so there exists great coverage margin in system. The longitude differences of direct and converse orbits should be $\Delta_1'=47.2°$ and $\Delta_2'=38.4°$ which had been optimized by STK.

2.2 Inclined Circular Orbit Constellation Design

In order to make comparison of coverage performance between polar and inclined circular orbit constellations, we assumed that there were also 32 satellites in inclined constellation with 4 orbit planes. The optimal phase of satellites in contiguous orbits can be confirmed by STK and the result is $F=1$. Simulations progressed with different orbit obliquity in order to select the best orbit obliquity (see Table 2.).

Table 2. Coverage Statistics Of Constellations With Different Obliquity

	Scheme A	Scheme B	Scheme C	Scheme D
Altitude(km)	1450	1450	1450	1450
Satellite Sums	32	32	32	32
Orbit Planes	4	4	4	4
Orbit Obliquity	38	40	43	45
Average Coverage Rate (40°S~40°N)	99.86%	99.41%	98.60%	97.74%
Average Coverage Rate (50°S~50°N)	99.85%	99.38%	98.54%	98.07%
Average Coverage Rate (60°S~60°N)	99.46%	99.89%	99.01%	98.08%
Global Average Coverage Rate	87.12%	88.27%	89.67%	91.25%

Table 2 shows that scheme A has the optimal coverage performance, so the best orbit obliquity should be $i=38°$ and the Walker ID is i: $T/P/F=38°:32/4/1$.

The relation between global coverage rate and latitude of different constellations has been shown in Fig.2.

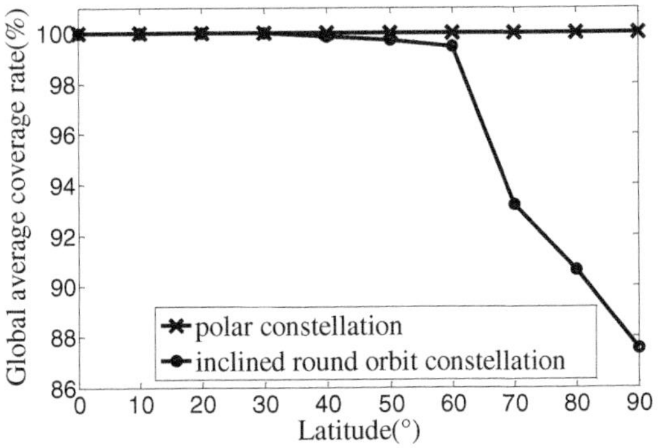

Fig. 2. Relations between Latitude and Global Coverage

The coverage rate toboggans with latitude rising in inclined circular orbit. The Walker constellation is not able to cover North and South Pole areas because of constrains of orbit obliquity which results in that it is more suitable to complete coverage in middle and low latitude zones. However, polar constellation dose not exit such problem because the satellites congregate in pole which enables the constellation could realize real none-gap global coverage.

3 Performance Simulations

Inter-Satellite Link, ISL, enables satellite communication system could be independent on ground communication network and supports communication between two or more users in any places on the earth (including ocean and low space) [7]. Azimuth, Elevation and Range are important parameters in ISL building conditions and link quality calculations; delay is a key factor to evaluate the satellite network performance; throughput and is significant criterion in multiple access performance evaluation; jitter reflects the stability of satellite networks [8].

In order to study the communication performance of polar and inclined circular orbit constellations designed above, AER of ISL, delay, jitter and throughput of satellite networks were simulated by STK and NS2.

3.1 Simulate Configuration

Simulation parameters in this work configured in NS2 were as follows: Beijing (39.9°N, 116.4°E) and Washington (38.91°N, 77.02°W) were selected to be ground terminals as data source, interval of data packets was set to be 30s, simulation period was 24h (86400s), latitude_threshold=75°, ground elevation E=8.2°, bandwidth of ISL bw_isl=25Mb, bandwidth of uplink bw_up=1.5Mb, bandwidth of downlink bw_down=1.5Mb, length of interface queue qlim=50. X label represents packet ID or time and Y label represents delay, jitter or throughputs of network in following figures.

3.2 Performance Simulations of Polar Orbit Constellation

Inter-satellite link changes between direct and converse orbits in polar constellation are very different; both AER performances of two ISL types in the constellation were simulated by STK with sampling interval 60s and results shown in Fig.3. The NS2 was used to validate the network performances; delay of network was shown in Fig.4 and throughput in Fig.5, jitter in Fig.6.

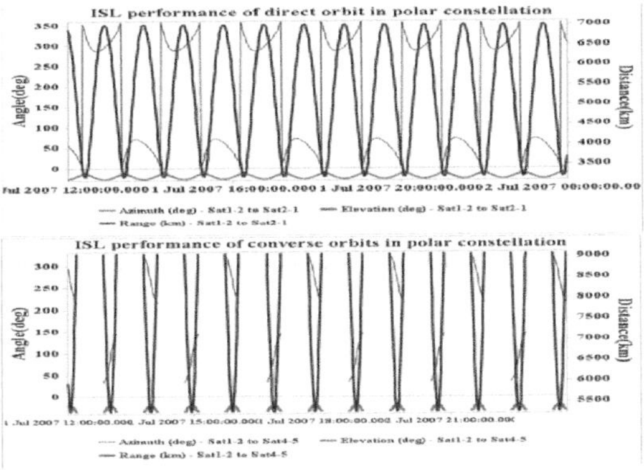

Fig. 3. AER Performance of Direct and Converse Orbits

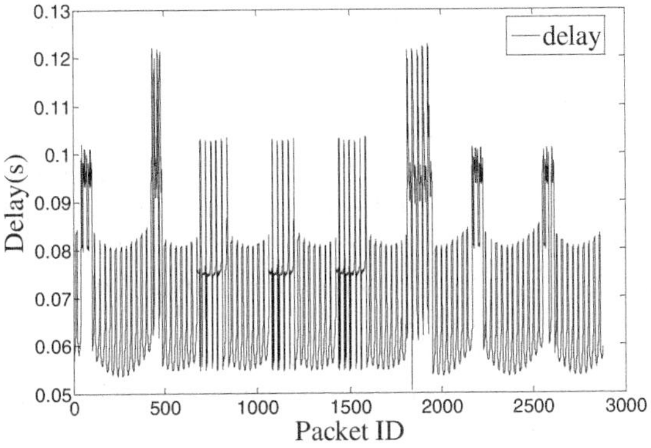

Fig. 4. Delay of Transmissions in Polar Orbit Constellation

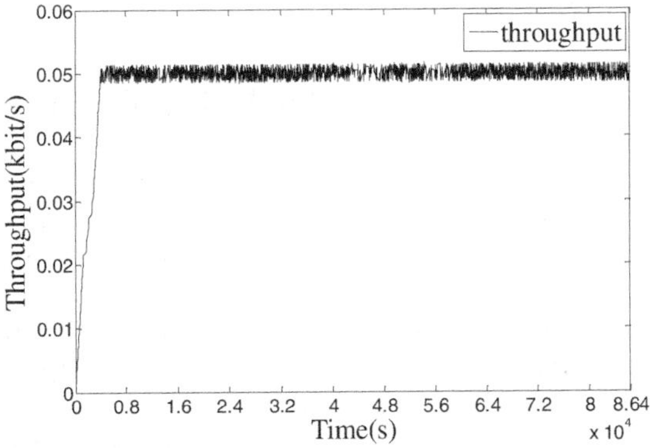

Fig. 5. Throughputs in Polar Orbit Constellation

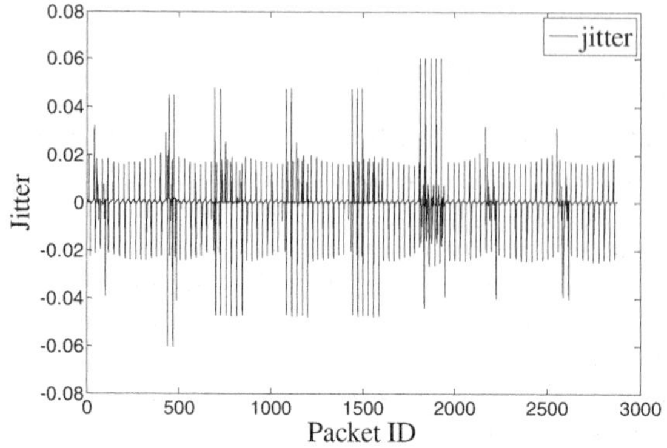

Fig. 6. Jitter in Polar Orbit Constellation

Fig.3 shows that the range of ISL length is 3125~6945km, range of elevation is -28°~12° and max difference of azimuth is 60° in direct orbit, which indicates rather few difficulties in ISL building. AER parameters in converse orbit change very rapidly which bring huge troubles in ISL building such as high requirements in stability of satellite, capturing and tracing ability of antenna on satellite etc.Fig.4~6 shows the maximum and minimum delay of satellite network are 0.125s and 0.055s, average delay is 0.07s; the throughput of network can be reach steady level quickly and keep stability at 0.05kbit/s; the range of jitter is -0.06~0.06 by center level 0, mainly between -0.02~0.02.

3.3 Performance Simulations of Inclined Circular Orbit Constellation

The AER performances of ISL simulated in inclined circular orbit constellation was chosen from Sat-1 with all contiguous satellites around it which was shown as Fig.6. The delay, throughput identity and jitter of network were simulated by NS2 and results were shown as follows.

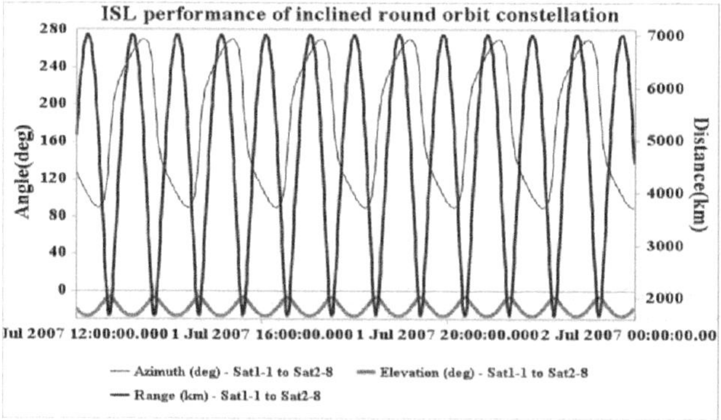

Fig. 7. AER Identity of Oblique Circular Orbit

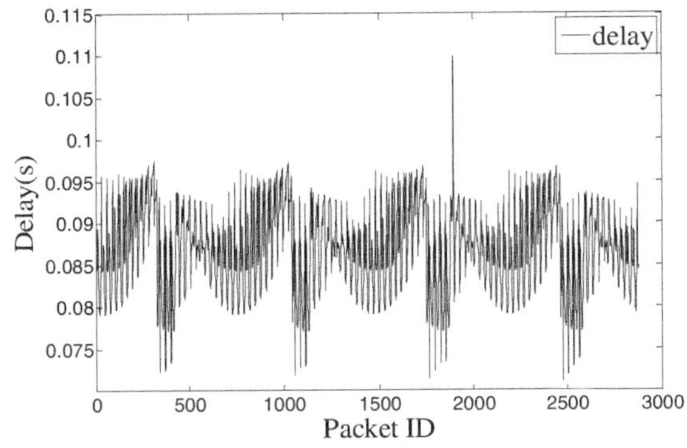

Fig. 8. Delay of Transmissions in Inclined Circular Orbit Constellation

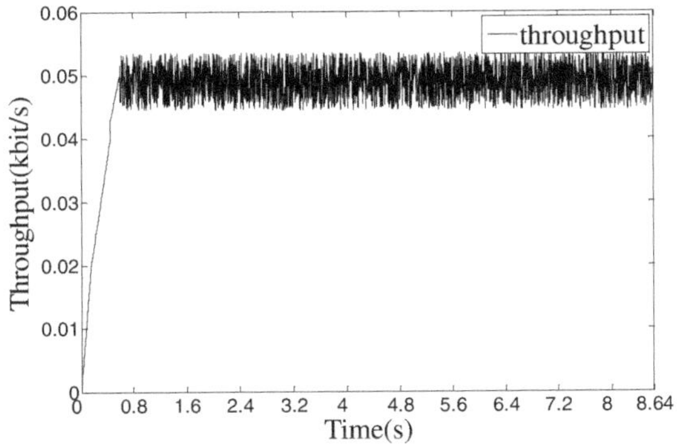

Fig. 9. Throughputs in Inclined Circular Orbit Constellation

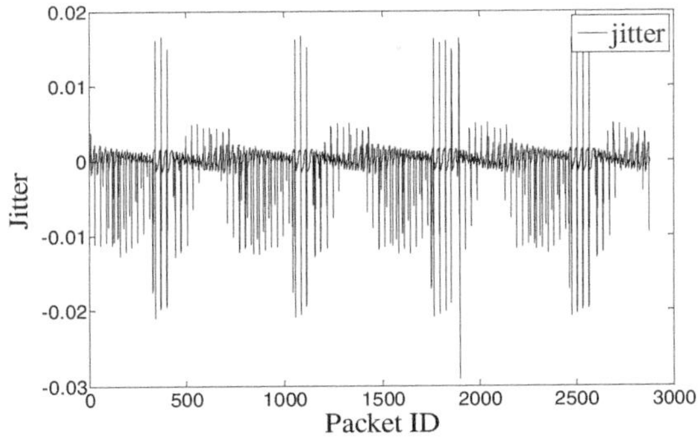

Fig. 10. Jitter in Inclined Circular Orbit Constellation

As shown in Fig.5, ISL in inclined circular orbit constellation could exit eternally and the range of ISL length is 1496~8480km, range of elevation is 78°~282° and range of azimuth is -33°~-6°. The fast changing of elevation leads high requirements of tracing ability of satellite antenna and stability technique of satellite stance, on the other hand, the big range of ISL length makes more difficulties in antenna design and lowers the quality of communication. From Fig.8~10 we can see that the maximum and minimum delay of satellite network are 0.112s and 0.072s, average delay is 0.085s, which is longer than polar orbit constellation; the throughput of network can be reach steady level in a short period and keep stability at 0.05kbit/s, but the scope of changing is greater; the range of jitter is -0.03~0.018 by center level 0, mainly between -0.012~0.003.

Synthetically considering, the polar orbit constellation is more suitable to build global LEO satellite communication system with ISL.

A design named orthogonal circular orbit constellation, was brought forward in paper [9]. With the same satellite sums and orbit plane numbers to realize global continuous coverage, orthogonal circular orbit constellation needs more satellites than polar orbit constellation, and furthermore, relative movements between contiguous satellites are so complex that it could hardly build global ISL network.

4 Conclusions

The global real-time communication ability is effectually to extend communication radius, improve remote operations ability of army and enhance air and missile defense power. The LEO constellation has vast advantages in real-time communication and anti-failure ability for its low orbit altitude and flexible networking ways, which makes it more propitious to build the global LEO satellite real-time communication system. However, the polar orbit constellation needs too many satellites and the cost of system will be high and the investment risk will be great if only depends on army, on the other hand, the whole resource of system used by military is very limited which will cause much waste, referring to the Iridium system operated by the US, the LEO satellite communication system of China should be constructed in the way of army-civilian cooperation in order to decrease the cost and risk of investments but enhance the military satellite communication capacity of National Defense.

References

1. Zahariadis, T.B., Vaxevanakis, K.G., Tsantilas, C.P., et al.: Global roaming in next-generation networks. IEEE Communications Magazine 40(2), 145–151 (2002)
2. Liu, G.: Research on the Key Technologies of Networking in the NGSO Satellite Mobile Communication Systems. Ph D Thesis. University of Electronic Science and Technology of China, Chengdu (2004)
3. Li, W.-h., Yuan, J.-p., Zhao, Y.-s.: Study on the Optimization of the LEO Small Satellite Constellation. Flight Dynamics 19(4), 41–42 (2001)
4. Jiang, H., Yin, Z.-s.: Gong Wen-bin etc, Brief Introduction of Constellations for Global Satellite Communications. GNSS World of Chin 40(3), 42–43 (2007)

5. Geng, L., Wu, S.-q.: Constellations Design and the Establishment of the Inter-Satellite Links. Chinese Space Science and Technology 27(6), 62–64 (2002)
6. Wang, Z.-y.: Architecture Design and Analysis of Multi-layer Satellite Networks, Ph D Thesis. Harbin Institute of Technology, Harbin (2006)
7. Marzano, F.S., Montopoli, M., Rossi, T.: Flower Constellation of Millimeter-Wave Radiometers for Tropospheric Monitoring at Pseudogeostationary Scale. IEEE Transaction on Geoscience and Remote Sensing 47(9), 3107–3108 (2009)
8. Zhang, Y.-m., Chen, S.-w., Xia, L.-y.: Research of the Wireless Network Simulation Based on NS2. Communications Technologies 59(3), 58–60 (2007)
9. Wu, T.-y.: Research on the Key Technologies of Networking in the NGSO Satellite Mobile Communication Systems. Ph D Thesis. University of Electronic Science and Technology of China, Chengdu (2008)

Study on the Current Situation of Soil Lead Pollution and Health Risk in Harbin

Peng Zhang, Xueping Zhang[1,*], and Lei Wang[2]

[1] School of Geography Science, Harbin Normal University
Key Laboratory of Remote Sensing Monitoring of Geographic Environment, College of Heilongjiang Province
Harbin, Heilongjiang Province, China
[2] School of Architecture and Civil Engineering
Heilongjiang Institute of Science and Technology
Harbin, Heilongjiang Province, China
zhangpeng_9988@126.com

Abstract. Soil health risk assessment is used to formulate soil environmental quality standard, and it is a new environment management method. Case study in urban center of Harbin, rural-urban fringe zone,primary toll gate in Songbei high way and Harbin Aircraft Manufacturing Company, full-scale study of Harbin soil lead pollution situation was made by the method the literature material and field survey, we calculated the lead exposure for the children from the soil in Harbin using biological dynamic model of MARCUS, and made the risk assessment of MMR incidence. The results showed that there was different of the levels of lead pollution in different function zones of Harbin, the both sides of road > around of factory > the banks of Songhua River > suburban farmland > residential areas. The content of lead obviously accumulated in 20 meters on both sides of the road, and the content of lead in industrial zone was of great difference for human factors, without obvious rule. Furthermore, the lead exposure of children aged from 0 to 6 from the soil in Harbin was between 7.96x10-5 and 1.98x10-4mg/d, and the health risk of MMR was 20.307 per 1,000.

Keywords: soil lead pollution, blood lead, health risk assessment, Harbin.

1 Introduction

Lead pollution of urban soil may cause lead poisoning in human body, and brings potential health hazard. Urban lead pollution was investigated in many cities of China. In 1991, the condition of lead pollution in 21 large cities was investigated and cluster analysis of lead pollution level was made by Changchun Geological Institute, Chinese academy of sciences [1]. In 2003, the survey made by maternity and child care centers in Haikou found that amount of blood lead of 53.8% infants and 56.1% students was more than 100μg/L [2], GuoPing [3] found the situation of soil lead

* Corresponding author.

D. Zeng (Ed.): ICAIC 2011, Part II, CCIS 225, pp. 228–235, 2011.

pollution in Changchun city was serious, RenHuiMin found soil lead pollution was serious on the highway in Shenyang three-ring [4]. Environmental health risk assessment took risk degree as evaluation index, connected the environmental pollution and human health, and made a quantitative description of pollution on human health hazard risk [5], the results of health risk assessment was used to formulate environmental standards and regulations, in order to protect the public from the effect of contaminants. The jobs of health risk assessment were interiorly done one after another, and the surveys about health risk assessment of heavy metal pollution in water, vegetable field, and polluted sites were in the majority, but only study of health risk assessment based on soil lead pollution of the city was still at the primary stage. Soil lead pollution situation of Harbin city was comprehensively discussed with the method of experimental research and literature, and health risk assessment of the children of soil lead exposure was made, which provided the scientific basis to lead pollution control and management in Harbin.

2 Data and Methods

2.1 Data and Pre-processing

In this study, 25 soil samples were collected in rural-urban fringe zone in Harbin, and the sampling points were chosen within the Third Ring Road. Soil was captured from plow layer, where was more than 30 meters away from the road, and depth was 0-20cm, then soil samples were taken back to the laboratory after packaged with polyethylene plastic bag and been numbered. After dried by nature, soil samples were filtered with 100 mesh nylon sieve, then they were digested and constant volume with the method of HF-HNO3-HCLO4, at last, the content of Pb element was determined by atomic absorption spectrometer (TAS-986, Beijing Persee Corporation).

In order to study the distribution regularities of lead pollution near the highway, sampling points were choosen near Songbei highway. Speed of motor vehicles was low in this section, and southwest wind dominated all year round. Samples could not be taken at the west of toll gate, as the land had been exploited. Samples were taken from eight lands planted corn at the east of Songbei highway, far away from the road 20 meters, 30 meters, 50 meters, 75 meters, 100 meters, 130 meters, 170 meters and 200 meters. Sampling and experimental methods were as above.

The research of lead pollution in industrial districts took Harbin Aircraft Manufacturer as an example, and sample of soil was taken from 22 places near the district. Sampling and experimental methods was as above.

2.2 Method

1) Exposure assessment.In Biological dynamics model of Marcus was adopted in this paper to calculate the exposure lead content intaken by Harbin children through hand - mouth way. Formula is as follows,

$$\boldsymbol{U}_{soil}=\boldsymbol{C}_{soil}\cdot\boldsymbol{Q}_{soil}\cdot\boldsymbol{A}_{soil} \tag{1}$$

In formula, $\boldsymbol{U}_{soil}$ is the exposure of medium i, $\boldsymbol{C}_{soil}$ is tested lead concentration in study area, $\boldsymbol{Q}_{soili}$ is the amount of soil that the children intake, and $\boldsymbol{A}_{soil}$ is children absorption factor from soil lead. This study adopted the children exposed parameters recommended by American EPA. $\boldsymbol{Q}_{soil}$ calculated by RenHuiMin, equaling to 0.093g/d, which is acceptable daily intake of children from soil. $\boldsymbol{A}_{soil}$ is lead absorption factor of children from soil, equaling to 0.030 [6].

2) Risk calculation-calculation of the risk of MMR.The formation of In method guide of disease burden caused by lead, WHO put forward that the level of intelligence quotient (IQ) and body blood lead levels were negatively correlative, and the level of IQ decreased with the rise of blood lead value. 1.3 points would decrease in IQ caused by per 5μg/dl rise of blood lead, within the range of 5~ 20μg/dl; As the blood lead was more than 20μg/dl, the level of IQ maintained at 3.5 points [7].

The value of IQ between 50 to 70 points is defined as mild mental retardation (MMR). According to the method of calculation given by WHO, with 5 μg/dl for interval, the formula calculating of MMR incidence caused by blood lead is as follows:

$$I(MMR)=AR\times[(R_1\times P_1)+(R_2\times P_2)+(R_3\times P_3)+(R_4\times P_4)] \tag{2}$$

Among them, AR is correction ratio. In China region the value of AR equals to 3.03; R_1, R_2, R_3, R_4 representes the proportion of children whose blood lead levels is in four interval, 5~ 10μg/dl, 10~ 15μg/dl, 15~ 20μg/dl and 20μg/dl above; p_1, p_2, p_3, p_4 representes the proportion of people, whose IQ level is in 70~ 70.65, 70~ 71.95, 70~ 73.25 and 70~ 73.5, respectively, equaling to 0.0024 ,0.008, 0.0145 and 0.00159 [8].

3 Result and Analysis

3.1 The Situation of Soil Lead Pollution in Harbin

The urban environment pollution of lead was mainly derived from painting, mining, smelting, casting and other industrial production activities and explosion-proof fuel of automobile. From a global perspective, automobile was the most serious lead pollution source. Harbin, being one of the important old industrial basis in China, whose industrial types include aircraft manufacturing, mechanical manufacturing, textile, medicine, metallurgy, chemical industry. Harbin is also the traffic hub in Heilongjiang province, as the national and local auto preferential policies come on, the consumption of family cars of residents enters into a period of rapid growth, and the automobile exhaust emission also brings residents health hazard.

1) The situation and distribution characteristics of urban soil lead pollution.The content of lead in land surface in Harbin was as follows, the content of lead in topsoil ranged in 4.67~ 176.11 mg/kg, and the average value was 48.11 mg/kg, being 1.10 times more than the soil background value of Harbin, being lower than 54.18 mg/kg, which was the average lead content of soil in Changchun [9].

In the five different functional areas, in both sides of highways and around the factories, where the content of lead was much more than that the background value of lead content, and in the two functional areas the average lead content was 2.5 times more than the background value of lead content in Harbin, close to 3 times more than the background value on both sides of road, being the zone where the content of lead was the most. The average value of lead content in both sides of Songhua River and farmland areas was in the medium level, being about 1.5 times more than the background value, while the average value of lead content in residential areas was at least level, being less than 1.5 times of background value. Clearly, industrial emissions and the urban traffic emissions from vehicle exhaust were the main sources of lead pollution. Sampling sites along the two sides of Songhua River have long been exploited, and they undertook important functions, such as the travel, transportation hubs, and the market, where it was crowded and traffic was heavy. Soil samples of suburban farmland were also taken from both sides of suburban road mainly, where the passenger cars, trucks, farm vehicles must be passed through. It was the most important reason that lead pollution level in these two functional areas was higher than others. Most urban residential areas were enclosed by buildings, forming semi-closed environment, external source of lead pollution was relatively hard to enter, so the lowest level of lead pollution was in this zone.

Table 1. Statictics of characteristics value of pb in farmland soil of rural-urban fringe zone in haerbin

Functional zone	Residential areas*	Factories around*	Two sides of road*	Banks of Songhua River*	Suburb farmland
Scope	4.67-68.45	26.35-108.25	26.64-176.11	20.47-78.32	13.93-101.49
Mean	28.56	64.25	70.95	39.29	37.51
I	1.12	2.52	2.78	1.54	1.47

Note: * quoted by XiaoFeng Ma [9]; I is the ratio of the average of lead content in soil and background values(lead background value in Harbin is 25.5).

2) The situation of soil lead pollution in rural-urban fringe zone. Statistical characteristics value of Pb in farmland soil of rural-urban fringe zone in Harbin was as follows, samples was 25, the minimum value was 13.95, the maximum value was 101.49, the mid-value was 30.50, the average value was 37.51, standard deviation was 21.85, skewness was 1.41, and the coefficient of variation was 0.58. The results showed that the content of soil lead in urban and rural fringe zone in Harbin did not exceed secondary standards of the state soil environment quality, but the average content of lead was higher than the background value of lead in Harbin, accounting for exogenous Pb. The maximum appeared in the farmland planted cabbage in front of Majiagou River (household waste was dumped into Majiagou River in quantity before treatment), this result accounted for that lead pollution was very serious. If the vegetables developed from this land be supplied to urban residents, lead could enter the body through diet ways and seriously endanger health of human body.

3) The situation of soil lead pollution in the east of primary toll station of Songbei highway. Fig. 1 showed the changing curve of lead content. The average value of the content of soil lead on the east side of primary toll gate of Songbei highway was slightly below the background value, and this may relate to less traffic flow before Songbei district developed. Great difference of lead content existed in those sample spots, and the peak of lead content appear in 20 meters and 200 meters far away from the high way, while minimum of lead content exists in 30 meters far away from highway, illustrating that lead pollution brought by the settlement from automobile exhaust pollutant accumulated obviously within 20 meters away from the highway. The variation of lead content at the position more distance far away from the highway

may be caused by man-made factors, such as pesticide sprays or fertilizer containing lead. Overall, the soil in the farmland on the east side of primary toll gate of Songbei highway has not been polluted seriously by lead.

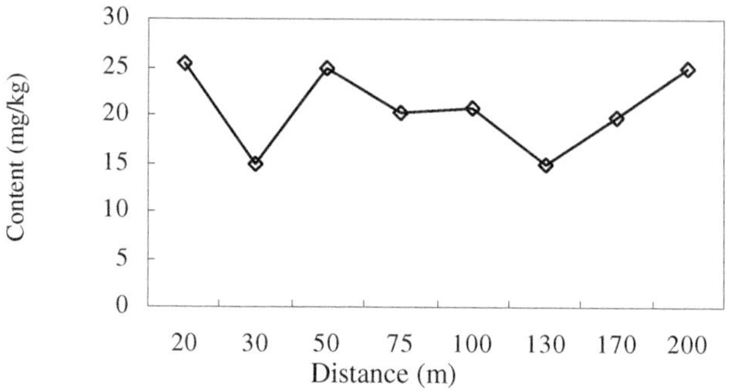

Fig. 1. Variation of soil lead content near the Songbei highway with different distance

4) The situation of lead pollution around industrial districts. Statistical results showed that the average value and maximum value of lead content was the highest in the northeast of Harbin Aircraft Manufacturer. The average value of lead content in the farmland was the lowest in the northwest. In general, the concentration of lead content was of nonuniform, which was related to annual dominant wind direction, meanwhile, it may be the influence of human factors and random disturbance in local area. Only the average value of soil lead content in northwest was less than the background value in Harbin, and lead content in other directions was about 4~ 5 times more than background value, which showed that soil lead pollution was serious.

Table 2. Statictics of characteristics value of pb around haerbin aircraft manufacturer

Element	Direction	Samples	AVG	MAX	MIN
	northwest	7	21.08	29	17
	southwest	6	96.50	111	87
Pb	northeast	4	141.75	171	77
	southwest	5	106.63	123	87

3.2 Health Risk Assessment of Lead Pollution

Environmental health risk assessment characterized the process of potential health effects caused by environmental pollution, which mainly assessed impacts and damage to human health caused by pollution within the field or region, to ensure the types and levels of environmental risk, to predict the scope of pollution and the degree of harm, and to provide scientific basis and technical support for risk management[10]. In 1983, National Academy of Sciences proposed the definition and the frame of health risk assessment, and four footwork of risk assessment, including harm judgment, doses-effect assessment, exposure assessment and risk characterization [10].

1) Exposure assessment. Statistical results showed that the average value and The average value and each parameter of lead content in different functional zones was substituted to the formula (1), obtaining the results of children lead exposure from the medium of soil, as is shown in the table 3. Table 3 showed the children intake of lead through the hand-mouth way, the most was around factories and on two sides of highway, and the least was in residential areas.

Table 3. Children Lead Exposure of Different Functional Zones(MG/D)

Residential areas*	Factories around*	Two sides of road*	Banks of Songhua River*	Suburb farmland
7.96×10^{-5}	1.79×10^{-4}	1.98×10^{-4}	1.10×10^{-4}	1.05×10^{-4}

2) Risk calculation-calculation of the risk of MMR. According to the formula (2) and data of table 4, the risk value of MMR of per 1,000 children in Harbin was calculated as follows:

$I(MMR)=AR\times[(R1\times P1)+(R2\times P2)+(R3\times P3)+(R4\times P4)]=3.03\times[(383\times0.0024)+(253\times0.0008)+(148\times0.0145)+(216\times0.0159)]=20.307$ (people/1000 children)

Results showed that incidence of MMR of children in Harbin equaled to 20.307, 20.307 children were likely to be suffered from MMR every 1,000 children. In 2000, WHO estimated the disease burden caused by lead in China, South Korea, the Philippines and east Asian countries, obtaining the result that the value of risk for MMR in these areas was between 3.4‰ ~ 14.6‰[11]. It was not difficult to find that the risk of sufferring from MMR of children in Harbin was much more than the upper limit value given by WHO, illustrating the health risk of residents in Harbin caused by soil lead pollution was serious. In addition, proper measures must be taken to avoid invasion of foreign lead pollution, at the same time, the treatment of lead adsorption should proceed effectively, to eliminate the hidden danger caused by lead pollution.

Table 4. Nunber of children that having the health risk of lead

Scope of blood lead(μg/dl)	Proportion of the children having the risk	Health effect	Ratio of children per 1000	Standard number of people in scope of IQ(%)
5~ 10	0.383	0.65 points decreased in IQ	383	0.24
10~ 15	0.253	1.95 points decreased in IQ	253	0.8
15~ 20	0.148	3.25 points decreased in IQ	148	1.45
> 20	0.216	3.50 points decreased in IQ	216	1.59

Note: quoted from Shuying Dong [12]etc.

Due to the lack of data resources, this study using the model calculation was quoted from the literature material content in 1999, spatial and temporal variation was great, the method of the calculation of MMR risk was according to the method given by WHO, parameters used in this calculation were based on western children, all of them resulted in error exist in the result.

3.3 Control Measures of Urban Soil Lead Pollution

1) Cutting off the source of lead pollution was the fundamental measures of preventing lead pollution. In Harbin, automobiles should be strictly restricted with using gasoline additive - tetraethyl lead, making sure that non-lead gasoline overall. Meantime, lead-based factories, enterprises and lead-acid battery industry should be rectified and reformed; backward equipment and technology should be sifted out; drainage or recycling treatment system should be improved; advanced technology of unleaded should be popularized; new lead-free green products should be developed, to minimize lead pollution [13].

2) Investigating blood lead for residents regularly. Children blood lead monitor net should be established, giving first place to maternity and child care, medical institutions of children and disease control, checking blood lead of children regularly, and intervening the children with high value of blood lead. Meantime, checking for adult blood lead also should not be ignored, strengthening investigating blood lead of the factory workers, people living near manufacturer, residents living near the main arteries and pregnant women.

3) Energetically strengthening propaganda and education, and improving self-protection awareness of public, especially the children's parents. To the greatest extent to avoid children playing near the high way or street sides, helping children to correct the habit of sucking hand, getting children into the good health habit of washing their hands frequently. All of them are necessary to effectively prevent children from lead poisoning.

4) More flowers and plants should be planted to absorb lead pollutants.

4 Conclusion

First, great difference of lead pollution level existed in the fringe zone of urban and rural within three-circle, its upper and lower limit was between 13.93 ~ 101.49 mg/kg, some of them was serious in lead pollution.

Second, the research of lead pollution of the two sides of the highway was based on the farmland on the east of primary toll gate of Songbei highway, results showed that the lead content was influenced by the distance from roads and personal factors, and lead pollutants accumulated obviously within 20 meters from road.

Third, on the whole, the lead content in heavy industrial districts was more than that in other regions, and serious lead pollution was in heavy industrial districts. Lead distribution surrounding Harbin Aircraft Manufacturer was related to the direction of prevailing wind, meanwhile, it was at random.

Fourth, the distribution of the lead content in functional zones of urban center showed two sides of highway > surrounding region of the factory > banks of Songhua River >suburban farmland > residential areas.

Fifth, the lead exposure of children in functional zones of urban center showed two sides of highway > surrounding region of the factory > banks of Songhua River >suburban farmland > residential areas.

Sixth, the health risk of MMR of children in Harbin was 20.307 per 1,000.

Acknowledgment. The authors would like to express gratitude to the research grant support kindly provided by National Science Foundation of China (No.41071033), the Natural Science Foundation of Heilongjiang Province, China (No. G200812D), the Doctoral Fund of Ministry of Education of China (No.200802310001) and Foundation of Heilongjiang Educational Committee(No. 11531224).

References

1. Wang, D.X., Fu, D.Y., Zhang, X.L., Wang, J.D., Meng, K.: Study on Lead Pollution of Large Cities in China. Urban Environment and Urban Ecology 4(2), 16–20 (1991)
2. Wang, Z., Shao, Z.Q.: Soil Lead Pollution and Management Measure. Agricultural Technology and Equipment 01B(158), 6–8 (2009)
3. Guo, P., Xie, Z.L., Li, J., Zhou, L.F.: Specificity of Heavy Metal Pollution and the Ecological Hazard in Urban Soils of Changchun City. Scientia Geographica Sinica 25(1), 108–112 (2005)
4. Ren, H.M., Zhang, J.D., Zhang, X.L.: Spatial Distribution of Pb in Soils and its Risk Assessment in Shenyang City. Advance in Earth Sciences 19(suppl.), 429–433 (2004)
5. Chen, H.: The Research of Environmental Health Risk Assessment Method. Science and Technology Information 34, 113–114 (2009)
6. Ren, H.M., Wang, J.D., Zhang, X.L., Wang, C.M.: Exposure Assessment of Environment Lead for Children in Shenyang City. Acta Scientiae Circumstantiae 25(9), 1236–1241 (2005)
7. Wu, J.G., Wang, Z.B.: Methods for Assessment on Environmental Burden of Disease. Environment and Occupation Medicine 25(1), 87–89 (2008)
8. Shi, J., Liu, M.: Investigation on Health Risk by Environmental Lead Pollution in Children. Chinese Journal Industry Medicine 22(1), 31–34 (2009)
9. Ma, X.F.: The Status and Ecological Risk Assessment of Soil Contaminated by Heavy Metals in Harbin. Northeast Forest University, Harbin (2009)
10. Li, Z.B., Luo, Y.M., Song, J., Zhao, Q.G., Liu, Z.Q.: Study on Soil Environmental Quality Guidelines and Standards II. Health Risk Assessment of Polluted Soils. Acta Pedologica Sinica 43(1), 142–151 (2006)
11. Jarosinska, D., Biesiada, M., Muszynska-Graca, M.: Environmental Burden of Disease Due to Lead in Urban Children from Silesia Poland. Science of the Total Envionment 367(1), 71–79 (2006)
12. Dong, S.Y., Zhen, X.Z., Li, P.C., Zhao, S.Y., Zhang, D.S., Guan, L., Liu, J.R.: Investigation on Blood Lead Levels of Preschool Children in Harbin and Jixi. Journal Environment Health 20(1), 28–29 (2003)
13. Jia, L.X., Song, W.B.: The Impact of Lead Pollution in Urban on Human Health and Measures Preventing. Studies of Trace Elements and Health 24(6), 38–41 (2007)

Study on Iterated Local Search Algorithm for Permutation Flowshop Problem with Total Flowtime Objective*

Xingye Dong[1], Houkuan Huang[1], and Ping Chen[2]

[1] School of Computer and Information Technology
Beijing Jiaotong University, Beijing, China
[2] TEDA College, Nankai University, Tianjin, China
xingming662@sohu.com

Abstract. Iterated Local Search (ILS) is a simple and efficient algorithm, which has been used to solve many combinatorial optimization problems. However, the effect of different search sequences in the local search procedure has not been studied. In this work, we used the ILS algorithm to solve the permutation flowshop scheduling problem with total flowtime objective, and observe the effect of different search sequences on this problem. Experimental results on benchmarks show that the search sequence does have some effect, though sometimes the effect may be little. If the iteration number is not so large, then the performance is better by changing the search sequence to the best solution found in the search process when the search process is trapped into local optimum. If the iteration number is rather large, then the performance is better by changing the search sequence to the best solution found in the search process immediately. Comparisons on benchmarks also show that ILS is comparable to the state of the art metaheuristics.

Keywords: Scheduling, Iterated local search, Search sequence, Flowshop, Total flowtime.

1 Introduction

Flowshop problems have attracted great attention of researchers since Johnson's pioneering work [1]. As one of the best known production scheduling problems, the permutation flowshop sequencing problem (PFSP), a simplified version of flowshop problem, has a strong engineering background and has attracted most researchers' attention. In the literature, the minimization of total flowtime is one of the common objectives, because this objective means to stabilize the consume of resources and minimize the work-in-process inventory, especially in the dynamic production environment nowadays [2]. The PFSP with total flowtime criterion is also a notorious combinatorial problem for its NP-completeness even with two machines [3].

* This work is supported by The Fundamental Research Funds for the Central Universities (Project Ref. 2009JBM018, Beijing Jiaotong University).

D. Zeng (Ed.): ICAIC 2011, Part II, CCIS 225, pp. 236–245, 2011.

Recently, the PFSP with total flowtime criterion has attracted many researchers' attention, and several metaheuristic algorithms were proposed to solve this problem. These algorithms include ant-colony algorithms by Rajendran and Zeigler [4, 5], particle swarm optimization algorithm by Tasgetiren et al. [6], and more recently, iterated local search algorithm by Dong et al. [7], hybrid genetic local search algorithm by Tseng and Lin [8, 9] and hybrid genetic algorithm by Zhang et al. [10]. These metaheuristics provide the state of the art in this field.

Iterated local search (ILS) is a simple and efficient metaheuristic, which provides a frame for improving the performance of local search methods. It has attracted much attention of researchers for its simplicity, effectiveness and efficiency, and it has been successfully applied to traveling salesman problem [11, 12], job shop scheduling [13], vehicle routing problem [14, 15] and PFSP with makespan criterion [16]. Recently, Dong et al. [7] proposed an ILS algorithm to solve the PFSP with total flowtime objective. They showed that the algorithm performed better than several state of the art metaheuristics, including two ant-colony algorithm M-MMAS and PACO by Rajendran and Zeigler [4], ant-colony algorithms ACO2 by Rajendran and Zeigler [5] and a particle swarm optimization algorithm PSO_{VNS} by Tasgetiren et al. [6].

In general, iterated local search is composed of three components, i.e., a local search procedure, a perturbation strategy and an acceptance criterion. Many works have shown that the local search procedure has a significant effect on the performance of the ILS. However, the search process in the local search was often neglected and few research was reported on it. Even though it is the case, there are some researchers have given a few discussion on this problem, e.g., in the ILS illustrated by Dong et al. [7], the search process is always changed according to the best schedule sequence when a new best schedule is found; also in the work of Laurent and Hao, the neighborhood of the local search is sampling in a random way [15]. Then, questions are risen: what is the effect of different search process? Can we improve the performance of the ILS by optimizing the process? In this paper, we do some effort on these issues.

The remainder of this paper is organized as follows. In Section 2, the formulation of PFSP with total flowtime criterion is presented briefly. In Section 3, the framework of general ILS algorithm is given, then the framework of ILS by Dong et al. [7] and a new one are illustrated; in addition, four different search sequences are discussed, and the effect of different search sequences are tested and analyzed. Then, in Section 4, the best variant of ILS is compared with several state of the art metaheuristics. Finally, the last section summarizes the main ideas of this work.

2 Problem Formulation

A set of jobs $J = \{1, 2, \ldots, n\}$, which are available at time zero, need to be processed on m machines, where $n \geq 1$ and $m \geq 1$. Each job has m operations, each of which needs an uninterrupted processing time. The processing time of the ith operation of job j is denoted by p_{ij}, where $p_{ij} \geq 0$. The process sequence of these m operations is that the first operation needs to be processed on the first machine, the second operation needs to be processed on the second machine, and so on. An operation of a job starts to be processed only if the previous operation of the job is completed and the requested machine is available. Each machine processes these jobs in the same order and only one operation can be processed at a time. Using the widely accepted triple notation [17], this problem can be denoted by $F_m|prmu|\sum C_j$, where C_j denotes the completion time of job j. Let π denote a permutation, which represents a job processing order, on the set J. Let $\pi(k)$, $k = 1, \ldots, n$, denote the kth job of π, then the completion time of job $\pi(k)$ on each machine i can be computed easily through a set of recursive equations:

$$C_{i,\pi(1)} = \sum_{l=1}^{i} p_{l,\pi(1)} \qquad i = 1, \ldots, m \tag{1}$$

$$C_{1,\pi(k)} = \sum_{l=1}^{k} p_{1,\pi(l)} \qquad k = 1, \ldots, n \tag{2}$$

$$C_{i,\pi(k)} = max\{C_{i-1,\pi(k)}, C_{i,\pi(k-1)}\} + p_{i,\pi(k)} \quad i = 2, \ldots, m; k = 2, \ldots, n \tag{3}$$

Then $C_{\pi(k)} = C_{m,\pi(k)}$, $k = 1, \ldots, n$. The total flowtime is $\sum C_{\pi(k)}$, i.e. the sum of the completion time on machine m of all the jobs. The PFSP with total flowtime criterion is to minimize $\sum C_{\pi(k)}$.

3 Iterated Local Search for PFSP

As aforementioned, iterated local search has been applied to several combinatorial optimization problems successfully. Its performance is comparable with other metaheuristics, such as simulated annealing, tabu search, genetic algorithm and ant-colony algorithm. Unlike some sophisticated metaheuristics, its frame is very simple. The pseudo code is illustrated in Fig. 1, where s^* denotes the best solution found in the search process.

1. Generate an initial solution s; let $s' = s$; let $s^* = s$;
2. Generate $s'' = LocalSearch(s')$; update s^* if a better solution is found in the process;
3. Let $s = Accept(s, s'')$ (this function can be a simulated-annealing one or some other types);
4. If the termination criterion is not satisfied, then generate $s' = Perturb(s)$, go to step 2; otherwise, output s^*.

Fig. 1. Pseudo code of an ILS frame

3.1 Two Variants of ILS

Dong et al. [7] proposed an efficient ILS algorithm to solve the permutation flowshop problem with total flowtime criterion. The framework of the ILS is as Fig. 2.

1. Set max_iter, $pert$, $iter = 1$, $cnt = 0$, $i = 1$;
2. Generate an initial permutation π, set $\pi^* = \pi$;
3. FOR $iter := 1$ TO max_iter DO
   BEGIN
4.   FOR $i := 1$ TO n DO
       BEGIN
5.   Find k, which satisfies $\pi(k) = \pi^*(i)$;
6.   Insert job $\pi(k)$ into other $n-1$ positions in π, respectively, and let π' be the best one among the $n-1$ generated permutations;
7.   If π' is better than π, then set $\pi = \pi'$, $cnt = 0$; otherwise, set $cnt = cnt + 1$;
8.   If π is better than π^*, then set $\pi^* = \pi$;
9.   IF$cnt \geq n$ THEN
           BEGIN
10.   Perturb π^* by $pert$ times random adjacent pair-wise interchange to generate a permutation π;
11.   If π is better than π^*, then set $\pi^* = \pi$;
12.   Set $cnt = 0$;
           END
       END
   END
   Output π^* and stop.

Fig. 2. Pseudo code of the ILS proposed by Dong et al. [7]

In this framework, the meanings of some parameters are listed as follows (while the others are omitted, because the meanings of them are obvious):

max_iter	maximum iterations allowed
$pert$	perturbation strength
cnt	used to count the consecutive moves without improvement in the current search process
π^*	the best solution found in the search process

Obviously, the search sequence in the local search is determined by the current best solution π^* (according to step 5). For π^* is changed when a better permutation is found (according to step 8 and 11), the searching sequence will be changed frequently at least at the early stage of the search process.

For the initial method, the H(2) algorithm proposed by Liu and Reeves [2] is chosen by Dong et al. [7], and so the time complexity of this algorithm is $O(mn^3)$ with a coefficient max_iter. More details about the time complexity of this algorithm can be referenced to Dong et al. [7].

In the literature, neighborhood structures are basically generated by two types of moves, so called SWAP and INSERT. The SWAP is swapping the jobs at the ith position and the jth position in the permutation, $i \neq j$. If an adjacent pair is swapped, then it is called API (Adjacent Pairwise Interchange). The INSERT is removing the job at the ith position from the permutation and then inserting it into the jth position, $i \neq j$. In step 6, the INSERT type move is used, as it has been shown good performance in the literature, e.g., the job-index-based local search proposed by Rajendran and Zeigler [4]. In step 10, the API is used to perturb permutation π^* in order to avoid disturbing π^* excessively. According to a group of experiments on benchmarks, the perturbation strength *pert* in step 1 is set to 6 [7].

This ILS algorithm performs quite well according to Dong et al. [7]. However, the search sequence in the local search procedure may be changed quite frequently in the early searching stage for the current best solution may be improved quite frequently. This may be not so good for exploiting good solutions.

In order to avoid this, a new framework, denoted by ILS2, is proposed as Fig. 3. In this framework, the search sequence in the local search is determined by $\hat{\pi}$, which is set to a specific sequence. The time to change this sequence is different from that in the ILS proposed by Dong et al. [7]. In the ILS2, the search sequence is only changed when the search is trapped into a local optimum, i.e., when the parameter cnt reaches to n (step 13).

In the ILS, Dong et al. [7] did not discuss the effect of different search sequences. Also, the search sequence is not discussed yet in the literature. In order to evaluate the effect of different search sequences, we will test four different search sequences in this paper for both ILS and ILS2. The search sequences are as follows: keep the sequence unchanged, i.e. the search sequence is always kept the initial sequence; the current best solution π^*; the reverse sequence of the current best solution, e.g., in the ILS2, we have $\hat{\pi}(i) = \pi^*(n-i+1), i = 1, 2, \ldots, n$; and a randomly generated sequence. In order to differentiate the algorithms with different search sequence, we use ILS-B to denote the ILS with setting the search sequence to the current best solution, ILS-RB to denote the ILS with setting the search sequence to the reverse of the current best solution and ILS-R to denote the ILS with random search sequence. For the ILS2, the variants are denoted by ILS2-B, ILS2-RB and ILS2-R, respectively. If the search sequence is kept unchanged, then the variants of ILS and ILS2 are the same, so it is only denoted by ILS-U.

```
1. Set max_iter, pert, iter = 1, cnt = 0, i = 1;
2. Generate an initial permutation π, set π* = π,
   π̂ = π;
3. FOR iter := 1 TO max_iter DO
   BEGIN
4.     FOR i := 1 TO n DO
       BEGIN
5.         Find k, which satisfies π(k) = π̂(i);
6.         Insert job π(k) into other n − 1 positions in
           π, respectively, and let π' be the best one
           among the n − 1 generated permutations;
7.         If π' is better than π, then set π = π', cnt
           = 0; otherwise, set cnt = cnt + 1;
8.         If π is better than π*, then set π* = π;
9.         IFcnt ≥ n THEN
           BEGIN
10.            Perturb π* by pert times random adjacent
               pair-wise interchange to generate a permu-
               tation π;
11.            If π is better than π*, then set π* = π;
12.            Set cnt = 0;
13.            Set π̂ to a specific sequence;
           END
       END
   END
   Output π* and stop.
```

Fig. 3. Pseudo code of the proposed ILS2

3.2 Analyses of ILS and ILS2

In this section, the variants of both ILS and ILS2 are compared and analyzed on benchmarks in order to observe the effect of different search strategies. The benchmarks are taken from Taillard [18], which consist of 120 particularly hard instances with 12 different sizes. For each size, a sample of 10 instances is provided. The scale of these problems varies from 20 jobs and 5 machines to 500 jobs and 20 machines. In the literature, many metaheuristics, such as the M-MMAS, PACO [4], PSO_{vns} [6], ILS [7] and Hybrid GA [8] were only tested on the first 90 benchmarks, so does it in this paper.

The settings for variants of both ILS and ILS2 are as follows: the initial solution is generated by H(2) and the perturbation strength *pert* is set to 6; also, in the local search procedure, the INSERT type move is used, while perturbing the current best solution, the API type move is used.

All the algorithms are implemented in C++, running on a PC with an Intel Core2 Duo processor (2.99 GHz) with 2G main memory. All the algorithms are compared by RPD (relative percentage deviation), which is calculated for each instance by Eq. (4):

$$RPD = (F - F_{best})/F_{best} \times 100 \tag{4}$$

where F is the result found by a specified algorithm such as ILS and ILS2, and F_{best} is the best result provided by BEST(LR) [2], M-MMAS and PACO [4]. Here, BEST(LR) denotes the best performing heuristic among many heuristics provided by Liu et al. [2]. Then the performance is compared by ARPD, where ARPD denotes the average RPD over all the instances.

All the algorithms are run five independent times for each benchmark instance. The best permutation is chosen as the final output. The maximum iterations max_iter is set to 120, 1000 and 5000, respectively in each run. The results are shown in Fig. 4.

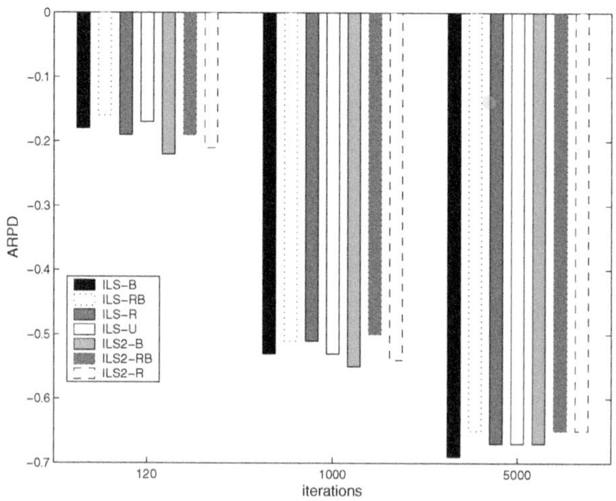

Fig. 4. Performances of ILS variants with different iterations

It can be seen from Fig. 4 that there is only slight difference among these variants, and this indicates that the search sequences have little effect on the performance. However, different search sequences still have different effect on the performance. When the variants are run 120 iterations, the ILS2-B, ILS2-RB and ILS2-R show better performance than their counterparts of the ILS, e.g., the average RPD of the ILS-B is −0.18, while that of the ILS2-B is −0.22; the average RPD of the ILS-RB is −0.16, while that of the ILS2-B is −0.19. This indicates that changing the search sequence only when the search trapped into local optimum is better than changing it much frequently such as ILS. However, when the variants are run much more iterations, the case is changed, e.g., when they are run 1000 iterations, the performance of these variants is almost the same: the RPDs for ILS2-B and ILS2-RB are −0.55 and −0.50, respectively, while they are −0.53 and −0.51 for ILS-B and ILS-RB, respectively; when run 5000 iterations, the performances of the ILS's variants are not worse than the ILS2's variants, e.g., the RPDs of ILS-B and ILS-RB are −0.69 and −0.65, respectively, while they are −0.67 and −0.65 for ILS2-B and ILS2-RB, respectively. Also, both the randomly generated search sequence strategy and the keeping the sequence unchanged strategy can achieve quite good results. All these indicate that though the search sequence has some effect on the performance of the algorithm, the effect is limited; and when the iteration number is large enough, the ILS proposed by Dong et al. [7] performs better.

4 Compare with Other Algorithms

From Fig. 4, it can be seen that the ILS-B performs the best when it is run 5000 iterations, so the results drawn by it with 5000 iterations are compared with other state of the art algorithms.

Recently, Zhang et al. [10] proposed a hybrid genetic algorithm for the discussed problem, and showed good performance of the algorithm. Also, Tseng and Lin [8, 9] proposed two genetic local search algorithms for the discussed problem. Though both algorithms proposed by Tseng and Lin performed quite well, the results reported in [8] are better not only in the term of total flowtime but also in the term of CPU time. As the result, the ILS-B is only compared with the hybrid genetic algorithm by Zhang et al. [10] and the hybrid genetic local search algorithm proposed by Tseng and Liu in [8]. Also, the algorithm proposed by Zhang et al. [10] is denoted by HGA-Z, and the algorithm proposed by Tseng and Liu [8] is denoted by HGA-T.

The HGA-Z was coded in JAVA and run on a PC with a P4 CPU (2.93GHz) and 512M main memory; the HGA-T was coded in C++ and run on a PC with an AMD K7 CPU (1.83GHz) and 512M main memory. The comparison results are listed in Table 1, where the RPDs are computed by Eq. 4, but the best solutions for HGA-Z are drawn from 5 independent runs, the best solutions for HGA-T are drawn from 10 independent runs, while the best solutions for ILS-B are drawn from five independent runs. Also, the $\bar{t}$ columns indicate the average CPU time per instance for running once.

From this table, it can be seen that the performances of these algorithms has little difference, though the ILS-B is slightly worse than the HGA-T and a little better than HGA-Z. More specifically, the HGA-T performs the best on 50 jobs instances and 100 jobs and 5 machines instances, while the ILS-B performs the best on 100 jobs and 10 or 20 machines instances. As for the CPU time, though the CPUs are different, the CPU time consumed by ILS-B is significantly less than the others, especially the HGA-Z. Though for the 20 jobs instances, the ILS-B needs more CPU time than HGA-T, it needs much less CPU time than HGA-T on larger instances, and on the average, the HGA-T needs 119.4 CPU seconds, while the ILS-B only needs 79.2 CPU seconds. So, the ILS-B is comparable to the HGA-T.

Table 1. Comparisons between HGA-Z, HGA-T and ILS-B on Taillard's benchmarks

	HGA-T[a]		HGA-Z[b]		ILS-B[c]	
$n\|m$	RPD	$\bar{t}$	RPD	$\bar{t}$	RPD	$\bar{t}$
20|5	-0.17	0.1	-0.17	2.2	-0.17	0.7
20|10	-0.04	0.2	-0.04	4.1	-0.04	1.6
20|20	-0.07	0.4	-0.07	7.6	-0.07	3.3
50|5	-0.84	16.6	-0.71	40.9	-0.69	10.0
50|10	-1.27	40.3	-1.04	111.7	-1.18	23.7
50|20	-1.08	82.2	-0.94	190.0	-1.02	49.4
100|5	-0.84	94.2	-0.74	461.3	-0.61	71.0
100|10	-1.04	251.6	-1.04	826.8	-1.08	176.2
100|20	-1.12	588.9	-1.30	2015.0	-1.35	376.6
Avg.	-0.72	119.4	-0.67	406.6	-0.69	79.2

[a] Best solutions are drawn from 10 independent runs on a 1.83GHz AMD K7 CPU with 512M RAM.
[b] Best solutions are drawn from 5 independent runs on a 2.93GHz P4 CPU with 512M RAM.
[c] Best solutions are drawn from 5 independent runs on a a 2.99GHz Intel Core2 Duo CPU with 2G RAM.

5 Conclusions

Iterated local search is a very effective algorithm framework. It has been applied to various problems and has shown powerful optimization ability. However, for the search sequences in the local search procedure, the effect of different search sequences have not been discussed yet. In this paper, we discussed it, and tested several search sequences through seven ILS variants on Taillard's benchmarks [18]. The results show that the search sequence has little effect on the performance of ILS, and the ILS performs quite well even with random generated search sequences. However, when the iteration number is not so large, the ILS can achieve a little better performance by changing the sequence to the best solution found in the search process only when the search is trapped into a local optima; when the iteration number is large enough, the ILS can perform better by changing the sequence immediately to the best solution found in the search process when a better solution is found. Comparisons with two hybrid genetic local search algorithms on Taillard's benchmarks show that the ILS is comparable to the state of the art metaheuristics.

References

1. Johnson, S.M.: Optimal two and three-stage production schedule with setup times included. Naval Research Logistics Quarterly 1(1), 61–68 (1954)
2. Liu, J.Y., Reeves, C.R.: Constructive and composite heuristic solutions to the $P//\sum C_i$ scheduling problem. European Journal of Operational Research 132, 439–452 (2001)
3. Garey, M.R., Johnson, D.S., Sethi, R.: The complexity of flowshop and jobshop scheduling. Mathematics of Operations Research 1, 117–129 (1976)
4. Rajendran, C., Ziegler, H.: Ant-colony algorithms for permutation flowshop scheduling to minimize makespan/total flowtime of jobs. European Journal of Operational Research 155, 426–438 (2004)
5. Rajendran, C., Ziegler, H.: Two ant-colony algorithms for minimizing total flowtime in permutation flowshops. Computers and Industrial Engineering 48, 789–797 (2005)
6. Tasgetiren, M.F., Liang, Y.-C., Sevkli, M., Gencyilmaz, G.: A particle swarm optimization algorithm for makespan and total flowtime minimization in the permutation flowshop sequencing problem. European Journal of Operational Research 177, 1930–1947 (2007)
7. Dong, X.Y., Huang, H.K., Chen, P.: An iterated local search algorithm for the permutation flowshop problem with total flowtime criterion. Computers and Operations Research 36, 1664–1669 (2009)
8. Tseng, L., Lin, Y.: A hybrib genetic local search algorithm for the permutation flowshop scheduling problem. European Journal of Operational Research 198, 84–92 (2009)
9. Tseng, L., Lin, Y.: A genetic local search algorithm for minimizing total flowtime in the permutation flowshop scheduling problem. Int. J. Production Economics 127, 121–128 (2010)
10. Zhang, Y., Li, X., Wang, Q.: Hybrid genetic algorithm for permutation flowshop scheduling problems with total flowtime minimization. European Journal of Operational Research 196, 869–878 (2009)
11. Martin, O., Otto, S.W., Felten, E.W.: Large-step markov chains for the traveling salesman problem. Complex Systems 5(3), 299–326 (1991)
12. Martin, O., Otto, S.W.: Combining simulated annealing with local search heuristics. Annals of Operations Research 63, 57–75 (1996)
13. Lourenco, H.R.: Job-shop scheduling: computational study of local search and large-step optimization methods. European Journal of Operational Research 83, 347–364 (1995)
14. Hashimoto, H., Yagiura, M., Ibaraki, T.: An iterated local search algorithm for the time-dependent vehicle routing problem with time windows. Discrete Optimization 5, 434–456 (2008)
15. Laurent, B., Hao, J.K.: Iterated local search for the multiple depot vehicle scheduling problem. Computers and Industrial Engineering 57, 277–286 (2009)
16. Stützle, T.: Applying iterated local search to the permutation flow shop problem. Technical Report, AIDA-98-04, FG Intellektik, TU Darmstadt
17. Pinedo, M.: Scheduling: theory, algorithms, and systems, 2nd edn. Prentice-Hall, Englewood Cliffs (2001)
18. Taillard, E.: Benchmarks for basic scheduling problems. European Journal of Operational Research 64, 278–285 (1993)

Robust Control of High Speed High Precision Linear Motion System

Chi Zhang[1], Xulei Yang[2], Yong Xiao[3], and Guangzhou Zhao[1]

[1] Ningbo Institute of Technology, Zhejiang University, Ningbo, China
[2] School of EEE, Nanyang Technological University, Singapore
[3] School of Information Engineering, Shenyang Uni. of Chemical Tech.
Shenyang, China
chi-z@in.com

Abstract. This paper presents an improved robust control of a high speed high precision XY-table linear motion system used in semiconductor wire bonding machine. H_∞ robust control is implemented in both inner velocity loop and outer position loop to achieve high and consistent control performance even in the presence of considerable resonance uncertainties and external disturbances. Auto-tuned feed-forward compensation based on fuzzy logic method is employed to obtain satisfactory dynamic and settling performance. High close loop bandwidth has been yielded, and high acceleration of 20g (1g = 9.81m/s^2), high speed of 0.9m/s can be achieved with the dynamic error of 2 μm, settling error of 1.2μm and static error of 0.2μm.

Keywords: Robust control, linear motion, XY-table positioning system, feed forward compensation, fuzzy logic.

1 Introduction

With the rapid development of semiconductor industry, microelectronics and bioengineering, mechanical systems such as machine tools, microelectronics manufacturing equipment, mechanical manipulators, and automatic inspection machines are intensively implemented with the requirements of robust, high speed, and high-accuracy positioning performance. Linear Permanent Magnet Synchronous Motors (LPMSM) has inherent advantages of direct drive, simple structure, high thrust density, high acceleration/deceleration capability and almost maintenance free [1,2]. Therefore it is particularly suitable for linear motion system where high speed & high precision are required.

In semiconductor wire-bonding machine application, high-speed operation is often required to yield high productivity, and the precision requirement becomes more and more stringent because less than 30 um bond pad pitch bonding capability is required in the industry. High-speed LPMSM direct-drive XY-table positioning mechanism is therefore implemented in the wire-bonding machine to achieve the required performances. In the XY-table positioning mechanism, Y-table is located on X-table and the bonding mechanical system that composes of bond head, optics, wire-clamps

D. Zeng (Ed.): ICAIC 2011, Part II, CCIS 225, pp. 246–254, 2011.

and cable bracket are mounted on the Y-table. The complicated mechanical design results in severe resonances, which vary with different operating conditions and from machine to machine due to manufacturing tolerance. The uncertain resonances cause difficulties in feedback control system design. It is difficult for traditionally designed proportional–integral–derivative (PID) controller, in combination with notch filters, to uniquely deal with a wide range of mass-produced machines with multi-varying resonance modes.

Intensive studies on the control of high-speed high precision linear motion systems have been conducted in many literatures. A disturbance observer was developed in [3] and used for table positioning systems [4,5]. An adaptive robust position control design was proposed for linear motor drives [6]. Composite nonlinear feedback control is proposed to achieve good tracking performance [7]. These designs, which are based on the mathematical modeling of low-frequency plant dynamics, may not effectively take the mechanical resonance uncertainty into consideration. Robust control design including velocity/position dual control loop and feed-forward compensator is presented in [8]. However, the acceleration of the XY-table is only 5g, the bandwidth is low and the high frequency resonances are not eliminated enough for close-loop frequency response, which will cause vibrations during settling.

In this paper, an improved robust control design is proposed for a high-speed high precision XY-table positioning mechanism. This improved H_∞ robust control design is employed in both velocity and position control loop to achieve high bandwidth that can reach 450Hz. Feed-forward compensation based on fuzzy logic control is implemented to obtain satisfactory dynamic and tracking performances. High acceleration of 20g is yielded with such design. Satisfactory performance has been achieved in the implementation of mass production.

2 Mechanical Structure and Dynamic Modeling

2.1 Mechanical Structure

Fig. 1 (a) shows the mechanical structure of the high-speed XY- table positioning mechanism mounted on a wire-bonding machine. Y table is mounted on top of the X table. A bonding mechanical system including a bonding head (BH) and its assembly accessories, such as bonding capillary, wire clamp and cable bracket are placed on the Y table. Each table is driven by a linear permanent magnet motor. The driving motors are composed of a base-mounted permanent magnet and one or two supporting carbon fiber brackets where the three-phase adjoining start-connection coils are embedded. The coil bracket shown in Fig. 1(b) is sandwiched between 2 permanent magnet mounted bases with air-gaps and acts as the moving part of the motor. By feeding appropriate three-phase current to the coils, the interaction between the permanent magnets and the coil bracket will generate the thrust force acting on the coil bracket and drive the table through a screw-tightened connection. X and Y tables are designed with cross-roller way guides to achieve extremely fast acceleration/deceleration. They are synchronized to move in the X and Y directions. X table has the largest moving mass and therefore is designed with 2 set coil bracket and permanent magnet stator, which can generate double moving force to drive the mechanical combination of Y table and bond head. High accuracy optical encoders are attached on XY table with an encoder resolution of 0.05 um to achieve high control accuracy.

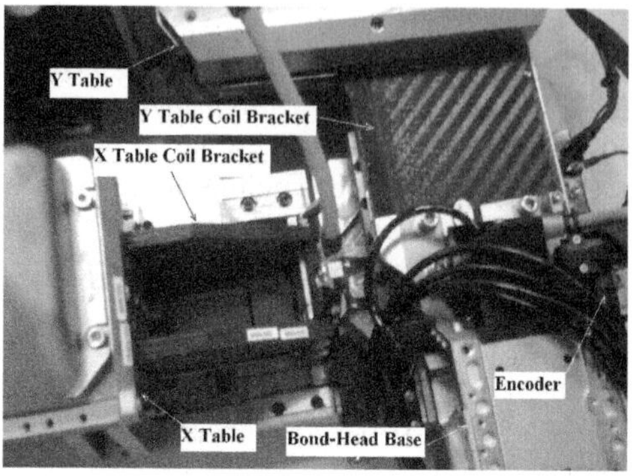

(a) XY table mechanical assembly

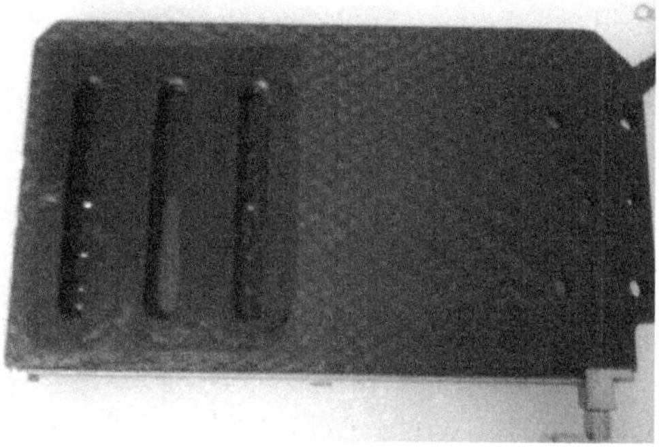

(b) Y table coil bracket

Fig. 1. Mechanical Structure of XY table

2.2 Dynamic Modeling

In general, the dynamics of the voltage-controllable LPMSM can be described as

$$Mx'' + Dx' + F_d = F_m \tag{1}$$

$$K_e x' + L_a \frac{dI_a}{dt} + R_a I_a = u \tag{2}$$

$$F_m = K_f I_a \tag{3}$$

Where x is the motor position; M, D, F_m and F_d are mass of the moving part, viscosity constant, generated force, and system disturbance, respectively; u, I_a, R_a, and L_a denote the electrical parameters, i.e., time-varying motor terminal voltage, armature current, armature resistance, and armature inductance, respectively; K_f is an electrical–mechanical energy conversion constant; and K_e is the back electromotive force (EMF) constant.

However, the dynamic model can only reflect the low frequency characteristics of the motor. It can not describe the high frequency resonances caused by attached mechanical system. Pseudo-random binary sequence (PRBS) has been employed to get the frequency response of the system by injecting PRBS singles to the current shown in (3) and reading the feedback encoder position signals shown in (1). The velocity frequency response that depicts the relationship between the input current and the velocity derived from the observed position can be obtained as shown in Fig.2, resonances can be observed from 600 Hz. High order model can be fitted based on the frequency response to design the robust controller.

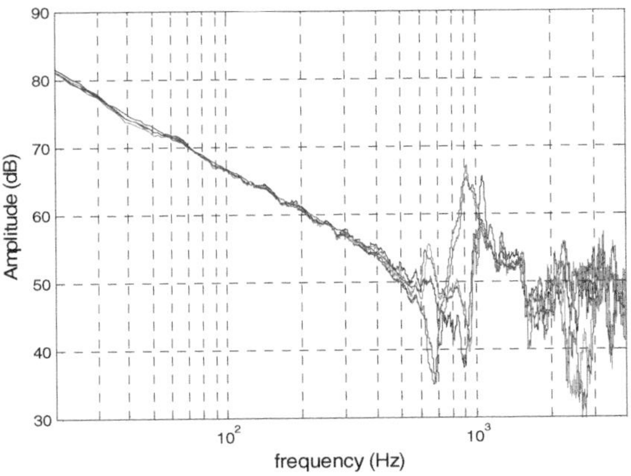

Fig. 2. Velocity Frequency Response of X Table

3 Robust Control Design of the System

3.1 Control Strategy

Fig. 3 shows the block diagram of the motion control system. The system consists of a hybrid control structure, which has two digital feedback control loops, inner velocity loop and outer position loop, plus feed forward compensator (FFC). Both velocity and position controllers are designed by employing H_∞ robust control. Fuzzy logic method is applied to auto-tune the FFC parameters based on the dynamic and settling performances. These parameters are altering along with the changes of the traveling distance, maximum acceleration and speed. Optimal motion performance can be achieved by employing the fuzzy tuned FFC parameters. The position, velocity, acceleration and jerk reference commands are produced on the basis of position motion profile with minimal time realization. The position error is generated with the comparison of the feed-back encoder position and the command position. The position controller output acts as velocity reference component and adds on the velocity command reference to compare with the encoder velocity. Generated velocity error is sent to velocity controller to produce current reference component. The sum of this component and the FFC output forms the current reference.

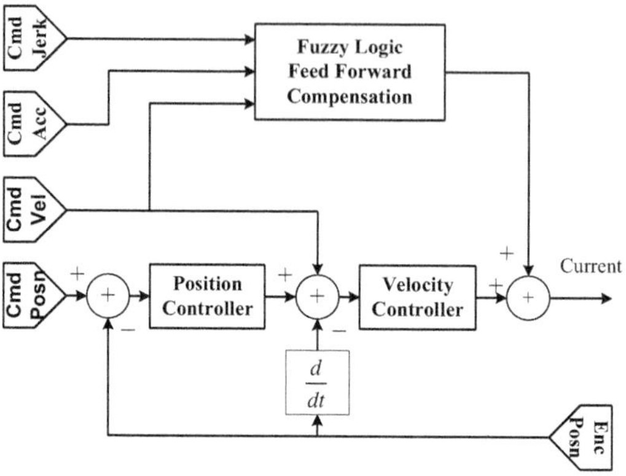

Fig. 3. Control Block Diagram

3.2 Robust Controller Design

Robust control is good at handling the uncertainty by bounding it rather than expressing it in the form of a distribution. The control can achieve results that meet the control system requirements in all cases by applying a bound on the uncertainty. Therefore robust control theory might be stated as a worst-case analysis method rather than a typical case method. It is thus a good choice for the complicated mass production system like XY-table motion mechanism with many uncertainties and high frequency resonances.

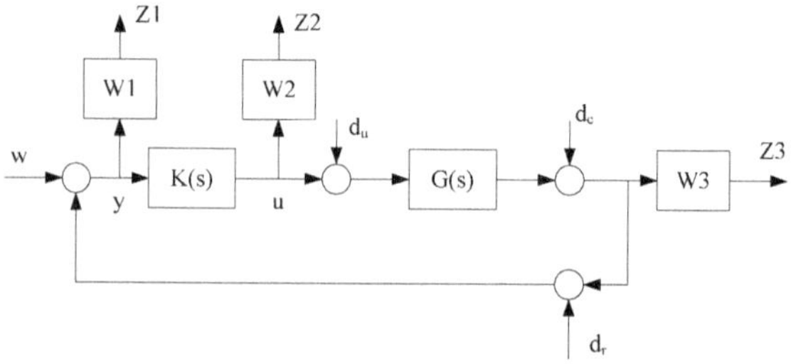

Fig. 4. Control Diagram with Weighting Functions

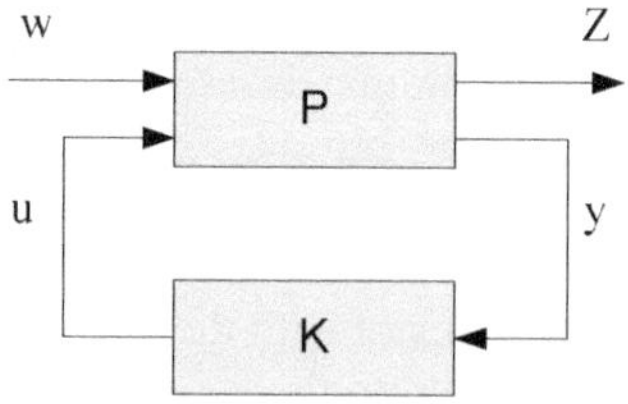

Fig. 5. Standard Problem Control Diagram

Fig.8 shows the feedback control diagram with three weighting functions that are control error, control input and control output respectively. The control structure can be transformed to the standard control problem as shown in Fig.9. Where, w is input signal, u is the control signal, y is measurement output signal and Z is the control output signal that can be defined as

$$Z = [Z_1 \quad Z_2 \quad Z_3]' \tag{4}$$

The augmented plant P can be derived as

$$P = \begin{bmatrix} W_1 & -W_1G \\ 0 & W_2 \\ 0 & W_3G \\ I & -G \end{bmatrix} \tag{5}$$

And the close-loop transfer function of the control system is shown as

$$T_{Zw} = \begin{bmatrix} W_1S \\ W_2R \\ W_3T \end{bmatrix} \tag{6}$$

Where, S, R and T are the sensitivity, control sensitivity and complementary sensitivity shown below

$$S = (I + GK)^{-1} \tag{7}$$

$$R = K(I + GK)^{-1} \tag{8}$$

$$T = GK(I + GK)^{-1} \tag{9}$$

The entropy equation is then be written as

$$Entropy = -\frac{\gamma^2}{2\pi}\int_{-\infty}^{\infty} \ln\left|\det I - \gamma^{-2}T_{zw}(jw)'T_{zw}(jw)\right|\left[\frac{s_0^{\,2}}{s_0^{\,2} + \omega^2}\right]d\alpha \tag{10}$$

The optimal γ can be derived by minimizing the entropy, and the sub-optimal stabilizing controller will be found to satisfy the weighting equation below

$$\|T_{zw}\|_\infty < \gamma \tag{11}$$

The velocity and position controller shown in Fig.3 can be designed by using the robust control method mentioned above. And the resulted velocity and position close-loop frequency responses, i.e., complimentary sensitivity functions are shown in Fig. 6 (a) and (b) respectively. The bandwidths of these two frequency responses can reach 450 Hz and 350 Hz. The gain margin of high frequency resonance is larger than 15dB, which ensure the stability of the system.

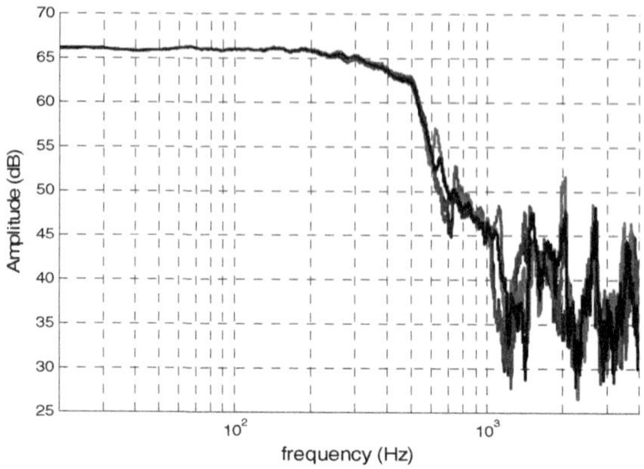

(a) Velocity Close-loop Frequency Response

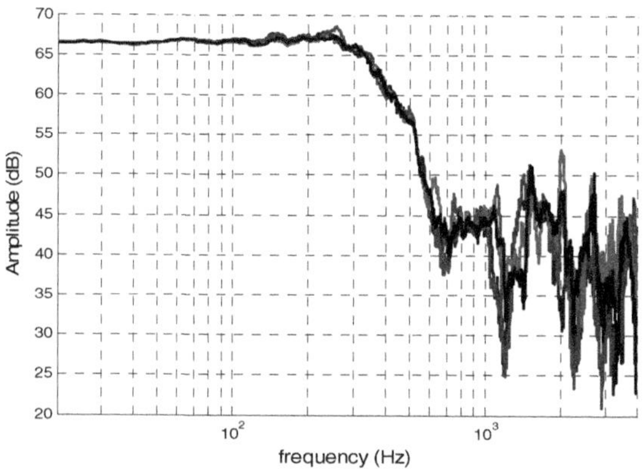

(b) Position Close-loop Frequency Response

Fig. 6. Close-loop Frequency Response of Velocity and Position Loop

4 Implementation and Test Results

The implementation results of profile motion are shown in Fig. 7. The position profile is generated to move the table with a travel distance of 3 mm, The peak velocity can reach 0.9 m/s while the maximum acceleration is around 20 g. The maximum dynamic error is about 1.5 μm, the maximum over-shoot is around 1μm, maximum under-shoot is about 0.5μm, The settling time, which is defined here as the time needed from the end of position profile command to the position error falling down within 1μm, is 2 ms. The maximum static tracking error is about 0.2μm.

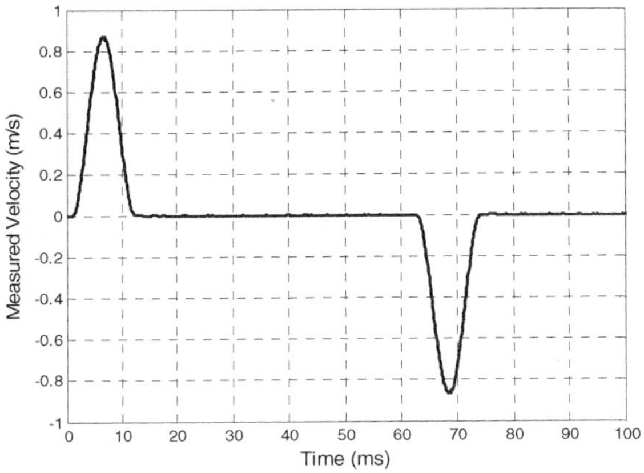

(a) Measured Velocity

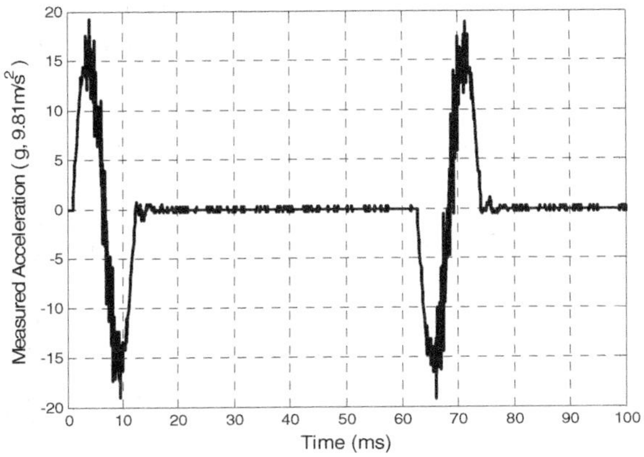

(b) Measured Acceleration

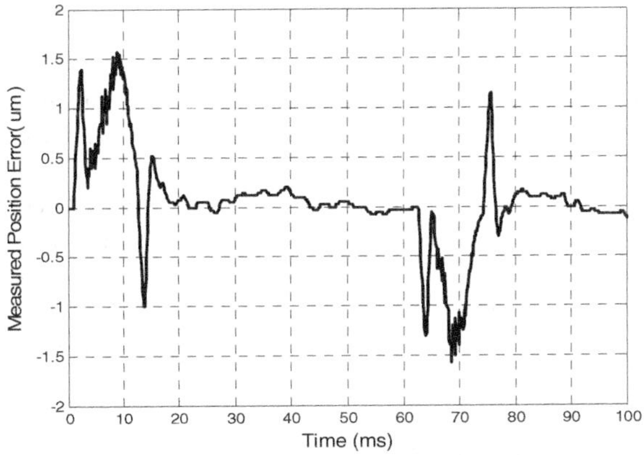

(c) Measured Position Error

Fig. 7. Motion Performance with 20 g Acceleration

5 Conclusions

This paper has introduced a hybrid H_∞ robust control strategy for high speed high precision XY-table positioning system used in wire bonding machine. Both inner velocity loop and outer position loop are controlled by H_∞ based robust controllers for the sake of consistent and robust control performance. Auto-tuned fuzzy logic feed-forward controller has been implemented to achieve high dynamic and settling performances. Experimental results demonstrate satisfactory motion performance with the proposed hybrid robust control system.

References

1. Boldea, I., Nasar, S.A.: Linear Electric Actuators & Generators. Cambridge University Press, Cambridge (1997)
2. Tam, M.S.W., Cheung, N.C.: A high speed high precision linear drive system for manufacturing automation. Applied Power Electronics Conference and Exposition (2001)
3. Umeno, T., Hori, Y.: Robust speed control of DC servomotors using modern two degree-of-freedom controller design. IEEE Trans. Ind. Electron. 38(5), 363–368 (1991)
4. Kempf, C.J., Kobayashi, S.: Disturbance observer and feedforward design for a high-speed.direct-drive positioning table. IEEE Trans. Contr. Syst. Technol. 7(5), 513–526 (1999)
5. Lee, H.S., Tomizuka, M.: Robust motion controller design for high accuracy positioning systems. IEEE Trans. Ind. Electron. 43(1), 48–55 (1996)
6. Xu, L., Yao, B.: Adaptive robust precision motion control of linear motors with negligible electrical dynamics theory and experiments. IEEE/ASME Transactions on Mechatronics 6(4), 444–452 (2001)
7. Cheng, G.Y., Peng, K., Chen, B.M., Lee, T.H.: Improving Transient Performance in Tracking General References Using Composite Nonlinear Feedback Control and Its Application to High-Speed XY-Table Positioning Mechanism. IEEE Trans. Ind. Electron. 54(2), 1039–1051 (2007)
8. Liu, Z.Z., Luo, F.L., Rahman, M.A.: Robust and precision motion control system of linear-motor direct drive for high-speed X–Y table positioning mechanism. IEEE Trans. Ind. Electron. 52(5), 1357–1363 (2005)

A Process Model with Event-Rule Mechanism

Xin Li

Department of Computer Science, University of Shantou
Shantou, Guangdong Province, China
li_xin1234@sohu.com

Abstract. Process systems are sorted according to the association form between process logics and business logics in this paper. The term of process service is defined with the view of system service. Then, a new process model is presented according to the goal of process service. The composition and the structure of process service model are different form those of workflow. All of the elements have strict formalized definitions. Business process can be expressed more clearly and flexibly now. In the end, an application instance is used to illustrate the function of process service model.

Keywords: Process Service, Workflow, Process Model, Dynamic Association, Event-Rule Mechanism.

1 Introduction

Process system has become more and more popular presently. A process is divided into several steps. Steps are organized with specific forms such as sequence, logic relationships and other conditions. Workflow is an important technique in process system. In workflow, a business process is regarded as the composite of two associated parts. One is called process logics and the other is called business logics. Process logics represent the relationships among steps and business logics represent concrete applications. The idea of workflow is the separation of process logics from business logics. The foundation of WFMC (Workflow Management Coalition) indicated that the development of workflow technique began to tend towards criterion. WFMC has brought forward a process model and most workflow products are constructed with the reference to WFMC criterion.

Although great progress has been made in workflow realm, problems and difficulties shouldn't be neglected either [1]. The development of process system has been a heavy-laden work all along. Therefore, providers of workflow products usually provide a complete set of solutions of process applications for the customers rather then a simple WFMS (Workflow Management System) to support independent development on it. It reveals that the process is still bound to the business, i.e. the idea of the separation of process logics from business logics hasn't been realized in process applications. It's considered that the issue has much to do with present process model [2]-[6]. In order to solve the troubles in process systems, a new process model is presented with a new conception of process service.

D. Zeng (Ed.): ICAIC 2011, Part II, CCIS 225, pp. 255–263, 2011.

2 Process System and Process Service

Process service is a kind of system service, which is constructed on net service (also called web service). Express a business process with the form as follows,

$$BP =< Process, Business, Association > .$$

Here, process denotes process logics, business denotes business logics and association denotes the association degree between process logics and business logics. Process systems are classified by association degree in this paper. There're four kinds of process systems.

2.1 Special Process System

Process section and business section are mixed together. Namely, process logics and business logics are highly coupled. Lots of Financial systems, official systems and management systems of product line are of this sort.

2.2 Workflow System

Business processes based on WFMS (Workflow management system) are of this sort. Process logics and business logics are partitioned in a certain extent. For example, the process can be abstracted as flow graph. Yet, the process in flow graph doesn't include business elements. So, such process is a kind of pure process. It's hard for it to deal with the process which is coupled with the business.

2.3 Flexible Workflow System

It's a kind of advanced workflow system. The ability to deal with the association of process logics and business logics is enhanced. However, the enhancement is limited. The process model is still from WFMC. Some complements are added into WFMC model to settle particular flexible problems. There aren't universal model for flexible workflow system and the flexible abilities of different systems are quite various either.

2.4 Process Service System

The association of process logics and business logics is dynamic. Business information can be embedded in the expression of process logics. There's a uniform model for the process, which is unlike the flexible workflow system where the static association is based on WFMC model and the flexible parts depend on extra models.

Now, a new conception called process service is presented. The accurate definition is given as follows,

Definition. Process service is a kind of net service which is process-oriented. The process model of process service is independent on the business model. The association between process logics and business logics is dynamic and simple reconstruction of the process is supported, which is mainly accomplished with the form of resetting the process templates.

In this definition, two factors are emphasized,

1) Dynamic association between process logics and business logics.
2) Simple reconstruction of the process.

The goal of process service has exceeded that of workflow, which is more a system service than a component to support the application. Such service will overlay net service (or called WEB service). It will become another system layer after OS and virtual machine such as Java VM and CLR.

3 Process Model for Process Service

WFMC model is the base of various workflow systems. The composition of process elements and their relationships are described in the model [7]. The model is shown in Fig. 1.

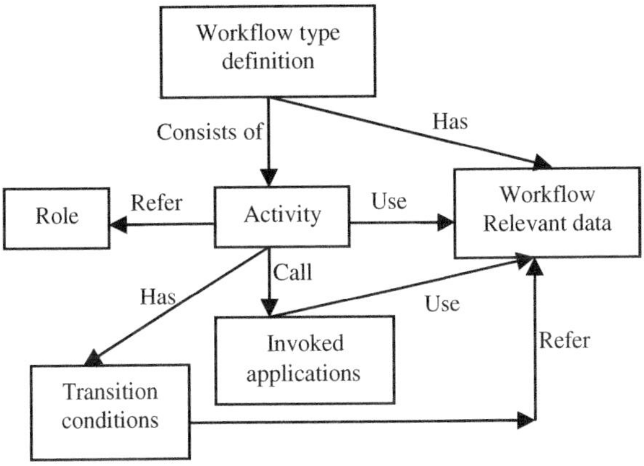

Fig. 1. WFMC model

Workflow relevant data in the model means the data which is associated with both process logics and business logics. The concrete disposal of the association is listed as follows,

1) Workflow relevant data is defined in workflow template.
2) Activities and applications both use workflow relevant data.

3) Transition conditions can be included in the activity, which is used to deal with questions about process logics.

4) Workflow relevant data may be referred by transition conditions.

However, WFMC model hasn't given the criterion to above contents. First, lack the criterion for the form of workflow relevant data. Second, lack the criterion about how to use and refer workflow relevant data. Third, lack the criterion for the form of transition condition. So, the disposal of the association between process logics and business logics is usually quite random in workflow systems, as leads to the result that process logics and business logics are tightly coupled together. The researches about process logics have been bound to pure process mainly.

Process service adopts a new model which can support dynamic variation and gives the criterion to embed business information into process logics. In the model, there exist some new elements which aren't in WFMC model. They are event, state and rule. These elements have appeared in flexible workflow [8]-[12]. But there, they are only regarded as complements to WFMC model.

3.1 Event

The occurrence of the behaviors in process system is abstracted as event. For example, the submitting of a document can be expressed as an event. There're two kinds of events in process service system: system event and business event. System behaviors such as starting an activity can trigger system events. Analogously, business behaviors such as submitting an order form can trigger business events. Usually, system events are predefined in process system. Yet business events can be self-defined by users according to the requirements. Two kinds of events are expressed with same form in the system.

3.2 State

Process, activity and role all have states. State isn't a basic element in WFMC model though there's the conception of the state in that model. However, in process service system, state is a normal element. The expression and the disposal to it both are regulated strictly, which will be narrated in the next section. Like event, states are also classified to two sorts: system state and business state and the latter can be self-defined.

3.3 Rule

In process service, process logics and some business logics are represented with rules. The form of the rule is like: P -> Q. P denotes the condition and Q denotes the results. Here, P is the composition of states and Q is the composition of operations. Particularly, the rule in process service isn't expressed with code form, but with data form, for example, being expressed with XML form.

The process model for process service is shown in Fig. 2.

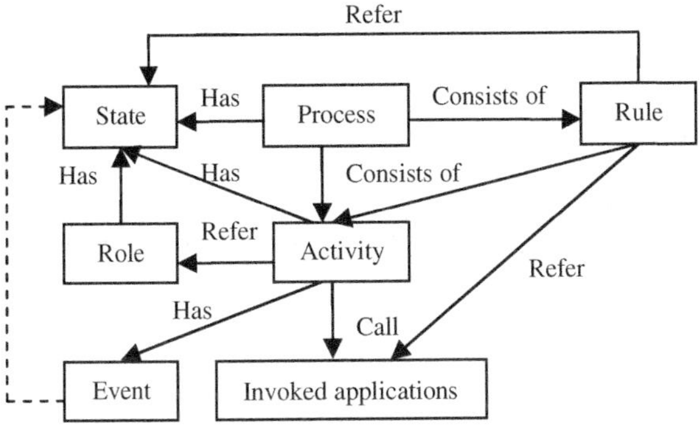

Fig. 2. Process model for process service

In this new model, workflow relevant data and transition conditions are eliminated. The functions of them are held by event, state and rule. The new elements have regular forms which is radically different from the old ones. At the same time, the contents of process and activity here aren't completely same as those in WFMC model because of the variation to the conception of the process. All the elements of process model constitute a framework which can support dynamic requirements for the process well.

4 Formalized Definition for the Elements of Process Model

Lots of terms may be expressed to different meanings in different workflow systems, even the term of workflow itself. It causes quite a few problems in workflow realm, for example modeling problems, implement problems, compatibility problems and portability problems, etc. To a great extent, it's because of lacking strict definitions in WFMC model. Unlike workflow, the elements in process service model all have strict formalized definitions, which are listed as follows.

4.1 Process and Process Instance

$$\begin{aligned} Process =< & ProcessID, ProcessName, ActivitySet, \\ & ProcessRoleSet, ProcessEventSet, \\ & ProcessStateSet, RuleSet > . \end{aligned}$$

Here, ProcessID is the identity of the process. ProcessName is the name of the process. ActivitySet is the set of activities which are included in the process. ProcessRoleSet is the set of roles who participate in the process in process layer. ProcessEventSet and ProcessStateSet are the sets of events and states which are in process layer. RuleSet is the set of rules which are related to process logics.

$$\begin{aligned} ProcessInstance =< & ProcessInstanceID, ProcessID \\ & ProcessInstanceState > . \end{aligned}$$

Here, ProcessInstanceID is the identity of the process instance. ProcessID is the identity of the process which the instance belongs to. ProcessInstanceState is current state of the process instance.

4.2 Activity and Activity Instance

$$Activity =< ActivityID, ActivityName, ActivityRoleSet, \\ ApplicationSet, ActivityEventSet, ActivityStateSet >.$$

Here, ActivityID is the identity of the activity. ActivityName is the name of the activity. ActivityRoleSet is the set of roles who participate in the activity. ApplicationSet is the set of application programs which can be called by the activity. ActivityEventSet and ActivityStateSet are the sets of events and states which are related to the activity.

$$ActivityInstance =< ActivityInstanceID, ActivityID, \\ ProcessInstanceID, ActivityInstanceState >.$$

Here, ActivityInstanceID is the identity of the activity instance. ProcessInstanceID is the identity of the process instance which includes the activity instance. ActivityInstanceState is current state of the activity instance.

4.3 Role and Executor

$$Role =< RoleID, RoleName >$$

Here, RoleID is the identity of the role. RoleName is the name of the type of the role.

$$Executor =< ExecutorID, ExecutorName, RoleID, \\ LogonTime, LogoffTime, Address, \\ ExcutorState, ProcessInstanceState >.$$

Executor is an individual who possesses some right of a role. Here, ExecutorID is the identity of the executor. ExecutorName is the name of the executor. RoleID is the role which the executor belongs to. LogonTime is the time when the executor has logged on. LogoffTime is the time when the executor has logged off. Address is the address where the executor has logged on. ExecutorState is current state of the executor. ProcessInstanceID is the identity of the process instance which the executor is dealing with.

4.4 Event

$$Event =< EventID, EventName, EventInfo >.$$

Here, EventID is the identity of the event. EventName is the name of the event. Eventinfo is the information of the event, including the description of the event and the method which related to the event. Two kinds of events have been narrated above: system event and business event. Common methods are related to system events and self-defined methods are related to business events. However, the modes of handling both kinds of events, which is from capturing the event to calling the method, are consistent. Process service system provides a common framework for it.

4.5 State

$$State =< StateID, StateName, StateType, StateValueSet >$$

Here, StateID is the identity of the state. StateName is the name of the state. StateType is the type which the state belongs to, such as ProcessInstanceState, ActivityInstanceState and ExecutorState. StateValueSet is the set of values which the state can be evaluated with. Two kinds of states have been narrated above: system state and business state. Like the event, Process service system also provides a common framework for the state, as workflow system lacks.

4.6 Rule

$$Rule =< RuleID, Conditon, \mathrm{Re}\,sult > .$$

$$Condition =< ConditionLogic(State*) > .$$

$$\mathrm{Re}\,sult =< \mathrm{Re}\,sultLogic((ActivityID, ActivityInfo)*) > .$$

Here, RuleID is the identity of the rule.ConditionLogic and ResultLogic are both compositions of logic operations and other operations. * denotes multiple. If the values of the states satisfied the condition of the rule, methods of activities in the result of the rule will be called with concrete logic defined in ResultLogic. ActivityInfo is the information of the activity related with rule, including the name of the method to be called.

5 Instance

An instance of medical treatment is used to illustrate the function of process service model. A section of medical process is shown in Fig. 3.

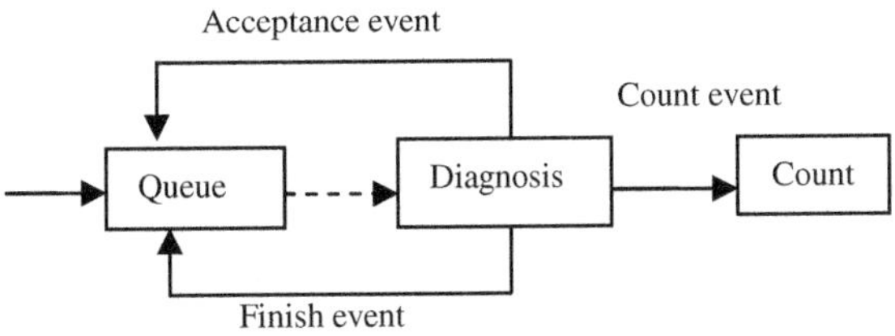

Fig. 3. A section of medical process

The contents of the process are explained here,

1) Patients wait with their numbers.

2) The queue program will call a patient according his number when a doctor is free.

3) A doctor will send acceptance event when he accepts a patient, as will change his state and the queue.

4) A doctor may send counting event when he gives the recipe or the examination to the patient.

5) A doctor will send finish event when he has finished the diagnosis, as will change his state.

The events, such as acceptance, count event and finish event, are of business logics. But now, they can be disposed of in a unified framework of process service. Some parts of the model for the process are shown as follows.

5.1 The Definitions of Business States and Role States

Only business states and Role states require self-definition because system states have been predefined in the system.

$$DiagnosisState =< StateID = 10, StateName = Diagnosis, \\ StateType = ActivityIns\tan ceState, \\ StateValueSet = \{Diagno\sin g, List\} > .$$

$$DoctorState =< StateID = 20, StateName = Doctor, \\ StateType = RoleState, \\ StateValueSet = \{Free, Busy, Absent\} > .$$

5.2 The Definitions of Rules

$$DiagnosisState.Statevalue = Diagno\sin g \rightarrow \\ QueueActivity.requeue .$$

$$DiagnosisState.Statevalue = List \rightarrow CountActivity.count .$$

$$DiagnosisState.Statevalue = Finished \wedge \\ Exists(DoctorState.Value = free) \rightarrow \\ QueueActivity.redistribute .$$

For reasons of space, other parts of the model for the medical process aren't listed here. It can be seen that process logics is well separated from business logics by event-rule mechanism in process service. It has finer granularity and more flexibility compared with WFMC model. At the same time, the reconstruction of the process also becomes much easier.

6 Conclusions and Future Work

WFMC provides an excessively loose framework to workflow realm, as result in large numbers of disagreements in workflow systems, even the term of workflow itself. On the other hand, one of the core issues, how to solve problems in the association of process logics and business logics, has been laid aside for a long time. It's considered the bottleneck in the development of process system in this paper.

So, a new process model is presented here. Some elements such as event, state and rule, are added in the process model and all of the elements in the model are formally defined. With the model, an executive mechanism based on event-rule form is presented. Now, a framework is constituted to support dynamic association of process logics and business logics. The software architecture of process server system will be researched in the next step, which includes the division of the modules, the design of the interface, the structure of the information and etc.

Acknowledgment. This work is supported by Youth Fund of university of Shantou (No. YR07004) and Science & Technology Project in Shantou (No. 36).

References

1. Luo, H.-b., Fan, Y.-s., Wu, C.: Overview of workflow technology. Journal of Software 11(7), 899–907 (2000)
2. Li, J.Z., Wang, H.Y.: Object-oriented event modeling method of workflow., Journal of Software (suppl.), 184–189 (2001)
3. Sun, R.-z., Shi, M.-l.: A meta-model supporting dynamic changing workflow. ACTA Electronica Sinica 30(12A), 2052–2056 (2002)
4. Hu, H., Wu, S.: Management system of dynamic workflow based on event-response. Journal of Huazhong University of Science and Technology (Natural Science Edition) 32(4), 177–184 (2004)
5. Wang, Y., Yao, S.-Z.: Research of an evolutional workflow model based on translating Petri net to ECA rules. Journal of System Simulation (suppl.), 131–135 (2005)
6. Zhou, J.-t., Shi, M.-l., Ye, X.-m.: State of arts and trends an flexible workflow technology. Computer Integrated Manufacturing Systems 11(11), 1051–1151 (2005)
7. Workflow Management Coalition, The workflow reference model. Technical Report, WFMC-TC00-1003, Hampshire: Workflow Management Coalition (1995)
8. Jang, J., Alan, F.: An event-driven workflow engine for service-based business systems., In: Proceedings of 10th IEEE International Enterprise Distributed Object Computing Conference, pp.233–242 (2006)
9. Wang, Y., Li, M.: ECA rule-based workflow modeling and implementation for service composition. IEICE Transactions on Information and Systems E89-D(2), 624–630 (2006)
10. Wang, Y., Li, M.: An ECA-rule-based workflow management approach for web services composition. In: Proceedings of 4th International Grid and Cooperative Computing Conference, pp. 143–148 (2005)
11. Chen, L., Li, M.: An ECA rule-based workflow design tool for Shanghai Grid. In: Proceedings of 2005 IEEE International Conference on Services Computing, pp. 325–328 (2005)
12. Xin, L.: The architecture of workflow on event mapping mechanism. In: Proceedings of The 6th International Conference on Distributed Computing and Applications for Business, pp. 392–395 (2007)

Conjugate Gradient Method Neural Network for Medium Resolution Remote Sensing Image Classification

Denghui Zhang[1] and Le Yu[2]

[1] College of Information and Technology, Zhejiang Shuren University, Hangzhou, China
[2] Department of Earth Sciences, Zhejiang University, Hangzhou, China
zhangdh8789@163.com

Abstract. Neural network (NN) classification has found wide use in remote sensing applications. There are a large number of ANN types available, and each focus on improving different classification performance. The conjugate gradient method is one of the efficient and low memory requirement methods. In this paper, a neural network using conjugate gradient method classifier is employed to classify three components derived by using principal component analysis to original six bands Landsat TM images. Comparison with a conventional classifier shows this NN performs better in both visualization inspection and quantitative evaluation.

Keywords: neural network, conjugate gradient method, remote sensing, classification.

1 Introduction

The artificial neural network (ANN), is a mathematical model inspired by the structure and/or functional aspects of biological neural networks; it consists of an interconnected layers (input, hidden, output) to processes information using a connectionist approach to computation. ANN has found wide use in remote sensing application (e.g. Jensen, 2005; Gopal & Woodcock, 1996; Weiss & Baret, 1999; Menzies et al., 2007). There are a large number of ANN types available, and each focus on improving different classification performance. Several studies investigated fast training techniques to find weights of neural networks by solving a set of linear equations (Barton, 1991; Rohani et al., 1992), methods including conjugate gradient algorithm (Chen & Manry, 1993) and SVD etc.

This paper use conjugate gradient algorithm neural network for a classification task on Landsat image. The experimental results show this type of NN performances better than conventional classification approach.

D. Zeng (Ed.): ICAIC 2011, Part II, CCIS 225, pp. 264–270, 2011.

2 NN Using Conjugate Gradient Method

The ANN model is a complex and nonlinear model. In this model, input neurons ($input_1$, ..., $input_n$), hidden neurons ($hidden_1$, ..., $hidden_m$), and p output neurons ($output_1$, ..., $output_p$). Let i, j, and k be the indices representing input, hidden, and output layers, respectively. The NN model can be noted as:

$$y_k = g_a\left(\sum_{j=1}^{h} z_j \beta_{jk} + \varphi_k\right) \tag{1}$$

$$z_j = f_a\left(\sum_{i=1}^{n} x_i w_{ij} + \mu_j\right) \tag{2}$$

Where g_a and f_a are activation functions. φ and μ are the bias for neuron. β and w are the weights.

The training of an ANN involves finding the optimal weight vector for the network. Many training techniques are available, they are generally compromising between computation cost and performance, and the most popular is the back propagation algorithm (Rumelhar et al., 1986).

The conjugate gradient method is an algorithm for the numerical solution of particular systems of linear equations, whose matrix is symmetric and positive-definite; it is an iterative method can be applied to sparse systems and solve unconstrained optimization problems such as energy minimization. For a large number of weights in NN, many conjugate gradient methods are efficient, and the memory required by these algorithms is proportional to the number of weights.

3 Experiments and Results

An experiment on using NN approach is conducted on a satellite remote sensing image – Landsat TM. The Landsat program is the longest running enterprise for acquisition of imagery of Earth from space and the instruments on the Landsat satellites have acquired millions of images. The Thematic Mapper (TM), which a major instrument on Landsat, is an advanced, multispectral scanning. Bands specifications for TM are listed in table 1.

Table 1. Landsat TM bands specifications

Band Number	Wavelength range (μm)	Resolution
1	0.45-0.52	30 m
2	0.52-0.60	30 m
3	0.63-0.69	30 m
4	0.76-0.90	30 m
5	1.55-1.75	30 m
6	10.4-12.5	120 m
7	2.08-2.35	30 m

In this experiment, only six bands, namely band 1, band 2, band 3, band 4, band 5 and band 7.

3.1 Test Image

The test image is shown in Figure 1, covered by several surface categories, there are vegetation, snow, soil, water, farmland, and shadow.

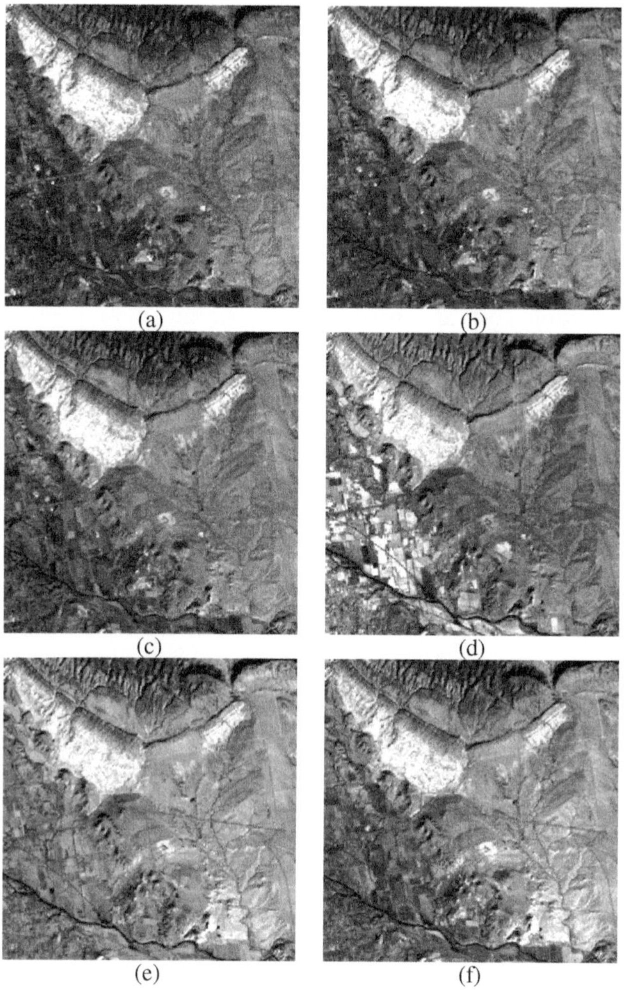

Fig. 1. Test image

A principal component analysis feature extraction process was conducted before classification. The results after PCA (Figure 2) shows the first three components tells most information (the variance contributions of increased PC are 53.94%, 74.35%, 87.56%, 92.14%, 96.12%, 100.00%, the first three components contributed 87.56% total information).

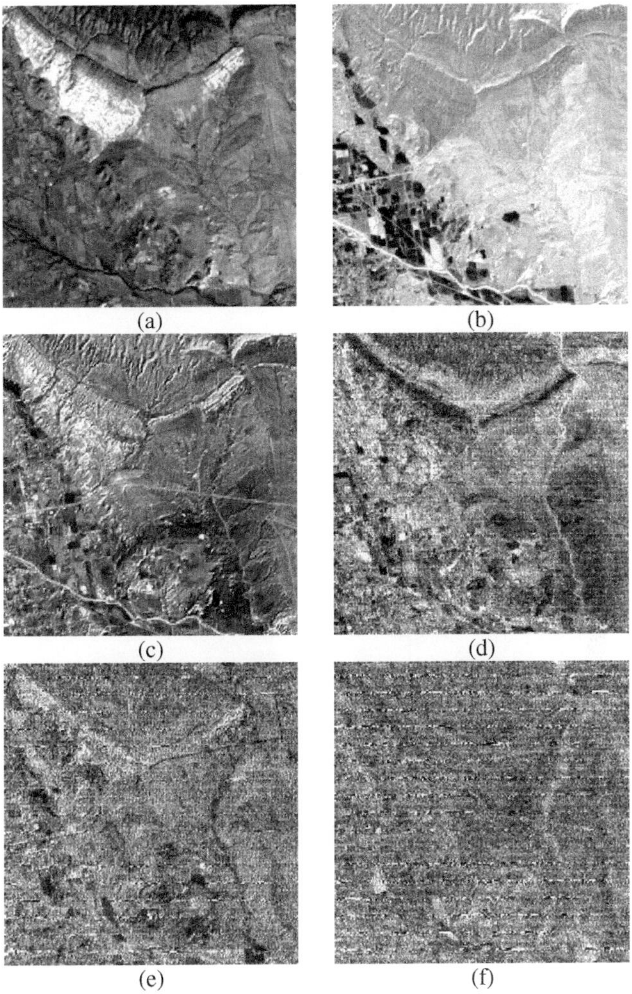

Fig. 2. Principal components

3.2 Classification Result

Classifications are conducted on the first three principal components. The classification results using NN and maximum likelihood (ML) are compared in Figure 3 and Figure 4.

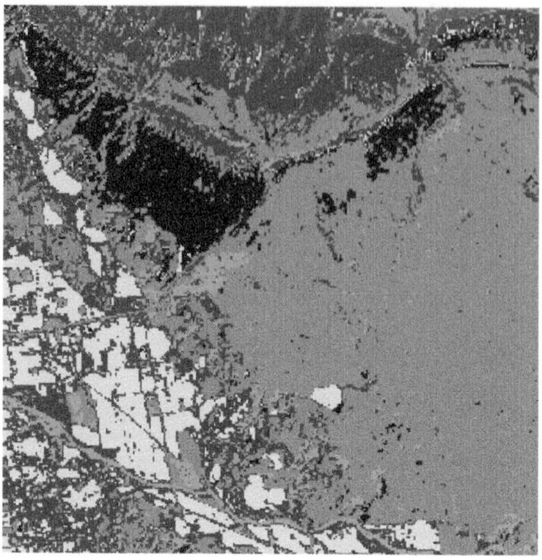

Fig. 3. Classification result using NN, Red: Vegetation, Green: Soil, Blue: Snow, Yellow: Farmland, Cyan: Water, Magenta: Shadow

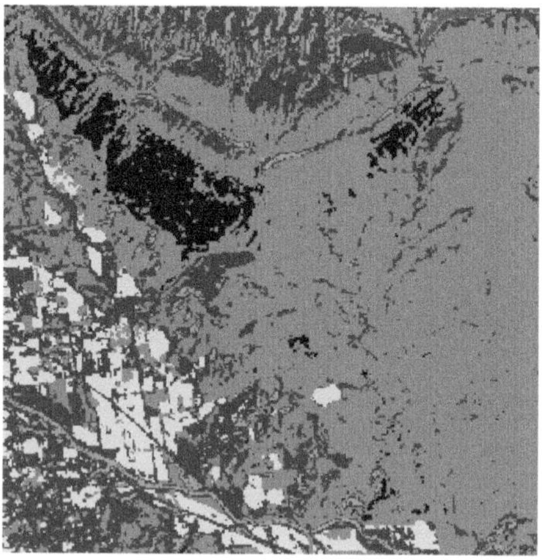

Fig. 4. Classification result using ML, Red: Vegetation, Green: Soil, Blue: Snow, Yellow: Farmland, Cyan: Water, Magenta: Shadow

Visual comparisons indicate that (1) both NN and ML are capable to classify those pre-defined classes, and (2) NN is a bit better than ML, especially for the vegetation (red) and snow (blue) area.

3.3 Quantitative Classification Accuracies Analysis

Validation samples are selected to quantitatively evaluate the classification results. Table 2 and Table 3 shows the classification accuracies: the overall accuracy of NN is about 4% higher that of ML.

Table 2. Classification Accuriecy of NN Method (C1: Vegetation, C2: Soil, C3: Snow, C4: Farmland, C5: Water, C6: Shadow)

	C1	C2	C3	C4	C5	C6
C1	100.00	0.00	0.00	0.00	0.00	0.00
C2	0.00	100.00	0.00	0.00	0.00	0.00
C3	0.00	0.00	100.00	0.00	0.00	5.88
C4	0.00	0.00	0.00	100.00	0.00	0.00
C5	2.67	0.00	0.00	0.00	85.29	24.53
C6	0.00	0.00	0.00	0.00	8.82	75.47

Overall Accuracy = 94.0722%.
Kappa Coefficient = 0.9286.

Table 3. Classification Accuracy of ML Method (C1: Vegetation, C2: Soil, C3: Snow, C4: Farmland, C5: Water, C6: Shadow)

	C1	C2	C3	C4	C5	C6
C1	93.33	0.00	0.00	0.00	8.82	0.00
C2	4.00	100.00	0.00	0.00	0.00	0.00
C3	0.00	0.00	100.00	0.00	0.00	0.00
C4	0.00	0.00	0.00	100.00	0.00	0.00
C5	2.67	0.00	0.00	0.00	76.47	32.08
C6	0.00	0.00	0.00	0.00	14.71	67.92

Overall Accuracy = 90.2062%.
Kappa Coefficient = 0.8821.

A detailed comparison on each class in terms of producer's accuracy and user's accuracy also indicate that NN method wins in almost all the values (only failed in the user accuracy for C2: soil).

Table 4. Accuracy comparision (C1: Vegetation, C2: Soil, C3: Snow, C4: Farmland, C5: Water, C6: Shadow)

	Prod. Acc.		User Acc.	
	ML	NN	ML	NN
C1	93.33	100	92.11	100
C2	100	100	95.83	94.52
C3	100	100	100	100
C4	100	100	100	100
C5	76.47	85.29	73.24	81.69
C6	67.92	75.47	78.26	86.96

4 Conclusion

In this paper, a neural network using conjugate gradient method classifier is used to classify three components derived by using principal component analysis to original six bands Landsat TM images. Comparison with a conventional classification method, namely maximum likelihood classifier, NN produced better result in both visualization inspection and quantitative evaluation.

Acknowledgements. This work is supported by the Zhejiang Provincial Natural Science Foundation of China (Grant Y107446), the Priority Theme Emphases Project of Zhejiang Province, China (Grant 2010C11045).

References

1. Jensen, J.R.: Introductory Digital Image Processing: A Remote Sensing Perspective, 3rd edn. Prentice Hall, Upper Saddle River (2005)
2. Gopal, S., Woodcock, C.E.: Remote sensing of forest change using artificial neural networks. IEEE Transctions on Geoscience snd Remote Senisng 34, 398–404 (1996)
3. Weiss, M., Baret, F.: Evaluation of canopy bioghysical variable retrieval performances from the accumulation of large swath satellie data. Remote Senisng of Envrionment 70, 293–306 (1999)
4. Menzies, F., Jensen, R.R., Brondizio, E., Maran, F., Mausel, P.: The accuracy of neural network and regression leaf area estimators in the Amazon Basin. GISience and Remote Sensing 44(1), 1548–1603 (2007)
5. Barton, S.A.: A matrix method for optimizing a neural network. Neural Computation 3(3), 450–459 (1991)
6. Rohani, K., Chen, M.S., Manry, M.T.: Neural subnet design by direct polynomial mapping. IEEE Transactions on Neural Networks 3(6), 1024–1026 (1992)
7. Chen, M.S., Manry, M.T.: Conventional modeling of the multi-layer perceptron using polynomial basis function. IEEE Transactions on Neural Netorks 4(1), 164–166 (1993)
8. Rumelhar, D.E., Hinton, D.E., Williams, R.J.: Learning internal representations by error progagation. In: Rumelhart, D.E., Mc-Clelland, F.L. (eds.) Parallel Distributed Processing: Exploartions in the Micriostructure of Cognition, MIT Press, Cambridge (1986)

Research on Real-Time Multi-dimensional Rendering Method of Fluid Phenomenon in Large-Scale Virtual Scene

Zhengwei He[1] and Kehua Su[2]

[1] School of Navigation,Wuhan University of Technology
Wuhan, Hubei, China
[2] Computer school,Wuhan University
Wuhan, Hubei, China
he_zhengwei@sina.com

Abstract. The real-time visualization of multi-dimensional spatial data of fluid phenomenon has always been a big issue. As to these temporal data, this paper provides a multi-dimensional rendering method for grid data of regular three-dimensional space--- dynamic partition projection. This method generates fluid space LOD dynamically through combining the idea of direct volume rendering and imposter technology, and then realizes the real-time scheduling of temporal space data rendering that based on temporal series. Experimental comparison shows that the method in this paper can reflect the relation between project and its order correctly and efficiently, and it can achieve real-time multi-dimensional visualization of fluid phenomenon in large-scale virtual scene.

Keywords: Real-time multi-dimensional rendering, fluid phenomenon, large-scale virtual scene, dynamic LOD generation, contaminative gas diffusion simulation.

1 Introduction

Generating smoke, contaminative gas, running water and other dynamic phenomena and objects in virtual three-dimensional scene has always been one of the most challenging research directions. The shapes of fluid phenomenon are irregular and uncertain, so they always have vague boundaries, and it is difficult to use a fixed formula to make qualitative description of their external shapes. Furthermore, since they are changing constantly, the complex variation makes the classical Euclidean geometry useless in their description, and rendering based on point, line, surface, volume (2.5 D) and other primitive in traditional computer graphics can not describe their evolution effectively.

In the field of spatial information, the representation of fluid phenomena is also based on the location representation, or using 2D to show its spatial area coverage [1]. There are no three-dimension or even multi-dimension visualized representation existing. In the research field of Computer Graphics and Visualization in Scientific Computing, there are lots of research fruits in three-dimension or multi-dimension

D. Zeng (Ed.): ICAIC 2011, Part II, CCIS 225, pp. 271–279, 2011.

rendering of the smoke, running water, cloud and other phenomena based on volume rendering algorithm[2-5]. In the field of virtual reality, the realization of visualization of smoke and fog is mainly based on the particle system [6-8]. However, most of them only describe the phenomena of small range with large calculated amount, which have serious influence on the real-time rendering. Many problems will emerge in the combination of single experimental environment and large-scale virtual scene. In the virtual environment, the real-time interaction between scene and other objects will be involved. Furthermore, it should meet the requirements on visual correctness (such as blocking relation), observation in any angle, real time, etc. Besides, the large scale of spatiotemporal data in fluid phenomena brings some difficulties in the limited hardware environment and real-time rendering.

Taking contaminative gas as an example, the new rendering proposed by this paper---dynamic partition projection does the multi-dimension (three dimensions + time) dynamic rendering addressing temporal order data of three-dimensional space field in contaminative gas diffusion process. It will show not only the internal spatial attribute of the contaminative gas, but also its evolutionary process, realizing the multi-dimensional dynamic visualization of real-time contaminative gas diffusion phenomena in virtual environment.

2 DYNAMIC Multi-dimensional Rendering of Real-Time

For the regular three-dimensional spatial data field rendering, there are 4 typical and frequently-used algorithms: ray casting, splatting, shear-warp, and the three-dimension texture mapping based on hardware. Volume rendering technique is suitable for the rendering of cloud and fog, providing the solution for vivid visualization of fluid phenomena. When it is adopted in the rendering of cloud and fog, the volume rendering technique is conducted in the targeting experimental environment in most cases. And in the virtual scene environment, the real-time interaction between scene and other objects will be involved. Furthermore, it should meet the requirements on visual correctness (such as blocking relation), observation in any angle, real time, etc. The existing rendering still has some problems in these aspects. Therefore, we put forward the dynamic partition projection to realize the real-time multi-dimension dynamic visualization for the fluid phenomena in virtual environment. Later, we will conduct the visualization experiment in contaminative gas diffusion by normal rendering and dynamic partition projection, and make comparisons between them.

2.1 The Basic Idea of Dynamic Partition Projection

Drawing from the combination of volume rendering algorithmic thoughts and imposter technique, the dynamic partition projection organizes the three-dimensional spatial data field to draw the visualization rendering in line with regular 3D grid way. The basic idea of it is that: first, redraw the data field in accordance with screen resolution or range of visibility (as Fig. 1), readjusting and generating dynamically the detail layering of the Fig.s; second, set an imposter to every subspace after redrawing, then project every voxel in the subspace to the imposter respectively according to the snow-ball throwing method (as Fig. 2), cumulatively calculating the voxel impacts gained from every pixel in imposter to get its color; finally, rendering these imposters in the graph rendering pipeline of the scene to get the final image.

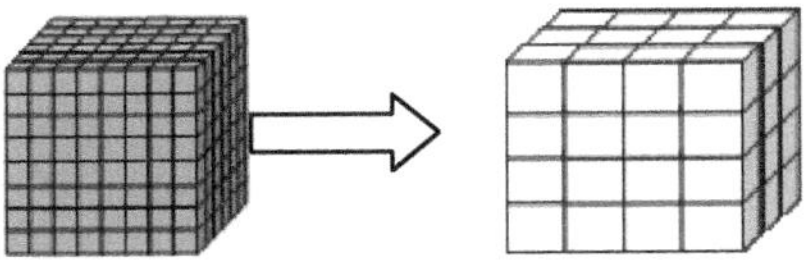

Fig. 1. Dynamic partition schematic

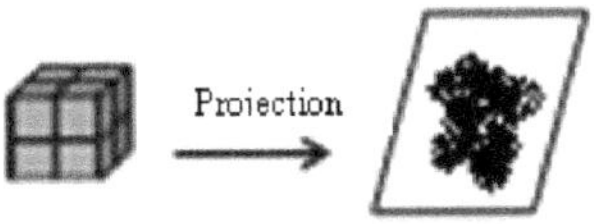

Fig. 2. Imposter projection schematic

2.2 The Algorithm Realization of the Dynamic Partition Projection

The specific processes of rendering by dynamic partition projection are as follows:

First, according to the requirements on the generated graph definition, spatially divide the data field in line with screen definition or range of visibility. The divided subspaces are multiple times to the single grid space of the former data field. Through the dynamic adjustment of the space subdivision, the different-detail-layering graphs, which are similar to LOD, can be generated. Based on the perspective projection principle, the derivation of space division standard (or LOD generation determination) by 3D spatial data field is as follows:

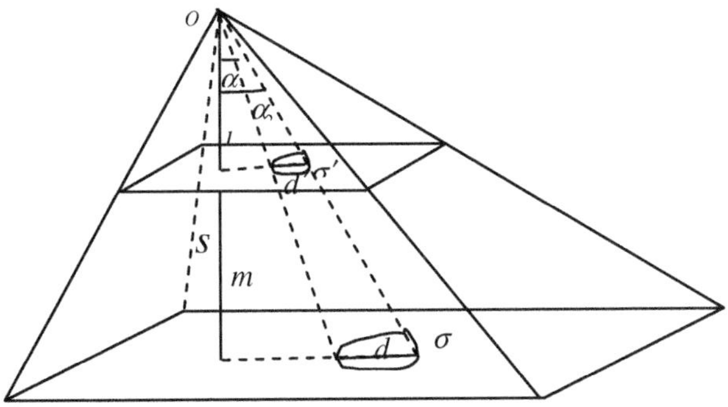

Fig. 3. Visual projection schematic

$$d' = d \cdot \frac{l}{l+m} = d \cdot {}^{l}\!/_{s}$$

$$\sigma' = \sigma \cdot \left(\frac{l}{l+m}\right)^2 = \sigma \cdot \left(\frac{l}{s}\right)^2$$

$$\sigma_{pixel} = \sigma' \left(\frac{L_{pixel\_num} \cdot W_{\text{pixel\_num}}}{Length_{frutum} \cdot Width_{frutum}}\right)$$

For the area of every space (single grid unit) projection on the screen (such as single area), we use a surface of the grid cube to do the approximate calculation.

$$\sigma_{pixel} = k \cdot \sigma_{grid} \cdot \left(\frac{l}{s}\right)^2 \cdot \left(\frac{L_{pixel\_num} \cdot W_{\text{pixel\_num}}}{Length_{frutum} \cdot Width_{frutum}}\right)$$

$$\sigma_{pixel} = \sigma \cdot \left(\frac{l}{s}\right)^2 \cdot \left(\frac{L_{pixel\_num} \cdot W_{\text{pixel\_num}}}{Length_{frutum} \cdot Width_{frutum}}\right)$$

In it, k is constant, as an accommodation coefficient

When σ_{pixel} is smaller than the threshold σ^i, the higher definition LOD should be rendered, meaning subdivision rendering of its grid. For i level of rendering, LOD_i should meet the following conditions:

$$\sigma_{pixel} \in \left(\sigma^{i+1}, \sigma^{i}\right] \tag{1}$$

In it, l, k, L_{pixel_num}, $W_{\text{pixel\_num}}$, $Length_{frutum}$, $Width_{frutum}$ are all constants.

Formula (1) is used to find out whether the subspace division can meet the visual threshold condition of its corresponding LOD_i.

Second, blank the divided subspaces, and estimate whether they are visual and are in need of rendering. If they are visual, generate an imposter for the subspace. In it, the normal of imposter is directing at the view point. The size of the imposter is that of the parallel projection area along line sight of the subspace. The center of imposer can be set as the geometric center of the subspace. The blanking and clipping are based on the grid joints. The specific methods are as follows:

The equations of the six surfaces of frustum respectively are:

$$A_i x + B_i y + C_i z + D_i = 0$$

Sight coordinates are $\left(X_e \quad Y_e \quad Z_e \quad W_e\right)$, M is viewing matrix.

$$\left(A_i \quad B_i \quad C_i \quad D_i\right) \mathrm{M}^{-1} \begin{pmatrix} X_e \\ Y_e \\ Z_e \\ W_e \end{pmatrix} \geq 0$$

In it, $i = \{0,1,\ldots,5\}$

The method of judgment is that: use the corresponding 8 vertices in the grid cube in the divided space to calculate the 6 equations. If 8 vertices all cannot meet the above 6 equations, this grid cube is invisible. It is necessary to blank the corresponding imposter of the grid space; otherwise, the imposter should be rendered.

Third, generate imposter. Project the grid in the data field, which is the voxel of the subspace, along the sight line to the imposter respectively. The corresponding grid or subspace "splatters" to do Splatting operation with the Westover [9] method. The impact and distribution of the voxel in the imposter image will be defined by Gaussian function to find out its contribution to the color and opacity of the imposter image (as Fig. 4). Then, project these Gaussian functions to imposter surface to generate images in the light of the calculated and sight-line Gaussian function integral stored tracking LUT.

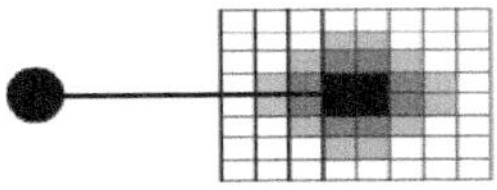

Fig. 4. The contribution of the single voxel (vexel) to the VOP

Supposing the color contribution of single voxel to imposter is σ_i , the color and texture compound result of imposter is:

$$\text{Tex}_{imposter} = \bigcup_{i=1}^{l} \sigma_i$$

l represents the basic voxel (or bottom grid) amount in the divided subspace.

The image obtained from the compound contributions of all basic voxel (l as the amount) VOP in the current subspace.

Fourth, visualize scene. After the generation of all the imposters in the visible subspace, in the scene pipeline, render all the imposters in the way of surface rendering and in line with the rendering order of objects in the scene and Z-Buffer depth. Finally, the scene graph containing all necessary visualized 3D data field can be obtained.

Fifth, generate the animation. The 3D spatial data generated in the process of evolution of fluid phenomena is the time sequence. In the rendering process, according to the time course, render the corresponding data field in the light of the above methods, and the four-dimensional (3D+time) visualized scene in the evolution process of the fluid phenomena can be gained. In the animation and cycle rendering refreshing, if the viewpoint or perspective projection environment do not change, all needed is to calculate the generated imposter texture. Otherwise, the projection space should be re-divided to calculate the imposter array and to form the new LOD level. The algorithm flow chart is as Fig 5.

3 Experiment Comparison and Discussion

In the virtual scene, the view angle and location of the observers may change constantly. Shearing-transforming algorithm has limitation on the observation angle, so it is not suitable for the real-time application in the virtual scene. Light projection algorithm and snow-ball throwing algorithm both rely on voxel to calculate the screen pixel. In the single experiment, both methods can generate the final image. However, in the virtual scene rendering, other objects exist in it. As a result of this, dodging, inserting, containing and other situations may happen between the objects. These two methods can not reflect the above relation between the project and its order, leading to scene error. After the modification, these two methods can be used in the virtual scene rendering---supposing that the viewpoint is close to volume data field, firstly, generate "screen" pixel by parallel projection in line with the algorithm along the sight line; then, taking "screen" image as a part of the scene, render it with other objects to get the final result.

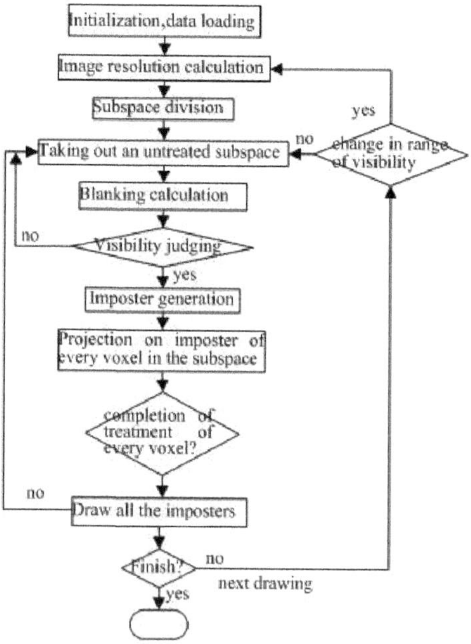

Fig. 5. Algorithm flow chart

Taking us for example, conduct the space visualization on volume data field. Based on the contaminative gas diffusion density field data, this paper uses light projecting, 3D texture projection based on GPU software and dynamic partition projection respectively to conduct visualization experiments in contaminative gas diffusion process. The experiments are conducted in PC, and the experiment environment is:

CPU:AMD 3000+ ,1.84G
DDR: 512M
GPU: NVIDIA GeForce 6600GT,128M VRAM, BUSTYPE - AGP 8X
GPU 3D texture maximum treatment size: 512×512×512
Graphics Engine: Microsoft DirectX9.0
System Software: Microsoft Windows XP Professional

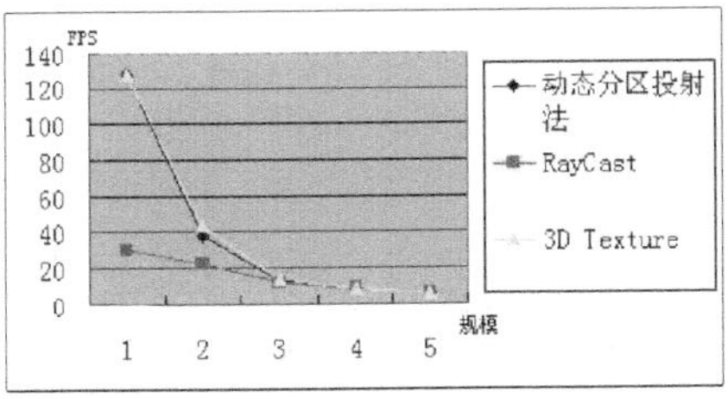

Fig. 6. FPS comparison of three algorithms

Fig. 7. Generated scene graph based on GPU software method

Fig. 8. Generated scene graph by dynamic partition projection

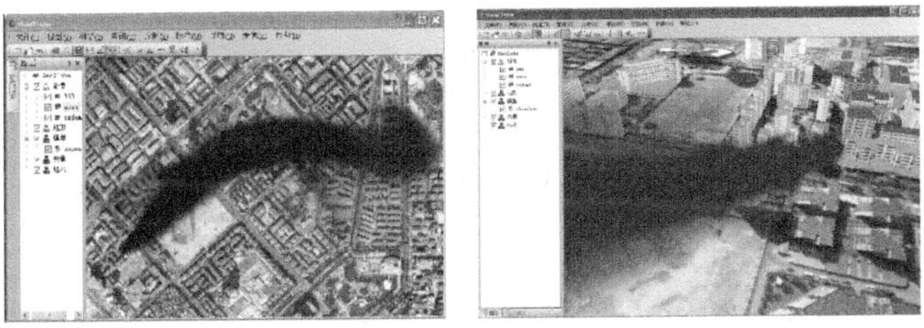

Fig. 9. Real-time visualized experiment rendering in the contaminative gas diffusion process

The experiments are conducted without any extra software speeding. Fig. 6 is the experiments of three methods in different scales: 1.16×8×8; 2.16×16×16; 3.64×16×16; 4.32×32×32; 5.64×32×32.

It can be seen from Fig.6 that in these three methods, light projection has the worst FPS. In the range of processing capability of GPU, the FPS of 3D texture projection is a little higher than that of dynamic partition projection. In the rendering scale, the two has similar overall FPS. The reasons for those are that: the light projection should have the integral along the light direction. The CPU should execute many calculations, which gives huge pressure to CPU, affecting FPS in a large extent. In the 3D texture projection, generate the 3D texture by GPU by means of cutting, and then, give the corresponding texture to every slice. In the 3D texture projection, GPU finishes the calculation similar to light integral. Furthermore, the calculation speed based on hardware is high, sharing the pressure of CPU to a large extent. So the FPS is higher. The imposter generation of dynamic partition projection is done by CPU, but it is similar to particle system in the rendering. After the projection, the calculation of light integral along the observation direction in the graph rendering pipeline is done by video card, transferring large amount calculation to graph software and balancing CPU pressure. The FPS of dynamic partition projection is much higher than light projection.

Fig.7 is the rendering scene graph of contaminative gas diffusion based on 3D texture projection. Fig.8 is a rendering scene graph of contaminative gas diffusion based on dynamic partition projection. Fig.9 is the real-time contaminative gas diffusion experimental rendering in large-scale scene ($1000\times1000m^2$), with average frame rate of 11.53 fps.

3D texture projection relies mainly on the processing capacity of GPU. Besides the requirement of the 3D texture processing capacity in the hardware environment, there is another limitation, which is the limitation on GPU processing scale. For example: the 3D texture maximum scale of NVIDIA GeForce 6600GT is 512×512×512. The texture who are larger than this cannot be processed. Now, because of the limitation on the GPU hardware development level, the large-scale fluid phenomena rendering cannot be done by this method. On the contrary, dynamic partition projection is not limited by this. In the rendering of large-scale fluid phenomena, apart from the update on hardware processing capacity, many software speeding technologies can also be adopted.

4 Conclusions

Directing at the multi-dimension visualization of spatial date field in fluid phenomena, the paper proposes the dynamic partition projection and makes comparisons among the methods of realizations of them. In the same experiment scale, dynamic partition projection can conduct real-time rendering in the complex virtual scenes to fluid phenomena. Moreover, without the limitations on hardware conditions, it can overcome the disadvantages of this method, handling correctly the scene blocking and relation between projects and order. It can realize the correct and vivid visualization in the virtual scene of contaminative gas diffusion process. As the hardware environment in the experiment is the low end PC, in the corresponding application and hardware conditions, it is possible to realize the real-time multi-dimension rendering in the large-scale scene of fluid phenomena.

With the development of modern technology, people, who are in the information age, is more and more expecting to show all the information related to the survival environment in a more direct and vivid way. The research of this paper can provide the exploration and solution to the real-time application and research in the field of virtual reality, simulation, games, etc.

In the future research, lighting effects and other realistic visualization research will be strengthened. It is possible to consider adopting parallel computing, grid computing, GPU and other ways to realize the real-time imitation and rendering of complex graphs by relying on net resources and software and hardware environment, upgrading the scale and visual effect of the virtual scene.

Acknowledgement. We acknowledge the financial supports of "the Fundamental Research Funds for the Central Universities" No. 2010-IV-034 and No.211275703) for this work.

References

1. Su, F.Z., Zhou, C.: Process Geographic Information System Frame Foundation and Prototyp Building. Geographical Research 25(3) (2006)
2. Rasmussen, N., et al.: Smoke simulation for large scale phenomena. In: ACM SIGGRAPH 2003 Papers. ACM, San Diego (2003); Californi Initialization, data loading Initialization, data loading a
3. Dobashi, Y., Yamamoto, T., Nishita, T.: A Controllable Method for Animation of Earth-scale Clouds. In: Proc. CASA 2006 (2006)
4. Schafhitzel, T., Falk, M., Ertl, T.: Real-Time Rendering of Planets with Atmospheres. In: WSCG International Conference in Central Europe on Computer Graphics (2007)
5. Bruneton, E., Neyret, F.: Precomputed Atmospheric Scattering. Computer Graphics Forum 27(4), 1079–1086 (2008)
6. Harris, M.J., Lastra, A.: Real-Time Cloud Rendering. Computer Graphics Forum 20(3), 76–84 (2001)
7. Angelidis, A., et al.: A controllable, fast and stable basis for vortex based smoke simulation. In: Proceedings of the 2006 ACM SIGGRAPH/Eurographics Symposium on Computer Animation. Eurographics Association, Vienna (2006)
8. Liu, S., et al.: Real time simulation of a tornado. The Visual Computer 23(8), 559–567 (2007)
9. Westover, L.: Footprint evaluation for volume rendering. Computer Graphics 24(4), 367–376 (1990)

Seismic Event Labelling in the Google Earth by Interactive User Interface Operation and by Program*

Hanming Huang[1,**], Haijun Zhou[1], Yinyan Chen[1], and Yinju Bian[2]

[1] College of Computer and Information Engineering
Guangxi Normal University
Guilin, Guangxi, China
Tel.: 13597338735
hhmmentor@gmail.com
[2] Institute of Geophysics
China Seismological Bureau
Beijing, China

Abstract. Google Earth has received popular acknowledgment due to its exuberant, dynamic and very low access overhead authorized geographic information. It also has been incorporated into many fields of earth science especially geographic information systems (GIS). Intense earthquakes may incur heavy damage of human life. More frequently occurring weak earthquakes may be the precursors of some strong earthquakes. Hence, the drawing and timely updating of historically and newly occurring earthquake epicentre distribution graph is one of the indispensable routine works for seismologists. Labelling earthquake or explosion's epicentre in Google Earth renders people more information about the quake and its surrounding area. And labelling observatory station locations of selected events is certainly helpful for the event's wave display, comparison, analysis and further researches. This paper achieves the programming drawing of epicentre distribution graph and introduces the selective labelling of observatory station locations which retrieve from event wave file (.evt format) on Google Earth. Finally, the potential application prospects of this paradigm in seismological researches are also discussed.

Keywords: Google Earth, KML, Earthquake Epicentre Distribution, Observatory Station.

1 Introduction

Google Earth which is the world's first desktop system to browse the global high-precision satellite images and three-dimensional data is commercial software but also opens free access to public [1]. It is launched by Google Company in 2005. It maps the Earth by the superimposition of images obtained from satellite imagery, aerial

* This work is financially supported by China Seismological Bureau on contract No. 200808003.
** Corresponding author.

D. Zeng (Ed.): ICAIC 2011, Part II, CCIS 225, pp. 280–286, 2011.

photography and geographic information systems in a three-dimensional earth model. Because Google Earth is a very good map platform, it has a huge and exciting prospect. It can not only give free access to the worldwide high-resolution satellite images, but it also provides the functions of processing superior image, controlling a map display based on API, operating map features and completing a certain task of spatial analysis. With appropriate code, we can make a GIS systems meet routine application and also scientific research. China is a country that frequently occur earthquakes. In order to make the seismic information more intuitively displayed, it is helpful to incorporate Google Earth with the seismic information processing system. It would have unparalleled advantages when releasing of the seismic information and dealing with emergency seismic events.

Google Earth is used in three-dimensional ways to view information, so the requirements of computer configuration are relative high. The minimum configuration of the software's running environment declared by the official website for Google Earth is as follows:

Operating System: Windows 2000, Windows XP, Vista, Windows7, Mac OS X, Linux
CPU: Pentium III, 500MHZ
Hard Disk: 400 MB free space
RAM: 128 MB
16 MB 3D-capable graphics card
Resolution of 1024x768, 16-bit High colour
Network Speed: 128 Kbps.

2 Development Approaches of Seismic Information in the Google Earth

There are two development approaches [2] in the Google Earth. The first is based on Google Earth API. It is mainly used to control the Google Earth view, operate the elements and achieve animation. Another approach is based on Keyhole Markup Language (KML). It's mainly used to generate geographic element files and achieve dynamic updating of elements. In most cases, two different ways can achieve the function of each other. Because it is easy to call the API provided by Google Earth, we need to combine these two different development methods in the seismic information processing system.

2.1 The Approach Based on Google Earth API

Any language that supports COM specification can write COM components and call their interfaces. Google Earth API also uses this principle. Program developers can use these languages for it, such as C#, Visual C++, Java, Delphi.

Several API interfaces been currently exposed by Google Earth can realize the functions that are required by the practical development and application of seismic information processing system. These functions include loading the KML files, controlling seismic information display, operating station and seismic interactive labelling and achieving dynamic display earthquake, etc.

IApplicationGE is the most important interface of Google Earth API. A number of functions can be realized through this interface, such as controlling the Google Earth program start-up and shutdown, operating the Google Earth window handle, accessing the camera field of view for the map, loading factor and converting coordinates. When we use other API interfaces, it is inevitably to apply IApplicationGE interface. Google Earth API [3] is shown in Table 1.

Table 1. Google Earth Api Interfaces

IApplicationGE	Operate the program and the handle
ICameraInfoGE	Control camera
IFeatureGE	Operate element
ITimeGE	Control the time and date
IAnimationControllerGE	Operate animation
IFeatureCollectionGE	Define the elements set
IViewExtentsGE	Control the range of coordinates
IPointOnTerrainGE	Access screen information
ISearchControllerGE	Search control interface
ITimeIntervalGE	Control time period
ITourControllerGE	Roaming interface

2.2 Another Approach Based on KML

The main task of interactive labelling is marking epicentres which read from earthquake catalog file or marking observatory station locations which extract from the seismic event file in the EVT format into the KML file [4], and then label these epicentres or station locations on the Google Earth. There are two types of placemark files in Google Earth. One is the KML file, the other is the KMZ file. KML which is an XML descriptive language has similar syntax and document structure with XML, built upon the features of XML about powerful data description and platform-independent. KML defines lots of the properties of space elements, such as point, line and surface. They can be explained and finally displayed on the Google Earth. KML file can be manually edited in the format of text file or generated by the program. Therefore, the use of KML file is very convenient to develop the Google Earth application. KMZ which is another form of KML file is the default output file format for Google Earth. When a KMZ file is unzipped, a single KML file is decompressed along with icon images referenced in the KML geographic map [5]. No matter what format you choice, it has little effect on programming. This paper uses KML file for labelling.

All classes of KML inherit Object which is the abstract base class. Several sub-classes like Feature, Geometry, ColorStyle, TimePrimitive and some abstract entity classes are derived from Object. KML mainly consists of element, style, perspective and animation, etc. The detailed class structure of KML2.2 version is shown in Fig.1.

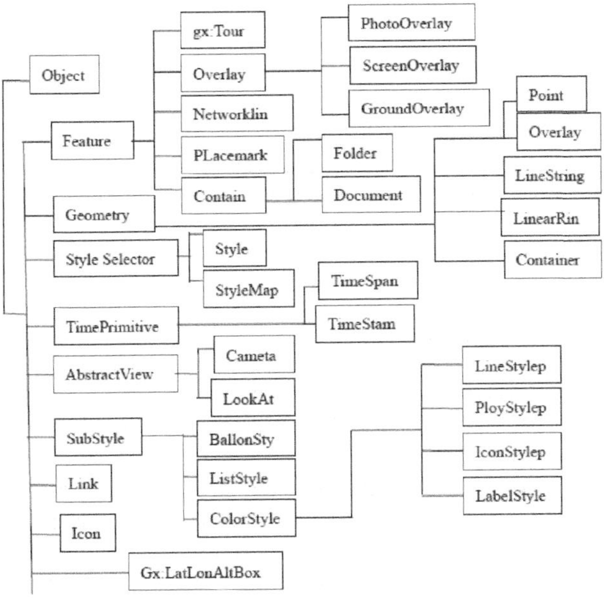

Fig. 1. The class structure of KML2.2 version

3 The Example of Seismic Application

The main purpose of applying Google Earth in seismic information system is to supervise the actual situation of earthquake events, placement situation of observatory stations and other seismic information, for timely understanding of the background of earthquake and taking the following measures for disaster preventing. In present implemented system, the main task is to label the placemarks and important seismic information into Google Earth, such as earthquake epicentre distribution, stations placement.

3.1 The Problem of Description of Seismic Events Interactive Labelling

Seismic events that have occurred are commonly having been recorded in some earthquake catalog, so the time of seismic occurrence, epicenter location, hypocenter depth and scale magnitude are read from earthquake catalog file. Seismic event files include seismic waveforms, locations (latitude, longitude and depth) of observatory stations, and some other seismic information. Since the vast majority of recorded seismic events are weak events, the observatory stations usually can only record near neighbor area events' seismic waves for weak events. In other words, the monitoring scope of observatory stations is very limited for weak event. Therefore, the three-dimensional map interface of Google Earth can be used to infer intuitively epicenters by stations with significant seismic wave recordings. So the newly occurring earthquake which has been recorded by some observatory stations but has not been localized yet can also be directly and interactively marked on Google Earth.

When we mark the earthquake epicentre and the selected event stations on Google Earth, the scope of the latitude and longitude need be firstly determined. These values which read from the earthquake catalog or the EVT file can be calculated. The KML placemark file which is generated by the earthquake epicentre or the location of coordinates of each station can be loaded into Google Earth with a program (such as C # code). The program also provides the interface labelling, removing the label information and other functions.

3.2 The Realization of Seismic Events Interactive Labelling Program

For referring to Google Earth in Visual Studio 2008, one should open "Project" → "Add Reference" → "COM" and add "Google Earth 1.0 Type Library". The file is named “googleearth.exe”. Then one should add “using EARTHlib” in the C# class file. After that, the interfaces and classes which are provided by Google Earth can be called ordinarily.

Google Earth application can be started through the program that is “ApplicationGEClass GeApp = new ApplicationGEClass ()”. If one wants to achieve visualization of the Google Earth window in the C#, the map window should be operated. In this way, Google Earth would start itself independently. So it needs to hide timely the main window of Google Earth. The main window handle of Google Earth can be acquired by the function of “GetMainHwnd” and the Google Earth interface can be controlled by the function of “SetWindowPos”. The handle of the map window can be got by the function of “GetRenderHwnd”, and the parent window can be changed by the function of “SetParent” which come from dynamic link library. The handle of “Form” will be set as the parent of the map window handle of Google Earth. The signature of “SetParent” function is as follows:

```
// set the map window of Google Earth as the parent window of C#
[DllImport("user32")]
public extern static IntPtr SetParent(IntPtr hWndChild, IntPtr hWndNewParent);
// “hWndChild” expresses the handle of map window of Google Earth;
// “hWndNewParent” expresses the handle of new parent window;
```

Although Google Earth does not provide access to get the screen coordinates, the function of “GetPointOnTerrain- FromScreenCoords” which offers an opportunity to operate the screen can change pixel coordinates of screen to latitude and longitude coordinates. In a certain range of latitude and longitude, the contents of the screen can be manipulated by adding another chart layer in Google Earth interface; meanwhile, Windows API can also capture elements of the map window.

This paper implements the dynamic labelling by means of KML file. Under normal circumstances, the latitude and longitude coordinates and some important seismic information tagging on Google Earth can be obtained from the seismic document. This information must be converted into a mode which is accepted by Google Earth. The nature of the KML file is a text file, so this system directly writes the information into a text file and changes the suffix into KML. A set of latitude and longitude information is stored as a KML file. It is benefit for the whole system to maintain and manage the system at any time. According to realistic need, the placemarks can be easily deleted and added.

3.3 Interactive Labeling of Seismic Events and Observatory Stations Placement

The distribution of stations and epicentre of the seismic event 20061106172613-1.evt is shown in Fig.2, where triangle represents the station and the circular is on behalf of the seismic event. At the same time, the size of the seismic graphic label represents the level of the earthquake. According to the actual graphics, labelling ratio of the size of the seismic graphic label is ScaleMode, as in (1). SM is on behalf of the earthquake magnitude. ScaleMode is on behalf of the radius of circle. The latitude and longitude of station are read from EVT files, and the event has recorded 111 seismic stations. These stations are distributed over the capital seismic zone. Fig.3 shows actual distribution of 37 natural earthquake epicentres in the capital seismic zone from 2003 to 2007.

$$\text{ScaleMode} = 1 + \text{SM} * 0.2 \tag{1}$$

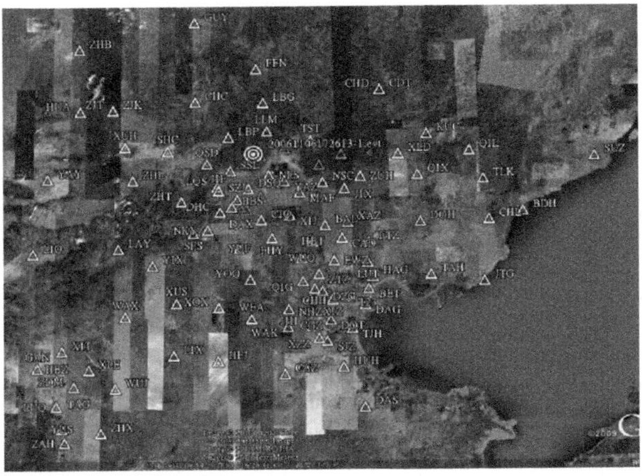

Fig. 2. Placement of observatory stations and epicentre of the seismic event 20061106172613-1.evt

Fig. 3. The natural earthquake epicentre distribution graph in the Beijing seismic zone from 2003 to 2007

3.4 Summary of Interactive Labelling Procedure

The location of earthquake epicentres and observing stations are included in the seismic event files. This information can be marked by KML on Google Earth, while the placemark is also described with some basic information about the earthquake. Especially in some areas where earthquakes have frequently taken place, a new earthquake epicentre will be marked on Google Earth in time, when the new seismic event file does not recorded by observatory stations. This will be an important role in the earthquake relief.

In the three-dimensional seismic information display system, Google Earth can achieve the earthquake dynamically display the location of ground stations and the epicentres. It avoids the original single display and improves the efficiency of workers in the earthquake research. But now the numbers of interfaces provided by Google Earth are limited. To a large extent, the lack of dynamic editing of three-dimensional data restricted its application in the seismic data processing.

4 The Perspectives of Potential Application

When the seismic information (epicentre distribution, station location map, etc.) is marked on Google Earth, it is effective to use the geographic image to facilitate the development of the system. We can further improve the system in some ways. Each station of the seismic events records waveforms in three directions. According to the actual needs, we can display and compare different seismic waveform in different stations. The earthquake epicentres can be located by the wave records and observatory station. We can also evaluate intensity map after knowing magnitude and estimate earthquake losses. With the constantly updating the Google Earth version, it will introduce more useful interfaces to the programmer and provide better opportunities to make seismic information system. The further development of seismic information system is very likely to promote the maturation of Google Earth application.

References

1. Qian, M.A.: Intellectual Map: the Secret of Google Earth / Maps / KML Core Development Technology. Electronic Industry Press, Beijing (2010)
2. Lu, B., Liu, Y.: Program Realization of Exploration Spot's Display in Google Earth. Resources Environment & Engineering 21(4), 450–454 (2007)
3. Jiang, X., Qin, F.: Research of Virtual Campus Based on Google Earth Technology. Computer Knowledge and Technology 36(5), 272–275 (2009)
4. Brown, M.C.: Hacking Google Maps and Google Earth. Wiley Publishing, Chichester (2006)
5. Sun, E., Nieto, A., Li, Z.: GPS and Google Earth Based 3D Assisted Driving System for Trucks in Surface Mines. Mining Science and Technology, 138–142 (January 2010)

Performance Analysis of GPS, GLONASS, GALILEO and Integrated GPS-GALILEO in China and Its Neighboring Area*

Shiwei Tian, Guangxia Li, Jiang Chang, Yangzhi Li, and Xiang Tian

Institute of Communication Engineering,
PLA University of Science and Technology,
Nanjing 210007, China
S.W_Tian@hainan.net

Abstract. This paper focuses on the performance of GPS, GLONASS, GALILEO and Integrated GPS-GALILEO in China and its Neighboring Area in order to provide some references to the design and construction of COMPASS. The four satellite navigation systems are briefly reviewed. Positioning with pseudo-range is analyzed. The performance is studied by software simulation. The results of simulation are presented using the number of visible satellites as a measure of availability, Position Dilution of Precision (PDOP) as a measure of accuracy with spatial and temporal variations. The results could be beneficial for the design and construction of Chinese COMPASS system.

Keywords: GPS, GLONASS, GALILEO, COMPASS, PDOP.

1 Introduction

Currently, there are mainly four Global Navigation Satellite System (GNSS) systems, including the United Stated (US) Global Positioning System (GPS), the Russian GLONASS System, the future European GALILEO System and Chinese COMPASS System.

GPS is a line-of–sight, all weather, world-wide continuously available satellite-based radio frequency system that is capable of providing three-dimensional position, velocity and time to users with an appropriate receiving equipment [1]. The nominal GPS constellation consists of 24 satellites, operated in six orbital planes. The orbital radius is about 26,560km and the inclination is 55° [2]. In each plane there are four operational satellites and the satellites are not evenly distributed within the planes. At present the constellation actually consists of 30 operational satellites. For the moment

* This work is supported by National High Technology Research and Development Program of China ("863" Program) (2008AA12Z307).

D. Zeng (Ed.): ICAIC 2011, Part II, CCIS 225, pp. 287–293, 2011.

we assume that the nominal 24-satellite constellation will be maintained. Currently, GPS transmits on the L1 and L2 frequency. Future satellites will also transmit on the L5 frequency [1].

GLONASS is a satellite navigation system operated by Russia's Ministry of Defense. The system consists of a 21 active satellites, operated in 3 orbital planes and nominally inclined at 64.8° to the equator [2]. Each orbital plane also contains a spare constellation, thus there are 8 satellites in each orbital plane placed orbiting at an altitude of 25500Km from the mean surface of the earth, with a period of approximately 11 hours 15 minutes.

GALILEO is a joint initiative of the European Commission and the European Space Agency. GALILEO will be put into operation by 2013 under the agreement signed by the EU transport ministers in November 2007. The designed GALILEO constellation consists of 30 satellites, and the satellites are evenly divided within three orbital planes with an orbital radius of 29,994 km and an inclination of 56° [3]. However, we will only consider the 27-satellite constellation here. GALILEO signals will be transmitted on the E2-L1-E1, E5a-E5b and E6 frequencies.

The COMPASS (or beidou) navigation system is China's entry into the realm of GNSS. The current design is to have a system comprised of 5 Geostationary Earth Orbit (GEO) satellites and 30 Non-GEO satellites (Inclined Geosynchronous Earth Orbit satellites and Medium Earth Orbit satellites) [4]. The Medium Earth Orbit (MEO) satellites will operate in six orbital planes to provide global positioning coverage.

In this paper we focus on the performance of GPS, GLONASS, GALILEO and Integrated GPS-GALILEO in China and its Neighboring Area. Positioning with pseudo-range will be analyzed and the measures for performance analysis will be described. The performance of GPS, GLONASS, GALILEO and Integrated GPS-GALILEO will be show in terms of visible satellites, Position Dilution of Precision (PDOP) with spatial and temporal variations.

2 Positioning with Code and PDOP

The linearized code measurement is given by [5]

$$\delta P = G\begin{bmatrix}\delta x\\ \delta b\end{bmatrix} + \tilde{e}_p \tag{1}$$

Where δP is the vector of predicted minus actual code measurement. δx is the three position offset of the user from the linearization point, and δb is the offset of the user time bias. The geometry matrix G is a $(m \times 4)$ matrix characterizing the satellite-receiver geometry; G only depends upon the line–of-sight effect. m is the number of measurements, and $\tilde{e}_p$ is the residual error vector.

The dilution of precision (DOP) is a measure for the geometrical strength of the observation model. Different types of DOP values are distinguished. Here only PDOP is considered. The DOP values depend on the cofactor matrix, $Q = (G^T G)^{-1}$. The PDOP is defined as

$$PDOP = \sqrt{Q_{11} + Q_{22} + Q_{33}} \tag{2}$$

The estimate of the rms value of the three position errors, $rms(x)$, may thus be given as [6]

$$rms(x) = \sigma_{UERE} \cdot PDOP \tag{3}$$

Where σ_{UERE} is the accuracy of the range measurements. The achievable position accuracy depends both on the accuracy of the range measurements and the satellite-receiver geometry. The lower the DOP, the more favorable the geometry will be, the higher the quality of the position estimate generally. In this paper, the PDOP is applied to evaluate the performance of positioning with code.

3 Performance Analysis and Simulation

In this section, the expected performance of GPS, GLONASS, GALILEO and Integrated GPS-GALILEO in China and its Neighboring Area is studied using software simulations [7, 8, 9]. The performance is evaluated spatially as well as temporally by analyzing the spatial and the temporal variations of the number of visible satellites and PDOP.

The number of visible satellites and the PDOP of the GPS, GLONASS, GALILEO and Integrated GPS-GALILEO were simulated respectively. Two simulations (spatial simulation and temporal simulation) were conducted.

In the spatial simulation, the receiver-satellite geometries were simulated in China and its neighboring area. In the temporal simulation, the receiver-satellite geometries were simulated at Beijing, from July 1, 2007, 12:00 to July 2, 2007, 12:00 with a sampling interval of 300 s. The spatial simulation and temporal simulation yielded spatial variations and temporal variations of the number of visible satellites and PDOP, in the case of GPS, GLONASS, GALILEO and Integrated GPS-GALILEO respectively.

3.1 Spatial Variations

Before considering the temporal performance of GPS, GLONASS, GALILEO and Integrated GPS-GALILEO, the spatial performance is analyzed.

Table 1 summarizes the spatial variations of the number of satellites, PDOP in the case of GPS, GLONASS, GALILEO and Integrated GPS-GALILEO.

Table 1. Spatial Variations of Pdop and the Number of Visible Satellites

		GPS	GLONASS	GALILEO	Integrated GPS and GALILEO
PDOP	MIN	1.50	2.02	1.24	1.03
	MAX	7.74	10	2.79	2.17
	MEAN	2.48	7.76	1.96	1.39
Number of Visible Satellites	MIN	5	2	7	12
	MAX	11	5	13	21
	MEAN	8	3.1	9.5	17.5

Fig. 1. shows the spatial variations of PDOP for GPS, GLONASS, GALILEO and Integrated GPS-GALILEO. The minimum PDOP of GPS, GLONASS, GALILEO and Integrated GPS-GALILEO are 1.50, 2.02, 1.24 and 1.03, and the average PDOP of the four systems are 2.48, 7.76, 1.96 and 1.39.

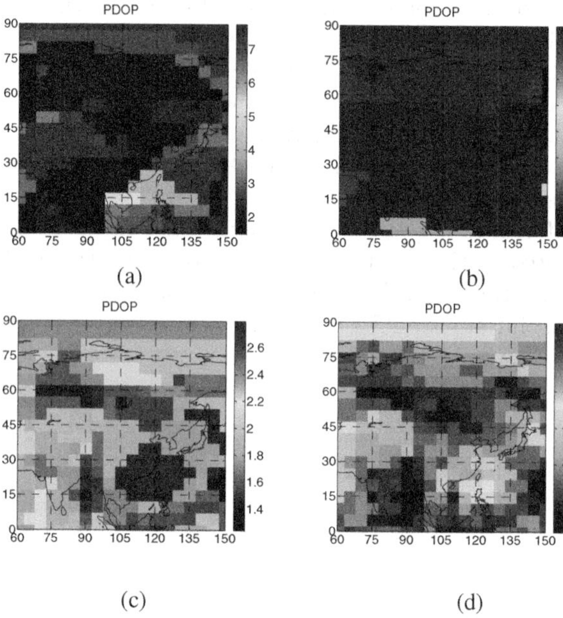

Fig. 1. Spatial variations of PDOP. (a) GPS. (b) GLONASS. (c) GALILEO. (d) Integrated GPS-GALILEO.

Fig. 2. shows the number of visible satellites as a function of geographic location in the case of GPS, GLONASS, GALILEO and Integrated GPS-GALILEO. The maximum number of the four systems is 11, 5, 13 and 21, and the average number of visible satellites of the four systems is 8, 3.1, 9.5 and 17.5.

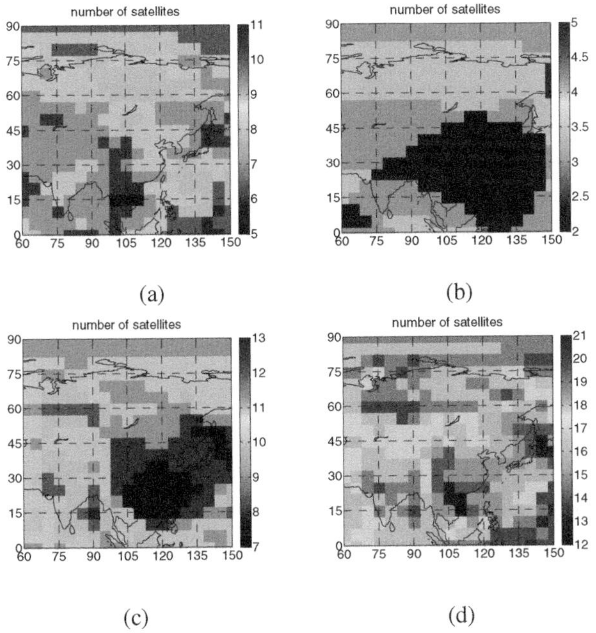

Fig. 2. Spatial variations of the Number of Visible Satellites. (a) GPS. (b) GLONASS. (c) GALILEO. (d) Integrated GPS-GALILEO.

3.2 Temporal Variations

Fig. 3. shows the PDOPs of GPS, GLONASS, GALILEO and Integrated GPS-GALILEO respectively as a function of time from July 1, 2007, 12:00 to July 2, 2007, 12:00 in Beijing. The minimum PDOPs of GPS, GLONASS, GALILEO and Integrated GPS-GALILEO are 1.41, 2.8, 1.29 and 0.96, and the average PDOPs of the four systems are 2.6, 9.54, 1.93 and 1.35.

Fig. 4. shows the temporal variations of the number of visible satellites in the case of GPS, GLONASS, GALILEO and Integrated GPS-GALILEO respectively from July 1, 2007, 12:00 to July 2, 2007, 12:00 in Beijing. The maximum number of GPS, GLONASS, GALILEO and Integrated GPS-GALILEO is 11, 5, 12 and 21, and the average number of visible satellites of the four systems is 7.87, 2.33, 8.6 and 16.47 respectively.

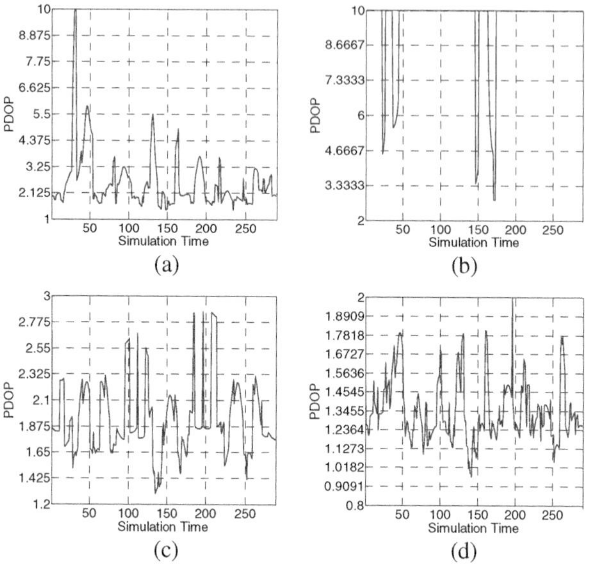

Fig. 3. Temporal variations of PDOP. (a) GPS. (b) GLONASS. (c) GALILEO. (d) Integrated GPS-GALILEO.

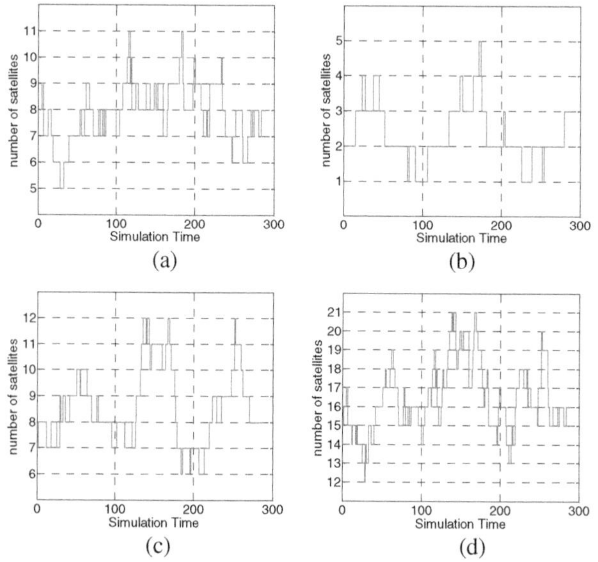

Fig. 4. Temporal variations of the Number of Visible Satellites. (a) GPS. (b) GLONASS. (c) GALILEO. (d) Integrated GPS-GALILEO.

The results show that GALILEO can provide a more favorable accuracy as well as a more favorable availability than GPS and GLONASS, and Integrated GPS-GALILEO will not only improve the number of visible satellites but also offer the best PDOP in the four systems.

4 Conclusion

This paper focused on the performance of GPS, GLONASS, GALILEO and Integrated GPS-GALILEO in China and its neighboring areas. The three satellite navigation systems have been briefly reviewed. Positioning with pseudo-range has been discussed, and the measures for performance evaluation have been described. The expected performance of the GPS, GLONASS, GALILEO and Integrated GPS-GALILEO has been obtained using software simulations and described by the spatial and temporal variations of the number of visible satellites and PDOP. The results show that GALILEO can provide a more favorable accuracy as well as a more favorable availability than GPS and GLONASS in China and its neighboring regions, and the results will be beneficial to the design and construction of Chinese COMPASS system.

Acknowledgment. The authors would like to acknowledge professor Jing Lv and professor Gengxin Zhang for discussions on GNSS. The authors would also like to acknowledge Ms. Sandra Verhagen at the faculty of Aerospace Engineering of the Delft University of Technology, the Netherlands, for software support.

References

1. Tran, M.T., Dao, T.H.D.: Global Navigation Satellte Systems (GNSS): An Overview
2. Kaplan, E.D., Leva, J.L., Pavloff, M.S.: Understanding GPS: Principles and Applications. Artech House, Boston (1996)
3. Kuusniemi, H.: User-Level Reliability and Quality Monitoring in Satellite-Based Personal Navigation, Tampere University of Technology, Finland. Publication 544 (2005)
4. `http://www.Beidou.gov.cn`
5. Misra, p., Enge, P.: Global Positioning System, Signals, Measurements, and Performance. Gange-Jamuna Press (2001)
6. Carnbinca, C.: Regional to global satellite based navigation systems. In: IEEE Position Location & Navigation Symposium, pp. 25–33 (1988)
7. Verhagen, S.: Studying the performance of Global Navigation Satellite Systems-A new software tool. GPS World 13(6), 60–65 (2002)
8. Wu, F., Kubo, N., Yasuda, A.: Performance Evaluation of GPS augmentation using Quasi-Zenith Satellite System. IEEE Transactions on Aerospace and Electronic Systems 40(4) (October 2004)
9. Verhagen, S.: Performance analysis of GPS, Galileo and Integrated GPS-Galileo. In: Proceedings of the 15th International Technical Meeting of the Satellite Division of the Institute of Navigation, Portland, OR, September 24-27, pp. 2208–2215 (2002)

Thread Schedulability in Windows CE

Xiaochun Wang and Hongwei Sun

Department of Computer Engineering, Shenzhen Polytechnic
Shenzhen, Guangdong, China
wang_xc307@sohu.com

Abstract. Real-time performance is very important to a real-time operating system (RTOS). It is also very important to determine schedulability of thread scheduler before we use the RTOS in our products. There are a lot of methods to value thread schedulability in practical application. Rate_Monotonic Scheduling Algorithm is an optimal static-priority processor scheduling algorithm; it is widely used in RTOS. We discuss the RM algorithm used in Windows CE , and give some useful conclusion.

Keywords: schedulability, rate monotonic, thread scheduling, Windows CE.

1 Foreword

Windows CE is used in both real-time environments and environments that are not real-time. To improve performance in a real-time environment, Microsoft suggest user to avoid using components that are not real-time. They also suggest user should only use functionality such as demand paging, power management, and device I/O in an appropriate manner because these components and functionality are not specifically designed for real-time application.

Real-time performance is defined in Windows CE as follows: [5]

Guaranteed upper bound on high-priority thread scheduling — only for the highest-priority thread among all the scheduled threads.

- Guaranteed upper bound on delay in executing high-priority interrupt service routines. The kernel has a few places where interrupts are turned off for a short, bounded time.
- Fine control over the scheduler and how it schedules threads.

Generally, Windows CE used as a real-time operating system in different application, such as mobile handhelds, telecommunication devices, or medical instruments. A real-time application is an application designed to manage time-critical systems, the unique characteristic of a real-time application is that it not only provides the correct response, but also responds within a specified period.

So, it is important to ensure threads scheduler will meet all threads deadlines in Windows CE operating system. Rate_Monotonic scheduling algorithm (RM) is an optimal static-priority algorithm. That is, if any static-priority algorithm can produce a

D. Zeng (Ed.): ICAIC 2011, Part II, CCIS 225, pp. 294–298, 2011.

feasible schedule, so can RM.[1] Here we will discuss how to use RM algorithm to determine thread schedulability in Windows CE.

2 Thread Schedulability in Windows CE

2.1 Rate_Monotonic Scheduling Algorithm

Given a set of n periodic threads, S={t1,t2,... ,tn}. Thread in the set is described as ti=(Ti,Ci,Di,Pi,Ui)(i=1,... ,n).

The notation used above is: Ti: period of thread ti, Ci: execution time of ti, Di: absolute deadline of thread ti, Pi: priority of thread ti, Ui: processor utilization of thread, that is Ui=Ci/Ti.

For threads set S={t1,t2,... ,tn}, we add an additional notation:

$$Wi(t) = \sum_{j=1}^{i} (Ci)[\frac{t}{Tj}]$$

$$Li(t) = Wi(t)/t$$

Wi(t) is the total amount of work carried by threads t1,t2,… tn, initiated in the interval [0,*t*].

RM Scheduling Algorithm:

1) If thread ti is RM-schedulable, when and only when:

$$Li = \min_{\{t \in Ti\}} Li(t) \le 1$$

2) If the entire set S is RM-schedulable, when and only when:

$$L = \max_{\{1 \le i \le n\}} Li \le 1$$

We can use RM algorithm to determine thread schedulability in an easy way. If the total utilization of the threads is no greater than n(21/n-1), where n is the number of threads to be scheduled, the RM algorithm will meet time demand of all threads. That is, all threads can be scheduled before their deadlines come.

We can also use RM Algorithm in Windows CE. But there are a lot of conditions before we use RM Algorithm, we will discuss them in details.

2.2 Sporadic Threads Released in the System

In RM algorithm, we think that all threads are periodic. In practical application, there are some threads which are released irregularly. These threads are called sporadic threads.

We consider the worst situation happened. In every thread period, there is a sporadic thread released. In thread set S, there are n periodic threads to be scheduled, and there are also n sporadic threads released at the same time. When thread ti is released, sporadic thread sti is also released corresponding with the thread ti, but the priority of thread ti is higher than sti, so sporadic thread sti runs first.

Example1. There are three threads, t1, t2, t3, the periods are T1, T2, T3, the execution times are C1, C2, C3, the priority are P1, P2, P3, and P1<P2<P3. We assume that each thread is released at time zero. There will be three sporadic threads released at

the same time, they are st1, st2, st3, execution times are SC1, SC2, SC3, and their priority are higher than the thread released at the same time.

For thread t1: Because of sporadic thread st1, to ensure thread t1 can be scheduled, the condition must be satisfied is $C1+SC1 \le T1$.

For thread t2: In the interval [0,t], thread t1 has been released [t/T1] times, and st1 has been released [t/T1] times too. To ensure that thread t2 can be scheduled properly, thread t1 and sporadic threads st1 must be iterated [t/T1] times, and there must be some time left to finish thread t2 and sporadic thread st2. That is:

$$t=(C1+SC1)\cdot t/T1+(C2+SC2)$$

We must find a t to meet the demand that $t \ge (C1+SC1)\cdot t/T1+(C2+SC2)$ and $t \le T2$.

For thread t3: In the interval[0,t], We must find a t to meet the inequality $t \ge (C1+SC1)\cdot t/T1+(C2+SC2)\cdot t/T2+(C3+SC3)$ and $t \le T3$. If this t can be found, the thread t3 can be scheduled.

For threads set S:

For threads set S={t1,t2,... ,tn}, in each thread period, there will be a sporadic thread released, sporadic thread queue is ST = {st1,st2,st3, ... ,stn}. Sporadic thread sti will be released at the same time with thread ti, we must find a t to meet:

$$t \ge \sum_{j=1}^{i}(Cj+SCj)[\frac{t}{Tj}] \text{ and } t \le Ti,$$

Then the thread ti can be scheduled.

So, in the worst situation, when each thread is executed, there is a sporadic thread released. We must meet the conditions described as:

The thread ti($1 \le i \le n$) is RM-scheduling, when and only when:

$$\min_{\{t\in Ti\}}\{\sum_{j=1}^{i}(Cj+SCj)[\frac{t}{Tj}]\Big/t\} \le 1$$

$$\max_{\{1\le i\le n\}}\{\sum_{j=1}^{i}(Cj+SCj)[\frac{t}{Tj}]\Big/t\} \le 1$$

In practical application, we must calculate the execution time of each sporadic thread, and analyze the total amount of threads execution time to ensure every thread can be scheduled.

2.3 Priority of the Thread

There are some assumptions when we use RM algorithms. One of them is the priority of a task is inversely related to its period, if thread ti has a smaller period than thread tj, ti has higher priority than tj.

In Windows CE, the priority of thread can be defined by programmer. So, the thread priority isn`t inversely related to its period, can we use the RM algorithm in Windows CE?

Liu and Layland proved that any threads set, which can be scheduled in any other static-priority processor scheduling algorithm, can also be scheduled by RM algorithm. Proof is: We assume that there is a thread set S, thread ti and thread tj are two threads that their priorities are adjacent, Ti>Tj and Pi<Pj. Now, we exchange the priority of this two threads, we can prove that this threads set can also be scheduled. According this, we can prove that any other static-priority processor scheduling algorithm can be changed to RM algorithm.

So, RM algorithm can be widely used in any case of static-priority scheduling algorithm. [3] Based on this conclusion, RM algorithm can also be used in thread scheduling in Windows CE .

2.4 Priority Inversion

Priority inversion occurs when a mutex, critical section, or semaphore resource held by a lower-priority thread delays the execution of a higher-priority thread when both are contending for the same resource.

To correct this situation and free the higher-priority thread, Windows CE enables the lower-priority thread to inherit the more critical thread's priority and run at the higher priority until it releases its use of the resource. [5]

But, priority inheritance can lead system to deadlock. There is an easy way to avoid deadlock, which is priority ceiling protocol. Windows CE doesn't support this protocol. And in practical, we must calculate the ceiling of the threads associate with the semaphore resource first, this may affect the execution efficiency of the system.

An easy way to resolve the deadlock of the priority inheritance is changing the way to get the critical resource. The detail is that each thread must apply the critical resource in a fixed way. For example, one thread needs two critical resource, r1 and r2. The thread must apply the r1 first, and then r2. To apply resource r2 first will not be permitted. This is a rule to all threads that may apply the critical resource r1 and r2. It is a good way to avoid the deadlock of the priority inheritance.

This is very useful to engineer to develop their products. We must list the critical resource first, and then list all threads. Then we can determine the sequence of critical resources that may be used by threads.

Microsoft also recommends that engineers avoid priority inversion by keeping all threads waiting on the critical resource at the same priority.

3 Conclusion

Windows CE has become more and more useful in embedded application system. Make sure that every thread can be released before their deadline comes is a important work to do. In practical development, we must consider different conditions that may affect the performance of the system, we discuss several factors here to give engineer some useful advices.

References

1. Krishna, C.M.: Real-time Systems. Tsinghua University Press, Beijing (2001)
2. Berger, A.: Embedded System Design. Publishing House of Electronics Industry, Beijing (2002)
3. Liu, J.-X., Wang, Y.-J.: An Improved Rate Monotonic Schedulability Test Algorithm. Journal of Software 15, 799–814 (2004)
4. Luolei: Real-time embedded operating system and application. Beihang Press, Beijing (2005)
5. Microsoft Windows CE5.0 Documents

The Compensation Control of Cutting Tool Position for Non-axisymmetric Optical Surface Machining*

Changping Song[1] and Yuanliang Zhang[2]

[1] Department of Mechanical Engineering Technology, Dalian Vocational Technical College
Dalian, Liaoning Province, China
[2] School of Mechanical Engineering, DaLian University of Technology
Dalian, Liaoning Province, China
songcp5484@163.com

Abstract. To eliminate the hysteresis error of piezoceramic servo-tools during axial asymmetric optical surface machining, three feedforward controlling mathematical models were established. These methods include voltage increasing models, the peak voltage dropping models and arbitrary point voltage dropping models. Based on these models, a series of feedforward controlling compensation experiments show that the maximal displacement error is 0.04μm. This result indicates that, by employing the feedforward controlling mathematical models, the displacement error compensation could greatly reduce the hysteresis error in machining.

Keywords: Feedforward controlling model, Axial asymmetric optical surface, Piezoceramic tools.

1 Introduction

As the rapid development of the electro-optic technology, high accuracy of optical instrument is required[1]. There is some limitation on improvement of the optical instrument accuracy because of the inevitable aberration of the lens and mirror. The wavefront corrector composed of the axial asymmetric optical surface can correct some aberration, and this can greatly improve the optical instrument accuracy[2]. If the corrector is introduced into the large telescope system, it will lower the system cost greatly.

The axial asymmetric optical surface machining is a new technology with high precision in this field, and it will become popular in the future. During performing the axial asymmetric optical surface machining, the intrinsic nonlinearity and hysteresis of the piezoceramic tools can affect the accuracy seriously[3]. How to control the position of the piezoceramic tools precisely becomes the key to assure the accuracy of the axial asymmetric optical surface machining[4].

* This study was supported by the "NSFC" Project under the contract No. 58975236.

D. Zeng (Ed.): ICAIC 2011, Part II, CCIS 225, pp. 299–305, 2011.

2 The Principle of Axial Asymmetric Optical Surface Machining

Each point on the axial asymmetric optical surface is asymmetric about its axis. The axial size z is not only the function of radial size x, and also a function of angle θ, that is Z=f(x,θ). During machining the axial asymmetric optical surface, while the spindle rotates and the turret fed in radial direction, the tool must be axially fed rapidly as the variation of the radial dimension and angle. Fig.1 illustrates the feeding principle of the axial asymmetric optical surface machining.

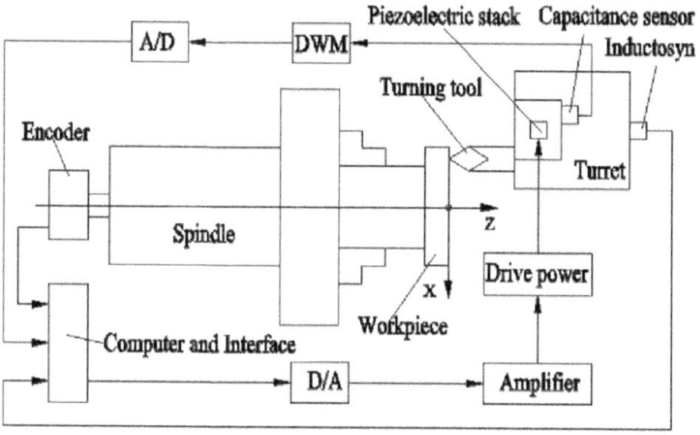

Fig. 1. Schematic diagram of machining principle

The axial rapidly micro-movement of diamond turning tool is driven by a pile of piezoceramics, while the pile of piezoceramics is controlled by a synchronous sign of the spindle angle and the turret movement in radial direction. In a moment, computer works out the axial coordinate of the point based on a specific mathematical model (as Zernike aberration equation), according to the data tested by the encoder and the inductosyn. Through D/A conversion, amplifying circuit and driving power make piezoceramics flex, the tool can make axial micro-movement, and then produce the axial asymmetric optical surface.

3 Position Compensation of Piezoceramic Tools

Surface machining precision of workpiece is affected by cutting tool axial position compensation according to the principle of axial asymmetric optical surface machining. Nonlinearity and hysteretic feature of piezoceramics will cause cutting tool displacement error. Fig.2 illustrates voltage increasing and voltage dropping in practical measurement measurement.

The mathematical model of the voltage increasing and voltage dropping curve can be got through the experiment for nonlinear of one-way stroke (voltage increasing or voltage dropping). Then piezoceramic tools will be controlled by the mathematical model. But it cannot eliminate the hysteresis phenomenon which will affect the cutter position accuracy. Because each telescopic range (or rises, step-down trip) is different in the machining process, voltage increasing and voltage dropping are different (as Fig.3 shows).

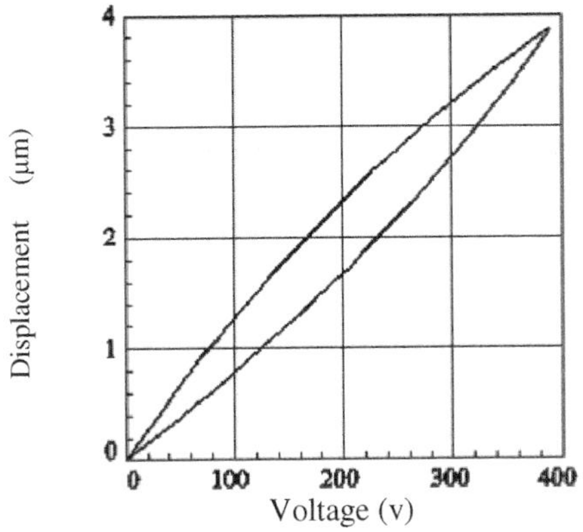

Fig. 2. Displacement of piezoceramic tools——Voltage Curve

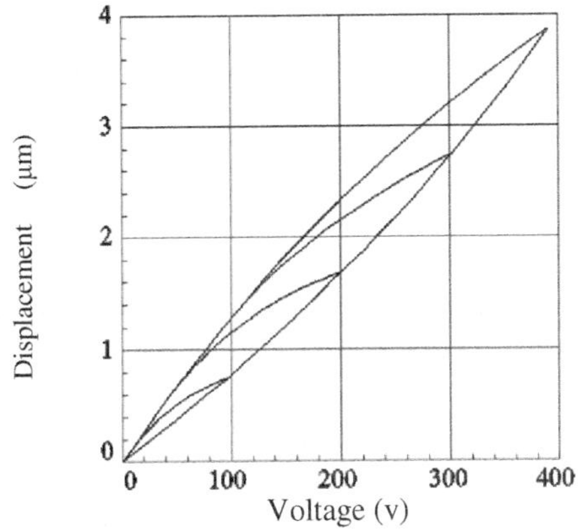

Fig. 3. Displacement of different stroke ——Voltage Curve

This makes the effects of piezoceramic tools position to the input voltage unpredictable. If the problem can't be solved, it will seriously limit the application of piezoceramics institutions. At present, many scholars engaged in this field, and put forward all kinds of error compensation control methods. To sum up, the main methods are as following:

3.1 Average Curve Method

We fit out the actual average tools through the experiment. Cutting tool position will be controlled according to the average curve model.

The advantage of this method is that mathematical model is simple and convenient. The disadvantage is imprecise. The authors also made some experiments by this method, and the maximum displacement error value is 0.2 μm.

3.2 The Self-learning Control

Among all sorts of self-learning control methods, P integral control is one of the most simple control methods, and you needn't to know model of the controlled object. The equation below is P integral controller:

$$u_j(\mathrm{k}) = u_1(\mathrm{k}) + k_r \sum_{i=1}^{j-1} a^{j-1-i} e_i(\mathrm{k}+\mathrm{d}) \tag{1}$$

In the formula, j is repetitions, u_j (k), u_j (k) is the output of controller for the first time and j times, e_i(k) is error sequence of the i times control process, a^{j-1-i} is index weight coefficient, a is called forgetting factor.

For P integral control, the precision of the process control can be improved by learning in less knowledge or unknown conditions. However self-learning control is based on the assumption that the control process can be repeated. For axial asymmetric optical surface of workpiece, the process of tools control is not repeated. Therefore, P integral control is not suitable for this processing.

3.3 Closed-Loop Control

Closed-loop control is that the actual displacement of cutting tools is detected by displacement sensor signal and computer will control the difference between the actual positions.

The controlling depends on displacement signal, not on piezoceramics hysteresis loop mathematical model. Theoretically speaking, it is very precise. The authors also did some experiments. It proved that in closed-loop control, the requirement for the transmission speed and accuracy is very restricted, especially high control precision of the sensor drift in the experiment. Therefore, the closed-loop control is not effective.

3.4 Abbreviations and Acronyms

Feedforward control is to compensate for the position of a piezoceramic tools in open-loop control system by specified magnetic hysteresis loop model.

Experiment results show that if increase the input voltage from zero (shown in fig.5), the voltage increasing curve will be uniquely determined. Expressed in polynomial as:

$$u_f(\mathrm{v}) = \sum_{k=0}^{m} \varphi_k v^k \tag{2}$$

In which, u_f(v) is the voltage dropping displacement of the piezoceramics tool, φ_k is coefficient of polynomial, determined by the experiment, m is the order number, v is voltage applied to piezoceramics.

Experiment also show that if voltage drop from saturation state (the maximum output state of piezoceramic tools), (see fig.5), the voltage dropping curve will also be uniquely determined. Expressed in polynomial as:

$$u_r(\mathrm{v}) \sum_{k=0}^{m} p_k v^k . \tag{3}$$

In which, u_r (v) is the voltage dropping displacement of the piezoceramic tools; p_k is coefficient of polynomial, determined by the experiment.

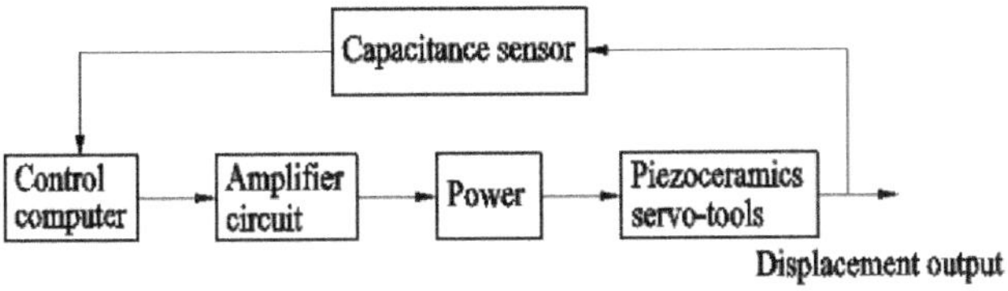

Fig. 4. Closed-loop control diagram

If voltage does not drop from saturation state, the voltage dropping curve will not be uniquely determined (see fig.5). As shown in fig.5, the nearer to the refraction points, the more compensation Δμ will be got. (The deviation between the actual displacement with arbitrarily input voltage and that of saturation voltage drop curve is called compensation). The maximum amount of compensation is:

$$\Delta u(v_r) = u_r(v_r) - u_f(v_r). \tag{4}$$

In which, u_r (v_r) and u_f (v_r) is the displacement of voltage dropping curve and voltage increasing curve corresponding to break over voltage v_r input.

As voltage drops on zero, Δu becomes nearly zero too. According to experimental results, Seung-Bae Jung, from Korea, deduced the following polynomial:

$$\Delta u / \Delta u(v_r) = \sum_{k=0}^{m} \psi_k [(v_r - v)/v_r]^k . \tag{5}$$

In which, ψ_k is coefficient of polynomial, determined by the experiment. The compensation of arbitrary tool position Δu can be obtained through equation (4) and (5).

Equation (2), (3) and (5) express the desired feedforword control model.

The above models are validated in our experiment. The maximum displacement of piezoceramic tools we used is 4.0 μ m, the voltage is 400v. The error of tool position is less than 0.04μm. The polynomial coefficient fitted by least square is shown in table 1.

4 Cutting Experiment

Cutting experiment is carried out on modified ultra-precision ultra-precision disk lathes. Industrial control computer is selected. Cutting tools is the natural single crystal diamond tools. The mathematical model of the workpiece surface is Zernike equation:

$$\begin{aligned} Z &= Z4e^2\cos(2\theta) + Z5e^2\sin(2\theta) \\ &+ Z6(3e^2 - 2)e\cos(\theta) + Z7(3e^2 - 2)e\sin(\theta). \end{aligned} \tag{6}$$

Z4, Z5, Z6, Z7 are Zernike coefficients. In the experiment, respectively, Z4 =1.7, Z5= 1.2, Z6 = 1.8, Z7 = 0. e=x/R, x is actual radius (radial dimensions) R is maximum surface radius. θ is angle. External diameter of the machining workpiece is 60mm, inner diameter is 10mm, feed is 5 um/r, the biggest cutting depth is 5 um, the workpiece material is brass.

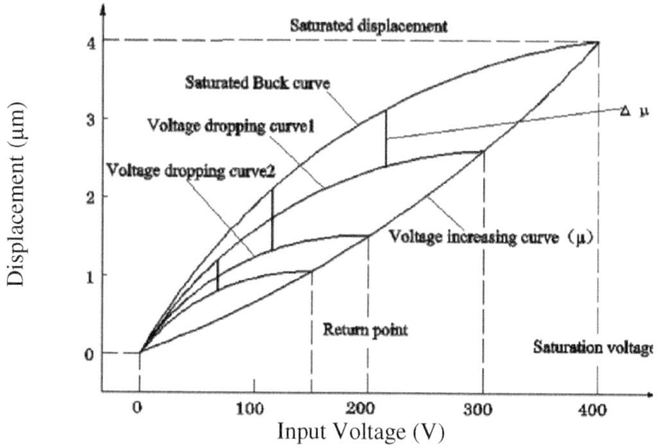

Fig. 5. Curve schematic diagram of limit state (Zero and Saturated state)

Table 1. Polynomial coefficient

$u_f(v)$:	φ_0=-5.5×10^{-3}	φ_1=8.699×10^{-3}	φ_2=3.0737×10^{-6}
$u_r(v)$:	ρ_0 =1.46×10^{-2}	ρ_1 =1.03955×10^{-2}	ρ_2 =-1.3976×10^{-6}
Δu / (v_r):	ψ_0=0.97	ψ_1=-2.426	ψ_2=3.215

In order to save time, the equation 6 is changed, some constants and trigonometric function are calculated using C language. The calculated data will be listed in memory. When the workpiece is processed, simple calculation with the transverse and angular displacement signal and the data from memory is enough. The computer control speed is increased greatly.

After processing, the workpiece is measured in radial and axial direction with high precision roundness measuring instrument and roughness profiler, and then the data is processed using multiple regression analysis method. The result in table 2 shows that the relative error is below 9.5% which is satisfactory.

Table 2. Zernike polynomial (wavelength) of wave front corrector

Zernike	input	output	absolute error	relative error %
Z4	1.7	1.59	-0.11	6.5
Z5	1.2	1.14	-0.06	5
Z6	1.8	1.97	+0.17	9.5
Z7	0.0	0.04	+0.04	

5 Conclusions

A. The feedforward controlling mathematical model is feasible for position compensation controlling of piezoceramic tools in open loop system.

B. The way of compensating the displacement error based on the feedforward controlling mathematical models can dramatically increase the machining accuracy.

Acknowledgment. We are grateful to National Natural Science Foundation of China (project No.58975236) for financial support.

References

1. Dale, S.R., Love, G.D., et al.: Wavefornt correction using a self-referencing phase conjugation system based on a Zernike cell. Opt. Commun. 191, 31–38 (2001)
2. Machowski, J.M., Pinard, L., Doqnin, L., Ganau, P., Lagrange, B., Michel, C., Morgue, M., et al.: Wavefront control. Opt. Quantum Electron. 31(5), 507–514 (1999)
3. Bifano, T., Bierden, P., Zhu, H., Cormelissen, S., Kim, J.H.: Megapixel wavefront correctors. In: Bifano, T., Bierden, P., Zhu, H., Cormelissen, S., Kim, J.H. (eds.) Proceedings of SPIE - Int. Soc. Opt. Eng. Advancements in Self-adjusting Optics, vol. 5490(3), pp. 1472–1481 (2004)
4. Zhou, Z.M., Zhang, Y.L., et al.: Application of single linear blade diamond tools to cutting of unitary wave-front corrector. In: 2009 International Forum on Information Technology and Application, vol. 3, pp. 69–71 (2009)

Positively Impelling Financial Innovation in China's Rural Areas

Liwen Wang

School of Economics and Finance, Xi'an Jiaotong University
Shaanxi Xi'an, Nostate 710061, China
liwen_lee@fastmail.fm

Abstract. Lack of financial innovation in China's rural areas has restricted rural economic development seriously. This thesis analyzes the characteristics of China's rural finance, the features and current situation of innovation, as well as the factors hindering innovation. Furthermore, based on foreign exemplary, countermeasures and suggestions are proposed on advancing rural financial innovation in China, which include constructing financial innovation mechanism oriented by the market, impelling multi-level financial innovation, actively cultivating talents of financial innovation, creating favorable environment for innovation, and so on.

Keywords: rural financial innovation, system innovation, innovation in vocational work.

1 Introduction

Innovation is the essence of progress of a country or a nation. Lack of financial innovation in China's rural areas has severely affected the development of "three *nong*" (agriculture, rural areas and farmers) and social stability. At present, in developing agricultural economy, increasing farmers' income, and constructing new countryside, capital is not only needed for massive investment, but also to be reasonably allocated and efficiently used. Lack of financial innovation in China's rural areas directly affects the ability of rural financing industry to raise funds, and also affects efficiency in allocating rural financial resources. At the same time, since the transitional period after entry into WTO has already expired, it is significant to confront challenge in international market by innovating in rural finance, supporting and protecting agriculture by financial means.

2 The Characteristics of Financial Innovation in Rural China

A. The rural financial innovation is basically a government-led, or led the financial regulatory authorities, rather than market-driven, customer involvement is not enough. Financial innovation has not yet become effective financial institutions reduce costs, increase profits, an important means to improve competitiveness.

D. Zeng (Ed.): ICAIC 2011, Part II, CCIS 225, pp. 306–310, 2011.

B. Imitative innovation in rural financial innovation and more, mainly for the expansion of the quantity, quality is poor. Focus on innovation and easy to learn, easy to operate, low-tech construction of the external form, such as financial institutions and the creation, expansion and other financial services, and in the operating mechanism of innovation is obviously insufficient, the real initiative by the Chinese, with Chinese characteristics of the original innovation rare.

C. Ignore the costs and benefits of innovation. Innovation is often biased in favor of disorderly competition in the market share, regardless of cost effectiveness do not speak. The lack of competitive, innovative products, leading the rural financial market "price war" smoke constantly, eroding the profitability of financial institutions and viability.

D. Rural financial innovation is not perfect in the environment and the context. For example, the monopoly of the financial system, financial control is still serious (including interest rates, business scope, capital markets, etc.), which have allowed financial institutions to innovation space is very restricted.

3 Lack of Financial Innovation in Rural China Mainly

China's rural financial innovation on the whole still in its infancy, the low level of rural financial development, overall development is not perfect, resulting in imbalance between capital supply and demand in rural areas, has seriously hampered the rapid development of rural economy. Especially with the deepening of banking reform, withdrawal of state-owned commercial banks in rural and remote areas; rural credit cooperatives reform is slow, a heavy historical burden, business strength is not strong; China Agricultural Development Bank of the narrow scope of operations, etc., limits its "three Agriculture "in support. Lack of formal innovation in rural financial development, causing symptoms including private fund raising, non-governmental cooperation will, of bidding clubs, underground banking and other non-formal financial system, and its potential to play an active role at the same time retain a greater risk. Lack of financial innovation in rural China the main features:

A. Institutional innovation is still in the surface state. Financial innovation is combined with the system, financial innovation must be interpreted from the systems have a real effect. Financial Innovation in China developed a great relationship with the legal system, inadequate rural financial system innovation in rural financial system in short supply.

B. The business of product innovation and low efficiency. Few rural financial services products, savings and loan exchange is basically a business, investment and wealth management products are almost none. Lack of product innovation and dynamic business, nor the procedural steps in accordance with the implementation of scientific innovation, making the business of product innovation and inefficient. When the business benefits of product innovation, cost, lack of consideration, the introduction of new products, new business from customers in, which in turn dampened the incentive to innovate.

C. The weak foundation of technological innovation. Rural Finance severe weak foundation of technological innovation, financial electronic spread slowly, based on electronic, information technology, network-based application of modern technology in the rural financial seriously behind urban finance, online banking, bank card business in rural development is slow, some extent, caused by high operating costs of rural financial institutions, falling profits and declining competitiveness.

D. Regulatory innovation lags behind. Separate supervision of financial institutions under the framework of the regulatory focus on compliance, the real risk of regulatory system has not yet formed. Coordination of separate supervision combined with the cost of larger, resulting in some lack of supervision of financial services, some businesses are in control, difficult to improve the efficiency of supervision. And serious market barriers to competition in the existing financial institutions, inadequate, causing financial institutions to ignore the market, cost and customer, slow growth at low levels.

4 Solutions and Suggestions on Boosting China Rural Financial Innovation

Using foreign experiences on rural financial innovation for reference, I suggest we adopt the following measures to push up the innovation in China.

A. Construct a market-oriented mechanism for financial innovation

Firstly, build up and improve the mechanism of rural financial innovation and development. This means to program the innovation from an overall perspective. Innovation and R&D groups need to be launched, and the creation and development should be consistent with the principles, features, process and steps required by innovations. Different development mechanisms need to be built up according to different hierarchies of financial innovations so that the costs are reduced and efficiency increased. Secondly, establish and promote the implementation mechanisms on rural financial innovation. This means to apply the creative ideas in practice, transform them into practical production. This process should be prudential with all possible results sufficiently evaluated and advantages and disadvantages traded off. The implementation of institutional and systematic innovation should first take the pilot process in specific regions and institutions and enlarge the range after gaining some experiences and becoming relatively stable step by step until fully implemented. Last but not least, up build and improve the incentive mechanisms for rural financial innovation. This needs measures to encourage, support and protect the innovation. The measures could include providing moderate subsidies for necessary investments, launching venture funds for business and products creation to transfer and reduce risks, adopting protective policies for innovation fruits, putting creative experts into important positions and premiums for successful innovation figures and allowing them to deduct a percentage from the benefits. The construction of an innovation culture and a sound innovation-encouraging atmosphere is also suggested and should be enhanced.

B. Promote financial innovations through multiple hierarchies

Above all, the institutional and systematic innovation should be boosted. We can push the financial originality from the perspective of macroeconomic development, which includes the creation of new institution on rural finance so that the enthusiasms of all financial institutions could be brought into play as much as possible. For example, apply the rural financial system of a mixed business and supervision model with banking, insurance and security integrated together so as to sufficiently take advantage of economical benefits on human resources, branches and techniques; Further improve and reform rural financial supervisory system, lower the threshold of market admission and exit approval to provide more flexibility and self-determination for financial institutions and enhance their vigor; Encourage civil finance to develop based on risk-controllable practices so that normal and supplementary institutions support each other and develop in harmony. Secondly, the origination of management mechanisms should be boosted. A self-developing and self-improving managerial system is needed for rural financial institutions. We should allow them to self-manage and self-finance within its business spectrums. Special policies such as subsidies, favorable conditions and tax breaks could be implemented for rural credit cooperatives and the newly-categorized three sorts of institutions to support the development of "villages, farmers and agriculture". Set up the incentive and restraint mechanisms for postal savings to flow back to rural regions and direct banks, insurers, security companies and financial holding corporations to invest in and support the development of Agriculture, Rural areas and Farmers with more efforts. Thirdly, promote products and business line innovations. We should direct financial institutions to enhance the development of rural financial business lines and products and build up mechanisms to stimulate the innovation. We can also subsidize the necessary investment for innovation, launch the venture funds for business and product creation to transfer and mitigate the risks. Fourthly, push up the technical service innovation. Financial institutions could get help on the setup of technical service centers with personnel, funding, wares and necessary technical support. Meanwhile, we should direct the technical sharing and cooperation among financial institutions, conduct joint projects or technical abutments for large and medium-sized technical programs so as to transform techniques to realized production, promote the upgrade and update of financial technical systems and the development of businesses. Finally, boost the supervisory innovation. We should implement integrated business and supervisory models, focusing on risk-based supervision supplemented by compliance overseeing. The supervisory conceptions should also be updated and methods and measures improved to conduct targeted supervision over rural financial institutions. It is also important to strengthen the supervisory cooperation among jurisdictions, continuously improve the supervision practice in the wake of the changes in international situations.

C. Devotedly Cultivate group of persons with special abilities on financial innovations and compose sound environment in favor of creations

Firstly, we should cultivate experts on rural financial innovations through diversified methods. We can organize innovation teams by introducing outside experts and sending inside members out to study. Incentive mechanisms should also be built up to inspire enthusiasm of creation. The propaganda, education and business training on financial origination should be strengthened as well to pilot staff to build up the conception that innovation equals effectiveness and even existence so that creations are made actively and bravely. Secondly, the construction of credit environment and protection of intellectual property rights should be enhanced. To forge a favorable eco-system for rural financial innovation is essential to the preservation of the fruits of financial originations. The research and development, try-out and popularization should be conducted as a connected sequence to speed up the conversion from creative ideas to implementation. We should also stress the culture of creation, cultivate an innovation-encouraging atmosphere so that the rural financial innovation could step to higher and deeper levels and broader spectrums.

5 Conclusion in Brief

The financial innovation in China rural areas is developing in an imperfect environment. It applies a government-led innovation pattern, focusing on the form and enlargement of quantity. The innovations are rather based on imitations than originations. The neglect on cost and benefit analysis is popular in innovation activities, which have not become important tools for financial institutions to lower their costs, increase earnings and enhance their competence. Thus the innovations in this field are insufficient and ineffective. It is an inevitable decision on boosting financial creations in the countryside to learn from abroad to form a marked-led mechanism. The devotion to advance diversified hierarchies of rural financial innovations must starts with issues on institutions and system, management mechanisms, technical services and products or business lines.

References

1. Molyneux, P., Shamroukh, N.: Financial Innovation. China Renmin University Press, Beijing (2003)
2. Zhang, J.: The Financial System in China's Rural Areas: the Structure, Changes and Policies. China Renmin University Press, Beijing (2003)
3. Zhang, Q.: The Analysis on China's Rural Financial Status and Polices[J/OL]. Yifu Lin Development Forum (2004.2.5), http://jlin.ccer.edu.cn
4. Li, Y.: The Thinking on Several Issues about the Improvement of Rural Financial System and Strengthening the Support for Finance in the Field of "Villages, Farmers and Agriculture". Journal of Financial Research 11 (2005)
5. Han, Z.: The Supply of Rural Financial Resources and Its Effects as the Rural Economy Grows. Doctoral Dissertation of Southwest University (June 2006)
6. The Research Group in the Shangrao Central Sub-branch of the Peoples' Bank of China. The Review of the Reality Lies in China's Rural Finance and the Systematic Restructure. Journal of Financial Research 1 (2006)

Innovation Design on Walking Structure of Diameter-Changeable Wheel Robot*

Dapeng Feng

School of Mechanic and Electric Engineering, Huangshi Institute of Technology
Huangshi, Hubei Province, China
fengdp555@yeah.net

Abstract. Walking structure is one of the significant research projects among some of the autonomous mobile robots studies. The content of the studies include the maneuverability of veer , the magnitude of driving force, the capability to adapt to the situation of the roadway and to cross the obstacles. A design was proposed in this paper that the diameter of wheel was changeable. The robot can adjust itself to different situation of the roadway through the change of the diameter of the wheels. The design has a lot of utility value. For example, the robot with changeable diameter is faster than the quadruped one, and it is stronger than the track mobile robot on the ability of cross barriers.

Keywords: robot, variable-diameter wheel, walking structure, innovation design.

1 Introduction

The definition of a robot isn't unified. But in general we can be understood as a kind of machine which has ability of autonomous or controlled, and can do some work instead of human. The development of robot has been gone a long time from the puppet of beginning to the service robots, entertainment robots, and special operations robot of today. Walking structure is one of the significant research projects among some of the autonomous mobile robots studies. The content of the studies include the maneuverability of veer , the magnitude of driving force, the capability to adapt to the situation of the roadway and to cross the obstacles. Most walking mechanism of robot are wheeled, foot type, crawler. Their advantages and disadvantages are clear. A design is proposes in this paper that the diameter of wheel is changeable. The robot can adjust itself to different situation of the roadway through the change of the diameter of the wheels. The design has a lot of utility value. For example, the robot with changeable diameter is faster than the quadruped one, and it is stronger than the track mobile robot on the ability of cross barriers.

* This work is supported by the Key Discipline in Hubei Province (Mechanical and Electronic Engineering) and the Outstanding Young Innovation Team Plans of Universities of Hubei Province (Research and Application of the Distributed Virtual Experiment Technology).

D. Zeng (Ed.): ICAIC 2011, Part II, CCIS 225, pp. 311–315, 2011.

2 Overview of the Robot Walking Mechanism

Most of the wheeled robots walking mechanism are foot type, crawler, etc., and we also use them together. They also have their own advantages and disadvantages; the ability to crossing obstacles is closely related to the diameter. It is obvious that the ability to crossing obstacles is greater which has the larger diameter, but it needs more space. The foot-type institutions crossing the barrier were markedly stronger than the wheel body, but its movement is slow, and the control is relatively complex. Crawler institutional has the characteristics of crossing obstacles with a simple but strong control of the larger driving force, but it lacks of flexibility, it has a larger contact area with the ground, when increasing the driving force, which still increases the friction when turning force, and it needs more power [1~2].

Crawler is also a special kind of wheels, which is a deformation of the ordinary. In fact, a two-legged walking mechanism is equivalent to a side length equalling to the rolling step polygon, when the step is small enough, than, the polygon is equivalent to a wheel.

3 Variable-Diameter Wheel

3.1 Design Ideas

Wheel walking structures are most widely used today such as cars, trains, boom crane and so on. They have a lot of advantages including mobility, controllability and stability. But they are designed basing on special circumstances (road network, rail network and truss). The superiority of the wheel is particularly prominent in these particular environments. But many robots are walking in the complex environment where traffic is the random variable. So the walking robots require very strong ability to adapt itself to the environment in order to stabilize its body. Therefore, the general wheel can not meet the requirements of complex environments. We need design a new wheel.

3.2 The Superiority of the Variable Diameter Wheel

The robot used Ordinary wheel as its walking mechanism when driving on the more complex road can avoid some of the bigger obstacles through the sensor, but some small objects (such as small stones on the ground) cannot be avoided. Because it is limited by the sensitivity of the other sensors (such as infrared sensors, ultrasonic sensors, etc.).Besides, the sensor cannot be installed too close to the ground, or it will hinder the movement of the robot, on the other hand, the sensor itself can also be easily damaged. If the robot's chassis can be always raised and lowered, ,then it can climb over small obstacles without being blocked. The design that raising the robot chassis has many species, for example, using suspension structure, if necessary, we can increase the distance of suspension and chassis to increase the distance between the chassis and the ground, as shown in Fig. 1.

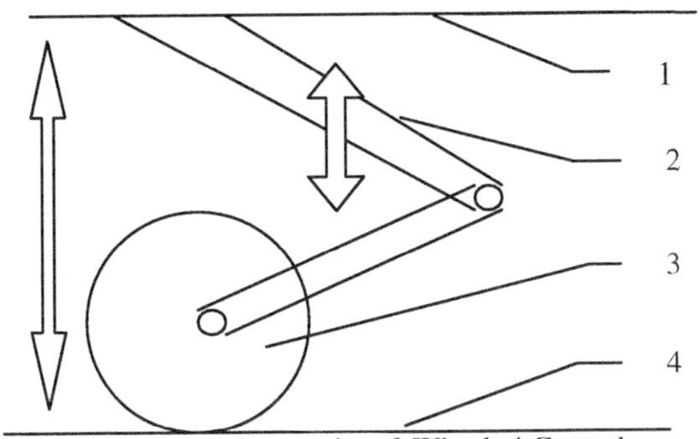

1 Chassis; 2 Suspension; 3 Wheel; 4 Ground

Fig. 1. Increasing the chassis and suspension structure diagram of the ground distance

In this method, the wheels away from the center and the ground has not changed, the axle is still can be obstructed by the obstacles. This design of the diameter can change the wheels, when increasing the diameter, the distance between the wheels and the ground also increases, so the ability of the robot crossing the obstacles would be increased. In the straight road, when minimizing the diameter of the wheels, the robot's center of gravity will lower, which can move steady and rapidly .In a more complex road, you can get a effective across by changing the diameter.

The control of Variable-diameter wheels is far more simpler then the foot-type running gear, the controls of the robot's direction are exactly the same, by controlling a DC motor reversing, which can control the wheel diameter[3~4].

3.3 The Specific Design Variable Diameter Wheel

1) Design of tire

Generally, there are two types of tire common, one is inflated tires (such as car tires), the other is non-pneumatic tire, that is, solid tires. Solid tire structure is relatively simple , and small robot use solid tires in general.The wheel diameter is variable, the tire circumference of the wheel diameter will change with changes. Therefore, the tire is divided into several segments, the tire with variable diameter choose to use tires wheels.

Tires is mounted on thin flexible steel, the steels can slide along an arc, which achieves a change in the tire circumference. Specific design is shown in Fig. 2.

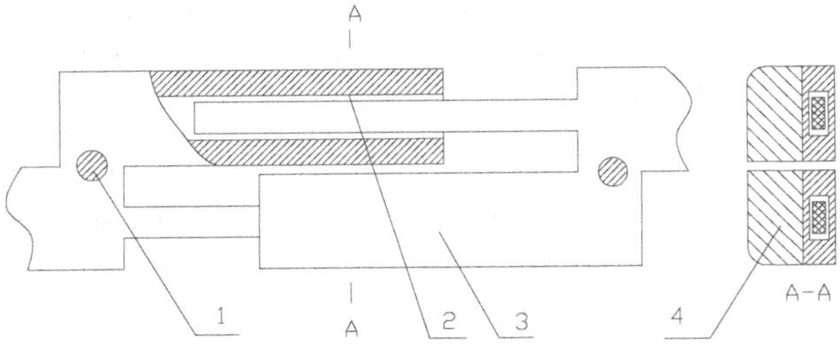

1 Connection with the spokes 2 Slide 3 Flexible steel 4 Solid tire

Fig. 2. Schematic diagrams of the tire's circumferential changes

The shape of the tire chip is selected as the “__⌐‾‾", whose structure is simple and its intensity is greater. The centrosymmetric structure makes the tires stress evenly. And each slice can combines easily. With the "__⌐‾‾__⌐‾‾__⌐‾‾" shape, the structure is compact.

2) Hub and Spoke Design

The diameter of the wheels is determined by the length of spokes, therefore, let all of the spokes change at the same speed, the diameter of the wheels can change uniformly and can be maintained circular all the time.

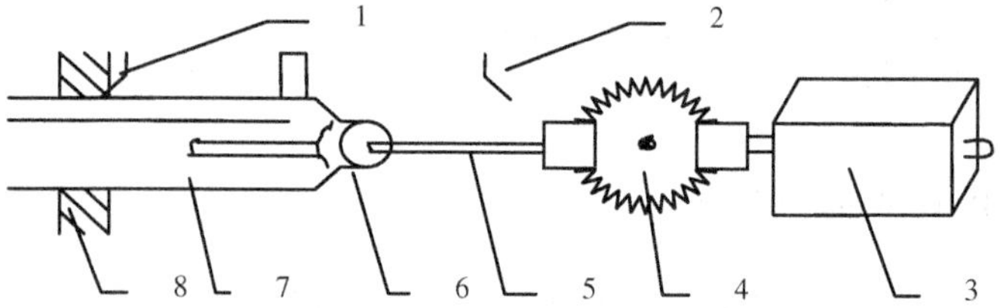

1~2 Limit switch; 3 Geared Motor; 4 Gear wheel; 5 Screw; 6 Nuts; 7 Spoke; 8 Sliding bearings

Fig. 3. Schematic diagrams of flexible spokes

The working principle: shown as Fig.3, driving the conical gears by the motor, and the conical gears pass transmits the drive to screw, changing circular motion into the spokes of the radial motion by the relative movement of screw and nut, and the bearings plays the role of guiding and fixing. Bearings and motor and gear box are fixed on the same chassis. This makes the change of the wheel diameter into reality. Installing two limit switches in the bottom can be used to limit the displacement of spokes in an appropriate limitation.

General assembly of the wheel shown in Fig. 4, the motor is mounted on the back, which isn’t shown in the figure.

Fig. 4. Schematic diagram of variable-diameter wheels general assembly

4 Conclusions

The minimum radius of the wheel "r", the minimum diameter d=2r, Diameter range $d < \Delta d < 2d$, the minimum circumference of the wheel $C = \pi \cdot d$; Changes in circumference: $\Delta C = \pi \cdot \Delta d$. "n" is the Wheel speed. By the above equation: When the diameter increases double, the perimeter increases double, and the line speed increases too.

1) With controlling the wheel speed "n" and the wheel circumference ΔC , we can control the speed of the robot more easily.

2) Wheel diameter increases with the height "r" doubling, the ability to climb obstacles increases.

3) When the left and right sides of the wheel diameter range, the shift can be very easy implemented. At the same time the centre gravity of the robot will move to the small diameter side of the wheel offset, so that when you can balance the center of gravity shift by the centrifugal force when the robot turns, which will increase the steering qualities by a wide margin.

References

1. Wang, W.: Mechanical basis (mechanical transmission and hydraulic drive, pneumatic drive). China Machine Press, Beijing (2005)
2. Ning, Y.: Machinery Basic. Harbin Institute of Technology Press, Harbin (2002)
3. Wang, L.: Robot Innovation Production. Tsinghua University Press, Beijing (2007)
4. Dennis, Clark, Zong, G., Huihui (translation): Robot design and control. Science Press, Beijing (2004)

A Two-Stage Decision Tree Algorithm on Constructing Hyper-Plane

Siping Zhang[1], Wei She[1], Mei Wang[1], and Zhimin Duan[2]

[1] Department of Computer, Hunan College of Information
Changsha, Hunan Province, China
[2] Software Institute of Hunan Vocational College of Science and Technology
Changsha, Hunan Province, China
sq_zhang123@qq.com

Abstract. How to construct the "appropriate" split hyper-plane in test nodes is the key of building decision trees. In this paper, we re-explain the process of building test nodes in terms of geometry. Based on this, we propose a method of learning the hyper-plane with two stages. The first stage searches for appropriate normal direction based on unsupervised methods (e.g. PCA, ICA etc), or supervised methods (e.g. neural network). The second stage detects the intercept of the hyper-plane in the normal direction according to some criterions, such as Gini of CART and information gain ratio (GainRatio) of C4.5. The experimental results conform that TSDT can improve the accuracy of univariate trees and that it is quite comparable to functional tree proposed by Gama, J. 2004, yet, needing much less learning time.

Keywords: decision tree, TSDT, hyper-plane.

1 Introduction

Decision trees, which encompass many diverse applications including diagnostics for linear regression[1], hierarchical multi-label classification[2]etc, have been among the most wide-spread models in machine learning and data mining. Some decision tree classifier's algorithms, such as ID3 [3], C4.5[4], CAR[5] etc, are often considered as the benchmark of evaluating the performance of other classifiers. Its popularity can be attributed to its simple form, fast computation in estimation, easy interpretation, and high accuracy.

The common way to build decision trees is by top-down partitioning, starting with the full training set and recursively finding a univariate split that maximizes some local criterion, such as Gini, until the class distributions in the leaf partitions are sufficiently pure. The tree obtained by this process is restricted to find the hyper-plane that is orthogonal to the features' axes. Brodley et al. pointed out that this maybe can not represent the space of regions in the instance space succinctly[6]. For example, the univariate decision tree has to approximate oblique boundary (e.g. $x + y \leq 8$) with a series of orthogonal splits. Therefore, multivariate (oblique) decision tree algorithms have been studied widely by investigators.

Since Breiman et al. first suggested a method of inducing oblique decision trees[5], there has been very much further research on such trees. Brodley et al. presented

D. Zeng (Ed.): ICAIC 2011, Part II, CCIS 225, pp. 316–323, 2011.

comprehensive studies for building oblique decision tree[6] by training a linear machine[8] on each test node. The work could be seen as the extensions of PT2[9] and LMDT[10]. LTree[11] and FTInner[12] belong to the type too. Though theses methods can find the optimal linear discriminates based on specific goodness measures, the training time of the linear program grows very fast with the number of instances and the number of attributes. Murthy et al.[13] employed hill climbing with two forms of randomization to find oblique hype-plane. The means could find "appropriate" hypeplane in much less time, yet, the learning process is difficult to understand. Liu et al.[14] used feature transformation via neural network, and then constructed univariate tree like C4.5 in transformed feature space. In the instance space, the hype-plane induced in this way would be oblique; however, it is orthogonal to the features' axes in transformed instance space.

In this paper, we re-investigate the course of constructing hyper-plane in terms of geometry, and then present TSDT algorithm which use two stages to construct split hyper-plane for each test node. The first stage detects the "appropriate" normal direction of the hyper-plane based on supervised or unsupervised method, and the second stage detects the intercept of it in the normal direction according to some criterions. Especially, when all the "appropriate" normal directions in test nodes are the directions of some features of instance space, the decision tree degenerates into univariate (normal) decision tree (e.g. tree induced by C4.5).

The major contributions of this paper are as follows.

1. We re-explain the process of building test nodes in terms of geometry and, based on that, propose TSDT algorithm using two stages to construct split hyper-plane. TSDT could incorporate supervised and unsupervised method together easily.
2. We give an in-depth study of the impacts of pruning methods and combining methods to the TSDT. The two methods impact decision tree deeply in accuracy and model size.
3. Experimental study suggests that TSDT outperforms univariate trees, even multivariate decision trees in accuracy with compromise training time, and that the impact of pruning and combination methods to TSDT is similar to other univariate decision tree such as C4.5, yet pruned TSDT algorithm and TSDT based combination methods are more competitive to other corresponding algorithms.

The rest of this paper is organized as follows. Sec.2 introduces the necessary background and related work. In sec. 3, we first re-interpret the process of constructing hyper-plane for test nodes, and then, present TSDT algorithm. In addition, we discuss the methods of selecting normal direction and give the corresponding reasonable explanations. Sec.4 evaluates the proposed algorithm on a random selection 36 benchmark data sets from the UCI repository using WEKA and studies the impact of pruning and combination methods to TSDT. The last section offers conclusions of the paper and presents the directions for future research.

2 Two-Stage Decision Tree Algorithm

Many algorithms have been applied to learn decision tree. In this section, we propose a two-stage decision tree constructing algorithm, which could explain the process of constructing hyper-plane of each test node from a new point of view.

2.1 Two-Stage Constructing Hyper-Plane

The central problem of building multivariate decision tree is how to learn the split hyper-plane function

$$w_1x_1 + w_2x_2 + \cdots + w_nx_n + threshold = 0 \tag{1}$$

in each test node for continuous attributes. Where X=(x_1, x_2... x_n, 1) is a pattern vector consisting of a constant and the n features that describe the instance. $W^T = (w_1, w_2, \ldots, w_n, w_{n+1})$ is a vector of parameters of X, also known as weight (we assume W^T is a unit vector in this paper). To find the appropriate function, traditional algorithms train a linear threshold unit [7] for two-class tasks and learn a linear threshold or a linear machine for multiclass tasks at each test node such as LMDT [10], FT [12] et al. If the tree induced in this way is stump, the classifier degenerates into a linear threshold unit or a linear machine. In fact, we can re-examine this issue from another point of view.

To re-examine the process of constructing the hyper-plane in each test node of a decision tree, we first adjust equation 1

$$w_1x_1 + w_2x_2 + \cdots + w_nx_n = threshold \tag{2}$$

The weight vector $W^T = (w_1, w_2 \ldots w_n)$ could be seen as the normal direction of the hyper-plane with the function 2. Then we can partition the process of searching for the hyper-plane function 2 into two stages: first detect the normal direction W^T, and then detect the parameter threshold. Let more than one parameter in W^T is not equal to 0, the hyper-plane we get is oblique to the features' axes. Let only one parameters in W^T is not 0, such as $W^T = (0, 0, \ldots, w_i, \ldots, 0)$, the hyper-plane is orthogonal to the features' axes. Obviously, if only one parameter in the W^T of each hyper-plane is not 0, the constructed decision tree would degenerate into univariate tree. For the purpose of further study of the problem, we first give definition 1.

Definition 1. Let $V=(v_1, v_2 \ldots v_n)$ (Unit vector) is a direction vector in instance space P, $a=(a_1, a_2 \ldots a_n)$ is a point in P, $\forall a$, we say a' is *V—component* of a if $a' = \sum_{1\le i\le n} a_i v_i$.

From definition 1 we know a' is the value on the axis V if we regard V as standard axis.

Proposition 1. Let H is the split hyper-plane with function 2. Suppose the normal—component of the intersecting point of A and H is v, then v = threshold.

Proof. Let $a=(a_1, a_2, \ldots, a_n)$ be a point in instance space, $\forall a \in P$, the normal—component b of a is

$$b = \sum_{1\le i\le n} w_i a_i$$

Let $a' = (a'_1, a'_2, \ldots, a'_n)$ be the mapped point from a onto normal axis we get

$$b = \sum_{1\le i\le n} w_i a_i = \sum_{1\le i\le n} w_i a'_i \tag{3}$$

Let $t=(t_1, t_2, \ldots, t_n)$ is intersecting point of A and P, since W^T is the normal of P, then $t=a'$. Combined with 3, we get

$$b = \sum_{1 \le i \le n} w_i a_i' = \sum_{1 \le i \le n} w_i t_i = v$$

Since equation 2, we have $v = threshold$.

If only one parameter in WT is not 0, such as WT = (0, 0, , wi, …,0), the normal direction in proposition 1 is exact a feature of the instance space. Therefore univariate decision tree satisfies proposition 1. From this point of view, our framework is the extension of univariate decision tree. In addition, once having found the normal direction, we can simply solve the threshold of the hyper-plane: calculate the normal—component of each instance as one-dimensional space value, and then searching for the best split value as the threshold of function 2 based on some criterion (e.g. gini).

2.2 TSDT Algorithm

Through the analysis in 3.1, the process of searching for the hyper-plane function can be partitioned into two stages. Based on this, we introduce TSDT algorithm which constructs the hype-plane for each test with two stages.(Fig 1). There is no different from other decision tree algorithm except steps 2 and 3. In step 2 (the first stage), normal list of candidate hyper-planes is constructed using some learning function. Many well-known methods can be applied directly here to search for the normals' direction such as PCA, ICA et al. Step 3 (the second stage) has two phases. In the first phase, the threshold for each candidate hyper-plane is detected based on some pure criterion (e.g. information gain ratio and gini), that has been discussed in section 3.1. The phase is similar to univariate decision tree algorithm in detecting the split point of continuous attribute. In the second phase, the model selects the best split hyper-plane from the candidate list based on the criterion.

```
Algorithm 1: GrowTree(Dataset, Learner, Criterion)
Input: Dataset    — the train dataset
   Learner   — the method of learning the normal direction of hyper-plane
   Criterion— the criterion for learning threshold of a hyper-plane.
Output: A well grown tree.
1. IF Stop Criterion(Dataset)
        Return a Leaf with class label
2. Construct candidate normal's set using Learner on Dataset
3. For each candidate i in normal's set
        Calculate hyper-plane i's threshold which split Dataset purest using
Criterion and the        corresponding pure gain purityGain_i
   Selecting the hyper-plane whose purityGain is largest, as the split hyper-plane
4. For each partition i of the Dataset using the split hyper-plane
        Tree_i = GrowTree(Dataset_i, Constuctor, Criterion)
5. Return a Tree, as a test node based on the split hyper-plane, containing
   descendants Tree_i
```

Fig. 1. Two stages to building a decision tree (TSDT)

As section 3.1 mentioned, if only one parameter of normal direction vector is not 0, such as WT = (0, 0, …, wi, …,0), the normal would be just the feature axis of instance space. If we just consider the normal of each hyper-plane in decision tree had the form WT = (0, 0, …, wi, …,0), the induced decision tree would be a univariate tree. So Algorithm 1 could be regard as the extension of univariate decision tree algorithm (e.g. CART).

Without loss of generality, we give the control algorithm of building TSDT in figure 2. Many algorithms can only handle a specific data set. For example, PCA and ICA require the continuous attribute values. To simplify the analysis complexity of the problem, step 1 preprocesses the input dataset. More study about data preprocessing is presented in section 3.3. After preprocessing dataset, step 2 constructs a TSDT using algorithm 1 (refer to Fig 2). Once a decision tree constructed, it is pruned back. Two measures should be estimated for each test node in pruning phase: an estimate of some index (e.g. error rate) of the subtree below the test node and the index of the test node if it is a leaf. If the former and the latter fulfill some condition (e.g. error rate of former is not less than the latter), the whole tree whose root is the node, is replace by a leaf. Different algorithm adopt different pruning index. Quinlan uses error rate estimated based on estimating statistic bounds[4], Breiman et al.use cost-complexity estimated based on both error rate and the scale [5] (measured by leaf node number) of a tree, Zhong et al. adopt standard deviation of error rate[15]. We just adopt EBP of C4.5 and CCP of CART to test the pruned TSDT performance and the impact of pruning algorithm, although all of them can be used.

```
Algorithm 2:LearnTree (Dataset, Learner, Criterion, Pruning)
Input: Dataset    —the train dataset
   Learner   —the method of learning the normal direction of hyper-plane
   Criterion—the criterion for learning threshold of a hyper-plane.
   Pruning   —indicate pruning tree or not.
Output: A well grown tree.
1.  Training-Set = Preprocessing(Training-Set)
2.  Tree = GrowTree(Dataset, Learner, Criterion) Refer to algorithm GrowTree()
    in figure 2
3.  IF Pruning is true
        Prune(Tree)
4.  Return Tree
End Function
```

Fig. 2. Construct decision tree with or without pruning

2.3 Data Preprocessing

Different type attribute and missing value for some attributes are commonplace in data mining that has greatly increased the difficulty of analysis of issues. To simplify problem's complexity, data preprocessing should be adopted.

For the problem, two different type attribute should be distinguished: nominal and continuous attributes. Some algorithms are able to handle both of them, yet some not. For example, univariate tree can deal with them with different approaches that we have discussed in section 2, however, PCA and ICA can not deal with norminal attribute and they are not sensitive to the dimension of different attributes. In this paper, all continuous attributes are normalized in [0, 1] and all nominal attributes are transformed to continuous attributes. In the latter, two candidate methods are applied. For a two-value attribute, we simply assign 1 to one value and 0 to the other. For a k-value attribute, each attribute-value is mapped to a new attribute which is 1 if the attribute has corresponding value and 0 otherwise.

For missing value, many approaches have been proposed: ignore the tuple, fill in the most probable value, etc (Han). In this paper, we replace missing value for nominal and numeric attributes with the modes and means. In PCA and ICA, the corresponding tuple has no contribution to the covariance matrix. This means that the tuple has no influence to PCA and ICA.

2.4 How to Select Normal Direction and Why

With respect to search for the appropriate normal direction, the underlying principle is that the constructed normal direction can reflect the patterns of the dataset. Several supervised and unsupervised methods could approach the purpose. It is easy to understand that supervised algorithms are used to find normal direction. Some processes in several algorithms are examples of the form although exist some differences, such as Gama constructs the split hyper-plane for test node[12] using linear machine, Liu et al. construct new features through two-layer feedforward neural network[14]. For unsupervised algorithms, Blockeel et al.do a very interesting experiment[16]. In the work, the authors present a hierarchic cluster algorithm with decision tree structure: the top-node corresponds to one cluster containing all data, which is recursively partitioned into smaller clusters such that intra-cluster variance is minimum and extra-cluster variance is maximum. They make a comparison between the algorithm and a supervised learners—TILDE (classification) and SRT (regression). The results show that the cluster tree is comparable to the supervised algorithms. From this point of view, unsupervised methods could be used in supervised algorithms. In deed, several unsupervised method have been adopted, such as rotation forests [17] etc. PCA which has been discussed in section 2 is particular unsupervised method. It can capture the maximum amount of variation that is just the eigenvector direction with largest eigenvalue (Theodoridis), in dataset. The split hyper-plane whose normal is the eigenvector, may maximize the variance between different sides of the hyper-plane. Besides, PCA is easy to implement in practice. Therefore, we apply it to detect the normal direction of split hyper-planes.

3 Conclusion and Future Work

In this paper, we first re-explained the process of building test nodes in terms of geometry and based on that, we proposed two-stage method to build the hyper-plane for each test node of decision tree. The first stage searches for the appropriate normal direction based on unsupervised methods or supervised methods. The second stage detects the intercept of the hyper-plane in the normal direction based on (according to) some criterions, such as gini and gain ratio. Then we presented TSDT algorithm. TSDT trees keep the interpretability of univariate tree. Experimental results show: (1) TSDT is competitive to other decision tree algorithms including univariate and multivariate decision tree in accuracy, training time and tree size;(2) In overall, the impact of pruning methods to TSDT is similar to other univariate decision tree in accuracy and model size, yet, the pruned TSDT is more competitive to other pruned decision tree in accuracy. (3) The same to other decision tree algorithm, combination algorithm could improve TSDT significantly. However, TSDT is more sensitive to combination algorithm so that TSDT based combination algorithm is much better than other tree based combination methods, such as C4.5 based combination methods.

In the future, we plan to apply more methods such as ICA, to search for the appropriate normal direction of each test node of decision tree and do more researches about the impact of combination algorithms to TSDT. Furthermore, we have started to apply TSDT to other problems (e.g. hierarchic class problems).

Acknowledgment. Owing to the instruction and assit of Professor She Wei, this paper could complete successfully. Profound knowledge, religious research style, open academic thought of teacher She influences and inspirits me profoundly. A thousand thanks for teacher She. Mei Wang and Zhimin Duan offered some materials and ideas for me., Thank their eagerness.

References

1. wSu, X.G., Tsai, C.-L., Wang, C.: Tree-structured model diagnostics for linear regression. Mach. Learn. 74, 111–131 (2009)
2. Vens, C., Struyf, J., Schietgat, L., Dzeroski, S., Blockeel, H.: Decision trees for hierarchical multi-label classification. Mach. Learn. 73, 185–214 (2008)
3. Quinlan, J.R.: Discovering rules by induction from large collection of examples. In: Michie, D. (ed.) Expert systems in the Micro Electronic Age, Edinburgh University Press, Edinburgh (1979)
4. Quinlan, J.R.: C4.5: Programs for Machine Learning. Morgan Kaufmann, San Mateo (1993)
5. Breiman, L., Friedman, J.H., Olshen, R., Stone, C.J.: Classification and Regression Trees. Chapman & Hall, New York (1984)
6. Brodley, C.E., Utgoff, P.E.: Multivariate decision trees. Machine Learning 19, 45–77 (1995)
7. Nilsson, N.J.: Learning machines. McGraw-Hill, New York (1965)
8. Duda, R.O., Hart, P.E.: Pattern classification and scene analysis. Wiley & Sons, New York (1973)
9. Utgoff, P.E., Brodley, C.E.: An incremental method for finding multivariate splits for decision trees. In: Proceedings of the Seventh International Conference on Machine Learning, pp. 58–65. Morgan Kaufmann, Austin (1990)
10. Utgoff, P.E., Brodley, C.E.: Linear machine decision trees (COINS Technical Report 91-10), Amherst, MA: University of Massachusetts, Department of Computer and Information Science (1991)
11. Gama, J.: Probabilistic linear tree. In: Fisher, D. (ed.) Proc. of the 14th International Conference on Machine Learning, pp. 134–142. Morgan Kaufmann, San Francisco (1997)

12. Gama, J.: Functional Trees. Machine Learning 55, 219–250 (2004)
13. Murthy, S., Kasif, S., Salzberg, S.: A system for induction of oblique decision trees. Journal of Artificial Intelligence Research 2, 1–32 (1994)
14. Liu, H., Setiono, R.: Feature Transformation and Multivariate Decision Tree Induction. Discovery Science. Springer, Heidelberg (1998)
15. Zhong, M.Y., Georgiopoulos, M., Anagnostopoulos, G.: A k-norm pruning algorithm for decision tree classifiers based on error rate estimation. Mach. Learning 71, 55–88 (2008)
16. Blockeel, H., De Raedt, L., Ramon, J.: Top-down induction of clustering trees. In: Proceedings of the 15th International Conference on Machine Learning, pp. 55–63 (1998)
17. Rodriguez, J., Kuncheva, L.: Rotation Forest: A New Classifier Ensemble Method. IEEE Transactions on Pattern Analysis and Machine Intelligence 28, 1613–1619 (2006)

Intelligent Component Library Design of 3D Industrial Configuration Software

Quangui Zhang[1,2], Pu Wang[3], and Yongsheng Qi[3]

[1] School of Electronic Information and Control Engineering
Beijing University of Technology
Beijing 100124, China
[2] School of Electronic and Information Engineering
Liaoning Technical University
Hu Ludao 125105, China
[3] School of Electronic Information and Control Engineering
Beijing University of Technology
Beijing 100124, China
zhangqg8123@yeah.net

Abstract. Analysed the advantages of 3D configuration software and the issues which need to be resolved. Proposed the design method of semantic-based intelligent component library. Using ontology to describe the graphics components of the 3D configuration software. Analysed the feasibility of using Ontology in the 3D configuration software. Proposed the architecture of the 3D intelligent components. And described the method of describing the 3D components using ontology. Depending on the support of the ontology library, it can reduce the complexity and difficulty to create the 3D monitor scene by the 3D configuration software using the intelligent components. This makes the automation engineers create 3D monitoring scene so easily who has no 3D experience in the 3D development.

Keywords: 3D configuration software, Intelligent component, Ontology.

1 Introduction

The Industrial configuration software provides users the express tools to construct the monitoring software of industrial automatic control system by a flexible way. Because of the characteristics of flexibility, simplicity, rapidity and universal, the configuration software has been widely applied in data acquisition, supervisory control and process control in power system, water supply systems, petroleum, chemical and many other fields. But the user interface of the configuration software is basically two-dimensional, which has limited its development and application. It can only express monitoring scene using simply 2D graphics. Its simulation effect is very limited. In addition, with the appearance of some new application fields, the 2D configuration software shows its limitation increasingly.

At present, 3D configuration software has become a hot issue both in industrial and academic circles. ICONICS company [1] has developed the first 3D configuration

D. Zeng (Ed.): ICAIC 2011, Part II, CCIS 225, pp. 324–329, 2011.

software-GENESIS64, and this is the first step in the industrialization process of the 3D configuration software. But, it has many issues to make the 3D configuration software widely applied. One of the issues is how to create the 3D monitoring scene rapidly. It is a very difficult thing to create a 3D scenario for most of the automation engineers who have no previous experience in 3D development using the 3D configuration software. This goes against the promotion application. So, how to realize the fast 3D scene configuration is a very meaningful issue.

First, this paper analyzed the relevant works on the fast creation 3D scene. And proposed a method of fast creation 3D scene. It used the ontology-based intelligent components library to accelerate the configuration process of the 3D configuration software.

2 Related Works on Fast Creation 3D Virtual Scene

At present, there is no unified standard on the research of fast creation 3D virtual scene in academic circles, but the universal methods used by most researchers is to integrate semantic information and constraint relations into scene objects. Users create scene at a high lever using these methods. Lode Vanacken et al [2] proposed a model-based user interface. In their works, they introduced the usage of semantic information during conceptual modeling of interaction for virtual environments represented using ontology. Félix Ramos et al [3,4] proposed the GEDA-3D project, the main purpose is to describe relations between the elements of environment to help to recreate the scenario and use the onlogy to describe the laws of the environment. Fabio Pittarello et al [5] proposed an alternate approach for associating semantic information to 3D worlds based on the integration of the X3D and the semantic web. Mario Gutierrez et al [6] considered every object participating in a VE application not only as a 3D shape, but as a dynamic entity with multiple visual representations and functionalities, and allowed for dynamically scaling and adapting the object's geometry and functions to different scenarios. Thomas Funkhouser et al [7] proposed the example-based modeling method.

The same is that they all integrate the semantic information into the virtual scene. And they all used the ontology to describe the scene. Learn more results from above, this paper describes the components in the 3D configuration software. And creates the intelligent component library. Each component is a agent, and it has some kinds of constraint relations.

3 Design of the Intelligent Component Library

3.1 Applies the Ontology in 3D Configuration Software

Ontology concept comes from philosophy. In the recent years, ontology has received wide application in artificial intelligence and knowledge engineering. In information system domain, Ontology is currently mainly used for knowledge representation, organization and management. The ontology's essential is sharing and reusing of the domain knowledge. Moreover, the standard and formalized domain ontology can provide tools for the high lever interactive among the information systems. Using the ontology in the 3D configuration software has two purposes: the first is to integrate semantic information into the 3D configuration software, and replace the geometric low-level configuration interaction with the high-level human-computer interaction, which will boost the configuration speed. Second, with the development of WEB 3D, there will be many free 3D graph libraries. And we can use the existing graph libraries to create 3D monitoring scene. We just need to integrate the web graph libraries with the existing ontology.

3.2 Architecture of the Intelligent Component Library

Similar to the traditional 2D configuration software, each monitoring interface element is a graph component. But the difference is that these graph components has intelligence. We describe each component and their relations using ontology. Each 3D component constitutes an agent. There are various relations in the agents. An agent has different performance in different environments.

We can make use of the component's structure information and the semantic information (Including the project properties, action sequences and other constraints information). For the integration of the scene graph (Geometry information) and semantic information, we need to associate all kinds of geometrical elements with the classes of the ontology.

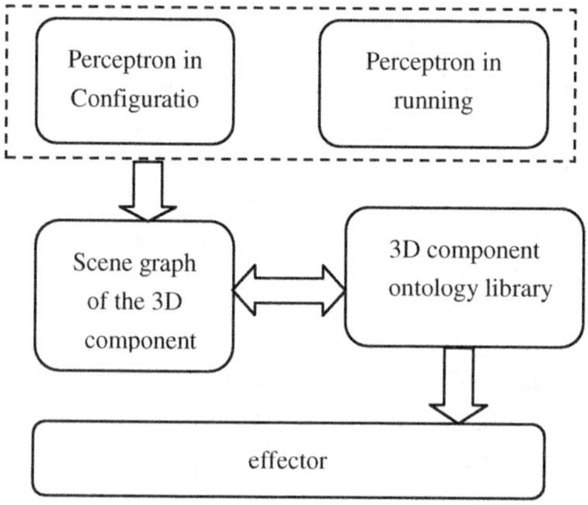

Fig. 1. Architecture of the intelligent component library

As shown in Fig. 1, the intelligent component library consists of five parts: perceptron in Configuration, perceptron in running, scene graph of the 3D component, 3D component ontology library and effector.

1) perceptron in Configuration: It perveives the user's input in the process of configuration, such as the actions of the keyboard and mouse. And sends the user's input information to the 3D component library. The component will judge the intent of the user interaction. Such as connectivity, translation, rotation, scaling, alignment and other operations.

2) perceptron in running: It perceives the input from environment, including input device such as the mouse, the keyboard and data cards, serial devices, net input devices et. al. And it sends the processed input information to the 3D component library, and produces the perception sequence.

3) 3D scene graph: 3D scene graph has two functions, first, it describes the geometric structure of the 3D component. As shown Fig.2. The intelligent boiler component consists of four parts: boiler body, water lever, temperature indicator, pipeline interface. These four parts correspond to classes in the ontology library. Second, it expresses the logic and spatial relations. It manages the instances of the ontology library.

4) 3D component ontology library: 3D component ontology library gives intelligence to 3D components, it describes the attributes and functions of each component. Further more, it defines the relations for the 3D components. All 3D components form a domain ontology.

Fig. 2. Monitoring scenario instance

3.3 Description of the 3D Component Ontology Library

Fig.2 describes the part scene of an boiler shop, It consists a boiler, two retorts, two pipelines, ground and wall. To create thus a simple 3D scene, we can use some commercial modeling softwares, such as 3Ds Max and Maya. But it will consume a long time to complete this simple scene, further more, the whole monitoring scene is usually very large. The time consume and complex operation will hamper the development of the 3D configuration software.

The intelligent component reduces the complexity of creating 3D monitoring scene for the automation engineers. To create the scene as shown in fig.2, first, users select "ground" component from the intelligent component library, then the ground will be placed into position by the description of the attribute and constraint relations in the ontology library. It can only be horizontally arranged. And it has the attribute of supporting, that, it can place other model on it. Also, For the boiler model, it has attributes and constraint relations in the ontology library. And it can be only placed on the models which has the supporting attribute. When we drag it out of the library, it judges which model has the supporting attribute in the current scene firstly. In the scene as shown in fig.2, it can only be place on the ground. We can't adjust it up and down, and it can only move along the ground. Also, for the creation of pipelines, after dragging a pipeline out of the library, we select the model which will be connected with the pipeline, then the pipeline will connect with the model and adjust the size of the connector automatically.

The configuration process above all used the ontology library. We select OWL [8] as the ontology language which is the modeling language of the semantic web. Due to its superiority and generality, it has been used in many fields where using ontology, not only be used in the semantic web.

For the ontology description of the scene above, we give a segment, as shown in fig.3.

```
<owl:Class rdf:ID="pipeline"/>
<owl:Class rdf:ID="water_level">
  <rdfs:subClassOf>
    <owl:Restriction>
      <owl:allValuesFrom>
        <owl:Class rdf:ID="boiler"/>
      </owl:allValuesFrom>
      <owl:onProperty>
        <owl:TransitiveProperty rdf:ID="is_part_of"/>
      </owl:onProperty>
    </owl:Restriction>
  </rdfs:subClassOf>
<owl:ObjectProperty rdf:ID="connect">
  <rdfs:domain>
    <owl:Class>
      <owl:unionOf rdf:parseType="Collection">
        <owl:Class rdf:about="#boiler"/>
        <owl:Class rdf:about="#pipeline"/>
      </owl:unionOf>
    </owl:Class>
  </rdfs:domain>
</owl:ObjectProperty>
<owl:DatatypeProperty rdf:ID="can_move_left_and_right">
  <rdfs:domain>
    <owl:Class>
      <owl:unionOf rdf:parseType="Collection">
        <owl:Class rdf:about="#ground"/>
        <owl:Class rdf:about="#boiler"/>
        <owl:Class rdf:about="#pipeline"/>
      </owl:unionOf>
    </owl:Class>
  </rdfs:domain>
  <rdfs:range
rdf:resource="http://www.w3.org/2001/XMLSchema#boolean"/>
</owl:DatatypeProperty>
```

Fig. 3. The segment of the ontology

4 Conclusion

This paper introduced the method of creation the intelligent component library. It can realize fast configuration of monitoring scene of 3D configuration software using the intelligent components. The validity of this method has been put into practice in our prototype system. With the related issues of the 3D configuration software solved, it's great expectations for the future development.

References

1. GENESIS64, `http://www.iconics.com/` (2010)
2. Vanacken, L., Raymaekers, C., Coninx, K.: Introducing semantic information during conceptual modelling of interaction for virtual environments. In: Proceedings of the 2007 Workshop on Multimodal Interfaces in Semantic Interaction, Nagoya, Japan, November 15, pp. 17–24 (2007)
3. Ramos, F., et al.: The Use of Ontologies for Creating Virtual Scenarios in GeDA-3D. In: Electronics, Robotics and Automotive Mechanics Conference (CERMA 2006), vol. 2, pp. 230–235 (2006)
4. Zaragoza, J., et al.: Creation of Virtual Environments through Knowledge-Aid Declarative Modeling. In: LAPTEC 2007, pp. 114–132 (2007)
5. Pittarello, F., De Faveri, A.: Semantic description of 3D environments: a proposal based on web standards. In: Proceedings of the Eleventh International Conference on 3D Web Technology, Columbia, Maryland, April 18-21 (2006)
6. Gutierrez, M., Vexo, F., Thalmann, D.: Semantics-based representation of Virtual Environments. International Journal of Computer Applications in Technology 23, 229–238 (2005)
7. Funkhouser, T., et al.: Modeling by example. In: ACM SIGGRAPH 2004 Papers, Los Angeles, California, August 8-12 (2004)
8. OWL Web Ontology Language Overview (2004), `http://www.w3.org/TR/owl-features`

Fault Diagnosis Model of Main Engine Water Cooling System Based on Attribute Hybrid Computing Network

Liu Nianzu[1], Xu Guanglin[1], and Liu Yongchang[2]

[1] School of Mathematics and Information,
Shanghai Lixin University of Commerce, Shanghai 201620
[2] College of Information Engineering,
Shanghai Maritime University, Shanghai 200135
minyoupp@tom.com

Abstract. On the basis of illustration of attribute hybrid computing network model, the paper gives the improving methods of training and learning of the model and applies it to fault diagnosis of the main engine water cooling system. The final results of simulation test verify the feasibility of methods.

Keywords: Attribute Hybrid Computing Network, Fault Diagnosis.

1 Introduction

The development of science and technology boosts scale of automation system, whose functions and complexity also improve dramatically. Any teeny fault may cause failure and stoppage of the entire system or even catastrophic results. The water cooling system is an important auxiliary system to main engine and proper and reliable cooling of main engine directly influences its normal operation. Therefore, for water cooling system, it is of utmost importance to monitor the system status, timely judge and detect the fault parts and forecast fault trend.

Currently, the main methods of fault diagnosis include signal possessing based method, direct analysis of detectable signal and expert system based method [1]. The signal possessing based method is to use signal model, e.g. correlation function, frequency spectrum and wavelet transform (WT), etc; the direct analysis of detectable signal is to diagnose fault by extraction and analysis of characteristics of variance, amplitude, frequency, etc; the expert system based method is generated by combination of diagnosis technology and artificial intelligence computing method, mainly including fault diagnosis methods based upon expert system, fuzzy mathematics, rough set and artificial neural network. All these methods are specially featured and some of them have been applied to fault diagnosis of main engine water cooling system with satisfactory results [2, 3]. In light of the advantages of expert system methodology, the paper utilizes the attribute hybrid computing network model to set up a fault diagnosis model to diagnose fault of the water cooling system of main engine.

D. Zeng (Ed.): ICAIC 2011, Part II, CCIS 225, pp. 330–336, 2011.

2 Fault Diagnosis Model Based on Attribute Hybrid Computing Network Model

2.1 Introduction of Attribute Computing Network

Generally speaking, attribute is the nature and characteristics of an object where nature is the internal property to differentiate objects while characteristics are the external signs of object's features. Since only the essential elements constituting object, also called the element differentiation or difference of profound structure can lead to diversity of object nature. The quality diversity will display characteristic difference at the surface of object. Attribute theory tries to perform qualitative and quantitative study of object's attribute where attribute computing network is mathematics abstract of the quantity-quality transfer regularity of the object's attribute and meanwhile a mathematics model for qualitative identification and classification.

Definition 1[4]: Supposing $\alpha_i(o)$ is the ith attribute of objective o, where i=1,...,n, $x_{ij}\in X_i$ is the quantitative characteristic value of $a_i(o)$, $p_{ij}(o)$ is the jth qualitative characteristics of $a_i(o)$, j=1,…,m, $[\alpha_{ij},\beta_{ij}]\subseteq X_i$ is the qualitative criterion of $p_{ij}(o)$, and then $\Gamma=\{[\alpha_{ij},\beta_{ij}]\}$ is the cluster of qualitative criterion, satisfying $[\alpha_{ij},\beta_{ij}]\cap[\alpha_{il},\beta_{il}]=\varnothing$, l=1,...,m,l≠j and $X_i = \bigcup_{j=1}^{m} [\alpha_{ij}, \beta_{ij}]$, if for any $x\in X$, there exists $[\alpha_v,\beta_v]\in\Gamma^n$ and the $[\alpha_v,\beta_v]$-based quality $p_v(o)\in P_o$, then:

$$\begin{aligned} &\tau\left((x_1,\cdots,x_n),\begin{pmatrix}[\alpha_{11},\beta_{11}] & \cdots & (\alpha_{1m},\beta_{1m}]\\ \vdots & (\alpha_{i_kj_i},\beta_{i_kj_i}] & \vdots\\ [\alpha_{n1},\beta_{n1}] & \cdots & (\alpha_{nm},\beta_{nm}]\end{pmatrix}\right) \\ &= \bigvee_{j_i=1}^{m}\bigwedge_{i_k=1}^{n}\{(x_1,\cdots,x_n)\in[\alpha_{i_1j_1},\beta_{i_1j_1}]\times\cdots\times(\alpha_{i_kj_i},\beta_{i_kj_i}]\times\cdots\times(\alpha_{i_nj_m},\beta_{i_nj_m}]\}\\ &= \bigvee_{j_i=1}^{m}\{\cdots\{\bigwedge_{i_k=1}^{n}\tau_{v(i_1j_1,\cdots,i_kj_i,\cdots,i_nj_m)}(x)\}\} \end{aligned} \tag{1}$$

Where:

$$\tau_{v(i_1j_1,\cdots,i_kj_l,\cdots,i_nj_m)}(x) = \begin{cases}1 & iff \quad x\in(\alpha_v,\beta_v]\\ 0 & iff \quad x\notin(\alpha_v,\beta_v]\end{cases} \tag{2}$$

Then (1) is a qualitative mapping to judge whether an object o with vector x has the quality pv (o), while (2) is the qualitative mapping with $[\alpha_v,\beta_v]$ as qualitative criterion, or the factor mapping of (1) that can also be recorded as: $\tau_p(x,[\alpha_v,\beta_v])$.

2.2 Attribute Hybrid Computing Network

In actual large-scale system, the operation status is reflected by huge amount of parameters, which may be quantitative or qualitative. Some parameters may be dominantly correlated. The attribute computing network standard model in formula (1) may encounter certain difficulty in solving the problems and thus needs further expansion, i.e. some hybrid input attribute computing network to solve the problem.

Definition 2: Under the prerequisite of compliance with the basic model of attribute computing network, if a certain layer has hybrid input of both initial input quantitative attribute x and qualitative attribute $p(x)$, such model is called hybrid input attribute computing network model.

The hybrid input attribute computing network model can be formed by several layers of interrelation. The one connecting with data input is the conversion layer where it firstly converts qualitative data, and then uniformizes the data by expansion or contraction of qualitative criterion. The arithmetic layer connecting conversion mainly perform conjunction calculation of qualitative attribute value $p_1(x),..., p_m(x)$ and quantitative attribute value x_j. The final layer is the output layer with final outcome of qualitative value p .

2.3 Training Model of Computing Hybrid Network

After establishment of attribute hybrid computing model, the next step is to find some reasonable mechanism that helps quick and accurate training and learning of structure and model of network.

Definition 3: Supposing (X,Γ) is the topological space of X and qualitative criterion cluster Γ, where $N_k(\xi_k,\delta_k)=[\alpha_k,\beta_k]\subseteq X\in P(X)$, which is the qualitative criterion of nature $p_j(o)$ after k (k=0,...,N) times of learning, where ξ_k and δ_k are the spherical center and radius of $N_k(\xi_k,\delta_k)$ respectively. $\tau_k(x,N(\xi_k,\delta_k))=p_k(x)$ is the result of judgment when $N_k(\xi_k,\delta_k)$ is applied.

$$e_k(x,N_k(\xi_k,\delta_k),p(x))=\tau_k(x,N_k(\xi_k,\delta_k)-p(x))=(p_k(x)-p(x))\in E(p) \tag{3}$$

The Formula (3) gives the error of such method, Supposing $\{z_{jk}(o),k=1,...w,...\}$ is the positive example cluster, $\{z_{jk}(0),k=1,...w,...\}\in[\alpha_j,\beta_j]\cap\{z_s(o)\}$, then judging from the nature of qualitative adjustment, the adjustment of qualitative criterion can be summarized as the following extreme process, i.e.:

$$\alpha_j = \lim_{k\to\infty}\{\min\{z_{jk}(o)\}\} \tag{4}$$

$$\beta_j = \lim_{k\to\infty}\{\max\{z_{jk}(o)\}\} \tag{5}$$

Just find the upper and lower limits from $\{z_{jk}(o),k=1,...w,...\}\subseteq\{z_s(o)\}$.

Some present learning methods of attribute network [5] are not perfect in terms of speed. So hereby take trigonometric function as the conversion degree function in this paper, which can greatly uplift calculation speed without impact on precision. The follows steps show the learning algorithm.

Supposing R is set of real numbers, j is sequence number of samples and k is sequence number of attribute quantity, where k may be omitted for single attribute quantity.

Step 1: Input the sample set with labels X={$z^{(k)}_j$, s_j | $z^{(k)}_j$ R ,∈ S_j ∈ S ,j=1,2…,k=1,2…}

Step 2: Preprocessing of sample set. Use Triangle conversion degree function (6) to treat there outcome so that the results fall within [0,1]; then the samples are turned into input sample set X={$z^{(k)}_j$| $z^{(k)}_j$ ∈ R, j=1,2…,k=1,2…} and output sample set Y={$r^{(k)}_j$| $r^{(k)}_j$ ∈ R, j=1,2…,k=1,2…}.

$$\eta_i(x) = 1 - \frac{|x - \xi_i|}{\delta_i} \tag{6}$$

Step 3: Based on output results, sequence the input sample set by insertion sort method; the adjustment times I=0; select two samples from sequence, one is close to zero and the other close to one as the initial criterion [$\alpha^{(k)}$,$\beta^{(k)}$].

Step 4:

(1) Learning α: if the result of sample is over zero, select a negative sample set $z^{(k)}_j$ against sequence and then turn to Step 5 or otherwise to Step 6;

(2) Learning β: if the result of sample is over zero, select a positive sample set $z^{(k)}_j$ in turn and then turn to Step 5 or otherwise to Step 6.

Step 5: if $z^{(k)}_j < \alpha^{(k)}$ and the output result makes $\alpha^{(k)} = z^{(k)}_j$, and if $z^{(k)}_j > \beta^{(k)}$ and | $z^{(k)}_j - \beta^{(k)}$ |>E that makes $\beta^{(k)} = z^{(k)}_j$ and I=I+1, turn to Step 4.

Step 6: output adjustment times i and qualitative criterion; finished.

2.4 Fault Diagnosis Model of Attribute Hybrid Computing Network

For fault diagnosis model, supposing there are several sample data that contain the known parameter values of all sensors and corresponding fault status. By the fault diagnosis model of attribute hybrid computing network, it just needs to learn the qualitative criterion of all sensor parameters corresponding to each kind of fault; records them in the qualitative criterion of each qualitative mapping and the structure of attribute network. By learning and training of attribute hybrid network, it can analyze the input parameters of sensors and judge the fault status of system.

3 Analysis of Fault Diagnosis System of Water Cooling System of Main Engine

Diesel engine cooling system is an auxiliary system of main engine system, generally using seawater compulsory cooling of freshwater and other heat-carrying fluid (e.g. lube oil, pressurization air, etc.) ,them the cooled freshwater goes on to cool down the main diesel engine. For system arrangement, the seawater system is of open recycling style while the freshwater and lube oil systems are of closed recycling style. According to actual conditions, there are six detectable technical parameters as follows:

Tf: Temperature of freshwater
Ts: Temperature of seawater
Pf: Pressure of freshwater
Ps: Pressure of seawater
Qf: Quantity of freshwater
Qs: Quantity of seawater

The above six parameters are the monitored input of fault diagnosis of cooling system. The common fault causes of system are as follows:

S0: Normal
S1: Cylinder overpressure
S2: Cylinder after-burn
S3: Closed or damaged freshwater valve
S4: Low pressure of freshwater pump
S5: Leakage of freshwater piping system
S6: High pressure of freshwater pump
S7: Small opening of freshwater bypass valve
S8: Block of freshwater piping system
S9: Block of cooler piping system
S10: Weather of sea area
S11: Low pressure of seawater pump
S12: Block of seawater
S13: High pressure of seawater pump
S14: Large opening of seawater bypass valve
S15: Block of seawater pump

The above 16 parameters serve as the output value for fault diagnosis system. According to references[6], the benchmarks of monitored input are:

T_f: High: $>80^\circ C$; Normal: $70^\circ C \sim 80^\circ C$, Low: $<70^\circ C$
T_s: High: $>50^\circ C$; Normal: $25^\circ C \sim 50^\circ C$, Low: $<25^\circ C$
P_f: High: >0.3Mpa; Normal: 0.2~0.3Mpa, Low: <0.2Mpa
P_s: High: >0.2Mpa ; Normal: 0.1~0.2Mpa, Low: <0.1Mpa
Q_f: High: >225m^3/hr; Normal: 75~225m^3/hr, Low: <75m^3/hr
Q_s: High: >450m3/hr; Normal: 150~450m3/hr, Low: <150m3/hr

According to the different values of the above six data, the possible faults may include:

Table 1. List of Failure

Monitored Data						Type of Failure
T_f	P_f	Q_f	T_s	P_s	Q_s	
N	N	N	N	N	N	F0
H	N	N	H	L	L	F1,F2
H	L	N	N	N	N	F3,F4,F5
L	H	N	N	N	N	F6,F7
H	N	L	N	N	N	F8,F9
L	N	H	N	N	N	F6
L	N	N	H	N	N	F10
H	N	N	H	N	N	F10
H	N	N	N	L	N	F11,F12
L	N	N	N	H	N	F13
H	N	N	N	N	L	F14,F15

4 Construction of Fault Diagnosis Model of Main Engine Water Cooling System Based on Attribute Hybrid Computing Network

Pursuant to analysis of main engine cooling system, it needs to set up a triple-layer input attribute hybrid computing network model. Shown in Figure 1, the first layer covers six qualitative criterions, i.e. low, normal and high; the second layer is the pool of rules after sample learning; the third layer is to judge fault types of cooling system by input.

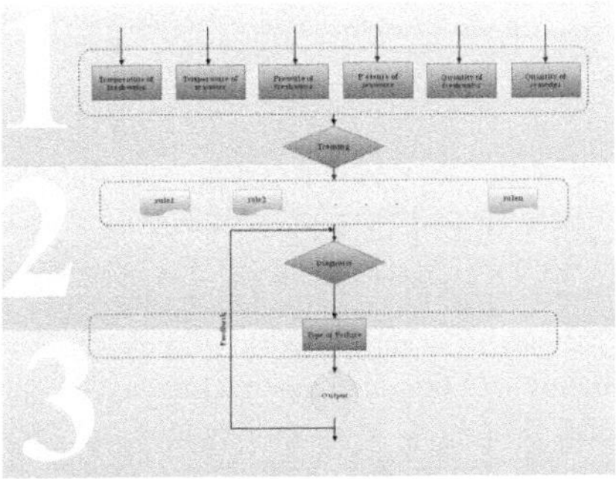

Fig. 1. Fault Diagnosis Attribute Computing Network Model of Ship's Main Engine Cooling System

The system firstly inputs 1,000 learning samples for training of attribute computing network and then randomly generates 1,000 data for test. Figure 2 shows the operation and results of system, among which Figure 2 (b) shows 100% of fault identification rate of system and the operation time is 0.2 seconds.

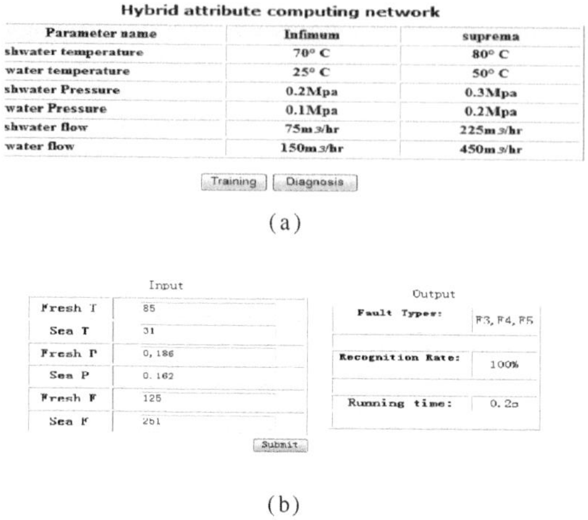

Fig. 2. Operation Results of Cooling System Based upon Attribute Hybrid Computing Network

5 Summary

This paper establishes the attribute hybrid computing network model and relevant learning methodology of it, them applies the model in the failure diagnosis system of main engine water cooling system. The results of simulation test fully testify the potentials of the model in fault diagnosis.

References

1. Chen, Y., Shu, S., Weng, Z.: General Comments on Failure Diagnosis Methods of Dynamic System. Chemical Automation and Instruments 28(3), 1–14 (2001)
2. Zhang, J., Feng, J.: Attribute Petri Net Modeling and Its Application. Computer Engineering 32(17) (2006)
3. Wu, G., Ren, X.: Dynamic modeling and simulation of cylinder jacker water cooling system for main marine diesel engine. Journal of Harbin Engineering University 24(4) (2003)
4. Feng, J.: Qualitative Mapping Orthogonal System Induced by Subdivision Transformation of Qualitative Criterion and Biomimetic Pattern Recognition. Chinese Journal of Electronics, Special Issue on Biomimetic Pattern Recognition 85(6A), 850–856 (2006)
5. Feng, J., Xu, G.: Study on Mixed Input Attribute Computing Network Model and Boundary Study Algorithm. Computational Intelligence in Decision and Control
6. Zhou, D., Ye, Y.: Modern Fault Diagnosis and Fault-tolerance Control, vol. 20. Qinghua University Press, Beijing

Design and Implementation of a Wireless Healthcare System Based on Galvanic Skin Response

Tao Huang, Jiping Xiong*, and Lifeng Xuan

College of Mathematics, Physics and Information Engineering
Zhejiang Normal University
JinHua, Zhejiang Province, China
huang_tao556@sina.cn

Abstract. Electrocardiography response (ECG) and Electroencephalography response (EEG) have been often applied in modern medical field. With the development of modern medicine, Galvanic skin response (GSR) also has becoming a reliable medical diagnosis. This paper presents the design and implementation of a novel medical healthcare system which use GSR as the relaxation factor. The system adopts wireless RF module PTR8000 to transmit the measurement data between the primary stage and the subordinate stage. The subordinate stage collects the GSR signal and converts analog signal into a digital signal, then sends it away; the primary stage receives the data, then stores and displays them. The paper illustrated the principle of wireless GSR healthcare system, its main components, and the flow chart of the system. Experiments have demonstrated that our system is reliable and useful in the healthcare field.

Keywords: PTR8000, Galvanic Skin Response, wireless data transmission.

1 Introduction

Galvanic skin response (GSR) is one of the methods to measure the skin conductance. As was indicated by researches, GSR can reflect the changes of sympathetic nervous system which was not controlled by subjective consciousness[1]. In the year of 1935, Wilhelm Reich confirmed the relationship between biological current and vegetal pleasure float in his institute. Similar to ECG and EEG, GSR can also provide a reliable criterion in clinical symptoms, which is a beneficial attempt in the process of pulse-taking diagnosis in the field of the Chinese traditional medicine[2]. Due to the ECG and EEG response detectors in the hospital are relatively large and heavy at present, patients have to receive treatments by carrying such an inconvenient monitor. However, it is a inexorable trend that the medical monitor go toward to miniaturization, embedded and network-oriented development, thus small and portable monitor is becoming a favor to most of the families[3-4].

* Corresponding author.

D. Zeng (Ed.): ICAIC 2011, Part II, CCIS 225, pp. 337–343, 2011.

In this paper, we describe the design and implementation of a wireless GSR healthcare system based on the PTR8000 module. This wireless healthcare system consists of the primary stage and the subordinate stage. The subordinate stage is mainly in charge of collecting the GSR signals from each user and then sending the signals away after converting the analog signals into digital signals. The primary stage is in charge of receiving, storing and displaying these signals. These two parts can communicate with each other by PTR8000 wireless module. After sampling and collecting, the diagnosticians can acquire the GSR signals from users, which will display on the primary stage.

2 System Architecture and Operating Principle

For the widely distributed group of patients, we propose the hierarchical structure of the healthcare monitor, and the entire system encompasses three layers: the primary stage, the PTR8000 wireless module and the subordinate stage. The microcontrollers are different, AT90S8515 for the primary and ATMEGA8 for the subordinate. Between them, the PTR8000 acts as a hinge of signal sending and receiving. Figure 1 depicts communication topology between primary stage and the subordinate stage.

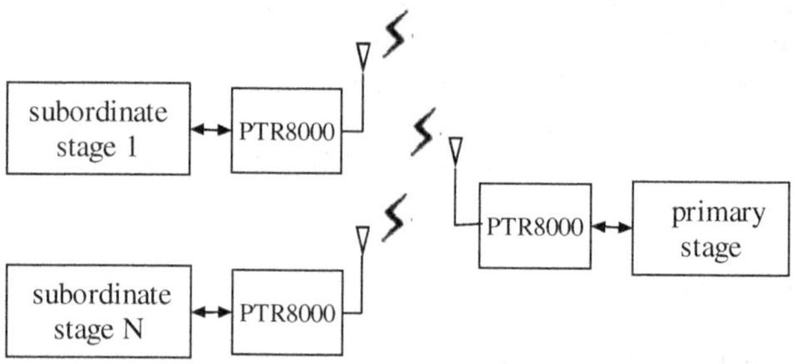

Fig. 1. Communication topology between the primary stage and the subordinate stage

The subordinate stage, including an amplifying circuit and a PTR8000 module, and fixing on patient body or fingers, samples the users' GSR values wherever they go. The PTR8000 sends the digital signals to the primary stage by means of wireless pattern after amplifying and analog-to-digital converting. The operating principle of the subordinate stage is shown in figure 2.

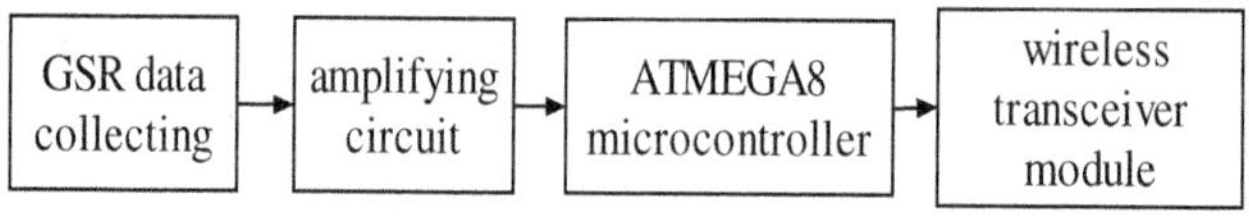

Fig. 2. Schematic of subordinate stage

The primary stage is mainly consisted of the PTR8000 receiving module and the PC server in medical institutions. The interface of receiving module and server is realized by a typical universal serial bus(USB). The operating principle of the primary stage is shown in figure 3.

Fig. 3. Schematic of primary stage

3 Hardware Architecture

Hardware in the monitor is mainly constituted by a GSR signal collecting circuit and a wireless module.

3.1 GSR Signal Collecting Circuit

The collecting circuit is realized by measuring the value of resistance or conductance between the patient's two fingers. The value collected from users directly is analog signal, and it will be transmitted to microcontroller interface after amplifying and A/D converting. A typical GSR signal collecting circuit is shown in figure 4. When a user is having a test, he may put two fingers into leather collars. The values of conductance are different when people in different states such as nervous or quiet. As is proved by experiments, the value of conductance will be increasing gradually when one's state changes from quiet to nervous. Such is the design of our collecting circuit, it is very convenient and of a low cost.

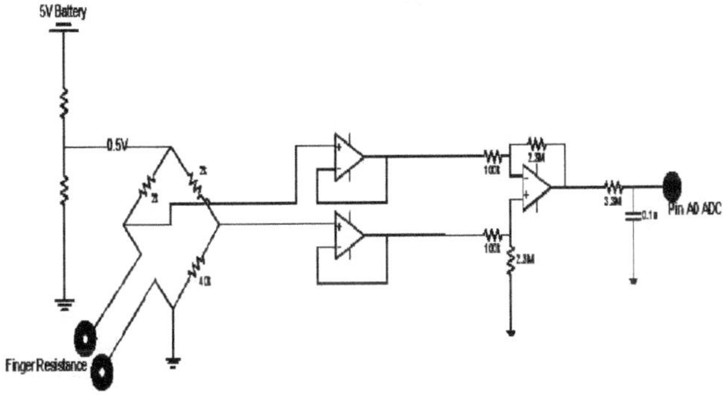

Fig. 4. GSR signal acquisition circuit

3.2 PTR8000 Module

PTR8000 is a highly embedded wireless module with multiple channels and frequencies. The operating voltage ranges from 1.9V to 3.6V and the standby power is 2uA. The nRF905 wireless transceiver is the key part of PTR8000 module, and it works at the ISM frequency of 433/868/915MHz, which belong to the international exploitation frequency that need not to apply for.

Table 1. PTR8000 electrical characteristics

Parameter	*Value*	*Unit*
Operating Voltage	1.9~3.6	V
Maximum Output Power	10	dBm
Maximum Data Transfer Rate	100	Kbps
Operating Current When -10dBm	11	mA
Operating Current in Receiving mode	12.5	mA
Operating Temperature	-40~85	°C
Sensitivity	-100	dBm
Operating Current when Power-down	25	uA

3.3 PTR8000 Hardware Interface

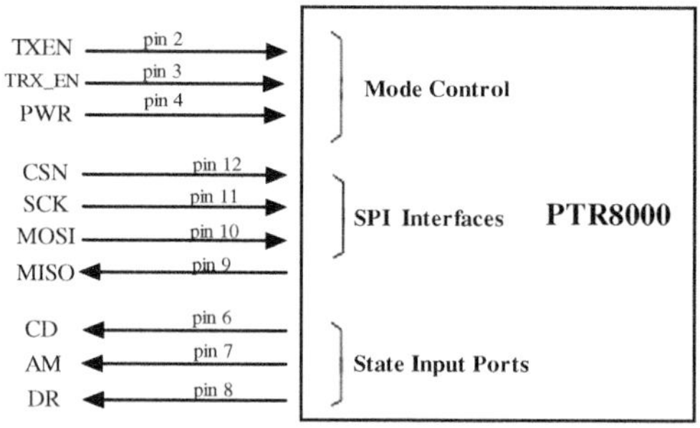

Fig. 5. PTR8000 hardware interface circuit

In figure 5, the overall PTR8000 interfaces are given. It is consisted of 10 digital input/output(I/O), and can be divided into three parts according to operating pattern.

1) Mode control

The three interfaces named TRX_CE, TX_EN, PWR control 4 operating modes which are power down and SPI programming mode, standby and SPI programming mode, sending mode and receiving mode. All of the modes are listed in table 2.

Table 2. PTR8000 control modes

PWR	*TRX_CE*	*TX_EN*	*Operating Mode*
0	X	X	Power-down and SPI programming mode
1	0	X	Standby and SPI programming mode
1	1	0	Receiving mode
1	1	1	Sending mode

2) SPI interface

The SPI interface is consisted of SCK, MISO, MOSI and CSN. Once it was configured, the microcontroller sets the operating parameter of PTR8000 by using SPI interface.

3) State output interface

This interface can provide a carrier detect (CD)input, an address matched(AM)output and a data ready(DR)output.

4 Software Architecture

The operating mode of PTR8000 is determined by the RF configuration register. The RF configuration register has a capacity of 10bytes which includes frequency, address, output power and effective bytes length. The situation of SPI interface determines all the configurations of the PTR8000. If the power was on, PTR8000 in the subordinate stage enters into the standby mode. Then it enters into the receiving state when the PTR8000 in the primary stage sending out commands. If the primary state has received a effective instruction from the subordinate stage, the microcontroller will output the GSR signals.

4.1 Event Management and Processing

When the PTR8000 is transmitting data, it will verify the data by adding the leading codes and CRC codes automatically and then complete the data packaging. The process of the transmitting program is shown in figure 6.

When the PTR8000 is in the state of receiving, it will be in charge of detecting the carrier signals, marching the address and decoding the data package constantly. When detected carriers keep in a constant frequency, the AM sets to high electrical level. At the same time, the PTR8000 conducts the operation of CRC detection. If the CRC detection was passed, it indicates that a right data package is received and then the DR sets to high level. After the SCM AT90S8515 has detected the DR high level, it will set the PTR8000 to standby mode. The SCM AT90S8515 reads the received data by the SPI interface. The process of receiving program is shown in figure 7.

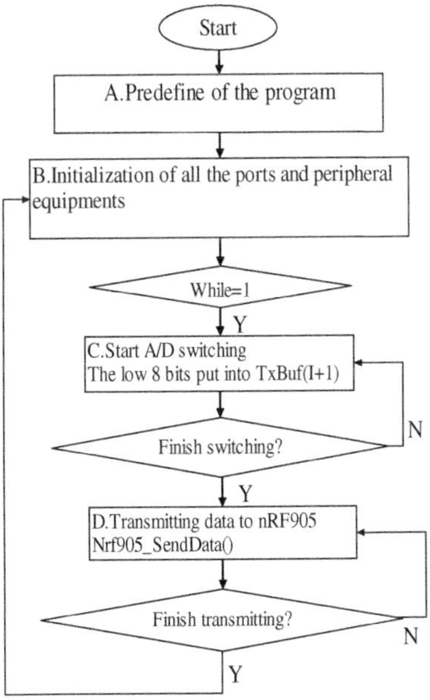

Fig. 6. Sending process flow chart

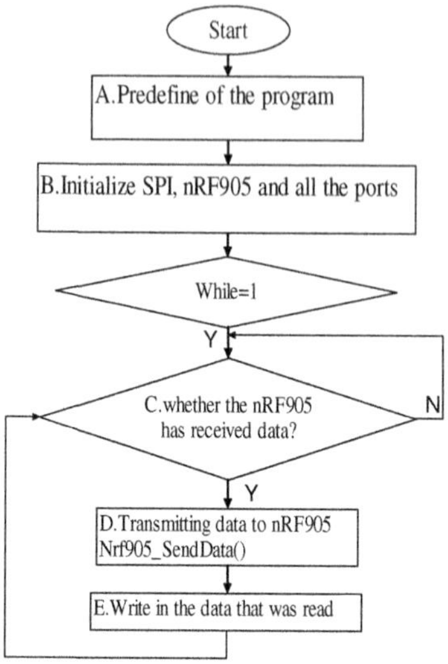

Fig. 7. Receiving process flow chart

4.2 GSR Displaying and Processing

With the development of object-oriented programming technology, various kinds of these languages, such as VB(Visual Basic) , VC(Visual C), are becoming useful instruments in users' software development gradually. In our project, the monitoring software of primary stage in the system is realized by using VB software and its interface can be programmed just like a serial port. In VB software, a MSComm control procedure can make the USB interface working as a virtual standard serial port[5]. The program includes two parts, the former one is USB interface program and the later one is applying program. At last, the collected GSR signal can be displayed by a visualized graph. Figure 8 shows the GSR signal graph when a tester relaxing himself by taking a deep breath. In the graph, x axis stands for time(T), y axis stands for conductivity (G). From figure 8, we can acquire a result that the mood loosens gradually when conductivity increases. So it has reached the design requirement.

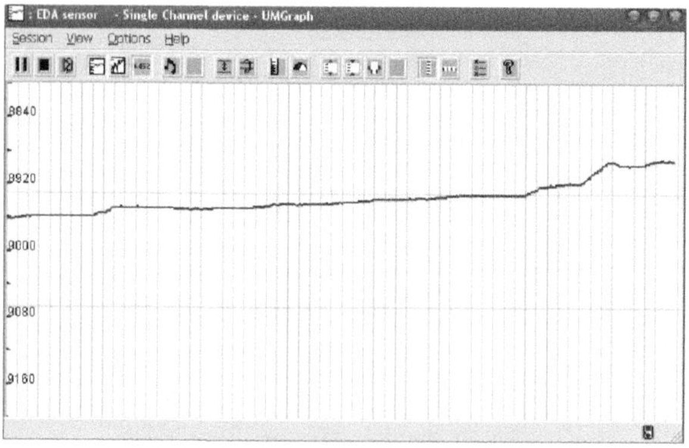

Fig. 8. Real-time GSR signals graph

5 Conclusion

In this paper, a novel wireless human healthcare system based on GSR from finger sensors is presented. The small and portable characteristics of the system will shift in the way people think about and manage their health. This shift will not only improve the quality of life, but will also reduce healthcare costs.

Our future work include:1)using more kinds of finger sensors to collect more GSR information;2) developing more accurate, real-time, reliable wireless healthcare system.

Acknowledgement. The project is sponsored by the XinMiao College Plan of Zhejiang province, No. ZC323010070.

References

1. http://en.wikipedia.org/wiki/Galvanic_skin_response
2. Reich, W.: Experimental Investigation of the Electrical Function of Sexuality and Anxiety, vol. 3(1&2) (November 1969)
3. Otto, C., et al.: System Architecture of a Wireless Body Area Sensor Network for Ubiquitous Health Monitoring. Journal of Mobile Multimedia 1(4), 307–326 (2006)
4. Wu, W.H., et al.: MEDIC: Medical embedded device for individualized care. University of California, Los Angeles
5. Westeyn, T., Presti, P., Starner, T.: ActionGSR: A combination galvanic skin response-accelerometer for physiological measurements in active environments. In: Proc. IEEE Int. Symp. Wearable Computers, vol. 06, pp. 11–14 (2006)

A New Motion Estimation Method and Its Application in Motion Compensation Deinterlacing Algorithms

Tao Zhang, Jianyu Wang, and Pengfei He

School of Electronic and Information Engineering,Tianjin University
Tianjin, P. R. China
zhangtao355@126.com

Abstract. In this paper, a new method of motion estimation is presented to replace the widely used method which calculates SAD to find the matching block. This method calculates the absolute difference between corresponding pixels of current block and reference block. Then according to the value of the absolute difference, different weights are assigned to these pixels. The level and value of these weights are set empirically. According to the sum of the pixels' weights, we can select motion vector and process motion compensation. This method avoids the error caused by few noise pixels when calculating SAD. The simulation shows this method can improve the quality of motion compensation in deinterlacing.

Keywords: Motion estimation, Motion vector, Deinterlacing, Motion compensation.

1 Introduction

Motion estimation is used in many video processing techniques to find the right motion vector. These techniques include MPEG-2 interframe predictive coding and motion compensation algorithms in deinterlacing process of video signal. The purpose of motion estimation is to find the matching block of the current macro block. The most widely used method of motion estimation is full-search (FS), and three-step search (TSS), four-step search (FSS), together with diamond-search (DS) are also commonly used fast algorithm [1-3]. These methods process every macro block (MB) by calculating the sum of absolute difference (SAD). The drawback of this method is the burst noise in the picture can cause dramatic influence on the SAD of this MB; therefore the outcome of motion estimation is undermined.

To solve this problem, this paper presents a new motion estimation method. Under this method, rather than summing the absolute difference of the pixels in the MB, a variant weight is assigned to every pixel, according to the value of its absolute difference. Then the weights of all pixels are added up in the block, and a larger sum of the weight (SW) indicates more likelihood of matching. This method avoids significant influence of burst noise and thus attains better effect and noise immunity.

D. Zeng (Ed.): ICAIC 2011, Part II, CCIS 225, pp. 344–351, 2011.

In television technology, interlaced scanning is a commonly used technique. We can obtain 50Hz field scanning rate using information of 25 frames, so we save bandwidth while meet the demand of human visual system's critical flicker frequency [4]. However, the size of modern television receiver is getting larger and larger, which makes the flicker caused by interlaced signal more and more obvious. What's worse, interlaced video brings difficulties when conducting video formats conversion and other processing [5]. So a large number of deinterlacing algorithms have been proposed.

Generally speaking, deinterlacing technology can be divided into 2 categories: non-motion compensation methods (non-MC) and motion compensation methods (MC). Non-MC methods include intra-field linear interpolation, inter-field linear interpolation, edge-dependent interpolation and motion- adaptive algorithms [5-12]. Motion compensation algorithms obtain best effect. Even though many motion compensation algorithms have been proposed, their effects largely depend on the correctness of motion vectors. Therefore, improving the correctness of motion estimation is key element of motion compensation deinterlacing. Applying the algorithm proposed in this paper to motion compensation deinterlacing process yields good effects. Furthermore, this method can be applied to almost all motion compensation algorithms' motion estimator. Evidently, our method can be used in other fields such as video coding, but this paper will focus on its application to deinterlacing algorithm.

2 PRIOR Works in Deinterlacing Algorithms and Motion Estimation Algorithms

Among deinterlacing algorithms, non-MC methods are comparatively simple and have lower hardware cost. Intra-field linear interpolation [7], such as line repetition and line average, causes deterioration in vertical resolution when high frequency details exist. Inter-field linear interpolation, such as field repetition, has good effect in stationary sequence. Yet serration appears at the profile of the objects when movement exists [8]. Edge-dependent deinterlacing, or edge-based line averaging (ELA), detects the presence of an edge, and the signal is interpolated in that direction [9-10]. Motion adaptive method (MA) switches algorithm according to the presence of motion, and can attain better outcome [11-12].

The motion compensation algorithm is the most advanced deinterlacing algorithm. It conducts interpolation along the motion trajectory and virtually converts a moving sequence into a stationary one [6], so inter-field interpolation yields good performance. Correct motion estimation is the key element of this algorithm. In earlier algorithms, Woods *et al.* [13] proposed the motion vector being extended to both previous and preprevious fields. Wang *et al.* [14] use both original pixels and interpolated pixels of previous field. New algorithms are more precise in determining the motion vector. Ohjae Kwon *et al.* [15], Kefei Ouyang *et al.* [16] adopt bi-directional motion estimation to make the motion vector smooth and avoid error caused by different sampling grid between odd field and even field. Yu-Lin Chang *et al.* [17] proposed four fields motion estimation. This method calculates two same parity field SAD values, which also avoids different sampling grid. Hossein Mahvash Mohammadi *et al.* [18] uses threshold to determine whether to use MC or ELA. Yu-Cheng Fan *et al.* [19] propose an algorithm to detect a wrong motion vector with multi-level method.

However, the algorithms mentioned above use same method when looking for matching block at the bottom level, which calculates the SAD between current block and reference block as

$$SAD_{(i,j)}(x,y) = \sum_{m=0}^{3}\sum_{n=0}^{7} | P_n(i+2m, j+n) - P_{n-1}(i+2m+x, j+n+y)| \quad (1)$$

where (i ,j) means the coordinate of the pixel at the upper left corner of current macro block (CB), every macro block is represented by a set of (i, j). Assuming CB's field number is n, (x, y) is the offset of reference block in previous field, which is virtually the motion vector (MV). An 8x8 macro block is used in our method. Due to interlaced format, the line number i is shifted in intervals of 2. The reference block which has the minimum $SAD_{(i,j)}(x, y)$ is the matching block (MB) and (x, y) is the correct motion vector. It is obvious that burst noise can cause dramatic influence in (1). For example, a black noise pixel in white background will contribute 255 to $SAD_{(i,j)}(x, y)$. In practice, we find most MB's $SAD_{(i,j)}(x, y)$ is less than 100. Therefore, a noise pixel will possibly change the result of motion estimation, which means distinguish the actually matching block as non-matching block due to the presence of few burst noise.

3 Proposed Motion Estimation Algorithm

3.1 Motion Estimation

In order to remove the error caused by calculating SAD, the absolute difference is not added up directly after calculating every pixel's absolute difference value. Rather, we assign a weight to every pixel in macro block (actually, it represents a pair of pixels in both current block and reference block) according to its absolute difference.

In detail, a 32x32 search window is adopted, so when it comes to MV=(x, y), x and y can be as large as ±12. Fig.1 shows the current block, search window, reference block and motion vector.

First, the absolute difference of every pixel is calculated.

$$AD_{(i,j)}^{(x,y)}(m,n) = | P_n(i+2m, j+n) - P_{n-1}(i+2m+x, j+n+y)| \quad (2)$$

In (2), (m, n) represents every pair of pixels in current block and reference block. Then, a weight is assigned to every set of (m, n) according to the value of $AD_{(i,j)}^{(x,y)}(m,n)$ and thresholds we set. The thresholds which decide the level and

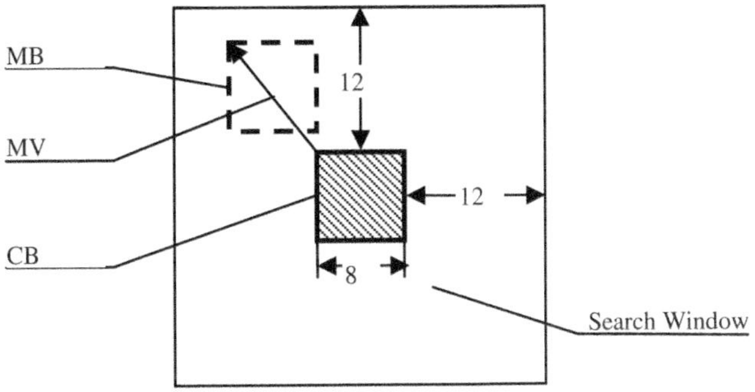

Fig. 1. Search Window

value of weight are set empirically. From the result of the simulation, multi-level weights are more reliable in motion compensation. After simulation and test, we set 3-level threshold as in (3).

$$W_{(i,j)}^{(x,y)}(m,n) = \begin{cases} 3, \text{if } AD_{(i,j)}^{(x,y)}(m,n) < TH1, & TH1 = 2 \\ 2, \text{if } TH1 \le AD_{(i,j)}^{(x,y)}(m,n) < TH2, & TH2 = 6 \\ 1, \text{if } TH2 \le AD_{(i,j)}^{(x,y)}(m,n) < TH3, & TH3 = 12 \\ 0, \text{if } AD_{(i,j)}^{(x,y)}(m,n) \ge TH3 \end{cases} \quad (3)$$

After calculating $W_{(i,j)}^{(x,y)}(m,n)$ of every pixel, the sum of $W_{(i,j)}^{(x,y)}(m,n)$ of every set of (m, n) in the macro block is calculated as in (4).

$$SW_{(i,j)}(x,y) = SAD_{(i,j)}(x,y) = \sum_{m=0}^{3}\sum_{n=0}^{7} W_{(i,j)}^{(x,y)}(m,n) \quad (4)$$

The matching block is the reference block with largest $SW_{(i,j)}(x, y)$ in 32 x 32 search window. A large $SW_{(i,j)}(x, y)$ indicates more matching pixels, so this reference block is regarded as more likely to be the matching block. If some burst noise exists in matching block, the influence is the decline of corresponding $W_{(i,j)}^{(x,y)}(m,n)$. The decline is at most 3 to 0, yet SW is generally larger than 50. Therefore, the influence of noise is limited. According to the simulation result, the decline has only slight influence on the search of MB. Another advantage of multi-level weights mentioned above is avoiding equal largest SW of a few blocks. This is due to that multi-level weights have more choices when determining pixels' weights and they are thus less likely to have equal sum value.

We should particularly point out that our motion estimation method can be applied to most kinds of motion compensation algorithms and other fields which need motion estimation. These algorithms can also be improved by using our method.

3.2 An Improvement Especially for Motion Compensation Deinterlacing

However, even using the new motion estimation algorithm, the motion vector cannot always be absolutely correct. To deal with those blocks which are not properly getting motion vectors, another threshold is utilized to decide whether the motion estimation is reliable. After the maximal SW is finded, it is compared with the threshold SUM_TH. If SW is larger than SUM_TH, the motion estimation is reliable and motion compensation is conducted in this macro block. If SW is smaller than SUM_TH, the motion estimation is regarded as unreliable. ELA is conducted at such block. The threshold SUM_TH is also selected empirically. We set it 90. By doing this, the block which fails to conduct motion estimation can be processed properly.

4 Simulation and Analysis

We apply our motion estimation algorithm to one kind of motion compensation deinterlacing algorithm, and then the result of motion deinterlacing is evaluated. Therefore the performance of motion estimation is evaluated indirectly. To make sure the performance is evaluated properly, we adopt both objective PSNR and subjective evaluation. We select 352×288 standard test sequences in CIF format. The interlaced sequences are processed with proposed method and previous method. PSNR values are calculated using the processed sequences and standard sequences.

In simulation, we apply our method to the algorithm proposed in [14]. Algorithm in [14] is a classical time-recursive motion compensation deinterlacing algorithm (TRMC). Many later algorithms are based on this classical algorithm. Therefore, using our method in TRMC can better prove its applicability. In TABLE 1, the PSNR results of different methods are listed. These methods include line average, ELA, TRMC, TRMC with proposed motion estimation method but without ELA, TRMC with proposed motion estimation method and ELA. We list both algorithms with ELA and without ELA because the performance of proposed motion estimation method can be better evaluated without ELA, but in practical application the algorithm combined with ELA is more useful.

From the column of TRMC and Proposed Method (without ELA) in TABLE 1, it is obvious that the performance of deinterlacing is improved by simply replacing the original motion estimator with proposed method. We should note that even though our PSNR values are lower than those of most advanced algorithms, which obtain 40dB to 50dB, our method can be used in those algorithms. Therefore, a lower PSNR does not mean our method is not useful. If our method is applied to advanced algorithms, their performance will also be improved. Furthermore, we can find out from the column of Proposed Method (with ELA) that combining ELA caused a decline in PSNR. However, when their performances are subjectively judged, combining ELA is helpful in enhancing the effect.

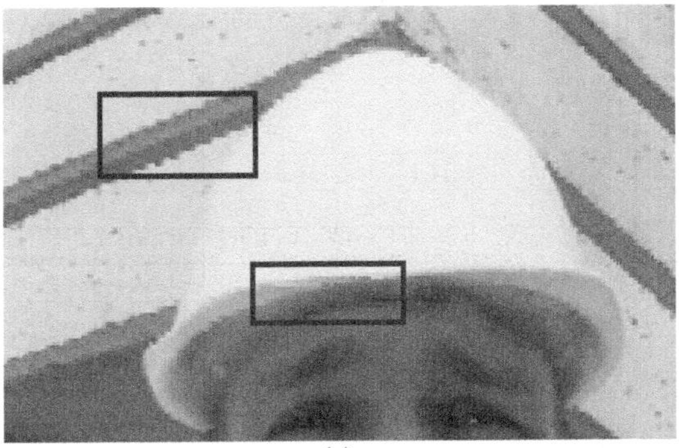

(a)

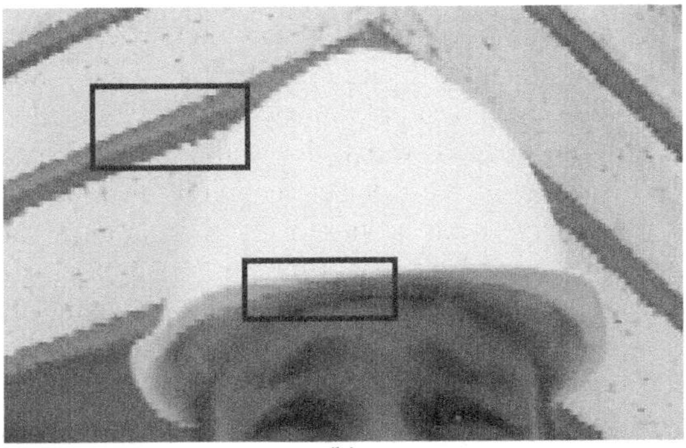

(b)

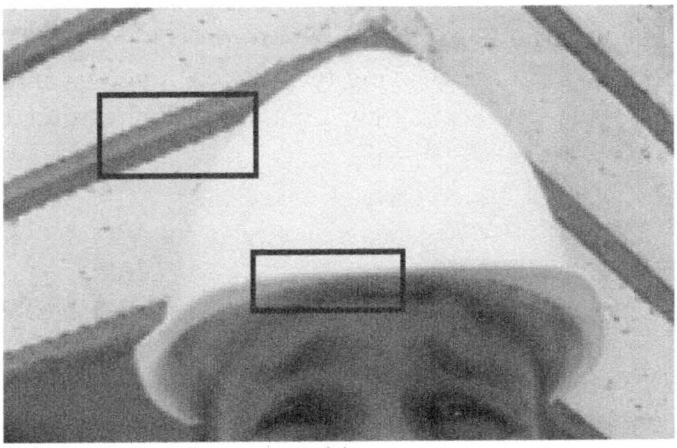

(c)

Fig. 2. (a) TRMC [14] (b) Proposed Method (without ELA) (c) Proposed Method (with ELA)

Fig. 2 shows that the picture processed by original TRMC has serration and blur at the oblique edge of building and helmet. After applying proposed algorithm, this effect is weaker. When combining with ELA, the serration has been mostly eliminated.

Table 1. psnr comparison

Name	Line Average	ELA	TRMC	Proposed Method (without ELA)	Proposed Method (with ELA)
Mother and Daughter	24.49573	34.13394	36.47232	39.22511	35.66899
Foreman	24.20272	29.39726	29.92898	30.85682	30.41310
Hall Monitor	24.98234	27.59323	30.32003	32.53923	27.7350
Silent	24.31665	29.32714	33.14730	34.57535	30.85473

5 Conclusion

In this paper, we propose a new motion estimation method to substitute the widely used SAD method. In our algorithm, the corresponding pixels in current block and reference block are utilized to calculate the absolute difference between them. Then a weight is assigned to this pixel according to the value of the absolute difference and empirically set thresholds. The sum of all pixels' weights decides the matching block. When applying to deinterlacing algorithm, the sum of weights can be compared to another threshold to determine whether the motion estimation is reliable. When the motion vector is not reliable, ELA is used in this macro block. The simulation test proved that the proposed method yield good performance when used in TRMC [14] algorithm. Furthermore, the proposed method can be used in other deinterlacing algorithms and video coding algorithms. Good performance can also be obtained in these fields.

References

1. Tekalp, A.M.: Digital Video Processing. Prentice Hall Signal Processing Serious, Upper Saddle River (1995)
2. Po, L.-M., Ma, W.-C.: A novel four-step search algorithm for fast block motion estimation. IEEE Trans. Circuits Syst. Video Technol. 6(3), 313–317 (1996)
3. Zhu, S., Ma, K.-K.: A new diamond search algorithm for fast blockmatching motion estimation. IEEE Trans. Image Process. 9(2), 287–290 (2000)
4. Yu, S., Li, W., Hou, Z., Feng, Q., Yang, Z., Lv, W.: Principles of Television. National Defense Industry Press, Beijing (2005) (in Chinese)
5. Fan, Y.-C.: Deinterlacing system and technology overview and discussion. In: IST 2009 - International Workshop on Imaging Systems and Techniques, pp. 287–290 (May 2009)
6. de Haan, G., Bellers, E.B.: Deinterlacing—An Overview. Proceedings of the IEEE 86(9), 1839–1857 (1998)
7. Bellers, E.B., de Haan, G.: De-interlacing—A key technology for scan rate conversion. Elsevier, Amsterdam (2000)
8. Smith, N.S., Walker, G.: Flexible upconversion for high quality TV and multimedia displays. In: 1996 IEEE Int. Conf. Consumer Electronics, pp. 338–339 (June 1996)
9. Lee, D.-H.: A new edge-based intra-field interpolation method for deinterlacing using locally adaptive-thresholded binary image. IEEE Transactions on Consumer Electronics 54(1), 110–115 (2008)

10. Lin, S.-F., Chang, Y.-L., Chen, L.-G.: Motion adaptiveinterpolation with horizontal motion detection for deinterlacing. IEEE Trans. on Consumer Electronics 49(4), 1256–1265 (2003)
11. Lee, G.G., Hsin-Te, Wang, M.-J., Lin, H.-Y.: Motion adaptive deinterlacing via edge pattern recognition. In: IEEE International Symposium on Circuits and Systems, ISCAS 2007, pp. 2662–2665 (May 2007)
12. Zhang, T., He, L., Ge, L., Li, H.: Motion adaptive deinterlacing with accurate motion detection and anti-aliasing interpolation filter. In: 25th PCS: Picture Coding Symposium 2006, PCS 2006, April 24-26 (2006)
13. Woods, J.W., Han, S.-C.: Hierarchical motion compensated de-interlacing. In: Proc. SPIE, vol. 1605, pp. 805–810 (1991)
14. Wang, F.M., Anastassiou, D., Netravali, A.N.: Time-recursive deinterlacing for IDTV and pyramid coding. Signal Process.: Image Commun. 2, 365–374 (1990)
15. Kwon, O., Sohn, K., Lee, C.: Deinterlacing using directional interpolation and motion compensation. IEEE Transactions on Consumer Electronics 49(1), 198–203 (2003)
16. Ouyang, K., Shen, G., Li, S., Gu, M.: Advanced motion search and adaptation techniques for deinterlacing. In: IEEE International Conference on Multimedia and Expo, ICME 2005, pp. 374–377 (July 2005)
17. Chang, Y.-L., Lin, S.-F., Chen, C.-Y., Chen, L.-G.: Video de-interlacing by adaptive 4-field global/local motion compensated approach. IEEE Transactions on Circuits and Systems for Video Technology 15(12), 1569–1582 (2005)
18. Mohammadi, H.M., Pierre Langlois, J.M., Savaria, Y.: A threshold-based deinterlacing algorithm using motion compensation and directional interpolation. In: 13th IEEE International Conference on Electronics, Circuits and Systems, ICECS 2006, pp. 459–462 (December 2006)
19. Fan, Y.-C., Lin, H.-S., Chiang, A., Tsao, H.-W., Kuo, C.-C.: Motion compensated deinterlacing with efficient artifact detection for digital television displays. Journal of Display Technology 4(2), 218–228 (2008)

An Application of the Particle Swarm Optimization on the Gasoline Blending Process*

Hui Cheng, Weimin Zhong**, and Feng Qian

Key Laboratory of Advanced Control and Optimization for Chemical Processes, Ministry of Education, East China University of Science and Technology
Shanghai, 200237, China
chenghui_123@sina.cn,
wmzhong@ecust.edu.cn

Abstract. In this paper, an optimization strategy on the basis of the particle swarm optimization (PSO) method is proposed to determine the optimal recipe offline for the gasoline blending process. An octane number model is proposed for optimization. Furthermore, the proposed strategy has been applied onsite at a gasoline production line in Nanjing, China. The results show that the optimized recipes are able to improve the first-time success rate for the blending process and significantly decrease the quality giveaways and blending cost. The stability of the gasoline production has been improved as well.

Keywords: Particle Swarm Optimization, Octane Number, Ethyl Model, Gasoline Blending.

1 Introduction

As the important product of the refineries, gasoline produces 60-70% of the refiner's income [1]. Therefore the gasoline blend process, which mixes different types of intermediate components into final product satisfying quality specifications, is critical for the refinery to operate economically.

In the traditional blending process (tank blending), different components are collected sequentially into the product tank, before the mixing operates. The recipe is calculated beforehand according to the offline property measurements of the components. Due to the fact that the component properties are not measured online, tank blending cannot handle the property variations of the components. This drawback usually results in the low first-time success rate, high blending cost and high quality giveaways, epically when the recipe is not calculated in a proper way.

---

* This work is supported by Shanghai PostDoc Fund (No. 10R21412200), National Science Fund for Distinguished Young Scholars (60625302), National 863 plans (2008AA042902), Shanghai scientific and technological project (09DZ1120400), the special funds of Shanghai College basic scientific research operating expenses, Shanghai Leading Academic Discipline Project (B504) and National Nature Science Foundation of China (No.60804029).

** Corresponding author. (Associate Professor)

D. Zeng (Ed.): ICAIC 2011, Part II, CCIS 225, pp. 352–360, 2011.

The studied process in this paper, a gasoline blending process in Nanjing, China, which is one of the most important lines for SINOPEC, produces gasoline by a typical tank blending way. During the production, the recipe is calculated with a linear model for the critical properties, such as octane number, with the trail error method. The success rate is very low because of the component variation. Therefore, the najor research aim of this paper is to use the modelling and optimization technique to improve the economic performance of the blending process.

2 Model for Gasoline Blending

Modelling of the blending process plays a critical role in the recipe optimization. Since dozens of specifications are applied on the final product, usually more than one property is considered for the recipe optimization. Among the properties, the octane number is the most critical one for the reason that it behaves nonlinearly and determines the performance of the gasoline in Engine.

According to the principle of the modelling for the octane number, the developed models can be categorized into two classes. Models belonging to the first class, usually predict the product octane number with the fixed structure constructed by the blending mechanism, while the parameters are identified with the experiment data. Ethyl model [2] is a typical example in this class. Ethyl model put extra nonlinear items on the basis of the linear blending model. The considered items include the effects of the component sensitivity, olefin content and aromatic content on the octane number of the product. The magnitudes of the nonlinear items are identified with the blending data. Due to its simplicity and generality, Ethyl model had the most applications among all the octane number models and are usually used as the benchmark to be compared with for new developed model [3].

In the second class, models predict the product octane number with the neural networks [4]. Being different from the model in first class, neural networks don't have the fixed model structure and physical explanation, while it can learn the model automatically from the blending data. Even though the neural networks model may have better accuracy for the training data, it lacks the generality and requires train the neural networks again after the components change. Therefore, the adaptive ability of the model is quite limited.

In this paper, an octane number blending effect model is proposed for the recipe optimization purpose. The basic idea is to quantify the blending effect of different components from the blending data first, and then use the quantitative information to compensate the component octane number. The weighted sum of the compensated octane numbers by the recipe is used to predict the octane number of the product. The advantage of this model is in accordance of the process knowledge from the site and accurate enough for optimization purpose, which improves the applicability of the proposed model.

First, it is assumed that n number of the components involved in the blending. Their octane numbers are compensated with the blending effect as following equation:

$$Qx(O_{j,i}) = O_{j,i} + pb_{j,i},\ j \in \{RON, MON\} \tag{1}$$

where $i = 1, 2, \cdots n$ is the number of the components, $j \in \{RON, MON\}$ is the research octane number (RON) and the motor octane number (MON), $pb_{j,i}$ is the blending effect for RON or MON of the i-th component, $o_{j,i}$ is the RON or MON for the i-th component, and $Qx(o_{j,i})$ is the compensation function.

Next, weighted sum is calculated for the compensated octane number in linear way,

$$o_j = \sum_{i=1}^{n} \left(r_i \cdot Qx(o_{ji}) \right) \tag{2}$$

where r_i is the recipe for i-th component, o_j is the prediction for the product octane number.

In above model, only the blending effect pb_{ji} is to be identified with the blending data, the number of which is defined by the number of components. Assume that m set of blending data are used for parameter identification. According to Equation (1) and (2), it is obtained as:

$$\begin{bmatrix} O_j^1 \\ O_j^2 \\ \vdots \\ O_j^l \\ \vdots \\ O_j^m \end{bmatrix} = \begin{bmatrix} r_{1,1} & r_{1,2} & \cdots & r_{1,n} \\ r_{2,1} & r_{2,2} & \cdots & r_{2,n} \\ \vdots & \vdots & r_{l,i} & \vdots \\ r_{m,1} & r_{m,2} & \cdots & r_{m,n} \end{bmatrix} * \left(\begin{bmatrix} pb_{j,1} \\ pb_{j,2} \\ \vdots \\ pb_{j,i} \\ \vdots \\ pb_{j,n} \end{bmatrix} + \begin{bmatrix} O_{j,1} \\ O_{j,2} \\ \vdots \\ O_{j,i} \\ \vdots \\ O_{j,n} \end{bmatrix} \right) \tag{3}$$

where O_j^l is the octane number for the l-th blending data, $\lfloor r_{m,1} \quad r_{m,2} \quad \cdots \quad r_{m,n} \rfloor$ is the recipe for the blending. If the recipe matrix in Equation (3) is denoted as $R_{m,n}$, then Equation (3) can be reformed as the linear regression form $R_{m,n}PB = Y$, where $PB = \lfloor pb_{j,1} \quad pb_{j,2} \quad \cdots \quad pb_{j,n} \rfloor^T$ and

$$Y = \begin{bmatrix} O_j^1 \\ O_j^2 \\ \vdots \\ O_j^l \\ \vdots \\ O_j^m \end{bmatrix} - \begin{bmatrix} r_{1,1} & r_{1,2} & \cdots & r_{1,n} \\ r_{2,1} & r_{2,2} & \cdots & r_{2,n} \\ \vdots & \vdots & r_{l,i} & \vdots \\ r_{m,1} & r_{m,2} & \cdots & r_{m,n} \end{bmatrix} * \begin{bmatrix} O_{j,1} \\ O_{j,2} \\ \vdots \\ O_{j,i} \\ \vdots \\ O_{j,n} \end{bmatrix} \tag{4}$$

Therefore, the parameter PB is identified as:

$$PB = \left(R_{m,n}^T * R_{m,n} \right)^{-1} * \left(R_{m,n}^T * Y \right) \tag{5}$$

Besides octane number, other properties, such as, Reid Vapor Pressure (RVP), sulphur content (SULF), density (SPGR), olefin content (OLE), aromatic content (AROM), and benzene content (BENZ), are taken into consideration for the recipe optimization. In this paper, for RON and MON, model in Equation (1)-(5) is used, while the anti-knock index (DON) is calculated by the average of the RON and MON, for RVP Chevron [5] mode is used, for the rest of the properties, linear blending model is chosen. On the basis of the model, the optimization algorithm and strategy are required to obtain the optimal recipe.

3 Particle Swarm Optimization for Gasoline Blending

The core technique in the recipe calculation is the optimization method. For the sake of easy use, linear programming (LP) methods, which only take linear blending models, are widely applied in blending process [6]. However, the major gasoline qualities, such as octane number and Reid vapour pressure, behave nonlinearly in the blending. Consequently, many applications of nonlinear programming (NLP) methods have been reported lately [7].

The deterministic optimization method, such as LP and NLP, usually suffers from being trapped in the local optima, which is not suitable to search the global optima for the multi-peak optimization problem. Recipe optimization is such problem that has more than one peak in the function surface, especially in the situation that additive MMT is used to boost the octane number. The particle swarm optimization (PSO) method is therefore selected to solve the recipe optimization problem, since the PSO has better ability to search the global optima.

PSO was proposed first by Kennedy and Eberhart in 1995 [8][9]. It mimics the social behaviour of the animals, for example bird flocking and fish schooling, to solve the combinational optimization problems. As an evolutionary and swarm intelligent method, PSO takes operation on a population of candidate solutions, named particles, through iterations to search the optima. The algorithm makes use of a velocity vector to update the current position of each particle in the swarm. The velocity vector is updated based on the history information gained by the swarm. And the positions of the swarm are updated to search for better positions according to the updated velocity vector [10][11] as following equations:

$$\begin{aligned} \boldsymbol{v}_i(t+1) = \omega \boldsymbol{v}_i(t) + c_1 \boldsymbol{r}_1(t)(\boldsymbol{p}_i(t) - \boldsymbol{x}_i(t)) \\ + c_2 \boldsymbol{r}_2(t)(\boldsymbol{p}_g(t) - \boldsymbol{x}_i(t)), \end{aligned} \tag{6}$$

$$\boldsymbol{x}_i(t+1) = \boldsymbol{x}_i(t) + \boldsymbol{v}_i(t+1) \tag{7}$$

for i=1, 2, …, s where s is the swarm size, c_1 and c_2 are the acceleration coefficients, ω is the inertia weight, $\boldsymbol{r}_1(t)$ and $\boldsymbol{r}_2(t)$~$U(0, 1)$, $\boldsymbol{x}_i(t)$ is the position of particle i at time t, $\boldsymbol{v}_i(t)$ is the velocity of particle i at time t, $\boldsymbol{p}_i(t)$ is the personal best solution of particle i at time t, and $\boldsymbol{p}_g(t)$ is the global best solution at time t.

The particle position $\boldsymbol{x}_i(t+1)$ is updated using its current value and the newly computed velocity $\boldsymbol{v}_i(t+1)$, which is determined by the values of $\boldsymbol{v}_i(t)$, $\boldsymbol{x}_i(t)$, $\boldsymbol{p}_i(t)$, $\boldsymbol{p}_g(t)$ and coefficients ω, c_1 and c_2. Since the method initializes and updates the particles randomly, PSO is able to jump out from the local optima and then to find the global optima.

The PSO method is then used to solve the following optimization problem for recipe calculation:

$$\underset{r_i}{Min}\left[w_c \cdot \sum_{i=1}^{n} c_i \cdot r_i + w_g \cdot \sum_{j \in S} sc_j \cdot g_j(P_j, s_j) \right] \tag{8}$$

where c_i is the specific cost of the component, sc_j is the specific cost of the product property, w_c and w_g are the weights for blending cost and quality giveaway respectively, S is the set of the property, $g_j(P_j, s_j)$ is the function to calculate quality giveaway based on the product quality prediction and the quality specification:

$$h_j(P_j, s_j) = \mathrm{sgn}(s_j) \cdot \left(\frac{\left(\begin{array}{l} P_j \cdot (targetvol - heelvol) \\ + P_j^h \cdot heelvol \end{array} \right)}{targetvol} - s_j \right) \tag{9}$$

$$g_j(P_j, s_j) = \max(h_j(P_j, s_j), 0)$$

where $targetvol$ is the target volume for blending, $heelvol$ is the volume of the heel in product tank, P_j^h is the property of the heel in product tank, $\mathrm{sgn}(s_j)$ is the sign function, whose output is +1 if the specification for the j-th property require to be bigger than certain value, otherwise, the output is -1.

The constrains for the optimization are

$$s.t.: \sum_{i=1}^{n} r_i = 1 \tag{10}$$

$$r_{i\_\min} < r_i < r_{i\_\max} \tag{11}$$

$$0 \le h_j(P_j, s_j) \tag{12}$$

where (8) is the material balance constrain, (9) is the constrain of the component storage, (10) is the constrain set by the quality specification.

By solving the optimization defined in Equation (8)-(12) with the PSO method, the optimal recipe can be obtained in the sense of the minimal blending cost and quality giveaways. The proposed blending model and optimization strategy have been applied onsite. Firstly, the onsite historical data for blending were collected for modeling with Equation (5). Secondly, the optimization problem was formulated and solved with PSO to get the optimal recipe, which is then used to produce gasoline onsite. The sample of the product is taken after the blending completes, and analyzed to check the result.

4 Application Results

In this paper, the proposed optimization strategy has been applied on a gasoline production line in Nanjing, China. In the application, the octane number model for 93# gasoline has been constructed, while the recipe for 93# gasoline production has been optimized and applied.

4.1 Results of the Blending Effect Model for Octane Number

90 sets of blending data from the production line for 93# gasoline have been collected to construct the RON and MON blending effect model, by Equation (5). With the

same data, the corresponding Ethyl models have been constructed for the comparison purpose. The results are shown in the following figures.

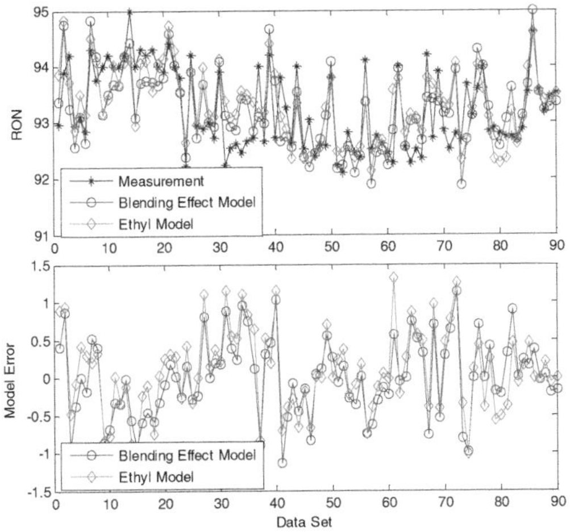

Fig. 1. Comparison of RON Models' Accuracy

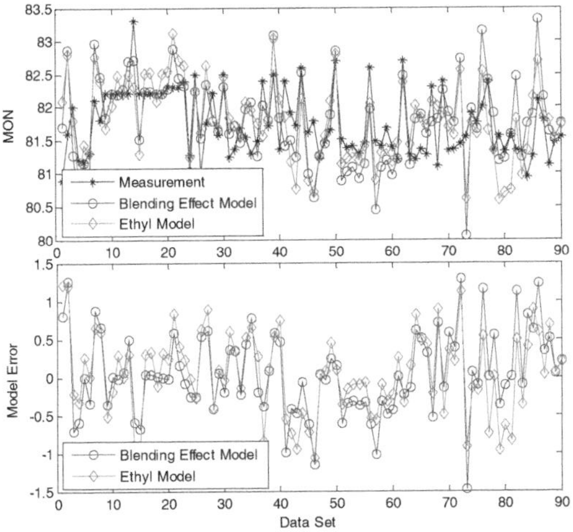

Fig. 2. Comparison of MON Models' Accuracy

From Fig. 1 and 2, it can be qualitatively seen that the blending effect model outputs for RON and MON are closer to the measurements than those of the Ethyl models. The quantitative comparisons between the blending effect model and the Ethyl model are listed in Table I. In the table, blending model has better accuracy in both criteria.

Table 1. Model Accuracy for Octane Number

	Average Absolute Error	Average Relative Error (%)
RON Ethyl model	0.4389	0.4711
MON Ethyl model	0.4411	0.5396
RON Blending effect model	0.4103	0.4397
MON Blending effect model	0.4213	0.5156

4.2 Results of Optimization

With the proposed strategy, the optimization problem was formulated first, which was then solved with the PSO method. The obtained recipe was applied to produce gasoline. The product was analyzed in the laboratory to verify the blending results, and the blending cost was calculated and compared with the original recipe given by the traditional recipe calculation system. In the application, 11 batches, around 60 thousand ton gasoline, have been produced with the optimized recipe. According to the laboratory measurements, 11 batches were all certificated by the national III specification. Table II lists the octane number results for the 11 batches in detail.

Table 2. Results of the Gasoline Product After Optimization

	RON	MON	DON(≥88.1)	DON give away
Set 1	93.80	82.4	88.1	0
Set 2	93.70	82.7	88.2	0.1
Set 3	93.60	82.7	88.15	0.05
Set 4	93.60	82.8	88.2	0.1
Set 5	93.50	82.8	88.15	0.05
Set 6	93.80	82.6	88.2	0.1
Set 7	93.80	82.6	88.2	0.1
Set 8	93.80	82.4	88.1	0
Set 9	94.00	82.3	88.15	0.05
Set 10	93.90	82.4	88.15	0.1
Set 11	94.10	82.3	88.2	0.1

Through the comparison between optimized 11 batches and the 17 batches without optimization, the benefit of the optimization strategy can be seen clearly in Fig. 3. Among the first 17 batches, large DON giveaways were produced in the 8th batch, while the 13th batch failed due to the low DON. Meanwhile, from the lower plot in Fig. 3, it can be seen that the RON of the product varied heavily from 93.2 to 94.4 and the quality giveaways in the upper plot varied from 0-0.8. On the contrary, after optimization, all the 11 batches succeeded with maximal 0.1 giveaways for DON. The variation of the RON decreased to 93.5-94.1. Therefore it can be concluded that the stability of the production has been improved.

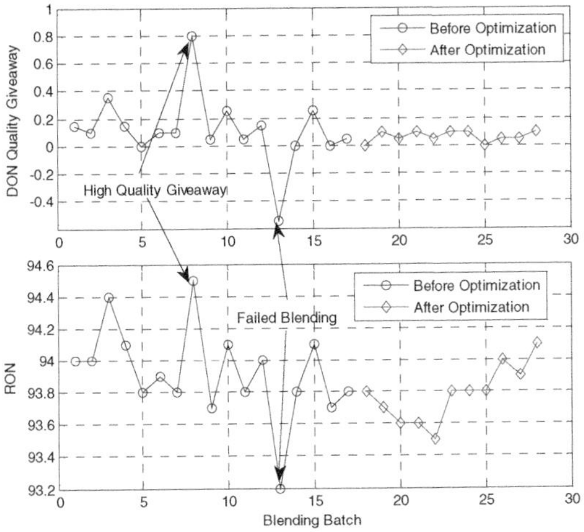

Fig. 3. Blending Results Comparison

Furthermore, the blending cost are calculated and compared. For the studied 11 batches, 42.5 thousand Yuan can be saved for the blending cost for each batch. According to the annual production of 93# gasoline for the studied process, 127.5 Million Yuan will be saved annually for the cost.

5 Conclusion

In this paper, an optimization strategy for gasoline blending process is proposed. Correspondingly, a blending effect mode is proposed to predict the octane number in nonlinear way. The formulated optimization problem is then solved by the PSO method in order to get the global optima for recipe.

With the industrial onsite data, the blending effect model is compared with the traditional Ethyl model for verification. Results show that proposed octane number model has better accuracy. To test the optimization strategy, 11 batches of gasoline were produced with the optimized recipe. The success rate is 100%, while the quality giveaways are controlled under 0.1. After the cost calculation, the optimization strategy has the potential to save 127.5 Million Yuan annually for the studied process.

References

1. DeWitt, C.W., Lasdon, L.S., Waren, A.D., Brenner, D.A., Melhem, S.A.: Omega: an improved gasoline blending system for Texaco. Interfaces 19, 85–101 (1989)
2. Healey, W.C., Maasen, C.W., Peterson, R.T.: A new approach to blending octanes. In: Proc. 24th Meeting API Refining Division, vol. 39, pp. 132–136 (1959)
3. Singh, A., Forbes, J.F., Vermeer, P.J., Woo, S.S.: Model-based real-time optimization of automotive gasoline blending operations. Journal of Process Control 10, 43–58 (2000)

4. Murty, B.S.N., Rao, R.N.: Global optimization for prediction of blend composition of gasolines of desired octane number and properties. Fuel Processing Technology 85, 1595–1602 (2004)
5. Gary, J.H., Handwerk, G.E.: Petroleum refining technology and economics, 3rd edn. Marcel Dekker, New York (1994)
6. Magoulas, K., Marinos-Kouris, D., Lygeros, A.: Instructions are given for building a gasoline blending LP. Oil and Gas Journal 4, 32–37 (1988)
7. Diaz, A., Barsamian, J.A.: Meeting changing fuel requirements with on-line optimization. Hydrocarbon Processing, 71–76 (1996)
8. Eberhart, R.C., Kennedy, J.: A New Optimizer Using Particle Swarm Theory. In: Proc. 6th Int. Symp. on Micro Machine and Human Science, pp. 39–43. IEEE Service Center, Piscataway (1995)
9. Kennedy, J., Eberhart, R.C.: Particle Swarm Optimization. In: Proc. 5th Int Conf. on Neural Networks, pp. 1942–1948. IEEE Service Center, Piscataway (1995)
10. Shi, Y., Eberhart, R.C.: A Modified Particle Swarm Optimizer. In: IEEE World Congress on Computational Intelligence, pp. 69–73 (1998)
11. Clerc, M.: The Swarm and the Queen: Towards a Deterministic and Adaptive Particle Swarm Optimization. In: Proc. ICEC, Washington, DC, pp. 1951–1957 (1999)

The Principle and Algorithm of Earthquake Alarm System Designed for Families

Ping He, Huiqi Sun, Wei Shang, and Pan Li

Department of Control Science and Engineering
Harbin Institute of Technology
Harbin, Heilongjiang Province, China
ping_he116@163.com

Abstract. Combined with a 51 SCM, AT89LV55, a three-axis accelerometer, ADXL335, and a digital-to-analog conversion chip, AD7708, a seismic alarm system has been designed, which can be placed and used in residents' homes or some public places. ADXL335 is used to measure the acceleration of three axis, X axis, Y axis and Z axis. The design, a combination of performance chip AT89LV55 and other interface functions, functioned with the relationship between seismic magnitude and acceleration, so that we can resort to the software enquiry to figure out the seismic magnitude. In addition, large variation on Z axis would trigger the alarm that earthquake is coming. The procedure has been simulated and debugged on the actual Printed Circuit Board and a summary was made. In fact, the system delivers the seismic message promptly with high accuracy and provides some additional functions such as voice prompt and calendar.

Keywords: 51SCM, Seismic alarm, digital-to-analog conversion chip, Acceleration measurement.

1 Introduction

Earthquake in China happens in high frequency with high magnitude and light focal, coupled with the characteristics of China's large population and poor seismic performance of buildings [1]. So earthquake often leads to serious disasters. Moreover, China and most of the other countries mainly depend on large earthquake information institutions which sometimes can not give accurate messages promptly, or may even give the false information. Suppose that each family has a flexible and independent earthquake alarm device, and the damage would be reduced in some way. Now China and other countries in the world rely mainly on the national earthquake prediction [2], while alarm equipments for earthquake are still seldom seen in the market. Therefore, it is necessary to do more in-depth seismic studies.

Ground motion would lead to variation in the ground acceleration. The magnitude has certain relationship with the variation of acceleration [3]. If the variation could be figured out by some way in high precision, the magnitude could be known. Then, the first problem is to find a proper method to ensure the detailed relationship. The second problem is how to put this method into reality and figure out the magnitude by procedure. Besides the function of detecting seismic magnitude, this system can also give early warnings. The estimation of strength of shaking at a site from the initial

D. Zeng (Ed.): ICAIC 2011, Part II, CCIS 225, pp. 361–368, 2011.

P-wave portion of ground motion is the key problems for shortening the alert time of the Earthquake Early Warning [4].

Real-time seismology deals with collection and quick analysis of the data after a seismic event. The information is used for the warning purposes or post-earthquake emergency response. The Earthquake Early Warning (EEW) systems utilize the time difference between the moments when the relatively small amplitude and rapidly propagating primary (P) waves and the damaging large-amplitude and slowly propagating secondary (S) waves would reach the site. The time difference, in some cases, may reach several tens of seconds. The recent advances in seismology, sensor, computer and telemetry technologies allow constructing the effective early warning system [5-6].

Based on these above theories, the principle and algorithm of earthquake alarm system designed for residents has been worked out, and seismic alarming system has been designed. To get the ground acceleration, an accelerometer is needed. Considering that the acceleration variation is so small, that a 16-bit ADC chip has been adopted. Among several kinds of mathematical methods, the interpolation has been found to have more superiority and has been finally applied to obtain the detailed relationship between magnitude and variation of acceleration. Additionally, the function of early warning has been added to this system. The variation of acceleration on Z axes would be processed separately to judge if P-wave has arrived. Certain related procedures and the module of software have been given.

2 The Principle of the Seismic Alarm System

When earthquake happens, the ground acceleration would change, and its variation reflects the intensity of the ground motion. Three-axis accelerometer ADXL335 is to convert three axes of accelerations into three voltage signals, then the voltage signals would be delivered into an A/D conversion chip AD7708 to be converted into some 16-bit binary codes. These codes would be processed by AT89LV55 to figure out the magnitude. Voice chip PM50S50 would choose different voice segments according to the magnitude, guiding people to make some preparations. Real-time clock chip DS1302 can provide the local time. Four buttons allows the user to interact with the alarm, such as resetting the system time and so on. All these information including the seismic magnitude, the calendar and the left battery power would be shown by LCD128*64. There are seven related modules, and the principle diagram is shown below.

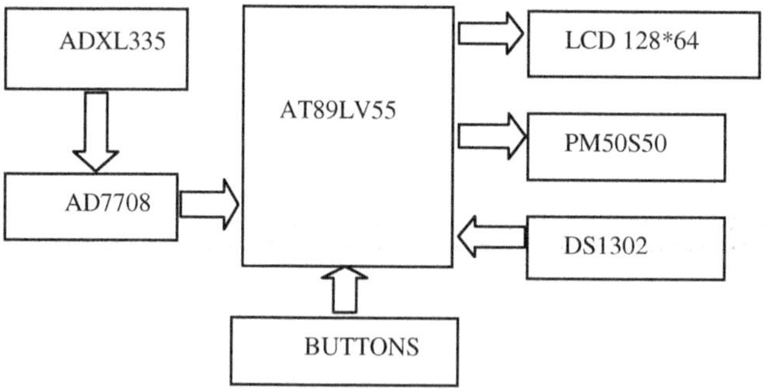

Fig. 1. The principle diagram of the alarm system

2.1 The Principle of Detecting Seismic Magnitude

Suppose the space is three-dimensional. In common, there is a constant downward acceleration of the ground, whose direction is defined as Z-axis positive direction. Other two axes are defined as X axes and Y axes, both of which are vertical with Z axes. And the values of the accelerations on the X axis and Y axis are zero, when there is no ground motion. When earthquake occurs, the ground acceleration will change. Greater acceleration variation would lead to higher seismic magnitude. This variation is the synthesis of changes of the three axis acceleration, and what we need is just its modulus. Some way should be found to get the seismic magnitude on the basis of the acceleration variation. Now, certain general relationship between the intensity in Japanese Meteorological Agency scale and the acceleration variation has been given. The Japan Meteorological Agency (JMA) seismic intensity has been transformed into the seismic magnitude that we usually see with some mathematical method, and the corresponding damages are given, just as the TABLE 1 show [2].

Table 1. The Relationship Between Magnitude and Variation

Seismic magnitude	Variation of acceleration (gal)	Human's feeling and the related damages
0	< 0.8	No feeling.
1	0.8 ~ 2.5	Very light shock.
2	2.5 ~ 8.0	Light shock.
3	8.0 ~ 25	Shock.
4	25 ~ 80	Serious ground shock. it is difficult to for people to walk.
5	80 ~ 250	Serious cracks would be found in the wall, with some collapses of buildings.
6	250 ~ 400	Less than 30% buildings collapse, and people may fall down to the ground.
7	> 400	More than 30% buildings collapse, and the ground would split.

The table just gives some general relationships and a series of variation ranges. In order to get the magnitude of each point (such as 2.3, 3.5, etc.) corresponding to the variation. Several mathematical methods are adopted and simulated, such as the curve fitting method and the interpolation method. To find the best choice, several aspects have been considered to comparing the advantages and disadvantages of them.

The curve fitting method has been given. The second order and the third order curves have been taken for example. According to the values of these original points, the coefficients of each corresponding function have been figured out. Simulate these functions, and the curves would be showed. The matlab procedure includes two steps, just as the below:

1) Calculate Coefficients: Calculate the coefficients of the second, the third and the fourth order of functions as the following procedure:

```
X = [0, 1, 2, 3, 4, 5, 6, 7];
Y = [0, 0.8, 2.5, 8.0, 25, 80, 250, 400];
```

A = polyfit (X, Y, 2) % A = 15.7042 -58.7923 26.7375;
B = polyfit (X, Y, 3) % B = 2.4255 -9.7636 7.9091 1.2697;

2) Simulation: Simulate the program according to the coefficient to get the graph of these curves. To make the simulation graph clear to see, only the second order curve-fitting and the third curve-fitting are simulated.

```
X = [0, 1, 2, 3, 4, 5, 6, 7];
xi = 0:0.1:7;
Y = [0, 0.8, 2.5, 8.0, 25, 80, 250, 400];
Y2 = 15.7042*xi.^2 -58.7923*xi+26.7375;
Y3 = 2.4255*xi.^3-9.7636*xi.^2+7.9091*xi+1.2697;
plot (X,Y, 'k.-',xi,Y2,'bo',xi,Y3,'r+')
legend ('The original curve', 'The second order curve-fitting', 'The second order curve-fitting')
grid on.
```

According to the procedure, the simulation graph is shown in Figure 2.

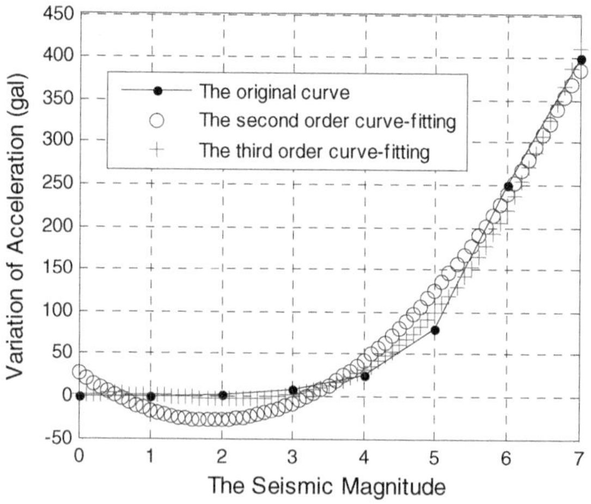

Fig. 2. Three fitting curves

The simulation results show that the precision is obviously low. Even the highest order fitting-curve has much difference with the original curve. To get enough accuracy, the order should be much high, which would occupy too much system resources. But, the advantage is that the procedure for this method is easy to realize. To solve this problem, several interpolation methods are introduced just as follows.

Now, the interpolation methods have been given. In order to make some contrasts and make the advantages or disadvantages clear to see, three interpolations have been given. The matlab procedure is shown below.

```
x = 0:7;
y = [0, 0.8, 2.5, 8.0, 25, 80, 250, 400];
xi = 0:0.1:7;
yi = interp1(x, y, xi, 'linear');
zi = interp1(x, y, xi, 'spline');
```

wi = interp1(x, y, xi, 'cubic');
plot (x, y, 'o', xi, yi, 'r+', xi, zi, 'g*', xi, wi, 'k.-')
legend ('The original curve', 'The linear interpolation', 'The three spline interpolation', 'The cubic polynomial interpolation')
grid;
The simulation graph is shown in Figure 3.

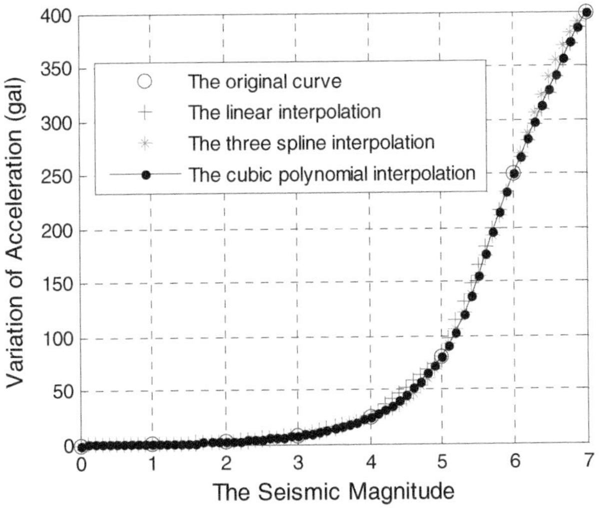

Fig. 3. Three interpolation curves

By contrast, the three interpolation methods, linear interpolation, three spline interpolation and cubic polynomial interpolation, are all very close to the original curve. Moreover, the value of each coordinate point can be directly obtained with the software, matlab. Therefore, interpolation has some superiority to curve-fitting. Anyone of these interpolations can be applied to get the detailed relationship of each point. However, the disadvantage is that the procedure may be somewhat complicated, because a certain amount of memory resources would be occupied to make a list, which stores these values of each coordinates.

From the above we can see that in respect of the precision, interpolation is obviously better than curve-fitting method with higher precision. In respect of the consumption on the system resources, the curve fitting method is simple and needs less RAM resources. Low order occupies less computing resources, but would bring in low accuracy. Higher accuracy needs higher order, but this would occupy too much computing resources. Although interpolation method needs a certain amount of RAM resources which would be used to make a list, it will bring in high accuracy. Moreover, in reasonable ways, it will not consume lots of computing resources. Thus, interpolation is feasible combined with the controlling chip AT89LV55 which has a 20K flash RAM, and enough program storage space. Therefore, the interpolation method was finally adopted to get the acceleration variation of each point.

2.2 Flow Diagram of Magnitude Measurement Module

The ADC chip AD7708 is the most important part in the process of calculating the seismic magnitude. ADXL335 is to convert acceleration into voltage signal, while AD7708 is to send these signals and some control signals into the microcontroller. Each action of the controller is based on the signals. Firstly, AD7708 needs some initialization and open the first channel. When the conversion is completed, the interrupt subroutine is triggered, and starts to process these data. In the terminal of the interrupt subroutine, the next channel should be opened, so that the channel1, channel2 and channel3 can be opened circularly. The first several times of conversions are used to obtain the original values. By comparing each new conversion with the original value of each corresponding channel, the variation would be figured out. The corresponding magnitude would be easy to obtain by tracing the list that has been made and stored before. Only after the three channels are all converted, can the magnitude be figured out once. The flow diagram of the interrupt subroutine is shown 4-6 in Figure 4.

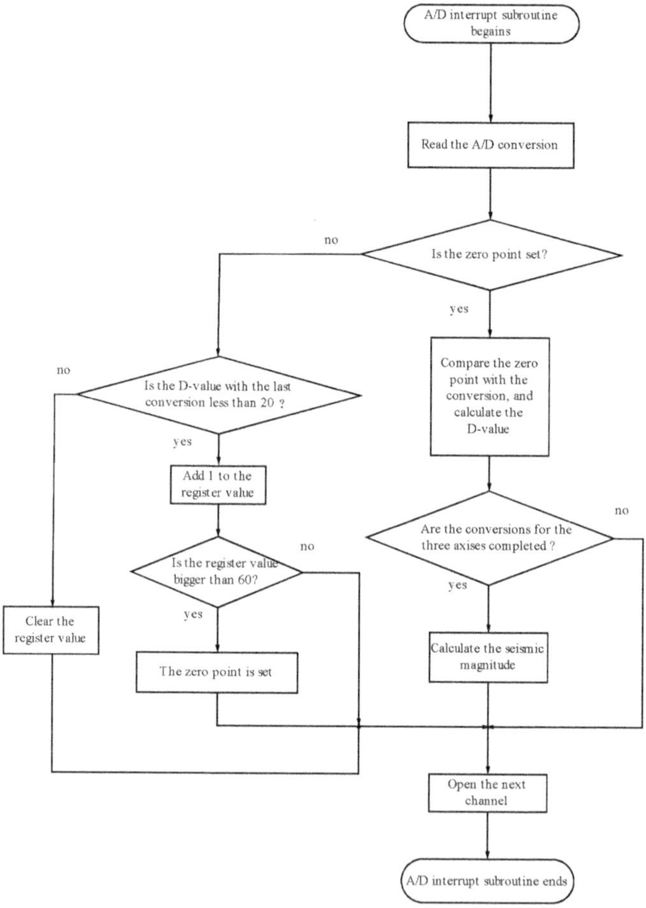

Fig. 4. The flow diagram of interrupt subroutine for AD7708

2.3 Magnitude Measurement Procedure Introduction

Each time AD7708 finishes its conversion, its pin RDY would be set high, triggering the interrupt subroutine. When the conversion result is read by the controller, the pin RDY would be set low and the next conversion should be started before the end of the subroutine. And, before calculating the magnitude, the zero point should be set. The zero point is regarded as the three reference values of the accelerations on these three axes. If it is not set, the first several conversions would be used to ensure the zero point. First, set a value which would be used to compare with the register value to judge when the seismic magnitude should be calculated, then compare the first two conversions. If the D-value between the two conversions is less twenty, just calculate their average, which would be compared with the next conversion, and the register value should be increased by 1. The initial value of the register is zero. That the D-value is more than 20 means unstable system environment, and the register value should be cleared. Keep working in this way, until the register value arrives to the value we have set before. From then on, each conversion should be compared with the zero point, and the next channel should be opened. When the conversions of the three channels are completed, the seismic magnitude would be figured out once.

In the step of calculating the seismic magnitude, the P-wave should be processed separately. Moreover, the P-wave refers to the variation of the acceleration on the Z axes. It often arrives some time ahead of large ground motions. Though this time varies with several complicated factors and is usually very short, this character should be applied properly to give early alarms, so that people can make some preparations. That high variation of acceleration on Z axis is detected would trigger the alarm to warn people.

3 Conclusion

The system functions in high quality of detecting ground motion. When outer motion was given to this system, it will response sensitively, sending voice signals and displaying the seismic information on the LCD. The quality of predictions of macro seismic intensity of the strongest part of shaking from the initial portion of earthquake ground motion depends on duration of the portion and availability of additional information about the earthquake characteristics (magnitude and distance). The time interval of 4~5s from the P-wave arrival may be considered sufficient. Since the interval is so precious and uncertain, people should take actions immediately when the alarm is triggered.

With flexibility, low costs, and uncomplicated design methods, this system is mainly to make up the shortcomings that large earthquake information institutions usually could not give accurate messages promptly, or may even give the false information. Because of its characteristics, it is available to realize the batch production, fill in the blank of civil seism products in the market and urge more institutions or enterprises to make further research and production. Furthermore, the results show that the application of the relations between the magnitude and the acceleration is feasible and the design has certain significance in the field of earthquake alarms.

References

1. Guo, L., Li, Y., Yang, G., Hu, X.: Characteristics of present-day active strain field of Chinese mainland. Acta Seismologica Sinica (English Edition) 6, 99–108 (2008)
2. National Center for Research on Earthquake Engineering, Taiwan Earthquake Loss Estimation System Handbook, pp. 29–31 (2008)
3. Sun, D., Sun, B.: Rapid prediction of earthquake damage to buildings based on fuzzy analysis, vol. 3, pp. 1332–1335 (2010)
4. Sokolov, V., Furumura, T., Wenzel, F.: On the use of JMA intensity in earthquake early warning systems. Bulletin of Earthquake Engineering 8, 767–786 (2010)
5. Aoi, S., Kunugi, T., Adachi, S., Nakamura, H., MorikawaH, H., Abara, K., Fujiwara, H.: Implementation of real-time capability of strong-motion seismographs network KiK - net. In: The 7th General Assembly of Asian Seismological Commission and the 2008 Fall Meeting of Seismological Society of Japan, pp. 24–27 (2008)
6. Hoshiba, M., Kamigachi, O., Saito, M., Tsukada, S., Hamada, N.: Earthquake early warning starts nationwide in Japan. Trans. Am. Geophys. Union 89(8), 73–80
7. Wooding, F.B., Peal, K.R., Colllins, J.A.: Broadband seismometry beneath the seafloor, vol. 2, pp. 1227–1231 (2000)
8. Jusoh, M.H., et al.: Earthquake Prediction Technique Based On GPS Dual Frequency System in Equatorial Region. In: 2008 IEEE International Rf and Microwave Conference Proceedings, December 2-4 (2008)
9. Yeh, C.-H., Loh, C.-H., Tsai, K.-C.: Overview of Taiwan Earthquake Loss Estimation System. Natural Hazards 37, 23–26 (2006)

Evaluation Study of Basic Public Service' Equalization Level on the Provincial Administrative Regions in China Based on the Wavelet Neural Network*

Zhiqiang Ma and Jiancheng Wang

School of Business Administration, Jiangsu University
Zhenjiang, Jiangsu Province, China
a.sree@tom.com

Abstract. Based on the pertinent researches and taking the current conditions of basic public service in the provincial administrative regions into consideration, this study proposed an evaluation indexes system to evaluate equalization level of basic public service in the provincial administrative regions from such four dimensions as foundation, security, development and environmental protection. This study also applied wavelet neural network (WNN) to evaluate equalization level of urban basic public service in 31 different provincial administrative regions in 2008 in a comparatively objective way. The results show that the evaluation model of basic public service's equalization level on the provincial administrative regions based on WNN is highly scientific, reasonable and feasible.

Keywords: basic public service, equalization, evaluation indexes, Wavelet Neural Network (WNN).

1 Introduction

Currently, China is at the transformation period to change its economic and social model. Establishing an effective public service system is the interior requirement of socialism marketing economy and the basic obligation of a service-oriented government. Because the basic public service is the most essential and critical part of public service, its equalization is the comprehensive policy goal of political, economic and social development ideal for a country. It is also a complicated and systematic social management project, not only related to value orientation of national public management, but also requiring objective understanding and evaluation of actual situation of public service according to reasonable standards and by scientific methods.

During recent years, many outstanding domestic researchers have done wide and profound studies on equalization of basic public service by theoretic study, investigation and empirical study and so on, proposing some suggestions of high

* This work is partially supported by National Statistics Research Projects Grant #2009LY051 to Zhiqiang Ma.

D. Zeng (Ed.): ICAIC 2011, Part II, CCIS 225, pp. 369–377, 2011.

significance. Most of the studies, however, mainly focus on qualitative analysis of current situations, realization paths or a certain profile of public service equalization. A reasonable and scientific evaluation system and method is absent. This study constructed an evaluation indexes system for equalization level of basic public services in the provincial administrative regions based on pertinent studies and evaluated the equalization level of basic public service in the provincial administrative regions in China by WNN, intending to provide reference for related theoretic studies and practical work.

2 Literature Review

The concept of public service is put forward by Adolf Wagner, who belonged to Germany Sociological School formed during the late 19th century. He emphasized particularly the social responsibility of government finance, regarding public service as a national fiscal requirement [1]. Presently, most scholars consider public service as a series of public activities to enhance equalization of social community's basic consumption by government, using public power or public resources. Basic public service is the most basic public service that the government shall provide to the community. Equalization of public service refers to that residents in different areas of the country are able to enjoy basic public service almost at the same level [2]. From the viewpoint of the development history of public services around the world, generally there are three models to realize equalization of basic public service: "equalization of financial capability per person (fiscal equalization)", "public service standardization" and "the minimum equity of basic public services" [3]. Theoretically, equalization of public service is unlikely to be achieved by automatic running of the marketing system. Therefore, evaluation of public service equalization is usually considered as evaluation of public policies. There has been an argument about the nature of public policies evaluation from the perspective of research method that whether the nature of this evaluation belongs to value judgment or to technical analysis [4]. Presently, the method applied by most researchers is technical evaluation, employing tools including maximum and minimum ratio, deviation, standard deviation (or mean difference), and coefficient of variation, etc. [5]. In recent years, some domestic scholars carried out scientific and reasonable analysis and objective evaluation for China's public service, achieving many valuable outcomes. An Tifu, Ren Qiang (2010) [6] constructed an index system to evaluate the equalization level of China's public service by comprehensive evaluation method, which consisted of four levels with 25 indexes in total. Liu Bin, Yu Xinghou et.al (2010) [7] studied the equalization level gap between western provinces and districts, western area and average level of the whole country, according to indexes selected based on comprehensive consideration of gap of regional economic development, local fiscal capability and public service level. Tan Yanhong (2010) [8] classified basic public service into 8 categories, referring to 40 indexes. Wang Yuerong (2010) [9] constructed an indexes system for basic public service of innovative cities in China.

In summary, domestic and overseas researchers produced plentiful research outcomes about basic public services in the provincial administrative regions with significant reference value. Due to the complexity of public service, however, the equalization level of public service is impacted by a variety of factors, beside fiscal input per person. Therefore, evaluation of public service's equalization level shall take regular value judgment into the evaluation process, constructing applicable evaluation indexes system and selecting scientific evaluation methods, to push the evaluation of public service equalization to develop into systematic comprehensive evaluation gradually.

3 Construction of Evaluation Indexes System for Equalization Level of Basic Public Services in the Provincial Administrative Regions

Based on the pertinent researches [6-9] and taking the current conditions of basic public services in the provincial administrative regions into consideration, this study classified the evaluation indexes system for the equalization level of urban basic public service into four categories: foundation, security, development and environmental protection. See Table 1 for further information.

Table 1. Indexes System for Equalization Level of Basic Public Services in the Provincial Administrative Regions

Level 1 indexes	Level 2 indexes	Level 3 indexes
Foundation U_1	Economic development level U_{11}	GDP per capita(Yuan)U_{111}
	Fiscal capacity U_{12}	Fiscal revenue per capita (Yuan)U_{121}
		Fiscal expenditure per capita (Yuan)U_{122}
	Infrastructure U_{13}	Road area per capita(m^2)U_{131}
		Ratio of population able to get access to water (%)U_{132}
		Ratio of population able to get access to natural gas (%)U_{133}

Level 1 indexes	Level 2 indexes	Level 3 indexes
Development U_3	Education U_{31}	Average population in primary school among every 100,000 people U_{311}
		Average population in junior middle school among every 100,000 U_{312}
		Teacher-student ratio in public primary school U_{313}
		Teacher-student ratio in public junior middle school U_{314}
		Illiterate ratio of population aged 15 and over (%)U_{315}
	Public education U_{32}	Average public library shared by one person U_{321}

Table 1. (*continued*)

Security U_2	Social security U_{21}	Occupancy ratio of unemployed insurance (%)U_{211}		Science technology U_{33}	The ratio of scientific research expenditure to total output value (%)U_{331}
		Occupancy ratio of basic hospitalization insurance (%)U_{212}			Granted quantity for the three types of patent application U_{332}
		Occupancy ratio of basic pension insurance (%)U_{213}	Environmental protection U_4	Environmental governance U_{41}	Treatment rate for discharged wastewater (%)U_{411}
	Public health U_{22}	Average quantity of hospitals/ health centers shared by one person U_{221}			Comprehensive utilization of the industrial solid wastes (%)U_{412}
		Average quantity of staff per health institution U_{222}			Treatment rate for discharged waste gas (%)U_{413}
					Harmless disposal rate for domestic waste (%)U_{414}

4 Construction of WNN-Based Evaluation Model of Basic Public Services' Equalization Level in the Provincial Administrative Regions

Put forward first by Qinghua Zhang in 1992, Wavelet Neural Network (WNN) is a new type of feed forward network based on WNN analysis, integrating good time-frequency localization property of wavelet and self-learning function of traditional neural networks, which equips it with good approximation and fault tolerance [10-11]. WNN applied here is BP neural network but in which the traditional non-linear Sigmoid function is displaced by non-linear wavelet bases. The fitting of the non-linear function is realized through linear superposition of non-linear wavelet bases. The non-linear function $y(t)$ can be fitted by wavelet base $\psi_{a,b}(t)$ as below:

$$\hat{y}(t)=\sum_{k=1}^{n} W_k \psi\left[\sum_{i=1}^{m} W_{ik} R_j(i)-b_k \Big/ a_k\right] \tag{1}$$

Whereof, $\hat{y}(t)$ is the fitting value sequence of the non-linear function $y(t)$; W_k represents the weight coefficient between the *Kth* unit of the input layer and the hidden layer; W_{ik} represents the weight coefficient between the *ith* node of the input layer and the *Kth* node of the hidden layer; $R_j(i)$ represents normalization data of the *ith* index of the input sample J; a_k and b_k are shift factor and extension factor of wavelet bases respectively; intriguing function of the hidden layer employs Morlet mother wavelet, that is,

$$\psi(t) = \cos(1.75t)\exp(-t^2/2) \tag{2}$$

Define WNN's error energy function as :

$$E = \frac{1}{2}\sum_{j=1}^{P}(y(t) - \hat{y}(t))^2 \tag{3}$$

whereof P is the sample size; the network E value is made minimum by adjusting the value of network parameters W_k, W_{ik}, a_j and b_j.

This study constructed a wavelet neural network structure for equalization level evaluation of basic public services in the provincial administrative regions (see Fig. 1 for details), applying WNN to equalization level evaluation of basic public services in the provincial administrative regions. A comprehensive evaluation model combined with qualitative and quantitative analysis which is more similar to human thinking model is to be constructed by applying dimensionless data of evaluation indexes based on universal index type first and then by acquiring tendency of evaluation expert's knowledge, experience, subjective judgment and importance degree of objectives through WNN' learning. During comprehensive equalization level evaluation of basic public services in the provincial administrative regions, this method realized combination of the qualitative analysis and quantitative analysis effectively by simulating evaluation expert's experience, knowledge and instinct. It can also guarantee objectiveness of the evaluation results of basic public services' equalization level in the provincial administrative regions. Automatic running, quick evaluation and decision support are achieved by this method.

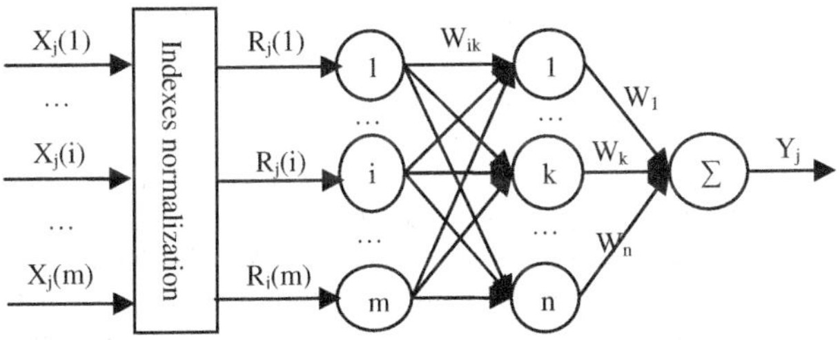

Fig. 1. WNN structure illustration

5 Model Running and Results Analysis

5.1 Data Collection, Preprocessing, and Indexes Weight Establishment

The value of single index for basic public services in the provincial administrative regions in 31 different provinces and municipalities are collected and received dimensionless treatment in this study (samples are sequenced randomly from 1 to 31). All the data are taken from China Statistical Annual 2009. Due to article length limitation, the data are not listed herein. At the same time, weights of all evaluation indexes are calculated by improved AHP method [12], and expertise evaluation value for each evaluation target is calculated by formula:

$$Z = \sum_{i=1}^{n} \sum_{j=1}^{m} \sum_{k=1}^{l} w_i w_j w_k x_k \tag{4}$$

Whereof, Z is expert evaluation value; w_i is weight of the *ith* level one index; w_j is weight of the *jth* level two index; w_k is weight of the *kth* level two index; x_k is the score of the *kth* level two factors.

Table 2. Evaluation Indexes' Weight of Basic Public Services' Equalization Level in the Provincial Administrative Regions

Level 1 indexes	Weight	Level 2 indexes	Weight	Level 3 indexes	Weight	Level 1 indexes	Weight	Level 2 indexes	Weight	Level 3 indexes	Weight
U_1	0.27	U_{11}	0.30	U_{111}	1	U_3	0.24	U_{31}	0.38	U_{311}	0.21
		U_{12}	0.33	U_{121}	0.5					U_{312}	0.21
				U_{122}	0.5					U_{313}	0.20
		U_{13}	0.37	U_{131}	0.32					U_{314}	0.20
				U_{132}	0.34					U_{315}	0.18
				U_{133}	0.34			U_{32}	0.32	U_{321}	1
U_2	0.27	U_{21}	0.52	U_{211}	0.32			U_{33}	0.30	U_{331}	0.52
				U_{212}	0.34					U_{332}	0.48
				U_{213}	0.34	U_4	0.22	U_{41}	1	U_{411}	0.27
		U_{22}	0.48	U_{221}	0.5					U_{412}	0.22
				U_{222}	0.5					U_{413}	0.27
										U_{414}	0.24

Table 3. Expert Evaluation Scores for Basic Public Services' Equalization Level in the Provincial Administrative Regions

No.	1	2	3	4	5	6	7	8	9	10	11
Experts evaluation score	0.52 139	0.42 963	0.46 312	0.35 626	0.30 058	0.48 508	0.29 431	0.30 997	0.57 336	0.70 730	0.40 410
No.	12	13	14	15	16	17	18	19	20	21	22
Experts evaluation score	0.41 390	0.46 827	0.45 671	0.43 074	0.75 777	0.37 402	0.25 605	0.37 257	0.28 105	0.40 644	0.09 696
No.	23	24	25	26	27	28	29	30	31		
Experts evaluation score	0.23 822	0.22 153	0.25 172	0.64 583	0.38 014	0.50 764	0.42 383	0.61 609	0.27 858		

5.2 Model Realization and Results Analysis

The software MATELAB7.6.0 was used in this study to accomplish the structure programming of WNN for equalization level evaluation basic public services in the provincial administrative regions. The first 25 groups of data are used to train the network as training sets while the rest 6 groups of data are used to test the feasibility of the network as testing samples. Rating standards for evaluation results are established to better analyse evaluation results of basic public services in the provincial administrative regions equalization level. See table 4 for details.

Table 4. Rating Standards for Evaluation Results of Basic Public Services' Equalization Level in the Provincial Administrative Regions

Rate	Excellent	Good	Average	Poor	Very poor
score	[1,0.8]	(0.8,0.6]	(0.6,0.4]	(0.4,0.2]	(0.2,0]

Analysis on results produced by different network data shows that training error reaches expected error when target error of network data being set as 0.0001, nerve cell of hidden layer being set as 15, learning rate of the network weight being set as 0.3 and training reaching 1053 steps. Convergence effect of the network is fine on these conditions. Network convergence diagram for 25 groups of learning training samples are produced as below:

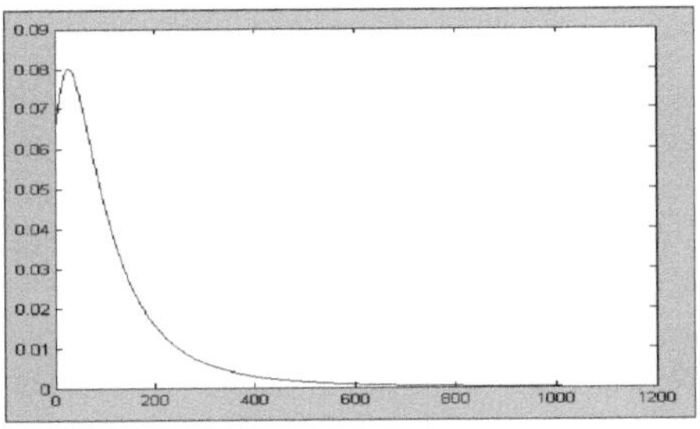

Fig. 2. Decrease curve of the network training error

Through learning and training, the WNN memorized weight coefficients and rules between different layers of the WNN. Expected value output was obtained through model verification for 6 groups of untrained testing samples. See Table 5 for the contrast results between the said expected output value and actual network output value. Approximation between those two sets of values indicates good generalization capacity of the model.

Table 5. Contract between Expected Output Value and Actual Output Value of the Network

NO.	Expected value	Actual value	Relative error
1	0.64583	0.64352	0.00231
2	0.38014	0.61462	0.00147
3	0.50764	0.50429	0.00335
4	0.42383	0.42042	0.00341
5	0.61609	0.37506	0.00508
6	0.27858	0.27453	0.00405

That the relative error between forecasting results of model evaluation results and expert evaluation results is less than 0.001 shows that the forecasting value is reliable. According to Table 4, the evaluation results of basic public services in the provincial administrative regions equalization level for 31 different provinces and municipalities are as follows: 0 excellent, 4 good, 13 average and 1 poor. From the perspective of area distribution, basic public services capacity decreases from eastern area, to central area, and finally to the western area. Regions with excellent basic public service are most in eastern area. Regions with average basic public service account for 53.85% of the eastern area, 30.77% of the central area, and 15.38% of the western area. Regions with poor basic public service account for 30.77% of the central area and 69.23% of the western area.

6 Conclusion

Based on study of pertinent researches, this paper constructed an evaluation indexes system for basic public services in the provincial administrative regions equalization level, consisting of 23 third level indexes classified into such 4 dimensions as foundation, security, development and environmental protection. This paper also evaluated comprehensively the overall equalization level of the basic public services in the provincial administrative regions in 31 different provincial administrative regions, reflecting comparatively objective conditions. Results show that the basic public services in the provincial administrative regions equalization level evaluation indexes system constructed in this study has high validity and feasibility. It also verified that such properties as self-organization, self-adaption, self-learning and error tolerance of the WNN are able to get rid of the impact of human acted subjectivity, enhancing objectiveness of evaluation results. Moreover, evaluation results show that Chinese basic public services in the provincial administrative regions capacity decreases from the eastern area to the central area and finally to the western area. The uneven problem of urban basic public service capability is obvious.

References

1. Guo, H.: Public service equalization: theoretical evaluation and practical application. Contemporary Finance & Economics 3, 29–33 (2008)
2. Liu, D.: Notions, systems and realization approaches of public service equalization: literature review. Shanghai Economic Review 4, 12–20 (2008)
3. Liu, W.: A review of the theoretical study of 'basic public services' in China. Review of Economic Research 16, 46–66 (2010)
4. Sen, A.K.: Inequality Reexamined, Translated by Wang Liwen, Yu Zhanjie, p. 307. Social Science Academic Press, Beijing (2006)
5. Chen, C., Cai, Y.: Public service in China: System Transformation and Local Comprehensive Evaluation, pp. 330–331. China Social Science Press, Beijing (2007)
6. An, T., Ren, Q.: Construction of indexes system for China public service equalization level – a quantitative analysis based on regional differences. Finance & Trade Economics 6, 79–82 (2008)
7. Liu, B., Yu, X., et al.: Research on the situation of the basic public service equalization among western areas. Journal of Chongqing Institute of Socialism 2, 32–35 (2010)

8. Tan, Y.: Basic public service equalization among ethnic areas: present situation evaluation and path selection – evaluation on comparison between Guangxi and other regions. Review of Economic Research 41, 56–61 (2010)
9. Wang, Y.: Appraisal of urban basic public service ability and equalization. Urban Problems 8, 10–16 (2010)
10. Zhang, Q.H., Benveniste, A.: Wavelet networks. IEEE Translated Neural Networks 3, 889–898 (1992)
11. Khayamian, T., Ensafi, A.A., Tabaraki, R.: Principle component -wavelet neural networks as a new multivariate calibration method. Analytical Letters 38(9), 1477–1489 (2005)
12. Huang, D., Xu, L.: Proportion criteria and method for building comparison matrices in the analytic hierarchy process. Control and Decision 17(4), 484–486 (2002)

An Empirical Research on Optimum—Scale of China's Foreign Exchange Reserve Format*

Xueping Ning

School of Finance, Shanghai Lixin University of Commerce
Shanghai, Nostate 201620, China
xueping_n@163.com

Abstract. The frequency of international financial crises showed that keeping abundant foreign exchange reserves and effective management of the reserves were an important measure to prevent the financial crises which brought about by extemal factors. China's foreign exchange reserves have amounted to $ 609.9 billion by the first three quarters of 2005. Gross reserve bas reached a historic high level. Its gigantic amount has been the focus of the debate among foreign and local researchers and professionals. This essay analyzes the moderation of China's current foreign exchange reserve quantity, and on this basis brings forward several suggestions for the current foreign exchange reserve management policies.

Keywords: Foreign Exchange Reserve, Demand Model, Regression Analysis.

1 Introduction

Since 1994 official rate of exchange for RMB was made uniform with the market rate, China's Foreign exchange reserves have been growing by 40 percent every year. According to State Administration of Exchange Control statisties, in 2003, China's Foreign Exchange Reserve reached $ 403.3 billion with 40.8 percent. Such growth reached as high as 51.3 percent in 2004. and the official reserve reached $ 609.9 billion. In the first three quarters of 2005, the quantity of Foreign Exchange Reserve hit $ 769.004 billion, 36.7 percent more than the same period of 2004 (Chart 1).

The foreign exchange Reserves are an important component of national economy. Sufficient reserve is of advantage to support a broad range of policy objectives, such as:

- Maintaining confidence in the monetary and exchange rate policies;
- Providing confidence to the international community
that the economy is able to meet itsexternal obligations;
- Maintaining foreign currency liquidity during periods of external shocks;
- Assisting the government in meeting foreign exchange obligations.

---

* This work is partially supported by #J51703 to Construction Program of Key Academic Subjects of Shanghai Education Committee.

D. Zeng (Ed.): ICAIC 2011, Part II, CCIS 225, pp. 378–383, 2011.

Given the above – mentioned objectives, the most commonly cited question is in the area of "what is the adepuate level of reserves?"

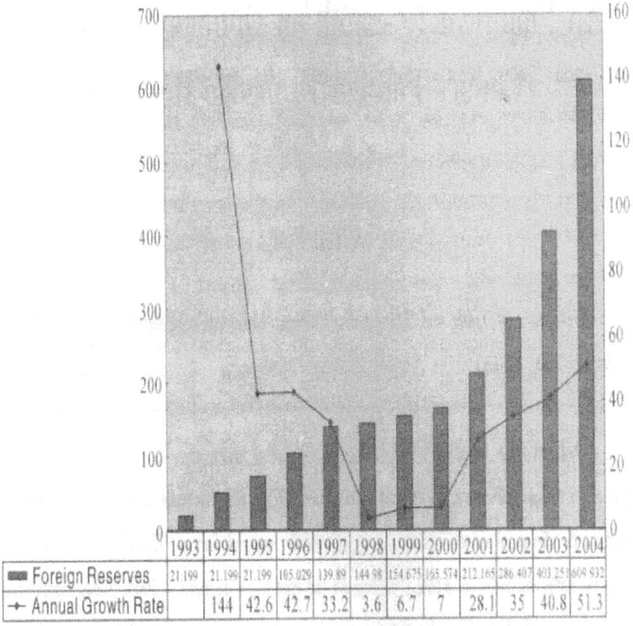

Chart 1 China's foreign exchange Reserve and growth rate, 1992-2004

Your goal is to simulate the usual appearance of papers in IEEE conference proceedings. For items not addressed in these instructions, please refer to the last issue of your conference's proceedings for reference or ask your conference Publications Chair for instructions.

2 A Brief Review of the Literature on Reserve

2.1 Several Theories to Foreign Exchange Reserve Scale

1) Method of proportion analysis

Proportion analysis is a conventional method that adopted proportional relations between reserve and other economic variable to measure one country's reserve. Proportional relations in everyday use have the ratio of reserve to import (R/M), the ratio of reserve to Gross National Product (R/GNP), the ratio of reserve to external debt (R/D) . R/M is the most popular among all. This method was put forward by American economist professor R. Triffin in "Gold and the Dollar Crisis: the Future of Convertibility". R. Triffin researched the change of reserve of twelve developed countries, and came a conclusion that the moderate ratio of reserve to import is 40%, less than 30% should take adjust measures, lowest can't less than 20%. The ratio of reserve to import over the year is 25%, then scale of reserve should satisfy three months to import.

2) Method of demand for reserve function

M. A. lyoha was of opinion that foreign exchange reserve scale of developing countries is dependent on their income of export (X^e), rate of change of import expenditure (8^2) , foreign exchange assets interest rate (r) and degree of economic openness (p) . Reserve and above – mentioned four factors present a direct correlation. Make use of the method of two times adjusting, establishing demand function for reserve as follows:

$$R = \alpha_0 + \alpha_1 X^e + \alpha_2 lgo^{-2} + \alpha_3 r + \alpha_4 P + \alpha_5 R_{-1} + \alpha_6 R_{-2} + \varepsilon$$

ε: *Random error*

3) Theory of Opportunity Cost

This theory was put forward by the end of 60's in 20th century by the economist J. P. Agaraual and H. R. Heller. They thought opportunity cost of the foreign exchange storage is the rate of return of the domestic investment (Agaraual,1971). One country holds excess reserves sacrifice that the part demand of the nations investment and consume. Hence one country demand of foreign exchange storage is decided by the marginal income of holding foreign exchange reservel. The appropriate reserve should be the balanced and the marginal income.

According to the balance principle of cost and benefit, Agaraual educed the calculation mode of optimum – scale of reserve in developing countries.

$R = B\ lgP\ (lgK + lgQ_2 - lgQ_1)$

R , optimum – scale of reserve

B , balance of payments deficit

P , probability of deficit occurrence

Q_1, the ratio of reserve to productive accumulation

Q_2, the ratio of reserve to output

4) Money supply

Monetarist, W. M. Brown and Herry Johnson, put forward the theory that the ratio of international reserve to the external payments balance and international reserve to domestic money supply is on the basis of the ratio of R. Triffin. This theory analyses the problem of suitable foreign exchange reserve scale from the point of view of money supply, thought disequilibrium of balance of payments is a kind of currency phenomenon essentially, being the domestic money supplly over the money demand, surplus of money will flow out, causing the decrease of cash balance. Therefore, demand for foreign exchange reserve is decided by increase or decrease of domestic money supply.

5)Qualitative analysis method

This theory is put forward by economist R. J. Carbaugh and C. D. Fan et al in the mid 70's in the 20th century. They thought that six factors influence one country's demand of reserve. First, one country's quality of reserve property; Second, the cooperation attitude of all countries' economic policy; Third, effect of one country balance of payments regulation mechanism; Fourth, the careful attitude that one country's government adopts to regulate measure; Fifth, the source of international liquidity on which one country depended and its measures of stability; Sixth, trend of balance of payments and economic condition. Though the factors which the method considered are more overall and practicable, most of them are difficult to be determined in quantity, they can't even reach a compatible conclusion, so the reference value of decision – making is meanless.

2.2 The Domestic Literature on Reserve

Whether the amount of foreign exchange is moderation or not becomes a popular problem, Many domestic scholars had two different opinions of analysis on this question. A kind of standpoint thinks Current Chinese foreign exchange reserves are ample, even a certain degree is surfeit. Hu Yuancheng (1997) analyzed China's reserve condition, the moderation scale model theories that he had put forward from both sides of the income and cost after comprehensive considerations. Wu Lihua (1997), Liang Di (2001) made use of J. Pgaowall model, making contrastive analysis of the income and cost of reserve, measuring and calculating the appropriate foreign exchange reserves level, and got a conclusion: the quantity of China's reserve is excessive, should carry on a valid management to the foreign exchange reserves property, in definite proportions investing excess reserve in the foreign government bond, depositing foreign exchange bank, investing the foreign real estate and stock, importing part of key technique and equipments that domestic need urgently, Wang Guolin (2001) made use of three metered index signs reckons that the foreign exchange stored in 1991-2000 of moderation value, getting our country's foreign exchange storage is ample, the scale is also appropriate. Shi Jianren, Guo Encai (2001) made use of international top usually adopt of three objective index signs to calculate the appropriate scale of China's foreign exchange reserve, result enunciation our country of foreign exchange storage consumedly high in nations generally accepted of reasonable storage level. Wu Nianlu (2003) thought our country current foreign exchange storage quantity, excessive, is to a kind of waste of the resources, put forward that our country should carry on a valid management to theforeign exchange storage property. Xu Chengming (2001) based – on Frenkel the foreign exchange storage need model foundation (1979), came to conclusion that; Since 90's in 20th centuries, our country actual foreign exchange storage. Mostly the age be placed in "opposite excess appearance". appearance". Another standpoint thinks currently our country foreign exchange reserve severity shortage, and can foreseereserve hardly to rise to the appropriate level in the future. Liu Bin (2000) founded on money supply determinism, thought the foreign exchange reserve severity shortage currently by comparison to the nation and the region in financial crisis of Asia. Guan Yuhua (2001) according to currently announce of national foreign exchange reserve data, came to conclusion that China's reserve outran the upper of appropriate scale consumedly, but according to the adjusted data reserve is far below the appropriate scale, this shows reserve is not more.

3 Regression Analysis Base on Statement of the Balance of International Payments

One country's Foreign Exchange Reserve is decided by its economical development level. These factors are popularly regarded as contributing factors on the volume of reserve, such as: scale of importation, balance of trade's fluctuation range, real FDI, the volume of external debts, change in balance of current account (CA) and the changes of exchange rate, etc. These factors whether have influence to the China's foreign exchange reserve, how are influence factors big or small, aiming at this problem, we carry on thorough analysis.

The data of regression equation hail from Nation Statistics of Bureau and Ministry of Commerce website.

Dependent Variable: R_t
Method: Least Squares
Sample: 1993-2004
Included observations: 12
(The analyses result is shown in the table right of the page)

4 Regression Explanation

According to above – mentioned regression analysis, we draw a conclusion. : of Export Variation of FDI and Variation of External Debt are these decision factors of Chinese for exchange storage quantity growth. Hence, these points have great weight with Chinese foreign exchange storage.

- Strict control foreign loan growth speed, reasonable arrangement foreign loan structure.
- keep the liquidity of the foreign exchange storage.
- Strengthening adjust the import and export industrial structure.
- The currency that impuoves the foreign exchange storage grows structure and mob structure.
- Build up exchange storage risk management system.

References

1. Di, L.: The Rethinking Based – on Our National Foreign Exchange Reserve Sca. Finance and Insurance (2001)
2. Yuhua, G.: Research the Appropriate Scale of Our National Foreign Exch Reserve. Statistic Research (2001)
3. Turnovsky, S.J.: A Determination of The Optimal Currency Basket:A Macroeconomice Analysis. Journal of International Economics 12, 333–354 (1982)
4. Yoshino, N., Kaji, S., Suzuki, A.: The Basket-peg,Dollar-peg,and Floating:A Comparative Analysis. Journal of the Japanese and International Economices 18, 183–217 (2004)
5. Black, S.W.: Exchange Policies forLessDevelopedCountriesin a World of Floating Rate. Princeton Essays in International Finance. No (1976)

Variable	Coefficient	Std. Error	t- Statistic	Prob.
M_t	-3.913851	1.372079	-2.852497	0.0463
X_t	4.868677	1.270464	3.832204	0.0186
$\Delta(X-M)$	-1.705774	1.030449	-1.655370	0.1732
ΔCA	1.139747	1.042578	1.093201	0.3357
ΔKA	2.164816	0.806298	2.684881	0.0549
ΔED	-2.602655	0.884171	-2.943612	0.0422
ΔFDI	3.802540	2.215058	1.716678	0.1612
β_0	-100.7280	20.77336	-4.848901	0.0083

R – squared	0.995835	Mean	197.3663
Adjusted	0.988547	dependent var	166.7168
R – squared	17.84207	S. D.	8.835717
S. E. of	1273.358	dependent var	9.158988
regression	-45.01430	Akaike info	136.6312
Sum squared	2.075564	criterion	0.000136
resid		Schwarz	
Log		criterion	
likelihood		F – statistic	
Durbin –		Prob (F –	
Watson stat		statistic)	

$$R_t = \beta_0 + \beta_1 M_t + \beta_2 X_t + \beta_3 \Delta(X-M) + \beta_4 \Delta CA + \beta_5 \Delta KA + \beta_6 \Delta FDI + \beta_7 \Delta ED + \varepsilon_t$$

R_t: of Foreign Exchange Reserve at t period

M_t: of Import at t period

X_t: of Export at t period

ΔCA: Variation of Balance of Current Account

$\Delta(X-M)$: Variation of balance of export and import

ΔFDI: Variation of FDI

ΔKA: Variation of Balance of Capital Account

ΔED: Variation of External Debt

ε_t: Stochastic disturbance term

Development of GE Series Motion Controller Utilized in Full Electric Plastic Injection Molding Machine

Hang-wei Zhang, Chan-juan Chen, and Ji-xian Dong

College of Mechatronic Engineering
Shaanxi University of Science & Technology
Xi'an, Shaanxi 710021, China
zhanghangwei_00@yeah.net

Abstract. This paper promotes a method to convert the molding machine's control system from a mechanical and hydraulic system into a motor-driven system. This conversion is done using the "PC + Motion Controller" architecture. It describes the components of hardware system based on the motion controller, and plots out software functions that match each of the machine component's roles. This work aims to produce a precise and low cost control system for customers.

Keywords: Motion Controller, Motor-driven, Plastic Injection Molding Machine.

1 Introduction

Fully electric plastic injection molding machines make up 75% of the market share, and this proportion increases every year [1]. *NISSEI Plastic Industrial Co. Ltd* developed the direct-press electric plastic injection molding machine with a mold clamping force up to 360 tons. It was the first of its kind in the field. In use, fully electric plastic injection molding machines are not as good as full hydraulic injection molding machines. However, the fully plastic machine is more environmental and energy efficient.

Fully electric plastic injection molding machines are driven by an AC servomotor with high power output. The machine can simultaneously perform mold clamping, die sinking and product ejection. It moves so fast that it could significantly shorten the process time and production cycle. The machine would be multi-level regulated by utilizing an AC servomotor with a more precise injection control system and greater revolution stability.

The electric plastic injection molding machine uses multiple low-power servomotors whereas the hydraulic machine uses a high-power motor and pump. The servomotors make much less noise. Unlike the motor and pump, the servomotors do not need a hydraulic operating unit, which produces environmental pollution when the hydraulic fluid leaks out. The servomotors also do not need a hydraulic cooling system, so their energy and water consumption is less. A significant disadvantage of the hydraulic injection molding machine is that it needs to be constantly supplied with

D. Zeng (Ed.): ICAIC 2011, Part II, CCIS 225, pp. 384–389, 2011.

electric power, regardless of whether the machine is being used or not. In contrast, the fully electric injection molding machine needs about 1/3 to 1/4 less electric power[2]. Compared with the hydraulic injection molding machine, the fully electric plastic injection molding machine uses less energy and is cheaper, more energy efficient, and more environmental.

The production cycle of the plastic injection molding machine can be divided into the following steps: clamping the mold, moving the injection seat forward, injecting the plastic, cooling the product and screw plasticizing, moving the injection seat out, sinking the die, and ejecting the product. The motion controller, the core of injection molding machine control system, collects on/off signals for sequential control. It gathers analog data, such as temperature, speed and pressure, for closed-loop control.

2 Ge Serias Motion Controller

The timing of the clamping mechanism in the plastic injection molding machine should be adjusted during the die sinking and mold clamping processes. The first reason for this is to reduce the length of the idle stroke to improve productivity. The second is to have a time buffer during the clamping mold process to prevent damage to the mold and product. The last reason is to reduce the impact of machine vibrations. Thus, the timing of clamping mechanism should be regulated. As well, the timing of the injection should be adjusted based on the variety of plastic, the product geometry and the die casting system being used. In addition, the motion of the ejector mechanism must be smooth and steady. The above timing adjustments were made to the *GE series motion controller* for the mold clamping, die sinking, injection and product ejection processes to improve the performance of the plastic injection molding machine.

The motion controller controls the position and velocity of mechanical moving parts to move them according to expected paths and prescribed parameters. The motion controller was initially developed for specialized, numerically controlled machines and other automated equipment. These machines and equipment had design related functional modules that matched process requirements in a certain industry. This motion controller was often unusable for different process requirements and could not be used in other industries. The motion controller could not be reconfigured based on user specified application requirements. It was necessary to develop a universal motion controller for the market. The market of universal motion controller is estimated to be more than 40 billion U.S. dollars, and the annual growth rate is expected to be 6.3% over the next five years[3].

The *GE-800-PV-PCI* motion controller is an 8-axis motion controller which uses *Digital Signal Processing* and *Field-Programmable Gate Array*. mainly with high performance computation[4]. The motion controller is able to perform complex control functions because the system offers a C language function library and Windows dynamic link libraries. For this study, the motion controller was connected to an industrial PC using an add-in card form, called the "PC+Motion Controller" mode. This combined the PC's information processing capability with the motion controller's open architecture. It therefore had a powerful information processing capability, and a high level of openness and versatility. A special application system meeting the requirements of the plastic injection molding machine was constructed. This system used application programs, such as the interface display and data processing, needed by control systems or control functions in the C language function library.

The *GE-800-PV-PCI* motion controller is a point-to-point motion controller[5]. Point-to-point motion controllers only deal with the final position, and have nothing to do with intermediate stages in a process, such as the path of motion.. Different control strategies were adopted when the point-to-point motion controller was making the machine accelerate or decelerate. In the acceleration stage, increasing the system gain and the acceleration were frequently used to accelerate the system to the desired velocity. At the end of deceleration stage, the S curve deceleration control strategy was utilized. Reducing system gain was useful for reducing vibrations when the system was in the right position[6]. Therefore, the point-to-point motion controller was able to perform modifications to control parameters, and acceleration or deceleration curves. Using these control strategies, the point-to-point motion controller was able to precisely control the speeds of the various machine components during the mold clamping, die sinking, injection, and product ejection phases.

3 Hardware of Control System

The motion controller controlled the fully electric plastic injection molding machine's six AC servomotors[7]. These were the mold clamping servomotor, injection servomotor, preplastication servomotor, product ejection servomotor, mold adjustment servomotor and injection seat forward servomotor. The motion controller also collected on/off signals for sequential control. The analog signals for the pressure, temperature and velocity at different positions were collected to complete the six closed-loop controls of servomotors[8].

The hardware was constructed using the following devices:

1) GE-800-PV-PCI Motion Controller, manufactured by Googol Technology Ltd. (Shenzhen)

2) Industrial PC or PC with PCI Slot

3) Driving Power Supply

4) +12V ~ +24V DC Power Supply (for Interface Board Supply)

5) Six different types of AC Panasonic Servomotors

6) AC Servomotor Drivers

7) Several Switches such as a change-over switch for inspecting the clamping mold on high and low speeds, a position switch for detecting whether injection seat was in the right position.

The steps taken to construct the fully electric plastic injection molding machine's control system were as follows. First, the motion control card was inserted into the industrial PC. Second, the communication driver was set up in the Windows operation system. Third, communication was established between the industrial PC and motion control card. Then, each individual servomotor was connected with its own driver. Lastly, the motion control card, servo drivers, terminal board's inputs/outputs were combined. The architecture of hardware system is depicted in figure 1.

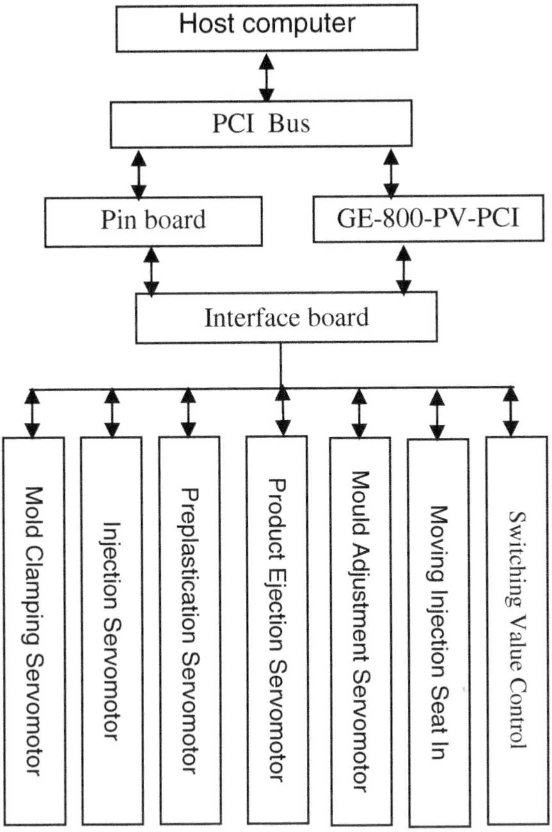

Fig. 1. Architecture of hardware

4 Software Functions

The control system of the fully electric plastic injection molding machine was based on the motion controller of a fore-background control system. The core of the control system was the *GE-800-PV-PCI motion controller*, which was the foreground level[9]. It was responsible for sequential control, close-loop or open-loop control for the six servomotors, input and output of on/off variables, input and output of analog variables, diagnosis, and receiving process information from the background control system. The industrial PC controlled the whole system. It was in charge of the man-to-machine interactions, parameters setting, dynamic simulation, the status display, and sending directives to the motion controller via the PCI.

The *GE-800-PV-PCI motion controller* in the foreground strictly had to control in real time. Therefore, interrupt priority was considered in programming. Some programs running in the background were indirectly related to physical devices, but and some of them were not. The requirement for real time control in the background was less strict than that in the foreground. Consequently, real time control in the background had less priority than that in the foreground. The man-to-machine interface developed for this system utilized control function library in the foreground and OOL (Object Oriented Language) C++ Builder Language in the background. Figure 2 shows the tasks that the software performed in the background.

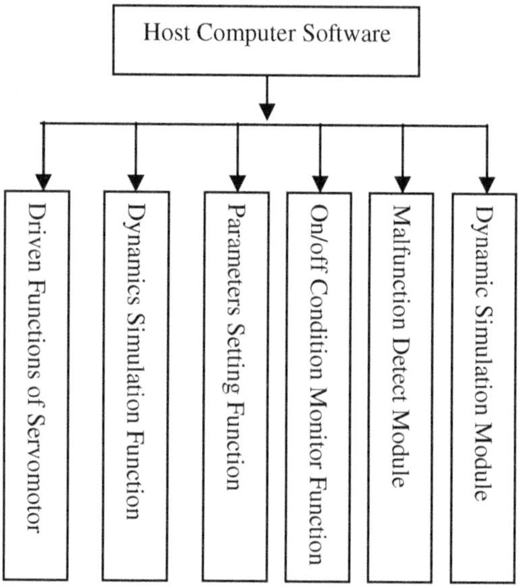

Fig. 2. Software Functions in Application Software

5 Summary

The technical reconstruction using the motion controller took full advantage of the original assemblies in injection molding machine. The mechanical and hydraulic plastic injection molding machine was successfully converted into a fully electric one. Using a general motion controller not only simplified the machine's configuration, but it also improved control precision and enhanced the product quality. It also led to products being made in a cheap, efficient, energy efficient, and environmental manner. Recently, open control systems tend to use motion controllers. The applications of motion controllers are becoming increasing broad.

Acknowledgments. The authors would like to thank Jun-sheen Liu, the researcher of the Molding Lab, for preparing the experiments; the College of Mechatronic Engineering in Shaanxi University of Science & Technology; Pro. Jia-hong Zhen, the Dean of Mechatronic Engineering Department, for providing technical support.

References

1. Wang, Q.: Rapid Popularity of Fully electric Plastic Injection Molding Machine in Japan. Manufacturing Technology & Machine Tool 8(2), 3 (2001)
2. Huang, B., Xu, Z.: Developing Trend of High-speed Precision injection Molding Machine Technology. China Media Technology 1(1), 47–49 (2003)
3. Xiang, P., Li, X., Du, Y.: Distributed Control System of Injection Molding Machine, vol. 5(6), pp. 37–39 (2007)
4. Jiang, S., Wu, H., Lu, S., Gong, X.: Present Status, Development and Application of General Motion Controller Technology, http://www.googoltech.com.cn/web/chi/institute.jsp
5. Google Technology Ltd., Technical Manual of GE Series Motion Controller, pp.10-16
6. Dong-sheen, D.: The Comparison of Several clamping Unit Structure in Electric Injection Molding Machine. Light Industry Machinery 24(2), 56–58 (2006)
7. Wang, X., Yang, H.: Process in Injection Molding Machine. China Plastics 15(8), 85–87 (2001)
8. Feng, B.: Design of Plastic Machine. Chemical Industry Press, CA (2008)
9. Xu, Y.: Design for Mechatronics System. Chemical Industry Press, CA (2005)

Classification of Multi-relational Databases

Xinchun Wang[1] and Sujuan Zhang[2]

[1] College of Science, Hebei United University
Tangshan, China
[2] Modern Technology and Education Centre, Hebei United University
Tangshan, China
x.c_wang@yahoo.com.cn

Abstract. Most existing datamining approaches look for patterns in a single table, Multi-Relational Data Mining(MRDM) approaches directly look for patterns that involve multiple tables (relations) from a relational databases. Multi-Relational Data Sorting(MRDS) is a important and basical technology, it has a extensive applications. The text analysed and comparede with classical sorting metheds, and put forward a expectation.

Keywords: decision-tree, star-pattern, snowflake-pattern, aggregation analysis.

1 Introduction

The original model from a single relationship of data mining method that is called propositional learning method or property - the value of learning method, which uses propositional logic to represent knowledge and model. And multi-relational data mining (Multi-Relational Data Mining, MRDM), is constituted by a number of relational tables in the relational database extraction patterns of a multidisciplinary field. It uses inductive logic programming (Inductive Logic Programming, ILP), KDD, machine learning and relational databases and so on, is committed to multi-relational data directly from the concentration of new methods of knowledge discovery, MRDM different mining methods based on single relations, it multi-relational data directly on the excavation, to avoid the complexity of the pretreatment process and the information loss due to compression problems [1].

Multi-database mining can be divided into three steps: First, the classification of multi-database, and then tap each database, and finally the same database mining to knowledge synthesis. The first step is to see data mining classification.

Classification is commonly used in data mining methods. Classification problem to categorize for an event or target . In use, this model can be used both existing data, you can also use it to predict future data. Classification goal is to establish a classification function or classification model (classifier), the model data in the database entries can mapped to a given category one. Classification can be used to predict, forecast data from the historical record of the purpose of automatically derived data on the promotion of a given description, which can predict the future data.

D. Zeng (Ed.): ICAIC 2011, Part II, CCIS 225, pp. 390–396, 2011.

2 Research Status of Classification on Multi Relational Database

In recent years, data mining has made a lot of classification algorithms, such as: decision trees, association rules, Bayesian, neural networks, rule learning, k-close, genetic algorithms, rough sets and fuzzy logic technology. However, these classification methods used in the field of relational databases and more research is still rare

3 Multi-relational Database Classification

3.1 Decision Tree Classification of Multi-relations

Decision tree is a tree structure similar to the flow chart, where each internal node represents a test on an attribute, each branch represents a test output, and each tree node represents class. In the CART method, based on the decision tree to overcome the limitations of language that brings, S-CART classification tree will be introduced to this method of statistical relational learning more than the field [2], which is the expansion of propositional learning method CART, S-CART to relational data directly from multiple tables and can easily use that knowledge background. S-CART system builds a tree, each node of the tree contains a text (text atoms) or a conjunction of the text, and each leaf node assigns a discrete value or data value. Inspired by the CART pruning method, S-CART is also added to the leaf nodes of the linear regression model to improve its performance [6]. Experiments showed that S-CART prediction accuracy comparable with other propositions methods, indicating that relations between classification and regression method can improve the comprehensibility of the rules and would not reduce the accuracy of the forecasts.

3.2 Spatial Classification Based on Multi-relations

Materialized join index by using the theme of space between the layers of neighbor links, and put them into a relational table that we can not directly use the relational schema predicate set classification. In the algorithm used in three tables: the target table, join index table and describes the properties of other topics neighbor layers [7]. Algorithm description:

Input: Target table: the object being analyzed
Neighbor: The neighbors of the object
Spatial join index table (SJI)
Target attribute: class labels
Output: binary decision tree

Step 1 decision tree to represent the training samples contain only a single node selection map S start.

Step 2 If the samples belong to the category, returns S as a leaf node.

Step 3 otherwise. Algorithm to the current selection graph S is optimized with the most information gain of optimizing operation R (R = optimal refinment (S)).

Step 4 best optimize operations through the implementation of R, spanning the two complementary branches of R (S),. Here, R (S) said that the selection graph S obtained by optimizing the choice of operator R maps indicated that the selection of Figure S implementation of R obtained by the complementary operation of the choice of map.

Step 5 use the same process for the two branches of the tree, recursive implementation of the 2 to step 4.

Step 6 Recursive division of one step only if the following conditions are established to stop:

① Select chart given on behalf of all the samples belong to the same class.

② not optimal operation can be used to further divide the sample, mark the node for all the samples of the most common class.

3.3 Methods Based on Probability – Bayesian Classification Multi Relations

Bayesian approach has a solid mathematical theory as the basis of prior information and data, and integrated the ability to sample information, is becoming the current machine learning and data mining research focus.

Naive Bayes algorithm is a simple and efficient classification algorithm, but the conditional independence assumption is not consistent with objective reality, which to some extent affects its classification performance. Weighted Naive Bayes is an extension of it.

1) Naive Bayes model: Bayesian classification is based on the Bayesian formula, namely:

where: P (C I X) conditional on X, C, posterior probability, P (C) as C prior probability, P (X I C) is Condition C, X, posteriori probability, P (X), said X a priori probability. For the convention: the variable with the capital letters, C variables that category, A, said property variable, assuming a total of m variables attributes, A = <A1,A2,…,Am>; with a lower case variable value, Val (C) = (C1, c2, ..., cl), Val (A) = (ai1, ai2, ..., aik) denote the type of the variable range of variables and attributes; said to be divided with the sample set X, x = <a1, a2, ..., am> that to be classified samples; said training sample set with a T, ti = <a1,a2,…,am,cl> that training example. In the naive Bayes assumption that the properties of conditional independence with respect to class, there are:

Thus the posterior probability formula: the test sample (E) was in the largest class posterior probability, because P (x) is a constant, then the naive Bayesian classifier model is:

2) Weighted naive Bayes model and the classification process: The more difficult in practice to meet the Naive Bayes conditional independence assumption, the properties can be assigned to different weights to different Bayesian be expanded, then the weighted naive Bayes model:, where wi Ai's right on behalf of property value, the greater the weight of property, the property the greater the impact on the classification. The key weighted naive Bayes problem is how to determine the weights of different attributes.

Weighted Naive Bayesian classification process:

Step1 Data Preprocessing: The classification of training samples and the samples were to be filled and discrete;

Step 2 Judgement: If the classification task, then switch to Step5, if the training task is transferred to Step3;

Step 3 Probability table and the weight parameter of the study:

Step 3.1 Probability parameter learning: Scanning all the training samples, calculate all of the prior probability P (aik | cl), cl is the class attribute Ai of the first k values of the probability of species; and P (cl), the value for the class cl probability;

Step 3.2 Weight parameter learning: Scanning all the training samples, calculate the weight of attribute Ai wi

Step 4 Generate weighted naive Bayesian probability table and a list of attribute weights, that the required weighted naive Bayesian classifier;

Step 5 Category: call the probability table and a list of attribute weights obtained classification results.

3.4 Nearest Neighbor Classification

In the field of data mining methods assume that the problem based on the distance between objects within each distance is calculated [9], distance-based methods, many data mining can be a simple and effective method to resolve. Extended to the case of multiple relations, the expansion of the concept of critical distance, Emde and ttschereck RIBL in the system presents a more defined relationship between the distance method. In addition, two defined using RIBL distance classification algorithm has been formed, namely RDBC and FORC. RDBC using hierarchical clustering and K means method using FORC.

3.5 Star Pattern Classification by Multi-dimensional Data

Star schema data can be classified, star schema [12] is a central fact table, connected to multiple dimension along the radius of the description [10]. Which: (1) the fact sheet: the relationship between store the actual business end-user data, and multiple dimension tables describe the keys, the combination of these keys form the primary key of the fact table; (2)-dimensional description: holds the key dimensions and describe the key of the other non-key attributes. The fact table contains the basic business data, which may include thousands of records; dimension table includes the SQL query can be used as business property. using star schema structure is conducive to the classification based on multi-dimensional analysis can regulate the connection SQL statement operation, reducing the level and complexity of connection, enhanced data query performance, multidimensional database does not support multi-table join, and support multi-table joins star structure is also prominent features.

3.6 Classification of Multidimensional Data Models by Snow

The multidimensional data model based on the relationship between the snow mode to the dimension table is decomposed into multiple tables with the dimension hierarchy. Snowflake star schema model is an extension of the process of decomposition of dimension table is on the dimension table further standardization process. Snowflake pattern in Table III-v increase the property sheet. The star schema dimension tables in most of the data described on the corresponding attribute table, dimension describes the main data table reservations only, of course, there is no dimension can be described in the table the corresponding property sheet. snowflake pattern has the advantage of using dimensional description is more standardized and easy to implement dynamic SQL generation. In the multi-dimensional analysis, the dimension is hierarchical, for example: geographical dimension can be divided into: national, regional, provincial, City, and several levels; product dimension can be divided into: brand, categories, subcategories, and specific standard models, etc.; customer dimension can be divided into: gender, age, occupation and so on. snowflake pattern reflects the hierarchical and easy to implement that is, from coarse to fine top-down data classification, and because the snow model dimension table are standardized, low particle size, increasing the flexibility of the application.

4 Multi-relational Database Comparing the Advantages and Disadvantages of Various Classifications

4.1 Decision Tree Classification of Multi-relations

Decision tree has the following advantages: first, fast: calculation is relatively small, and easily converted into classification rules. As long as the leaves along the roots all the way down, along the split conditions can uniquely identify a classification rule. Second, high accuracy: the classification rules mined with high accuracy. Third, easy to understand: the semantic type of classification rules is very easy to translate for people to understand explanation.

4.2 Spatial Classification Based on Multi-relations

Spatial data are highly correlated, so spatial data mining involves complex spatial relationships, space Between the objectives of data mining not only to consider the properties of the object itself, but also consider the properties of the target neighbor [8]; the same time, spatial data is often implicitly stored in multiple layers. Use of spatial relations and multi-relational data mining materialized the idea of simplifying spatial relations, make full use of the database structure information in the data model, and ultimately converted to create a binary decision tree.

4.3 Naive Bayesian Classification of Multi-relations

Naive Bayes algorithm is a simple and efficient classification algorithm, but the conditional independence assumption is not consistent with objective reality, which to some extent affects its classification performance. Weighted Naive Bayes is an extension of it. However, pretreatment of the original information system has changed more or less the same time the original information directly to a loss of how to value and continuous value attribute information system weighted naive Bayesian classifier needs further study.

4.4 Nearest Neighbor Classification

Case-based learning methods in the classification of all instances must be stored before, this requires a huge storage space; between the calculation of the distance between objects is not as simple as the Euclidean distance calculation, and only when calculating the distance to be classified, so does not require a higher classification rate for the task; the existing relationship between the three multi-method (RIBL2, RDBC, FORC) is only suitable for application to contain thousands of instances in the problem domain, and therefore also in the large-scale data intensive applications require more more research; In addition, the method of computation cost is very expensive.

4.5 Star Pattern Classification by Multi-dimensional Data

Star schema has the advantage of multi-table join, easy to divide the hierarchical relationship of each dimension to reduce the query to deal with the number of physical connections, greatly improving performance. Model view simplified and requires less effort to the data warehouse to be completed by the maintenance and development, expand the data warehouse can be flexible. drawback is that the central star schema fact table that the amount of data (record number) is huge.

4.6 Snowflake Pattern Classification by Multi-dimensional Data

Snowflake pattern is extended in the star pattern, snowflake pattern reflects the hierarchical top-down easy to realize that the data from the coarse to fine classification, and because snow model dimension table are standardized, low particle size, increasing the flexibility of the application. However, increasing snow mode the user must deal with the number of tables, increasing the complexity of some of the data query.

5 First Cluster and Then Proposes a Method of Classification

The basic idea of clustering is that the sample under study exists between the different degrees of similarity, and then be able to identify the specific measure of the similarity of the statistics, according to the statistics of some degree of similarity were clustered into one large sample class, the other a large degree of similarity between the sample and aggregated into another, until all the samples are clustering is completed. Clustering can also use the idea to determine the classification based on the actual number needed. Approach is to first construct a similar function to characterize the similarity between the library, according to the definition of similarity distance between the two libraries; a given distance threshold, within the threshold into one class; also given a good classification and bad standards, is the key to a good number of classifications classified as little as possible, the distance between the sum of all types are as small as possible. In accordance with the classification by the fact that relations between the database and closer and more reasonable.

6 Conclusion

More and more data mining applications require the type of complex structured data (such as gene sequence analysis, HTML and XML document analysis, etc.) for analysis, which requires the use of a more expressive mode of language; In addition, Based on a number of practical application of the original property - the value proposition of learning simply can not solve all of which contributed to the development of MRDM. In general, relational data mining methods can be extended to multi-relational database mining, many data mining methods to data classification is for the purpose. The existence of a classification is not yet suitable for a variety of features to the data, many scholars are trying to propose a broader range of applications of data classification.

References

1. On your other constant, Data Mining Algorithms. Its Applications, Phase II (2005)
2. Xu, G.: United States, and many more relational data mining method. Computer Applications 9 (2006)
3. Dzerosk, S., Lavrac, i.N.: Relational Data Mining. Springer, Berlin (2001)
4. Wrobe, S.: Inductive Logic Programming for Knowledge Discovery in Databases, pp. 74–101. Springer, Berlin (2001)
5. De Raedt, L., Blockee, H., Dehaspe, l.L., et al.: Three Companions for Data Mining in First Order Logic. Relational Data Mining, pp. 105–139. Springer, Heidelberg (2001)
6. Kramer, S., Widmer, G.: Inducing Classification and Regression Trees in First Order Logic. Relational Data Mining, pp. 140–159. Springer, Heidelberg (2001)
7. Liu, Y., et al.: Multidimensional data structures based on relational database model. Computer Engineering (September 2000)
8. Liu, W.s.: The relationship between space based on multi-classification algorithm. Yunnan University 28(S1), 158–163 (2006)
9. Ma, et al.: Context-based Web Deep Web database classification. CODEN RUXUEW Journal of Software 19(2), 267–274 (2008) ISSN 1000-9825
10. Liu, D.S.: Multidimensional data model and its implementation in a relational database. Microelectronics and Computer 2 (1999)

An Anti-JPEG Compression Image Perceptual Hashing Algorithm

Bin Zhang[1], Yang Xin[1], Xinxin Niu[1], Kaiguo Yuan[1], and Zhang Bin[2]

[1] Key Laboratory of network and information attack & defence technology of MOE
Beijing University of Posts and Telecommunications
Xitucheng Road, Haidian District, Beijing, China
[2] Department of Information Security Service Certification
China Information Security Certification Center
Chaowai Street, Chaoyang District, Beijing, China
zhang_bin345@163.com

Abstract. Perceptual hashing schemes were widely applied to image content authentication fields in recent years. However, most hashing algorithms were difficult to differentiate malicious attacks from normal signal processing operations, such as image compression, enhancement. As well known, digital images are JPEG compressed mainly by quantifying the DCT coefficients. The higher frequency DCT coefficients are removed, while the lower values are protected in this process. This paper will propose a novel image perceptual hashing method based on image DCT low frequency coefficients which not only can achieve authentication for the JPEG image, but also can detect some malicious manipulations effectively, including tampering, noise. Also, Theoretical and experimental results are used to demonstrate the effectiveness of this algorithm.

Keywords: Perceptual image hashing, Quantization JPEG compression, DCT.

1 Introduction

The developments of digital signal process technology and internet communication have made digital multimedia easier to forgery or modify than ever. To ensure information security, integrity and availability, many novel multimedia authentication technique, such as cryptography, digital signature, digital watermarking, were proposed in the past decades. The traditional cryptography method achieving information authentication is based on digital signature technology and hashing functions. Digital signature is to add version right or symbol into the protected multimedia carrier, and most signature systems are based on PKI algorithms. Digital watermarking schemes aim to verify the authenticity and integrity of the host carriers and provide tamper proofing by location and recovery.

All these methods can protect the multimedia data in a way, but have their drawbacks at the same time. For example, the traditional image authentication methods can not differentiate malicious attacks from this inevitable JPEG compression. Being a lossy compression method, JPEG images are widely distributed

D. Zeng (Ed.): ICAIC 2011, Part II, CCIS 225, pp. 397–406, 2011.

and used on the internet because it not only can guarantee the image quality but also occupies smaller memory than BMP and other format images. So, JPEG compression becomes an inevitable or even desirable image manipulation. Hence, we need a kind of image authentication method which is resilient to normal signal processing operations, while fragile enough to detect malicious manipulations. Several image authentication schemes have been proposed in this literature, and can be classified into two categories: watermark-based algorithm and hash-based algorithm.

As a novel image authentication method, image perceptual hashing (hash-based algorithm) which was first brought forward last century and has solved the many problems faced by traditional image authentication technology. Image perceptual hashing is a class of single way mapping from image presentations to a perceptual hash value in terms of their perceptual content. In contrast to the other methods, this technique is required to be robust to incidental image modifications. That is, the images being perceptually identical to the original image should be regarded to be authentic by applying the image hashing technique. Image hash algorithms have been widely applied in digital watermarking, content authentication and database search fields.

(1) Perceptual hashing model

A number of media-specific hash functions have been proposed for multimedia authentication. Multimedia hash can be regarded as a kind of content-based digital signature. To generate multimedia hash, a secret key is used to extract certain features from the image. These features are further processed to form the hash value. The hash is transmitted along with the media either by appending or embedding it to the original media data.

At the receiver side, the authenticator uses the same key to generate the hash values which are compared to the ones transmitted along with the data for verifying its authenticity. This process is illustrated in Fig 1.

(2) Research status

Based on the research achievement of digital image process, many image perceptual invariable feature extraction methods were applied to this field such as image block histogram, image edge features, DCT and DWT coefficients relation etc. Using the brightness histogram statistics characters as the hash values, Schneider M and Chang S. F[1] proposed a robust image content authentication method based digital signature. The most flaw of this method is: the attacker can change the image content without distorting the histogram. Kailasnathan[2] used image mean and variance as the statistical characters, but this method has the similar drawback as the former. Venkatesan et al.[3] designed an image hash function based on the statistical characters of the discrete wavelet transform (DWT), which believed that the mean of the DC sub-frequency band and the variance of the AC sub-frequency band were invariable with the image content. Fridrich[4] proposed a method that projected an input image onto DC-free (zero-mean) random smooth patterns. Swaminathan et al.[5] proposed an image hash scheme by using Fourier-Mellin transform's scale invariant property.

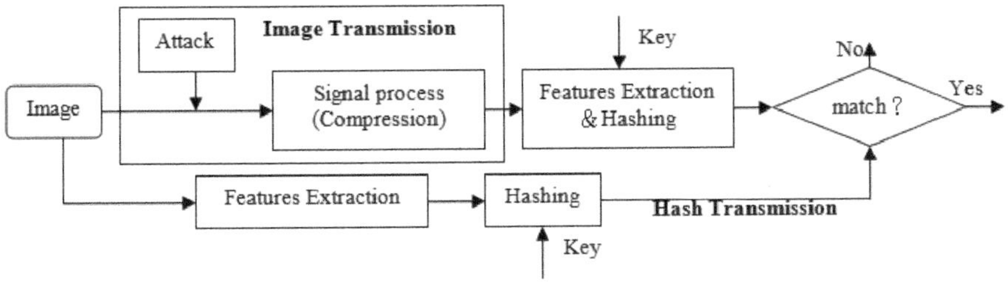

Fig. 1. Image hashing authentication procedure

Lin Chingyung and Chang Shihfu[6] proposed a technique for image authentication which can prevent malicious manipulations but allow JPEG lossy compression. The authentication signature is based on the invariance of the relationships between discrete cosine transform (DCT) coefficients at the same position in separate blocks of an image. These relationships are preserved when DCT coefficients are quantized in JPEG compression. Chien-chang Chen and Cheng-shian Lin[7] presented a robust image authentication approach which can distinguish malicious attacks from JPEG lossy compression. The authentication procedure calculates the relationships between important DCT coefficients in each pair of DCT blocks and predefined thresholds to form the authentication message, and then embeds the encryption of the authenticated message into other DCT coefficients. The message calculation and embedding procedure are based on two proposed quantization properties that always exist under different JPEG quantization tables. Mona F. M. Mursi, Ghazy M. R. Assassa et al.[8] proposed an image authentication system that is tolerant to JPEG lossy compression operations. A scheme for JPEG grey scale images is proposed based on a data embedding method that is based on a secret key and a secret mapping vector in the frequency domain. An encrypted feature vector extracted from the image DCT coefficients is embedded redundantly and invisibly in the marked image.

Fridrich proposed a robust watermarking technique for authentication[9] [10]. He divided images into small blocks and embedded quasi-VQ codes for each block by the spread spectrum method. This technique is robust to manipulations, but can not detect small area modification. The error between the extracted watermark and the reconstructed quasi-VQ codes is too large after JPEG compression. Therefore, this technique would be hard to distinguish malicious manipulations from JPEG compressions. Zou Jiancheng, Zhou Hongli and Deng Huanjun[11] proposed a secure and robust image hashing scheme based on V-system. It first modulated the image pixels to get an intensity-transformed version, then extracted coefficients after V transform to generate the hash value. Zhao Yuxin et al.[12] developed a robust hash scheme for image authentication. The image's feature matrix in the wavelet frequency domain is extracted by dithering quantization. Zhang Weike et al.[13] proposed a secure and robust image perceptual hashing algorithm. In this scheme, secure hash vector index is generated using perceptual invariance of image DCT low frequency coefficients to authenticate content of image that has been processed. A hash vector is generated for digital image using the technique of combining standardized DCT low frequency coefficients matrix with key based random matrix.

2 Theory of Algorithm

(1) JPEG compression introduction

The JPEG compression standard which was named ISO/IEC IS 10918-1 in 1995 includes four parts: basic JPEG standard, rules and software framework, standard extension, detail description. In general, JPEG compression includes two schemes: lossy compression and lossless compression. Lossy compression is achieved based on DCT, while lossless compression is by DPCM coding in space region. After being quantized, the DCT coefficients can not be recovered. The DCT framework of lossy JPEG compression is shown in Fig 2 [14]:

In JPEG compression standard, the 8×8 DCT coefficient should be quantized including the brightness and the color-difference using formula (1). $F(u,v)$ is the DCT coefficient, $Q(u,v)$ is the quantization matrix. So, the larger the quantization coefficient $Q(u,v)$ is, the more information of $F(u,v)$ may be lost in this process.

$$\tilde{F}(u,v) = Round(\frac{F(u,v)}{Q(u,v)}) \quad (1)$$

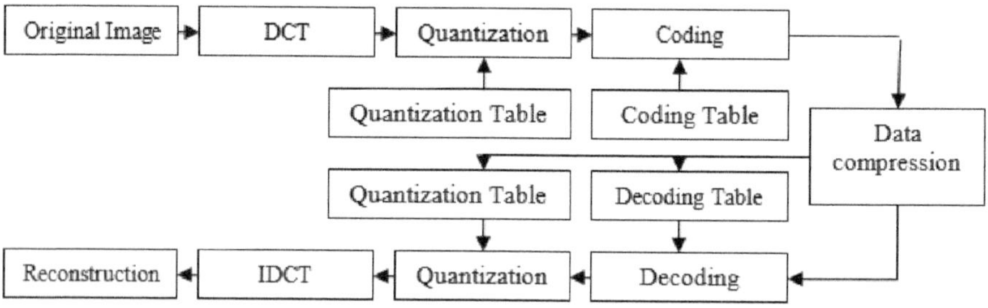

Fig. 2. Framework of JPEG compression

The quantization table is user-defined generally, but a recommended quantization table is given in JPEG standard.

$$Q_l(u,v) = \begin{bmatrix} 16 & 11 & 10 & 16 & 24 & 40 & 51 & 61 \\ 12 & 12 & 14 & 19 & 26 & 58 & 60 & 55 \\ 14 & 13 & 16 & 24 & 40 & 57 & 69 & 56 \\ 14 & 17 & 22 & 29 & 51 & 87 & 80 & 62 \\ 18 & 22 & 37 & 56 & 68 & 109 & 103 & 77 \\ 24 & 35 & 55 & 64 & 81 & 104 & 113 & 92 \\ 49 & 64 & 78 & 87 & 103 & 121 & 120 & 101 \\ 72 & 92 & 95 & 98 & 112 & 100 & 103 & 99 \end{bmatrix}$$

From the recommended quantization tables, we can find a rule that the quantization value of low frequency is smaller than that of high frequency. As well known, image profile is decided by low frequency coefficients, and the detailed information of the image is preserved in high frequency coefficients. So, the detailed information of the image would be lost, but the image profile is protected in compression process. This compression mechanism is mainly realized by these quantization tables.

(2) Feature extraction method

In this section, we will use an example to describe this feature extraction process in details. When the lightness values of a gray image (8×8) are as follows:

$$I(i,j)=\begin{bmatrix} 223 & 221 & 223 & 221 & 223 & 223 & 223 & 221 \\ 154 & 154 & 154 & 154 & 155 & 150 & 156 & 153 \\ 154 & 154 & 154 & 154 & 155 & 150 & 156 & 153 \\ 154 & 154 & 154 & 154 & 155 & 150 & 156 & 153 \\ 154 & 154 & 154 & 154 & 155 & 150 & 156 & 153 \\ 156 & 156 & 150 & 149 & 154 & 152 & 152 & 152 \\ 153 & 153 & 155 & 151 & 153 & 154 & 152 & 149 \\ 151 & 151 & 149 & 150 & 151 & 151 & 149 & 150 \end{bmatrix}$$

The DCT coefficient is:

$$F(u,v)=\begin{bmatrix} 1293.9 & 1.4838 & 1.4835 & 1.774 & -1.125 & 2.118 & -5.89 & 6.132 \\ 10395 & -3.662 & 0.6479 & -1.905 & -0.859 & 5.165 & -4.105 & 3.672 \\ 86.492 & -2.277 & -0.485 & 2.471 & -2.895 & 1.421 & 1.835 & -0.762 \\ 84.228 & -0.915 & 2.0355 & 3.024 & -2.576 & 0.61 & -0.429 & 1.145 \\ 68.625 & -3.032 & -0.084 & -0.437 & -0.375 & 2.663 & -1.183 & 0.460 \\ 56.129 & -0.663 & -0.828 & 1.165 & -3.074 & 1.688 & 1.108 & 1.023 \\ 36.018 & -0.878 & 2.585 & 2.112 & 0.5228 & -1.041 & -1.015 & -0.646 \\ 20.550 & -0.772 & -1.726 & -1.104 & -1.318 & 2.113 & 0.366 & 0.950 \end{bmatrix}$$

Using the JPEG recommended lightness quantization table and formula (1), the quantized coefficients table is:

$$\tilde{F}(u,v)=\begin{bmatrix} 81 & 0 & 0 & 0 & 0 & 0 & 0 & 0 \\ 9 & 0 & 0 & 0 & 0 & 0 & 0 & 0 \\ 6 & 0 & 0 & 0 & 0 & 0 & 0 & 0 \\ 6 & 0 & 0 & 0 & 0 & 0 & 0 & 0 \\ 4 & 0 & 0 & 0 & 0 & 0 & 0 & 0 \\ 2 & 0 & 0 & 0 & 0 & 0 & 0 & 0 \\ 0 & 0 & 0 & 0 & 0 & 0 & 0 & 0 \\ 0 & 0 & 0 & 0 & 0 & 0 & 0 & 0 \end{bmatrix}$$

After IDCT, we can get the JPEG compressed image. Using the first low frequency coefficient of the original image and JPEG image, we can calculate the compression ratio of the low frequency by $R(1,1) = F(1,1) / \tilde{F}(1,1)$. Also, we can get all the low frequency coefficients of the original image.

Take the first low frequency coefficient for example, the original low frequency coefficient is $F(1,1)=1293.9$, the quantization value is $Q(1,1)=16$, the compressed coefficient is $F(1,1)/Q(1,1)=80.8$, $\tilde{F}(1,1)=Round(80.8)=81$.

Using $\tilde{F}(1,1), F(1,1)$, we can get the approximate ratio $F(1,1)/\tilde{F}(1,1)=1293.9/81\approx 15.97$ which is very close to 16.

In return, we can get the new first low frequency coefficient $F'(1,1)=81\times 15.97=1293.57$. The error of round process is only 0.33.

Because we use high bit data as the hashing value, the error would not affect the authentication result.

3 Algorithm Procedure

(1) Features extraction

In this part, the image feature will be extracted from the low frequency coefficients of the image using the method described in 2.2. Then, we can get the hashing values using the Huffman coding algorithm.

Step 1: Read into the image file I, and M,N is the row and column of the image individually. Divide I into non-overlapping 8×8 blocks. If $M=a\times 8$ and $N=b\times 8$, then $I=\bigcup_{\varsigma=1}^{\Psi} I_{\varsigma}$, and $\psi=a\times b, a,b\in Z^{+}$. If $M\neq a\times 8$ or $N\neq b\times 8$, extend the row or column first until $M'=a\times 8, N'=b\times 8$, then $I'=\bigcup_{\varsigma=1}^{\hbar} I'_{\varsigma}$, and $\hbar=a\times b \;\; a,b\in Z^{+}$.

Step 2: Each 8×8 block I_{ς} is sent to the discrete cosine transform (DCT) sequentially and the results are preserved in a coefficient matrix $C=\bigcup_{\varsigma=1}^{\Psi} C_{\varsigma}$. Extract the low frequency coefficient from every C_{ς}, and compose a low frequency coefficient matrix $L=\bigcup_{\varsigma=1}^{axb} C_{\varsigma}(1,1)$. Transform $L(1,1)$ to unsigned 16-bit binary number and save it as H.

Step 3: Extract the higher bit value from the low frequency coefficient matrix L and save in L' using $L(i,j)'=\left\lfloor \frac{L(i,j)}{100} \right\rfloor$.Transform $L'(i,j)$ to binary value, extract the 4th bit and 3rd bit values of $L'(i,j)$ and connect them to a binary string S. This binary string is regarded as the image invariable feature in compression process and its length is $a \times b \times 2$.

Step 4: Using Huffman coding algorithm, S can be transformed into the hashing values $Hash_1$. H , $Hash_1$ are connected as the authentication data.

(2) Hashing Authentication

The hashing authentication is realized by comparing the hashing values and the extracted image feature from the JPEG image. The authentication approaches are as follows:

Step 1: Separate the received authentication data and get the former 16-bit data H and a hashing value individually. Transform H to decimalization.

Step 2: Make discrete cosine transform (DCT) of the image, and get the coefficient matrix $C^{'}$. Extract all the low frequency coefficients from $C^{'}$ and preserve it as $L^{'}$. Get the value of $L^{'}(1,1)$ and calculate the compression ratio $R = H / L^{'}(1,1)$. Using the compression ratio R, we can get another low frequency coefficient matrix $L^{''} = L^{'} \times R$ which is close to the original coefficient matrix L.

Step 3: Using the same method as 'Step 3 of Features Extraction Process', we can get another binary string $S^{'}$.

Step 4: Using Huffman compression and coding algorithm, $S^{'}$ can be transformed into the hashing values $Hash_2$. Calculate the distance of $Hash_1$ and $Hash_2$ using formula 2. Set a threshold $Thrd$, If $D < Thrd$, match successfully; else, not match.

$$D = NORM\left[\frac{|Hash_1 - Hash_2|^2}{NORM(Hash_1) \times NORM(Hash_2)}\right] \tag{2}$$

4 Experimental Results and Analysis

In this section, we will demonstrate the performance of the proposed algorithm from two aspects: robustness and vulnerability. We verify the robustness of the extracted feature and the hash value when the image is compressed. The vulnerability of the extracted feature can be authenticated by some malicious attack, such as tampering, noise and filtering.

(1) Robustness to JPEG compression

The main propose of JPEG compression is to decrease the redundancy of image by removing the high frequency characters and maintaining the low frequency information of the image. The profile is decided by the low frequency, and the details are saved in the high frequency. Because the hash value is completely extracted from the low frequency coefficients, the JPEG compression can not affect the authentication effectiveness. So, this algorithm has good robustness to JPEG compression. In simulation, we use the recommended quantization table of JPEG standard to compress the high frequency coefficients. Also, we get the compressed images and the hash values according to the different low frequency quantization coefficient.

The test image Lena (512×512, 8 bits per pixel) and the compressed image are shown in Fig 3. Image (b),(c),(d),(e),(f) are the compressed images of (a). Only the high frequency characters are removed in image (b). Besides the high frequency characters, the low frequency characters are quantized in image (c),(d),(e),(f). The extracted image feature keeps unchanged in this process.

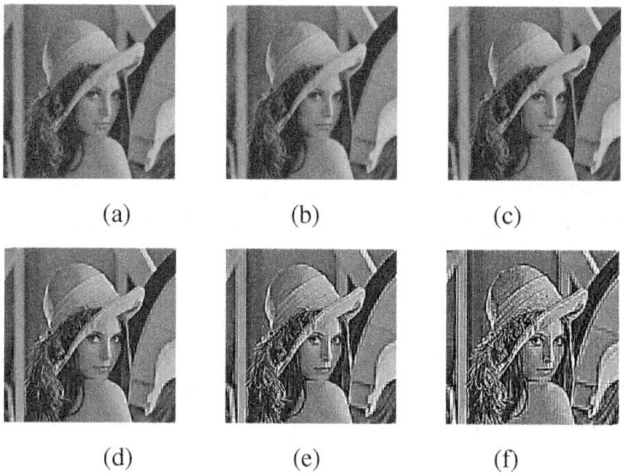

Fig. 3. (a) the original image (b) the low frequency quantization coefficient=1 (c) the low frequency quantization coefficient=16 (d) the low frequency quantization coefficient=32 (e) the low frequency quantization coefficient=64 (f) the low frequency quantization coefficient=128

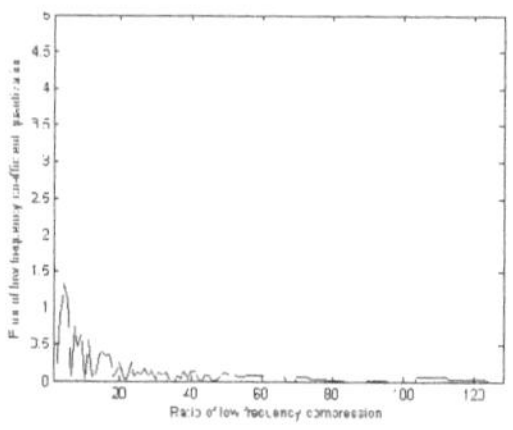

Fig. 4. The error of low frequency coefficients

When the quantization values of the low frequency coefficients are range from 1 to 128, the error between the original coefficient and the quantized coefficient are illustrated in Fig 4. Because most of the error is less 1 and the maximum error is 1.33, it does not affect the high bit of the extracted image feature.

(2) Vulnerability to malicious attacks

A: Tamper attack

From the algorithm procedure, we know the low frequency coefficients of the extracted feature only may change in a small range. And the image pixel value would change greatly when image is tampered or deformed maliciously even a pixel. It means that the hash value has the ability to detect malicious modification and the extracted image features can locate malicious modification. By comparing and calculating the distance of the hash values using formula (2), the proposed scheme can detect the tampered pixels location which is signed in Fig 5.

Fig. 5. (a) Image tampering (b) tampering detection and location

B: Noise & Filtering attack

In this section, we will calculate the account of changed pixels when the image is attacked by four types of noise and filtering, and verify whether the algorithm can detect malicious noise and filtering attacks or not.

TABLE 1 illustrates the mean ratio of the changed pixels when the image is attacked 100 times by noise and by filtering.

$$Ratio = P/(M \times N) \qquad (3)$$

Here, P and $M \times N$ represents the number of changed pixels by noise attack and the size of the extracted image feature.

Table 1.

Noise type	Gaussian	Salt & Pepper (0.02)	Poisson	Speckle
Ratio(by noise)	3.2%	2.6%	1.8%	3.4%
Ratio(by filtering)	4.1%	2.7%	3.1%	5.1%

So, the experimental results prove that the proposed algorithm can detect noise and filtering attack.

5 Conclusion

This paper firstly introduced the research progress of the image perceptual hashing algorithm and the principle of JPEG compression. Then, we present a novel image perceptual hashing algorithm based on the relative stability of the low frequency coefficients in JPEG compression. The simulation results proved that the proposed algorithm not only had good robustness to normal JPEG compression, but could detect some malicious attacks such as tampering, noise and filtering attack. In the future, more deeply research can be made in distinguishing different malicious attacks using the hash values.

Acknowledgment. This work is supported by National Basic Research Program of China (973 Program) (2007CB311203), National Natural Science Foundation of China (No. 60821001), National Natural Science Foundation of China (No. 61003284), Specialized Research Fund for the Doctoral Program of Higher Education (No. 20070013007) and the 111 Project (No. B08004).

References

1. Schneider, M., Chang, S.F.: A robust content based digital signature for image authentication. In: Proc of IEEE Conf Image Processing, Lausanne, Switzerland, vol. 3, pp. 227–230 (1996)
2. Kailasanathan, C., Naini, R.S.: Image authentication surviving acceptable modifications using statistical measures and k2mean segmentation. In: IEEE EURAS IP Workshop Nonlinear Signal and Image, Baltimore, Maryland, USA (2001)
3. Venkatesan, R., Koon, S.M., Jakubowski: Robust image hashing. In: International Conference on Image Processing, Vancouver, Canada, vol. 3, pp. 664–666 (2000)
4. Fridrich, J., Goljan, M.: Robust hash functions for digital watermarking. In: Proc. IEEE Int. Conf. on Information Technology: Coding and Computing, pp. 178–183 (2000)
5. Swaminathan, A., Mao, Y., Wu, M.: Image hashing resilient to geometric and filtering operations. In: Proc. IEEE Workshop on Multimedia. Signal Processing, Siena, Italy (2004)
6. Lin, C.Y., Chang, S.F.: A robust image authentication method distinguishing JPEG compression from malicious manipulation. IEEE Transactions on Circuits and Systems of Video Technology 11, 153–168 (2001)
7. Chen, C., Lin, C.: Toward a Robust Image Authentication Method Surviving JPEG Lossy Compression. Journal of Information Science and Engineering 23, 511–524 (2007)
8. Mursi, M.F.M., Assassa, G.M.R., Aboalsamh, H.A., Alghathbar, K.: A DCT-Based Secure JPEG Image Authentication Scheme. World Academy of Science, Engineering and Technology 53, 682–687 (2009)
9. Fridrich, J.: Image watermarking for tamper detection. In: Proc. IEEE Int. Conf. Image Processing, Chicago, IL (October 1998)
10. Methods for detecting changes in digital images. In: Proc. IEEE Workshop on Intelligent Signal Processing and Communication Systems, Melbourne, Australia (November 1998)
11. Zou, J., Zhou, H., Deng, H.: Secure and robust image hashing scheme. Application Research of Computers 26(6) (June 2009)
12. Zhao, Y., Liu, G., Dai, Y., Wang, Z.: A robust hash scheme for image authentication. Systems Engineering and Electronics 31(2) (February 2009)
13. Zhang, W., Kong, X., You, X.: Secure and robust image perceptual hashing. Journal of Southeast University 137 (September 2007)
14. Wang, Q.: Digital Image Process. Science Publication (February 2009)

Detection for Rotor Position of Brushless DC Motor Based on Pulse Injection Method

Yunhai Hou, Shihua Sun, and Ershuai Li

School Of Electric & Electric Engineering, Changchun University of Technology
Changchun, Jilin Province, China
Yunhai_Hou@163.com

Abstract. This paper presents a method to detect the rotor position of Permanent Magnet motors at standstill by using a suitable sequence of voltage pulses. Based on this method, a suitable sequence of voltage pulses is applied to the stator windings. We can estimate the magnets position by the current variation of the pulses. The obtained results show that this method is not only efficient but also need bit of motor parameters, it can omit motor internal examination parts and may be used on other type of motors.

Keywords: BLDC Motor, Rotor position, Pulses injection, Detection.

1 Introduction

In recent years, the non-position sensor brushless DC motor (BLDCM) is more and more widely applied for it has such advantages as good speed regulation characteristics, no switch fire, no radio interference, high efficiency, long life, reliable operation, easy maintenance, etc. Detecting the position of rotor is an important subject in study on BLDCM. The uncertainty of rotor position will directly lead to the motor started to fail or a reversal. At present, the start methods of BLDCM are hardware and software start. But the main disadvantage of hardware start is that the extra start circuit enlarges the size of motor, which reduces the reliable of it. The rotor pre-setting start method of software start needs to drive the rotor to a particular location and drive it after stopping swing [1]. As the rotor has inertia, so it will extends the drive time. This paper presents a method to detect the motor at standstill.

For the different rotor's position of Permanent Magnet Motor, the EM field of motor is different. So it is different to the influence of the each phase current of stator [2]. By using the change of current in the stator and sending a series of current pulses to the stator, the rotor position can be determined. This paper introduces the working principle, and carry out the rotor position detection experiments under static condition, which verify the reliability and feasibility of the method. The method is not only applicable to brushless DC motor, but also is effective to the PMSM and other permanent magnet motor. And which has no special requirements to the parameters of motor.

D. Zeng (Ed.): ICAIC 2011, Part II, CCIS 225, pp. 407–412, 2011.

2 Basic Principles

The Schematic diagram of permanent magnet motor is shown as Fig. 1:

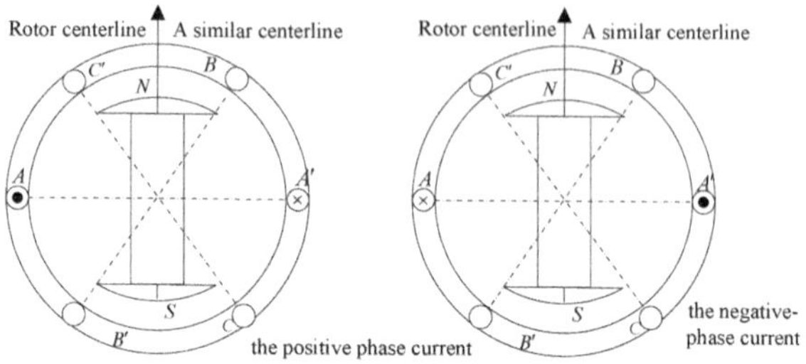

Fig. 1. Basic Principle

To A phase, for example, the internal magnetic field (φ_A) is formed of the magnetic field (φ_{AM})of permanent magnet (PM) and magnetic field (φ_{Ai}) of stator current, namely as:

$$\varphi_A = \varphi_{AM} + \varphi_{Ai} \tag{1}$$

If the phase current then becomes higher, the magnetic circuit will be saturated and the self-inductance coefficient of winding will be smaller. If the current direction changed, then the total flux can be:

$$\varphi_A = \varphi_{AM} - \varphi_{Ai} \tag{2}$$

Then, the flux will be out of saturation and the self-inductance coefficient of winding will become higher. With positive voltage pulse, digital controller continuously carried out A/D conversion to current by current sensors [3]. When the test pulse is over, a peak current can be obtained. Subsequently a negative voltage pulse be sent, the negative peak phase current can be obtained. Comparing with two amplified current value ,the current difference is:

$$\Delta i_A = i_{A1} - i_{A2} \tag{3}$$

The phase diagram of the test current is shown as follow in Fig. 1. Since the position of phase B and C are symmetrical relative to the magnet center axis, their current difference should be the same size. Because the magnetic field produced by phase A is exactly as same as magnet at this time, the magnetic field produced by positive and negative current is also the largest. And the resulting current difference is also the largest in theory.

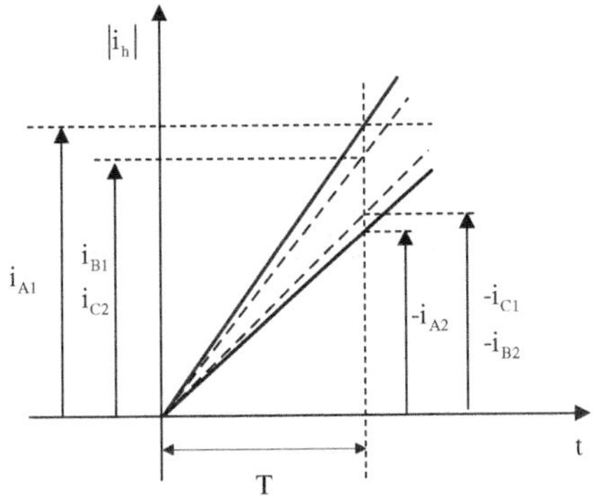

Fig. 2. Current Diagram

Similarly, for other phases, there be:

$$\Delta i_A = i_{A1} - i_{A2}$$
$$\Delta i_B = i_{B1} - i_{B2}$$
$$\Delta i_C = i_{C1} - i_{C2} \tag{4}$$

The Fig. 3 shows the schematic of current difference in the state of Fig. 1.

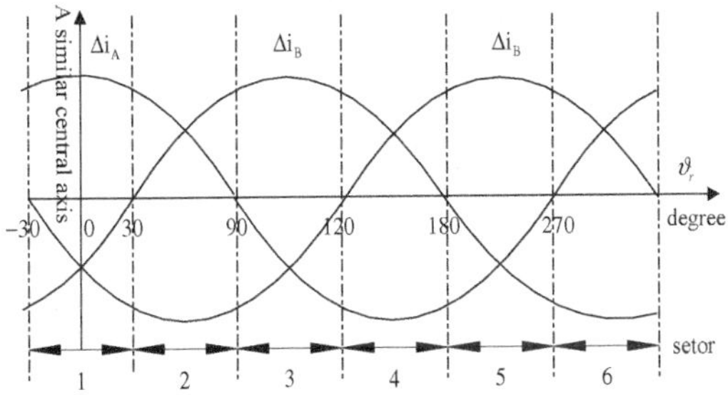

Fig. 3. Relationship Diagram of Current Difference and Rotor Position

Comprehensively analysis all three-phase current difference, the rotor position can detected [2]. Set the current difference changes continuously with the rotor position (set as change with sine wave shown as Fig. 3), the absolute position of rotor can be determined according to their different combination[4].

3 The Hardware Circuit of System

The design of system is based on TI's DSP320LF2407A. As mentioned earlier, the current pulses can be given in the order in Table 1.

Table 1. Given The Timing

	1	2	3	4	5	6
VT_1	+	–	–	+	–	+
VT_2	+	–	+	–	–	+
VT_3	–	+	+	–	–	+
VT_4	–	+	+	–	+	–
VT_5	–	+	–	+	+	–
VT_6	+	–	–	+	+	–

First of all, the PWM unit gives high voltage to VT1, VT2, VT6, the positive current of any phase can be detected. And then add the reverse current to that phase to force the current to be zero. Subsequently another high voltage is given to VT3, VT4, VT5, the negative current of the phase can be obtained. Based on equation (4), there will be the current difference. Similarly, the current difference of other phases can be got.

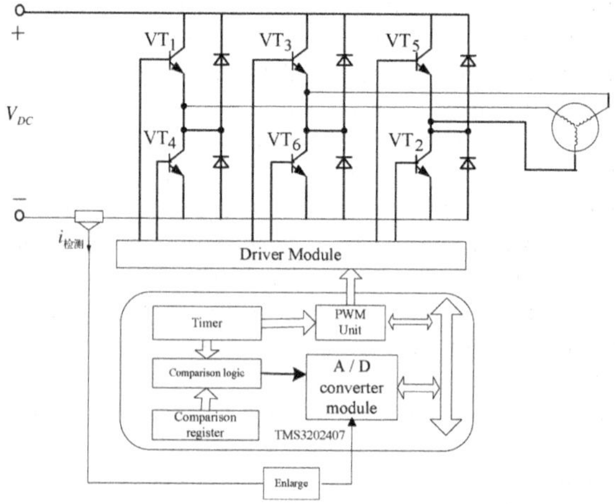

Fig. 4. Detection Circuit

In the experiment, two modules of DSP were used. First of all, the event management module A (EVA) produced a PWM signal, then the current signal from motor compared with the existing values to determine the rotor position after adopting the current difference through the analog-digital conversion block (ADC).

4 Experimental Methods

In this study, experiment was done combined with 70ZWN24-30 type motor of Yahequan company, which is a brushless DC motor with the internal sensor. So the exact rotor position can be obtained from the sensor, and then can compare with the value obtained in experiment. The parameters were shown in Table 2.

Table 2. Parameters of motor

Phase	Pole	Rated voltage (V)	Rated Current (A)	Rated Power (W)	Rated speed (r / min)
3	2	24	1.9	30	2450

When the sensor is sampling on current in each phase of the motor, the gain and offset error obtained are different. To avoid the problems it brings, we can installed a shunt resistor on DC bus (see Fig. 4). It will bring the same gain error and no offset error, which get the more precise current difference. Advantages of the design is that the phase current of a dc link in six working thyristors is equal to one of the phase currents or its opposite value precisely [5]. According to TABLE 1, we get:

$$\begin{aligned} i_{h'1} &= k \cdot i_{h1} + i_p \\ i_{h'2} &= k \cdot i_{h2} + i_p \end{aligned} \quad (5)$$

Here k and i_p are the current gain and offset error, in which k> 0. Computing the current deviation $i_{k,SAMPLE}$, we get:

$$\Delta i_{h'} = i_{h'1} - i_{h'2} = k \cdot (i_1 - i_2) = k \cdot \Delta i_h \quad (6)$$

Therefore, the adjustment error which may occur in the circuit measurement will not affect the accuracy of the calculated current difference. The gain error is same for the current difference of each phase, so it will not affect the judgment to rotor.

In order to enable the correct synchronization of the A/D converter and PWM waveform and improve the reliability of current sampling, PWM waveform need to delay 1us.

5 Results

The Fig. 5 below shows the current difference obtained from the above method. Ideal experimental results show that it can be enough to detect the current difference, and effectively prove the effectiveness of the proposed method. At the same time, it can ensure the motor to start in security mode.

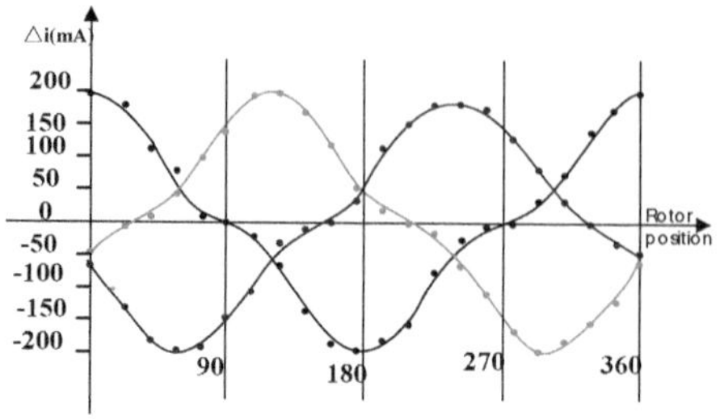

Fig. 5. Three-Phase Current Difference Relationship With The Rotor Position Changes

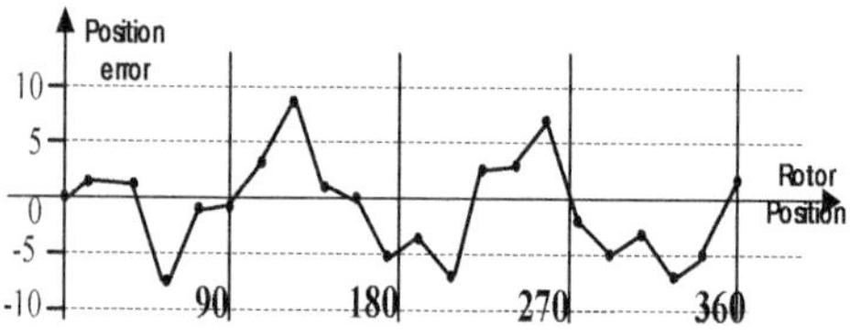

Fig. 6. Rotor Position Error

6 Conclusions

This paper presents a pulse injection method to detect the rotor position of permanent magnet motor in the static state. By using the changes of stator current when the magnet locates on different positions, the method can estimate the rotor position. It can also effectively avoid the transient reversal or start fails. At the same time, it is not only applicable to brushless DC motor but also to DC synchronous motor, and do not need to know the parameters of motor. Experimental results show that the estimation error is within the electrical angle of [-8 +9], so the motor can start in safe condition. As can be seen from Fig. 6, the estimation error is a certain regular. To continuously reduce the error, the program's algorithm can be further improved.

References

[1] Wang, R., Liu, Y.: Comparison and Research on Sensorless Brushless DC Motor. Micromotor (January 2003)
[2] Cardoletti, L., Dum, A., Jufer, M.: Indirect Position Detection at Standstill for Brushless DC and Step Motors. In: Proc. of EPE 1989, Aachen, pp. 1219–1222 (1989)
[3] Deng, C., Zhang, S.: A New Start Method of Brushless DC Motor. Micromotor (June 2002)
[4] Schmidt, P.B., Gasperi, M.L., Ray, G., Wijenayake, A.H.: Initial Rotor Angle Detection of Non-Salient Pole Permanent Magnet Synchronous Machine. In: Proc. IEEE Industry Application Society Annual Meeting, New Orleans, pp. 459–463 (1987)
[5] Parasiliti, F., Petrella, R., Tursini, M.: Low Cost Phase Current Sensing in DSP Based ACDrives. In: Proc.of the IEEE ISIE 1999, Bled, Slovenia, vol. 3, pp. 1284–1289 (1999)

Multi-objective Optimization of Crane Luffing Mechanism Based on Gray Fuzzy Optimal Model

Fei Ye, Zhu Tianen, and Yu Haiyang

School of Traffic & Mechanical Engineering
Shenyang Jianzhu University
Shenyang, China
feiye_33@fastmail.fm

Abstract. Aiming at the problem that the traditional multi-objective optimization method is sensitive to the shape of the Pareto frontier and defect that can not guarantee to obtain Pareto decision solution, bases on the idea of system decision, combines the grey system theory and fuzzy comprehensive evaluation, establish grey fuzzy optimal model of Pareto solution and make a math demonstration about its correctness, while it gives the corresponding genetic algorithm. Project examples show that the model is practical and effective.

Keywords: hoist crane, luffing mechanism, genetic algorithm, multi-objective optimization.

1 Introduction

Generally a good design must embody the principle of the greatest benefit and least cost. In fact, optimum design is a multi-objective optimization problem. It has no unique optimal solution but has more satisfactory solution sets called Pareto optimal solution set [1].

At present, there are lots of methods to solve multi-objective optimal problems, they are divided into two categories: first, unified objective function method, integrating multiple objects into a new objective function to solve, obtain decision solution of the problem directly, it is a solution model that optimization searching after decision-making; second, Pareto set generation method, namely selecting decision solution of the problem according to some criterion after finding Pareto solution set of multi-objective optimization problem, it is a solution model that decision-making after optimization searching. Methods on the representative of two categories are linear weighting method and genetic algorithm based on vector evaluating mechanisms[2][3]. Linear weighting method for multi-objective optimization problem is simple and practical, but it is sensitive to shape of Pareto frontier in multi-objective problem, sometimes it discards decision optimal solution of Pareto frontier[4]; Non-inferior solution set generation method results in a Pareto solution set, it must make decision by virtue of preference information can we obtain optimal solution of engineering, actually this algorithm is complicated, it has been limited in engineering.

D. Zeng (Ed.): ICAIC 2011, Part II, CCIS 225, pp. 413–421, 2011.

From the above analysis we know that each of two methods has respective advantage, but they also have dissatisfaction to multi-objective optimization of engineering.

Therefore, we can try to combine their advantages, design a simple and practical solving method which can be obtained arbitrary decision solution.

2 Pareto Solution Set of Multi-objective Optimization Problem

A. The Mathematical Description of the Problem

The multi-objective optimization problem is usually stated as equation (1):

$$\begin{aligned} &\min \ F(x) = (f_1(x), f_2(x), \cdots, f_n(x)) \\ &s.t. \ \ g(x) = (g_1(x), g_2(x), \cdots, g_r(x)) \le 0 \\ &\qquad x = (x_1, x_2, \cdots, x_k) \in X \end{aligned} \tag{1}$$

In this equation, x is design variable, F is object vector, $g(x) \le 0$ denotes the feasible field of objective function.

B. Pareto solution of multi-objective optimization problem

It has given definition of Pareto solution of model (1) in literature[5][6]:If $\forall j$, $f_j(u) \le f_j(v)$ and $\exists j$, $f_j(u) < f_j(v)$, $j = 1, 2, \cdots, n$,u, $v \in X$. So we say that u is better than v . If there is no such a feasible solution u , we name v as non-inferior solution, valid solution, Pareto solution, $f_j(v)$ is called non-inferior solution of objective space. The Pareto solution set of multi-objective optimization problem is constituted with all non-inferior solutions, there is no superior solution that is better than other solutions in all objects.

3 Pareto Solution Fuzzy Optimal Model Based on Grey Correlation Analysis

The convenience of linear weighting method for solving multi-objective optimization problem is making decision beforehand,then it can obtain engineering decision solution by virtue of single-objective optimization algorithm directly; the non-inferior solution set generation method can avoid losing decision solution effectively. Therefore, it integrates linear weighting thought into the non-inferior solution set generation method, and make decision during the process of searching the non-inferior solution set. It involves a mass of fuzzy factors,grey information, this paper constructs the engineering decision model of Pareto solution set by grey system theory and fuzzy optimal method.

A. The grey fuzzy member ship of optimal model

The system is divided into white system, black system and grey system which between the white and the black system in accordance with the clarity of information that it expresses. Grey correlation is the basic concept of grey system, it means uncertain correlation between things, between the factors or factors to main action, we call it as grey correlation [7]. Grey system theory thinks that it is impossible that any two behavior sequences are non-associated strictly. Therefore, grey system theory which adopts correlation analysis, compares and judges micro or macro geometry proximity to the behavior sequences, it is quantitative analysis for the dynamic development state of behavior sequences.

From vector angle and combining model (1), The Scheme F_i is defined by n objects, to facilitate and call it short as: $F_i = (f_{01}, f_{02}, \cdots, f_{0n})$.

After determining value of n objects, the advantages and disadvantages of the scheme are also determined, so the scheme optimization will be converted to compare the advantages and disadvantages of each object. The comparative sequence of grey correlation analysis is composed of sub-object values within scheme to be selected.

It must build a consensus sequence F_0 to compare the schemes, Scheme optimization has relative of comparison, therefore , it can define an imaginary ideal scheme, each attribute index of this scheme is optimal value of corresponding index which is in all the schemes to be selected. Namely characterizing ideal scheme's index sequence:

$F_0 = (f_{01}, f_{02}, \cdots, f_{0n})$, among:

$$f_{0j} = \begin{cases} \vee_j f_{ij} & \text{the index more greater the better} \\ \wedge_j f_{ij} & \text{the index more smaller the better} \end{cases} \tag{2}$$

By doing correlation analysis to comparative sequence and the ideal consensus sequence of the scheme, obtains the correlation coefficient that plan index relative to its ideal value [7]:

$$\gamma_{ij} = \frac{\Delta_{\min} + \zeta\Delta_{\max}}{\Delta_{ij} + \zeta\Delta_{\max}} \tag{3}$$

in this formula:

$$\Delta_{\min} = \min_{i\in m} \min_{j\in n} \left| f_{0j} - f_{ij} \right|$$

$$\Delta_{\max} = \max_{i\in m} \max_{j\in n} \left| f_{0j} - f_{ij} \right|$$

$\Delta_{ij} = \left| f_{0j} - f_{ij} \right|$,$0 \le \zeta \le 1$ as identification coefficient, generally take it be 0.5

The implication of type (3), after considering influence that each object in all schemes to be selected to i scheme. For j object, this scheme maybe reaches the degree of the ideal scheme. Therefore, it is also measurement which some objects belong to ideal and optimal,this paper calls it as grey fuzzy membership degree object. It does not only considers the impact of each object itself, also considers that the correlation object has influence on its optimal membership degree. So it embodies the scheme's integrity.

B. Gray fuzzy optimal model

Literature [8] proposed one kind of fuzzy optimal model based on the scheme of weight derivation, The basic idea is to convert attribute of the scheme into the corresponding membership degree, subsequently according to the different levels of influence that various attributes to the scheme to establish weight vector and form the judgment matrix of the scheme; optimal membership degree of scheme which to be selected is defined as the interval power of the most ideal optimal scheme. Be determined by use of extreme value method, obtains the optimum order relation of various schemes according to this, the optimal membership degree calculation model of this method is:

$$u_i = \frac{1}{1+[(1-I_i)/I_i]} \tag{4}$$

In the formula: u_i is optimal membership degree of scheme to be selected i relative to the most ideal optimal scheme; $I_i = \sum_{j=1}^{n} \omega_j r_{ij}$, ω_j is weight of attribute. r_{ij} is membership degree of various attribute of scheme which to be selected i relative to respective optimum index.

According to Literature [8], r_{ij} just depends on eigenvalue of objective, but noting to do with other objective, therefore it only reflects its own and partial information, This article uses the objective correlation coefficient obtained by type [3] as a gray fuzzy membership degree instead of membership degree of Literature [8], introduces the fuzzy optimal model type(4), namely becomes fuzzy optimal model based on the gray correlation analysis.

It should be noticed that different objective attribute to different universe of discourse generally, therefore, first we must eliminate impact of dimension and unify measure scale, namely carries on standardized processing to the objective value, guarantees the equivalence of the judgment, completion of its standardization by the following formula:

$$f_{ij}^0 = \begin{cases} f_{ij}/f_{0j} & \text{the index greater the better} \\ f_{0j}/f_{ij} & \text{The index smaller the better} \end{cases} \tag{5}$$

It denotes agreement that object relative to the ideal index, forms the standard ideal consensus sequence and comparative sequence, they are:

$$\begin{aligned} F_0^0 &= (1,1,\cdots,1)_{1\times n} \\ F_i^0 &= (f_{i1}^0, f_{i2}^0, \cdots, f_{in}^0) \end{aligned} \tag{6}$$

According to above can we compute gray fuzzy membership degree of various schemes objects, Form the judging matrix of schemes which to be selected $\gamma = (\gamma_{ij})_{m\times n}$ .thus, original multi-objective optimization problem(1) is converted into ask the biggest x^* of u_i in type(4), calls it as Pareto solution of question(1).

4 Based on Grey Fuzzy Multi-objective Genetic Algorithm Optimization

Optimization model (4) presented in this paper, the state of function is not continuous, traditional gradient optimization algorithm cannot be used. The genetic algorithm is a global optimization method which is developed based on the theory of biological evolution and the probability search technology in recent years[4], because of only using the evaluation information based on the objective function value during searching process, that it dose not requires derivative information of objective function offers possibility for the global optimization model(4).

A. Decision Operator

An important feature of the genetic algorithm is the operation of population. This makes it will appear a number of feasible solutions in each generation during the evolutionary process. Therefore, Grey fuzzy optimization model is introduced into a genetic algorithm as operator, so it is convenient to construct attributes sequence of scheme to be selected by population individual of every generation. Calculate membership degree relative to the ideal optimal scheme as the fitness function of the genetic algorithm, taking it as a navigation through repeated selection, crossover and mutation. Obtain x^* which make u the largest as Pareto solutions of the original problem. We call the operator as decision operator. This is the gray fuzzy optimization based on multi-objective genetic algorithm.

B. Algorithm Realization

(1) Calculate the optimal value under the constraints of single-object, to construct the ideal consensus sequence, and make standardized process to the sub-objective function in accordance with the formula(5);

(2) Coding: Using floating-point code;

(3) Initial population: Randomly generated; Scale m;

(4) Introduce operator, calculate membership degree u_i of different individuals of population according to equation(4);

(5) Selection: select individuals of the size m by use of roulette wheel;

(6) Crossover: Individual that is selected crossover according to crossover probability to generate a new individual;

(7) Variation: Individual that is selected crossover according to vary probability to generate a new individual;

(8) Introduce new generation individual which decision operator iteracts(5)(6)(7) to calculate relative membership degree;

(9) make a judgement of conclusion criterion, If satisfied then stop, or turn (4).

We proposed genetic algorithm based on gray fuzzy optimization model in this paper, As each generation of evolution select the best in all populations through decision operators, thereby avoiding the problem of Pareto isostatic surface convexity and concavity of multi-objective optimization problem, it can still obtain the decision optimal solution of Pareto isostatic surface though probability1.

5 Engineering Example-Multi-objective Optimization of Pulley Block Compensation of Crane Luffing Mechanism

A. Optimization model

Pulley block compensation device of crane demand lifting load move along the horizontal as possible in the process of luffing, decreasing additional inertial load in the elevation direction and variable amplitude energy consume. Therefore, the design of luffing mechanism of pulley block compensation can attribute to the following multi-objective optimization problem[9]:

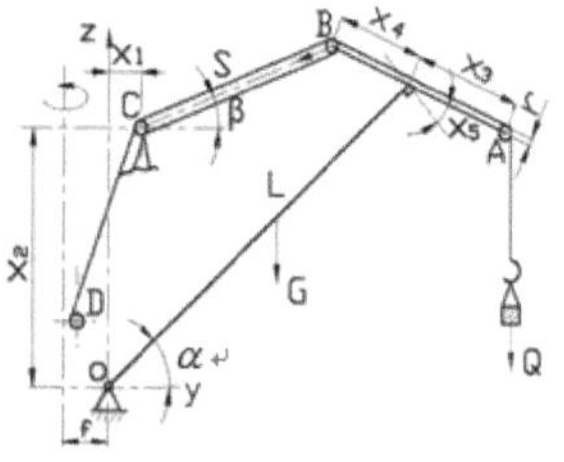

Fig. 1. Derricking mechanism use compensative pulley block

$$X = [x_1, x_2, x_3, x_4, x_5]^T$$

$$\min f_1(X) = \max \Delta h_\alpha - \min \Delta h_\alpha$$

$$\min f_2(X) = \max |d[\Delta h_\alpha] / dt|$$

$$\min f_3(X) = \int_0^T P_t dt$$

s.t. $-0.4 \le x_1 \le 0.5$

$3 \le x_2 \le 8$

$0.3 \le x_3 \le 0.8$

$0.4 \le x_4 \le 0.7$

$0 \le x_5 \le 25°$ $5.8 \le L\cos\alpha + x_3 \sin(\alpha + x_5) + f + r \le 12$

$\Delta h_\alpha = \Delta z_\alpha - \Delta l_\alpha$ $\Delta z_\alpha = L(\sin\alpha - \sin\alpha_1) + x_3[\cos(\alpha_1 + x_5) - \cos(\alpha + x_5)]$

$l_1 = \sqrt{a^2 + b^2}$ $l_\alpha = \sqrt{c^2 + d^2}$

$a = L\cos\alpha_1 - x_4 \sin(\alpha_1 + x_5) - x_1$ $b = L\sin\alpha_1 + x_4 \cos(\alpha_1 + \alpha_5) - x_2$

$$c = L\cos\alpha - x_4\sin(\alpha + x_5) - x_1$$
$$d = L\sin\alpha + x_4\cos(\alpha + x_5) - x_2$$
$$P_t = \omega Q[L\cos\alpha + x_3\sin(\alpha + x_5) + r]$$
$$-\omega Q(z_B\cos\beta - y_B\sin\beta)m_b / m_q + \omega GL\zeta\cos\alpha$$

G: boom deadweight, ζ is the distance between the boom center and down hinge of O boom and the ratio of arm length.

B. The solution of Optimization model

By virtue of genetic algorithm[10] based on structural parameters and related literature, we can get the three single-objective optimal solution. as if table 1, according to this can we construct the ideal scheme index sequence $F_0 = [0.01479, 0.00218, 2.946]$

Note: optimal single-object $f_1(x)$ is obtained from optimization for this paper, while optimal single-object $f_2(x)$ and $f_3(x)$ are from the literature [11].

According to grey fuzzy optimal model(4), optimizing by adopting adaptive genetic algorithm based on structure and parameters, three decision optimal solutions of objective optimization can be got.

In the literature [12], it turned the three-objects optimization into a single-objective problem, which by virtue of fuzzy nearness, and obtained decision solutions with improved genetic algorithm; It solved with game decision analysis in the literature [11]. This optimal solutions obtained by the three optimization methods are listed in the table 2:

Table 1. The optimal solution of single-objective optimization

objective function	Design variables					target value
	x_1/m	x_2/m	x_3/m	x_4/m	x_5/(°)	f(X)
f_1(X)	0.0566	4.1004	0.6179	0.3151	22.1735	0.01479 m
f_2(X)	-0.06318	4.10981	0.71116	0.57573	19.5424 2	0.00218 m/s
f_3(X)	0.00954	4.83460	0.77453	0.57673	17.8785 5	2.946 kJ

Table 2. The search result of different optimization methods

	Design variables					Optimal target value		
	x_1/m	x_2/m	x_3/m	x_4/m	x_5/rad	f_1/m	f_2/m/s	f_3/KJ
Literature [12]	0.02911	4.11177	0.63089	0.42633	0.42864	0.01973	0.0031 9	31.68 6
Literature [11]	0.12127	4.08723	0.40340	0.30002	0.43625	0.01773	0.0031 9	33.02 7
This paper	0.0555	4.1018	0.6218	0.3071	0.3902	0.01763	0.0031 2	31.94 8

According to the table 2, the solutions are obtained from algorithm in this paper are better than that in the literature [11] in three goals; compared to the literature [12] f_1 decreases by 10.64% and f_2 decreases by 2.19%, while the third objects f_3 increases slightly by 0.83%. Overall the solution in the paper is better.

6 Relevant Conclusions

This paper from the perspective of the system decision, combine fuzzy and gray system theory, establish a optimal model of multi-objective optimization question Pareto solution, according to this we proposed the genetic algorithm that fit to the engineering application, it has these advantages as following:

(1) With the aid of fuzzy theory to define this fuzzy concept that objects subordinate optimal, describe grey coupling relationship of sub-objects by use of grey system theory, make the model not only consider local characters of sub-objects, also reflect the overall characteristics of the object space, so it ensures validly coordination between sub-objective optimization and multi-objective optimization;

(2) This model has no special requests to the number and nature of sub-objects, it is generally applicable.

(3) For complex multi-objective optimization problem, the most effective method is the combination of human and computer, this model not only considers subjective preference of humans(use ω_j), also pay attention to the objective status of the problem itself(use γ_j), thereby it gives the proper interface of human-computer combination.

References

1. Cours, P.V.: D' Economic Politique, vol. I and II. F. Rouge, Lausanne (1896)
2. Ritzel, B.J., Wayland Eheart, J., Ranjithan, S.: Using genetic algorithms to solve a multiple objective groundwater pollution containment problem. Water Resources Research 30(5), 1589–1603 (1994)
3. Fujita, K., Hirokawa, N., Akagi, S., et al.: Multi-objective Optimal Design of Automotive Engine Using Genetic Algorithm
4. Zitzler, E.: Evolutionary Algorithms for Multiobjective Optimization: Methods and Applications. Swiss Federal Institute of Technology, Switzerland (1999)
5. Deb, K., Pratap, A., Agarwal, S., Meyarivn, T.: A Fast and Elitist Multiobjective Genetic Algorithm: NSGA-II. IEEE Transactions on Evolutionary Computation 6(2), 182–197 (2002)
6. Collo, C.A.C.: Handling Preferences in Evolutionary Multiobjective Optimization: A survey. In: 2000 Congress on Evolutionary Computation, vol. 1, pp. 30–37. IEEE Service Center, Piscataway (2000)
7. Sifeng, Guo, T., Dang, Y.: Grey system theory and application. Science, Beijing (1999)
8. Chen, S.: Project hydrology water resources system fuzzy set analysis theory and practice. Dalian University of Technology, Dalian (1998)
9. Sun, G.: Optimal design and application. China Communications, Beijing (2000)

10. Fei, Y., Li, N., Zheng, X.: Improved genetic algorithm based on structure and adaptive parameters. Shenyang University of Architecture 22(2), 338–340 (2006)
11. Xie, N., Fang, H., Bao, J.: The application of Game decision analysis in multi-objective design of compensative sheave block luff mechanism. Mechanical Strength 27(2), 202–206 (2005)
12. Xie, N., Wang, Q.: The Multi-objective fuzzy Optimization of pulley block compensation luffing mechanism based on genetic algorithm. Mechanical Transmission 27(1), 30–32 (2003)

Design of New Marine Environmental Data Acquisition and Transmission System*

Hailiang Zhang, Jingbiao Liu, and Haibin Yu

College of Electronic Information
University of Hangzhou Dianzi
Hangzhou, Zhejiang Province, China
zhanglin994@tom.com

Abstract. A new marine environmental data acquisition and transmission system that can monitor multiple environment parameters of an ocean fixed-point vertical profile is described. The system collects, archives marine profiling multi-parameter data and transmits them to Surveillance center through satellite communication. It consists of Buoy Electronic and Communication subsystem, Magnetic Coupling Communication subsystem, Underwater Data Acquisition and Storage subsystem. This paper mainly describes the detailed information about the key technologies, including multi-serial port communication technology, method of multi-source data acquisition, underwater inductive coupling data transmission technology and technology of low power consumption design, also information about system's hardware and software design is given. Marine experiment results show that this system can works well in less than 300m depth, due to the careful low power consumption design, the system can continuously work for more than 90 days, the magnetic coupling data transmission subsystem can reliably transfer data at rate less than 9600bps through Serial RS232.

Keywords: marine environment monitoring, multi-parameter data acquisition, magnetic coupling, low power consumption.

1 Introduction

At present, the situation of Chinese marine environment monitoring is not optimistic: the amount of ocean environmental data is small, the method that using data of small sea area forecast situation of large sea area is not reliable, and the accuracy of ocean forecast field is lower. The capability of real-time observation data collection, calculation, multi-source data transmission and comprehensive processing is not enough [1]. Most products that analyze and forecast marine information are mainly based on hydrometeorological information, lacking diversity in content. The means of underwater communication are mainly conventional wired communication which is actually not fit in water, or expensive sonar system, lacking lower cost and no-contact

* This work is supported by Science and Technology Department of Zhejiang Province as one project of technology priority topics (2008C13056).

D. Zeng (Ed.): ICAIC 2011, Part II, CCIS 225, pp. 422–429, 2011.

data transmitting technology that is suitable for water environment. Therefore, it is necessary to research and develop a new ocean environmental data acquisition and transmission system, it should have the following features: it can automatically, continuously collects data in a fixed point of marine area without the control of people in a long time, it can communicate with multiple sensors; supports multi-task operation; possess the ability of real-time acquisition and processing data; has the ability of underwater no-contact data transmission.

2 Overview of System

A. Overall Design Requirements

In view of the system's features described above and its special working conditions, the issues that need to be solved in system's hardware and software design can be summarized as below:

1) The system is designed to monitor multi-parameter of marine environment, so it should have the capability of connecting with multiple measuring sensors and large local storing capacity.

2) Realize reliable no-contact data transmission between underwater data acquisition subsystem and buoy platform communication subsystem.

3) Define and realize a high effective and reliable protocol for the underwater no-contact communication subsystem.

4) Reduce system's power consumption from hardware design and software development.

5) Select a proper communication means between ocean buoy station and Surveillance center.

When designed, the whole system is divided into three parts, underwater control module, underwater data transmission module and buoy platform control module. This method of modular design in hardware and software has many advantages, the whole system's structure and each subsystem's function is clearer; design is more simple.

B. Overview of System's Operating Principle

This system is applied in device of wave-driven vertical profiler for observations ocean elements. Its overall structure's schematic diagram is shown in figure 1. It is mainly composed of an underwater data collection and storage module, two magnetic coupling data transmission modules, several sensors for marine environment, a solar energy panel and a buoy platform of satellite communication subsystem.

The system's working process can be described as following. When system starts to work, the underwater data collection subsystem automatically goes down with the driven of wave-power. As it reaches seabed or depth that has been set before test, the unidirectional-switch is opened and the underwater instrument compartment goes up with the float's buoyancy force. During the process of the subsystem's floating up, the underwater subsystem collects sensors' data. After processed (get meaningful data from packaged data or acquire analog data and convert), all sensors' data is stored into SD card or U-disk. When the underwater device goes up near the fixed sea level, it stops the task of collecting data and tries to get contact with the buoy platform communication subsystem. During this process, it will first wake up the buoy platform subsystem which is in sleep mode before, and then the underwater subsystem fixes itself by closing unidirectional-switch and communicates with the buoy platform subsystem through the Magnetic Coupling Data Transmission Subsystem. After all data transferred, the subsystem will again automatically go down, repeating the above process. At the same time, the buoy platform communication subsystem transfers the sensors' data real-time to the monitoring department's superior machine through satellite communication.

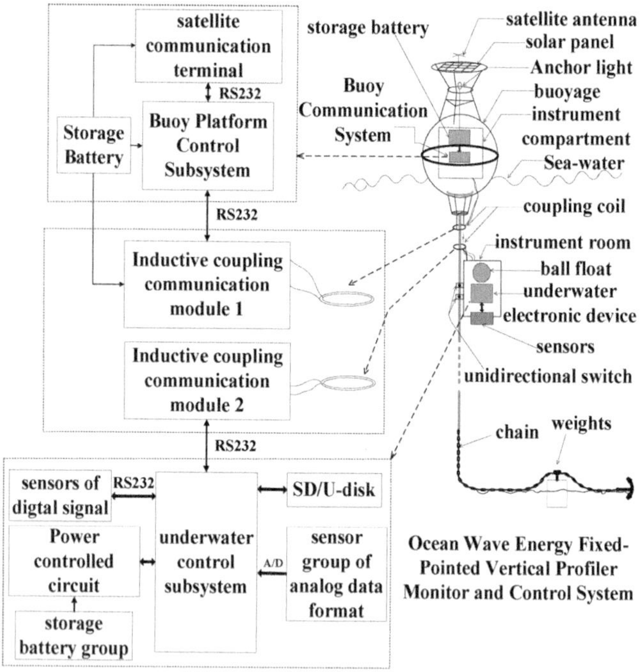

Fig. 1. Schematic diagram of system

3 Hardware Design

A. Data Collection and Storage Module

In order to monitor ocean environmental parameters as more as possible, the subsystem should connect with multiple measuring sensors (nutritive salt, temperature, salinity, turbidity, PH, chlorophyll, etc). The format of sensors' output data includes analog signal which is usually voltage and digital signal which uses serial communication. Besides we need one more serial port to connect with the magnetic coupling communication subsystem and a debugging serial interface. The system has to use a step motor to control the unidirectional-switch, it needs several I/O ports.

In the design, the underwater subsystem takes 32-bit RISC microprocessor AT91SAM9260 as controller core. Its processing speed reaches as high as 200MIPS when it works in 180MHz. It is rich in system resources and peripheral interfaces, including 7 UARTs, 4-channel 10-bit Analog-to-Digital converters, three 32-bit Parallel Input/Output Controllers and 96 Programmable I/O Lines. Besides it supports SD card interface without capacity limit and it also supports USB 2.0 full speed. This chip is very suitable for applications in certain occasions such as data collect, storage and procession. The structure of the data collection and storage subsystem is shown in figure 2.

The power manager circuit which adopts optical coupler greatly reduces power consumption. It controls the on/off of the sensors' power supply switch, when not collecting data, it shuts down the power supply which reduce lots of power consumption, also the subsystem can monitor voltage condition of Power Supply and packages the information together with acquired data and finally be send to monitoring station. It is a very important feature of this subsystem.

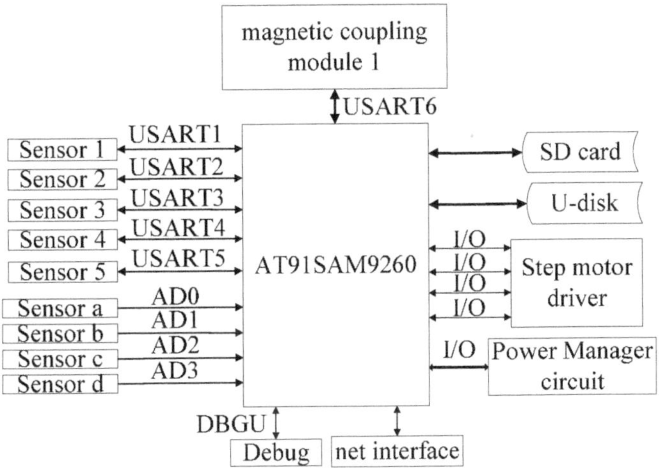

Fig. 2. Acquisition control module

B. Magnetic Coupling Communication Subsystem

This subsystem is mainly designed for small-sized underwater transmission system. Compared with the expensive sonar system and wired communication which is not fit for the marine environment, the magnetic coupling technology has lots of advantages, such as strong adaptability to environment, communication is reliable, cost is low, hardware circuit is simple and low power consumption etc [2]. It is composed of a MCU; ASK modulation-demodulation circuit and magnetic coupling coils. This module takes STM32F101C6XX as the microprocessor, it is based on high-performance ARM Cortex-M3 32-bit RISC core operating at a 36 MHz frequency, A comprehensive set of power-saving mode allows the design of low-power applications. It integrates two USARTs and it only has 48 pins which makes the circuit simpler. It can be seen from figure 1 that system needs two magnetic coupling communication modules. They are completely same in the hardware and software. The schematic diagram of hardware is shown in figure 3.

For the transmitting circuit, signal that comes from serial's output pin goes into the amplitude modulation circuit, then the signal is loaded into carrier wave of 200khz, after be amplified by the power amplifying circuit, the signal that be modulated will finally be send by the resonant coupling coil [3].

For the receiving circuit, the inductance coil and the circuit will generate voltage if the near electromagnetic fields make changes. The weak voltage signal will be amplified by the amplifier circuit. Then the useful signal will be separated from the mixed signal after it comes through ASK non-coherent demodulation circuit. Finally the data comes into the serial's input pin.

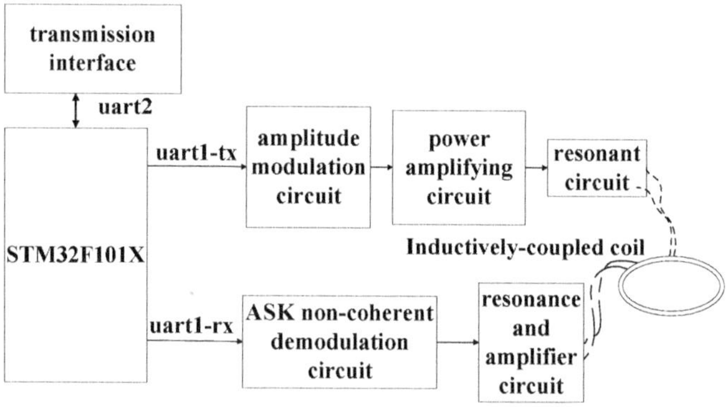

Fig. 3. Schematic diagram of Inductively-coupled communication

4 Software Development

A. Data Collection and Storage module Software development

The data collection and transmission module uses Linux as operation system. Linux requires a little resource of hardware, runs stably, high efficiency, and it can give full play to the advantages of hardware, Linux is very well-suited for embedded devices [4].

Before the software development, it needs to build the cross-compile development environment of embedded Linux. After finish developing applications, it can begin the transplanting work, put the compiled application program files into the root file system that have been constructed before, and then the root file system together with the compiled Linux Kernel is programmed to the flash of AT91SAM9260. Now the work of transplanting Linux is finished [5], after the ARM chip restarts, the system will run the applications.

The execution procedures of the data collection and storage module can be described as follow:

1) Initialization and configuration

2) In order to reduce power consumption, turn off the power of some sensors in the process of going down.

3) When reach the seabed or depth that has been set before test start, turn on the sensors' power supply, during the device's floating time, collect, process and store data.

4) As reach near the water level, stop the collection work, power off, contact with the buoy platform subsystem and transfer the data.

5) After transmit all data, the data collection and transmission will go down together with instrument room and then repeat the above process.

B. Develop Software of Magnetic Coupling Communication Module

As the resonance circuit of receiver and sender uses the same inductance coil, when sends data, the electromagnetic field will generate self-inductance on the inductance coil, it will influence the receiver circuit. In order to eliminate this Self-Sending and Self-Receiving phenomenon, this paper uses programmer to control the process of sending and receiving, simulates the half duplex communication mode. The direction of data transmission is as follow:

1) USART2-Rx→USART1-Tx →ASK modulation circuit.

2) ASK demodulation circuit →USART1-Rx →USART2-Tx.

The main program mainly does some initialization work, then configures the two serials, enables their receiving interrupts, finally it goes into closed loop. Data transmitting and receiving is done in the interrupt service program of serial. Figure 5 shows flow chart of the interrupt service program of serial one. It is same with flow chart of serial two. It can be seen from hardware and software that the Inductive Coupling Data Transmission module does no extra treatment to the data of transmission, it realizes transparent transmission for data.

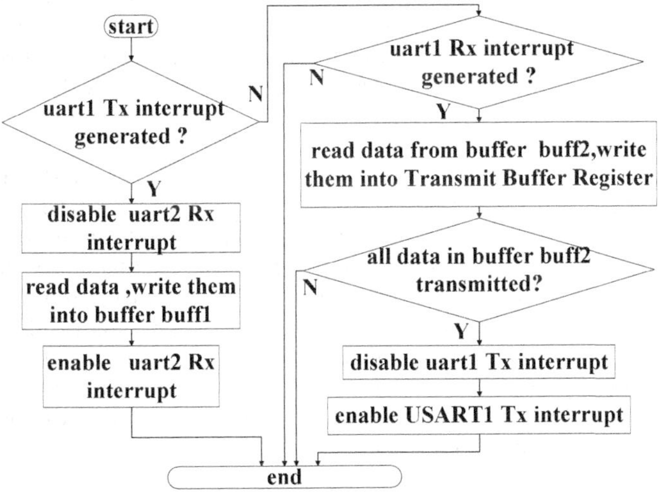

Fig. 4. Flow diagram of magnetic coupling module's Program

5 Results and Conclusion

After completed the instrument's production, system's initial debugging was performed in laboratory, including tested system's overall performance and measured its power consumption. Then from August 2010, the system's sea performance tests was started in an Ningbo's offshore area, where the water depth is about 150m, wave height is 25cm-35cm, flow velocity is 2m/s-3.5m/s. Before starting device, set up collecting depth ranges to 1m-100m, sampling frequency to 1s/times. There are three sensors, a SBE 49 FastCAT CTD (RS-232 Interface), a SBE-43 Dissolved Oxygen Sensor (voltage output) and a Seapoint Turbidity Meter (voltage output) that mounted to the system. Several groups of data that collected and processed by the system are given in Table 1.

Table 1. Sampling Data

Depth(m)	Temperature(°C)	Salinity(PSU)	DO(ml/l)	Turbidity(FTU)
5.065	16.9863	4.54377	2.9258	0.0901
10.847	17.0356	4.55507	2.9777	0.0802
15.006	17.0433	4.55626	2.9956	0.0801
51.394	17.0512	4.55892	2.9901	0.0811
93.062	17.0465	4.55998	2.9957	0.0802

Table 2 shows the system performance which comes from experimental results.

Table 2. System Performance

	Performance
Underwater Subsystem	5-channel available sensor RS-232 4-channel sensor A/D interfaces (10-bit) Data Storage: SD-card(U-disk) without capacity limited
Magnetic Coupling Communication Subsystem	half duplex no-error communication conditions: baud rate <= 9600bps coils' distance <= 0.2m axial angle <= 50°
Overall System	average power consumption: underwater subsystem 5.54w buoy platform subsystem 5.76w Depth Capability < 300m continuous work time > 90 days

In conclusion, this system has multiple RS-232 Interfaces and A/D interfaces available to sensors, it can mount various sensors according to actual situation, it adopts Iridium Satellite to transmit remote data and supports acquired data local preservation. The no-contact magnetic coupling data transmission can well meet requirement of system's underwater small-sized data transmission. The low power consumption design of system's hardware and software makes the device continuously work for long time as possible without maintenance. In addition, this system can be easily applied to other water environment monitoring device after a little modification on software and has good scalability and integration.

References

1. Ma, Y.: Overview of marine observation and forecast System for china. Marine Forecasts 25(1), 31–40 (2008)
2. Chen, Q., Bu, W., Liu, F.: Application and Research of ICL in the Offshore Wireless CPT Beneath Ocean. Science Technology and Engineering 8(12), 3110–3114 (2008)
3. Troyk, P.R., DeMichele, G.A.: Inductively-Coupled Power and Data Link for Neural Prostheses using a Class-E Oscillator and FSK Modulatio. In: Proceedings of the 25th Annual International Conference of the IEEE EMBS, Cancun, Mexico, September 17-21 (2003)
4. Huang, P.: Design and Implementation Of Linux Platform For an Embedded Industrial Controller System. University, Xian (2008)
5. Sun, P., Peng, P., Zhang, Y.: Linux Transplantation Based on The Processor S3C2440. In: 9th International Conference on Electronic Measurements & Instruments, ICEMI 2009 (2009)

A Key Management Scheme for Secure Communications of Advanced Metering Infrastructure

Jinshan Chen, Nian Liu, Wenxia Liu, and Hong Luo

School of Electric and Electronic Engineering
North China Electric Power University
Beijing 102206, China
j.s_che@yahoo.cn

Abstract. Key management for secure communications of Advanced Metering Infrastructure is an unsolved problem until now. The difficulties and their basic solutions for AMI to choose a key management scheme will be summed up after the studies of AMI system structure, business process and cyber security requirements, proposing a key management scheme based on a hybrid-tree graph for AMI. The cyber security, including key security, key freshness, forward & backward security, etc., can be satisfied by the presented scheme, while the overhead of the scheme is reasonable. Based on the analysis, the presented scheme is a possible solution for AMI.

Keywords: Advanced Metering Infrastructure, Hybrid-tree graph, Key Management, Smart Grid.

1 Introduction

As a new emerging technology for smart grid, Advanced Metering Infrastructure (AMI) provides an intelligent connection between consumers and power system operators. AMI is tightly integrated with distribution automation system and meter data management system. Besides, open network and information techniques have been introduced to AMI, increasing cyber security threats. Therefore, cyber security of AMI shouldn't be ignored for the construction and operation of smart grid.

In general, the security requirements of AMI include confidentiality, integrality and availability [1]. The confidentiality and integrality can be solved by encryption and authentication, which depend on the security of cryptograph key. Therefore, the key management is important to secure the communications of AMI. There have been many Key Management Schemes (KMSs) proposed for secure communications of power control systems, such as SCADA system [2-4]. But until recently, there is not any reported KMS for AMI systems. Because of the comparability both in system structure and process, the KMS used in general IT systems can not be directly applied in AMI.

According to the topology of key management systems, there are three types of KMS: centralized, distributed and hierarchical approach [5]. Modifications have been presented to improve the three original schemes, all of which depend on the specific system requirements.

D. Zeng (Ed.): ICAIC 2011, Part II, CCIS 225, pp. 430–438, 2011.

In order to ensure secure communications in AMI, a novel KMS based on a hybrid-tree graph is proposed in this paper. The rest contents of this paper will be organized as bellow. Section II studies the structure and business process of AMI, and briefly analyzes the difficulties for key management; In section III, a novel KMS for AMI is introduced in details; Performance of the presented KMS will be analyzed and calculated in section IV; An example is presented to deepen understanding in section V; Finally, a conclusion is given in section VI.

2 The Chosen of KMS for AMI

A. System Characteristics of AMI

AMI is a real-time network processing system, which can not only measure, acquisition, store and deal with users' load data, but can also manage demand-response (DR) projects.

(1) System Structure of AMI

AMI is composed by smart meter (SM), home area network (HAN), user gateway (UG), data concentrator(DC), data concentrator, communication network and metering data management system (MDMS) [6]. Fig. 1 shows the structure of AMI.

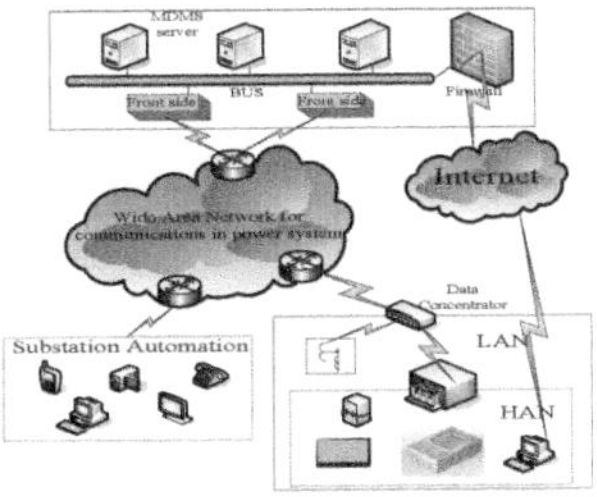

Fig. 1. The structure of AMI

SM is a typical embedded system; HAN is a kind of Local Area Network (LAN), which combines SM with UG and smart appliances; UG, generally deployed in other devices such as PC and SM, is the bridge across inner and outer networks.

Although the communication medium used in AMI is uncertain yet, the communication process is based on TCP/IP according to IEC.

(2) Function Layers of AMI

According to function characters, member devices in AMI can be attracted into three layers. The user side is the bottom layer, which contains SM, UG, SG, transformer meter, feeder meter and so on (all of these are denoted as SX for simple); Data concentrator (DC) is the middle layer; The topmost layer is MDMS, denoted as MS for simple.

(3) Businesses in AMI

Devices in the three layers transmit interactive electrical data through communication network. Business types are shown in table I with their data volume and time requirement.

All the transmitted businesses can be divided into two classes according to transmission mode: unicast and multicast, which are also shown in table 1.

Table 1. AMI business types

No.	Business Type	Transmission Mode	Data volume	Time Requirement
1	Metered data	Unicast	Large	Low
2	Electricity price	Multicast	Large	Low
3	DR distribution	Multicast	Large	Low
4	DR response	Unicast	Large	Low
5	Alarm	Unicast	Small	High
6	Control	Unicast	Small	High

(4) Communication Model and Characteristics of AMI

Till now we can abstract the communication model that is shown in Fig.2 from the communication function of AMI to simplify the analysis.

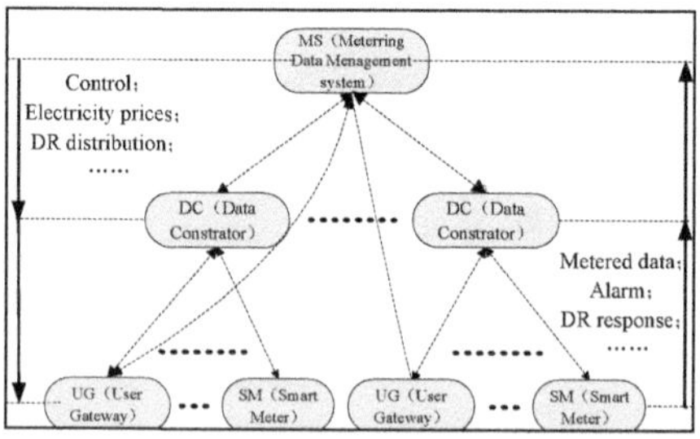

Fig. 2. AMI communication model

Besides, as is analyzed above, AMI system characteristics mainly incarnate:

1) The computing and storage ability of the nodes in bottom layer can't be equaled to a personal computer.

2) Both unicast and multicast businesses are required.

3) Some types of businesses require real-time transmission.

B. Suitable KMSs for AMI

Generally speaking, a KMS composes its key graph, key generating & refreshing policy, key distribution policy and storage method and so on.

What kind of KMS that is effective for AMI, is closely related to the specific architecture and business. In summary, problems as below may be met when choosing a KMS for AMI. Initial solutions will also be put forward.

(1) Key Graph

AMI sustains both unicast and multicast businesses, which have different requirements for information security. Forward and backward security is especially required for multicast services. KMSs suitable for AMI should meet the different requirements at the same time.

Key tree graph is quite suitable for both of them synchronously, in consideration of which, keys for multicast are easy to create and key-refreshing in nodes becomes very simple.

Besides, hierarchical key graph, which applies session keys and key-encryption keys at the same time, should be adopted to achieve well security and validity.

(2) Key Calculating

Key calculating is quite necessary both in key generating and refreshing policy. But it's a considerable limit for nodes in bottom layer to calculate keys.

On the premise of the security need satisfaction, cryptography algorithms that need less overhead should be preferred.

(3) Key Storing

According to a specific KMS, communication nodes need to store a certain number of keys and additional related information. Storage space of nodes in bottom layer is another considerable limit.

Taking calculating and storing ability into account, symmetric cryptographic algorithm is exactly the choice for AMI, security need being satisfied.

(4) Key Distribution

The traffic of key distribution should not affect the main businesses in AMI. On the other hand, the key distribution center (KDC) should not take too much distribution tasks, or it may easily become a single point of failure.

If some communication nodes can refresh keys themselves rather than receiving them from KDC, the main traffic will be less affected and KDC will less easily fail.

(5) Key Freshness

Key freshness is a very important measure to avoid replay attacks. In order to ensure the key freshness and forward and backward security, keys should be refreshed in time. Additionally, using old keys to calculate keys in use should not be successful.

Hash functions have a very good one-way feature and can be adopted for updating keys. Besides, the refresh period should be appropriate to ensure both system complexity and security at the same time.

3 A KMS Based on Hybrid Key Tree for AMI

In this paper, a novel KMS based on hybrid key tree has been proposed for AMI according to the analysis above.

In the presented scheme, a hybrid key tree which consists of a binary tree and multi-ary trees is built to support simultaneously unicast and multicast businesses, and to make key generating and refreshing become easy.

When there are keys need to be refreshed, some of them can be accomplished on the SX nodes' own without receiving from KDC. Therefore, the dispatching tasks can be reduced, preventing single point of failure and reducing the traffic in the network.

The presented KMS, which consists of these parts: the KMS architecture, key generating and renewing policies, will be introduced in details as below.

Table 2 shows the denotation and definition which will be used in the following context.

Table 2. Denotation and Definition

No.	Denotation	Definition
1	q	The number of DC nodes.
2	h	Height of the hybrid tree.
3	r	The number of SX nodes per DC nodes (r is a fixed value in this paper for simple).
4	N_i	The i-th node in the tree.
5	KP_{N_i}	The set of keys on the key path of N_i, from the parent node of N_i to the root.

A. Architecture of the Novel KMS

The hybrid tree, whose constructor method will be presented in details, is not only a key tree, but also a node tree and an additional value tree.

Firstly, all the AMI communication devices in the three layers can be constructed into a node set:

$$GM = \{N_1, \cdots, N_n\}, \text{ and } GM = MS \cup DC \cup SX$$

The number of SX nodes is ($n - q - 1$), and they are $SX = \{N_1, \cdots, N_{q*r}\}$; The number of DC nodes is q, they are $DC = \{N_{q*r+1}, \cdots, N_{n-1}\}$; The number of MS node in one area is only 1, it is N_n.

Secondly, a hybrid tree corresponding to GM can be constructed as shown in figure 4. $K_{0,1}$ corresponds to MS node; The leaves of the binary tree correspond to DC nodes; The leaves of the whole hybrid tree correspond to the SX nodes. All the keys on the tree are key-encryption keys.

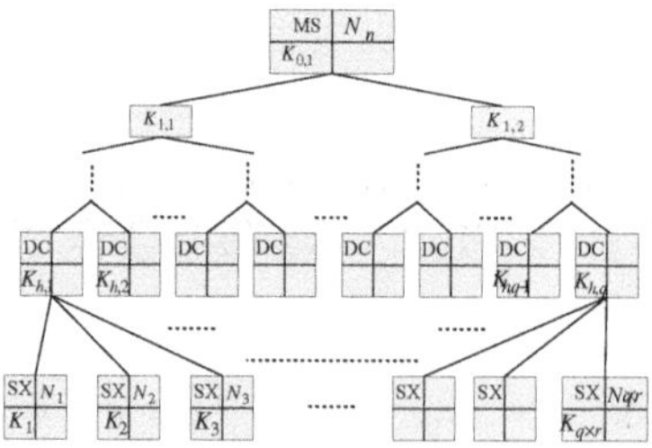

Fig. 4. Structure of the a hybrid key tree

Thirdly, the MS node stores all the keys on the hybrid tree; Every SX node stores all the keys on its key path. For example, K_1, $K_{h,1}$, $K_{h-1,1}$, ……, $K_{0,1}$ are stored in N_1.

B. Key Generation and Refreshing

In order to maintain key freshness and communication security of forward and backward, both the session keys and the key-encryption keys in use need to be updated in time.

C. Session Key Generation and Refreshing Policy

(1) Generation Policy

In the presented scheme, session keys' generation depends on a related key-encryption key (K) and a related additional value (C) using a hash function, that is:

$$SK = H(K, C)$$

One MS or SX node relates to one key-encryption key K and one additional value C which is a numeral with a fixed length. The same to key-encryption keys, MS stores the additional values of all nodes in the tree, and each SX stores the additional values of all nodes on its key path.

For unicast, we set the additional value as the sum of electric quantity in the former day and the number of message ever transmitted in the day; For multicast, the additional value C_c is set for business customers, C_i for industrial customers and C_p for residential customers.

When MS and node N_j need to communicate with each other, they generate session keys both in the two communication parties by formula $S_{0,j} = H(K_j, C_{K_j})$, in which K_j and C_{K_j} are related to N_j.

When MS wants to multicast messages to some SX nodes, the required session key are generated in both MS and $GM_{K_{i,j}}$ by formula $BS_{i,j} = H\left(K_{i,j}, C_{K_{i,j}}\right)$.

(2) Refreshing Policy

All session keys should be replaced after each session to guarantee the keys' freshness. The refreshing policy is as below:

For unicast business, the related additional value is set to be automatic refreshed.

For multicast business, the related additional value will be added 1 in all the communication parties respectively after each session.

D. Key-encryption Key Generation and Refreshing Policy

When a key-encryption key $K_{i,j}$ needs to be updated, all keys in $KP_{N_{i,j}}$ should be updated, too. The related work which should be done in each node is as below:

1) KDC updates the keys in $KP_{K_{i,j}}$;

2) The SX nodes that have keys in $KP_{K_{i,j}}$ should refresh them by themselves, or receive new keys from KDC.

(1) Generation and Refreshing Periodically

Key-encryption keys can be generated or periodically refreshed by $K' = Kgen(1^b)$, which is a b-bit security random key algorithm.

(2) Dynamically Refreshing

When there is a node want to join in or quit the group, all the keys on its key path should be refreshed, with the method which will be discussed in further researches.

4 Performance of the Presented KMS

A. Cost for Storage and Calculating in Nodes

Each SX node needs to store all the keys on its key path and their related additional values. The number of keys and values are both ($\lceil \log_2 q \rceil$+2).

If O_E denotes one time calculating cost for $E_K(D)$, O_g denotes a random key producing, and O_H denotes the cost for executing $H(D)$ for one time, calculating costs at different conditions are showed in table 3.

B. Cost for Key Distribution in the Network

Table 3 shows the performance of the algorithm, with the denotation of O_T for the cost of one time key transmission.

Table 3. Algorithm property of a key management scheme Based on a hybrid key tree

Resources used		KDC	SX node
Storage cost	Number of keys	$n+q-2$	$\lceil \log_2 q \rceil + 2$
	Number of additional value	$n+q-2$	$\lceil \log_2 q \rceil + 2$
Calculation cost	Session key producing	O_H	O_H
	Refreshing periodically	$nO_g + nO_E$	O_E
Transmission cost	Session key producing	0	
	Refreshing periodically	nO_T	

5 Case Study

A. An Example of the Novel KMS

There are one MS node, 12 DC nodes and 60,000 SM nodes in a city. Without loss of generality, the values of O_g, O_H, O_E and O_T are respectively 1ms, 10ms, 10ms and 256 bits in this example.

(1) Hybrid tree Construction

All the nodes in the city can be constructed into the node set:

$$GM = MS \cup DC \cup SX = \{N_1, \cdots, N_{60000}, N_{60001}, \cdots, N_{60012}, N_{60013}\}$$

The hybrid node tree graph is as shown in Fig.5.MS generates the key-encryption keys related to all the nodes using $K_i = Kgen(1^b)$, and distributes them in a secure way. In this paper, we consume that b=128.

Finally, all the additional values, the length of which is 56 bits, are randomly generated in MS node, and then distributed to their related nodes in a secure way.

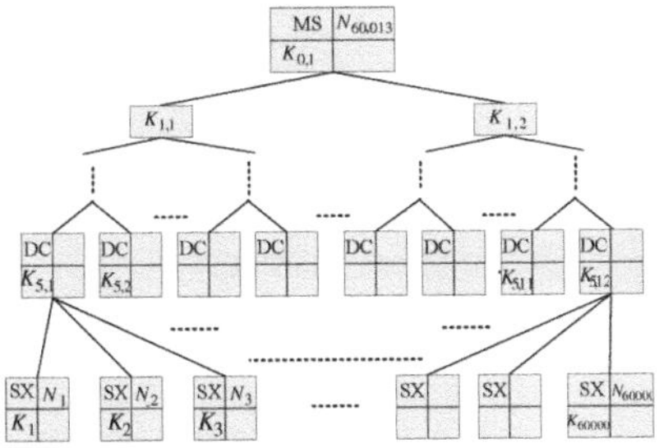

Fig. 5. The hybrid-tree graph

(2) Session Key Generation for Unicast

Consume the power volume of customer N_1 yesterday is 5 degrees. When MS and N_1 want to communicate with each other for the first time today, both of them generate a session key respectively:

$$S_{0,1} = H(K_1, C_{K_1}), \text{ where } C_{K_1} = 5+1=6.$$

(3) Session Key Generation for Multicast

When MS wants to distribute DR projects to residential customers located in $GM_{K_{1,1}}$, the session key should be created respectively as $BS_{1,1} = H(K_{1,1}, C_{K_{1,1}})$, $C_{K_{1,1}} = C_p + 1$.

(4) Key-encryption Key Refreshing Periodically

When it's time to refresh the whole key tree, MS generates all the new key-encryption keys using a random key algorithm and distribute them to the related nodes.

B. The Overhead of the Example

Based on the analysis in chapter IV, the overhead of the example is listed in table 4, from which we can see that the overhead is very reasonable.

Table 4. Overhead of the example

Resources used		KDC	SX node
Storage cost(kb)	Number of keys	10,083	1.008
	Number of additional values	10,083	1.008
Calculation cost(ms)	Session key producing	10	10
	Refreshing periodically	660,000	10
Transmission cost(kb)	Session key producing	0	
	Update periodically	15,360	

6 Conclusion

AMI plays an important role in smart grid. An appropriate KMS will guarantee the confidentiality and integrality of interactive information. A novel scheme based on a hybrid-tree graph has been presented in this paper.

Firstly, the use of hybrid tree graph makes it easy to create session keys and refresh key-encryption keys, result in that single point of failure may be avoided and the traffic will be reduced.

Secondly, the usage of functions $E_K(D)$ and $H(D)$ guarantees the independence between the keys in use and previous keys, ensuring the key security.

Finally, the usage of additional values which vary from different types of customer makes it efficiency to manage keys.

References

1. Cleveland, F.M.: Cyber Security Issues for Advanced Metering Infrastructure (AMI). In: Power and Energy Society General Meeting - Conversion and Delivery of Electrical Energy in the 21st Century. IEEE, Pittsburgh (2008)
2. Robert, D., Colin, B., Ed, D., Juan, M.G.N.: SKMA a key management architecture for SCADA systems. In: Proc. 4th Australasian Information Security Workshop, vol. 54, pp. 138–192 (2006)
3. Wong, C.K., Gouda, M., Lam, S.: Secure group communication using key graphs. IEEE/ACM Trans. on Networking 8, 16–30 (2000)
4. Mittra, S.: Lolus: a framework for scalable secure multicasting. ACM SIGCOMM Computer Communication Review 27, 277–288 (1997)
5. Yan, K., Xie, D.-q.: Scheme and Implementation of Key Management Based on LKH. Computer Engineer and Application 31, 145–148 (2006)
6. Luan, W.-p.: Advanced Metering Infrastructure. Southern Power System Technology 3, 6–10 (2009)

Virtual Simulation of Printing Machine Training System Based on Virtual Reality

Zhang Bin, Li Jiajia, Qi Yuansheng, and Wang Xiaohua

School of information and mechanical engineering
Beijing Institute of Graphic Communication
Beijing, China
zhang1232123@tom.com

Abstract. Virtual reality-based disassembly techniques place a unique role in new product development, product maintenance and operation training. So we give a virtual reality-based training system for the printing-machine. Given the functional model and system architecture, Three-dimensional modeling by Solidworks, analysis of the assembly model about printing machine, then finish the virtual environment interaction by the software called EON Professional 5.5; publish it through Visual Basic 6.0 at last.

Keywords: printing-machine, virtual reality, virtual assembly and disassembly, simulation.

1 Introduction

In recent years, the global manufacturing market is growing competition, and with the shortest time to develop and produce high-quality and low-cost products becomes the goal for each enterprise. From the beginning of birth, printing-machine improves and develops continuously. No matter the development and design printing-machine, or disassembly practice, we achieve the desired results by low-cost and high-speed. Virtual disassembly is actually the integration of multiple technologies, and its development is in the relevant with the development of product design methodology, visualization, simulation, decision theory, assembly and manufacturing technology, assembly and manufacturing equipment, etc. [1] It brings a new opportunity for the development of printing-machine by using the technology of virtual assembly and disassembly into printing-machine design, manufacture and assembly.

It is similar for the development and design between printing-machine and other industry design. People continuously explores and improves new design method (design, manufacture, assembly and disassembly in the virtual environment) in order to reduce costs and shorten the development cycle. This will get rid of dependence on the trial sample. In a virtual environment, pre-assembling three-dimensional model on the design of the product, can confirm and adjust the rationality of the design plan and

D. Zeng (Ed.): ICAIC 2011, Part II, CCIS 225, pp. 439–445, 2011.

interference before the actual processing. Apart from this, in order for learners to better understand and master the structure of the modern printing-machine, principles, process and maintenance theory, learners need to line the actual assembly and disassembly experiments for printing-machine. But the equipment is very expensive, it is impossible for students studying in repeated assembly and disassembly, especially for new models of printing-machine products. In this situation, we carry out a task which called "Virtual Simulation of Printing Machine Training System Based on Virtual Reality". Virtual Simulation Of Printing Machine Training System can show the inner complex workings process of the machine, operate those equipments which may cause damage due to error operation, and can be into those structures which cannot observe or touch. It won't limited by time and space, and has high simulation, strong combination and expanding easily. It can reduce the duplicated investment of major equipment and update soon. Those advantages meet the needs of professional and technical education, and especially fit the diversity and variability of professional training.

2 System Design

A. Function of the system model

This system mainly completed the function of showing the printing-machine's internal and external structure, demo works and the training of virtual assembly and disassembly. It contains four modules which we call user login module, component structure display module, display module works and the virtual disassembly training.

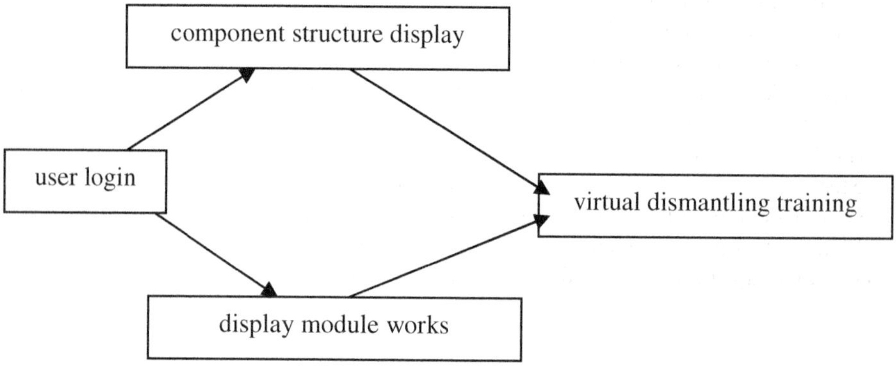

Fig. 1. System functional modules

B. System Architecture

This system provides designers with a real-time interactive virtual training environment by multi-mode interactive way. Figure 2 is the System Architecture.

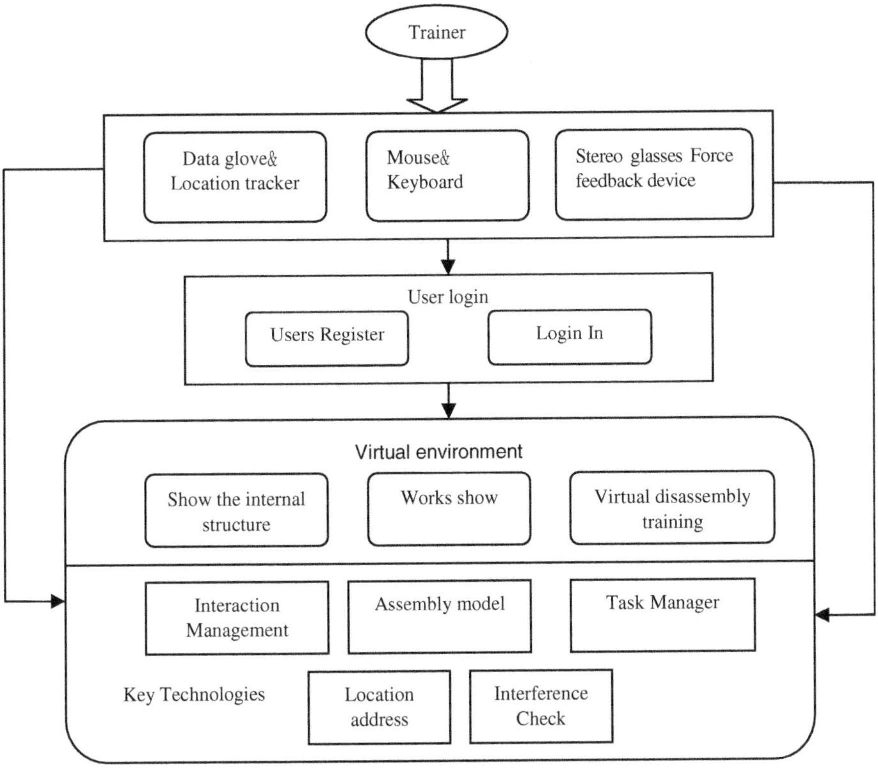

Fig. 2. System Architecture

1) User login module: After the first loading system runs into the module, and it has two sub-register and login. Legitimate registered users can be into this simulation training system, and non-registered users can access the temporary browser interface. After the user enters the system ,it allows to see personal information.

2) Component structure display module: Mainly learn and operate parts of the printing-machine. This module includes the following sections: equipment model showing the performance of real material, describing the model by text, pictures, etc; when display the internal structure, the external is translucent; supporting for multiple view switching and full observation by 360 degree.

3) Display how module works: Primarily simulating on the printing-machine's work theory of printing-machine simulation, this module includes the following sections: supporting multi-view demo and multiple presentations of real-time window switch; principles of simulation demo; explaining with text and sound.

4) Virtual disassembly training: This module is the core module of the system. This module mainly allows users to complete the virtual disassembly training, and it is divided into preview mode, guiding model, research model and assessment model. Users can choose different training mode by personal circumstances.

Model Features:

a) Preview mode

Demonstrate experimental steps and process mainly by the animation. Assist with experiments goals and the experiment instrutions. Each steps has prompted the corresponding audio and text display; Demo user can choose different angles to observe the experiment; when presentation is completed, the user can either choose to preview, or choose other module.

b) Guiding mode

Users operate the experiment in accordance with established procedures. It has appropriate action tips, and the action tips mainly have the following ways: Text,Sound,Direction icon (the corresponding model highlighting or flashing, etc.)

Operator errors have errors prompting (sound, text), after the error correction can be continued; after training you can choose to continue training in the model, also exit the training module and access into other training modes.

c) Research mode

This training mode is not the fixed disassembly procedures. The user explorers the disassembly training by himself, but simulation follows the objective law. There is none prompt during operation, only on error tips. After training you can choose to continue training in the model, also exit the training module and access into other training modes.

d) Assessment Mode

This training mode is mainly used on the user's operating procedures for evaluating. The operation does not need any tips. After the operation completed, the system must have the appropriate performance evaluation function; a high degree simulating operation of the field operations contains operational procedures.

3 The Realization of the System

A. The main interface of the system

Interface of the system is developed with Visual Basic 6.0, which can be login in. The function of system can be implemented by EON Professional 5.5, with the communication between each other, interaction will be accomplished.

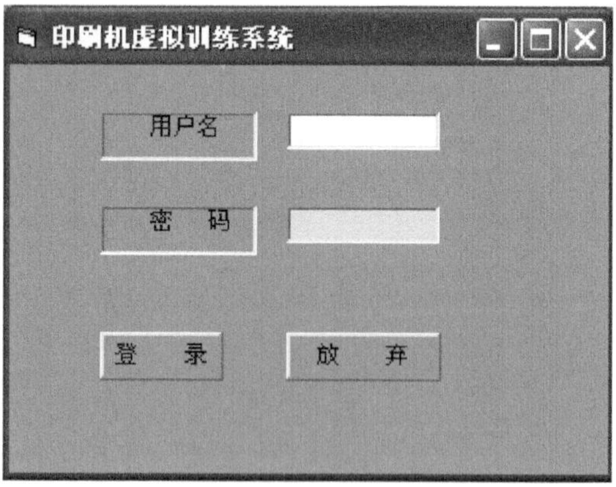

Fig. 3. Login interface

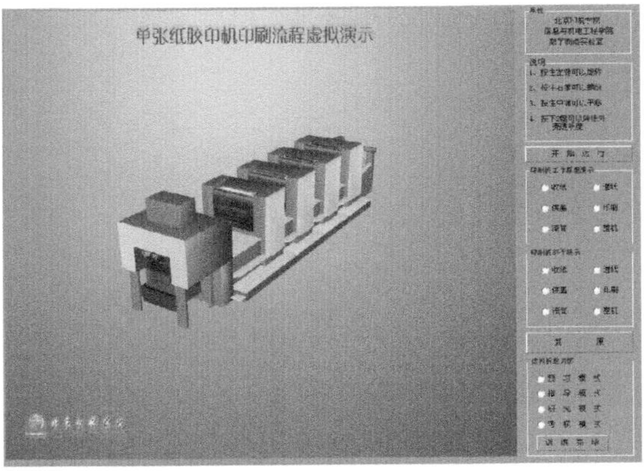

Fig. 4. Function interface

B. The implement of Assembly model

Assembly model of printing-machine is the vector of product design, and is the basic data structure, which provides information sources and access mechanisms for product design and assembly. The model of this system is made by SolidWorks modeling and assembling, then with data conversion, which is imported into the virtual environment in EON Professional. However, the model used in virtual environment, which is usually facet to describe collection object, is different from that used in 3D software. But, facet model loses parts of geometric topology and engineering design information, though it can meet the demand of real-time display and real-time collision detection [2].

C. Simulation to the vitual object

In order to carry out works of principle display and disassembly training on the printing-machine, it needs to build virtual simulation model of printing-machine in EON, update and manage the state of virtual printing-machine, simulate constrained motion of printing-machine and respond to the request of behavior of user actions, and then the real work environment can be reflected realistically.

1) Modifying imported 3D model. 3D model of printing-machine is imported into the system environment with the EON Import, then under the simulation conditions, the parts can be set on constraint, scaling, material, color, light and so on.

2) Setting interactions. To drag Nodes and Prototype given by software under specified things by Simulation Tree, to provide different parameter setting based on different states, then to set the logic relation between them in Routes window, so that interactions among parts and between man and components are implemented.

3) Issuing simulation program. After system debugging, the . Edz format document is output, which can be browsed directly and interact by EON Viewer, so, the Virtual Maintenance Training System can be operated without Eon software.

During the course of virtual disassembly, there is an important issue of preventing parts from penetrating when moving them that needs to be solved. Collision Node provided by EON Professional 5.5 is for the detection of collisions among components.

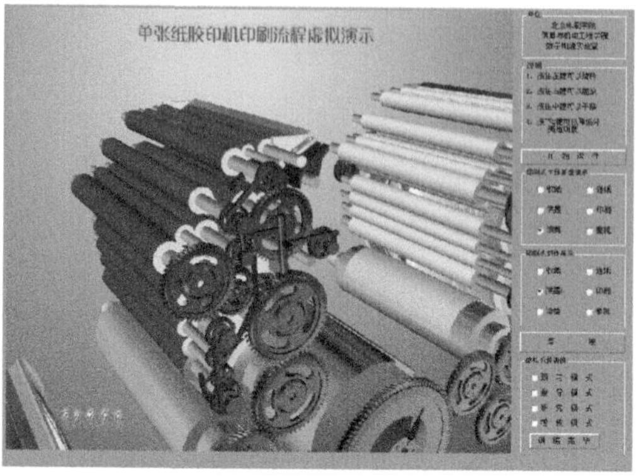

Fig. 5. Internal structure presentation

Fig. 6. Assessment Model

D. Planning to the path of virtual disassembly

Rational plan for path of disassembly and assembly based on the constraint relationship among components of printing-machine may guarantee disassembly and assembly are reversible each other. Assembly models represent mainly the message of the relationship among components, which can be described by Object-oriented approach. According to disassembly characters, the parts class can be built, the specific disassembly part just is the object in parts class, and the geometric information and a variety of physical information are the attributes of the parts class. Parts information obtained by the various methods of calling parts object which can offer information supports for the disassembly planning, as well as the other follow-up functions. Through reasoning and decomposing sub-body to generate all feasible disassembly sequence, disassembly and assembly in a virtual maintenance environment will be completed. [3]

E. Design and implement of interactive How-to

Interaction is the way for user to exchange with virtual environment, uses' selection, operate and control of simulation course to the object in the virtual environment must be accomplished through Human-Computer Interaction. Only building simple and effective Human-Computer Interaction, the users can finish training in immersion virtual environment more better. As immersion virtual training, the interaction of virtual training system of printing-machine mainly contains: User location tracking, navigation, manipulation to the objects in virtual environment (mainly grabbing, releasing, location), putting through between objects, system control and providing reminder message of intervene, collision and so on in the course of manipulation. These interaction relations may come true with EON Simulation Tree and logic settings window Routes. That provides representations for all defined flow charts, the connection between functional nodes is represented by line. Its executive flow begins with Out-Field of triggered functional nodes and ends with In-Field of corresponding.

4 The Conclusion

As a new mode, virtual simulation training can offer advanced experimental environment and simulation methods for training in equipment operation, maintenance, technical support ect., which has great role in improving training effectiveness, the quality of personnel training and scientific research. The virtual simulation training of printing-machine in this paper has the following features: first, providing a virtual practice platform. It breaks through the limit in range, ground and time of traditional training and change passive training into active participation; second, training in advance. The training model developed is consistent with the real engineering design and production exactly, before actual equipment used, the training may be conducted; third, reflecting the link of design, usage and message feedback. The users making design defects feedback during virtual training in time allowing for dynamic, parallel design become a reality; fourth, systematization of information. The fault type library integrated a number of theoretical and practical knowledge has large message, good expandability and diversified representation methods, so that, maintenance personnel can hold the training content in time, then avoid unilateral turn and singleness.

References

1. Li, B.M., Wu, Q.: The virtual disassembly and assembly system of a marine diesel engine based on SolidWorks. China Shiprepair 21, 27–30 (2008)
2. Peng, G.L., Liu, X.H.: Modular fixture assembly design system based on virtual reality. Computer Integrated Manufacturing Systems 14(5), 984–990 (2008)
3. Li, L., Li, H.J.: Virtual maintenance of missile based on EON Professional 5.0. Journal of Shandong Institute of Light Industry 21(4), 11–13 (2007)
4. Wang, Y., Jayaram, S., Jayaram, U.: Physically based modeling in virtual assmbly. The International Journal of Virtual Reality 5(1), 1–14 (2001)

Stable Analysis of a Predator–Prey SIRS Model with Saturation Infectious Force*

Xiaoying Chen, Wanzhen Zhu,
and Xiaowen Tang

Department of Computer and
Information Science
Fujian University of Technology
Fuzhou 350000, China
x.ychen@yahoo.com.cn

Abstract. By establishing a predator-prey automatic SIRS model with saturation infectious force and Holling II functions, we have obtained the conditions of the existence of various equilibriums. Furthermore, the global stability or the local stability of equilibrium has been proved. In this paper, we have obtained the existence and the asymptotical stability of the positive equilibriums when the saturation infectious force is large enough.

Keywords: saturation infectious force, predator-prey, SIRS model, stability of equilibrium.

1 Introduction

Mathematical models have become important tools in analyzing the spread and control of infectious diseases. In [7], it has studied a predator-prey SIRS model with saturation infectious and Holling II functions. However, the author did not consider the immunity period of the diseases. Based on [7], we improved the infectious SIRS models and the sufficient conditions of global asymptotical stabilities and local stabilities of the equilibrium points are obtained.

Our Model:

* This work is supported by the science and technology project of Fujian's Education Department #JBO8194.

D. Zeng (Ed.): ICAIC 2011, Part II, CCIS 225, pp. 446–454, 2011.

$$\begin{cases} \dot{x} = x(a-bx) - \dfrac{\alpha x}{1+\beta x} y \\ \dot{y} = y[-d + e\dfrac{\alpha x}{1+\beta x}] \\ \dot{S}_1 = a_1(S_1 + R_1) + a_1(1-\lambda_1)I_1 - (a_2+bx)S_1 \\ \quad -(\overline{\beta}_{11}I_1 + \overline{\beta}_{21}I_2)\dfrac{S_1}{x} - \dfrac{\alpha y}{1+\beta x}S_1 + \delta_1 R_1 \\ \dot{I}_1 = (\overline{\beta}_{11}I_1 + \overline{\beta}_{21}I_2)\dfrac{S_1}{x} + a_1\lambda_1 I_1 - (a_2+bx)I_1 - \dfrac{\alpha y}{1+\beta x}I_1 - \gamma_1 I_1 \\ \dot{R}_1 = \gamma_1 I_1 - (a_2+bx)R_1 - \dfrac{\alpha y}{1+\beta x}R_1 - \delta_1 R_1 \\ \dot{S}_2 = e\dfrac{\alpha x}{1+\beta x}(S_2+R_2) + e\dfrac{\alpha x}{1+\beta x}(1-\lambda_2)I_2 - dS_2 \\ \quad -(\overline{\beta}_{12}I_1 + \overline{\beta}_{22}I_2)\dfrac{S_2}{y} + \delta_2 R_2 \\ \dot{I}_2 = (\overline{\beta}_{12}I_1 + \overline{\beta}_{22}I_2)\dfrac{S_2}{y} + e\dfrac{\alpha x}{1+\beta x}\lambda_2 I_2 - dI_2 - \gamma_2 I_2 \\ \dot{R}_2 = \gamma_2 I_2 - dR_2 - \delta_2 R_2 \end{cases} \tag{1}$$

The above parameters are positive, $\lambda_i (0 < \lambda_i < 1)$ is the hereditary infective rate of species i , a_1, $a_2 + bx$ are the birth rate and the death rate respectively $(a = a_1 - a_2,), d$ is the death rate of the predator, γ_i, δ_i are the recovery rate and the rate of losing immunity of species i. $\overline{\beta}_{i1} \triangleq \overline{\beta}_{i1}(x) = \dfrac{\alpha_{i1}x}{1+\beta_{i1}x}$, $(\alpha_{i1} > 0, \beta_{i1} > 0)$ are the infectious rate of species i to the prey. $\overline{\beta}_{i2} \triangleq \overline{\beta}_{i2}(y) = \dfrac{\alpha_{i2}y}{1+\beta_{i2}y}$, ($\alpha_{i2} > 0, \beta_{i2} > 0$) are the infectious i rate of species i to the predator. Variables x、 y are the population size of prey and predator at time t . S_i, I_i, R_i denote the susceptible individuals number, the infectious individuals number and the removed individuals number at t, respectively. $i = 1$ denotes the prey, $i = 2$ denotes the predator.

Since $x = S_1 + I_1 + R_1, y = S_2 + I_2 + R_2$, system (1) can be write as system (2):

$$\begin{cases} \dot{x} = x(a-bx) - \dfrac{\alpha x}{1+\beta x} y \triangleq Q_1 \\ \dot{y} = y[-d + e\dfrac{\alpha x}{1+\beta x}] \triangleq Q_2 \\ \dot{I}_1 = (\overline{\beta}_{11}I_1 + \overline{\beta}_{21}I_2)\dfrac{S_1}{x} + a_1\lambda_1 I_1 - (a_2+bx)I_1 - \dfrac{\alpha y}{1+\beta x}I_1 - \gamma_1 I_1 \triangleq Q_3 \\ \dot{R}_1 = \gamma_1 I_1 - (a_2+bx)R_1 - \dfrac{\alpha y}{1+\beta x}R_1 - \delta_1 R_1 \triangleq Q_5 \\ \dot{I}_2 = (\overline{\beta}_{12}I_1 + \overline{\beta}_{22}I_2)\dfrac{S_2}{y} + e\dfrac{\alpha x}{1+\beta x}\lambda_2 I_2 - dI_2 - \gamma_2 I_2 \triangleq Q_4 \\ \dot{R}_2 = \gamma_2 I_2 - dR_2 - \delta_2 R_2 \triangleq Q_6 \end{cases} \tag{2}$$

2 Existence and Stability of Equilibrium Points of System (3)[7]

At first, let us consider the sub-system of (2)

$$\begin{cases} \dot{x} = x(a - bx) - \dfrac{\alpha x}{1+\beta x} y \\ \dot{y} = y[-d + e\dfrac{\alpha x}{1+\beta x}] \end{cases} \tag{3}$$

A. *Lemma 1*

Let $\varpi = e\beta(\frac{\alpha}{\beta} - \frac{d}{e})$.

1) When $\frac{d}{\varpi} \geq \frac{a}{b}$, or $\varpi < 0$, system (3) just has two equilibrium points $O(0,0), M_1(x_1,0)), O$ is a saddle point and M_1 is at first quadrant and is asymptotical stable.

2) When $0 < \frac{d}{\varpi} < \frac{a}{b}$, system(3) has three equilibrium points: $O(0,0), M_1(x_1,0)), M_0(x_0, y_0); O, M_1$ are saddle points.

For positive equilibrium points M_0, there are two cases:

When $(\frac{a}{b} - 2\frac{d}{\varpi})\beta \leq 1$, M_0 is at first quadrant and is asymptotical stable;

When $(\frac{a}{b} - 2\frac{d}{\varpi})\beta > 1$, M_0 is unstable.

Where $x_1 = \frac{a}{b}, x_0 = \frac{d}{\varpi}, y_0 = \frac{1}{\alpha}(a - bx_0)(1 + \beta x_0)$.

B. *Lemma 2*

The solutions of System (3) with positive initial conditions are bounded.

In fact, Let the initial point be $(\bar{x}_0, \bar{y}_0)$; $L_1 : x = \bar{x}_1$, $L_2 : y = l_0 - ex$, where $l_0 = \max\{\frac{1}{d}\max_{0 \le x \le \bar{x}_1}[ex(a - bx) + dex], \quad e\bar{x}_1 + \bar{y}_0\}$, $\bar{x}_1 = \max\{\bar{x}_0, x_1\}$ .Then the trajectory starting from $(\bar{x}_0, \bar{y}_0)$ stay in the area with bounds of x-axis, y-axis, L_1 and L_2.

3 Existence of Equilibrium Points of System (2)

Let $A(x, y) = a_2 + bx + \gamma_1 + \dfrac{\alpha y}{1+\beta x} - a_1\lambda_1, B(x, y)$

$$= d + \gamma_2 - e\frac{\alpha x}{1+\beta x}\lambda,$$

$$A(x_0, y_0) = A_0,\ B(x_0, y_0) = B_0, \quad A(x_1, 0) = A_1,\ B(x_1, 0) = B_1.$$

Let $C(x, y) = a_2 + bx + \dfrac{\alpha y}{1+\beta x} + \delta_1$,

$$C(x_0, y_0) = C_0,\ C(x_1, 0) = C_1, \quad R = \frac{\overline{\beta}_{21}(x_0)\overline{\beta}_{12}(y_0)}{[A_0 - \overline{\beta}_{11}(x_0)][B_0 - \overline{\beta}_{22}(y_0)]}.$$

It is not difficult to know that $\Omega = \{(I_1, R_1, x, I_2, R_2, y)^T \mid 0 \le I_i, 0 \le R_i, 0 \le I_1 + R_1 \le x, 0 \le I_2 + R_2 \le y, i = 1,2\}$ is positive invariant set of system (2). The trajectory starting from Ω is bounded.

A. *Theorem 1*

The existence equilibrium points of system (2) is shown as follows.

Conditions of existence			Equilibrium points of system(2)
I	II	III	
			$p_0(0,0,0,0,0,0)$, $p_1(0,0,x_1,0,0,0)$
	$0 < A_1 < \overline{\beta}_{11}(x_1)$		$p_2(I_{1E}, R_{1E}, x_1, 0,0,0)$, where: $I_{1E} = \dfrac{C_1 x_1}{\gamma_1 + C_1}\dfrac{[\overline{\beta}_{11}(x_1) - A_1]}{\overline{\beta}_{11}(x_1)}$, $R_{1E} = \dfrac{\gamma_1}{C_1} I_{1E}$.
$0 < \dfrac{d}{\varpi} < \dfrac{a}{b}$			$p_3(0,0,x_0,0,0,y_0)$
	$0 < A_0 < \overline{\beta}_{11}(x_0)$ $B_0 > 0$		$p_4(I_{1EE}, R_{1EE}, x_0, I_{2EE}, R_{2EE}, y_0)$ where: $R_{1EE} = \dfrac{\gamma_1}{C_0} I_{1EE}$, $R_{2EE} = \dfrac{\gamma_2}{d+\delta_2} I_{2EE}$ (I_{1EE}, I_{2EE}) is the intersection of $f_1(I_1)$ and $f_2(I_2)$.
$0 < \dfrac{d}{\varpi} < \dfrac{a}{b}$	or $0 < B_0 < \overline{\beta}_{22}(y_0)$ $A_0 > 0$		
	or $A_0 > \overline{\beta}_{11}(x_0)$ $B_0 > \overline{\beta}_{22}(y_0)$	$R > 1$	

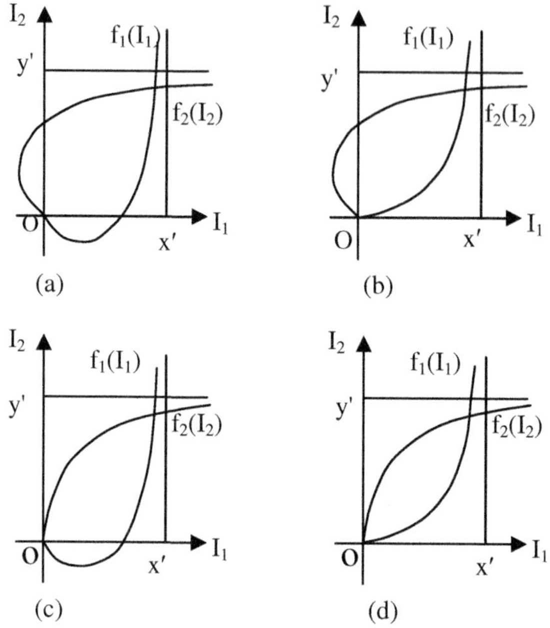

Fig. 1. (f_1 and f_2)

Note: Let

$$x_0' \triangleq \frac{x_0}{1+\dfrac{\gamma_1}{C_0}}, \quad y_0' \triangleq \frac{y_0}{1+\dfrac{\gamma_2}{d+\delta_2}}.$$

We just prove the existence of p_4.

B. Proof

Let $Q_3=0, Q_4=0$, $Q_5=0, Q_6=0$. Substituting (x_0, y_0) to system (2), we have

$$I_2=\frac{-\overline{\beta}_{11}(x_0)I_1\dfrac{S_1}{x_0}+A_0I_1}{\overline{\beta}_{21}(x_0)\dfrac{S_1}{x_0}}=-\frac{\overline{\beta}_{11}(x_0)}{\overline{\beta}_{21}(x_0)}I_1-\frac{A_0x_0}{\overline{\beta}_{21}(x_0)}+$$

$$\frac{A_0x_0^2}{(1+\dfrac{\gamma_1}{C_0})\overline{\beta}_1(x_0)}\cdot\frac{1}{x_0-(1+\dfrac{\gamma_1}{C_0})I_1}\triangleq f_1(I_1)$$

$$I_1=\frac{-\overline{\beta}_{22}(y_0)I_2\dfrac{S_2}{y_0}+B_0I_2}{\overline{\beta}_{12}(y_0)\dfrac{S_2}{y_0}}=-\frac{\overline{\beta}_{22}(y_0)}{\overline{\beta}_{12}(y_0)}I_2-\frac{B_0y_0}{\overline{\beta}_{12}(y_0)}+$$

$$\frac{B_0 y_0^2}{(1+\frac{\gamma_2}{d+\delta_2})\overline{\beta}_{12}(y_0)} \cdot \frac{1}{y_0-(1+\frac{\gamma_2}{d+\delta_2})I_2} \triangleq f_2(I_2)$$

$$\frac{df_1(I_1)}{dI_1} = -\frac{\overline{\beta}_{11}(x_0)}{\overline{\beta}_{21}(x_0)} + \frac{A_0 x_0^2}{\overline{\beta}_{21}(x_0)} \cdot \frac{1}{[x_0-(1+\frac{\gamma_1}{C_0})I_1]^2}$$

$$\frac{d^2 f_1(I_1)}{dI_1^2} = 2\frac{A_0 x_0^2}{\overline{\beta}_{21}(x_0)} \cdot (1+\frac{\gamma_1}{C_0}) \frac{1}{[x_0-(1+\frac{\gamma_1}{C_0})I_1]^3} > 0$$ where

$A_0 > 0,\ 0 < I_1 < \dfrac{x_0}{1+\frac{\gamma_1}{C_0}}$.

$$\frac{df_2(I_2)}{dI_2} = -\frac{\overline{\beta}_{22}(y_0)}{\overline{\beta}_{12}(y_0)} + \frac{B_0 y_0^2}{\overline{\beta}_{12}(y_0)} \cdot \frac{1}{[y_0-(1+\frac{\gamma_2}{d+\delta_2})I_1]^2}$$

$$\frac{d^2 f_2(I_2)}{dI_2^2}) = 2\frac{B_0 y_0^2}{\overline{\beta}_{12}(y_0)} (1+\frac{\gamma_2}{d+\delta_2}) \cdot \frac{1}{[y_0-(1+\frac{\gamma_2}{d+\delta_2})I_2]^3} > 0$$ where

$B_0 > 0,\ \ 0 < I_2 < \dfrac{y_0}{(1+\frac{\gamma_2}{d+\delta_2})}$.

And $f_1(0) = 0,\ \ \lim\limits_{I_1 \to \frac{x_0}{1+\frac{\gamma_1}{C_0}} - 0^-} f_1(I_1) = +\infty,\ \ (A_0 > 0)$,

$f_2(0) = 0,\ \ \lim\limits_{I_2 \to \frac{y_0}{(1+\frac{\gamma_2}{d+\delta_2})} - 0^-} f_2(I_2) = +\infty\ \ (B_0 > 0)$, then :

if $0 < A_0, 0 < B_0$, and $A_0 < \overline{\beta}_{11}(x_0)$ or $B_0 < \overline{\beta}_{22}(y_0)$, the graphs of $f_1(I_1), f_2(I_2)$ refer to $1(a),(b),(c)$;

if $A_0 > \overline{\beta}_{11}(x_0), B_0 > \overline{\beta}_{22}(y_0)$,

$$\frac{[A_0-\overline{\beta}_{11}(x_0)][B_0-\overline{\beta}_{22}(y_0)]}{\overline{\beta}_{21}(x_0)\overline{\beta}_{12}(y_0)} < 1,$$

the graphs of $f_1(I_1), f_2(I_2)$ refer to $1(d)$.

Therefore, under the above conditions, in the area

$$D:\begin{cases} 0<I_1<\dfrac{x_0}{1+\dfrac{\gamma_1}{C_0}} \\ 0<I_2<\dfrac{y_0}{1+\dfrac{\gamma_2}{d+\delta_2}} \end{cases},$$

there is one and only one intersection between $f_1(I_1)$ and $f_2(I_2)$: (I_{1EE}, I_{2EE}). If one the conditions is not met, there is no intersection between $f_1(I_1)$ and $f_2(I_2)$ in D.

4 Stability of Equilibrium Points of System (2)

Let

$$D_1 \triangleq \begin{pmatrix} \dfrac{\partial Q_1}{\partial x} & \dfrac{\partial Q_1}{\partial y} \\ \dfrac{\partial Q_2}{\partial x} & \dfrac{\partial Q_2}{\partial y} \end{pmatrix}, \quad D_2 \triangleq \begin{pmatrix} \dfrac{\partial Q_3}{\partial I_1} & \dfrac{\partial Q_3}{\partial I_2} & \dfrac{\partial Q_3}{\partial R_1} & \dfrac{\partial Q_3}{\partial R_2} \\ \dfrac{\partial Q_4}{\partial I_1} & \dfrac{\partial Q_4}{\partial I_2} & \dfrac{\partial Q_4}{\partial R_1} & \dfrac{\partial Q_4}{\partial R_2} \\ \dfrac{\partial Q_5}{\partial I_1} & \dfrac{\partial Q_5}{\partial I_2} & \dfrac{\partial Q_5}{\partial R_1} & \dfrac{\partial Q_5}{\partial R_2} \\ \dfrac{\partial Q_6}{\partial I_1} & \dfrac{\partial Q_6}{\partial I_2} & \dfrac{\partial Q_6}{\partial R_1} & \dfrac{\partial Q_6}{\partial R_2} \end{pmatrix}.$$

Then the eigenvalues of Matrix D_1, D_2 are the roots of derivation operator of system (2).

A. Theorem 2

The local stability of equilibrium points of system (2) is following:

1) $p_0(0,0,0,0\ 0\ 0)$ always exists,but is unstable;

2) If $\varpi<0$ or $\dfrac{d}{\varpi}>\dfrac{a}{b}$, when $0<A_1<\overline{\beta}_{11}(x_1)$, and $\alpha_{11}x_1<B_1$, $p_2(I_{1E},R_{1E},x_1,0,0,0)$ is locally stable;

3) If $0<\dfrac{d}{\varpi}<\dfrac{a}{b}$, $(\dfrac{a}{b}-2\dfrac{d}{\varpi})\beta\le 1$, when $0<A_0<\overline{\beta}_{11}(x_0), 0<B_0$, or $0<B_0<\overline{\beta}_{22}(y_0), 0<A_0$, or $A_0>\overline{\beta}_{11}(x_0)$, $B_0>\overline{\beta}_{22}(y_0)$ and $R>1$, $p_4(I_{1EE},R_{1EE},x_0,I_{2EE},R_{2EE},y_0)$ is locally stable.

We only prove the existence of P_4.

Proof:

$$D_2(P_4)=\begin{pmatrix} -\frac{I_{2EE}}{I_{1EE}}\overline{\beta}_{21}(x_0)(1-\frac{I_{1EE}+R_{1EE}}{x_0}) -\frac{1}{x_0}[\overline{\beta}_{11}(x_0)I_{1EE}+\overline{\beta}_{21}(x_0)I_{2EE}] & \overline{\beta}_{21}(x_0)(1-\frac{I_{1EE}+R_{1EE}}{x_0}) & -\frac{1}{x_0}[\overline{\beta}_{11}(x_0)I_{1EE}+\overline{\beta}_{21}(x_0)I_{2EE}] & 0 \\ \overline{\beta}_{12}(y_0)(1-\frac{I_{2EE}+R_{2EE}}{y_0}) & -\frac{I_{1EE}}{I_{2EE}}\overline{\beta}_{12}(y_0)(1-\frac{I_{2EE}+R_{2EE}}{y_0}) -\frac{1}{y_0}[\overline{\beta}_{12}(y_0)I_{1EE}+\overline{\beta}_{22}(y_0)I_{2EE}] & 0 & -\frac{1}{y_0}[\overline{\beta}_{12}(y_0)I_{1EE}+\overline{\beta}_{22}(y_0)I_{2EE}] \\ \gamma_1 & 0 & -C_0 & 0 \\ 0 & \gamma_2 & 0 & -(d+\delta_2) \end{pmatrix}$$

Under the assumptions, $D_2(P_4)$ has four eigenvalues with negative real parts. By lemma 1, P_4 is locally asymptotically stable.

B. Theorem 3

The global stability of equilibrium points of system (2) is following:

1) If $\varpi<0$ or $\frac{d}{\varpi}>\frac{a}{b}$, when $A_1>\overline{\beta}_{11}(x_1)$, $p_1(0{,}0,x_1{,}0$ $0\,{,}0)$ is globally asymptotically stable in Ω^+;

2) If $0<\frac{d}{\varpi}<\frac{a}{b}$, $(\frac{a}{b}-2\frac{d}{\varpi})\beta\le 1$, when $A_0>\overline{\beta}_{11}(x_0)$, $B_0>\overline{\beta}_{22}(y_0)$, $p_3(0{,}0,x_0{,}0{,}0,y_0)$ is globally asymptotically stable in Ω^+.

Where $\Omega^+=\{(I_1,R_1,x,I_2,R_2,y)|0<I_1+R_1\le x,$

$$0<I_2+R_2\le y, I_i>0, R_i>0, i=1,2\}.$$

We just prove the stability of P_1.

Proof: By Lemma 3.6.2 in [9], from third inequality of system (2) we have

$$\overline{\beta}_{21}(x_1)I_2^{\infty}\ge[A_1-\overline{\beta}_{11}(x_1)]I_1^{\infty},$$

By Lemma 1, $\lim_{t\to\infty}I_2(t)=0$.

Since $A_1>\overline{\beta}_{11}(x_1)$, we have $I_1^{\infty}=0$ and $\lim_{t\to\infty}I_1(t)=0$.

By Lemma 3.6.2 in [9], from fourth inequality of system (2) we have $\gamma_1 I_1^\infty \geq C_1 R_1^\infty$. Therefore, $\lim_{t\to\infty} R_1(t) = 0$.

By Lemma 1 again, $p_1(0,0,x_1,0,0,0)$ is globally asymptotically stable in Ω^+.

References

1. Dushoff, J., Huang, W., Castillo-Chavez, C.: Backwards Bifurcations and Catastrophe in Simple Models of Fatal Diseases. J. Math. Biol. 36(3), 227–248 (1998)
2. Venturino, E.: The Influence of Disease on Lokta-Volterra Systems. Rocky Mt. J. Math. 24, 381–402 (1994)
3. Han, L., Yuan, S., Ma, Z.: Preistence of an SIRS Epidemic Model of Two Competitive Species. Journal of Engineering Mathematics 27(2), 172–176 (2004)
4. Yuan, S., Jing, L.: An SIS Model with Nonlinear Saturation Infectious Force. Journal of Engineering Mathematics 18(4), 98–102 (2001)
5. Zhu, W., Xu, M.: SIS epidemic model of Two Competitive Species with Nonlinear Incidence. Journal of Zhejiang University of Science and Technology 16(2), 75–79 (2004)
6. Zhu, W.: A Qualitative Analysis of A SIS Model with Nonlinear Infection Rate When the Two Species Coexist. Journal of Jiangxi Institute of Education 25(3), 4–7 (2004)
7. Chen, X., Zhu, W.: Qualitative Analysis of Autonomous SIS Model of Predator-Prey System with the Holling II Functional Response. Journal of Xiamen University(National Science) 44(1), 16–19 (2005)
8. Chen, X.Y.: Modelng and Analysis of a Predater-Prey Model with Disease in the Prey. Math Biosci. 171, 59–82 (2001)
9. Ma, Z., Zhou, Y., Wan, W.: The Mathematical Modeling and Study of Epidemic Dynamics. Scientific Publishing Company, Beijing (2004)

Kalman Filter Based on SVM Innovation Update for Predicting State-of Health of VRLA Batteries

Li Chang and Jiang Xiaoluo

College of physics and electronic information engineering
WenZhou University
WenZhou, China
a1L3381@tom.com

Abstract. This paper describes the dynamic model of valve regulated lead-acid (VRLA) battery and presents application of state-estimation techniques for online prediction of the state-of-health (SOH) and state-of-charge(SOC) of VRLA batteries. Specifically, approach of Kalman filter (KF) based on the innovation obtained through support vector machine (SVM) is presented. In the proposed approach, factors related with the capacity of a VRLA battery have been taken into account as elements of the multi-dimensional input vector while time series is trained by SVM to predict future innovation. Then step by step prognosis is implemented through Kalman predictor with innovation obtained through SVM. Measurements using real-time data are used to verify the capability of the proposed method for estimating SOH and SOC of VRLA batteries. Experiment results show that our proposed methodologies are suitable for monitoring the SOC and SOH of the battery, with accuracy in determining the SOH and its trend of deterioration within 4%. Moreover, by accounting for the nonlinearities present within the dynamic cell models, the application of KF prediction based on SVM innovation update is shown to provide verifiable indications of SoH.

Keywords: Kalman Filter, state of charge, state-of health, SVM, VRLA battery.

1 Introduction

Due to advantages of maintenance-free, large capacity, low prices, VRLA batteries have been widely applied in various fields: power of communication systems, UPS system, power DC panel, hybrid-electric vehicles and so on. It will play a positive role if capacity of the batteries could be forecasted accurately and work-state of the batteries could be predicted exactly.

With accurate measurement of SOC and exact prediction of SOH, it has a special significance in the cases of ensuring the batteries to function properly; Only with the exact separation of batteries with deterioration or the failed ones from a battery bank, could the monitoring system play an efficient role to facilitate safely and operate correctly. The predicted available capacity of a battery is an important performance parameter in electrochemical industry, many laboratory methods and several commercial devices have been reported and reviewed to predict this SOC status [1].

D. Zeng (Ed.): ICAIC 2011, Part II, CCIS 225, pp. 455–463, 2011.

Method of load discharge test is usually an accurate and accepted way to measure SOC of a battery, which is recommended by most of the manufactures. However, this proposed method shows its own limitation if requirement of online supervision is taken into the consideration, which restricts the scope of its application. Methods of online test rely on measuring the transient voltage electric current and inner resistance or the combination of these quantities. Method of measuring the resistance can provide the accuracy better than 80% relatively since the electrolyte density of an aged battery is reduced and its resistance is increased [2].However, VRLA battery is a complex electrochemical system because the actual SOC and SOH of the tested batteries are co-vary with temperature, cycles of charge and discharge, depth of discharge, and many other factors. How to improve the prediction accuracy of the residual capacity and the efficiency in the use of battery becomes a focus of research in battery industry.

Different methods have been proposed in literatures for the SOH estimation [3]. These proposed methods for the SOH determination in literatures can be divided into tow categories:

The first one is to determine the SOH in laboratory by measuring the extrinsic parameters of a battery based on the relationship between the parameters and the residual capacity, which rely on the theory of the internal workings in VRLA battery. Due to the complexity of the theory, few methods works well enough while varieties of batteries with different bands are taken into the consideration.

The other one is to determine the SOH based on system identification and estimation of the model parameters, using the AC and DC impedances and conductance measurements. VRLA battery is viewed as a black box in these approaches and modeled as a nonlinear system which will be estimated by Kalman Filter neural networks and fuzzy logic algorithms. The SOH of a VRLA battery can be estimated accurately by measuring the voltage, appearing in the early stages of the battery discharge [4][5]. Combination of these intelligence algorithms becomes a trend in the research works on determining the SOH of VRLA batteries.

In this paper, expert system for status supervisory and fault diagnosis with real-time prediction for SOC of VRLA batteries is presented with a dynamic model based on the electrochemical operation of the battery and variations of terminal voltage and current. Support vector machine is introduced with Kalman Filter while the unknown model of the battery is estimated by means of system identification. Factors related with the capacity of the VRLA battery have also been considered while time series is trained by means of SVM to predict future innovation. Then step by step prognosis is implemented as Kalman predictor with the updated innovation obtained by means of SVM. One of the main advantages of the proposed method is that it can improve the accuracy of the prediction and identify the trend of the deterioration correctly as an online estimator for the SOH of these failed batteries.

This paper is organised as follows. Section 2 illustrates the model of VRLA batteries and model-based state-estimation techniques using Kalman Filter. Section 3 provides the scheme of support vector machine and its application in the estimation process while KF is employed to predict SOC and SOH of the batteries. Section 4 shows the experimental results. Section 5 gives a brief conclusion of the paper.

2 Model of VRLA BATTERIES AND Kalman Prediction

A. Equivalent model of VRLA Batteries and its state-space description

Fig 1 shows a simplified equivalent model of VRLA batteries. This model has been employed before to simulate the response of VRLA cells to constant current and dynamic power cycles. In this figure, R_t is the lumped resistance due to the cell interconnections. R_e and R_s are the surface resistance and end resistance, respectively, meanwhile, the capacitance C_b models the charge storability of the battery, C_s models surface capacitance and diffusion effects within the battery. The voltage across the bulk and surface capacitors are denoted V_b and V_s, respectively. Voltages and currents describing the characteristics of the net-work shown in Fig.1, are given by

$$V_O = IR_t + I_b R_e + V_b \tag{1}$$

$$V_O = IR_t + I_s R_s + V_s \tag{2}$$

After some algebraic manipulation, yields:

$$I_b(R_e + R_s) = IR_s + V_s - V_b \tag{3}$$

Since $I_b = \frac{dV_b}{dt} C_b$, (4) can be rearranged to give

$$\frac{dV_b}{dt} = -\frac{V_b}{C_b(R_e+R_s)} + \frac{V_s}{C_b(R_e+R_s)} + \frac{IR_s}{C_b(R_e+R_s)} \tag{4}$$

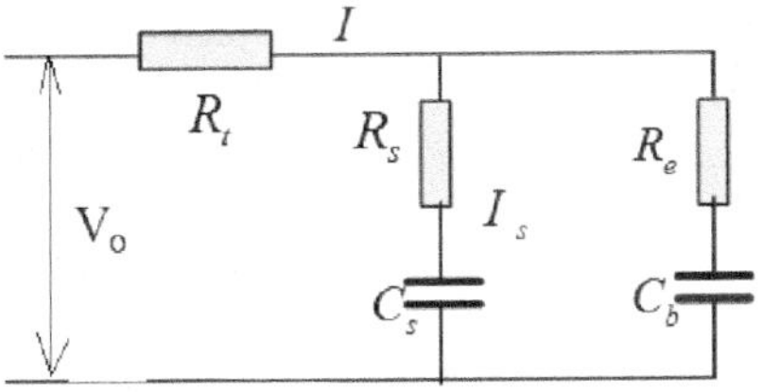

Fig. 1. Equivalent circuit of a VRLA battery

$$\frac{dV_s}{dt} = -\frac{V_s}{C_s(R_e+R_s)} + \frac{V_b}{C_s(R_e+R_s)} + \frac{IR_s}{C_s(R_e+R_s)} \tag{5}$$

A state variable description relating the voltages and currents is obtained as follows:

$$\begin{bmatrix} \frac{dV_b}{dt} \\ \frac{dV_s}{dt} \end{bmatrix} = \begin{bmatrix} -\frac{1}{C_b(R_e+R_s)} & \frac{1}{C_b(R_e+R_s)} \\ \frac{1}{C_S(R_e+R_s)} & -\frac{1}{C_s(R_e+R_s)} \end{bmatrix} \begin{bmatrix} V_b \\ V_s \end{bmatrix} + \begin{bmatrix} \frac{R_s}{C_b(R_e+R_s)} \\ \frac{R_e}{C_s(R_e+R_s)} \end{bmatrix} I \tag{6}$$

The output voltage, as a function of terminal current, is given as follows:

$$V_O = \begin{bmatrix} \dfrac{R_s}{R_e + R_s} & \dfrac{R_e}{R_e + R_s} \end{bmatrix} \begin{bmatrix} V_b \\ V_s \end{bmatrix} + \begin{bmatrix} R_t + \dfrac{R_e R_s}{R_e + R_s} \end{bmatrix} I \tag{7}$$

Since the rate of change of terminal current is negligible, it is obvious: $\dfrac{dI}{dt} \approx 0$, (6) can be rearranged to give:

$$\begin{aligned} \frac{dV_o}{dt} = & \left[-\frac{R_s}{C_b(R_e + R_s)^2} + \frac{R_e}{C_s(R_e + R_s)^2} \right] V_b \\ & + \left[\frac{R_s}{C_b(R_e + R_s)^2} - \frac{R_e}{C_s(R_e + R_s)^2} \right] V_c \\ & + \left[\frac{R_s^{\,2}}{C_b(R_e + R_s)^2} + \frac{R_e^{\,2}}{C_s(R_e + R_s)^2} \right] I \end{aligned} \tag{8}$$

The complete state variable description of the model can be obtained as follows:

$$\begin{bmatrix} \dfrac{dV_b}{dt} \\ \dfrac{dV_s}{dt} \\ \dfrac{dV_O}{dt} \end{bmatrix} = \begin{bmatrix} -\dfrac{1}{C_b(R_e + R_s)} & \dfrac{1}{C_b(R_e + R_s)} & 0 \\ \dfrac{1}{C_s(R_e + R_s)} & -\dfrac{1}{C_s(R_e + R_s)} & 0 \\ A_{3,1} & 0 & A_{3,3} \end{bmatrix} \begin{bmatrix} V_b \\ V_s \\ V_o \end{bmatrix} + \begin{bmatrix} \dfrac{R_s}{C_b(R_e + R_s)} \\ \dfrac{R_e}{C_s(R_e + R_s)} \\ B_{3,1} \end{bmatrix} I \tag{9}$$

$$\begin{aligned} A_{3,1} = & -\frac{R_s^{\,2}}{C_b R_e (R_e + R_s)^2} + \frac{R_s}{C_s(R_e + R_s)^2} \\ & + \frac{R_s}{C_b(R_e + R_s)^2} + \frac{R_e}{C_s(R_e + R_s)^2} \end{aligned}$$

$$A_{3,3} = \frac{R_s}{C_b R_e (R_e + R_s)} - \frac{1}{C_s(R_e + R_s)}$$

$$B_{3,1} = \frac{R_s^{\ 2}}{C_s(R_e + R_s)^2} - \frac{R_s R_t}{C_b R_e (R_e + R_s)}$$
$$+ \frac{R_t}{C_s(R_e + R_s)} + \frac{R_e R_s}{C_s(R_e + R_s)^2}$$

(9) can be simplified as follow:

$$\begin{bmatrix} \frac{dV_{Cb}}{dt} \\ \frac{dV_{Cs}}{dt} \\ \frac{dV_o}{dt} \end{bmatrix} = A \times \begin{bmatrix} V_{Cb} \\ V_{Cs} \\ V_o \end{bmatrix} + B \times I \tag{10}$$

B. Kalman prediction and its shortcomings

Assuming the applied input is the current I , a discrete-time equivalent model of the system is given by

$$\begin{aligned} \mathbf{x}(k+1) &= A\mathbf{x}(k) + B\mathbf{u}(k) + \boldsymbol{\tau}(k) \\ \mathbf{y}(k+1) &= C\mathbf{x}(k) + D\mathbf{u}(k) + \boldsymbol{\zeta}(k) \end{aligned} \tag{11}$$

$\boldsymbol{\tau}(k)$ and $\boldsymbol{\zeta}(k)$ denote the fault term and high order term in the state equation and output equation, respectively [6]. The former represents system disturbances and model inaccuracies caused by deterioration in the batteries. The latter represents the effects of measurement noise. However, disturbances caused by deterioration will lead the linear model to a non-linear one or a local linear one. Kalman Filter is used to estimate the approximate linear system as follows:

$$\begin{aligned} \hat{\mathbf{x}}(k+1) &= A\hat{\mathbf{x}}(k) + B\mathbf{u}(k) \\ \hat{y}(k+1) &= C\mathbf{x}(k+1) + D\mathbf{u}(k+1) \\ \varepsilon(k+1) &= y(k+1) - \hat{y}(k+1) \\ \hat{\mathbf{x}}(k+1) &= \hat{\mathbf{x}}(k+1) + \mathbf{K}(k+1)\varepsilon(k+1) \end{aligned} \tag{12}$$

As a matter of fact, the predicted innovation $\varepsilon(k)$ presented in formula (12) corresponds to model error caused by deterioration in the battery. Because the output in the future time is unknown due to the variation of the model, Kalman prediction will loss the function of the correction to model bias caused by the deterioration in the battery, the iteration in the KF algorithm is doomed to failure. It is necessary to modify the innovation to meet the variation of battery model caused by the deterioration while KF prediction is carried out.

3 Kalman Prediction Based on SVM Innovation Update

Since KF prediction should be modified to adapt to the variation of the model caused by the deterioration in the battery, we propose a new approach based on SVM. Innovation predicted through the SVM time series is used for forecasting the future values as the iteration in KF algorithm is modified, which makes it possible for KF prediction to correct the model bias.

A. Prediction of time series based on SVM

Support Vector Machine is a linear machine working in the highly dimensional feature space formed by the nonlinear mapping of the N-dimensional input vector x into a K-dimensional feature space (K>N) through the mapping function $\phi(x)$ [7] [8].

The key point in prediction of the time series using SVM is that how could sequence of the measured values could be fitted to the character of a nonlinear system. In the proposed SVM method, values in the past are collected as an input vector $\mathbf{x}(k)=[x(k),x(k-1,\cdots,x(k-p+1))]$, the output is the value that will be measured in the future $x(k+1)$. A mapping function $f:R^p \to R$ is adopted in the proposed regression based on SVM. The input learning data and the output data of the regression based on SVM are as follows:

$$\mathrm{X}=\begin{vmatrix}\mathbf{x}(p)\\ \mathbf{x}(p+1)\\ \vdots\\ \mathbf{x}(n-1)\end{vmatrix}=\begin{vmatrix}x(p) & x(p-1) & \cdots & x(1)\\ x(p+1) & x(p) & \cdots & x(2)\\ \vdots & \vdots & \ddots & \vdots\\ x(n-1) & x(n-2) & \cdots & x(n-p)\end{vmatrix},\mathrm{Y}=\begin{vmatrix}\mathbf{x}(p+1)\\ \mathbf{x}(p+2)\\ \vdots\\ \mathbf{x}(n)\end{vmatrix}$$

The prediction function obtained is as follows:

$$x(k)=f(\mathbf{x}(k-1))=\sum_{i=p}^{n-1}(\alpha_i-\alpha_i^*)K(\mathbf{x}(i),\mathbf{x}(k-1))+b \tag{13}$$

As we know, α_i is the lagrange multiplier [9], $K()$ is kernel function and b is the bias.

B. Kalman filter prediction of the failed batteries based on SVM innovation update

The basic idea of Kalman detection based on SVM innovation update is that: SVM time series model of the innovation is formed by the measured data at first, Kalman filter detection is combined with SVM innovation prediction before the SOC and trend of deterioration could be forecast. The prediction system can be divided into the following two main steps:

(1) Prediction: State and the output in the future time are predicted as follows:

$$\begin{aligned}\hat{\mathbf{x}}^*(k+1)&=A\hat{\mathbf{x}}(k)+B\mathbf{u}(k)\\ \hat{y}^*(k+1)&=C\mathbf{x}^*(k+1)+D\mathbf{u}(k+1)\end{aligned} \tag{14}$$

Innovation in the future can be obtained from time series measured in the past:

$$\tilde{\varepsilon}(k+1) = f(\varepsilon(k), \varepsilon(k-1), \cdots, \varepsilon(k-p+1)) \quad (15)$$

(2) Modification: the output and state variables are modified by the innovation as follows:

$$\hat{\mathbf{y}}(k+1) = \hat{\mathbf{y}}^*(k+1) + \tilde{\varepsilon}(k+1) \quad (16)$$
$$\hat{\mathbf{x}}^*(k+1) = \hat{\mathbf{x}}^*(k+1) + K\tilde{\varepsilon}(k+1)$$

As depicted in equation (15) and (16), the predicted innovation $\tilde{\varepsilon}(k+1)$ obtained from the sequence of the error values in the past by means of SVM regression is used in the Kalman filter, which enables the ability to fit with the deterioration of the battery.

4 SOH Estimation and Experimental Results

The designed estimation approach, explained in previous sections, is tested in the next step as an estimator. The proposed algorithm is applied in the Power Environments Monitor System made by Wenzhou Makepower Electrionics Co.,Ltd , in which voltage acquisition and online testing of the resistance are adopted to achieve a comprehensive monitoring of the battery. Data used for verifying the efficiency of the proposed approach is extracted from the database of the Monitor system which has been used in WenZhou Branch of China Unicom Co.,Ltd for about 4 years. Four batteries with different SOH are selected as our samples from the database in the experiment.

The extent of deterioration of a battery is defined as follows:

$$SOH = \frac{Q_{\max}(Aged)}{Q_{\max}(New)} \quad (17)$$

On the other hand, terminal voltage V_o and SOC of a battery observe the following linear principle during charging processes:

$$V_o = \alpha_1 SOC(t) + \alpha_0 \quad (18)$$

(18) shows that SOC can be estimated when terminal voltage is measured. Experiment results showed in fig.2 verify this phenomenon: the relationship between Q and V_o can be fitted to linear, while the slope of the line denotes the SOH of a battery. For a given battery, SOH is estimated through the proposed approach.Fig.2 shows that the estimated SOH during the charging and discharging processes is 76% and 72%, respectively. The actual SOH of this battery calculated by Eq. (1) is 73%.The accuracy of the proposed method is verified.

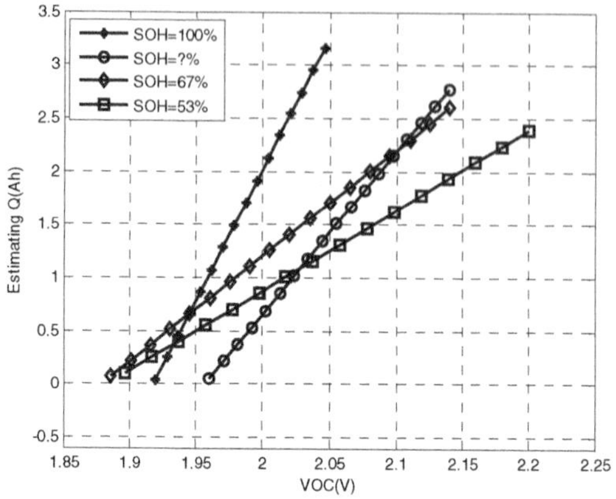

Fig. 2. Estimation line with different SOH through SVM-KF method

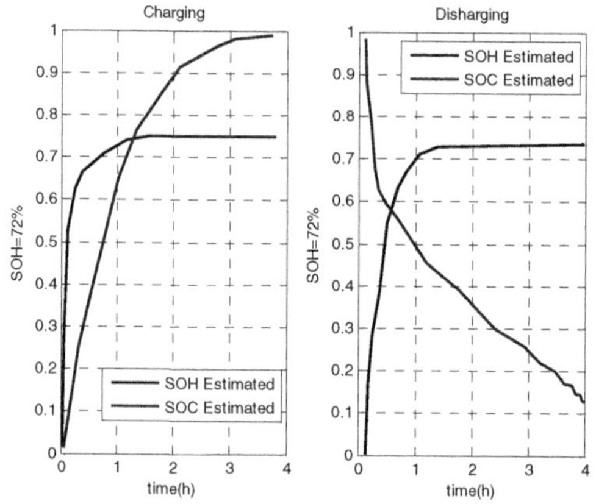

Fig. 3. Estimation of the SOH and SOC of a VRLA battery

5 Conclusions

In this paper, a novel method was proposed for on-line estimation of SOH and SOC of VRLA batteries. The proposed method is based on the KF prediction of an equivalent model of the battery. SVM is introduced while KF is employed to predict SOC and SOH of the batteries. Experimental results show accurate estimation of the SOH as well as SOC during the charging and discharging processes.

References

1. Pan, s., Qian, z., Lei, n.: A novel charge and discharge equalization scheme for battery strings. Journal of Zhejiang University 2(1), 56–59 (2000)
2. Johnson, V.H.: Battery performance models in ADVISOR. Journal of Power Sources 110, 321 (2002)

3. Coleman, M., Lee, C.K., Hurley, W.G.: State of Health Determination:Two Pulse Load Test for a VRLA Battery. In: Proc. 2006 IEEE Power Electronics Specialists Conference, Jeju, Korea, vol. 37, pp. 1–6 (2006)
4. Kaiser, R.: Optimized battery-management system to improve storage lifetime in renewable energy systems. Journal of Power Sources 168(1), 58–65 (2007)
5. Jossen, A.: Fundamentals of battery dynamics. Journal of Power Sources 154(2), 530–538 (2006)
6. Xiong, K., Zhang, H.Y., Chan, C.W.: Performance evaluation of UKF-based nonlinear filtering. Automatica 42(2), 261–270 (2006)
7. Haykin, S.: Adaptive filter theory (Fourth Edition). Prentice-Hall, Englewood Cliffs (2002)
8. Vapnik, V.: Statistical learning theory. Willey, N.Y (1998)
9. Zhang, J.F., Hu, S.S.: Nonlinear time series fault prediction based on clustering and support vector machine. Control Theory & Applications 24(1), 64–68 (2007)

Application of Multi-task and Real-Time Communication Technology Based on DM642

Kai Liang and Shuguang Dai

School of Optical-Electronical and Computer Engineering
University of Shanghai for Science and Technology
Shanghai, China
w.l.l_so@qq.com

Abstract. We have completed a project named "car navigation by machine vision technology", which was consisted of several smaller projects, and one of them was to schedule all the tasks in a real-time manner. This article tries to give some views about how to use the DSP / BIOS real-time operating system to cope with multi-task communication by solving the following problems: "how to achieve synchronization of image processing and car motion"," how to use the image threshold to control the state transition of car motion", and "how to arrange the required schedule of various states of car motion". Finally, we made the car move in a similar "8" shaped path periodically.

Keywords: DM642, DSP/BIOS, RTOS, Machine Vision.

1 Introduction

DM642 (TMS320DM642) is a TI company's 32-bit fixed-point DSP product, which is dedicated to multimedia applications. It uses TI's latest advanced structure of instruction word (VelociTI), enabling to process 8 instructions in one instruction cycle parallel. Thus, in the case of the maximum frequency up to 720MHZ, the processing speed can reach 5760MIPS. In addition, it has a large capacity of double Cache, 64 independent channels of EDMA controller and rich peripheral interfaces. These features constitute DM642's obvious advantages in the field of developing equipments of real-time image processing and multimedia communications.

CCS3.3, published by TI cooperation, based on windows operating system, is currently the best and most popular IDE for DSP software development. Its greatest feature is including RTDX (Real Time Data Exchange) technology, and DSP / BIOS real-time operating system (RTOS). The former enables data exchange between DSP program and other programs written by LABVIEW or C++, without interrupting the target system. The latter is mainly designed for multitasking real-time scheduling and synchronization, as well as the application of host / target communications and real-time monitoring system.

We have done a project based on DM642, which used real-time image processing technologies for car navigation. The main task was to guide the car moving in a similar "8" shaped path periodically. In the terms of software programming, there are

D. Zeng (Ed.): ICAIC 2011, Part II, CCIS 225, pp. 464–472, 2011.

four tasks: image process, forward, turn left, and turn right; as well as the synchronization and communication among them.

2 Basic Principles of Multi-task Scheduing

The target of synchronization and communication is to use some skills to enable more than one task to use the same data storage units in different time spots, so as to arrange the required schedule of these tasks.

Available data storage units, include: global variables, semaphore, locks, mailboxes, pipes, queues, and streams. DSP / BIOS provides four types of communication skills: SEM (semaphore manager), MBX (mailbox manager), QUE (atomic queue manager), and LCK (resource lock manager), in TCF (a configuration tool in CCS3.3). They have different functions, with which combined, specific tasks can be completed.

A. DSP/BIOS Real-time Operating System

The role of DSP / BIOS, in the whole process of DSP system development, is shown in Fig.1:

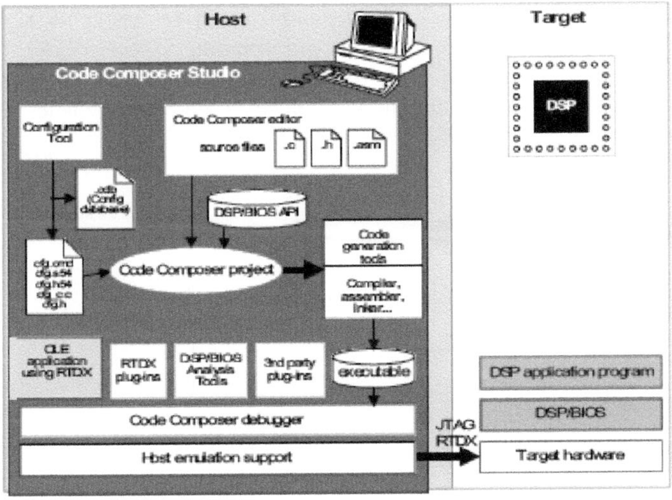

Fig. 1. DSP/BIOS Component

The same with windows operating system, DSP / BIOS operating system plays a role of basic platform for application programming, and it is actually a software package of 24 modules. It gives you a lot of API functions written in assembly language, which you can call directly in the preparation of an application; It provides you a DSP / BIOS configuration tool, with which, you can set parameters of DSP / BIOS module and chip support library, you can also establish DSP / BIOS objects. It supplies DSP / BIOS real-time analysis tools to visually observe the program's performance, and almost does not affect the operation of the application.

There are three steps for DSP / BIOS programming: configure your own DSP/BIOS system with the help of configuration tool; develop applications in the use of API functions; and then debug your applications online real-time analysis tools. You can check out TI cooperation's manual book TMS320C6000 DSP BIOS API Reference Guide for more information. In addition, TI also provides users with BIOS-based high-level application library: DDK (Driver developer's kit), for the development of device drivers; NDK (network developer's kit), used for the development of network transmission system based on DSP; RF5 (reference framework 5), for development of multi-task programs.

B. Tasks Scheduled By DSP/BIOS

Thread is the basic unit of DSP / BIOS, which refers to any independent instruction stream. A task can be composed of one or more threads. DSP / BIOS enables us to design the structure of applications by threads; each thread is supposed to complete a structural function. In multi-threaded programs, it's permitted that high-priority threads preempt lower-priority threads, as well as thread synchronization and communication. There a are four kinds of threads in DSP / BIOS: 1) *hardware interrupt (HWI)*: used to respond to asynchronous external events, such as the timer's clock-match interruption, the external sensor's threshold-match interruption, and it can't be stopped while running; *2) software interrupt (SWI)*: triggered by calling SWI functions, it can only be stopped by HWI while running; *3) task (TSK):* triggered by functions, and can be suspended while running; *4) background threads (IDL)*: it is an idle cycle, performing those functions without deadline. The priorities of the above threads are shown in Fig.2. HWI has the highest priority.

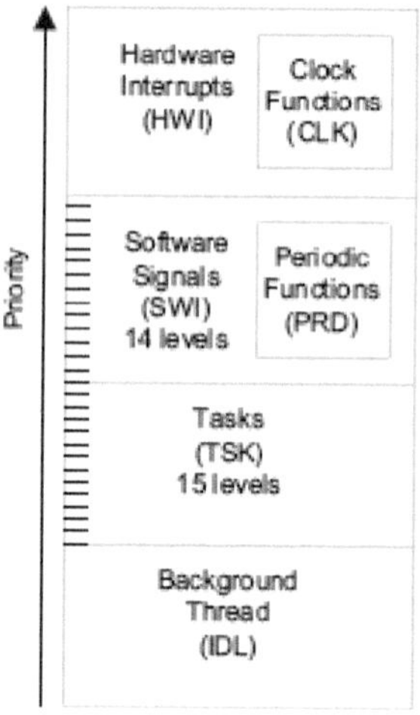

Fig. 2. The Priorities of Threads

A thread has four states during its life time: 1) *running*, get the cpu, and is running; 2) *ready*, once get the cpu, it runs; 3) *blocked*, run after something has happened;

4) *terminated*, run over. At a time point, only one thread can be running. So what can we do to make two or more tasks to run synchronously? The concept of time slice can solve this problem. Most of the time, the operating system will divide the executing period of a task into a lot of slices, the implementation of some tasks is based on the conditions created by the accomplishment of other tasks, so we can execute the time slices from different tasks alternate. So we can say that synchronization among different tasks is actually the reasonable arrangement of threads' different states, with the help of all kinds of API functions, and thus, we can achieve our goals finally. The tasks schedule is shown as fig.3.

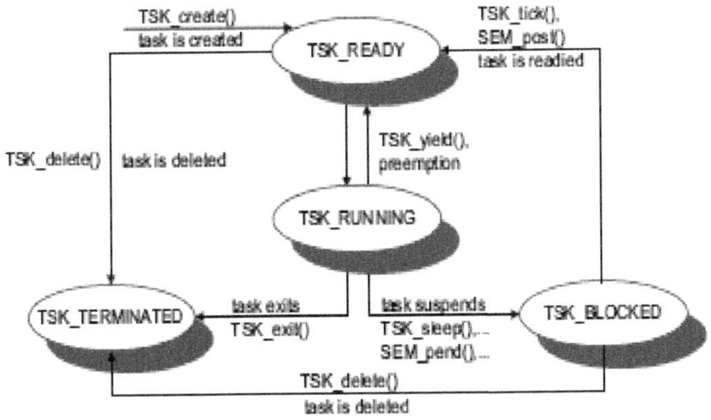

Fig. 3. Tasks Schedule

3 Specific Application

From the analysis above, we can see that, the fulfilment of the real-time and multi-task synchronization and communication is base on a hardware condition and a software condition: 1) hardware: the processor should be fast enough, so the 720MHZ clock speed and 5760MIPS processing speed of DM642 is competent; 2) *software*: to have a powerful real-time operating systems, and DSP / BIOS can make it, too. However, if we want to arrange the schedule of these tasks reasonably, so as to enable the car to go in a similar "8" shaped path, there is a lot of work to do in practice. The following are three difficult questions.

A. Synchronization of Image processing and Forward-motion

This question requires the system can process the image data in the process of car's going straight. We use semaphore as the storage unit and SEM as the communication skill to answer this question.

You can use TCF configuration tool to generate static semaphores, whose initial values can be set, but can not be changed in the programs; you can also create semaphores with the function Void SEM_new () in the programs.

When a task is pending (applying) for a semaphore, it's in the state of waiting for other tasks to release the semaphore; once the application is successful, it starts running. The flow chart is as Fig.4 shown (PROC_run means PROCESS is running, FORW_SEM means FORWARD is applying for the semaphore, and you can guess the meanings of FORW_run and PROC_SEM easily).

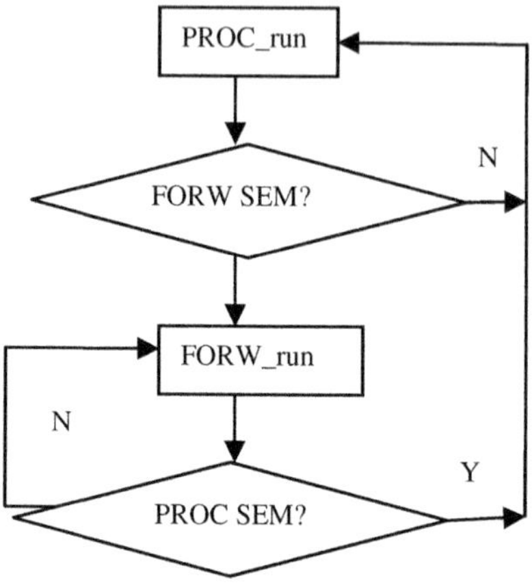

Fig. 4. Flow Chart

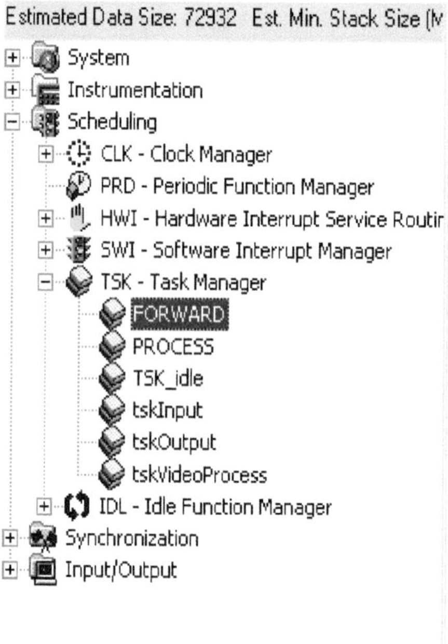

Fig. 5. Configuration tool

We can see from Fig.4 that the two tasks are running at the "same" time by competing with the same semaphore; in fact, they are not. However, because of the very fast processing speed of DM642 (5760MIPS), the result is that they are running almost at the same time.

The following is the steps to answer this question:

1) *Use TCF to create two tasks*: PROCESS (for image process) and FORWARD (for going straight), and two relevant functions: process () and forward (). In order to make PROCESS run before FORWARD, we have to set the former with a higher priority. As the Fig.5 shown.

2) *Use TCF to create a semaphore, named sem, whose initiate value is 1.*

3) *The referenced codes are as following*:

```
Void process ()
{
while (1)
{/*apply for the semaphore */
SEM_pend (sem, SYS_FOREVER);
IMG_process () ;  //the function of image process
SEM_post   (sem) ;   // release the semaphore
TSK_sleep (1); // give up cup for other tasks
}
}
Void forward ()
{
while (1)
{/*apply for the semaphore */
SEM_pend (sem, SYS_FOREVER);
Forward  () ;  //the function of forward
SEM_post   (sem)  ;  //release the semaphore
TSK_sleep (1); // give up cup for other tasks
}
}
```

B. Use Image Threshold to Control Car's Left Turning

We use queue as the storage unit and QUE as the communication skill to answer this question.

Queue is a data structure, with which the elements can be operated in a first in first out (FIFO) manner. Elements of the queue can be any structure, but its first member variable must be QUE_Elem type.

Following are the steps to solve the problem:

1) *Use TCF to create two tasks*: THRESH (for getting the threshold message) and LEFT (for left turning), and two relevant functions: threshold () and left ().
2) *Establish a structure used as a data storage unit.* typedef struct ScomBufChannels

```
{
   QUE_Elem queElem; //must be the first member
   int thre;   //threshold information
} ScomBufChannels;
```

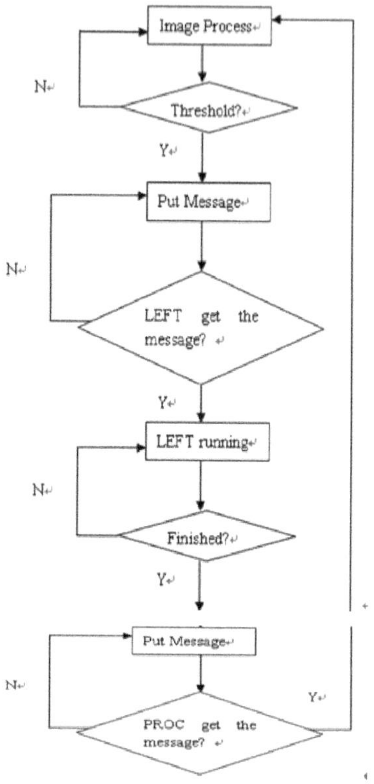

Fig. 6. Flow Chart

3) *Create a new queue for the transmission of information*, using the function SCOM_create ().For example: SCOM_create ("PROCTOLEFT", NULL).
4) *Flowing Chart*, as Fig.6 shown.
5) *The referenced codes*

```
ScomBufChannels message; //define a message
Void proc ()
{
int thr;
SCOM_Handle fromProctoLeft, fromLefttoProc; //define a queue handle
   fromProctoLeft = SCOM_open ("PROCTOLEFT ");
fromLefttoProc = SCOM_open ("LEFTTOPROC "); //connect the queue with its handle
   while (1)
{
thr= threshold () ; // get the result of image process
   if (thr>=THR)// When it reaches the threshold value
   {
message.thre=thr;
SCOM_putMsg (fromProctoLeft, &message); // put the message into the queue
```

```
TSK_sleep (2000); // give up cpu
  SCOM_getMsg (fromLefttoProc, SYS_FOREVER);
  // when get the message from LEFT, rerun the function threshold ()
  }
  }
  }
  Void left ()
  {
  SCOM_Handle fromProctoLeft, fromLefttoProc; // define a queue handle
   fromProctoLeft = SCOM_open ("PROCTOLEFT ");
   fromLefttoProc = SCOM_open ("LEFTTOPROC "); // connect the queue with its
handle
   while (1)
  {
  SCOM_getMsg (fromProctoLeft, SYS_FOREVER);
  //get the message from proc, when it comes, begin running following codes.
  Turn_left () ;
  SCOM_putMsg (fromLefttoProc, NULL);
  // after the task Turn_left () is finished, send message to proc for next loop
  }
  }
```

C. Arrange the Required Schedule of Various Motion State

We just need a global variable, named "flag", to accomplish this task. The initial value flag is 0, when the condition "thr>=THR&&flag==0" is met, turn left for the first time, and flag plus 1 in the body of function left turning. When the condition "thr>=THR1&&flag==1" is met, go forward for the second time. And so as to complete the required order of the motioning states.

4 Summary

It has such advantages as high efficiency, fast, reliable, simple debugging, and so on, to develop DSP applications with DSP / BIOS real-time operating system. The process of programming is actually a process of creating several tasks and integrating them with a few communication skills, and the latter is the key point as well as the difficult point. We have to master a variety of basic communication methods, but don't rigidly adhere to the existing variety of forms, on contrary; we need to define the specific carrier to transmit information according to the specific issues, so as to achieve the goal of multi-task-and-real-time operations.

References

1. Tian, L.Y., He, P.K., Zhu, M.Y.: TMS320C6000 DSP programming tools and guidelines. Tsinghua University Press (September 2006)
2. Texas Instrument Incorporated, DSP/BIOS User's Guide (Rev. B) (2009)

3. Texas Instrument Incorporated, TMS320C64x Image/Video Processing Library Programmer's Reference (Rev. B) (2009)
4. Texas Instrument Incorporated, TMS320C6000 DSP/BIOS API Refenerce Guide (Rev. F) (2009)
5. Texas Instrument Incorporated, TMS320C6000 Programmer's Guide (Rev. G) (2009)

Substation Intelligent Monitoring System Based on Pattern Recognition*

ShuMin Liu[1] and XuJuan Xu[2]

[1] Nanchang campus Jiangxi University of Science and Technology, NanChang city, east street of shuanggang, NO.1180, jiangxi province, China
[2] Educational administration, Schools of Civil Affairs of Jiangxi Province, NanChang city, east street of shuanggang, NO.1180, jiangxi province, China
shui.yang@sohu.com

Abstract. Pattern recognition technology is a new intelligent science with a wide range of applied. It can be applied in a variety of intelligent control and monitoring areas. According to the technical characteristics of pattern recognition, it can be used in substations video surveillance system. Tests show that this control method can effectively deal with the increasing workload of the substation monitoring of tasks, but can also greatly improve the control accuracy and reduce cost. This paper briefly discusses the basic principles of pattern recognition, and then introduces some commonly used pattern recognition method, And applies them to the specific substation control and inspection process respectively.

Keywords: pattern recognition, unattended, Substation Monitoring, illegal intruder.

1 Introduction

In recent years, To meet the growing demand for electricity in our country, the power system must continuously expand. In order to improve the automation level of production for power system, many unmanned substation and communication station are building. In the meantime, numerous electric power company are building centralized Remote Image Monitoring System in order to improve the security and dependability of Electricity production system. The feature of "remote viewing", as a new function of power system , with "remote measuring", "remote control", and "remote regulating", have achieved a variety of important detection for the system. For example, the main control room, high-pressure chamber, breaker and isolation switch can be monitored with real-time visualization. Now , available Video Surveillance System have only the function of video monitoring and recording, unable to complete the goal of intelligent control analysis to monitoring objectives.

In the base of tradition monitoring system, using image processing and pattern recognition technology to automatically identify the monitoring target in order to

* This work is worded for pattern recognition in the field of Substation Intelligent Monitoring System.

D. Zeng (Ed.): ICAIC 2011, Part II, CCIS 225, pp. 473–478, 2011.

make more effective use of existing video surveillance system, and equipment failures can be reported to the police and inform the location and cause of the malfunction of staff when the equipment fails or a security risk situation.

2 Image Processing and Pattern Recognition+

Image processing and pattern recognition are two key areas of intelligent control technology, it can be used effectively in substation remote video monitoring system. The Image processing and recognition structure shown in Figure 1.

1) Get image information: after sampling the analog video signals getting by a dedicated camera video capture card into digital video and quantify the signal. During the conversion process, in order to save computer resources and transmission time, the image must be compressed by a compression video capture card.

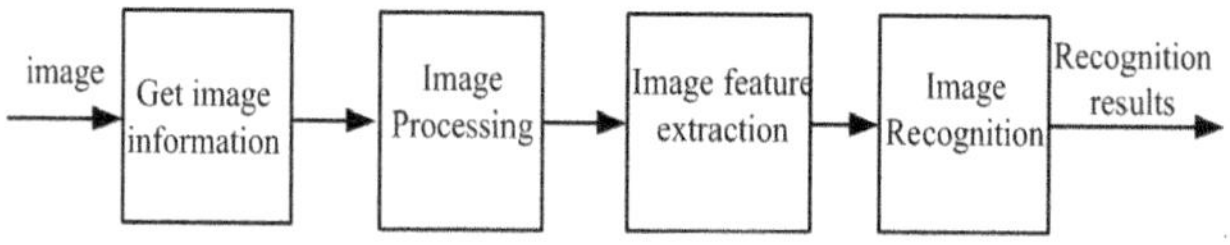

Fig. 1. The Image processing and recognition structure

2) Image processing: Image processing including image enhancement, image segmentation, edge detection, image restoration and other process. In order to remove the interference, noise and differences, we must first preprocess the original image, and converting it into a suitable form so that computer to complete the image feature extraction easily. There are several image preprocessing methods, such as histogram, gray level transformation, image smoothing, image sharpening and so on.

3) Image feature extraction: In image recognition, have to be image feature extraction, because an image contains a very large amount of information, and we do not need to know all the information. The role of the image feature extraction is driveing useful information from the filtering data, and from a number of features to find the best value that reflect the nature of the image, for example, the texture features, shape features, color characteristics, etc, in order to reduce the difficulty of follow-up process.

4) Image recognition: the image contains a variety of information. Now, There is no general method to identify different objects. But there are several popular methods: statistical pattern recognition methods, fuzzy identification, and neural network methods, etc.

3 The Methods of Pattern Recognition

Application of pattern recognition is a very wide range of subjects. It can be used in speech recognition, face recognition, fingerprint recognition, character or signature recognition, fault detection. Pattern Recognition diagram shown in Figure 2, It is a deeper level of image recognition method on the basis of image processing, which contains the following types of identification.

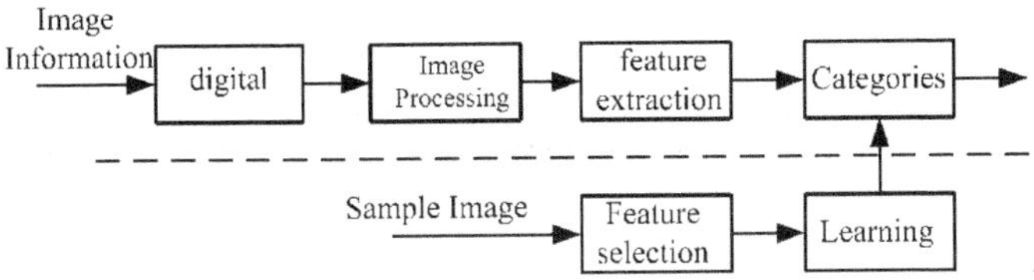

Fig. 2. Pattern Recognition diagram

1) Statistical Recognition

When the statistical properties of image feature is known or can be derived, you can use statistical pattern recognition methods. Commonly used method is the Bayesian classifier.

Here, $\{\omega_1, \omega_2, \ldots, \omega_n\}$ represent a limited set of categories n, Vector X represents a d-dimensional feature vector, $p(X/\omega_j)$ is the conditional probability density function based on ω_j, and $P(\omega_j)$ is the prior probability for ω_j. Therefore, the Bayes theorem can be expressed as:

$$p(\omega_j / X) = \frac{p(X/\omega_j)P(\omega_j)}{p(X)} \tag{1}$$

In the formula,

$$p(X) = \sum_{j=1}^{c} p(X/\omega_j)P(\omega_j)$$

Work rules for the Bayesian classifier: If there exists i $\neq$ j, and $P(\omega_i / X) > P(\omega_j / X)$, then is sentenced to ω_i.

Bayesian classifier has been shown to be optimal by literature 1 in the condition of minimum classification error rate. But sometimes the smallest error rate is not necessarily the best standard. Taking into account the different degree of error may be caused by the consequences of different circumstances, literature 2 Proposed the classification criteria based on minimum risk.

In addition, the image histogram can reflect a lot of information, such as mean, variance, standard deviation, Mean difference, this is the most basic features of statistical values can be calculated by computer

When the image features obvious or easily inferred, the statistical pattern recognition method is the simplest, and we can objectively describe the image features by it. But this method has two shortcomings: First, if the image is not obvious or can not be inferred, it is easy classification error; Second, if the vector used to represent the characteristics is large dimensions, there will be "dimension disaster" problem.

2) Template matching

Template matching method is a relatively high recognition accuracy and faster method[3] Equation (2) as the standard template matching algorithm used in the relevant metric function:

$$D(i,j)=\sum_{m=1}^{M}\sum_{n=1}^{N}[I_{ij}(m,n)-T(m,n)]^2 \tag{2}$$

In the formula, $I_{ij}(m,n)$ representative sample of the image; $T(m,n)$ represents the standard template image; and M, N, respectively, expressed the image width and height

The working principle of template matching is to calculate minimum variance between the sample image and the standard template, and then determine the type of sample according to the matching principle of minimum variance between the two image. The results of this method are more accurate, but it is difficult to determine the match value. And the matching results are vulnerable to the environment.

3) Fuzzy Identification

In the natural sciences or social sciences, there are many definitions of less stringent or fuzzy concept. In order to process, analyze the "fuzzy" data, then we need to use the knowledge of fuzzy theory. During the fuzzy identification, first, the typical sample parameters be entered into the computer to learn, and then become a standard pattern for recognition process. Then, respectively, after image processing, feature extraction, and get characteristic parameters for target image, and finally establish the corresponding similarity matrix by which we can recognize the objects finally.

Fuzzy identification methods include: the principle of maximum membership method, the principle of selection methods and recent fuzzy clustering method. This paper briefly describes the first method.

For x, if

$$u_{\overline{A}_i}(x)=\max\{u_{\overline{A}_1}(x),u_{\overline{A}_2}(x),....,u_{\overline{A}_n}(x)\} \tag{3}$$

Then, the element x belongs to class $\overline{A}_t$,

In the formula, $\overline{A}_1, \overline{A}_2$, ..., $\overline{A}_n$ is a n-dimensional domain of fuzzy sets composed by object Recognition. Fuzzy identification method can make the computer closer to human intelligence, has a high adaptability. The disadvantage is that: ① the selection of threshold is difficult, if selected properly, false acceptance and rejection easily.② difficult to accurately determine the membership function

4) Neural network recognition

There are many types of neural network model, in which BP (error back-propagation algorithm) network model is one of the most extensive network. BP network identification method is a supervised learning algorithm, can self-study, and has a fast recognition speed and classification speed. It has the characteristics of slow convergence in the larger schema mapping, easy to fall into local minimum, inaccurate judgments and other defects. BP network learning as follows: ① Initialize network weights, the weights and the node threshold is set to a small random value;② training network; ③ according to the given input training pattern, calculate the network output mode and compared with the desired output, if there are errors, perform ④, otherwise return to ②; ④ back propagation, correction weights. Weight correction formula is:

$$W_{ij}(t+1)=W_{ji}(t)+\eta\delta_{pj}o_{pj}+\alpha[W_{ji}(t)-W_{ji}(t-1)] \quad (4)$$

In the formula, η is the learning factor; o_{pj} is the actual output of unit j. BP network as easy to fall into local minimum, the training may lead to failure, it must be optimized and improved. literature 4 proposes an optimized method of BP network, that is, on the basis of the traditional BP algorithm, Proposed an improved BP learning algorithm.

This method requires to make the dynamic adjustment for learning step, to avoid learning too slowly, and repeated shocks and other issues in traditional learning. Experiments show that the algorithm can reduce network complexity to some extent, and has a faster convergence speed than the traditional algorithm.

Each pattern recognition method has its own characteristics and applications, meanwhile, there are complementarities between the various methods. It can significantly improve recognition accuracy if combined with a variety of methods.

4 Pattern Recognition in Substation

Pattern Recognition in the substation video system, which can effectively identify and monitor its internal environment and equipment. The Identify objects include: the status of lights, 7-segment digital meters, indicators and meters, and so on.

1) The state of signal recognition

Literature 6 proposed a recognition method for red, green, yellow three states of the lights in substation using the color characteristic value reflected by the histogram. Control Figure 1, we can recognize the state of the lights just by extracting the characteristic value of color signal. We need a lot of statistics on the scene after the order to obtain correct results.

Template matching method can also be used for signal recognition. First, we get the standard sample of signal light and compare it with the specified area in the video image for substation, and then find the sample that has a highest level match value , which can reflect the the state of the signal light.

2) Identification of 7-segment digital instrument

There are a lot of digital instruments in substation. Literature 6 proposed a recognition method of 7-segment digital meters: After images were gray-scale transformation, thresholding, median filtering and expansion of pre-form, and then split it to the character, as long as determine whether the black pixels in the corresponding paragraph, we Can identify the 7-segment numeric value.

3) Identification of a pointer insturument

Substation power system has many pointer instruments, such as: voltage meter, ammeter, barometer, thermometer and so on. literature 5 proposed a meter pointer angle recognition algorithm based on Hough (Hough) transform principle.

The algorithm uses a pointer subtraction algorithm to determine the center according to the position pointer difference between the reference position and the cursor , and then complete the point of recognition by finding the the pointer position.

4) Identification of illegal intruder alarm

Environmental testing for unattended substation is also very necessary. The system Should automatically recognize and alarm when the illegal intruder. Literature 7 presents a moving target detection method combining the background subtraction and temporal entropy. The method first construct a background image, and get the difference image by the current frame and background subtraction, then calculate the entropy space of difference image, finally, we can detect moving targets.

5 Conclusion

The pattern recognition technology for remote substation monitoring system not only saves manpower, and overcome the possible human error. For the substation to provide the conditions to achieve unattended, but also for substation automation, intelligence provides the conditions. Image processing and pattern recognition technology can not only realize the substation equipment and environmental monitoring and identification, but also can be used for fingerprint identification, product testing in the industrial manufacturing, military reconnaissance, weather analysis and forecasting of natural disasters, etc.

References

1. Bian, z.: Pattern Recognition (Second edition). Tsinghua University Press, BeiJing (2002)
2. Wald, A.: Contributions tO the theory of statistical estimation and testing of hypotheses. Annals of Math-ematieal Statistics 10, 299–326 (1939)
3. Wang, Q., et al.: A fast template matching target recognition algorithm[J]. Computer Engineering and Applications 6, 42–43, 49 (2000)
4. Meng, B., et al.: Feedforward neural networks in an improved BP algorithm. Southeast University press(Natural Science) 31(4), 40–42 (2001)
5. Liu, Y., You, Z.: For Image Object Recognition and Neural Network Applications Vehicle Identification. Computer engineering 29(3), 30–32 (2003)
6. Zhang, J., Yang, S., et al.: Pattern Recognition in Remote Video Monitoring System Substation Applied Research. Power Systems 26(12), 50–53 (2007)
7. Chang, X., Feng, X.: Based on background subtraction and temporal entropy new method of moving target detection. Computer Simulation 25(3), 235–238 (2008)

The Design and Application of Sewage Treatment Automation System Based on S7-300 PLC

Bai Jing-Jing, Zhao Zi-Xian, Bao Chun-Hong, and Feng Shi-Guang

College of Electronic Engineering
Inner Mongolia University
Hohhot, China
J.J_Bai@yahoo.cn

Abstract. For the demand of domestic sewage treatment in county-level city, a plan of Distributed Control System in municipal wastewater treatment automation system is proposed. The main function of which is to supervisory control each sub-system of municipal wastewater treatment via the configuration software WinCC, Siemens S7-300 PLC. The feature of the system is mainly as follows, collection of PLC data, protocol conversion, the transmission of data through fiber, the realization of projects in configuration software, the real-time data storage in database and the release of each kind of report forms, curves, alarm events in display and so on. The idea of the system designation and the detailed characterization of the system are also presented in this paper.

Keywords: sewage treatment, Distributed Control System, Siemens S7-300 PLC, WinCC Configuration software.

1 Introduction

Along with the rapid development of economy in China, environmental protection has been a prominent issue. Sewage Treatment is one of the most important parts of environmental protection. As the computer technology and sewage treatment technology developed rapidly, it is necessary to monitor the wastewater treatment process with advanced control technology and equipment.

The sewage treatment project of Inner Mongolian Huhhot Tumd Left Banner is the typical project among the county-level city of Hohhot, using Cyclic Activated Sludge System (CASS) to treat wastewater. For this project, we designed a DCS system, whose control system is Siemens S7-300 PLC and monitoring software is Siemens WinCC 6.2.

2 CASS Process

CASS biochemical treatment system is a complex system of multi-parameter (such as liquid level, water quality, flow rate, pressure, etc.), multi-tasking (such as water-in,

D. Zeng (Ed.): ICAIC 2011, Part II, CCIS 225, pp. 479–486, 2011.

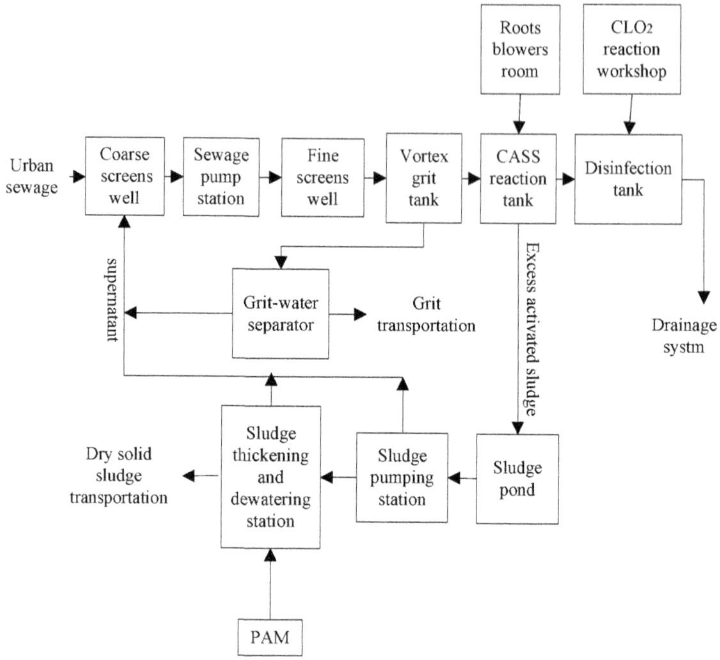

Fig. 1. The flow chart of sewage treatment

aeration, sedimentation, drainage, etc.), multi-equipment (such as water decanter, blower, regulator etc.). It is the core of the whole biochemical disposal system. The main reaction zone of CASS pool works according to "aeration, idle, sedimentation, and decanter" cycle procedures. Among them, the control of dissolved oxygen in CASS pool, anaerobic segment and aerobic segment is the core.

3 The Structure of Control System

The structure of automatic sewage treatment control system is centralized management and decentralized control. The control system employs the control method combined with remote automatic control, remote manual control and field automatic control.

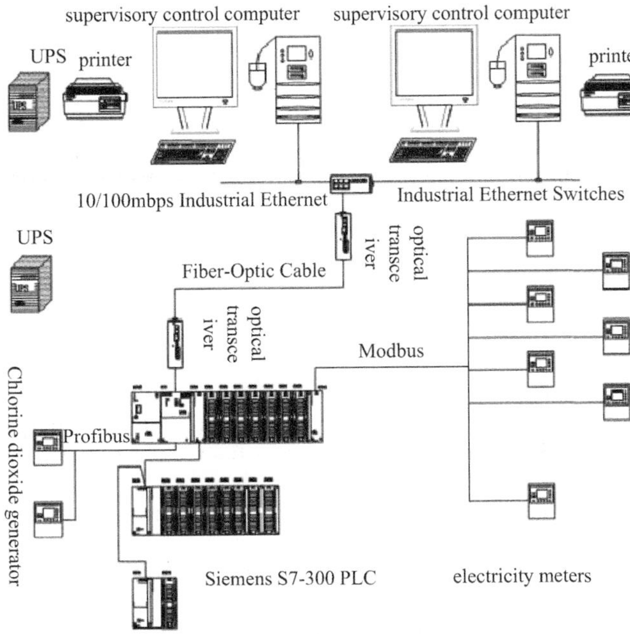

Fig. 2. The structure of control system

In order to ensure sewage treatment systems well running, the master station and slave station of PLC are independent. If one of the sub-systems breaks down, it will not affect the normal running of other system devices. Even if the whole control system malfunctions, it also can adjust itself to field manual control mode in time to ensure the normal operation of the sewage treatment system. The PC in the central control room uses the model of dual-redundant to realize real-time monitoring. Two servers standby each other to jointly perform monitoring tasks. If one server conk out, the other one can undertake the tasks, automatically ensure the system can continue to provide services in the case without manual intervention.

4 Data Communication

S7 series PLC communications network mainly includes the following: multi-point interface (MPI) protocol of data communications, PROFIBUS, industrial Ethernet, point-to-point connections, through AS-I process of communication. This project has PROFIBUS, used in industrial Ethernet, point-to-point connections.

A. The communication between S7-300 and WinCC

PLC and PC communicate with each other through industrial Ethernet, PLC transmit the data of real-time production to the real-time database of configuration software of

PC via optical fiber. PC can monitor the field device from far away. Because the PLC selected in this system is CPU 315-2PN/DP, which has integrated PROFINET interface. Communication between S7-300 and WINCC uses TCP/IP network protocol. TCP/IP network protocol is a de facto standard about communication protocol in IT services. Although the response time is about 100 ms magnitude, for application of factory control, the response time is enough.

B. The Communication between master station and slave station of PLC

PLC master station uses Siemens S7-300 PLC. The PLC of ClO_2 workshop uses Siemens S7-200 PLC; Communication between master station and slave station of PLC uses Profibus. Profibus, is internationalized and open, do not depend on the standard fieldbus of equipment manufacturer. It is a fieldbus which is used for control of factory automation workshop-level and data communications, and control of field device-level. The transmission speed of Profibus is at 9600 bauds ~1.2×107 bauds, and when the main system startup, all devices connected to the general line should be set at the same speed. This can achieve the dispersed digital control and scene communications network from field device-level to workshop-level, so as to provide a viable solution for factory to realize the comprehensive automation and field device intellectualized. The processor module of Siemens S7-300 PLC is CPU 315-2 PN/DP, which has its own Profibus interface. The S7-300 PLC Profibus network is used as the master station, S7-200 PLC Profibus network is used as the slave station. S7-200 PLC has no Profibus communication interface itself, so it has to communicate with master station through EM277.

C. The communication between PLC and power equipment

According to the design requirements, the communication between PLC and power equipment need to be established, for the current, voltage and power in the power supply system should be monitored. Siemens PLC make the point-to-point communication link with the power equipment through the communication module CP341. The power equipment provides RS485 communication interface in half-duplex asynchronous serial communication, and uses MODBUS RTU communication protocol.

MODBUS protocol is one of the network protocols in the industrial control unit. The control equipments produced by different manufacturers can be linked as industrial network by MODBUS protocol, and that equipments linked in the different networks can make the data communication. The MODBUS network can connect multiple equipments on one line, and each of the equipments can set communication address itself. Each controller can distinguish the devices by the address. Also, it decides what to do with the message send by the address. If response is needed, the controller will create feedback information and send it by MODBUS protocol. On the other network, there are frame and package constructs converted by the message in the MODBUS protocol. The conversion also extends the way to solve the problems such as address, routing paths and mistake detection in the specific network.

5 Siemens S7-300 Hardware

The S7-300 is a modular PLC. It consists of Rack, CPU Module; Signal Module, Function Module, Interface Module, Communications Processor, Power Module and

Programming Module. The PLC system of the project contains three racks, the main rack communicated with Expansion rack by IM360/361. The Phase I of Sewage Treatment Phase has 23 analog input points, 3 analog output points, 222 digital input points and 68 digital output points. In this PLC hardware configuration, 15% of all reserve margins is set aside, which is leaved for the phase II. The S7-300 PLC system hardware diagram is shown in Figure 3.

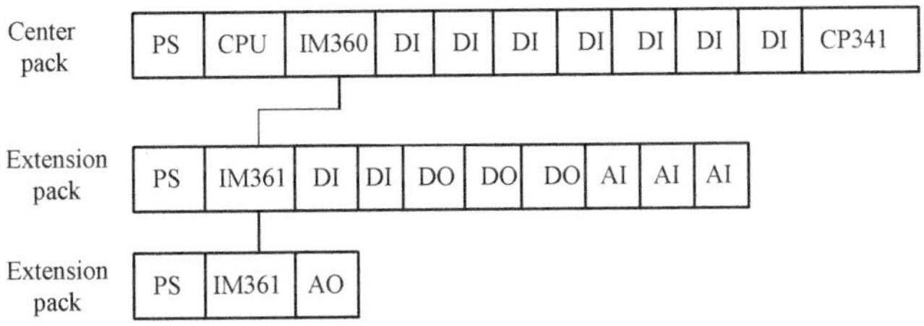

Fig. 3. S7-300 PLC hardware diagram

6 Composition of PC Conitoring System

This paper takes the Chinese version WinCC 6.2 as systematic configuration software to realize the configuration function of supervision appearance. It provides a variety of drivers for PLC, which makes the connection of PLC and PC much easier and the variable list configured in STEP7 can be used directly when connected with WinCC. The functional block diagram for PC monitor system is shown in Figure 4.

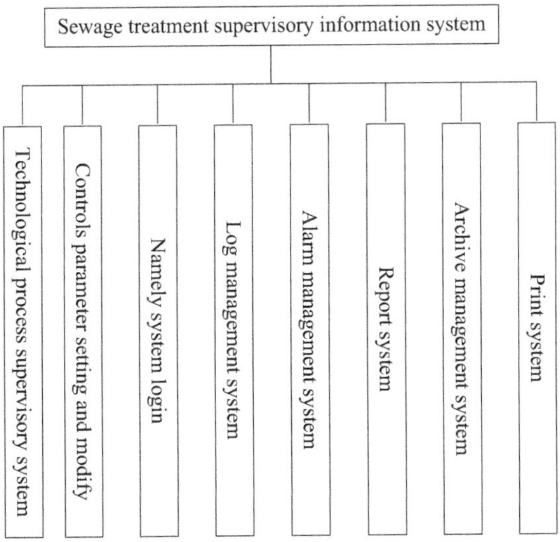

Fig. 4. The functional block diagram for PC monitor system

"PC monitoring system" functional description:

1) Control function: to realize the real-time control of the on-line device in the main control room, such as equipments start/stop, analog output regulation, PLC operating parameters setting online, etc.

2) Display function: to show the running state of the device being controlled via the graphics and animation, and display the real-time operating parameters of the field device.

3) Archive management: to set up a production database and save the original data of the production for calculation and analysis, according to the change speed, the importance of operating parameters, the real-time and historical running state.

4) Alarm function: to realize different alarm grades according to different needs, when an analog value (such as current, pressure, water level, concentration, etc.) measured exceed the given range, or a digital value (such as motor start/stop, valves on/off) is out of order.

5) Print function: to realize the real-time print of reports, graphics, various events and alarm. The print way can be divided into timed print, event trigger print and user-defined print.

6) Log management: to record the information for users to log in and out, in order to search the use interval for different users.

7) Rights management: to set corresponding visit protection for different users, and prevent the users unauthorized to visit the monitor system.

7 PLC Control Program

The system uses modular design to realize the program of PLC system via the STEP7 software. The program flow chart is shown in Figure 5.

The control system adopts modularization program, and falls into 10 subprogram according to the workshop in the whole technical procedures. In the main program, it can call each subprogram in proper order. First, writting an analog acquisition subprogram in the orgnization block 35, and the analog data will be refreshed every 100 millisecond. Then, the control program which is classified as remote manual control and remote auto control can switch through the bit flags between the two ways. Remote manual control can artificially control the devices start/stop, and remote auto control can automatically adjust the state of the device by the change of on-the-spot parameters. So the program can execute automatically and timely for every cycle, and all the devices can auto-stop to control when there are cascading failures. Modbus uses multi-site polling mode to communicate with power meters.

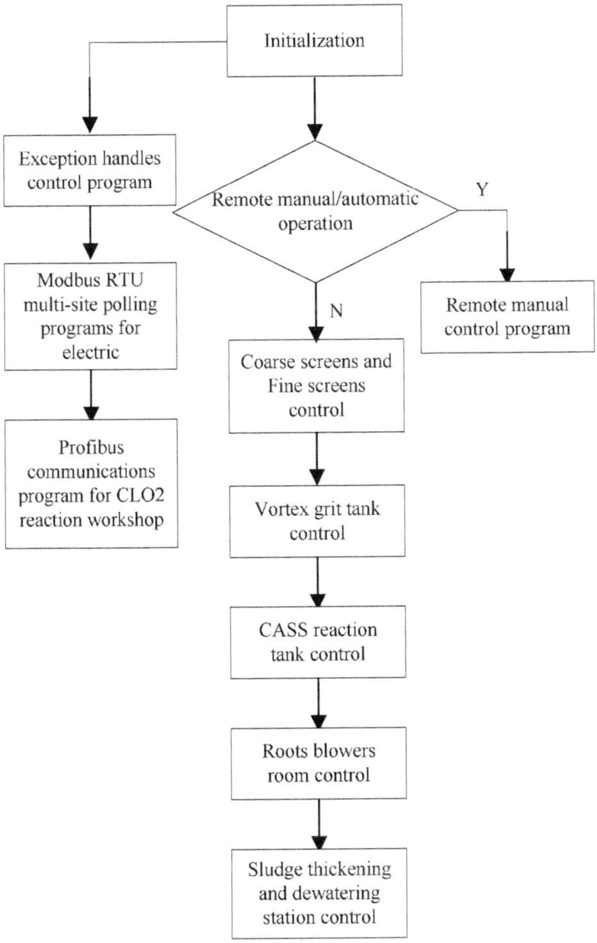

Fig. 5. The flow chart of PLC control program

8 Conclusions

The automation system used in the sewage treatment is worthy of popularization, for it can meet the requirements of modern sewage treatment, reduce the day-to-day maintenance management, decrease the cost of sewage treatment, and ensure the operation and water quality of effluent. With the development of engineering technology, the automation system in the sewage treatment plant will surely show its obvious advantages. The project was tested in 2010.10. The result turned out that the system is an advanced, reliable and simply operated control system, which can better resolve some problems such as low automation, low reliability and large time delay and some other exist problems in the past sewage treatment.

References

1. He, X.Z., Guo, Y.J.: The Online Monitoring System in Wet Zinc Metallurgy Leaching Process Based on Fieldbus. In: Proceedings of the 2009 Second International Symposium on Computational Intelligence and Design, vol. 2, pp. 262–265 (2009)
2. Tian, P., Gao, M., Zhou, P., Li, P.: The Intelligent Control System of Flocculation Process of Sewage Treatment Based on Wavelet Neural Networks. In: Proceedings of the Sixth International Conference on Intelligent Systems Design and Applications, pp. 219–224 (October 2006)
3. Li, C.-y., Chen, H.-g., Lv, F.-c.: Research on Cost Control of Sewage Treatment Engineering Based on Project Financing. In: Proceedings of the 2009 International Conference on Information Management, Innovation Management and Industrial Engineering, vol. 4, pp. 160–165 (December 2009)
4. Sun, X., Hou, Z., Zuo, J.: Real-time Control with Low DO Concentration in SBR Process. In: Proceedings of the 2010 International Conference on Challenges in Environmental Science and Computer Engineering, vol. 1, pp. 350–353 (March 2010)
5. Ling, Z., Jie, C., Lixin, X.: The Research and Application of Freeport Communication of SIEMENS PLC. In: Proceedings of the 2008 International Seminar on Future BioMedical Information Engineering, pp. 318–321 (December 2008)
6. Li, J., Peng, X., Ma, S., Yang, Y., Han, X., Liu, L., Zhang, Y.: Treating Domestic Sewage by Using CASS Technique. In: Proceedings of the 2009 International Conference on Environmental Science and Information Application Technology, vol. 1, pp. 395–397 (July 2009)
7. Ding, J., Shen, Q.: Automatic Control of a Headstock Gear Based on PLC for Water Conservancy. In: Proceedings of the 2010 International Conference on Measuring Technology and Mechatronics Automation, vol. 1, pp. 711–713 (March 2010)
8. Liu, Y., Chen, X.: Design of Traffic Lights Controlling System Based on PLC and Configuration Technology. In: Proceedings of the 2009 International Conference on Multimedia Information Networking and Security, vol. 2, pp. 561–563 (November 2009)
9. Budha, M., Thapa, D., Park, S.C., Wang, G.-N.: Generation of PLC Ladder Diagram Using Modular Structure. In: Proceedings of the 2008 International Conference on Computational Intelligence for Modelling Control & Automation, pp. 1194–1198 (December 2008)
10. Feng, L.-h., Gui, W.-h., Yang, F.: Application of Communication Optimization Strategy Based on Cascade PLC MODBUS in Fire Water System of Hydropower Station. In: Proceedings of the 2009 Second International Conference on Intelligent Computation Technology and Automation, vol. 4, pp. 45–48 (October 2009)

BP Neural Network Algorithm for Multi-sensor Trajectory Separation Based on Maximum Membership Degree*

Chao Sun, Qinquan Gao, Xinhui Li, Zhifeng Luo, and Xiaolin Liu

Department of Automation
Xiamen University
Xiamen, Fujian Province, China
xiaoqin_ma@yahoo.cn

Abstract. Aiming at the problem of trajectory separation which belongs to the data fusion technology, the multi-sensor trajectory separation algorithm of BP neural network based on the maximum membership degree is presented in this paper. The trajectory points can be predicted by establishing the trajectory prediction model which is based on the BP neural network, and the new radar data can be judged whether they belong to the prescriptive trajectory; Based on the prediction of the BP neural network, the multi-trajectory separation algorithm of fuzzy clustering of maximum membership degree is added to improve the effectiveness and accuracy of the trajectory separation. The experimental tests show that the algorithm which is presented in this paper effectively improves the efficiency and accuracy of the trajectory separation, and has a better application value.

Keywords: BP neural network, maximum membership degree, fuzzy clustering, mean square error.

1 Introduction

With the development of information technology, the technology that a single sensor is applied to capture and track has increasingly obvious shortcomings, while the multi-sensor data fusion technology not only expand by several times the advantages of capturing and tracking spatial and temporal coverage, but also reduce information ambiguity, enhance credibility, improve the detection performance and so on, multi-sensor data fusion technology has been widely applied [1].

The multi-sensor data fusion can be divided into the data preprocessing, trajectory correlation, trajectory fusion and so on. If a radar station at any time receives the electromagnetic waves which are reflected from the sky, it will record the relevant data and calculate them, and a set of data can be obtained as follows: the longitude, the latitude, the altitude, the velocity in the direction of longitude, the velocity in the

* This work is partially supported by the Aeronautical Science Foundation under the grant number of 20080768004.

D. Zeng (Ed.): ICAIC 2011, Part II, CCIS 225, pp. 487–495, 2011.

direction of latitude, observation time and so on. An important link in the multi-sensor data fusion technology is that how to accurately separate the different trajectory which consists of discrete points [2].

The complex factors which influence the target trajectory are fully considered in this paper, the trajectory points can be predicted by training and correcting the history trajectory and establishing the trajectory prediction model which is based on the BP neural network, and the new radar data can be judged whether they belong to the prescriptive trajectory through repeatedly train and correct the historical trajectory point and establish the trajectory prediction model which is based on the BP neural network; Based on the prediction of the BP neural network, the multi-trajectory separation algorithm of fuzzy clustering of maximum membership degree is added to improve the effectiveness and accuracy of the trajectory separation[3].

2 BP Neural Network Algorithm

In view of simple hierarchy, easy implementation and convergence, BP neural network is usually used as the supervised classifier[4]. According to the trajectory characteristics, a 3-layer back-propagation BP Neural Network with multi-input units, multi-hidden units and multi-output units is adopted to forecast the trajectory. The basic structure of BP network is shown in the Figure 1.

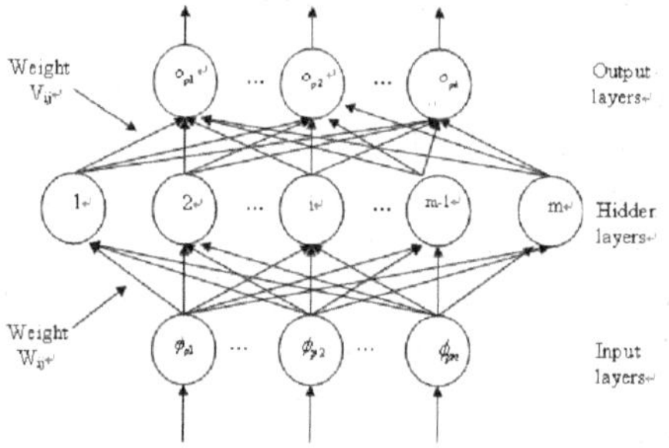

Fig. 1. The basic structure of BP Neural Network

In the Figure 1, n and m represents the number of Neuron of input layer and hidden layer respectively, while ϕ_{p1}, ϕ_{p2} ,......, ϕ_{pn} are the input values of Neural Network which represent the attribute value of the Pth sample model to forecast the trajectory[5].

3 Fuzzy Clustering Based on Maximum Membership Degree

A. Fuzzy mathematical foundation

Fuzzy set is the characterization of fuzzy phenomenon or fuzzy concept. The universe of discourse u is given, and then a value of membership function $u_{\lambda}(x)$ have been appointed for arbitrary $x \in u$, the ordered pair is as follows[7]:

$$\tilde{A} = (u_\lambda(x) / X), \forall x \in u, u_\lambda(x) \in [0,1] \quad (1)$$

B. Membership function

Membership function is used to describe the ambiguity and uncertainty of the middle transition of the difference between objective things. The common membership function which is the normal type has been selected in this paper, the concrete form is as follows[8],where $b > 0$:

$$\mu(x) = \exp[-(\frac{x-a}{b})^2] \quad (2)$$

C. The principle of maximum membership degree

Suppose that X is a set of all the objects to be identified, and $A_i \in F(X)(i = 1, 2...n)$ represent n fuzzy patterns. For any element x in X, if the element x should be identified which model it belongs to, it can be judged by the following principle [9]:

$$\mu_A(x) = \max\left\{\mu_{A_i}(x)\right\} 1 \le i \le n \cdot \quad (3)$$

4 BP Neural Network Hybrid Algorithm Based on the Maximum Membership Degree

A. Fuzzy mathematical foundation

Because the kinetic equation of the target which is flying in the sky is unknown, the future position of the corresponding target can not be calculated according to the state equation. Therefore, the trajectory is predicted by applying the BP neural network in this paper.

1) Definition of input vector and target vector

The radar observation time t, the velocity in the direction of longitude V_x and the velocity in the direction of latitude V_y are the input neurons of neural network, which the input vector matrix $P = (t, V_x, V_y)$; The longitude x, latitude y and altitude h of the target are the training set goals, which constitute the output vector matrix $T = (X, Y, Z)$ in this paper. The position vector of target trajectory of future time point can be represented by defining the input vector $P_c = (t_c, V_{xc}, V_{yc})$ which will be predicted and calculate the output vector $T_c = (X_c, Y_c, Z_c)$ with the neural network. In this paper, the trajectory is predicted by applying the BP neural network which is the three layers error back propagation and has three input units, multi-hidden layer units, and three output units.

2) Parameter Setting of BP neural network

There are 3 input layer nodes, representing the radar observation time t, the velocity in the direction of longitude V_x and latitude V_y respectively; 3 output layer nodes, representing the longitude x, latitude y and altitude h of the object respectively.

The number of hidden layer nodes of neural network l is usually designed according to the experience of the predecessors and the repeated experiments. According to the analysis of the literature references, the initial value of hidden layer nodes can be determined by the following two formulas [11]:

$$l = \sqrt{0.43mn + 0.12n^2 + 2.54m + 0.77n + 0.35 + 0.51} \tag{4}$$

Where, m and n represent the number of input nodes and output nodes respectively[12].

3) Neural network training and target trajectory prediction

According to the six groups' data which are passing back from the radars and the existing historical trajectories of the radar fusion center, the future trajectory is predicted by using BP Neural network. According to the historical radar trajectory data, the trajectory data of one of the targets that selected for training and testing are listed in the following table 1[13].

Table 1. The characteristic values of historical trajectory of a target

t	V_x	V_y	x	y	z
50778	318.9	42	121.6789	25.5575	0
50843	315.6	34	121.8047	25.70417	0
50850	317.8	27	121.8117	25.72611	0
50887	308.6	30	121.8558	25.81	0
50912	297.5	31	121.8878	25.85139	0
50931	292.5	32	121.9153	25.88472	0
50949	291.4	33	121.9456	25.92139	0
50971	292.2	34	121.9872	25.96861	0
51011	291.1	37	122.0592	26.04889	0
51019	292.2	39	122.0797	26.06528	0
51053	293.1	42	122.1508	26.13278	0
51101	292.5	41	122.2439	26.22722	0
51137	291.7	41	122.31	26.29861	0
51164	290.3	41	122.3561	26.34972	0
51197	287.8	40	122.4086	26.41	0
51266	284.4	38	122.5217	26.53861	0
51307	281.9	38	122.5853	26.61083	0

So, the trajectory prediction (x, y, z) of the goal in the next 5 seconds can be obtained in the following Table 2.

Table 2. The forecasted trajectory of a goal in the next 5 seconds

t	V_x	V_y	x	y	z
51308	281.4	38	122.6211	26.64972	0
51309	281.7	38	122.6636	26.69333	0
51310	277.2	39	122.7394	26.77445	0
51311	272.8	39	122.7933	26.83306	0
51312	271.7	39	122.8319	26.87361	0

And then, the forecasted trajectory which is simulated in the MATLAB can be shown in the Figure 3.

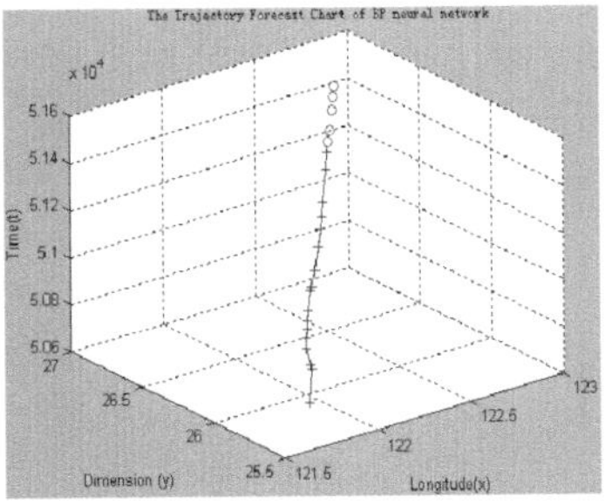

Fig. 2. The forecast trajectory map by using BP neural network

B. Trajectory clustering based on the maximum membership degree

The trajectory features of objective which are in the data provided by the radars are as follows: observation time t, the velocity in the direction of longitude V_x , the velocity in the direction of latitude V_y , longitude x , latitude y ,altitude h and so on, which also represent the characteristics of the space coordinate information and the state of motion changes of different objectives[14].

Assuming that the number of the existing historical system trajectories in the fusion center is n , and the Universe $A=\{A_1, A_2,......, A_n\}$ with the existing trajectories of the fusion center is established. The factors set that affect the cluster is defined as $X=\{X_1, X_2, X_3, X_4, X_k\}=\{x, y, z, V_x, V_y\}$ ={ longitude x , latitude y, altitude h, the velocity in the direction of longitude V_x, the velocity in the direction of latitude V_y }[13].

Supposed that characteristic parameters of trajectory A_i are $X_i=\{x_i, y_i, z_i, V_{x_i}, V_{y_i}\}$, the target in the moment k is defined as

$P^k = \{x^k, y^k, z^k, V_x^k, V_y^k\}$. The following membership function $\mu_i(A)$ of forecasted trajectory is designed with the normal type membership function algorithm [15]:

$$\mu_i(A) = \begin{cases} \mu_i(x) = \exp[-(\frac{x^k - x_i}{b_1})^2] \\ \mu_i(y) = \exp[-(\frac{y^k - y_i}{b_2})^2] \\ \mu_i(z) = \exp[-(\frac{z^k - z_i}{b_3})^2] \\ \mu_i(V_x) = \exp[-(\frac{V_x^k - V_{x_i}}{b_4})^2] \\ \mu_i(V_y) = \exp[-(\frac{V_y^k - V_{y_i}}{b_5})^2] \end{cases} \tag{5}$$

Where, $(b_1, b_2, b_3, b_4, b_5)$ are the radar observation errors of different characteristics.

Then it can obtain the following decision matrix:

$$U = (\mu_{ij})_{5\times n} = \begin{bmatrix} \mu_{11}(x) & \mu_{12}(x) & \cdots\cdots & \mu_{1n}(x) \\ \mu_{21}(y) & \mu_{22}(y) & \cdots\cdots & \mu_{2n}(y) \\ \mu_{31}(z) & \mu_{32}(z) & \cdots\cdots & \mu_{3n}(z) \\ \mu_{41}(V_x) & \mu_{42}(V_x) & \cdots\cdots & \mu_{4n}(V_x) \\ \mu_{51}(V_y) & \mu_{52}(V_y) & \cdots\cdots & \mu_{5n}(V_y) \end{bmatrix} \tag{6}$$

Where, μ_{ij} is the membership degree of the single factor to the target trajectory A_j, which is the fuzzy set of trajectories set A. n is the number of different membership function of the same parameter characteristic which is caused by the different radar.

And then, increase the weights to the membership degree of each factor, which is defined as follows.

$$W = \begin{bmatrix} w_{11} & w_{12} & w_{13} & w_{14} & w_{15} \\ w_{21} & w_{22} & w_{23} & w_{24} & w_{25} \\ \cdots & \cdots & \cdots & \cdots & \cdots \\ w_{n1} & w_{n2} & w_{n3} & w_{n4} & w_{n5} \end{bmatrix} \tag{7}$$

Where, the weight W can be obtained by the following formulas:

$$\omega_j = \frac{c_j}{\sum_{j=1}^{m} c_j} = \frac{\sum_i \mu_{ij}}{\sum_{j=1}^{m} \sum_i \mu_{ij}} (j = 1, 2, ..., m)$$
$$c_j = \sum_i \mu_{ij} \tag{8}$$

Therefore, the decision-making model can be established which is based on the maximum membership degree of the trajectory separation:

$$\max z(\mu_j) = \sum_{i=1}^{5} w_{ji}\mu_{ij},\ j = 1, 2, \ldots\ldots n; \tag{9}$$

For a new trajectory point $P^k = \{x^k, y^k, z^k, V_x^k, V_y^k\}$, this algorithm can judge it belonging to the trajectory which has the greatest degree of members. Therefore, it can achieve the trajectory separation.

C. Implementation of BP neural network trajectory separation algorithm based on maximum membership degree

1) The Algorithm Steps

Step 1. The data which are detected by the multi-sensor are preprocessed. And then, the data which are obtained from the different radars are separated according to the number of sensor. Finally, the data are sorted according to the observation time t.

Step 2. Judge the current trajectory point whether exceed the previous trajectory point more than 5 minutes time interval, if they have exceeded, a new trajectory is generated.

Step 3. For each radar, BP neural network is used to predict the future trajectory point. And then, the prediction point $T_c = (X_c, Y_c, Z_c)$ is used as the center of a circle, the spatial distance R is used as radius to set the space wave door (In this paper, by repeated simulation test, when R is valued as 0.45 meters, it can obtain the best trajectory separation). When the new observation point data fall into the space wave door, it is believed that the current observation point may be the extension of real trajectory, but it do not immediately judge that it is the trajectory. Because the point maybe also fall into the space wave door of trajectories.

It should firstly save all the trajectories that a new observation point falls into their predicted space wave doors, and eliminate the impossible trajectories. If a new observation point doesn't fall into wave door of any existing tracks, build a new trajectory.

Step 4. Then, by using the maximum membership degree of fuzzy clustering method, take the longitude x, latitude y and altitude h, the velocity in the direction of longitude V_x and the velocity in the direction of latitude V_y as the feature to calculate Gaussian membership degree of each trajectory, and give different weight to the feature. After repeated adjusting, we finalize the weight vector of (0.2,02,0.25,0.25,0.1), which has a better effect. And then, using fuzzy membership degree, cluster the point with the trajectory which has the greatest membership degree.

Step 5. By prediction and fuzzy clustering of each point, it finally takes the collection of every radar trajectory point, which is a separated trajectory.

Step 6. For the trajectory of only a single point, which is caused by the observational error, determine it to be an invalid trajectory. So it should be removed.

2) The flowchart of the Algorithm

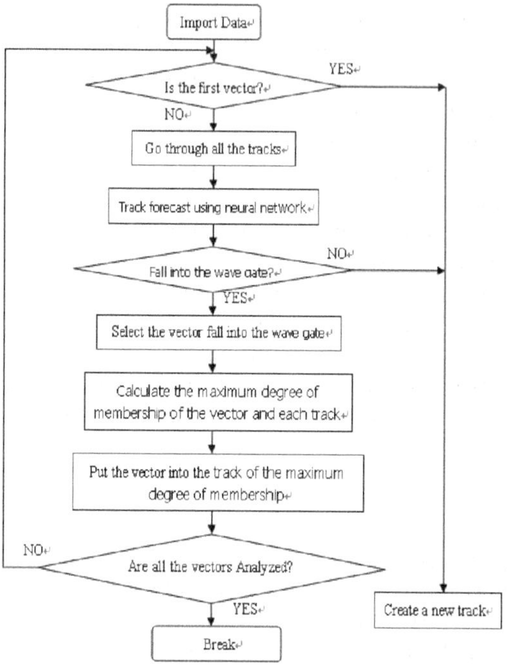

Fig. 3. The flowchart of BP neural network algorithm based onthe maximum membership degree

3) The extraction of trajectory result

The trajectories of 6 radars are respectively separated by the algorithm which is implemented by the C++ program, it is shown in the following Table 3.

Table 3. The trajectories of all the radars

Radar label	The number of effective track	The number of invalid track
1	2	0
2	22	13
3	7	0
4	18	6
5	31	0
6	137	16

5 Conclusion

The method of the BP neural network trajectory separation algorithm based on the maximum membership degree is proposed in this paper in order to effectively separate the multi-sensor data and complex trajectory characters, which improves the efficiency and accuracy of the trajectory separation. The complex motion law has been converted into the training and output of the neural network with BP neural network, which have avoided the internal complex process of model building. The algorithm converges quickly by constantly modify the hidden layers and the intermediate training parameters of neural networks, which have a better prediction for the future trajectory, and judge whether they belong to the prescriptive trajectory with the specific wave gate. The algorithm can be applied in the clustering which has the unknown clusters number, and separate successfully the target trajectory which is monitored by the multi-sensor.

Acknowledgment. The authors of this paper would like to thank the anonymous reviewers of this paper for their carefully reading of the manuscript as well as their many helpfully suggestions and corrections. The work of this paper is financially supported by the Aeronautical Science Foundation(20080768004).

References

1. Besada-Portas, E., Lopez-Orozco, J.A., Besada, J.A.: Multisensor Out of Sequence Data Fusion for Estimating the State of Discrete Control Systems. IEEE Transactions on Automatic Control 54(7), 1728–1732 (2009)
2. Wang, Q., Nie, W.: Distributed Multi-sensor data fusion. Sensor Technology 16(5), 8–10 (1997)
3. Haykin, S., McMaster: Neural Networks and Learning Machines. China Machine Press, Beijing (2009)
4. Adams, R., Davey, N., Kaye, P.: Gene sequence data sets analysed using a hierarchical neural clusterer. International Journal of Systems Science 36(14), 877–885 (2005)
5. Wang, Z., Li, H.: The fracture parameter prediction based on BP neural network algorithm. Chinese Journal Of Applied Mechanics 26(2), 379–382 (2009)
6. Yu, Q., Zhao, H.: Nenural network prediction model based on differential evolution algorithm and its application. Computer Engineering add Applications 44(14), 246–248 (2008)
7. Chen, C.L., Hsu, S.H., Hsieh, C.T.: A simple method for identification of singleton fuzzy models. International Journal of Systems Science 36(13), 845–854 (2005)
8. Liang, B., Cao, D.: Fuzzy Mathematics and Its Applications. Establishing of Science Press, Beijing (2007)
9. Fang, C., Hang, X., Wang, X.: Target Threat Evaluating and Sequencing Method Based on the Maximum Degree of Membership. Systems Engineering and Electronics 25(1), 50–52 (2003)
10. Geng, F., Zhu, X.: Research on Neural Network-aided Multi-target Tracking Data Fusion. Fire Control and Comm. and Control 33(9), 51–54 (2008)
11. Yu, Q., Zhao, H.: Track Fusion with NN Compensated in a Multi-sensor Environment. Journal of Shanghai Jiaotong University 44(14), 246–248 (2008)
12. Ribes-Gomez, E., Mcloone, S., Irwin, G.: A taxonomy for wavelet neural networks applied to nonlinear modelling. International Journal of Systems Science 39(6), 607–627 (2008)
13. Ruan, X., Bien, Z., Park, K.H.: Iterative learning controllers for discrete-time large-scale systems to track trajectories with distinct magnitudes. International Journal of Systems Science 36(4), 221–233 (2005)
14. Bar-Shalom, Y., Campo, L.: The effect of common noise on the two-sensor fused-track covariance. IEEE Transactions on Aerospace and Electronic Systems 22(6), 803–805 (1986)
15. Pei, L.: Fuzzy comprehensive evaluation on radar track fusion effectiveness. Electronics Optics and Control 13(6), 18–20 (2006)

Research on Modeling Cognitive Behaviors of Pilot Processing Alarm in Cockpit*

Luan Yichun and Xue Hongjun

College of Aeronautic
Northwest Polytechnical University
Xi'an, Shaan xi Province, China
l.y.now@qq.com

Abstract. A general process about modeling the perception, attention, decision-making behaviors of cockpit pilot processing alarm was proposed. Based on information process stage model and ACT-R theory, modeling was combined with description about modeling tasks of processing alarm, cognitive behavior of pilot and operational components. A serial of fundamental experiments of specific operational tasks were developed to support modeling. Validity and exactitude of the model was proved through matching the model perform those tasks with pilot practice. The modeling aim was to improve flying safety.

Keywords: modeling cognitive behaviors, information process stage model, ACT-R, process alarm.

1 Introduction

Warning system is one of the important factors influenced flying safety in cockpit. Modeling cognitive behaviors of cockpit pilot processing alarm was named cognitive modeling, aimed to explore theirs mind-action mechanisms especial perception, attention and decision-making courses in an emergency. Based on information process stage model and ACT-R theory, combined with description about modeling tasks of processing alarm, cognitive behavior of pilot and operational components, a model of cognitive behaviors of pilot was built. Experimental and simulating methods were used to study pilot perception course involved information content, alarm display mode, process logic and display devices layout of alarm, pilot decision-making course involved information fused, analyzed, diagnosed and action. Base on above research, cockpit warning system was optimized to improve flying safety. We can't explore the brain's actual operation like detect other body parts, so we could check whether the model was validity and exactness from how it keep cognitive theory, the reality and accurate of its prediction about what happen to tasks.

---

* This work is partially supported by National Basic Research Program of China (973 Program), 2010 CB734101.

D. Zeng (Ed.): ICAIC 2011, Part II, CCIS 225, pp. 496–503, 2011.

2 The Information Process Stage Model

Cognitive modeling mostly use information point of view and information process theory to study how human attend, select, comprehend and memorize, and make decision to act according to information[1]. Information was emphasized to study system functions by information process method and system was thought through information required, transfer, transform and process to act. Information process of system was made qualitative and quantitative analysis to build the model which represent operational of system. By this, cognitive behaviors was divided many stages or modules including percept, consciousness, attention, memory, decision-making, thought and action. Information was inputted from visual receptor, aural receptor, feeling receptor, proprioceptor or other receptors, and was processed by stages such as consciousness, attention, memory, decision-making, thought, at last outputted action. Figure 1 was the information process stage model [2].

Based on information process theory, the information process stage model drew some good ideas from control theory, information theory and system theory. It made researchers analyzing problem through information flow from the environment to mind then to the environment, and made mind study based on a great deal of experiments and become more rigorous and theoretical consistent, but mechanical defect was outstanding. In addition, single modules' function, work mechanism, information configuration and representation had more studied, but the connection and coupling between modules had few studied.

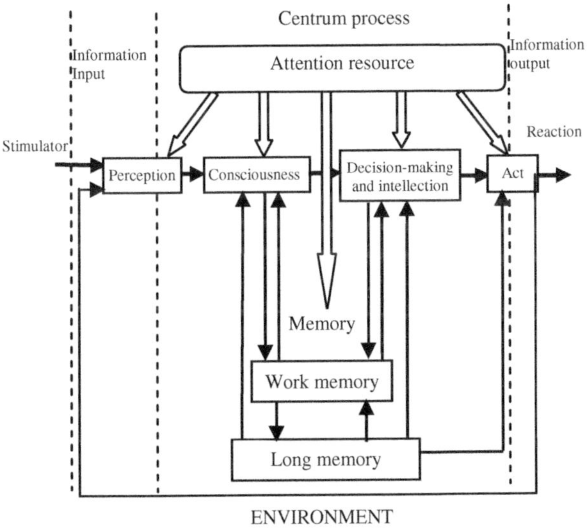

Fig. 1. Human information process stage model

There was a gap between the information process stage model in cognitive psychology and modeling cognitive behaviors on computer. Cognitive psychology and the information process stage model was semantic description, but modeling cognitive behaviors must transform those semantic relations into mathematic formulas and parameters which implemented on computers. So a great deal of study had to do to actualize that transformation.

3 Modeling Cognitive Behaviors of Pilot Processing Alarm in Cockpit

A. Orientation of the Model

Modeling cognitive behaviors of cockpit pilot processing alarm was focus on brain action mechanism in information process, and had a distinct difference with two kinds of intelligence system which is expert support system and other is brain model for product's creative design. Expert support system such as fault diagnosis expert system of the civil aviation aircraft mainly built a database from a great deal of fault information and got some reference information by index words to support assist for human. Brain model for product's creative design was focus on creativity. The outstanding point of cognitive behaviors of cockpit pilot processing alarm was:

- ✧ Strongly goal-directed processing;
- ✧ Extreme time pressure;
- ✧ A rich body of declarative knowledge;
- ✧ A great deal of practice;
- ✧ Grow into master.

Pilots could become masters from greener by a great deal of training and practice. Their knowledge acquired and skilled enhance went through declarative stage, knowledge compile stage and procedural stage. So cognitive behaviors of pilot processing alarm may show many views, but dynamic, growing, and auto-learning features were outstanding. There were distinct differences in response time, accuracy, etc. between novice and expert.

B. a General Process of Modeling Cognitive Behaviors of Pilot Processing Alarm in Cockpit

Modeling cognitive behaviors of pilots was a huge systems engineering which integrated with experiments, models and theory, and engineering applications. Experiments included fundamental ergonomical experiments and verification experiments to prove model whether was exact and validity. Model and theory was based on cognitive psychology, cognitive science, and ACT-R theoretical system. Engineering applications were cockpit pilot processing alarm of civil aircraft. The associated architecture of those parts was in Figure 2.

1) Object: Modeling cognitive behaviors mostly aimed at exploring human brain action when implementing some specific tasks or solving problems, not a general or single cognitive behavior. So it must have an object that was cockpit pilot processing alarm information. The course of pilot processing alarm includes fault diagnosis, intelligence estimate and operational process. The cognitive behaviors of pilot on commercial airplane could assist to establish the alarm diagnostic database and operational process database. Fault diagnosis or intelligence estimate was the pilot's mental process and couldn't be obviously observed. We should combine with the alarm information content, alarm display mode and refer the pilot questionnaire to get those. Pilot questionnaire on the significance of that was this.

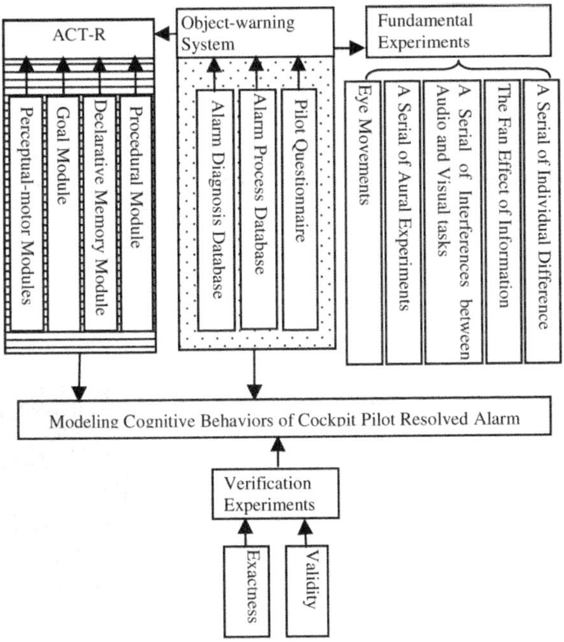

Fig. 2. A general process modeling cognitive behaviors of pilot processing alarm in cockpit

2) Theory and model: A unified model structure based on information processing stage model was established. Like the information processing stage model, the model will comprise several modules. Working mechanism, information representation, information input and output of each module would be established. Information conversion and coupling between two modules would be importance. There were two types of cognitive models, the local architecture and unified cognitive model. Local architecture used computers to simulate one particular aspect of brain model; however, unified cognitive model could explain how all the components of the mind worked to produce coherent cognition. Wegman put forward the concept of Unified Architectures in cognitive science[3]. So far there were three well-known unified architectures, two of which were symbolic processing framework, SOAR(state, operator and results) named by Newel[4] and ACT-R(adaptive control of thought-rational) put forward by Anderson[5]; Another connectionist architecture was, a parallel distributed processing (PDP) by Rumelhart and McClelland. Bogdan summarized the framework of classical cognitive science as "ICM Method", that "information, capacity and mechanisms to approach", in order to understand information processing mechanism for how to run, people must first understand the operation of the program, the ability to define; in order to understand a process, one must identify in advance the information of the program execution tasks, as well as in the implementation and problems encountered. Therefore, This approach to proceed from top to bottom: from the information and task to the ability of implementing tasks and then to run the program function mechanism [6].

ACT-R was cognitive system which was to develop continuously for exploring human cognitive behavior and widely used in many fields[7][8]. It was based on the study of the cognitive psychology and cognitive psychology; a large number of experiments were designed to support it. Its' resource management tasks were successfully implemented by linguistic, conceptual manners. Finally some theories of cognitive psychology which reflected one particular aspect of brain model as local architectures were proved in unified brain; the exactness and validity of the model was verified through matching human implement practical tasks and model simulation[9]. ACT-R should be a powerful tool for simulation human cognitive behaviors and prediction of performance.

Model consisted of two kinds of manipulative modules, symbolic module and subsymbolic module. Symbolic module was the statement of facts, based on extensive statistics derived from ergonomic experiments. It was important concepts and theories of cognitive psychology and information processing stage model representing. Subsymobolic module included learning and performance mechanisms, declarative mechanisms and procedural mechanisms which manipulated through many time-related formulas parallel processing. Symbolic work was controlled by subsymbolic module. Most subsymbolic module was used to simulate human learning process[10]. Learning was achieved by knowledge cells collection and transformation. These knowledge cells could be combined to produce a complex cognitive process.

Fig. 3 was the architecture of ACT-R model and it liked the information process stage model. Table 1 was function and modules match between two models. The environment played an importance role in those models and formed the main. In ACT-R model, the environment established the structures of problems and played an important role in the learning process. These structures could assist learning of chunks, and promote the formation of production rules, reemphasized the key step in understanding of human cognition-the importance of analyzing essential characteristics of the environment.

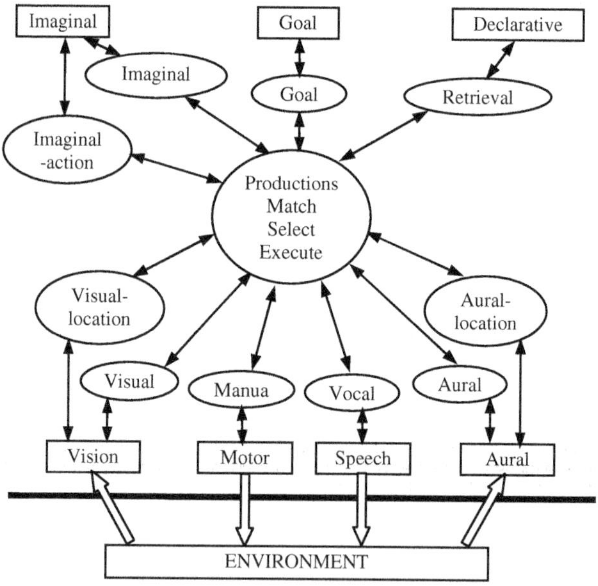

Fig. 3. Architecture of ACT-R model

Table 1. Match modules between information process stage model and ACT-R model

Information process stage model	ACT-R modular structure
Perception(vision, aural, esthesia, taste, proprioceptor)	Vision, aural
Instantaneous memory	Visual-location buffer, visual buffer, aural-location buffer, aural buffer
Consciousness(contex relevant information to current task)	Imaginal module
Attention(intent, motivation)	Goal module
Work memory	Imaginal buffer, imaginal-action buffer, goal buffer, retrieval
Long memory	Declarative modules
Decision-making and intellection	production
Motor(speech, manual, other body action)	Speech module, motor module

Modeling cognitive behaviors of pilot processing alarm would use ACT-R theory to simulate pilot psychological process and exterior actions. Psychological process would include goal, imaginal, memory and intellection; Exterior actions would include visual, aural, speech, manual motor. The scenes of alarm processing tasks were built by programming language, then we used built-in cognitive theory in ACT-R and coupled with necessary assumption and knowledge description of alarm processing tasks to establish ACT-R cognitive architecture model. The comparison of the model operational results and experimental results will verify the validity of the model; then the model compatible with human cognitive was used to instruct study, and achieved the prediction, guidance and control purposes of tasks.

Modeling cognitive behaviors of pilot processing alarm was based on the descriptions of tasks of warning information processing, model of cognitive behaviors of pilots and cognitive description of control components. That included four sections: procedural knowledge representation and content, declarative knowledge representation and content, operational process and goals about alarm processing. Alarm information processing tasks consisted of a great deal of elements of cognitive tasks or operating rules which were no longer decomposed. These smallest units were closely consecutive and formed sequences. The database of typical cognitive behaviors of processing alarm was based on rules and sequence steps of information diagnosed and manipulated which were extracted from courses of pilot processing alarm and pilot's interview, those were divided into procedural knowledge and declarative knowledge about alarm processing. The perception control theory [11][12], means-end method were used to create the goal hiberarchy of pilot processing alarm, then the goal hiberarchy instructed to build operational procedural rules jointing means and control procedural rules process. Response time of some specific rules could get by basic experiments, most of the operating rules used interior parameters of ACT-R model which were obtained by experimental means early. Finally response times of steps were added up and compared with reaction times of pilots practice on alarm tasks.

When pilot began training, they followed every step in the instructions. After some periods of time, some strategies were generated in their mind and could integrate some separated declarative and procedural knowledge into chunks, simplify by eliminating redundant steps, shorten response time and heighten veracity. The model could do so. Those production rules which often were employed could be strengthened, united and integrated; those production rules which few were employed could be weaken and abated. The ultimate results could be showed reaction time and accuracy. The change of reaction time can be compared with pilots' to verify whether the model was exact.

3) Experiments: Modeling cognitive behavior of pilot processing alarm required a lot of basic experimental support to obtain operational time of some specific minimum tasks. Those experiments include:

- A serial of experiments of eye movement. Several factors influencing the divert time of pilot's point of view would be obtained by statistical analyzing experimental data; the influencing law of formula for reaction time would be explored and fit for instructing optimizing layout of alarm displays in cockpit.
- A serial of experiments of aural distinguish. Experiments on the pilot' memory and retrieval capacity of sounds were carried out. Reaction time of pilot disposing specific alarm sounds and alarm speech were tested and compared.
- A series of experiments of audio-visual alarm dual-task interference.
- Information associated with activation strength test.
- A series of experiments of individual differences.

In addition, the complex alarm experiments would be designed to verify the accuracy of the model. Pilots would process those typical and importance alarms information in test platform. Processing course and the important parameters, such as reaction time, were recorded to compare with the course and parameters of model processing alarm information to verify the accuracy of the model. The difference between the novice and experts which had trained and long-term actual flight was clarified to explore internal psychological mechanism of expert converting knowledge into skill and provide experimental guidance for pilot training to quickly and accurately grasp the alarm information processing skills.

4 Conclusion

Modeling cognitive behavior of pilot process alarm was a huge systems engineering which included experiments, models and theory, and engineering applications. In this paper, the general process of modeling cognitive behavior of pilot had put forward with various contents working logistically and supporting each other. Basic experiments supported theoretical models; experiments which pilot tested in the experimental platform could be verified the accuracy and exactness of the model; the database of pilot process alarm information could support for establish the structure and task scenes of model; at last the model would provide guidance for optimal design of the cockpit warning system and pilot training program.

Acknowledgment. This research was supported by National Basic Research Program of China (973 Program), 2010 CB734101. I would like to thank Su Run'e for discussions relevant to preparation of the last section of this article and Zhang Yugang for his comments on the manuscript.

References

1. Shi, Z.: Intelligence Science. Tsinghua University press (2005)
2. Zhu, Z.: Engineering Psychology Tutorial. People's Education Press (2003)
3. Wagman: Cognitive psychology and artificial intelligence: theory and research in cognitive science, p. 20. Praeger Publishers, Connecticut (1993)
4. Newell, A.: Unified Theories of Cognition. Harvard University Press, Cambridge (1990)
5. Anderson, J.R.: The architecture of cognition. Harvard University Press, Cambridge (1983)
6. Bogdan, R.J.: Grounds for cognition: How goal-guided behavior shapes the mind. Erlbaum, Hillsdale (1994)
7. Liu, Y.: Researches on driver behavior modeling. Zhejiang University (2007)
8. Liu, Y., Wu, Z.-h.: Driver behavior modeling in ACT-R cognitive architecture. Journal of Zhejiang University 40(10), 1657–1662 (2006)
9. Anderson, J.R., et al.: An Integrated theory of the mind. Psychological Review 111(4), 1036–1060 (2004)
10. Anderson, J.R., et al.: Information-Processing Modules and their Relative Modality Specificity, pp. 185–217 (2007)
11. Powers, W.T.: Behavior: the Control of Perception. Aldine, Chicago (1973)
12. Hendy, K.C., Beevis, D., Lichacz, F., Edwards, J.L.: Analyzing the cognitive system from a perceptual control theory point of view. In: Cognitive systems engineering in military aviation environments: Avoiding cogminutia fragmentosa: A report produced under the auspices of the Technical Cooperation Programme Thechnical Panel HUM TP-7 Human Factors in Aircraft Environments. Human Systems IAC, Wright-Patterson AFB, OH (2002)

Research on over Frequency Load Adding and Its Application to Hainan Power Grid

Yu Jiaxi, Wei Guoqing, Li Xian, Hu Jianchen, and Xing Tao

Power Dispatching and Communication Center
Hainan Power Grid
Haikou, Hainan Province, China
yumessenger@hotmail.com

Abstract. In allusion to the frequency increase problem caused by under frequency load shedding in Hainan power grid, over frequency load adding, which is a complementary scheme to guarantee stable operation, is presented in the paper. The general condition of Hainan power grid is introduced. The idea of over frequency load adding is elucidated. Present relative research and application are summarized. The schemes of over frequency load adding are established by detailed calculation and analysis according to the model of Hainan power grid. Over frequency load adding has not only improved the stability of Hainan power grid to some extends, but also brought remarkable economic benefits.

Keywords: excessive load shedding, frequency stability, Hainan power gird, load adding.

1 Introduction

Power system is to provide customers with reliable and qualified power supply [1]. Frequency is one of the most important indexes of power quality. Hainan Island, the province located at south end of China, is often hit by typhoon. With the weak power network and unreasonable distribution of power sources, Hainan power grid has suffered a lot from the unstable operation. The blackout of 2005 caused by DaWei Typhoon is a typical incident. Hainan power system is a small one containing generators with large unit capacity. Frequency would drop quickly with the trip of large generator. And blackout in large scale would happen in serious accidents. Though a 500kV AC submarine cable links Hainan with Guangdong, the frequency problem still may occur when the cable is in maintenance and failure.

In order to avoid the rapid frequency drop with the trip of large generator, some relevant loads should be cut off. In most cases a little more loads is shed to insure the effect of Under Frequency Load Shedding (UFLS) schemes. The paper presents a complementary measure to amend the excessive load shedding according to the current stability control measures of Hainan Power Grid. When the frequency increases after the load shedding, some load would be restored with the Over Frequency Load Adding (OFLA) devices. Frequency increasing would be restrained and the stability of power grid is improved with the load adding schemes.

D. Zeng (Ed.): ICAIC 2011, Part II, CCIS 225, pp. 504–512, 2011.

2 Brief Condition of Hainan Power Grid

By the end of 2009, the total installed unit capacity of the generators dispatched by the Power Dispatching and Communication Center of Hainan power grid is 3040MW. There are 8 coal-fired units, 12 gas-fired units, 10 water units. The unit with the maximum capacity, which is 350MW, is in DongFang Power Plant. And the minimum total load of Hainan is less than 1000MW. There would be frequency stability problem with tripping failure of one large unit, especially the 350MW generator. The 220kV AC power networks have been built around Hainan. Besides one circle tie-line in the eastern part of the island, there are double tie lines in the rest parts. There are thirteen 220kV substations and thirty-six 220kV transmission lines with total length of 1777km. The geographical diagram of Hainan power gird is shown as Fig.1.

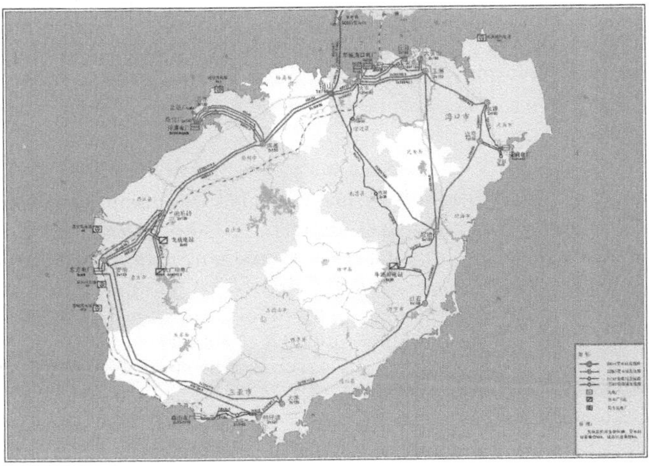

Fig. 1. The geographical diagram of Hainan power grid

The main automatic safety control systems of Hainan Power Grid are UFLS, security and stability control system, over frequency generator splitting system and under frequency generator starting system. It is difficult to make those devices and systems coordinate with each other. Excessive load shedding is a general problem bothering Hainan power grid.

3 General Introduction of over Frequency Load Adding

A. Definition of Over Frequency Load Adding

According to the regulations of *Guide for Power System Security and Stability*, the sequence and amount of loads split by the under frequency load shedding system should be arranged reasonably [2, 3]. When power system lacks of active power, some minor loads could be cut off in sequence corresponding to the frequency dropping condition. Thus, the operation of power system is safe and the power supply of important customers is not interrupted.

Over frequency load adding is an automatic reclosing device whose action is set with the frequency changing of power system. After some loads are cut off by the UFLS devices when there is active power lacking, the frequency of power system would recover gradually. Once the frequency is beyond the value expected, some lines that have been cut off would be reloaded. The load loss caused by the UFLS is reduced as much as possible.

B. Contemporary Research On Over Frequency Load Adding

Some research and application on over frequency load adding has been done in Russia [4-6]. Russian scholars call it as frequency automatic reclosing, which has been a representative characteristic of Russian power grid. After the actions of Under Frequency Load Shedding I (UFLSI) that is fast and Under Frequency Load Shedding П (UFLSП) that is slow, the frequency of power system would recover to 48.5~49.5Hz or a little higher value. To remove the lacking of active power, the reserve capacity of the whole power system is employed. The power grid operators would put the divided systems reunited. Usually the frequency is 0.1~0.2Hz beyond the designated value. In most cases, the frequency recovers to an acceptable level with the help of automatic devices.

For the isolated power systems or the ones probably divided from the main system for a long time, the starting frequency is designated as 49.2~50Hz. And the starting time is 10~20 seconds. In some cases, the starting is postponed to 90~120 seconds, especially for isolated power systems. The reclosing sequence is determined by the importance of the customers. For example, the most important customers would be the last one shed from power grid and the first one recovered with power supply. The unattended substations would be the first one shed when there is lacking of active power. We have not found similar application of OFLA except Hainan power grid in China.

C. Basic Principles of Over Frequency Load Adding

OFLA is mainly employed to get rid of excessive load shedding. To shorten the outage time and improve the reliability of power supply, some basic principles are demanded for OFLA as following.

1) Rapidness. The OFLA should be started quickly to recover some important customers. It is dangerous for power system to operate with high frequency in a long time. And the frequency should recover to a reasonable value rapidly after accidents. The starting frequency of OFLA need correspond to that of the generator high frequency protection. Thus, the remaining large steam turbines would not trip and further aggravate the accident. The frequency should not be less than 49.5Hz when the OFLA is started. On the other side, the frequency of power system is not allowed beyond 51Hz. Some delay measures are taken to insure the power supply of the sub important customers.

2) Sequence. The sequence of load recovered by OFLA is determined by the importance of customers.

3) Cooperation. OFLA should coordinate with over frequency generator splitting system. Before the startup of over frequency generator splitting system, OFLA is supposed to be triggered. The starting value of OFLA should be less than that of over frequency generator splitting system of small hydropower generators.

4 Analysis of the Effect of OFLA on Frequency Stability of Hainan Power Grid

In this section, the effect of OFLA on frequency stability of Hainan power grid is analyzed in detail. The status of Hainan power grid in 2008 is taken as an example for calculation.

A. Calculation Condition

The models of the main apparatuses in power grid are built. For generators, E_q'' and E_d'' are variables; the effect of automatic exciting regulator is considered; the influences of governor and power system stabilizer are neglected; the damping coefficients are designated as 0. For load, the static ZIP model is taken. And the proportion of constant-impedance load, constant-current load and constant-power load is 40%, 30% and 30% respectively. The frequency characteristic of load is taken into account. The frequency factors are as following.

$$\frac{dP}{df}=2, \quad \frac{dQ}{df}=-2$$

B. Criterion of Frequency Stability

The criterion of frequency stability is that power system could recover to normal operation status after accidents. There is no frequency collapse during the course of accidents with the help of generator and load shedding.

C. Main Factors Affecting Frequency Recovery

The main factors affecting frequency recovery after the action of UFLS are the configuration of device, action times, the setting value and the load amount shed after triggering etc. The principle of frequency recovery after the action of OFLA is similar to that of UFLS. Whether the frequency could come back to normal mainly depends on three factors, that is, the frequency after accident, the load amount recovered by OFLA and the setting value of OFLA. The primary research of the paper focuses on the measures adapting the frequency to preset range, that is, 49.5~50.5Hz. The load amount added and the starting frequency of OFLA are the crucial factors, which would be calculated and analyzed in the next section.

D. Calculation of Load Amount Added and Starting Frequency

Based on the calculation data of four typical operation modes, *KD Mode*, *FD Mode*, *KX Mode* and *FX Mode*, the two crucial values of OFLA, load amount added and Starting frequency, are obtained.

1) KD Mode

The whole load of Hainan power grid is 1540MW in *KD Mode*. The calculation results are shown in Table 1 below.

Table 1. The calculatoin results of kD mode

Capacity of generator trip (MW)	Load shed by UFLS (MW)	The highest frequency after accidents (Hz)	Load added by OFLA (MW)	Starting frequency (Hz)	Frequency after recovery (Hz)
100	77	49.52	-	-	-
200	181	49.48	-	-	-
300	285	49.23	-	-	-
400	389	49.21	-	-	-
500	586	51.65	10	50.1	51.377
			10	50.3	51.377
			10	50.5	51.377
			20	50.1	51.104
			20	50.3	51.104
			20	50.5	51.103
			30	50.1	50.843
			30	50.3	50.842
			30	50.5	50.840

When there are excessive loads shed after generator trip, the system frequency recovers to different values in the above four cases, that is, without any OFLA measures, the amount of load added by OFLA is 10MW, the amount is 20MW and the amount is 30MW (The starting frequency is 50.1Hz). From Table I, we find that 30MW load should be added at least to ensure the frequency stability of Hainan power grid after the trip of 500MW generating capacity in *KD Mode*. The curves showing the process of frequency recovery are as following.

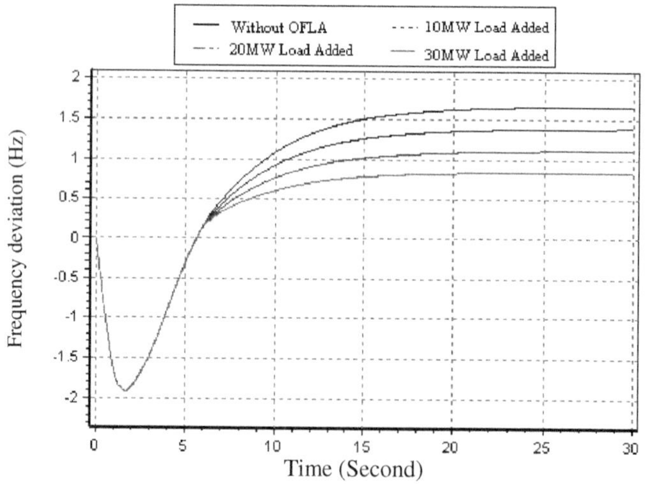

Fig. 2. The curves of frequency recovering process when the loads added by OFLA are 10MW, 20MW, 30MW respectively

When the load added by OFLA is 30MW, the frequency of power system reaches to a more reasonable value.

Here some further research is done. The amount of load added by OFLA is divided into two rounds. 30MW load is added when the frequency is 50.1Hz at the first round. 10MW, 20MW and 30MW loads are added separately when the frequency is 50.3Hz. The case of 500MW generator trip is calculated as an example, which is shown in Table 2.

Table 2. Further calculatoin results of kD mode

Capacity of generator trip (MW)	Load shed by UFLS (MW)	The highest frequency after accidents (Hz)	Load added by OFLA (MW)	Triggering frequency (Hz)	Frequency after recovery (Hz)
500	586	51.65	30	50.1	50.574
			10	50.3	
			30	50.1	50.308
			20	50.3	
			30	50.1	50.035
			30	50.3	

The calculation results show that two or more rounds load recovery can be adopted by OFLA if the amount of the excessive load shed is much bigger. And the frequency of power system will approach 50Hz. The relevant curves showing the course of frequency recovery are as bellowing.

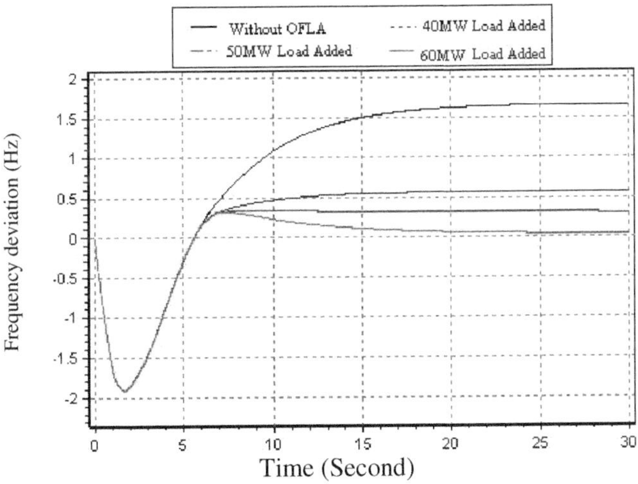

Fig. 3. The curves of frequency recovering process when the loads added by OFLA are 40MW, 50MW, 60MW respectively

2) FD Mode

The whole load of Hainan power grid is 1500MW in *FD Mode*. The calculation results are shown in Table 3 below.

Table 3. The calculatoin results of FD mode

Capacity of generator trip (MW)	Load shed by UFLS (MW)	The highest frequency after accidents (Hz)	Load added by OFLA (MW)	Triggering frequency (Hz)	Frequency after recovery (Hz)
100	62	49.26	-	-	-
200	204	49.72	-	-	-
300	350	50.71	10	50.1	50.46
			10	50.3	50.47
			10	50.5	50.48
			20	50.1	50.22
			20	50.3	50.23
			20	50.5	50.25
			30	50.1	49.96
			30	50.3	49.97
			30	50.5	49.99
400	458	50.86	10	50.1	50.60
			10	50.3	50.60
			10	50.5	50.60
			20	50.1	50.33
			20	50.3	50.34
			20	50.5	50.34
			30	50.1	50.08
			30	50.3	50.08
			30	50.5	50.08
500	555	50.83	10	50.1	50.54
			10	50.3	50.54
			10	50.5	50.54
			20	50.1	50.24
			20	50.3	50.25
			20	50.5	50.25
			30	50.1	49.95
			30	50.3	49.95
			30	50.5	49.95

3) KX Mode

The whole load of Hainan power grid is 970MW in *KX Mode*. The calculation results are shown in Table 4 below.

Table 4. Further calculatoin results of kD mode

Capacity of generator trip (MW)	Load shed by UFLS (MW)	The highest frequency after accidents (Hz)	Load added by OFLA (MW)	Triggering frequency (Hz)	Frequency after recovery (Hz)
100	116.3	50.3	-	-	-
200	185.6	49.3	-	-	-
300	322.4	50.59	10	50.1	50.194
			10	50.3	50.194
			10	50.5	50.192
			20	50.1	49.821
			20	50.3	49.805
			20	50.5	49.800

4) FX Mode

The whole load of Hainan power grid is 850MW in *FX Mode*. The calculation results are shown in Table 5 below.

Table 5. Further calculatoin results of kD mode

Capacity of generator trip (MW)	Load shed by UFLS (MW)	The highest frequency after accidents (Hz)	Load added by OFLA (MW)	Triggering frequency (Hz)	Frequency after recovery (Hz)
50	49	49.9	-	-	-
100	93	49.5	-	-	-
150	158	50.0	10	50.1	50.01
			10	50.3	50.03
			10	50.5	50.03
			20	50.1	49.57
			20	50.3	49.58
			20	50.5	49.60

Some conclusions could be gotten from the above calculation and analysis as following.

a) Generally, only the trip of large unit would cause excessive load shedding and frequency problem. Frequency seldom exceeds the upper limit when the trip capacity is less than 200MW.

b) The starting value of OFLA should be bigger than 50Hz.

c) The starting value of OFLA would affect the recovery frequency, which would be a little smaller if OFLA has been triggered earlier.

d) The amount of load added has a great impact on the recovery frequency, which would get back to a relatively lower value if the load added is bigger. In *KD Mode*, the starting load is 30MW; in *KX Mode*, the starting load is 10MW.

e) Another round of load adding could be taken if the frequency has not recovered to the expected range. Most of the time, the starting frequency is designated as 50.3Hz and the delay time is less than 0.2 second.

Since 2008, the OFLA has been put into use in Hainan power grid. The loss reduced by OFLA has been more than 4 million Yuan.

5 Conclusions

To alleviate the excessive load shedding problem that would bring frequency instability to Hainan power grid, the paper proposed the idea of over frequency load adding. OFLA not only improves the frequency stability, but also enhances the reliability of power supply in practice. With the help of OFLA, great economic benefit has been achieved in recent years.

References

1. Yuan, J.: Power System Security and Stability Control, 1st edn. China Electric Power Press, Beijing (1996)
2. Li, X.: Power System Automation, 1st edn. Hydraulic and Electric Power Press, Beijing (1986)
3. Yang, G.: Theory of Power System Automatic Devices, 1st edn. Hydraulic and Electric Power Press, Beijing (1989)
4. Qin, M., Yang, X.: Measures to reduce over-shedding caused by under-frequency load shedding project. Power System Technology 26(3), 83–86 (2002)
5. Chen, J., Hong, L., Deng, Z.: Investigation and application of under-frequency loading shedding. Relay 35(14), 79–82 (2007)
6. Qin, M., Yang, X.: Research of the frequency urgent control in Enshi power system. Huazhong Power System 16, 10–14 (2003)

Dynamic Electronic Forensics Based on Plug-in

Yong Li[1] and Wenqi Wang[2]

[1] College of Educational Information Technology, Anyang Normal University, Anyang, Henan Province, China
[2] College of Computer Science, Zhongyuan University of Technology, ZhengZhou, Henan Province, China
gqliuwhu62@sohu.com

Abstract. Considering the characteristics of judicial forensics as well as diversity types of network forensics, this paper proposed a forensic framework based on XML and plug-in technology. The framework maintains scalability of forensic analysis and ensures the stability of forensic process simultaneity. Specified host forensics and hacker intrusion forensics were implemented to proof the validity of the framework.

Keywords: Forensic framework, Plug-in technology, Intrusion forensics.

1 Introduction

With the development and in-depth application of information technology, computer-related cases, such as e-commerce disputes, computer-related crimes, are emerging. The motivation of research on computer forensics technology is to solve the problem that how to get electronic evidence timely and effectively[1-3]. Computer Forensics technology adopt a proactive approach to collect evidence of the invasion, to reproduce the invasion process, to analyze attacks, and finally, to prevent hackers effectively. As with the traditional evidence, electronic evidence must be true, accurate, complete, compliance with legal and acceptable for the court.

Computer Forensics technology can be divided into static forensics and dynamic forensics. Static forensics technology is referring to data extraction and analysis from static storage medium after the incident. The shortcoming of the static forensics is that electronic evidence for criminals may be damaged and cannot be restored. Network-based dynamic forensics technology, through analysis of criminal information had been acquired, adopt real-time data acquisition and tracing techniques to capture and preserve evidence without the awareness of criminals. This is typically a crime in progress or prior evidence. The former one is still the mainstream of computer forensics technology, while with the increasing popularity of network application, network-based forensics technology is gradually becoming a hot spot, though there are still many problems to be solved[4-5].

D. Zeng (Ed.): ICAIC 2011, Part II, CCIS 225, pp. 513–522, 2011.

2 An Extensible Framework

According to the existing literature, current studies on dynamic forensics technology are mostly focusing on the evidence against hacker attacks. However, practically the types of dynamic forensics technology are varied, such as acquire evidence from specified IP, specified E-Mail and instant messaging software, etc. Forensics technology is extensible. While taking into account the judicial characteristics of electronic forensics, evidence collection process should be controllable, verifiable, stable, and even with judicial authentication. That is to say, the dynamic extensibility and stability should be considered simultaneously[6-7].

The extensibility of Network-based forensics lies in the diverse types of network-based crime and so the contents of network crimes are varied. Therefore, the analysis of network data is complex, dynamic and extensible. Extensibility is also reflected in real-time analysis of the new cyber crime. The forensic framework requires a secondary development capacity.

The stability of the network-based forensics is mainly reflected in the handling of the raw data as well as the overall framework of the program. Raw data processing should comply with the requirements of judicial evidence. That is to say original data is not changed, or change process is reproducible. This can be implemented through encrypted signature. By doing this, it can prevent the original data from being tampered as well as ensure that forensics officers is liable for the original data. While the stability is also reflected in the framework of program need to be certificated. The certification is of vital importance because it can proof that the program has no back door or other illegal acts during forensics process.

Based on the above considerations, XML-based plug-in technology is adopted to achieve the separation of evidence analysis and data processing. XML extensibility feature is used to achieve various types of information analytic processing[8]. Plug-in technology is used to ensure the process stability as well as to realize the analysis of multiple types of evidence event. Forensics system with plug-in technology has a secondary development capacity.

To this end, the extensible framework of dynamic forensics is designed, which is shown in Figure 1: XML parser has two Features: 1) parsing the configuration files, which consists of plug-in loading information and database information, so that plug-ins can be called by the main frame program 2) analytical forensics command, the forensic information are diverse. Information need to analysis also have various types. So it is impossible to design a unified command format. In order to ensure its extensibility, XML format is used. Therefore, the evidence-gathering procedures have good scalability, and able to adapt to a wide variety of evidence types.

XML parser was designed based on free open-source development tool libxml2. Libxml2 is an XML-based C-language version parser. It was originally developed for the Gnome project, and it is a free open source software compliance with MIT License. In addition to supporting C-language, it also supports C++, PHP, Pascal, Ruby, Tcl and other languages. It runs on Windows, Linux, Solaris, MacOsX and other platforms. Because of its cross-platform feature, libxml2 makes the main program easy to port to Linux and other system environments.

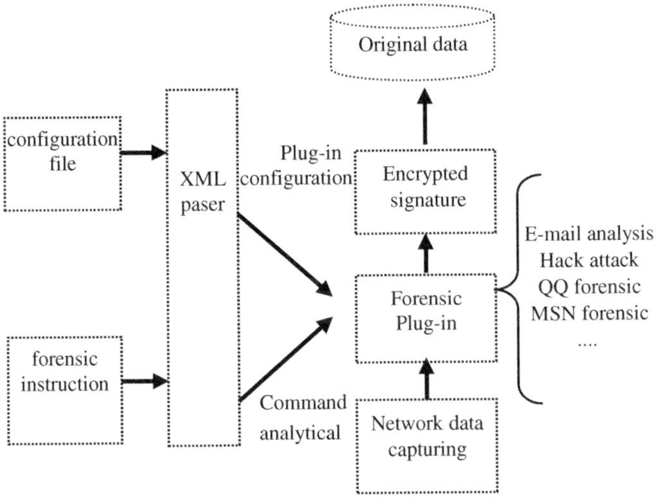

Fig. 1. Forensic framework

Currently there are a variety of plug-in technologies; ActiveX controls and Microsoft's COM technology are two of them, but they are of poor security. Considering the complexity of network data analysis via plug-in technology, a dll technology is adopted. Different types of network data analysis procedures are implemented in Dlls. Registration of different plug-ins are implemented in configuration files. The main program calls the appropriate plug-in by reading the configuration files. Two export functions must be implemented in plug-in, the one is command analytic function, and it's responsible to resolve XML-format forensics messages sent from main function; and the other is network data analysis function. The function analysis network data packets captured by the main program to determine whether forensics procedure should be active. Plug-in (if possible) can also produce forensics command export function. The function is usually a dialog box with the main function calls to generate forensics command, and it returns with a package of XML format forensics message. As analytic forensics command and encapsulate forensics command are both based on the XML formatted messages, XML parser is necessary in plug-in, so XML parser must be open.

The main functions of main program include loading plug-in information according to configuration files, database maintenance, storing forensics instructions into database, network data capture, as well as data encryption, certification and preservation according to the forensic analysis results.

When register plug-in in configuration file. The information relevant to forensics types, plug-in's dll files and the parameters of forensics instructions must be included. For instance, a plug-in registration information of IP-specified host forensic analysis:

```
<IPForensic MenuName="IP traceing" dllName="IP_Forernsic.dll" TableName="IPTrace">
    <IPAdress>varchar(14)</IPAdress>
    <Port>varchar(14)</Port>
```

```
<IPKeyword>varchar(50)</IPKeyword>
<Subject>varchar(200)</Subject>
 </IPForensic>
```

Forensics instruction in XML format must include the type of forensics, the specific settings parameters. For instance, there is a host holds an IP address 192.168.0.3, the following XML message shows forensics message format for that host:

```
"<?xml version="1.0"?>
<forensic>
  <IPForensic IPAdress="192.168.0.3"
Subject="&#x53D6;&#x8BC1;IP&#x5730;&#x5740;">
<IPForensiccon IPAdress="" Subject=""/>
 </IPForensic>
</forensic>"
```

From the above description, we can see that the main program on the one hand can receive forensics instruction events generated by the local plug-in; it also can receive the correct forensics message sent from remote. By doing so, remotely control is realized; the system has distributed forensics capabilities. With correct configuration and export function, users can generate a new type of network forensics plug-in, which also makes forensics system has secondary development capability.

The following describes several different types of network data forensics technology.

3 Forensics of Specific Events

Mainly used for a particular IP, a specific port number, and network packets those contain specific keywords forensic. Forensic analysis of the data for specific events is relatively easy, there is resolve forensic events export functions, resolve the host IP address, port number, and specific keywords and other information, according to analysis of forensic command parameters, in network data analysis export functions, IP address in network layer and TCP port in transport layer are resolved from network data packet, and then determine whether forensic procedure should be active. Content-based keyword forensic can use mature search matching algorithm to match the network data package, such as forensic for all packets that contain "Fa lun Gong", search matching algorithm can search network packets and match with the keyword "Fa lun Gong" . If the packets need to be collected as evidence based on forensic instruction, the exported DLL function returns TRUE. Then, the main program will store the entire session of the network data as evidence.

4 E-Mail Forensics

E-mail forensic refer to various e-mail message forensic. E-mail forensic achieve the following specific items forensic: the sender, recipient, subject keywords, content, keywords and other information of the evidence. If forensic of a particular sender, all messages that contain the specified sender should be obtained as evidence. The feature of E-mail forensic is that forensic keywords can be extracted from the network data which contain E-mail message.

There are two protocols of E-mail transmission, 1) Simple Mail Transfer Protocol(SMTP), such as OutLook, FoxMail and other SMTP mail client using SMTP to send messages, and receive messages using the POP3 protocol. 2) HTTP protocol, when using browser to send and receive e-mail, because the data in web page is presented as HTTP protocol packets, the format of these messages depends on the individual e-mail service providers. Service providers and each has its own style, such as XML, plain text format. It brings some difficult to identify the sender, recipient, subject and other keywords. Therefore special algorithm need to be designed to obtain forensic keywords.

HTTP and SMTP are both application layer protocol, so forensic information cannot be obtained completely through a single network packet. All network data must be restored to the application layer to form a complete application layer session.

E-mail forensic algorithms based on two different protocols are described as follows:

A. E-Mail forensic based on SMTP Protocol (Figure 2)

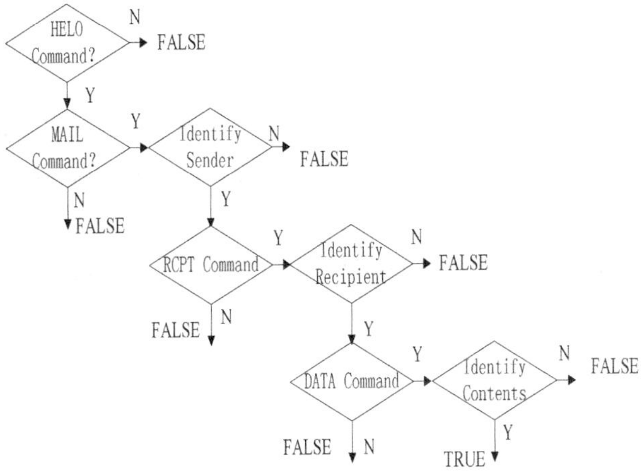

Fig. 2. SMTP-based E-Mail Identify

Network packet received from main program:

1) Determine the destination port number is 21, if not return FALSE, otherwise the next step;
2) Detect whether is HELO or EHLO command, if not return FALSE;
3) Detect the correctness of SMTP protocol network data format, and whether there is MAIL command, If has, extract e-mail address. If the address is the current sender forensic address or the current sender forensic address is null, go to the next step, otherwise return FASLE;
4) Detect the correctness of SMTP protocol network data format, and whether there is RCPT command, If has, extract e-mail address. If the address is the current recipient forensic address or the current recipient forensic address is null, go to the next step, otherwise return FASLE;

5) Detect the correctness of SMTP protocol network data format, and whether there is DATA command, If has, and content-based keyword forensic has configured in current mail forensic instruction, then keyword searching and matching algorithm becomes active. If search results are not null return TRUE, otherwise return FASLE;
6) Detect the correctness of SMTP protocol network data format, and whether there is QUIT command, If has, and content-based keyword forensic has configured in current mail forensic instruction and the search result is null return FALSE, otherwise return TRUE;

B. E-Mail forensic based on HTTP Protocol

The same as other types of electronic documents, e-mail content must follow a certain format, in order to let a variety of mail handling procedures can analyze and extract the sender, recipient, subject, attachments and other information. The basic format and the specific details of E-Mail content are documented by RFC822 and MIME protocol.

RFC822 gives a basic specification of HTTP-based E-Mail. The format described as follows: Each mail has two parts: header and body. Header part field can be divided into two types. One is generated by the client program, and the other is added by SMTP server which the mail goes through.

Table 1 lists the RFC822 specified mail header field which can be controlled by user's E-Mail program. But not all fields are required. You can ignore to form the header, and only to send the body, then SMTP server will add the minimum required fields.

Table 1. The meaning of Mail header fields specified in RFC822

Header Field	Meaning	Header Field	Meaning
From	Mail Author	Subject	Subject
Sender	Sender	Comments	Comments
Reply-To	Reply Address	Keywords	Keywords
To	recipient Address	In-Reply-To	E-Mail IDs replied by the current Mail
CC	CC recipient address	References	Sameas In-Reply-To
BCC	recipient address	Encrypted	Encryption type of encrypted E-Mail
Date	Sending date		

RFC822 only specify e-mail document format. When transmit e-mail message via HTTP protocol, as browser client and server-side are controlled by service provider, the actual transmission protocol is not entirely comply with the RFC822. Such as using Microsoft's website to send Hotmail e-mail message, the sender key words is "fFrom", and using sina's website to send mail, the mailbox use name has been recorded, and the session is stored by appropriate cookie, so send a message does not provide sender's mailbox. The web-based e-mail also has different format. E-mail format has plain text-based, such as Sina, Hotmail, also has XML-based format, such as Yahoo.

Because of this diversity, it becomes more difficult to analysis and forensic mail message.

Though, different mail server provider adopt different text format, the message itself has certain semantic descriptions (such as sender, receiver, etc.), so keywords identify different format e-mail message are approximate.

In addition, in HTTP protocol, the client browser send mail through "POST" command, and request to download from server through "GET".

Based on the above analysis, a fuzzy algorithm for e-mail forensic is designed.

Algorithm is described as follows:

1) HTTP application layer data recovery. Firstly, the current HTTP data, including data packet length, data type, and the command type, are extracted from HTTP header information. The session time span and session compression algorithm also can get from HTTP header. Then, a complete session data can be received according to session time span, and the corresponding decompression algorithm can be applied to decompress the messages.

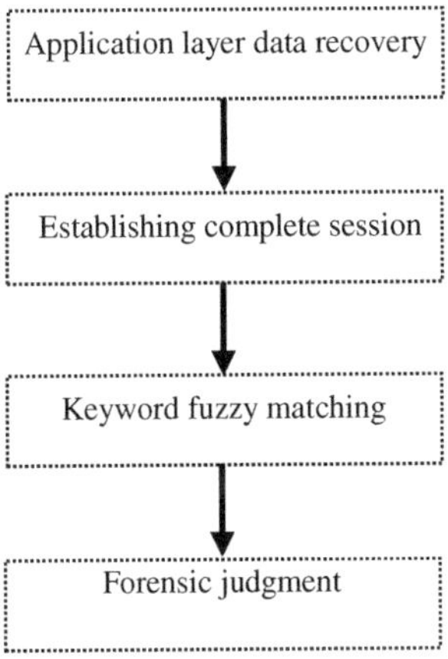

Fig. 3. E-Mail forensic algorithm

2) E-Mail message format parsing. Different algorithms are applied to different message formats. Encapsulated XML parser is used to parse the forensic field for XML-based message format. Keywords searching and matching algorithm is used to search keywords for the plain text format message.

3) Keyword matching using fuzzy matching algorithm. Collect as much e-mail keywords information as possible to form a fuzzy searching space. Then, keywords can be matched through approximate fuzzy matching method in searching space. For example, for different E-Mail server provider like Sina, Sohu, 163, Hotmail, Yahoo, The keywords representing email subject are "subj", "fSubject", "subject" and so on. The keywords representing email contents are "fMessageBody", "editmessagearea", "msgtxt", "content" and so on. A matching space can be formed by extensive collect keywords identifying email message. Thus, forensic information (such as sender, recipient, etc.) can be matched according to the forensic instructions.

4) According to the matching result to determine whether need to collect evidence, if need returns TRUE, otherwise returns FALSE.

5 Hacker Attack Forensics

Hacker attacks are of variety of ways and features, so it's impossible to set up forensic commands for every single type of hacker attack. Currently, many researchers are focusing on how to analyze hacker forensics, and then make it equivalent to the intrusion detection system. In fact, analyze the behavior characteristics of hacker attack is the main study area of intrusion detection system. Hacker attack's forensics is focused on the preservation of original attacking evidence. So according to the intrusion detection results, the corresponding network data can be stored. Therefore, network electronic forensics and intrusion detection system are not the same. That is to say, when you need to collect the hacker attack evidence, you can rely on the results of intrusion detection system.

To do this using the following methods: The hacker attack behavior is detected by the intrusion detection system. When the behavior has detected, corresponding XML-based forensic message is constructed. Then, the message sent to forensic system via network. The forensic system parses forensic command and then save corresponding evidence of network packets.

To verify the hacking forensic system, free open-source snort is used as hacking forensics intrusion detection system.

Snort is a powerful, cross-platform, lightweight network intrusion detection system, which is packet sniffer and logging tools based on the libpcap. Snort can run on Windows, Linux, OpenBSD, FreeBSD, Solaris, and other UNIX systems. Snort uses the form of plug-ins to facilitate the administrator to call a variety of plug-in module. Including input and output plug-ins, in which the output plug-in is mainly used to output detected alarms in a variety of ways, Snort's output plug-in will run on condition that alarm and recording subsystem is called. Through output plug-in, the results can output to mysql, sql and other databases in XML, tcpdump and other formats.

In accordance with the output plug-in specifications, users can write your own plug-ins to deal with the alarm and thus to respond in your own way. So that Snort has a very good scalability and flexibility.

Based on the characteristics of Snort, Its output plug-in is rewritten here. The detected invasion information and network data packets which need to be collected as evidence are encapsulated in XML format, and then sent over the network to forensics system. The configuration file Snort.conf also need to be modified, configure the output to the override plug in, and according to the Snort's configuration rules, configure other detecting information (such as detecting network, detecting rule sets, etc.). When forensic system gets the message, forensic procedure is active.

Hacker attack forensics process described as below (Figure 4):

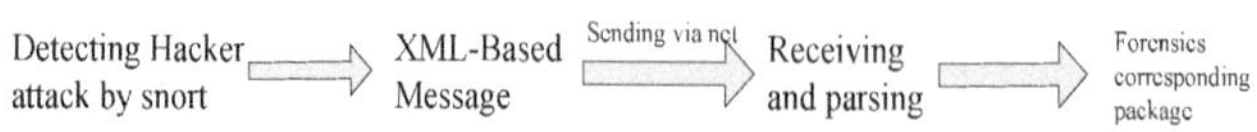

Fig. 4. Hackers attack forensics process

1) First, start Snort to detect hacker attacks.

2) When the attack is detected, construct XML messages and then send to forensic system. The message contains network data packet to be collected, such as source address, source port, destination address, destination port and so on.

3) After got the forensic message, the forensic system will parse the message, and then according to the analysis results, collecting corresponding network data as evidence.

Currently, a major flaw of intrusion detection system is that it has large number of repeated alarms, especially scanning event, it needs to send a large number of abnormal data packets. Using a simple Nmap scan can cause dozens of repeated alarms by Snort. Repeated alarms may produce repeated forensic instructions. This on the one hand reduce the efficiency of forensic, on the other may mislead the judgment of real forensic event. So an algorithm should be designed to normalize the repeated alarms.

Generally speaking, repeated alarms caused by network intrusion detection system typically detect only one specific attack. While, in a period of time, one hacker attack may contains different kinds of attacks. That's why there are a large number of repeated alarm events. So there can have the following assumptions: in a certain time t, alarm events of network attacks with the same source and destination IP address can be seen as repeating alarm.

Based on the above assumptions, intrusion forensic algorithm is designed as shown in Figure 5.

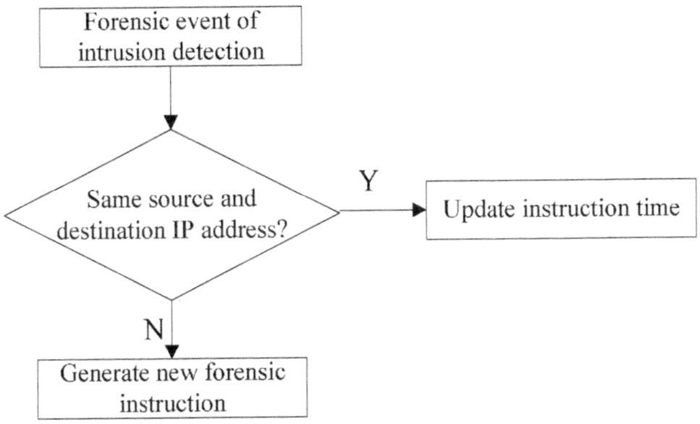

Fig. 5. Intrusion forensic algorithm

When a forensic event is received from intrusion detection system, first check the current forensic event whether there is the same source and destination IP address of the two events. If not, generate a new forensic instruction and record instruction time, otherwise only update instruction time.

A timer is set to check the current forensic event periodically, if forensic event time over a threshold T, then terminate the forensic event.

Experiment show that, like NMAP scan events, the threshold T can be set to 60s.

6 Conclusion

Network forensics is a hot spot of current forensics research. Network electronic forensics should comply with judicial originality and certification of data. In this paper, an XML and plug-in based dynamic network electronic forensic framework was proposed. Meanwhile, specified host forensics and hacker attack forensics were implemented to proof the validity of the framework. The framework maintains stability of forensic process and scalability of forensic analysis.

References

1. Zhao, X.M., Chen, Q.: The research sotuanon and expectation of computer forensics. Computer Security 10, 78–81 (2003)
2. Haber, S., Stornetta, W.S.: How to time stamp a digital document. Journal of Cryptology 3(2), 99–111 (1991)
3. Palmer, G.: A road map for digital forensics research. Report From the First Digital Forensics Research Workshop, DFRWS (2001)
4. Reith, M., Carr, C., Gunsch, G.: An Exanmination of Digital Forensic Model. International Journal of Digital Evidence 1(3), 1–12 (2002)
5. Honing, A.: Adaptive model generation:an architecture for deploy of data mining based intrusion detection systems [EB/OL] (2007-1-12), `http://citeseer.ist.psu.edu/`
6. Yasinsac, A., Manzano, Y.: Honeytraps:A network forensic tool. In: Proc. of the 6th Word Multiconference on Systems,Cybernetics and Informatics (SCI 2002), Orlando, Florida, USA, pp. 1–6 (2002)
7. Thompson, E.: MD5 collisions and the impact on computer forensics. Digital Investigation 2(1), 36–40 (2005)
8. Stevens, M.W.: Unification of relative time frames for digital forensics. Digital Investigation 1, 225–239 (2004)

Empirical Study on the Relationship between Tacit Knowledge Creation and the Evolution of Core Competence

Xiaojing Chen

School of Finance, Shanghai Institute of Foreign Trade, Shanghai, P.R. China, 201620
chen.xiaojing@yahoo.cn

Abstract. In recent years, according to many scholars' view, core competence is in essence the tacit knowledge. Then, does tacit knowledge creation have impact on the evolution of core competence? The current literature is based on qualitative analysis instead of quantitative analysis. In this paper, on the basis of the questionnaire surveys of hundreds of senior managers mainly from Shanghai, Zhejiang, Beijing and Jiangxi, the empirical study is carried out on the relationship between tacit knowledge creation and the evolution of core competence. The result shows that tacit knowledge creation has significant impact on the evolution of core competence. Finally, some suggestions are put forward for the enterprises to conduct tacit knowledge creation.

Keywords: Tacit knowledge, Knowledge creation, Core competence.

1 Introduction

In recent years, scholars both at home and abroad have carried out a lot research on knowledge creation and the core competence. Some scholars point out that corporate core competence is in essence the specific, inimitable tacit knowledge which can bring about the value to the customers, such as the systematic software developing technology of Microsoft Company, the sales & logistics system of Walmart. Since core competence is tacit knowledge, then does tacit knowledge creation have the significant influence on the evolution of core competence? What is the micro-perspective mechanism of tacit knowledge creation? How to assess the evolution of core competence? After an exhaustive review of the literature, little was found on the key to the previous questions. Via questionnaire surveys, this paper carries out the empirical study on those questions to perfect relevant theories, thus providing guidance to the corporations.

2 Literature Review and Research Hypothesis

A. Tacit Knowledge Creation

Polanyi (1996) classified human knowledge into two categories [1]. "Explicit" or codified knowledge refers to knowledge that is transmittable in formal, systematic

D. Zeng (Ed.): ICAIC 2011, Part II, CCIS 225, pp. 523–530, 2011.

language. On the other hand, "tacit" knowledge has a personal quality, which makes it hard to formalize and communicate. Tacit knowledge is deeply rooted in action, commitment, and involvement in a specific context.

Tacit knowledge and explicit knowledge are the key factors in knowledge creation (Nahapiet & Ghoshal, 1998)[2]. A Firm's innovation capability has been linked to tacit knowledge transfer (Cavusgil et al., 2003)[3]. This is 'learn-by-doing' knowledge and is difficult for competitors to buy in the marketplace.

Scholars have done a lot research on knowledge creation and proposed many models, among which, Nonaka & Rui Mingjie pioneer in this field. Nonaka et al.(2000) viewed that knowledge creation is a continuous, self-transcending process through which one transcends the boundary of the old self into a new self by acquiring a new context, a new view of the world, and new knowledge[4]. To understand how organizations create knowledge dynamically, Nonaka et al proposed the SECI process, the process of knowledge creation through conversion between tacit and explicit knowledge.

On the basis of evaluating Nonaka's SECI model, Rui Mingjie et al.(2004) proposed the high-tech enterprise knowledge-creating model[5]. They pointed out that the knowledge creation falls into 6 stages, that is, knowledge acquisition, knowledge selection, knowledge integration, knowledge creation, knowledge diffusion and knowledge sharing.

The typical organizational learning models stressed the importance of knowledge acquisition, knowledge sharing and knowledge integration (Dixon, 1995; Rui Mingjie, 2004)[6, 5]. Tsai (2001) asserted knowledge acquisition and knowledge sharing can promote organizational performance and innovation [7]. Knowledge absorption is increasingly becoming a managerial priority. New knowledge provides the basis for organizational renewal and sustainable competitive advantage. Innovation results from the firm's effective development and use of new technologies and/ or knowledge about market opportunities.

Hence, this study views that the tacit knowledge creation mainly covers 4 stages: knowledge absorption, knowledge selection, knowledge transmission and knowledge sharing.

B. The Evolution of Core Competence

Researchers often assume that descriptors of core capabilities such as 'unique,' 'distinctive,' 'difficult to imitate', or 'superior to competition' render the term self-explanatory, especially if reference the term self-explanatory, especially if reference is also made to 'resource deployment' or 'skills'. A few authors include activities such as 'collective learning' and explain how competence is and is not cultivated (Prahalad and Hamel, 1990) [8]. Teece, Pisano and Shuen (1990) provided one of the clearest definitions: 'a set of differentiated skills, complementary assets, and routines that provide the basis for a firm's competitive capabilities and sustainable advantage in a particular business' [9]. Barton (1992) adopted a knowledge-based view of the firm, defined core capability as the knowledge set that distinguishes and provides a competitive advantage [10]. There are four dimensions to this knowledge set. Its content is embodied in (1) employee knowledge and skills, which encompass both firm-specific techniques and scientific understanding; and embedded in (2) technical systems. This knowledge constitutes both information (e.g. a data base of product tests conducted over decades) and procedures (e.g. proprietary design rules).The processes of knowledge creation and control are guided by (3) managerial systems, which represents formal and informal ways of creating knowledge (e.g. through sabbaticals, apprenticeship programs or networks with partners) and of controlling knowledge (e.g. incentive systems and reporting structures). The fourth dimension is (4) the values and norms associated with the various types of embodied and embedded knowledge creation and control. Thus a core capability is an interrelated, interdependent knowledge system.

Core competence is the unique knowledge owned by the corporation, which is formed through hard-won knowledge, reflecting the corporate feature and leading to the business opportunities of the corporation (Rui Mingjie, Yuan Anzhao, 2005)[11].

As to a corporation, knowledge acquisition, knowledge innovation and implementation play an important role in evolving corporate core competence (Grant, 1996; Kogut and Zander, 1992) [12-13].

The core competence can be featured by brand, know-how, service, innovation management, market net-work and so on (Rui Mingjie, Yuan Anzhao, 2005)[11]. With the enhancing of corporate core competence, the performance, scale as well as the sales volume will rise (Shi Dongming, 2002)[14].The evaluation index of core competence includes market control capability and profitability (Huang Jigang, 2004)[15]. Customer satisfaction is an important index to measure the competitiveness in European countries (Xu Zhaoyuan, Bao Wei, 2001) [16]. All in all, this paper wants to harness market share, assets return ratio, average product price, overall product quality, overall customer service level, profit and customer satisfactory to assess the evolution of corporate core competence.

C. The Proposition of the Hypothesis

Based on the previous literature review, the following hypotheses have been proposed and are to be tested:

Hypothesis 1: Tacit knowledge innovation has significant positive influence on the evolution of core competence.

Hypothesis 1a: Knowledge absorption has significant positive influence on the evolution of core competence.

Hypothesis 1b: Knowledge selection has significant positive influence on the evolution of core competence.

Hypothesis 1c: Knowledge transmission has significant positive influence on the evolution of core competence.

Hypothesis 1d: Knowledge sharing has significant positive influence on the evolution of core competence.

3 Research Method

A. Sample and Data Collection

Data were collected with the senior manager as the key informant (Kumar et al., 1993) [17]. The key informant approach has been widely used in empirical studies (Morgan and Hunt, 1994; Sen and Egelhoff, 2000) [18-19]because of the key informants' knowledge of the firm, access to strategic information, and familiarity with the environment of the firms (Aguilar, 1967)[20]. A Limited pilot study was undertaken to ensure that respondents had no difficulties in completing the questionnaire (Greenley, 1995) [21].

The senior managers mainly from Shanghai, Beijing, Jiangxi and Zhejiang have been surveyed. 570 questionnaire surveys have been sent out. Follow-up phone calls were made to all potential respondents who had not returned the surveys after four weeks. A comparison of the early-responding firms with the late-responding firms showed that these groups did not differ in terms of number of employees, sales revenue, and years in business. As a result, 307 usable questionnaires were returned, resulting in an efficient response rate of 53.86%. This response rate is similar to others who have surveyed top management (e.g. Geletkanyez, 1997) [22]. All variables were measured on a five-point scale to ensure the uniform scale width.

B. Measures and Validation

All scales used a five-point scoring format ranging from 'Strongly Disagree' to 'Strongly Agree'. Some items were adapted and reworded to fit the present context. The measures in this study were drawn on several sources. The items to measure tacit knowledge creation were consisted of eleven items, which were adapted from Ikujiro Nonaka (2000) [4], Rui Mingjie (2004)[5], Baetz (2003)[23], Ma Hongjian (2005) [24]and so on. The items to measure the evolution of core competence were drawn from the studies by Shi Dongming (2002)[14], Huang Jigang (2004)[15], Xu Zhaoyuan (2002) [16]which were composed by seven items, namely, market share, assets return ratio, average product price, the overall quality of the product, the overall customer service level , profit and the ratio of customer satisfactory.

To assess the discriminate validity of multi-item measures, factor analysis was conducted. Table 1 shows the result of this analysis. It also shows that almost all measures have a Cronbach's alpha above 0.80, indicating a good evaluation of reliability of these measures.

4 Analysis and Results

To test the hypotheses, correlation analysis was performed. Table 1 presents the results of correlation analysis.

Table 1. The Correlation between Each Item of Tacit Knowledge Creation and the Evolution of Core Competence

		Knowledge absorption	Knowledge selection	Knowledge transmission	Knowledge sharing	The evolution of core competence
Knowledge absorption	Pearson Correlation	1	.721(**)	.633(**)	.644(**)	.374(**)
	Sig. (2-tailed)	.	.000	.000	.000	.000
	N	307	307	307	307	307
Knowledge selection	Pearson Correlation	.721(**)	1	.685(**)	.616(**)	.304(**)
	Sig. (2-tailed)	.000	.	.000	.000	.000
	N	307	307	307	307	307
Knowledge transmission	Pearson Correlation	.633(**)	.685(**)	1	.689(**)	.317(**)
	Sig. (2-tailed)	.000	.000	.	.000	.000
	N	307	307	307	307	307
Knowledge sharing	Pearson Correlation	.644(**)	.616(**)	.689(**)	1	.390(**)
	Sig. (2-tailed)	.000	.000	.000	.	.000
	N	307	307	307	307	307
The evolution of core competence	Pearson Correlation	.374(**)	.304(**)	.317(**)	.390(**)	1
	Sig. (2-tailed)	.000	.000	.000	.000	.
	N	307	307	307	307	307

** Correlation is significant at the 0.01 level (2-tailed).

Table 1 shows that knowledge absorption, knowledge selection, knowledge transmission and knowledge sharing have positive relationship with the evolution of core competence, which are 0.374, 0.304, 0.317 and 0.390 respectively, significant at the 0.01 level. Thus, hypotheses 1a, 1b, 1c and 1d receive solid support from the analysis.

Through the regression analysis of the relationship between tacit knowledge creation and the evolution of core competence, their B-value is 0.408, t-value is 7.813, which is very significant under 0.001level, so hypotheses 1 receive strong support.

5 Conclusion

A. Managerial Implications

The findings of this study offer several insights for managers of corporations. Today's business environment requires consistent growth in order to remain competitive and sustain corporate growth. This leaves the ability to innovate as the primary mechanism for organizations to sustain growth in the long run. In particular, tacit knowledge is crucial for innovation, great efforts are necessary to promote an environment for tacit knowledge sharing between individuals through socialization processes (Nonaka, 1994) [25]. In order to nurture corporate core competence, the corporations should pay attention to the construction of the following four factors:

Firstly, pay attention to knowledge absorption. Tacit knowledge absorption plays a vital role in cultivating corporation competitive advantage (Lubit, 2001) [26]. Knowledge absorption is the activity of accepting a unit of knowledge from the external environment and transforming it into a representation that can be internalized, and/or used within an organization. Corporate capabilities are closely related to their knowledge assets, identification of corporate prior knowledge, its capacity to absorb new knowledge, and its transition from its current form of knowledge to another form, is necessary to build corporate dynamic competitiveness. As to the corporation, it should actively absorb knowledge, establish solid and multi-channel data bank. The absorption of knowledge can be via selection and learning. The results of this study strongly support the notion that managers and corporations need to prepare themselves for knowledge absorption through learning to create a facilitating culture in an effort to foster and sustain the corporate core competence. Managers should encourage organizational learning by emphasizing the organization's ability to learn and considering learning as an investment, rather than an expense.

Secondly, emphasize on knowledge selection. It is essential to select new knowledge, because some knowledge acquired is just the accumulation in quantity instead of the accuracy, which is useless to the corporation. Therefore, knowledge selection must be carried out to exclude the useless information. Knowledge selection can greatly reduce the cost of knowledge management and improve the knowledge application efficiency. The mode of knowledge selection includes personnel analysis and computer-based automatic selection. Hence, the corporation should take measures to select important knowledge.

Thirdly, lay emphasis on knowledge transmission. Knowledge, once acquired, must be quickly disseminated to all parts of the corporation. There are formal and informal channels in transmitting knowledge. The formal channel depends on the technological basis and rules, while the informal channel is through the establishment of personnel ties and sound negotiation capability. Much latest tacit knowledge is uneasy to acquire through formal channel, but one can understand the essence of tacit knowledge through language, body gestures and feelings, so as to share those tacit knowledge. The individual network is crucial in transmitting the knowledge. In order to establish human relationship network, the corporation should establish communication mechanism. For example, shift the staff from one sub-corporation to another to provide the career opportunity to them, thus making them gaining the experience of working in different countries. Therefore, in order to acquire tacit knowledge which can improve corporate competitive advantage, the corporation should constantly recruit the staffs who are the master of a certain know-how, and harness the informal network (such as the communication with suppliers, customers and rivals) to spur the transmission of knowledge.

Fourthly, stress on knowledge sharing. The aim of knowledge sharing is to break down the barrier among team members and harness knowledge within a certain extent. Knowledge sharing can promote the adequate flow of knowledge, minimize repeated input of knowledge so as to reduce the cost of knowledge acquisition, thus facilitating knowledge application and innovation. Knowledge sharing is the source of innovation. Only through dialogue and co-transformation, can corporate knowledge be shared by the staffs in the corporation. So the corporation should adopt all kinds of measures to enhance the connection among staffs, thus sharing information, ideas and exchanging mental mode. In order to share the knowledge, the corporation should establish the effective incentive mechanism, encourage the staff to learn from each other and share good experiences. In addition, the trust-based corporate culture should be set up, thus promoting negotiation and understanding among staffs.

B. Limitations and Further Research

The findings of this study should be interpreted in the light of its inherent limitations. Firstly, we haven't made an in-depth analysis of the tacit knowledge creation mechanism. Thus, the future research should cover this area. Secondly, this study does not control for the corporate ownership and origin. The ownership (e.g. independent and corporate) and origin (joint venture and privately owned) could affect tacit knowledge creation. Hence, future research should control for these variables. Thirdly, regional sampling does limit the generality of the results. Our samples are mainly from Shanghai, Beijing, Zhejiang and Jiangxi in China, it may be instructive for future research to extend to other regions or other transitional economies.

Acknowledgment. The author thanks the grant by Key Principle Project of Shanghai Municipal Education Commission (J51201), Key Innovation Project of Shanghai Municipal Education Commission (09ZS187), Shanghai Natural Science Foundation (09ZR1420000).

References

1. Polanyi, M.: The Tacit Dimension, pp. 16–18. Routledge &v Kegan Paul, London (1996)
2. Nahapiet, J., Ghosal, S.: Social Capital, Intellectual Capital, and the Organizational Advantage. The Academy of Management Review 23(2), 242–266 (1998)
3. Cavusgil, S.T., Calantone, R.J., Zhao, Y.: Tacit Knowledge Transfer and Firm Innovation Capability. Journal of Business & Industrial Marketing 18, 6–21 (2003)
4. Nonaka, I.: SECI, Ba and Leadership: a Unified Model of Dynamic Knowledge Creation. Long Range Planning 33, 5–34 (2000)
5. Mingjie, R., Xin, L., Hongbo, R.: On Knowledge Innovation Mode of High-tech Corporations: the Amendment of the Model of Nonaka. Foreign Economies and Management 5(26), 8–12 (2004)
6. Dixon, N.: The Organizational Learning Cycle. McGRAW-HILL Book Company, Europe (1995)
7. Tsci, W.: Knowledge Transfer in Intra-organizational Networks: Effects of Network Position and Absorptive Capacity on Business Unit Innovation and Performance. Academy of Management Journal 44(5), 996–1004 (2001)
8. Prahalad, C.K., Hamel, G.: The Core Competence of the Corporation. Harvard Business Review 68(3), 79–91 (1990)

9. Teece, D.J., Pisano, G., Shuen, A.: Firm Capabilities, Resources and the Concept of Strategy., Consortium on Competitiveness and Cooperation Working Paper # 90-9, University of California at Berkeley, Center for Research in Management, Berkeley, CA (1990)
10. Leonard-Barton, D.: Core Capability and Core Rigidities: A Paradox in Managing New Product Development. Strategic Management Journal 13, 111–127 (1992)
11. Rui, M., Yuan, A.: Theories and Operations of Modern Corporations. Shanghai University of Economics & Finance Press (2005)
12. Grant, R.M.: Toward a Knowledge-Based Theory of the Firm. Strategic Management Journal 17, 109–122 (1996)
13. Kogut, B., Zander, U.: Knowledge of the Firm, Combinative Capabilities, and the Replication of Technology. Organization Science 3, 383–397 (1992)
14. Shi, D.: On Core Competence of Enterprises: Construction of International Competitiveness. Beijing University Press (2002)
15. Huang, J.: Dynamic Management of Core Competence. Economics & Management Press (2004)
16. Xu, Z., Bao, W.: Empirical Study on the Indexes of American Customer Satisfaction. World Standardization and Quality Management 2, 10–17 (2002)
17. Kumar, N., Stern, L.W., Anderson, J.C.: Conducting inter-organizational Research Using Key Informants. Academy of Management Journal 36(6), 1633–1651 (1993)
18. Morgan, R.H., Hunt, S.D.: The Commitment-trust Theory of Relationship Marketing. Journal of Marketing 58(3), 20–39 (1994)
19. Sen, F.K., Egelhoff, W.G.: Innovative Capabilities of a Firm and the Use of Technical Alliances. IEEE Transactions on Engineering Management 47(2), 174–183 (2000)
20. Aguilar, F.: Scanning the Business Environment. Macmillan, NY (1967)
21. Greenly, G.: Market orientation and company performance: empirical evidence. British Journal of Management 40, 33–46 (1995)
22. Geletkanyez, M.: The Salience of "Culture's Consequences": the Effects of Cultural Values on Top Executive Commitment to the Status Quo. Strategic Management Journal 18(8), 615–634 (1997)
23. Baetz, W.G.: Organizational Learning Practices. State University of New York (2003)
24. Ma, H.: The Theories and Empirical study on Knowledge Management of High-tech Enterprises in China. Fudan University, Shanghai (2005)
25. Nonaka, I.: A Dynamic Theory of Organizational Knowledge Creation. Organization Science 5(1), 14–37 (1994)
26. Lubit, R.: Knowledge and Knowledge Management: The Keys to Sustainable Competitive Advantage. Organizational Dynamics 4, 164–178 (2001)

An Organization-Extended RBAC Model and Its Application in Workflow System*

Zhang Cheng[1] and Lai Zulong[2]

[1] School of Arts and Communication, China University of Geosciences, Wuhan, Hubei Province 430074, China
[2] Faculty of Information Engineering, China University of Geosciences, Wuhan, Hubei Province 430074, China
304858424@qq.com

Abstract. Access control and management of organization-role are the keys to workflow system. A new role-based access control model, OE-RBAC, which extends the concept of role and organization and is based on RBAC model, is proposed. Organization information and roles are managed uniformly in the new model. This mechanism enables the organizations and roles to have the same authority to assign permissions, and enhances the organization-role management of workflow. The model provides the mechanism of uniform management of hierarchical organization-role, which makes the application more suitable for large-scale deployment.

Keywords: security model, workflow, OE-RBAC, role-based access control.

1 Introduction

Access control is the core of security management in workflow system, and its mission is to ensure that information system resources are not illegally used or unauthorized access. The basic idea of RBAC has been applied in some applications [1-3]. But, there are still some problems in large-scale workflow systems. First, cross-boundary system of the business is hard to deploy. All administrative departments are relatively independent in the daily management, but there are some nodes in the flow of the business which needs assistance from the same or superior departments. Every department is autonomic, so hierarchical management authority is imperative[4-5]. Second, e-government system based on workflow may be complex, and there may be large numbers of users. If all authorization is assigned by one super administrator, the workload of authorization is large in this case. It is recognized that the work of creating the user account and managing the user's right is becoming harder and harder[6-8]. Third, the structure of the organization in workflow system is multifarious and variable. The workflow system is a practical system, which combines with the department's organization closely. The structure of department's organization must be fully considered. The workflow system is used widely in many fields, and administrative departments and companies must adapt to the rapid development of human society, and

* This work is supported by the project of "special fund of basic scientific research of Education Ministry of China (No. CUGW090233)".

D. Zeng (Ed.): ICAIC 2011, Part II, CCIS 225, pp. 531–537, 2011.

the organization has to be changed correspondingly. So the workflow system should be flexible, adaptable and manageable.

Workflow is an essential infrastructure of e-government system. At the same time, workflow and organization-role are inseparable from empowerment. So, it is propitious to assign authority in e-government system that workflow is combined with RBAC, which enhance the management of organization-role.

Facing these problems, the organization information of administrative department is introduced to traditional RBAC model. Organizational structure and roles participate in access control at the same time, and the enhanced access control model is combined with workflow. It provides a solid foundation for e-government. Hierarchical definitions of role and organization coincide with hierarchical management in reality. So, this paper put forward an organization enhanced RBAC (OE-RBAC), in which organization information can participate in access control. The e-government system can be authorized not only by one super administrator, but also by hierarchical administrators in the model. The model is fit for large-scale users' cooperation in work intuitively.

2 Improved RBAC

A. RBAC Summary

Role is introduced as intermediary in RBAC[9]. Administrator defines all roles according to application's demand, and grants proper access permission to roles. So role and user are the components of the authorization logically. Users can be assigned with different roles and get different access permissions. RBAC simplifies the operation of authorization and security in distributed environment. Thus, the access procedure is divided into two parts: relevance of access rights and roles, relevance of roles and users. RBAC successfully separates user and access right, which makes management of authorization easy. RBAC model[9] is shown as Figure1:

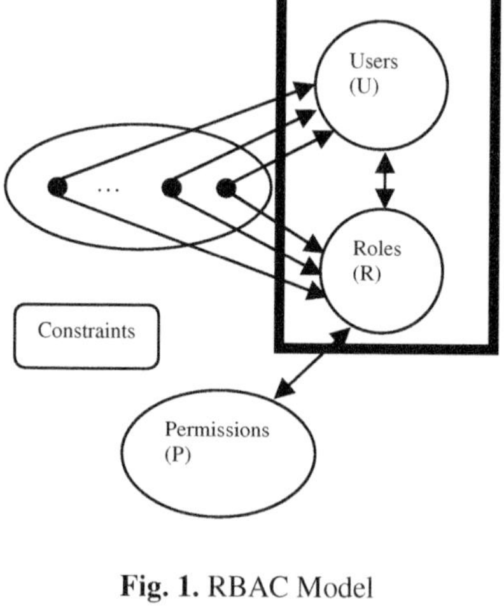

Fig. 1. RBAC Model

RBAC includes: user, role, permission and session. Its basic principle is that the permission of access to resource is granted to roles, and the role is assigned to user according to user's duty[1]. Role is the bridge of permission and user. The whole process is divided into role definition and permission assignment, between which the relation is shown as Figure1.

B. OE-RBAC

OE-RBAC allows organization information participate in allocation and management of permission, and enhances the model that only uses role to grant permission. Organization and role are used in one workflow system complementarily, and strengthen the ability of authorization. The conceptual model of OE-RBAC is shown as fugure2.

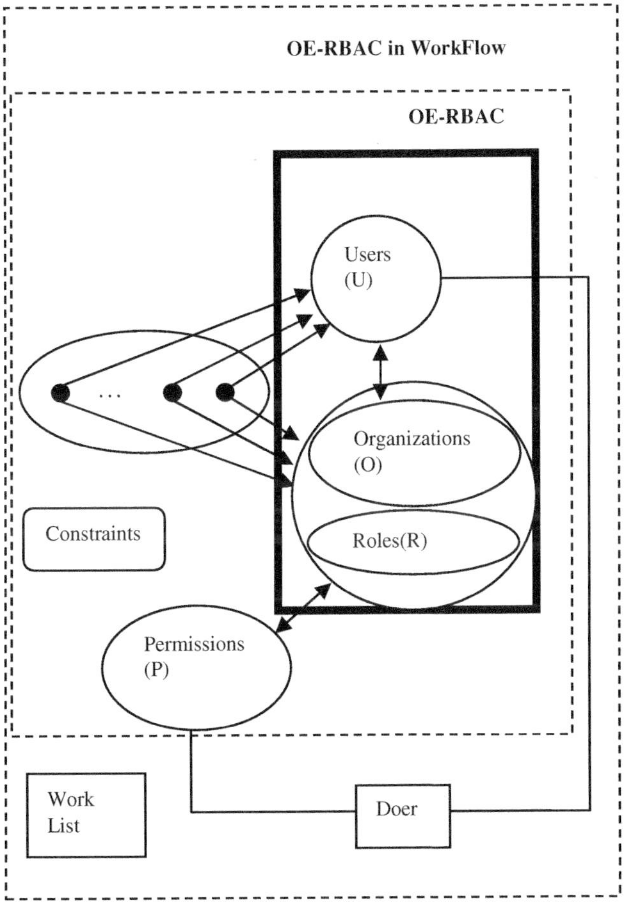

Fig. 2. A Conceptual model of OE-RBAC and OE-RBAC in Workflow

Obviously, an organization set is added to RBAC, OE-RBAC emphasizes integrated management of organization, role and user.

O: organization set, it can be granted permission like role set.

$\forall$ o $\in$ O has the same ability of authorization like R. So, the OE-RBAC can grant permission to O or R according to demands. RBAC, which only uses R to assign permission, is strengthened.

Organization set is not only used in management of organization, but also has the same ability to assign permission as R. Organization has become a carrier of permission like role. So, the expressions in RBAC can be adjusted as the following:

UA:UA U$\times$(O+R), UA is the set of relationship between users and (organizations and roles).

PA:PA (O+R)$\times$P, PA is the set of relation between (organizations + roles) and permissions.

RH:RH (O+R)$\times$(O+R), RH is the description of the relationship of (role or organization) inheritance, for example: (o1, role2) $\in$ RH represent that role2 inherits from o1.

C. The tree structure of organization and role in OE-RBAC

In OE-RBAC, organization, role and user are integrated. Organization and role have the same ability to assign permission. Organization is fixed in specific application, but role is defined according to the demand of authorization with flexibility. Role can exist in organization tree, and can exist in isolated role tree. But, organization can not be in role tree.

Any system is designed to serve a particular organization. Current organizations are mainly structured like a tree. In the management of an organization, one or more roles will be designed inside the organization in order to complete a particular task, while roles are likely to be designed among several organizations to complete some special tasks.

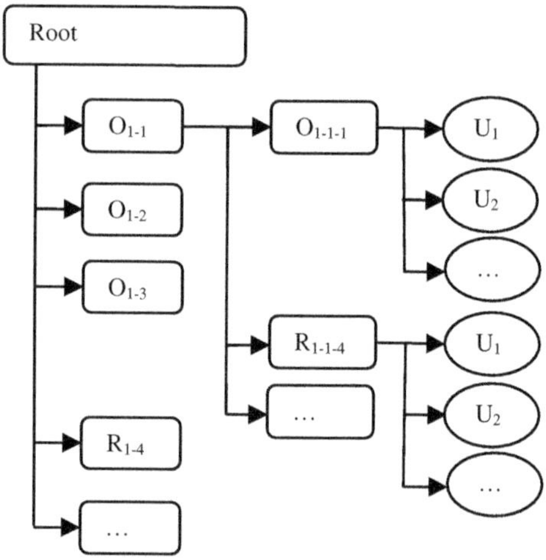

Fig. 3. Organization-role-user tree structure

The above figure shows that role in RBAC is replaced by organization and role, in which tree-structure visual management is more human-centered and conforms to reality. To personnel, organization is a form to organize people, and role is also a form to organize people. The model of hierarchical organization and role has characteristics:

First, OE-RBAC can reflect the real organization structure of application system, which is very beneficial to administrator to maintain organization and user information. Administrator can respond rapidly to organization restructured and user changed.

OE-RBAC can organize users in their real existence. So, whenever it is at the beginning of application construction or in the process of system operation and maintenance, the organization or role is very similar to the real existence, even identical.

Second, organization has the same authorization ability like role in RBAC to assign permission properly. If the permission is based on organization, what to do is to adjust the personnel in organization, and permission will be adjusted automatically. If the permission is based on role, it must be sure that corresponding user should be in right role.

Third, OE-RBAC is shown in tree which is familiar to every person, so organization is easy to construct at the beginning of the application based on OE-RBAC.

Role can exist in organization tree and can exist in isolated role tree. Those roles which are specialized set for one branch can be set under the branch node. Some roles which are set for cross-branch must be set under the node of isolated role tree. The organization-role tree is abstract in e-government for authorization, and is also a powerful tool to manage organization and user.

When the organization is being abstracted into tree, there may be some persons belonging to several branches. To solve this problem, the person must have only one identity which also exists in corresponding branches. The mixed management of organization and role makes it true that any single user could be managed by sub-administrator of his own department.

D. Hierarchical Authorization Management in OE-RBAC

Permission assignment is generally the duty of administrator, but because of the scope of work, administrator hardly has comprehensive understanding of all duties of all posts. Those large-scale systems may involve many departments which may be located in different places and have a large number of employees. It is fit to grade in tree, as permission under any node of the organization-role tree can be granted to different sub-administrator. Every sub-administrator is responsible for authorization under the node in the tree, which makes the duty of sub-administrator clear and accurate.

For small applications, an administrator can manage the authorization competently. For large applications, across multiple departments which are located far apart, and each department's management is basically autonomous, the corresponding node should be assigned administrative rights within the department.

Hierarchical authority management is definitely not only just the permission level of the inheritance hierarchy, but rather: 1, It represents the organization's classification; 2, In some branches of organization, personnel managements are relatively independent, their own business is managed by themselves, and their responsibilities are obviously in process.

E. OE-RBAC is Flexible

With changes in social conditions and increasingly rapid pace of social life, organizations and roles of enterprises and administrative departments will be adjusted accordingly. So, OE-RBAC must be flexible to adapt to these uncertain changes, otherwise, the life cycle of application system will be greatly shortened. The outstanding characteristics of OE-RBAC model are more flexible, more practical and more suitable for large organizations than ever. The module of organization-role management gives the facility to personnel adjustments, and organization adjustments.

3 Application of OE-RBAC in Workflow

Workflow management technology aims at the automated support and coordination of business processes in order to reduce costs as well as increase the effectiveness and efficiency of an organization. Workflow management systems allow for the definition and enactment of business processes. Workflow and OE-RBAC are inseparable. So, the combination of workflow system and OE-RBAC and the enhanced organization management of the user's role will contribute more to the attribution of authority in the application system.

B. Combination of OE-RBAC and Workflow

OE-RBAC has been discussed above, and workflow management system also needs permissions to user management which can be seen in International Workflow Reference Model (as shown figure4). Workflow and access control are very important infrastructures of application system. The combination of OE-RBAC with workflow management can enhance the ability of authorization based on organization and role.

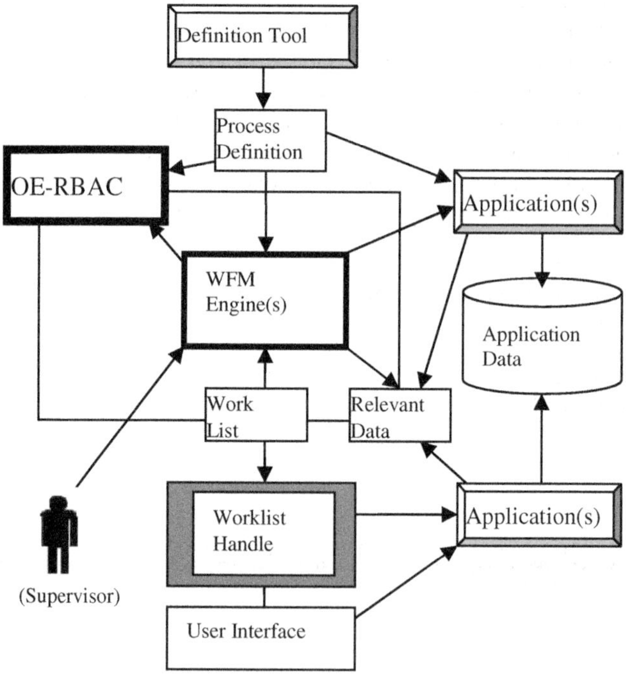

Fig. 4. Workflow Product Structure with OE-RBAC

4 Conclusion

This paper studies the access control model of application system, gives OE-RBAC model unified administration of organization, role and user, and enhances the traditional RBAC model which only uses roles to assign authorization. OE-RBAC model is put forward, which can be applied in workflow system, and strongly supports the application systems such as the e-government system and makes it meet the demands of large-scale hierarchical management. For instance, the OE-RBAC workflow management system has been in application in the rent management system of Shenzhen City. There are many organizations that scatter in different places with more than 10000 users, so it is nearly impossible for a super administrator to manage vertically so many users and authorities. Through the actual application in the management system, it is proved that the OE-RBAC model is very suitable for large-scale e-government authorization management.

References

1. Kang, L.-z., Huang, Q.-s., Liu, L.-j.: An Improved Role-Based H ierarchicalAuthorization and Access Control M odel. Journal of Kunming University of Science and Technology (Science and Technology) 34, 39–42 (2009), doi: CNKI:SUN:KMLG.0.2009-01-010
2. Zhu, Y.-q., Li, J.-h., Zhang, Q.-h.: A New Dynamic Hierarchical RBAC Model for Web Services. Journal of Shanghai Jiaotong University 41, 783–787 (2007), doi: CNKI:ISSN:1006-2467.0.2007-05-025
3. Zhao, J., Liu, X., Feng, H., Zhou, P.: Research and Application of Function Field-Based RBAC Model. ComputerApplications and Software 26 (April 2009), doi: CNKI:SUN:JYRJ.0.2009-04-043.
4. Wang, S.: A Brief Researeh on the Double-Ievel Teaching Administration from the University and Colleges in High School. Journal of Social Science of Jiamusi University 24, 80–81 (2006), doi: cnki:ISSN:1007-9882.0.2006-04-035
5. Su, P., Cao, B.: A Implementation of Classified Rights Management Scheme. Information Technology and Informatization 5, 59–69 (2006), doi: CNKI:SUN:BGJS.0.2009-01-007
6. Yi, K., Zheng, M., Qian, X., Gu, Z.: A Dynamic Hierarchy Access Control Model Based on Roles. Baosteel Technology, 19–22 (January 2009)
7. Le, Y., Choi, Y.: A Flexible Access Control Model for Dynamic Workflow Using Extended WAM and RBAC copy. In: Shen, W., Yong, J., Yang, Y., Barthès, J.-P.A., Luo, J. (eds.) CSCWD 2007. LNCS, vol. 5236, pp. 488–497. Springer, Heidelberg (2008)
8. Oh, S., Park, S.: Task-role based access control (T-RBAC): An improved access control model for enterprise environment. In: Ibrahim, M., Küng, J., Revell, N. (eds.) DEXA 2000. LNCS, vol. 1873, pp. 264–273. Springer, Heidelberg (2000), doi:10.1007/3-540-44469-6_25
9. Sandhu, R., Coyne, E., Feinstein, H., Youman, C.: Role-based access control models. IEEE Computer 29(2), 38–47 (1996)

Semi-supervised Co-training Algorithm Based on Assisted Learning*

Hong-li Wang and Rong-yi Cui**

Intelligent Information Processing Lab.
Department of Computer Science and Technology
Yanbian University
Yanji, Jilin, China
w.l@myself.com, cuirongyi@ybu.edu.cn

Abstract. The classification performance of the learner is weakened when unlabeled examples are mislabeled during co-training process. A semi-supervised co-training algorithm based on assisted learning (AR-Tri-training) was proposed. Firstly, the assisted learning strategy was presented, which is combined with rich information strategy for designing the assisted learner. Secondly, the evaluation factor was calculated, and noise was eliminated from unlabeled example set by using the assisted learner and the evaluation factor. Finally, three single learners were trained using labeled examples, wrong-learning examples on validation set and less noise unlabeled examples. The experimental results on application to voice recognition indicate that AR-Tri-training can compensate for the Tri-training shortcomings and the average classification accuracy is increased by 15%. As can be drawn from the experimental results, AR-Tri-training not only removes the mislabeled examples in training process, but also takes full advantage of the unlabeled examples and wrong-learning examples on validation set.

Keywords: Semi-supervised Learning, Rich Information Strategy, Assistant Learning Strategy, Tri-training.

1 Introduction

In the traditional setting of supervised learning, a large amount of labeled training examples should be available for building a model with good generalization ability. However, in many real applications labeled ones are fairly expensive to obtain because they require human effort; while the unsupervised learning can not get higher learning accuracy due to using only the unlabeled examples. As a consequence, semi-supervised learning that exploits unlabeled examples in addition to labeled ones has become a hot topic [1].

* This work is supported by Sci. & Tech. Department of Jilin Prov. Grant#20050703-1.
** Corresponding author.

D. Zeng (Ed.): ICAIC 2011, Part II, CCIS 225, pp. 538–545, 2011.

Many current semi-supervised learning algorithms are very interested in how to utilize unlabeled examples for training learners. A prominent achievement in this area is the Co-training paradigm proposed by Blum and Mitchell [2], which trains two classifiers separately on two different views, i.e. two independent sets of attributes, and uses the predictions of each classifier on unlabeled examples to augment the training set of the other. The standard Co-training algorithm requires two sufficient and redundant views. Unfortunately, such a requirement can hardly be met in most scenarios. So, Goldman and Zhou proposed an algorithm which does not exploit attribute partition. However, it requires using two different supervised learning algorithms that partition the instance space into a set of equivalence classes, and employing time-consuming cross validation technique to determine how to label the unlabeled examples and how to produce the final hypothesis [3]. The Tri-training algorithm proposed by Zhou and Li is a new co-training algorithm and not constraint in the attribute sets and supervised learning algorithm, therefore, it has a broader scope [4].

However, the unlabeled examples are often incorrectly labeled and the mislabeled examples are accumulated in the iterative learning process, which is the same with other semi-supervised learning method, so the learning performance is weakened [5][6], more particularly, this problem is more severe in Tri-training and other Co-training algorithm[7]. Therefore, in this paper, a new Co-training style semi-supervised learning algorithm named AR-Tri-training is proposed. In contrast to previous algorithms, i.e. Tri-training, AR-Tri-training not only can initiatively identify and remove possible mislabeled examples, but also improves the utilization of unlabeled examples.

2 Related Work

Tri-training generates three learners from the original labeled example set. These learners are then refined using unlabeled examples in the Tri-training process. In each round of tri-training, an unlabeled example is labeled for a learner if the other two learners agree on the labeling, under certain conditions. The final hypothesis is produced via majority voting.

Let **L** denote the labeled example set with size $|\mathbf{L}|$ and **U** denote the unlabeled example set with size $|\mathbf{U}|$. These learners h_1, h_2, h_3 are initially trained from **L**, the training process is in the following. If h_2 and h_3 agree on the labeling of an example x in **U**, then x can be labeled the prediction of h_2 for h_1, so $\mathbf{S}_1$ is formed which can be denoted $\mathbf{S}_1=\mathbf{L}\cup\{x|x\in\mathbf{U} \text{ and } h_2(x)=h_3(x)\}$. Similarly, the training sets of h_2, h_3 are also extended to $\mathbf{S}_2$, $\mathbf{S}_3$, and then three learners are retrained, repeatedly the iteration, until h_1, h_2, h_3 are invariant.

It is obvious that in such a scheme if the prediction of h_2 and h_3 on x is correct, then h_1 will receive a valid new example for further training, otherwise h_1 will get an example with noisy label. Zhou and Li get the sufficient condition for iteration which can reduce the classification error rate, but Tri-training can not take full advantage of unlabeled examples.

As mentioned above, Tri-training not only fails to discard the noise, but also limits the number of unlabeled examples added to the new training set, so the assisted learning strategy is proposed in our work.

3 Ar-Tri-Training Algorithm

A. Learning Strategies

The sampling strategy is necessary for machine learning. The rich information is a special sampling strategy, that is, if an example can't be correctly learned, the next weak learner will focus on learning it [8]. Inspired by Wang, the method of selecting examples with rich information is as follow. If an example is inaccurately classified by the current weak learner, it is selected with probability 1 and added to the training set of component learners; otherwise, it is selected with probability 0. In this way, learners can pay close attention to more difficult examples which are prone to be misclassified, so learners get a better learning result and improve the classification accuracy.

The labeled examples are re-sampled using the rich information strategy in the training process, yet a large number of unlabeled examples are employed in the semi-supervised learning, and unlabeled examples may be wrongly marked by the machine automatically. Therefore, how to remove the mislabeled examples is of great significance in the AR-Tri-training process. To address this problem, the assisted learning strategy with guidance is proposed and the corresponding assistant learner is designed in this paper. The basic idea of the strategy is as follow. A learner is trained using labeled examples including the labeled example set and misclassified examples on validation set which have a certain influence on this learner, then the learner generates evaluation factor which can make decision on whether unlabeled examples are correct labeled or not, so achieving the purpose of removing noise. In other words, the learner and unlabeled examples are regarded as a teacher and students respectively, the teacher can guide their students after studying a particular area of knowledge, and then judging the results of students' actions. The learner plays an important role in guiding unlabeled examples, so the learner is defined as the assisted learner.

In specific implementations, consider the consistent unlabeled example *ConsistentU* in t-th iterative training. Firstly, the assisted learner H is trained using the initial training set **L** and the misclassified examples *Fdata* on validation set **V**, and meanwhile, the classification accuracy *OldTestRate* on **V** is computed. Secondly, H is retrained using **L**, *Fdata*, *ConsistentU*, and the classification accuracy *NewTestRate* is recomputed. Finally, compare *OldTestRate* and *NewTestRate*, if *NewTestRate*≥*OldTestRate*, *ConsistentU* is correctly labeled and added to the training set of component learner, otherwise *ConsistentU* is discarded. Note that H is trained using the total misclassified examples including the 1st, 2nd, …, t-th in t-th iterative process with the running of AR-Tri-training algorithm, which not only embodies the rich information strategy, but also makes the learner H be trained fully.

In summary, the AR-Tri-training algorithm adopts the rich information strategy and the assisted learning strategy, so this method can improve the utilization of unlabeled examples.

B. Description of AR-Tri-training Algorithm

The basic settings of AR-Tri-training algorithm consist of labeled examples **L**, validation set **V**, unlabeled examples **U**, test set **T**, which contain N class data entirely, and three component learners are SVM1 (polynomial kernel function), KNCN, SVM2 (RBF kernel function), and the assisted learner SVM3 (linear kernel function). The algorithm terminates after the integration testing arrives at a steady state on the validation set **V**. The pseudo-code of AR-Tri-training and Rnoise are presented in Tab.1 and Tab. 2 respectively.

Table 1. Pseudo-code of AR-Tri-training algorithm

```
AR-Tri-training(L, U, V, T, Learn)
 Input: L: Original labeled example set
          U: Unlabeled example set
          V: Validation set
          T: Test set
          Learn: Learning algorithm
   for i ∈ {1... 3}
      h_i←Learn(L)
   end of for
   Fdata←Ensembletest(V)
   H←Learn(L∪Fdata)
   L' ←L∪Fdata
   repeat until the ensemble testing to a steady
state
          for i ∈ {1... 3}
              RUdata←RNoise(h_j, h_k, H, V, U,
L' ) (j, k ≠ i)
              h_i←Learn(L∪Fdata∪RUdata)
          end of for
          Fdata←Ensembletest(V)
          H←Learn(L' ∪Fdata)
          L' ← L' ∪Fdata
   end of repeat
   Output: Totalrate←Ensembletest(T)
```

Table 2. Pseudo-code of RNoise algorithm

```
RNoise(h1, h2, H, V, U, L′ )
   Input: h1, h2: Component learners
          H: Assisted machinery
          V: Validation set
          U: Unlabeled set
          L′ Training set
   CUdata←NULL
   RUdata←NULL
   OldTestRate←Test(H, V)
   for every x∈U do
      if h1(x)=h2(x)
      then CUdata←CUdata ∪ {x, h1(x)}
   end of for
   for every ConsistentU∈CUdata
      H←Learn(L′ ∪ConsistentU)
      NewTestRate←Test(H, V)
      if NewTestRate≥OldTestRate
      then RUdata←RUdata ∪ ConsistentU
   end of for
   Output: RUdata
```

4 Experimental Results and Analysis: Voice Recognition Using Ar-Tri-Training Algorithm

A. Experimental Data

Our data set comprise 150 audio data, including voice examples, music ringtones, animal sounds, environmental sounds and transport sounds. All the audio data are digitized at 8kHz sampling frequency and 16bit quantization accuracy, their lengths are 4s. In the experiment 14-dimensional features are extracted including the MFCC mean of layer 3th wavelet low-frequency coefficients, centroid variance, and spectral entropy variance. These features are calculated on 256-sample frames of audio, 80-sample overlap, pre-emphasis coefficient 0.97, and the number of Mel filters is 24.

B. Experimental Results and Analysis

The performance of AR-Tri-training is compared with Tri-training, R-Tri-training only using the rich information strategy, and the experiments are based on the same validation set and test set, having 30 and 50 examples respectively. For each experiment, in order to simulate different amount of labeled example set **L** and unlabeled example set **U**, three different |**L**|, i.e. 5, 10, 20 and seven different |**U**|, i.e. 1, 10, 20, 30, 40, 50, 60, are investigated here respectively. The classification accuracy of the compared algorithms are depicted in Fig.1 to 3 for test set.

Fig.1 to 3 show that AR-Tri-training can effectively improve the classification accuracy. Tri-training uses some criterions in training process, so the unlabeled examples can't be fully used. R-Tri-training which is added a validation set can pay attention to wrong-learning labeled examples. However the noise examples still exists in both algorithms. In contrast to Tri-training and R-Tri-training, the proposed AR-Tri-training algorithm considers that correctly identifying the current unlabeled example plays a supplementary role in the next unlabeled example, so it designs the additional learner by using the assisted learning strategy and rich information strategy. As a result, AR-Tri-training not only focus on wrong-learning labeled examples, but also removes the mislabeled examples.

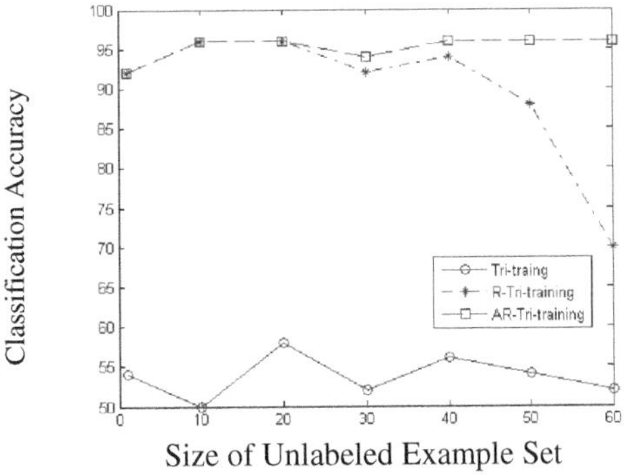

Fig. 1. 5 labeled examples

In the case of the same labeled examples, labeled examples play an important role in Tri-training, and with the increase in unlabeled examples, the role of labeled examples is reduced, noise is accumulated seriously, so the classification accuracy is decreased gradually in Tri-training. R-Tri-training takes advantage of wrong-learning examples on validation set, so an equivalent classification accuracy is obtained. But AR-Tri-training achieves the purpose of removing the mislabeled examples and makes full use of unlabeled examples. As a consequence, the classification accuracy is further improved and this approach reflects the essence of semi-supervised learning.

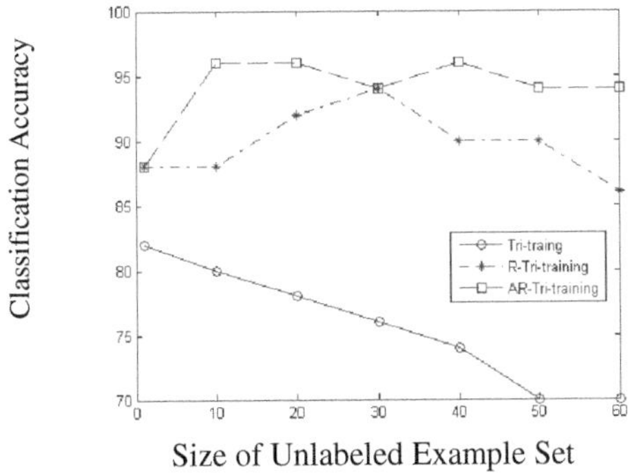

Fig. 2. 10 labeled examples

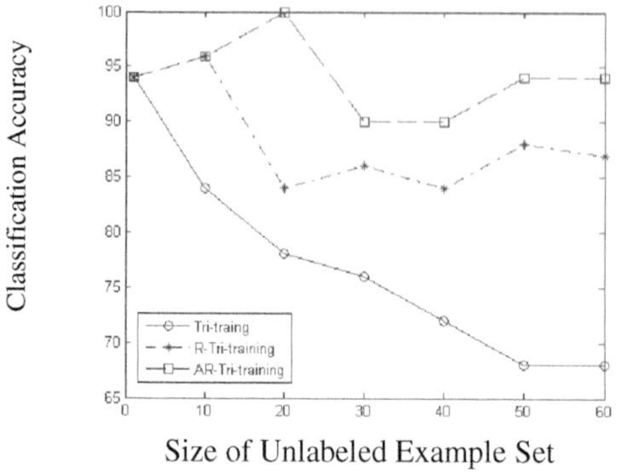

Fig. 3. 20 labeled examples

Meanwhile, in order to demonstrate the ability to remove noise in AR-Tri-training, the number of noise which appears in the training process for AR-Tri-training and R-Tri-training are counted. The statistical results are shown in Tab. 3, Tab. 4 for |**L**|=10, |**L**|=20, |**U**| is 10, 20, 30, 40, 50, 60 respectively.

Table 3. Noise Examples in Training Process for |L|=10

Num. of Unlabeled Examples	AR-Tri-training	R-Tri-training
10	1	3
20	1	6
30	2	8
40	2	10
50	4	13
60	4	18

Table 4. Noise Examples in Training Process for |L|=20

Num. of Unlabeled Examples	AR-Tri-training	R-Tri-training
10	1	1
20	0	3
30	4	7
40	4	8
50	6	12
60	7	15

It can be drawn from Tab. 3 and Tab. 4 that AR-Tri-training algorithm can remove a lot of noise in training process, compared with R-Tri-training. It is no doubt that supervised learning can improve the performance of learner by only using labeled examples, so the assisted learner is designed based on this idea in AR-Tri-training. But it is very important that the assisted learner pay attention to wrong-learning examples particularly, which is different from supervised learning. As mentioned above, the assisted learner plays an important role in removing noise by using the assisted learning strategy and rich information strategy for AR-Tri-training.

5 Conclusion and Next Step

In this paper, the AR-Tri-training algorithm is proposed in order to tackle two difficult problems in Tri-training including noise accumulation and low utilization of unlabeled examples and is applied to voice recognition. AR-Tri-training adopts the rich information strategy which makes a weak learner focus on learning wrong-learning examples and improves the learner's classification accuracy and the assisted learning strategy which can remove the accumulated noise in the training process and improve the utilization of unlabeled examples. Experimental results show that AR-Tri-training is effective on removing noise.

Co-training and sample selection have different goals, their selection methods focus on different criteria: Co-training typically favors selecting accurately labeled examples, while sample selection typically favors selecting examples with high training utility, therefore the selection strategy of unlabeled examples is an important future work.

References

1. Li, K.-L., Zhang, W., Dai, Y.-N.: Semi-supervised SVM based on Tri-training. Computer Engineering and Applications 45(22), 103–106 (2009) (in Chinese with English abstract)
2. Blum, A., Mitchell, T.: Combining labeled and unlabeled data with Co-training. In: Proceedings 11th Annual Conf. on Computational Learning Theory, Wisconsin USA, pp. 92–100 (1998)
3. Goldman, S., Zhou, Y.: Enhancing supervised learning with unlabeled data. In: Proceedings 17th Annual Conf. on Machine Learning, California, USA, pp. 327–334 (2000)
4. Zhou, Z., Li, M.: Tri-training: Exploiting unlabeled data using three classifiers. IEEE Transactions on Knowledge and Data Engineering 17, 1529–1541 (2005)
5. Nigam, K., Mccallum, A., Thrun, S., Mitchell, T.: Text classification from labeled and unlabeled documents using EM. Machine Learning 39, 103–104 (2000)
6. Blum, A., Chawla, S.: Learning from labeled and unlabeled data using graph mincuts. In: Proceedings 18th Annual Conf. on Machine Learning, Williamstown, MA, pp. 19–26 (2001)
7. Deng, C., Guo, M.-Z.: ADE-Tri-training: Tri-training with adaptive data editing. Chinese Journal of Computers 30(8), 1213–1226 (2007) (in Chinese with English abstract)
8. Bi, H., Liang, H.-L., Wang, J.: Resampling methods and machine learning. Chinese Journal of Computers 32(5), 862–877 (2009) (in Chinese with English abstract)

Estimating the Number of People in Crowded Scenes Based on Energy in Frequency Domain*

Guo-qing Yang and Rong-yi Cui**

Intelligent Information Processing Lab.
Department of Computer Science and Technology
Yanbian University
Yanji, Jilin, China
y.p.yu@sohu.com, cuirongyi@ybu.edu.cn

Abstract. A method of estimating the number of people in crowded scenes was proposed based on energy of the foreground image in the frequency domain. Firstly, the foreground image including the target area was obtained by background subtraction. And furthermore, with the second order wavelet decomposition of the foreground image, the energy of high frequency sub-band was computed. Finally, by analyzing the relation between population size and high frequency sub-band energy, the mathematical model was established and the model was solved by the least square method. The experimental results show that the proposed method is highly effective for estimating the number of moving people in crowded scenes.

Keywords: estimation of number of people, wavelet decomposition, energy in frequency domain, the least square method.

1 Introduction

With the development of human civilization and the advancement of society, crowd density estimation has become an important topic in the field of public safety. Recently, much attention has been paid to crowd monitoring and pre-warning in automatic surveillance systems. The traditional artificial way to safeguard crowd safety is using Closed Circuit Television (CCTV) by which specific objects and their behaviour could be monitored for a long period of time. However, a human observer might lose some information because monitoring crowds through CCTV is very laborious and cannot be performed for all the cameras simultaneously, and furthermore, may cause irreparable consequences.

In recent years, various methods for intelligent monitoring have emerged. Early work on crowd monitoring using image processing was reviewed by Davies et al. in 1995[1] by mainly analysing the crowd size and crowd motion estimation. Counting the number of pixels in foreground area and in edge of foreground, the "optimal"

* This work is supported by Sci. & Tech. Department of Jilin Prov. Grant#20050703-1.
** Corresponding author.

D. Zeng (Ed.): ICAIC 2011, Part II, CCIS 225, pp. 546–554, 2011.

estimation of crowd density was achieved using linear regression training model and Kalman filter. Ref. [2] proposed Bayesian-Poisson regression for crowd counting and Ref. [3] adopted the foreground pixel area and edge pixels of foreground, and verified that there exists a linear relation between them, by which the crowd density was estimated through the least squares method. Nowadays, the most methods for density estimation are mainly researched in space domain, while some frequency-domain-dependent methods just simply accumulate the values of amplitude [4].

In this paper, under complicated background, background subtraction was adopted to obtain the moving target groups, and furthermore, the energy of the high frequency sub-band details in the wavelet transform was extracted as the feature. In our work it was found that there exists a linear relation between the number of people and the energy of the high frequency sub-band details. Finally, the least square was employed to fit the linear equation and estimate the number of crowd. Experimental results show that the proposed method is highly effective for estimating the number of moving people in crowded scenes.

2 Moving Object Extraction

Moving object detection and extraction is an important field of computer vision research, and is the basic part of the later tasks such as moving target tracking, identification, target behaviour analysis and understanding. Moving object detection methods are mainly divided to three types [5,6]: frame difference, optical flow and background subtraction. In view of the moving objects detected by background difference is more complete, we use background subtraction to extract the moving object. Extraction process is as shown below:

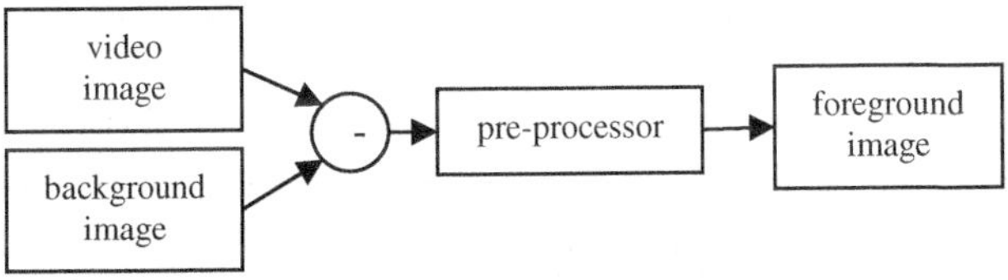

Fig. 1. The process of moving object detection

In the process of background reconstructing, available background models are median background, single Gaussian model and Gaussian mixture models, and the simplest model is the time average image. In our work, we adopted the Kalman-filter-based background updating algorithm (see Fig. 2).

As shown in Fig.3, the moving object detecting is always interfered by shadow. Using color information may improve the accuracy of shadow detection. Instead of RGB space, HSV color space is closer to human vision system in which the

Fig. 2. The background image

Fig. 3. Result after removing background

chromaticity and intensity of pixel are distinguishable more easily than in RGB space, so that shadow and foreground targets may be separated conveniently in HSV space. Hence, in our work the shadow was removed in HSV space. Shadow model [7] is described as a shadow mask *SP* in (1), where $I_k^V(x,y)$, $I_k^S(x,y)$ and $I_k^H(x,y)$ describe the *V*, *S* and *H* components in HSV space respectively for the pixel of foreground image at coordinate (*x*, *y*) in the *k*-th frame. Same as above, $B_k^V(x,y)$, $B_k^S(x,y)$ and $B_k^H(x,y)$ denote corresponding components for background image at the same position in the same frame. In (1), the constants α, β are less than 1, and α depends on light intensity, the higher light intensity, the less the value of α; variable β could avoid misidentification of shadow caused by background pixels which are interfered by noise. The image after removing shadow is shown in Fig. 4. When the color of moving object is close to that of background, empty areas may appear in the region of moving object, and this may be solved by morphological processing.

$$SP(x,y)=\begin{cases}1 & \text{for } \alpha \le I_k^V(x,y)/B_k^V(x,y) \le \beta \text{ and} \\ & I_k^S(x,y)-B_k^S(x,y) \le T_S \text{ and} \\ & I_k^H(x,y)-B_k^H(x,y) \le T_H \\ 0 & \text{otherwise}\end{cases} \quad (1)$$

Fig. 4. Result after removing shadow

3 Wavelet Decomposition

A. Wavelet Decomposition of Image

With wavelet decomposition of image we can obtain the high frequency sub-graph and the low frequency sub-graph. The high frequency sub-graph retains the edge information of original image, and the low frequency sub-graph retains the original image details [8].

From the perspective of image processing, wavelet decomposition has some advantages as follows:

1) Wavelet decomposition can cover the entire frequency domain;

2) By selecting the appropriate filter, Wavelet transform can greatly reduce or remove the extracted correlation between different features;

3) The wavelet transform has a "zoom" feature, high frequency resolution and low temporal resolution may be used in low frequency band, low-frequency resolution and high temporal resolution may be used in high frequency band.

As a tool for video processing, wavelet transform has the following advantages:

1) Wavelet transform maintains the spatial characteristics of the original image, which allows us to extract the high frequency information of the image;

2) Wavelet can remove correlation and related energy;

3) Wavelet components have orientation selectivity, and can be divided into horizontal, vertical and diagonal direction;

4) Proper wave filter can improve the bad image quality and gain the better video reconstruction.

Thus, selecting high frequency sub-graph has two advantages: firstly, we can get the marginal information accurately; secondly, dimensionality reduction of image can reduce the computing amount for feature extracting. The *db2* of Daubechies wavelet radical is employed for decomposition in this paper.

B. Energy

We can get low frequency image and high frequency image after wavelet decomposition. High frequency image includes horizontal detail sub-band, vertical detail sub-band and diagonal detail sub-band. Because the high frequency sub-graph retains the marginal information of original image, and the marginal information of image represents the basic feature of image, which is invariant for changing illumination [9]. So we used high frequency sub-graph to extract marginal information. In the wavelet transform domain, the amplitude of factor represents change rate of original image gray scale under this resolution, so the energy can represent the definite feature of original image as possible as it can.

The energy of high frequency sub-graph in this paper adopts the definition given in Ref. [10]:

$$E = \sum_{i=1}^{M} \sum_{j=1}^{N} |f(i,j)| . \tag{2}$$

where M and N are the size of high frequency sub-graph, $f(i,j)$ is the image of high frequency sub-graph. Because there are three images of high frequency sub-graph,

Table 1. Number of People And The energy of High Frequency Sub-Band

Actual number of people	Energy of high frequency sub-band	Actual number of people	Energy of high frequency sub-band
5	149.3141	13	367.6555
7	163.4963	17	449.5771
9	216.3634	24	846.8471
11	279.2682	29	907.0659

so the energy in this paper is the sum of three detail sub-bands calculated by Eq.(2). Tab.1 presents the number of crowd and corresponding energy of high frequency sub-band. It can be seen from the table that the energy of high frequency sub-band increases with the increase of number of people.

4 Curve Fitting

A. Principle of Least Square

The least square method—a very popular technique—is used to compute the estimations of parameters to fit given data. As a mathematical optimization technique, it aims to find the best matching function for given data by minimizing the error sum squares. The unknown data can be easily calculated by the least square method, which produces minimum squared error between actual data and calculated data [11].

Curve fitting can be roughly divided into the following steps:

1) Draw the scatter diagram and choose suitable curve type.

2) Variable transformation

$$\begin{cases} x' = g_1(x) \\ y' = g_2(y) \end{cases}. \tag{3}$$

The method of variables replacement is adopted in (3) so that there exists linear relation between the two new variables.

3) The experimental data are used to seek for undetermined coefficient according to the least square principle.

4) The linear formula is transformed to the function about original variables x, y.

B. Correlation Coefficient

The correlation coefficients stand for degree of the pertinence between variables. Let $r \in [-1,1]$ stands for correlation coefficient of samples, so that larger value of $|r|$ corresponds to smaller error, which represents higher correlative linear degree; on the other hand, if the value $|r|$ is close to 0 then the error becomes large, the correlative linear degree is lower. The two variables may be considered highly correlative in the case of $|r|>0.75$. The equation for calculating correlation coefficient is as follow:

$$r = (\sum_{i=1}^{n}(x_i - \bar{x})(y_i - \bar{y})) / \sqrt{\sum_{i=1}^{n}(x_i - \bar{x})^2 \sum_{i=1}^{n}(y_i - \bar{y})^2} . \tag{4}$$

where, x_i , y_i are sample variables and $\bar{x}$, $\bar{y}$ stand for average value of sample variables and n stands for sample size.

5 Experimental Results and Analysis

A. Experimental Procedures

We choose 40 frames of images (there are different number of people in different frame) randomly from video to form training image base. In order to reduce the error caused by the incompleteness of the foreground images, the training images are divided into 4 groups, each group has 10 frames of images, and the numbers of people in these groups are 5, 7, 9 and 11 respectively. Experimental training process is as follows:

Step1: Count the number of each frame manually and store the number into array y;

Step2: Execute background subtraction for training image base, and obtain moving object, i.e., foreground image;

Step3: Wavelet decompose the foreground image, and store the high frequency sub-band energy into array x;

Step4: Draw scatter diagram according to statistical data;

Step5: Calculate correlation coefficient of the data;

Step6: Ascertain curve type according to scatter diagram and correlation coefficient, compute undetermined coefficient and obtain curve equation.

The experimental process chart is as follows:

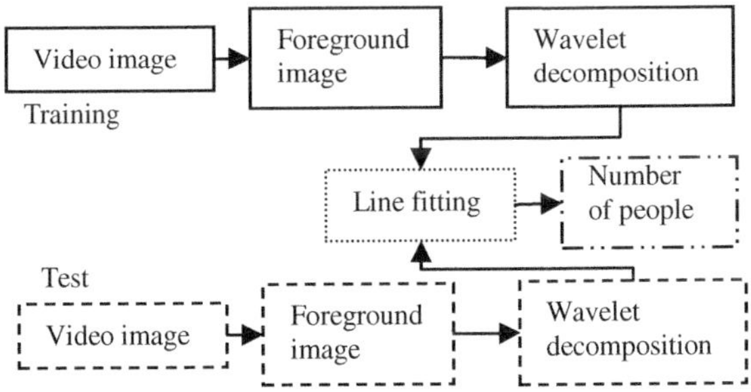

Fig. 5. Experiment process

B. Experimental Results and Analysis

After wavelet decomposition of all foreground images and calculation of the energy of their high-frequency sub-bands, curve can be fitted with all the energy and respective number of people. Part of energy for foreground images in the training base is indicated in Tab.II, and the experimental result is shown as in Fig.6.

Table 2. The energy of frame difference image

Foreground image	Energy	Number of people	Foreground image	Energy	Number of people
1	141.75	5	6	168.48	7
2	149.31	5	7	216.72	9
3	136.05	5	8	213.09	9
4	161.21	7	9	278.78	11
5	171.99	7	10	289.13	11

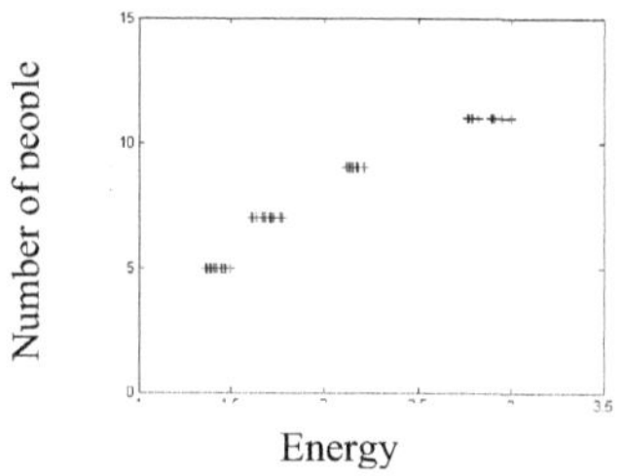

Fig. 6. Scatter of experimental data

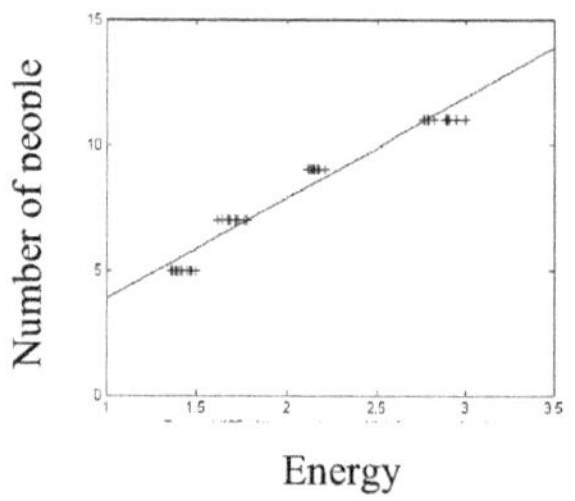

Fig. 7. Result of linear fitting

According to the scatter diagram, the data distributed intensively with the shap of strap. Conclusion can be drawn through experimental analysis that the correlation coefficient between the number of people and the energy of the high frequency sub-bands of the foreground images in the crowd scenes is 0.9759, which indicates that there exist a strong linear relation between the number of people and the energy. Hence, the method of linear fitting can be adopted, in other words, estimation of the number of people in the crowd scenes can be described in the form of linear equation $y = ax + b$, where a, b are undetermined coefficients, y stands for the number of people in the crowd scenes, and x stands for the energy of high frequency sub-bands of the foreground images. Fig.7 is the resunlt of linear fitted by the least square method.

We use the relative error to measure the results, namely:

$$error = \left| N_t - N_c \right| / N_t \times 100\% . \tag{5}$$

where N_t is the actual number, N_C is the test value.

Tab. 3 is the testing results, and Fig.8 is the result of comparison between test value and real value, where '*' indicates the actual number of people, '□' stands for the tested number of people. According to the experimental result, the maximum relative error of the test result is 12.50%, and the maximum difference number of people is 2. Tab. 4 and Fig. 9 are the results of comparisons between proposed methods and the method of Canny operator edge detection in Ref. [3]. As shown in the figure, our result has a higher preciseness rate than the method proposed in Ref. [3], especially when the number of people increases, its estimation becomes closer to the actual situation.

Table 3. Testing results

Test image	Test value	Actual value	Relative error(%)
1	18	20	10.00
2	19	20	5.00
3	30	30	0.00
4	7	8	12.50
5	14	16	12.50
6	30	28	7.14
7	27	28	3.57
8	13	13	0.00
9	4	4	0.00
10	15	15	0.00

Table 4. Results of comparison

Test image	Test value	Test value in Ref. [3]	Actual value
1	18	22	20
2	19	22	20
3	30	34	30
4	7	34	8
5	14	17	16
6	30	34	28
7	27	32	28
8	13	15	13
9	4	4	4
10	15	17	15

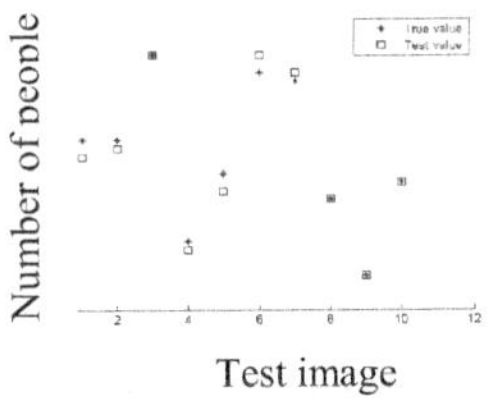

Fig. 8. Comparison between test and real value

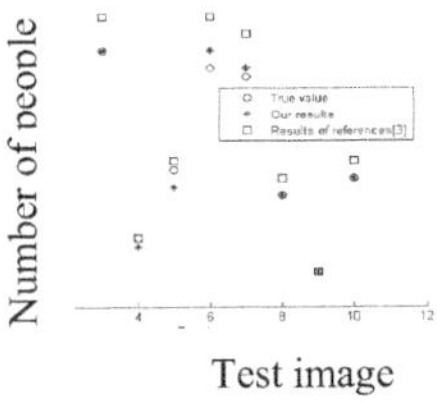

Fig. 9. Comparison with Ref.[3]

6 Conclusion

In this paper, moving objects were captured by background subtraction, and through wavelet decomposition of foreground images, the energy of high frequency sub-bands was extracted as the feature for estimation of number of people. And furthermore, with the least square method, the linear equation for number of people and energy was found. Finally, the estimation for number of people in the crowd scenes was conducted. The experimental results show that simple and fast method of linear equation proposed in this paper is very effective to estimate the number of people in the crowd scenes.

Sometimes moving objects are incomplete, which affects the subsequent work greatly. The incompleteness of moving objects is mainly due to the similar color between background and moving objects. So extracting the complete moving objects is a problem to be solved during the work of algorithm improvement in the future.

References

1. Davies, C., Yin, J.H., Velastin, S.A.: Crowd monitoring using image processing. IEE Electronic and Commnications Engineering Journal 7, 43–45 (1995)
2. Chan, B., Vasconcelos, N.: Bayesian poisson regression for crowd counting. In: IEEE Int'l Conf. on Computer Vision, Kyoto, Japan, pp. 1–7 (2009)
3. Du, J.: Crowd's density and motion estimation, pp. 21–25. Dalian University of Technology, Dalian (2006) (in Chinese with English abstract)
4. Liu, X.R., Zhou, J.L., Li, X.H.: Grey level dependent matrix based crowd density estimation in Fourier spectrum. Control & Automation 24, 310–311 (2008) (in Chinese with English abstract)
5. Wang, S.L., Kang, M.N., Deng, Z.H., et al.: Object separating algorithm based on background alternation and dynamic expanding filter. Computer Engineering and Applications 46, 246–247 (2010)(in Chinese with English abstract)
6. Shi, Z.L., Chen, Y.M., Xu, J.: Enhanced human motion detection algorithm and its implementation based on video surveillance. Computer Engineering and Applications 44, 206–207 (2008)(in Chinese with English abstract)
7. Cucchiara, R., Grana, C., Piccardi, M., et al.: Improving shadow suppression in moving object detection with HSV color information. In: Proc. IEEE Intel. Transp. Syst. Conf., Oakland, USA, pp. 335–336 (2001)
8. Xue, L.Y., Pan, J.J.: Edge Detection Combining Wavelet Transform and canny operator based on fusion rules. Proceedings of the 2009 International Conf. on Wavelet Analysis and Pattern Recognition, Baoding, China, pp. 324–325 (2009)
9. He, Y., Wang, L.: Image stitching algorithm based on feature block and wavelet transform. Computer Engineering and Design 31, 1959–1960 (2010) (in Chinese with English abstract)
10. Yang, D.Q., Xiao, S.P., Jiang, J.F.: Texture image recognition based on differentia evolution probabilistic neural network. Computer Engineering and Applications 44, 180–181 (2008) (in Chinese with English abstract)
11. Abdi, H.: Least squares. In: Lewis-Beck, M., Bryman, A., Futing, T. (eds.) Encyclopaedia for Research Methods for the Social Sciences, pp. 559–561. Sage, Thousand Oaks (2003)

Video-Based Recognition of Walking States*

Guang-yu Xu and Rong-yi Cui**

Intelligent Information Processing Lab.
Department of Computer Science and Technology
Yanbian University
Yanji, Jilin, China
xusongxiaorong@sina.com, cuirongyi@ybu.edu.cn

Abstract. An approach to analyzing human walking states based on multi-viewing-angle was proposed in this paper. Firstly, the moving person was detected using self-adaptive background updating algorithm. Then, the moving period of the centroid in the upper part of the human body was extracted by 4-direction chain code and the next motion of person was predicted. Finally, fuzzy set of walking states was obtained by the training set of period, and three kinds of walking states including promenade, walking and running were analyzed by membership function. The experimental results show that the movement velocity and the walking states are exactly analyzed by the proposed method, and furthermore, the method is reasonably robust in background noise and varying view angles.

Keywords: video sequence, moving person detection, centroid extraction, velocity estimate, membership function.

1 Introduction

Gait recognition has received growing interest within the computer vision community recently for it is the only one of biometric recognition method which is suitable for long distance apperceiving [1]. In recent years, increasing attention has been paid to this field, though there is less gait-feature-based analysis on moving states [2].

Shi Chen et al. designed invariant histogram-based descriptor, the gait appearance characteristics are described as a sequence of shape distributions, and recognition capability is illustrated by an 87.59% CCR on Soton database [3]. ChangYou Chen et al. proposed an iterative prototype algorithm based on tangent distance for gait recognition, experiment results show a promising results in accuracy compared with the GEI's [4]. Kim D et al. presented a gait recognition system using infra-red images, and the experimental results improve the highest accuracy of 90.6% [5]. However, human activity is one of the most important ingredients of context information. Activities such as walking, standing and sitting were inferred from data provided by body-worn acceleration sensors by Huynh T [6]. Tabbone S, Wendling L, Salmon J P

* This work is supported by Sci. & Tech. Department of Jilin Prov. Grant#20050703-1.
** Corresponding author.

D. Zeng (Ed.): ICAIC 2011, Part II, CCIS 225, pp. 555–562, 2011.

presented a novel approach to identify complex shapes based on the Radon transform, and obtained higher accuracy based on action recognition [7]. Aggarwal et al. processed image sequences for action recognition and proposed high-level recognition schemes with domain knowledge [8]. There is no doubt that research work on walking states analysis would be useful to implement a system that reports and analyzes human behaviors and activities in detail.

In our work, moving person was detected, and the walking states were analyzed based on multi-viewing-angle. The proposed algorithm is robust in background noise and varying view angles.

This paper is organized as follows. Motion area is detected by advanced background subtraction method in Section 2. Section 3 analyzes moving states and predicts the next motion. In Section 4 our experimental results are showed with explanation. Finally, conclusions and future research directions are described in Section 5.

2 Motion Detection

Segmentation of moving object in video sequences is a basic task in many computer vision analysis applications, for instance, video surveillance, indexing for multimedia, people detection and tracking, perceptual human computer interface. Accurate moving object segmentation will greatly improve the performance of object tracking, recognition, classification and activity analysis [9]. It is still a difficult problem to detect moving objects precisely, because of varying illumination, small movements of background, moving shadows, and occlusion.

We detect moving object by advanced background subtraction method [9], which estimates the intensity variation of each pixel by integrating all the variations exhibited by other pixels with the same intensity value, labeled as 'static' in the last frame, then the shadow is deleted by comparing the correlation exhibited between regions selected from the reference image and the current one (see Fig. 1).

a) Image of 63rd frame b) Background frame c) Detected motion in (a)

Fig. 1. Demonstration of moving object detection

Because the background is not fixed, we used morphological operation after motion detection in order to eliminate moving noise, then we obtained moving person which was enveloped by rectangle according to the ratio between length and width of human body.

3 Moving States Analysis

There are more human gait recognition methods based on the side-away view angle while less frontal view angle, meanwhile, less upper information of person are used. In our work, the period of motion was extracted based on the upper body information, then the moving state was recognized.

A. Trail Extraction

The centroid and area of moving person will be change with as time go on. There are three main methods of trail extraction: physiological method, Fourier-transform-based method and physical method. We obtained the trail and recognized the moving states of person by physical method.

The motion period can be detected in 90 degree view angle or near to 90 degree by physiological method and Fourier-transform-based method [10], but the trail curve is clutter in 0 or near to 0 degree view angle and the two above methods didn't work well.

The centroid of a physical object is its mass centre, and a moving person can be treated as an abstract point. Human walking has periodic property, so centroid can be used to estimate the moving states of person. The centroid coordinate $x_c(t)$ of person in vertical direction at t-th second can be computed as follows:

$$x_c(t) = \sum_{i=1}^{M}\sum_{j=1}^{N} i \cdot S^t(i,j) \Big/ \sum_{i=1}^{M}\sum_{j=1}^{N} S^t(i,j) \tag{1}$$

where i is the row of the image, and $S^t(i,j)$ is the gray value at t-th second in the i-th row and j-th column.

Because of the difference in gait extents and gait sequences for varying persons, there exist noises in moving trail period. We only extracted the centroid upper body (see (2)) which is more stable in different view angles.

$$x_c'(t) = \sum_{i=1}^{M/2}\sum_{j=1}^{N} i \cdot S^t(i,j) \Big/ \sum_{i=1}^{M/2}\sum_{j=1}^{N} S^t(i,j) \tag{2}$$

The mean smoothing was adopted in our method to eliminate the noises in moving trail.

B. Period Estimation

4-direction chain code can be used to detect moving trail because the moving trail is only related to the moving period. Connected edge points can be associated by chain code, and chain code can represent a digital curve. One of connected edge points but endpoints has exactly one 4-neighbor point. The position of 4-neighbor point can be represented by a chain code $a_i \in \{0,1,2,3\}$, which denotes the direction of 4-neighbor point (see Fig. 2(a)).

There are many 3, 0 serial numbers or 1, 0 serial numbers in declivous lines (see Fig.2 (b)). Because the speed of a running person can not exceed 5 steps in one second, 7 continuous frames are chosen to extract the wave trough of moving period in (3).

$$\{x_c'(t-3), x_c'(t-2), x_c'(t-1), x_c'(t+1), x_c'(t+2), x_c'(t+3)\} > x_c'(t) \tag{3}$$

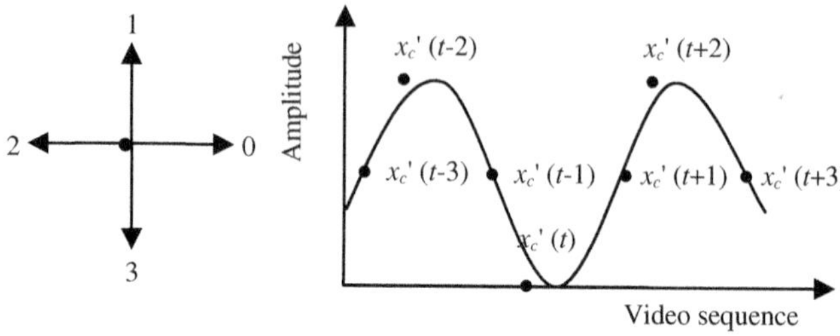

(a) (b)

Fig. 2. Period detection for moving object: (a) serial number of 4-direction chain code, (b) image of alterant amplitude

We estimate the motion velocity of person by following algorithm:

Step 1: Detect $Length_{checked\_code}$ of valley floor in (3);

Step 2: If $(Length_{checked\_code}/2+1)^{th}$ ~ $(Length_{video}-Length_{checked\_code}/2)^{th}$ frames are detected, the total number frames is $Length_{video}-Length_{checked\_code}/2$;

Step 3: If the code sequences satisfy (3), a wave trough $r(i)$ has been found;

Step 4: The continuous frames $D_{dist}(i-1)$ and frequency $f(i-1)$ will be computed if two or more wave troughs have been found;

Step 5: Iterate Step 2~Step 4 until all frames have been processed.

The jitter noise can be eliminated effectively by above algorithm.

C. Moving States Recognition

The moving states can be classified by the maximum membership principle. We adopt the membership function described in (4).

$$\mu_{\underset{\sim}{A}}(x)=e^{-(\frac{x-\alpha}{\sigma})^2}\ (\sigma>0) \tag{4}$$

where α is mathematical expectation, and σ is mean-square value of normal distributions.

For any set **X**, a membership function on X is a mapping from X to the real unit interval [0, 1]. Membership functions on X represent fuzzy subsets of **X**. The membership function which represents a fuzzy set $\underset{\sim}{\mathbf{A}}$ is denoted by $\mu_{\underset{\sim}{A}}$. For an element x of **X**, the value $\mu_{\underset{\sim}{A}}(x)$ is called the membership degree of x in the fuzzy set $\underset{\sim}{\mathbf{A}}$. The membership degree $\mu_{\underset{\sim}{A}}(x)$ quantifies the grade of membership of the element x to the fuzzy set $\underset{\sim}{\mathbf{A}}$. Value 0 means that x is not a member of the fuzzy set, and value 1 means that x is fully a member of the fuzzy set. The values between 0 and 1 characterize fuzzy members, which belongs to the fuzzy set only partially.

There are three moving states considered in this paper, so three membership function was obtained by the training samples.

D. Moving Trend Prediction

The moving states of person must accordant to the criterion of the given environment, and the period of the upper body centroid can tell the purpose, moving trend and mood of the person. We take the moving period into account to predict the trend of the moving person. When there are some continuous periods are large, we can determin that the person is moving slowly, otherwise he/she is moving quickly. If the periods become larger or smaller in a short time, the moving is difined as abnormality.

4 Experiments and Analysis

Three kinds (there are 50 samples for each kind) of moving states were recognized in our experiments (25 *fps*, 240×320 *pixels*). Firstly, motion was detected by advanced background subtraction method from video sequence and the centroid of the upper part was obtained; then, moving period of the centroid was detected by 4-direction chain code; finally, the moving states were recognized by membership function, and motion trend was predicted. In this paper, four experiments were designed to verify our approach, and the average processing frame rate of our system is 32 *fps*.

1) Experiment 1: Trail extraction

In order to recognize moving states accurately, we take 10 view angles into account, and 7 kinds of methods are used to validate our method (see Fig.3(b)~(h)) in the second experiment, then the moving period of the centroid of the upper body is extracted by 4-direction chain code.

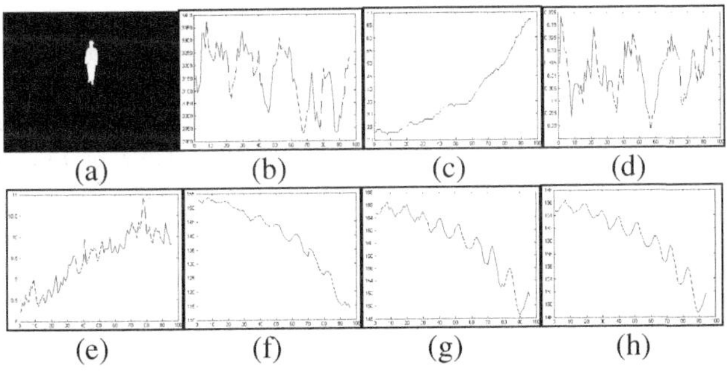

Fig. 3. Moving trail extraction for 0 degree view angle: (a) current frame, (b) the trail of area, (c) the trail of width, (d) the trail of ratio between length and width, (e) the trail of average amplitude in frequency domain, (f) the trail of centroid, (g) the trail of centroid of upper body, (h) the smoothed trail of centroid of upper body

It is shown that our method is the best one to detect moving period. The experimental results show that the period can be extracted accurately by the proposed method.

The trail images in other views are shown in Fig.4.

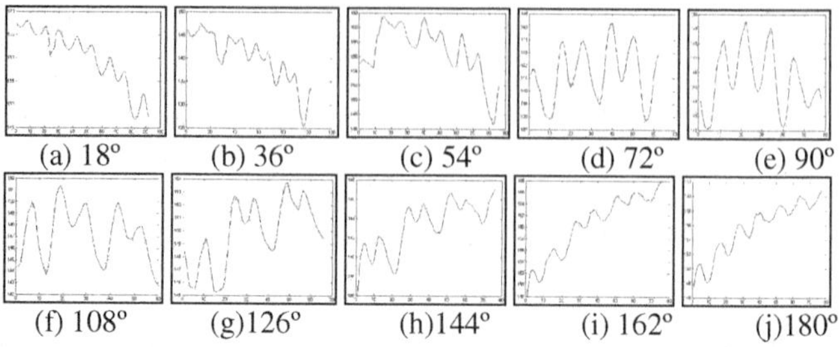

Fig. 4. Moving trail extraction for different view angles

2) Experiment 2: Period estimation

By plotting the trail-time curve of the centroid of the upper body (see Fig.5) the wave trough motion curve based on 4-direction chain code is extracted accurately in 0 degree view angle, and furthermore, by computing the distance between two neighboring wave troughs the time consumed by each step is obtained.

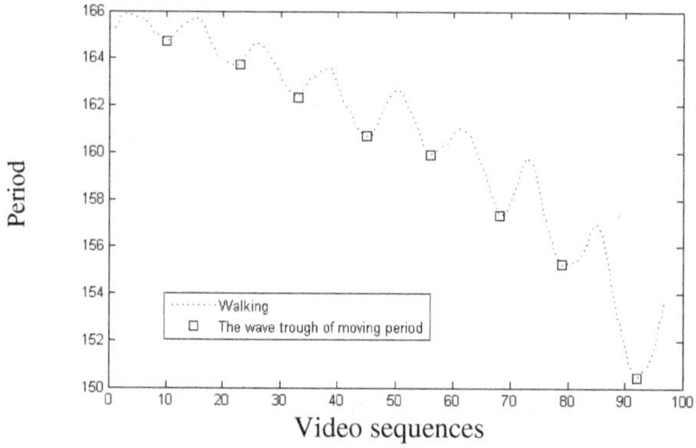

Fig. 5. Motion period estimation

3) Experiment 3: Moving states recognition

Three kinds of moving states including promenade, walking and running are recognized by normal membership function. We obtain 3 groups of membership function as described in equation (5) and shown in Fig.6, and the recognition results are shown in Tab I, which shows that the average recognition rate is 92.00%, and the rates for various moving states are stable.

$$\begin{cases} \underset{\sim}{A}_{running} = e^{-\left(\frac{x-0.2905}{0.0478}\right)^2} \\ \underset{\sim}{A}_{walking} = e^{-\left(\frac{x-0.4643}{0.0289}\right)^2} \\ \underset{\sim}{A}_{promenade} = e^{-\left(\frac{x-0.6514}{0.0459}\right)^2} \end{cases} \tag{5}$$

The accuracy of promenade is lowest and running is highest, because the motion of running is vehement and the centroid period is most obvious. From the results we can see that gait period can be used to estimate the moving states.

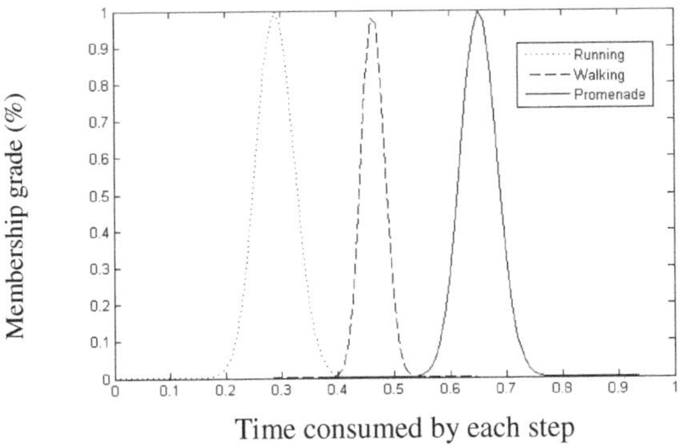

Fig. 6. Curves of membership function

4) Experiment 4: Human Motion Analysis

We take the ratio between two neighboring periods into account, and plot the histogram in Fig.7. From Fig.7 we can see that when the ratio is greater than 1, the human speed becomes more and more slowly, otherwise the human speed becomes faster and faster, and if the ratio deviates from 1 largely, the human motion is decided as abnormal behavior.

Table 1. Recognition Rate

	Kinds of moving states			
	Promenade	Walking	Running	
Total number in each state	50	50	50	
Number of recognition	45	46	47	
Accuracy(%)	90.00	92.00	94.00	
Average accuracy(%)				92.00

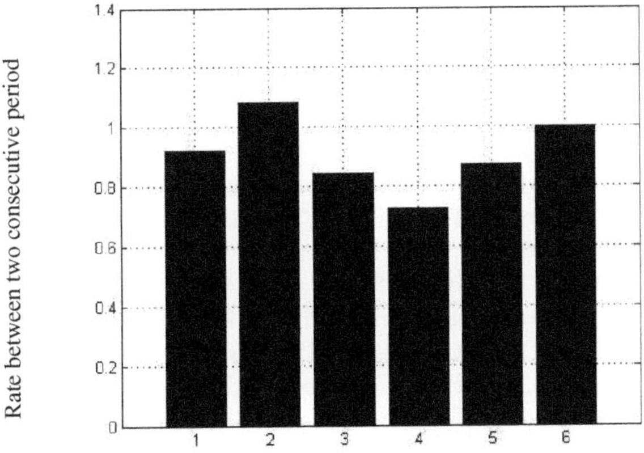

Fig. 7. Period of motion sequence

5 Conclusions

In this paper, we implemented 3 moving states recognition according to moving period and predicted the motion trend.

1) We only take the centroid in the upper part of the human body into account, so our approach is robust in varying view angles.
2) 4-direction chain code can be used to effectively detect the moving period.
3) Motion trend can be predicted by the change between continuous periods.
4) Moving states can be recognized by membership function with acceptable accuracy rate.

The future work includes analyzing multi-person states, and using the detail information of motion for state recognition and analysis.

Acknowledgment. We would like to thank Graduate School of Yanbian University for financial support.

References

1. Wang, K.J., Hou, B.B.: A survery of gait recognition. Joumal of Image and Graphics 12(7), 1152–1160 (2007) (in Chinese with English abstract)
2. Su, H., Huang, F.G., Hong, W.: Automatic human identification method based on gait analysis. Journal of System Simulation 18(7), 1292–1296 (2006) (in Chinese with English abstract)
3. Chen, S., Ma, T.J., Gao, Y.X.: Gait recognition using distributions of silhouette feature. Journal of Computer Research and Development 46(2), 295–301 (2009) (in Chinese with English abstract)
4. Chen, C.Y., Gao, J.P.: An iterative gait prototype learning algorithm based on tangent distance. Journal of Computer Research and Development 45(7), 1177–1182 (2008) (in Chinese with English abstract)
5. Kim, D., Lee, S., Paik, J.: Active Shape Model-Based on Gait Recognition Using Infrared Images. Communications in Computer and Information Science 60(1), 275–281 (2009)
6. Huynh, T., Schiele, B.: Analyzing Features for Activity Recognition. In: Proc. Soc EUSAI 2005. ACM Int. Conf. Proceeding Series, pp. 159–163. ACM Press, Grenoble (2005)
7. Tabbone, S., Wendling, L., Salmon, J.P.: A new shape descriptor defined on the Radon transform. Computer Vision and Image Understanding 102(1), 42–51 (2006)
8. Aggarwal, J.K., Park, S.: Human Motion: Modeling and Recognition of Actions and Interaction. In: Proceedings of the Second International Symposium on 3D Data Processing, Visualization and Transmission, Thessaloniki, Washington, DC, pp. 640–647 (2004)
9. Spagnolo, P., D'Orazio, T., Leo, M., Distante, A.: Advances in background updating and shadow removing for motion detection algorithms. In: Gagalowicz, A., Philips, W. (eds.) CAIP 2005. LNCS, vol. 3691, pp. 398–406. Springer, Heidelberg (2005)
10. Zhao, G.Y., Xiang, S.M., Li, H.: Gait sequences recognition based on reflective symmetry. Journal of Computer-Aided Design & Computer Graphics 17(10), 2239–2244 (2005) (in Chinese with English abstract)

The Accelerated Power Method for Kernel Principal Component Analysis*

Weiya Shi[1] and Wenhua Zhang[2]

[1] School of Information Science and Engineering, Henan University of Technology, Zhengzhou, Henan Province, 450001, China
[2] Computer Engineering Department, Luohe Vocational and Technical College, Luohe, Henan Province, China
xiao.m.li@sohu.com

Abstract. When faced with the large-scale data set, Kernel principal component analysis (KPCA) is infeasible because of the storage and computational problem. To overcome these disadvantages, an accelerated power method of computing kernel principal components is proposed. First, the accelerated Power iteration is introduced to compute the first eigenvalue and corresponding eigenvector. Then the deflation method is repeatedly applied to achieve other higher order eigenvectors. The space and time complexity of the proposed method is greatly reduced. Experimental results confirm the effectiveness of proposed method.

Keywords: KPCA, Large-scale, power, deflation.

1 Introduction

Kernel Principal Component Analysis (KPCA) [1] is the nonlinear generalization of Principal Component Analysis (PCA) [2]. The standard KPCA generally needs to eigen-decompose the Gram matrix [3], which is acquired using the kernel function. It must firstly store the Gram matrix of all data, which takes the space complexity of $O(m^2)$, where m is the number of data samples. In addition, it needs the time complexity of $O(m^3)$ to extract the kernel principal components. But traditional kernel function is based on the inner product of data vector, the size of kernel matrix scales with the number of data points. When faced with the large-scale data set, it is infeasible to store and compute the kernel matrix because of the limited storage capacity.

In order to solve the problem of the large-scale data set, some approaches have been proposed to compute kernel principal components. Zheng [4] proposed to partition of the original data set into several small-scale data sets and handle them separately. Approximation algorithms were also proposed to extract the nonlinear components [5][6]. The major difference between these methods lies in the sampling

* This work is partially supported by NSF of Henan Educational Committee under contract 2010B520005 and Doctor Fund of Henan University of Technology under contract 2009BS013.

D. Zeng (Ed.): ICAIC 2011, Part II, CCIS 225, pp. 563–570, 2011.

way. One disadvantage is that these methods will lose some information in the sampling process.An iterative procedure is proposed to estimate the kernel principal components by kernelizing the generalize Hebbian algorithm [7]. But the convergence is slow and cannot be guaranteed. We also proposed an iterative Algorithm [8] based on Gram-power matrix.

We have given an iterative power method of computing Kernel Principal Component for large-scale data set [9]. In this paper, the accelerated power method is introduced to improve the converge speed. The two methods in numerical analysis are used, which are the accelerated power iteration and the deflation method. First, the accelerated power iteration [10] is introduced to compute the first eigenvalue and corresponding eigenvector. Then the deflation [11] method is repeatedly applied to achieve other higher order eigenvectors. In the process of computation, the kernel matrix needs not to compute and store in advance. Instead, each row of kernel matrix is computed when used in the computing process. Thus, the kernel principal components can be iteratively computed without the eigen-decomposition.

The space complexity of the proposed method is only O(m) and its time complexity is also greatly reduced. More important, it still can be used even if traditional eigen-decomposition technique cannot be applied when faced with the extremely large-scale data set. Experimental results demonstrate the effectiveness of the proposed method. Its performance is also compared with the standard method and our proposed power method in the paper [9].

The rest of this paper is organized as follows: section 2 gives a short review and analysis of the KPCA. Then, we describe the proposed method in section 3. The experimental evaluation of the proposed method is given in section 4. Finally we conclude with a discussion.

2 Review and Analysis

Let $X = (x_1, x_2, ..., x_m)$ be the data matrix in input space, where $x_i, i = 1, 2, ..., m$, is a d-dimensional vector and m is the number of data samples. There exists a mapping function ϕ, which projects the data into high-dimensional (even infinite dimensional) Reproducing Kernel Hilbert Space (RKHS).

$$\begin{aligned} \phi: \quad & \Re^d \rightarrow \mathbb{F} \\ & x_i \mapsto \phi(x_i) \end{aligned} \tag{1}$$

Using mapping function ϕ, we can get the projected data set $\Phi(X) = (\phi(x_1), \phi(x_2), ..., \phi(x_m))$ in feature space. A positive definite kernel function $\kappa(.,.)$ is used to calculate the dot product between mapped data set, where $\kappa(.,.)$ is given by $\kappa(.,.) = \kappa(x_i, x_j) = \phi(x_i)^T \phi(x_j)$. In mapping feature space, the covariance matrix is given as follows:

$$C = \frac{1}{m} \sum_{i=1}^{m} \phi(x_i) \phi(x_i)^T \tag{2}$$

It accords with the following equation:

$$Cv = \lambda v \tag{3}$$

Where v and λ are corresponding eigenvector and eigenvalue of covariance matrix. Using the Representer theorem, the solution v can be expanded by all the data set $\Phi(X) = (\phi(x_1), \phi(x_2), ..., \phi(x_m))$ as:

$$v = \sum_{i=1}^{m} \alpha_i \phi(x_i) \tag{4}$$

By substituting Eq.3, Eq.4 into Eq.2, we can get the following formula:

$$K\alpha = m\lambda\alpha \tag{5}$$

Where α is a vector of span coefficients, K is Gram matrix denoted as $K = \Phi(X)^T \Phi(X)$. The entry of Gram matrix is $k_{ij} = k(x_i, x_j)$ .To compute the kernel principal components, the traditional method is to diagonalize Gram matrix K using eigen-decomposition technique. After having gotten the eigenvector α, we can achieve the kernel principal components V using Eq.4. For a test sample x, the nonlinear feature is given:

$$(v, \phi(x)) = \sum_{i=1}^{m} \alpha_i (\phi(x_i) \cdot \phi(x)) = \sum_{i=1}^{m} \alpha_i k(x_i, x) \tag{6}$$

In the process of deduction, it is assumed that the data have zero mean, if it is not, we can get the matrix $\tilde{K} = K - 1_m K - K 1_m + 1_m K 1_m$, where $1_m = (1/m)_{m \times m}$.

3 Extracting the Kernel Principal Components Using the Accelerated Power and Deflation Method

In order to solve the problem and introduce the proposed method, we firstly give some introduction about power method in numerical analysis. Then, the accelerated power and deflation method is given in detail.

A. power method.

Power method is one of these methods, which can compute the largest eigenvalue and corresponding eigenvector. Its main procedure is to repeatedly multiply an arbitrary nonzero vector with the matrix K, as much as to say finally multiplying the initial nonzero vector by higher powers of the matrix K.

Assuming that $m \times m$ matrix K has the eigenvalues $\lambda_1 > \lambda_2 > \cdots > \lambda_m$, with corresponding eigenvectors $v_1, v_2, \cdots, v_m$ The starting nonzero vector x_0 can be expressed as the linear combination of all eigenvectors $v_1, v_2, \cdots, v_m$, which means $x_0 = \sum_{j=1}^{m} \alpha_j v_j$. The procedure of power iteration can be illustrated as follows:

$$\begin{aligned} x_i &= Kx_{i-1} = K^2 x_{i-2} = \cdots = K^i x_0 \\ &= K^i \sum_{j=1}^{m} \alpha_j v_j = \sum_{j=1}^{m} \alpha_j K^i v_j \\ &= \sum_{j=1}^{m} \alpha_j \lambda_j^i v_j = \lambda_1^i (\alpha_1 v_1 + \sum_{j=2}^{m} \alpha_j (\lambda_j / \lambda_1)^i v_j) \end{aligned} \quad (7)$$

For $\lambda_1 > \lambda_2 > \cdots > \lambda_m$, $|\lambda_j / \lambda_1| < 1$. As a result, $(\lambda_j / \lambda_1)^i \xrightarrow{i \to \infty} 0$, only the corresponding term V_1 is left with the successively iterations. Considering that the components of vector x_i may be geometric growth or decay at each iteration. In order to avoiding the situation, the iterative vector x_i is generally rescaled to normalization (the vector ∞-norm is typically used). The algorithm of normalized power iteration is summarized in Algorithm 1:

Algorithm 1: The method of power iteration

1. Set arbitrary nonzero vector x_0
2. For i=1,2, …

$$y_i = Kx_{i-1}$$
$$x_i = y_i / \| y_i \|_\infty \quad (8)$$

3. end

B. the accelerated power method

But the power method generally converges very slow, the Rayleigh quotient power method can be used to accelerate the computation procedure.

For an nonzero vector x, its definition of Rayleigh quotient can be given as follows: $R(x) = \dfrac{(Ax, x)}{(x, x)}$.

The iteration procedure for the Rayleigh quotient power method is outlined in Algorithm 2:

Algorithm 2: The method of Rayleigh quotient iteration

1. Set initial nonzero vector $x_0 \neq 0$ and error limit $\varepsilon = 1 \times 10^{-6}$
2. For i=1,2, …do following steps

$$y^{i+1} = Kx^i ; m^{i+1} = \frac{(Kx^i, x^i)}{(x^i, x^i)}$$
$$x^{i+1} = \frac{y^{i+1}}{m^{i+1}}$$
$$\textit{if} \quad | m^{i+1} - m^i | < \varepsilon \quad \textit{then} \quad \textit{exit}$$

3. end

C. the deflation method

But the method of power iteration can only compute the largest eigenvalue λ_1 and corresponding eigenvector v_1. We must consider how to produce other higher order eigenvalues and corresponding eigenvectors. The deflation method gives the solution.

Without loss of generality, we assume $v_1 = (1, v_{21}, ..., v_{m1})^T$ and define the first row of K as $k^T = (k_{11}, k_{12}, ..., k_{1m})$. The new array can be given as follows:

$$K_1 = K - v_1 k^T$$
$$= \begin{pmatrix} 0 & \cdots & 0 \\ k_{21} - v_{21}k_{11} & \cdots & k_{2m} - v_{21}k_{1m} \\ \vdots & \vdots & \vdots \\ k_{m1} - v_{m1}k_{11} & \cdots & k_{mm} - v_{m1}k_{1m} \end{pmatrix} \tag{9}$$

It is clear that K_1 is an array of degree reduction and has zero eigenvalue. It is easy to get:

$$\begin{aligned} K_1 v_1 &= (K - v_1 k^T) v_1 = K v_1 - (K^T v_1) v_1 \\ &= (\lambda_1 - K^T v_1) v_1 \end{aligned} \tag{10}$$

The formula $K^T v_1 = \lambda_1$ can be easily verified because of $K v_1 = \lambda_1 v_1$. According to the Eq.10, The eigenvector as to zero eigenvalue of K_1 is also v_1

It is assumed that the eigenvector $v_2 = (1, v_{22}, ..., v_{m2})$ as to the second largest eigenvalue λ_2 of array K_1 and $k^T v_2 = \lambda_2$ is also true. As a result,

$$\begin{aligned} K_1(v_1 - v_2) &= (K - v_1 k^T)(v_1 - v_2) \\ &= K v_1 - K v_2 - (k^T v_1) v_1 + (k^T v_2) v_1 \\ &= \lambda_1 v_1 - \lambda_1 v_1 + \lambda_2 v_1 - \lambda_2 v_2 = \lambda_2 (v_1 - v_2) \end{aligned} \tag{11}$$

That means λ_2 is also the eigenvalue of array K_1 and its corresponding eigenvector is $(v_1 - v_2)$. Because the first row of $(v_1 - v_2)$ is zero, the first column of array $K_1(v_1 - v_2)$ has nothing with the first column of array K_1. We could remove the first row and column of the array K_1 and get the array B of m-1 order. The array B has the same eigenvalues $\lambda_2 > ... > \lambda_m$ as the array K except the eigenvalue λ_1. We can use the power method to calculate the largest eigenvalue λ_2 and its corresponding eigenvector. The method is called deflation method. It can be proven that the eigenvector $v_2 = v_1 - cz = v_1 - (\lambda_1 - \lambda_2)/(k^T z) z$, where $z = (0, \overline{v}_2)^T$ and $\overline{v}_2$ is the eigenvector as to eigenvalue λ_2 of array B. Similarly, the deflation method can be iterative used to compute other eigenvalues and corresponding eigenvectors.

We can use the accelerated power iteration and deflation method to compute the eigenvalues and eigenvectors in iterative manner. In the computation, the Gram matrix K is repeatedly multiplied by an arbitrary nonzero vector. Because the Gram matrix is positive semi-definite, we need not to store the whole matrix and its row is computed when iteration occurs. By some iteration, the eigenvalues and corresponding eigenvectors can be quickly achieved.

4 Experimental Results and Discussion

To demonstrate the effective of the proposed method, we do some experiment to validate it performance. We also compared with our proposed power method in the paper [9]. In our experiments, the Gaussian kernel function $k(x,y)=\exp(-\frac{\|x-y\|^2}{2\sigma^2})$ is only used if it is explicitly stated (σ is kernel parameter which needs to be adjusted by cross-validation). The convergence condition is when the norm of adjacent iterative vector is less than a predefined parameter $\xi < 1\times10^{-3}$.

A. Toy examples:

We firstly use 2-dimensional toy problem to demonstrate the effectiveness of proposed method. There are 3 clusters, the center of which is [-0.5 -0.2; 0 0.6; 0.5 0]. Each cluster consists of 30 samples, which is chosen from a normal distribution with mean zero, standard deviation 0.1. The kernel parameter $\sigma = 0.1$

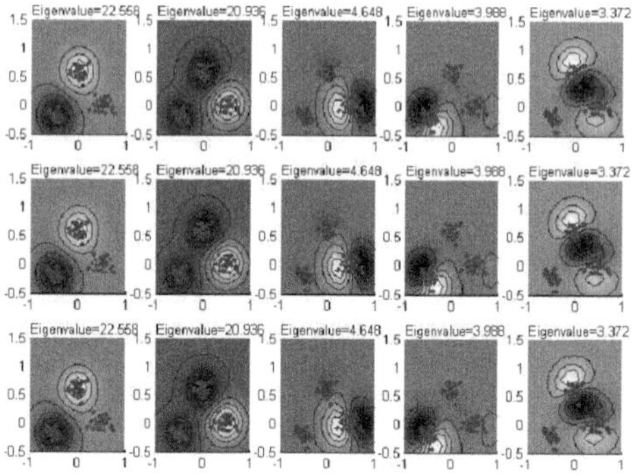

Fig. 1. Contour image of first 5 principal components obtained from standard KPCA (the top row), the power method (the middle row), the accelerated power method (the bottom row)

Table 1. The iterative numbers(INs) of first 5 principal components(PCs) obtained from the proposed method. INs(1)and INs(2) are respectively corresponded to the power method and the Rayleigh quotient accelerated power method.

PCs	First	Second	Third	Fourth	Fifth
INs(1)	97	11	57	71	32
INs(2)	78	8	41	60	29

The experiment results are given in Fig.1. It gives contour lines of constant value of the first 5 principal components using the standard method, the power and accelerated power method, where the gray values represent the projected feature value. From the result, the proposed method can get comparable performance with standard KPCA. To validate the convergence speed, Table 1 also gives the iterative numbers of first 5 principal components obtained from the proposed power and accelerated power method. It is clear that the eigenvectors acquired by the proposed method have quickly approached the objective ones obtained from power method in less iteration.

B. USPS examples:

We also test the proposed method on real-world data. The US postal Service (USPS) data set[1] is 256-dimensional handwritten digits. It consists of 7291 training samples and 2007 testing samples.

We firstly use the extracted features to compute the pre-image and denoise the noisy image. Then we projected the denoising image on the first 64 principal components. In the experiment, the kernel parameter $\delta = dc$, where d is the image dimension and equals to twice the average variance of data. The standard KPCA and the proposed method are trained with 3000 randomly choosing training samples, respectively. Then, the testing samples were added additive Gaussian noise with zero mean and standard variance 0.5. Fig.2 gives the original testing image, noising image and the de-noising result obtained by standard KPCA, the proposed power method and accelerated power method, respectively. It can be found that proposed methods achieve the similar de-noising image.

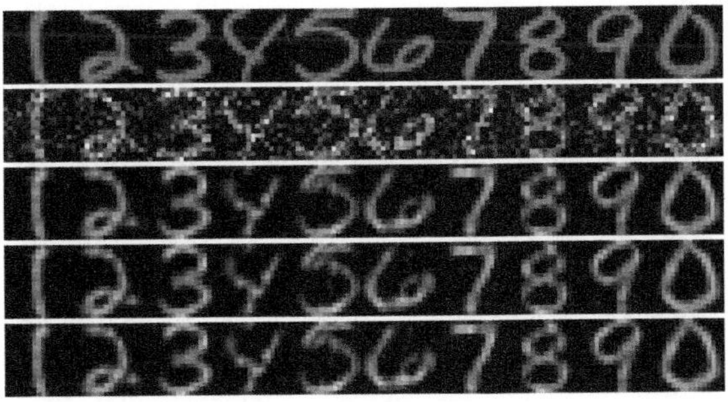

Fig. 2. De-nosing results. First row: original image; Second row: noisy image; Third row: de-noising result by batch KPCA; Fourth row: de-noising result by power method; Fifth row: de-noising result by accelerated power method.

[1] Available at http://www.kernel-machines.org.

Next, we use all the training samples to extract the nonlinear feature. The polynomial kernel is used to compute the Gram matrix. Because the size of Gram matrix is 7291*7291, it is impossible for standard KPCA algorithm to run in normal PC. But the proposed method still works well. Table.2 gives the error rate of testing sample of USPS data set. We can see that the proposed method can achieve the classified performance even the eigen-decomposition technique cannot work out.

Table 2. Error rate of 2007 testing sample of USPS data set

Degree d	2	3	4	5	6
Error rate(%)	5.98	6.43	6.78	7.13	7.57

5 Conclusion

In this paper, we propose an efficient iterative method to compute kernel principal components for large-scale data set. The method improves the converge speed of our proposed power method. We do not need to eigen-decompose the Gram matrix like tradition method, which is infeasible for large-scale matrix. The space and time complexity is also greatly reduced.

Acknowledgment. This work was supported in part by Natural Science Foundation of Henan Educational Committee under contract 2010B520005 and Doctor Fund of Henan University of Technology under contract 2009BS013.

References

1. Scholkopf, B., Smola, A.: Learning with Kernels: Support Vector MAchines, Regularization, Optimization and Beyond. MIT Press, Cambridge (2002)
2. Fukunaga, K.: Introduction to Statistical Pattern Recognition. Academic Press, London (1990)
3. Scholkopf, B., Smola, A., Muller, K.R.: Nonlinear Component Analysis as a Kernel Eigenvalue Problem. Neural Computation 10, 1299–1319 (1998)
4. Zheng, W.M., Zou, C.R., Zhao, L.: An Improved Algorithm for Kernel Principal Components Analysis. Neural Processing Letters 22, 49–56 (2005)
5. Smola, A., Cristianini, N.: Sparse Greefy Matrix Approximation for Machine Learning. In: International Conference on Machine Learning (2000)
6. Williams, C., Eeger, S.M.: Using the Nystrom Method to Speed up Kernel Machine. In: Advances in Neural Information Processing Systems (2001)
7. Kim, K.I., Franz, M.O., Scholkopf, B.: Iterative Kernel Principal Component Analysis. IEEE Trans. Pattern Anal. Mach. Intell. 27(9), 1351–1366 (2005)
8. Shi, W.Y., Guo, Y.F., An, X.Y.: efficient kernel principal component analysis algorithm for large-scale data set. Journal of Software 20(8), 2153–2159 (2009)
9. Shi, W.Y., Guo, Y.F., Xue, X.Y.: The power and deflationmethod based Kernel Principal Component Analysis. In: International Conference on Intelligent Computing and Intelligent Systems (2010)
10. Michael, T.H.: Scientific Computing–An Introductory Survey, 3rd edn. The Johns Hopkins University Press, Baltimore (1996)
11. David, R.K., Cheney, E.W.: Numerical Analysis:Mathematics of Scientific Computing, 3rd edn. Brooks Cole Press, Pacific Grove (2001)

Research on Hierarchical Genetic Algorithm Optimized Based on Fuzzy Neural Network*

Yuan Hao[1], Zhaohui Ren[1], and Bingcheng Wang[2]

[1] School of Mechanical engineering and Automation, Northeastern University, Shenyang, Liaoning Province, 110004, China
[2] School of Mechanical engineering, Shenzhen University, Shenzhen, Guangdong Province, 518040 China
ZhaohuiRen025@163.com

Abstract. A new algorithm of fuzzy neural network learning is presented. It is based on combining genetic algorithm of hierarchical structure with evolution programming. This algorithm is used to optimize the structure and parameters of fuzzy neural network, reject redundant nodes and redundancy connections, and improve the treatment ability of the network. The results of analysis and experiment show that, by using this method the fuzzy neural network of mechanical fault diagnosis has good concise structure and diagnosis effect.

Keywords: Fuzzy neural network, Hierarchical genetic algorithm, Evolutionary programming.

1 Introduction

In the mechanical fault diagnosis, fuzzy logic theory and the neural network technology played a large role in the knowledge representation, knowledge storing and reasoning speed[1]. Fuzzy logic combined with neural network constitutes the fuzzy neural network fault diagnosis [2], which can have the advantages of fuzzy logic and neural network. It can express knowledge and also have strong knowledge of learning ability and the data processing ability, thus further improve mechanical fault diagnosis effect.

The design method common fuzzy neural network structure can be resolved into structure of "increment method" and "reduction method". However, these two methods of fuzzy neural network structure are easy into the local minimum value and the search space also just a tiny space in the structure of the space. The study of neural network training usually adopt BP algorithm to adjust the membership function parameters and connection weights, but the BP algorithm has the convergence on initial conditions, easily into the local minimum. After fuzzy neural network training, there will be a mass of redundant nodes and redundancy connections, therefore, to optimize fuzzy neural network, and reduce the number of the nodes and the network

* This work is financed by National Natural Sciencc Foundation of China #50975044 to #50875175.

D. Zeng (Ed.): ICAIC 2011, Part II, CCIS 225, pp. 571–578, 2011.

connection number to improve the concise degrees of the network, while further optimize the input space of membership functions and the connection weight between nodes. It's necessary to reduce the complexity of the fuzzy neural network and improve the efficiency in the use of network. Therefore how to optimize the structure and parameter of fuzzy neural network to improve the fault diagnosis effect of fuzzy neural network is the key.

Genetic algorithm [3] is a kind of global optimization algorithm, simulating the evolution process of nature, adopting the principle of the survival of the fittest, using genetic operator (copy, crossover and variation) to seek optimal individuals. Genetic algorithm provides a feasible method to further optimize fuzzy neural network [4]-[5]. Genetic algorithm can not only overcome the shortcomings of the BP algorithm, but also can optimize the rule layer's number of nodes in fuzzy neural network, the linking number and connection weight number of rule layer and condition layer and conclusion layer, the number of nodes in the input variables and membership function, which guarantees the most concise fuzzy neural network under the premise of the accuracy in the system to satisfy certain condition. And it greatly improved fuzzy neural network fault diagnosis effect. This paper presents a new algorithm of fuzzy neural network learning which based on combining a genetic algorithm of hierarchical structure [6] with evolution programming [7], using this algorithm to optimize the structure and parameters of fuzzy neural network. At last, through the examples show that the mechanical fault diagnosis of fuzzy neural network which based on this paper's method has concise structure and good effect of the mechanical fault diagnosis.

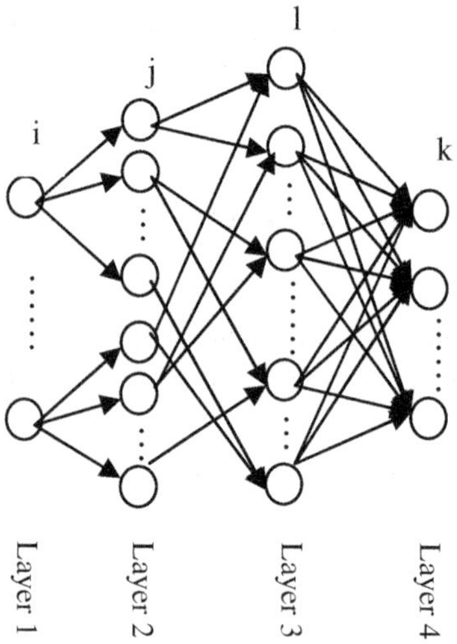

Fig. 1. Fuzzy neural networks

2 Fuzzy Neural Network Model

Fuzzy neural network model is shown in figure 1:

The first layer: $O_i = I_i = x_i$;

The second layer: $O_{x_i}^j = e^{-(\frac{I_i - m_{ij}}{\delta_{ij}})^2} = e^{-(\frac{x_i - m_{ij}}{\delta_{ij}})^2}$;

The third layer : $O_l = \prod_{i=1}^{n} I_{x_i}^j$;

The fourth layer : $O_k = \sum_{l=1}^{m} w_{kl} I_l$;

Among these four layers, the first layer is the input layer. The second layer is the rules and conditions layer which obfuscation dispose of input variables. Assuming that this network has n inputs and i have k fuzzy partition, the number of this layer's node is $k_1 + k_2 + ... + k_n$, and its function is the membership functions of each input to its fuzzy subsets. The third is variable rules layer, and each node represents a fuzzy rules. This layer node number uses genetic algorithm to optimize. The number is variable, and the most is $m = \prod_{i=1}^{n} k_i$. Generally each node of the third layer with one of each condition attributes fuzzy partition has one connection, so there is n connection. Because the connections exit redundancy, the number of the connections is less than n. There is one connection between each node of the third layer and fourth layer. The weight coefficient of the connection is w_{kl}. It is the support amount of each conclusion fuzzy partition to each rule. But the connections also exit redundancy, so the number of the connections is less than l. This paper adopts hierarchical structure. It optimizes the number of nodes in the third floor, the number of the connections between the second and third layer, the number of the connections between the third and the fourth layer, and the connection weight coefficients between them. The fourth layer is rule conclusion layer. Assuming the output conclusion is l, this layer node number is 1.

3 Using Hierarchical Genetic Algorithm to Optimize the Fuzzy Neural Network Learning New Algorithm Is Realized

This paper adopts the genetic algorithm of the hybrid coding hierarchical structure to optimize the structure and parameters of fuzzy neural network. Network structure optimization makes sure the number of the third hidden nodes, the number of the connections between the third and second, the number of the connections and the connection weight coefficients ω_{ij} between the third and the fourth layer. Network parameters optimization includes location parameter m_{ij} and width parameter δ_{ij} in the membership functions of the input variable.

A. coding

Using the genetic algorithm of the hybrid coding hierarchical structure, the network structure and the input membership function parameters constitute the individual. The network structure uses the hierarchical structure coding, and the input membership function parameters m_{ij}, δ_{ij} use real number coding. The fuzzy neural network to optimize the parameters has N = 2 $(k_1+k_2+\ldots+k_n)$. The network structure and its corresponding the code of network parameters constituting each, it is mixed chain of code whose genetic length is M + N. M is the length of the genes in network structure, and N is the length in network parameters. The coding of network structure uses hierarchical structure. The structure uses control gene to express the number of hidden nodes .It also uses structural gene to express the number of connections between the third and the second layer, and the number of connections and connection weights between the third and fourth layer. As shown in figure 2.

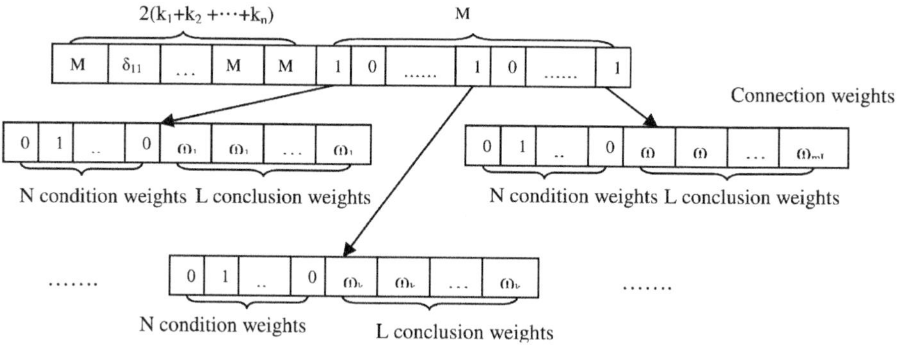

Fig. 2. Iterative structure based control gene and network weights

Coding hidden layer structure to a binary string, one byte of the string says a hidden node. As control gene, "1" says the hidden nodes having effect and "0" says the hidden nodes having no effect. So the number of "1" in the control genes is the number of network hidden nodes which have effect. Identify the most number of hidden layer nodes. If there is M = k_1, k_2, ...,K_n, the length of the string genes is m. Network connection weight structure includes n conditions weight numbers and a conclusion weight number. Conditions weight number using a binary string, "1" says that there have connections between conditions layer and rules layer, and "0" says that there is no connection. Conclusion weight number ω_{ij} using real number coding, is possible to be "0", which says there has no connection.

B. Select the initial population

The structure and parameters of fuzzy neural network is represented as a chromosome. The initial population contains the individual which corresponding to the most complete network structure and the input variables and the output variables evenly divide inkling within the scope of the variations. The rest of the individual randomly generated. If there is certain knowledge about the modelling system, you can elect the rule set which according to the experience and the vector corresponding to the input and output fuzzy partition into initial population. So it uses experience adequately meanwhile it ensures the final system better than experiential system.

C. The fitness function

In this paper, network error function uses mean-square error and network complexity uses the sum of node number between the second and third layer in network and the connection number of the second , the third and fourth layer to evaluate. Optimizing the network error and considering the complexity of control network .So that get the optimal fuzzy rules. The fitness function form of the network is shown:

$$f(i,t)=\alpha\frac{1}{E(i,t)}+\beta\frac{1}{H(i,t)}$$

Among them, f(i,t) says the fitness of the ith network individual in the t generation, E(I,t) says the error of it , and H(i, t) says the network complexity of it. α and β are greater than 0 constants, and $\alpha+\beta=1$. With such function as fitness function can guarantee the optimization of network weights and get a fit network structure. α and β can be ensured beforehand according to the actual problems, also can be adjusted in the training process adaptively.

D. Genetic operation

Genetic operations include copy, crossover and variation. Form the previous generation populations copy uses roulette to select individual according to adaptionswert. The algorithm uses heuristic algorithm which elect elite in this paper. It unconditionally keeps the optimal individual in each generation of population to the next in this algorithm. This way ensures asymptotic convergence in the algorithm.

In the process of genetic operations, crossover and variation rate of p_c and p_m have very big effect on the performance of genetic algorithm. The values of should change with adaptive value's changes. To the value of high fitness, take the lower p_c and p_m , and make the chance increased that enter into the next generation. And to the value of low fitness, take the higher value to eliminate it. When the mature convergence occurs, should increase p_c and p_m, in order to accelerate the produces of new individuals. This paper uses a kind of method that p_c and p_m are self-adaptive[8]. It uses individual fitness function to measure the convergence degree of the algorithm. The expressions are:

$$p_c=\begin{cases} p_{c1}-\dfrac{(p_{c1}-p_{c2})(f'-f_{avg})}{f_{\max}-f_{avg}}, f'\geq f_{avg} \\ p_{c1}, f'<f_{avg} \end{cases}$$

$$p_m=\begin{cases} p_{m1}-\dfrac{(p_{m1}-p_{m2})(f'-f_{avg})}{f_{\max}-f_{avg}}, f'\geq f_{avg} \\ p_{m1}, f'<f_{avg} \end{cases}$$

Among them: f_{max} and f_{avg} respectively express the maximum fitness and the average fitness. f' is the biggish fitness between the two individuals which to be crossed. the in the. p_{c1}, p_{c2}, p_{m1}, and p_{m2} are all less than 1, and all of them are constant. Crossover operation is only for hidden layer control gene. When the hidden layer control gene is doing cross operation, it will take away control network weights of each gene correspondingly. The crossover operation of hierarchical structure of hybrid coding makes not only the control gene recombined, but also network weights recombined.

Variation operations includes control gene and its control network weights, and the condition parameter m_{ij} and δ_{ij} which coding with control gene. According to the certain probability, the variation of control gene and condition weight number make the byte of the control gene string changed from 0 to 1, or from 1 to 0. The variation of m_{ij} , δ_{ij} and the connection weight number w_{ij} is solved with evolutionary programming in this paper. Their variations are shown:

$$m_{ij}' = m_{ij} + \alpha \frac{1}{f} N(0,1)$$

$$\delta_{ij}' = \delta_{ij} + \alpha \frac{1}{f} N(0,1)$$

$$w_{ij}' = w_{ij} + \alpha \frac{1}{f} N(0,1)$$

α is evolution rate .f is fitness of each individual. N (0,1) is random quantity of normal distribution when expected value is 0 and standard deviation is 1.

4 Fault Diagnosis Application Examples

Here quote the data of references [9]. Simulate three kinds of faults such as grinding machine imbalance, misalignment and oil whirl on the test-bed. Take (0.4, 0.5) x, 1 x, 2 x, 3 x and greater than 3 x of the turning frequency as objective parameter. Each input parameter is fuzzy into big, middle and small one. The initiating structure of fuzzy neural network is intended to be 5-15-243-3. Select five parameters as input training vector and three kinds of faults as corresponding training object vector. When process online learning and test, use normalization processing with data to make data on [0, 1]. There are 12 groups of the standard model fault samples. They are shown in table 1.

Table 1. Simulating different grinder faults of standard sample

number	fault	Input sample					Ideal output		
		(0.4~0.5)x	1x	2x	3x	>3x	F1	F2	F3
1	oil whirl	0.92	0.21	0.03	0.01	0.02	1	0	0
2		0.90	.020	0.05	0.02	0.02	1	0	0
3		0.85	0.25	0.06	0.02	0.01	1	0	0
4		0.82	0.28	0.05	0.04	0.03	1	0	0
5	imbalance	0.05	0.90	0.11	0.05	0.02	0	1	0
6		0.03	0.96	0.12	0.04	0.03	0	1	0
7		0.02	0.91	0.08	0.01	0.02	0	1	0
8		0.02	1.00	0.08	0.03	0.01	0	1	0
9	misalignment	0.01	0.52	0.40	0.32	0.10	0	0	1
10		0.01	0.40	0.47	0.35	0.18	0	0	1
11		0.02	0.45	0.42	0.28	0.29	0	0	1
12		0.01	0.48	0.48	0.36	0.20	0	0	1

Each input parameter is fuzzy into big, middle and small one. The initiating structure of fuzzy neural network is intended to be 5-15-243-3. Select the number of 1, 2, 5, 7, 8, 10, 11 and 12 as network input data. And these data make up the network training sample. In this paper's genetic algorithm, population scale is taken as 80, maximum evolution algebra is 4000, α is 0.9, β is 0.1, $P_{c1}=0.9, P_{c2}=0.6, P_{m1}=0.1$ and $P_{m2}=0.001$. Through using hierarchical genetic algorithm to optimize the fuzzy neural network training, fuzzy neural network structure is changed into 5-8-20-3 finally.

Select the number of 3, 6 and 9 in standard fault as network test fault. The test faults are shown in table 2. As can be seen from table 2, fuzzy neural network which is optimized with hierarchical genetic algorithm can identify the grinding machine fault accurately. And it also can achieve high accuracy. Comparing with the number 9 literature, this paper can get not only higher fault recognition accuracy, but also clearer semantic structure of fuzzy neural network. You still can get fuzzy rule which is provided for expert system to process with fault diagnosis.

Table 2. Recognizing results of simulating different grinder faults

Serial number / falut		3	6	9
		oil whirl	imbalance	misalignment
undetermimed identification sample	(0.4-0.5) x	0.85	0.03	0.01
	1x	0.25	0.96	0.52
	2x	0.06	0.12	0.40
	3x	0.02	0.04	0.32
	>3x	0.01	0.03	0.10
indentification result	F1	0.999995	0.000006	0.000000
	F2	0.000000	0.999998	0.000007
	F3	0.000004	0.000012	0.999924

5 Conclusion

This paper presents the new algorithm of fuzzy neural network learning which based on combining the hybrid coding genetic algorithm of hierarchical structure with evolution programming. You can use this algorithm to optimize the structure and parameters of fuzzy neural network. The fault diagnosis fuzzy neural network which uses this optimizing learning algorithm overcomes the shortcoming of BP algorithm such as falling into local minimum and containing a mass of redundant nodes and connections. It also can get higher mechanical fault diagnosis accuracy.

References

1. Yu, H., Han, Q., Li, S.: Equipment fault diagnosis engineering. Metallurgical Industry Press, Beijing (2001)
2. Liu, Z., Liu, L.: Skidding Scrape Diagnosis of Rolling Bearing Based On Fuzzy Neural Network. Transactions of the Chinese Society for Agricultural Machinery 33(1), 97–99 (2002)

3. Liu, M., Wu, J., Guo, J., et al.: Study on Matching Method of Mechanical Products with Multi-parameters Based on Genetic Algorithm. Transactions of the Chinese Society for Agricultural Machinery 37(1), 124–127 (2006)
4. Ishigami, H., Fukuda, T., Shibata, T.: Structure Optimization of Fuzzy Neural Network by Genetic Algorithm. Fuzzy Sets and Systems 71(2), 257–264 (1995)
5. Shi, Y., Russell, E., Chen, Y.: Implementation of Evolutionary Fuzzy Systems. IEEE Trans. on Fuzzy Systems 7(2), 109–119 (1999)
6. Feng, J., Diao, Z., et al.: Wavelet neural network optimization based on hybrid hierarchy genetic algorithm. Systems Engineering and Electronics 22(6), 1112–1115 (2004)
7. Tang, Y., Wang, M., et al.: Biology Multiple Sequence Alignment Algorithm Based on Genetic Annealing Evolution. Transactions of the Chinese Society for Agricultural Machinery 36(9), 116–119 (2005)
8. Gong, D., Pan, F.: The Theory and Research of Adaptive Genetic Alforithm. China University of Mining And Technology Press (2003)
9. Zhao, H., Xia, Z.: The application of optimizing the genetic neural network to the machinery fault diagnosis. Journal of China Institute of Metrology 15(2), 153–155 (2004)

New Fast-ICA Algorithms for Blind Source Separation without Prewhitening*

Jimin Ye and Ting Huang

Department of Applied Mathematics
Xidian University
Xi'an, 710071, China
ji.kam@qq.com

Abstract. This paper focuses on proposing a new fast-ICA algorithm without prewhitening. First, existing fast-ICA method is reviewed. then, by combing the separating vector in the existing fast-ICA algorithm with the prewhitening matrix, we propose a new separating vector, which is used to separate statistically independent component from the observed data without prewhitening. The iterative rule of new separating vector is developed. Finally, the effectiveness of this new algorithm is verified by computer simulations.

Keywords: blind source separation, fast-ICA, pre-whitening.

1 Introduction

Blind source separation (BSS) has been studied widely in recent years since it is a fundamental problem encountered in many practical applications such as wireless telecommunication and image denoising etc.(see[1] and [2]) In the instantaneous case, observed mixture of several signals can be express by the simplest mathematical model, in which noise is ignored :

$$\mathbf{x}_t = \mathbf{A}\mathbf{s}_t \tag{1}$$

where $\mathbf{A}$ is an $n \times n$ fixed but unknown mixing matrix with full column rank, $\mathbf{s}_t = (s_1, \cdots, s_n)^T$ is a $n \times 1$ vector composed of independent source signals and $\mathbf{x}_t = (x_1, \cdots, x_n)^T$ is a $n \times 1$ vector composed of observed signals. We assume that source signals are statically independent with zero-mean and unit variance.

The task of BSS is to construct a separating matrix $\mathbf{B}$ given just the observed signals, using the independence assumption of source signals. The result:

$$\mathbf{y}_t = \mathbf{B}\mathbf{x}_t \tag{2}$$

is an estimation of source signals. Since the amplitude and order of recovered signals are not exact, the overall transfer matrix $\mathbf{C} = \mathbf{BA}$ should converge to a permutation matrix with one and only one nonzero value at each row and each column. (see [3])

* This work is supported by National Science Foundation of China (Grant No. 61075117).

D. Zeng (Ed.): ICAIC 2011, Part II, CCIS 225, pp. 579–585, 2011.

A variety of approaches have been proposed for BSS. In Some methods, prewhitening is not necessary, such as the EASI algorithm [4] and natural gradient algorithms [5]. But other approaches such as NPCA algorithm [6] and exited fast-ICA algorithm [1], need to prewhiten the observed data first. In prewhitening cases, the final separating matrix should be $\mathbf{B} = \mathbf{W}\mathbf{V}$, where $\mathbf{V}$ is the prewhitening matrix and $\mathbf{W}$ is an orthonormal matrix.

Fast-ICA algorithm and its related prewhitening algorithm are offline methods. Although we could obtain a good prewhitening result, each sample will be calculated to get a prewhitening matrix. That is to say, prewhitening step may affect the convergence speed. Furthermore, the error of prewhitening will propagate to the separation layer. In this paper, we introduce the autocorrelation matrix $\mathbf{R}_{\mathbf{x}} = \mathbf{E}\{\mathbf{x}_t\mathbf{x}_t^{\mathrm{T}}\}$ into fast-ICA algorithm and propose a new fast-ICA algorithm which avoids prewhitening procedure and works also efficiently.

2 Fast-Ica Algorithm for Prewhitened Bss

A. *Prewhitening algorithm*

A zero-mean random vector $\mathbf{z}_t$ is said to be whitening if its elements are uncorrelated and have unit variances. In terms of the covariance matrix, this means that $\mathbf{E}\{\mathbf{z}_t\mathbf{z}_t^{\mathrm{T}}\} = \mathbf{I}$, where $\mathbf{I}$ is a unit matrix. Because whitening is essentially decorrelation followed by scaling, it can be done with a linear operation. We will find a linear transformation $\mathbf{V}$ which will be applied to observed data $\mathbf{x}_t$, in order to get whitened data $\mathbf{z}_t = \mathbf{V}\mathbf{x}_t$.

In fast-ICA algorithm, we use the offline whitening algorithm as follow: $\mathbf{R}_{\mathbf{x}}$ is the covariance matrix of $\mathbf{x}_t$, which can be computed from the samples of $\mathbf{x}_t$. Let $\mathbf{E} = (\mathbf{e}_1,...,\mathbf{e}_n)$ be the matrix whose columns are the unit-norm eigenvectors of the covariance matrix $\mathbf{R}_{\mathbf{x}}$. Let $\mathbf{D} = diag(\mathbf{d}_1,...,\mathbf{d}_n)$ be the diagonal matrix of the eigenvalues of $\mathbf{R}_{\mathbf{x}}$. Then whitening transform is given by

$$\mathbf{V} = \mathbf{D}^{-1/2}\mathbf{E}^{\mathrm{T}} \tag{3}$$

It is easy to show that $\mathbf{V}$ is indeed a whitening transformation. $\mathbf{R}_{\mathbf{x}}$ is written as $\mathbf{R}_{\mathbf{x}} = \mathbf{E}\mathbf{D}\mathbf{E}^{\mathrm{T}}$. With $\mathbf{E}$ an orthogonal matrix satisfying $\mathbf{E}^{\mathrm{T}}\mathbf{E} = \mathbf{E}\mathbf{E}^{\mathrm{T}} = \mathbf{I}$, it holds $\mathbf{E}\{\mathbf{z}_t\mathbf{z}_t^{\mathrm{T}}\} = \mathbf{V}\mathbf{E}\{\mathbf{x}_t\mathbf{x}_t^{\mathrm{T}}\}\mathbf{V}^{\mathrm{T}} = \mathbf{D}^{-1/2}\mathbf{E}^{\mathrm{T}}\mathbf{E}\mathbf{D}\mathbf{E}^{\mathrm{T}}\mathbf{E}\mathbf{D}^{-1/2} = \mathbf{I}$. The covariance of $\mathbf{z}_t$ is the unit matrix, hence $\mathbf{z}_t$ is white. (see [1])

B. *Fast fixed-point algorithm using kurtosis*

Assume that data vector $\mathbf{x}_t$ is distributed according to the model (1): $\mathbf{x}_t = \mathbf{A}\mathbf{s}_t$. Let us invert the mixing as $\mathbf{s}_t = \mathbf{A}^{-1}\mathbf{x}_t$. That is to say, estimating the independent components can be accomplished by finding the right linear combinations of the mixture data. Let us estimate one component as

$$y = \mathbf{b}^{\mathrm{T}}\mathbf{x} = \sum_i b_i x_i \tag{4}$$

where $\mathbf{b}^{\mathrm{T}}$ is just one row of separating matrix $\mathbf{B}$ to be determined, termed as separating vector. Note that

$$y = \mathbf{b}^{\mathrm{T}}\mathbf{x} = \mathbf{b}^{\mathrm{T}}\mathbf{A}\mathbf{s} = \mathbf{q}^{\mathrm{T}}\mathbf{s} = \sum_i q_i s_i \tag{5}$$

y is a certain linear combination of $s_i, i = 1,2,...,n$ We know that a sum of even two independent random variables is more gaussian than the original variables. So y is usually more gaussian than any of $\mathbf{s}, i = 1,2,...,n$ and becomes least Gaussian when it equals one of $\mathbf{s}, i = 1,2,...,n$. So maximizing the nongaussianity of $y = \mathbf{b}^{\mathrm{T}}\mathbf{x}$ gives us one of the independent components. For whitened data $\mathbf{z}$, we seek for a linear combination $y = \mathbf{w}^{\mathrm{T}}\mathbf{z}$ that maximizes nongaussianity.

Kurtosis is a classic measure of nongaussianity. The definition of nongaussianity of y is:

$$kurt(y) = E\{y^4\} - 3(E\{y^2\})^2 \tag{6}$$

Typically nongaussianity is measured by the absolute value of kurtosis. These measures are zero for a Gaussian variable and greater than zero for most nongaussian random variables. So let us select kurtosis as the contract function

$$J(\mathbf{w}) = |kurt(y)| = |kurt(\mathbf{w}^{\mathrm{T}}\mathbf{z})| \tag{7}$$

To maximize the contract function, we could start from some vector $\mathbf{w}$, compute the direction in which the absolute value of the kurtosis of $y = \mathbf{w}^{\mathrm{T}}\mathbf{z}$ is growing most strongly. The direction is called gradient. We can derive the gradient of (7) as follows:

$$\partial |kurt(\mathbf{w}^{\mathrm{T}}\mathbf{z})| / \partial \mathbf{w} = 4sign(kurt(\mathbf{w}^{\mathrm{T}}\mathbf{z}))[E\{\mathbf{z}(\mathbf{w}^{\mathrm{T}}\mathbf{z})^3\} - 3\mathbf{w}\|\mathbf{w}\|^2] \tag{8}$$

Estimation y should have unit variance, so for the whitened data $\mathbf{z}$ we have $\mathbf{E}\{(\mathbf{w}^{\mathrm{T}}\mathbf{z})^2\} = \mathbf{I}$, That is to say , we are optimizing this function on the unit sphere $\|\mathbf{w}^2\| = 1$. Therefore, we may project $\mathbf{w}$ on the unit sphere after every step.

If we move the vector $\mathbf{w}$ in its gradient direction, we get the gradient algorithm. Now we note that, at a stable point of the gradient algorithm, the gradient must point in the direction of $\mathbf{w}$. In this case, adding the gradient to $\mathbf{w}$ does not change its direction and we can have the convergence. So we get the fast-ICA algorithm if we equating the gradient of kurtosis with $\mathbf{w}$ as follow:

$$\begin{aligned} &\mathbf{w} \propto 4sign(kurt(\mathbf{w}^{\mathrm{T}}\mathbf{z}))[\mathbf{E}\{\mathbf{z}(\mathbf{w}^{\mathrm{T}}\mathbf{z})^3\} - 3\mathbf{w}\|\mathbf{w}\|^2] \\ &\mathbf{w} \leftarrow \mathbf{w} / \|\mathbf{w}\| \end{aligned} \tag{9}$$

3 New Fast-Ica Algorithm for Bss without Prewhitening

In existent fast-ICA algorithm, we get the whitened data $\mathbf{z}_t = \mathbf{V}\mathbf{x}_t$ by a linear transformation $\mathbf{V}$. Applying the separating transformation to $\mathbf{z}_t$ we obtain the estimation of source data $y_t = \mathbf{w}^{\mathrm{T}}\mathbf{z}_t = \mathbf{w}^{\mathrm{T}}\mathbf{V}\mathbf{x}_t$. Now we focus on a new algorithm without prewhitening. Let us combine the separating and whitening operation and denote

$$\mathbf{b}^{\mathrm{T}} = \mathbf{w}^{\mathrm{T}}\mathbf{V} \tag{10}$$

that is $\mathbf{b} = \mathbf{V}^{\mathrm{T}}\mathbf{w}$.

So we have

$$\partial\left|kurt(y)\right| / \partial\mathbf{b} = \mathbf{V}^{T}(\partial\left|kurt(y)\right|^{T} / \partial\mathbf{w}) \tag{11}$$

We know:

$$\begin{aligned} \mathbf{z} &= \mathbf{V}\mathbf{x} \\ \mathbf{w}^{\mathrm{T}}\mathbf{z} &= \mathbf{b}^{\mathrm{T}}\mathbf{x} \end{aligned} \tag{12}$$

Since whitened data has unit variance, we have $\mathbf{E}\{\mathbf{z}_t\mathbf{z}_t^{\mathrm{T}}\} = \mathbf{V}\mathbf{E}\{\mathbf{x}_t\mathbf{x}_t^{\mathrm{T}}\}\mathbf{V}^{\mathrm{T}} = \mathbf{V}\mathbf{R}_{\mathbf{x}}\mathbf{V}^{\mathrm{T}} = \mathbf{I}$. Then we can derive an useful conclusion as follows:

$$\mathbf{V}^{\mathrm{T}}\mathbf{V} = \mathbf{R}_{\mathbf{x}}^{-1} \tag{13}$$

By applying (8), (12), (13) to (11), we get a new form of gradient

$$\partial\left|kurt(y)\right| / \partial\mathbf{b} = 4sign(kurt(y))\mathbf{R}_{\mathbf{x}}^{-1}\mathbf{E}\{\mathbf{x}(\mathbf{b}^{\mathrm{T}}\mathbf{x})^{3}\} - 3\mathbf{b}^{\mathrm{T}}\mathbf{R}_{\mathbf{x}}\mathbf{b}\mathbf{b} \tag{14}$$

This gradient is the iterative rule of $\mathbf{b}$, Using (14) and similar derivation as in prewhitening fast fixed point algorithm, we get a new fast-ICA algorithm without whitening.

Take notice of that elements of estimation are mutually irrelevant. That is to say

$$\mathbf{E}\{(\mathbf{b}_i^{\mathrm{T}}\mathbf{x})(\mathbf{b}_j^{\mathrm{T}}\mathbf{x})^{\mathrm{T}}\} = \mathbf{b}_i^{\mathrm{T}}\mathbf{R}_{\mathbf{x}}\mathbf{b}_j = 0 \text{ when } i \neq j \tag{15}$$

Meanwhile the source signals have unit variance, therefore,

$$\mathbf{E}\{(\mathbf{b}_i^{\mathrm{T}}\mathbf{x})(\mathbf{b}_i^{\mathrm{T}}\mathbf{x})^{\mathrm{T}}\} = \mathbf{b}_i^{\mathrm{T}}\mathbf{R}_{\mathbf{x}}\mathbf{b}_i = 1 \tag{16}$$

Combining these two constraints we have

$$\mathbf{B}^{\mathrm{T}}\mathbf{R}_{\mathbf{x}}\mathbf{B} = \mathbf{I} \tag{17}$$

That is to say, the separating vectors corresponding the different independent should be weighted orthogonal with weighting matrix $\mathbf{R}_{\mathbf{x}}$.

We have many methods to achieve (17). In this paper we apply the generalized Gram-Schmidt algorithm as follows (see [3]):

$$\begin{aligned} &\mathbf{B}_t(:,1) = \mathbf{B}_t(:,1) / \sqrt{\mathbf{B}_t^{\mathrm{T}}(:,1)\mathbf{R}_{\mathbf{x}}\mathbf{B}_t(:,1)} \\ &i = 2,\ldots,d \\ &\tilde{\mathbf{B}}_t(:,i) = \mathbf{B}_t(:,i) - \sum_{j=1}^{i-1}[\mathbf{B}_t^{\mathrm{T}}(:,i)\mathbf{R}_{\mathbf{x}}\mathbf{B}_t(:,j)]\mathbf{B}_t(:,j) \\ &\mathbf{B}_t(:,i) = \tilde{\mathbf{B}}_t(:,i) / \sqrt{\mathbf{B}_t^{T}(:,i)\mathbf{R}_{\mathbf{x}}\mathbf{B}_t(:,i)} \end{aligned} \tag{18}$$

Combining the iterative rule with deflationary orthogonalization, we get the new fast-ICA algorithm as follows:

$$\begin{aligned}
&\mathbf{B}_t(:,i) \propto 4sign(kurt(\mathbf{B}_t(:,i)^{\mathrm{T}}\mathbf{x}))\times \\
&\qquad \mathbf{R}_{\mathbf{x}}^{-1}\mathbf{E}\{\mathbf{x}(\mathbf{B}_t(:,i)^{\mathrm{T}}\mathbf{x})^3\} - 3\mathbf{B}_t(:,i)^{\mathrm{T}}\mathbf{R}_{\mathbf{x}}\mathbf{B}_t(:,i)\mathbf{B}_t(:,i) \\
&\tilde{\mathbf{B}}_t(:,i) = \mathbf{B}_t(:,i) - \sum_{j=1}^{i-1}[\mathbf{B}_t^{\mathrm{T}}(:,i)\mathbf{R}_{\mathbf{x}}\mathbf{B}_t(:,j)]\mathbf{B}_t(:,j) \\
&\mathbf{B}_t(:,i) = \tilde{\mathbf{B}}_t(:,i)/\sqrt{\mathbf{B}_t^{T}(:,i)\mathbf{R}_{\mathbf{x}}\mathbf{B}_t(:,i)}
\end{aligned} \tag{19}$$

4 Computer Simulations

In order to verify the effectiveness of the proposed algorithm, we consider the separation of the following source signals (taken from [7]): (1) sign function signal: $\mathrm{sgn}(\cos(2\pi 155t))$; (2) high frequency sinusoid signal: $\sin(2\pi 800t)$; (3) low frequency sinusoid signal: $\sin(2\pi 90t)$; (4) phase-modulated signal: $\sin(2\pi 300t + 6\cos(2\pi 60t))$. The mixing matrix **A** is generated randomly. The signals are sampled at 10kHz. Taking 1000 samples, the source signals, mixed signals and separating signals are independently plotted from fig.1 to fig.3. We can see that the proposed algorithm can separate the mixed signals effectively without prewhitening. By counting the average iterative times for each component of 10 runs (see tableI), we can verify the proposed algorithm is faster than the existing fast-ICA algorithm.

Table 1. Iterative times for each component

	First component	Second component	Third component	Fourth component
Existing algorithm	31	75	12	2
Proposed algorithm	9	9	7	2

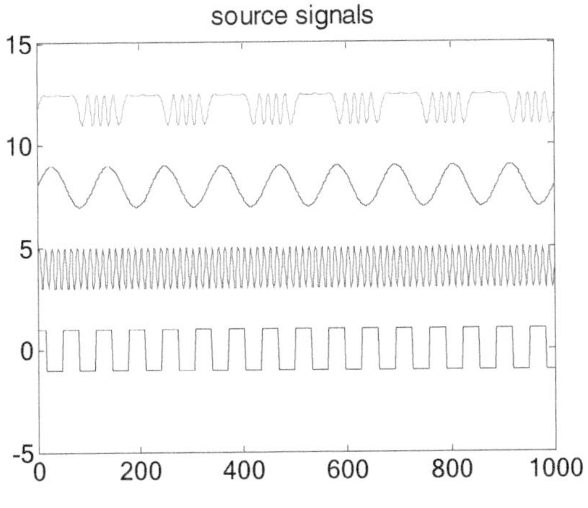

Fig. 1. Source signals

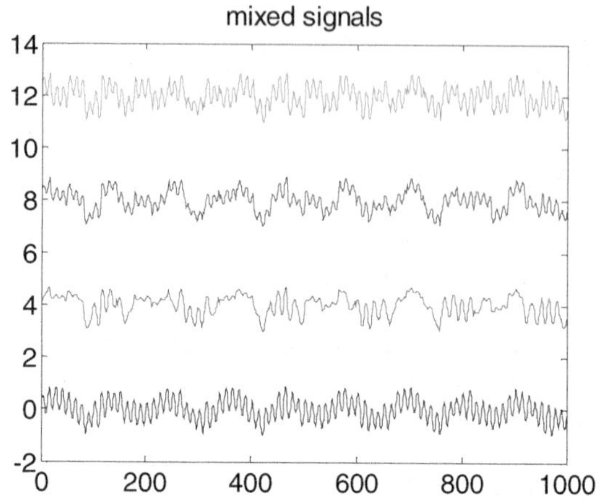

Fig. 2. Mixed signals

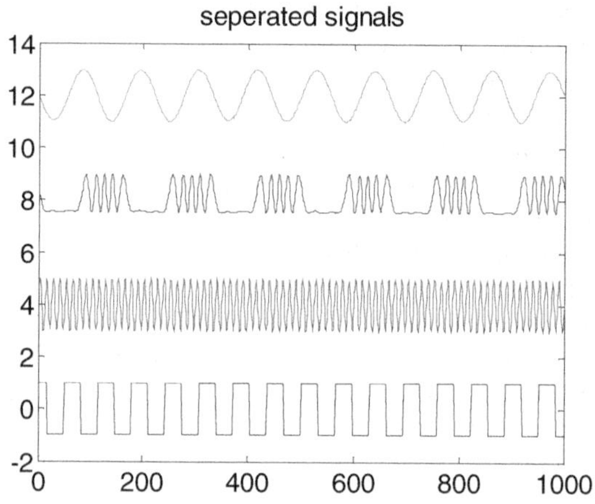

Fig. 3. Separated signals

5 Conclusions

Since the prewhitening process may affect greatly the convergence speed and the accuracy of the BSS methods, we promote a new fast-ICA algorithm that removes the prewhitening process. Gradient of the separating vector, which is applied to data without prewhitening, is obtained by multiplying the gradient of separating vector in [1] with a linear transform matrix. The autocorrelation matrix is taken in to account when we look for the orthonormal constraints. The computer simulation shows that this new algorithm works effectively.

References

1. Hyvarinen, A., Karhunen, J., Oja, E.: Independent Component Analysis. Wiley, New York (2001)
2. Girolami, M.: Self-Organizing Neural Networks: Independent Component Analysis and Blind Source Separation. Springer, London (1999)
3. Zhu, X., Zhang, X., Ding, Z., Jia, Y.: Adaptive Nonlinear PCA Algorithms for Blind Source Separation Without Prewhitening. IEEE Transactions on Circuits and Systems-I: Regular Papers 53(3) (March 2006)
4. Cardoso, J.F., Laheld, H.: Equivariant Adaptive Source Separation. IEEE Trans. Signal Process. 44(12), 3017–3029 (1996)
5. Amari, S., Cichocki, A., Yang, H.H.: A New Learning Algorithm for blind Source Separation. In: Advance in Neural Information Processing Systems, vol. 8, pp. 757–763. MIT Press, Cambridge (1996)
6. Karhunen, J., Pajunen, P., Oja, E.: The Nonlinear PCA Criterion In Blind Source Separation: Relations With Other Approaches. Neurocomputing 22(1-3), 5–20 (1998)
7. Zhu, X., Zhang, X., Ye, J.: Natural gradient-based recursive least-squares algorithm for adaptive blind source separation. Science in China Ser.F Information Sciences 47(1), 55–65 (2004)

Research on High-Speed Asynchronous Serial Transmission Based on 8b10b

Weigong Zhang, Yongqin Hu, Rui Ding, and Bo Yang

College of Information Engineering
Capital Normal University
Beijing, China
W.G_Zhang@TOM.com

Abstract. This paper points out the application limitations of the current synchronous transmission and asynchronous transmission, presents a new type of high-speed asynchronous serial transmission scheme based on 8b10b, introduces the principle of the scheme, and pointed out the program's features and key technologies.

Keywords: Asynchronous serial transmission, 8b/10b, Data recovery.

1 Introduction

The signal transmission is one of the most important parts of the communication system. The speed and quality of the signal transmission impact the validity and reliability of communications systems, therefore, the signal transmission scheme has been one of the hot subjects both at home and abroad. Signal transmission can be divided into synchronous and asynchronous mode. Synchronous transmission is accurate and efficient. A large number of signals can be transmitted at one communication process. However, separate clock line is necessary for the system, while the sending and receiving devices are more complex and costly. Asynchronous mode is simple and flexible but the transmission rate is too low. With the technology development and the increasing application demand, the traditional signal transmission scheme has been unable to meet the communication needs. In light of this, a kind of high-speed asynchronous serial transmission scheme based on 8b10b is presented.

2 Basic PRINCIPLE

The data is divided into several groups (8 bits per group) first, and then is encoded into a kind of new groups (one group per 10 bits) according to 8b10b encoding. The second step, the parallel data is converted to serial data, and then is transmitted to the receiver asynchronously. The third step, at the receiver, the clock information is recovered by the data recovery module. The fourth step, the serial data is converted to parallel data. The last step, the data is decoded according to the 8b10b decoding, as shown in Figure 1:

D. Zeng (Ed.): ICAIC 2011, Part II, CCIS 225, pp. 586–592, 2011.

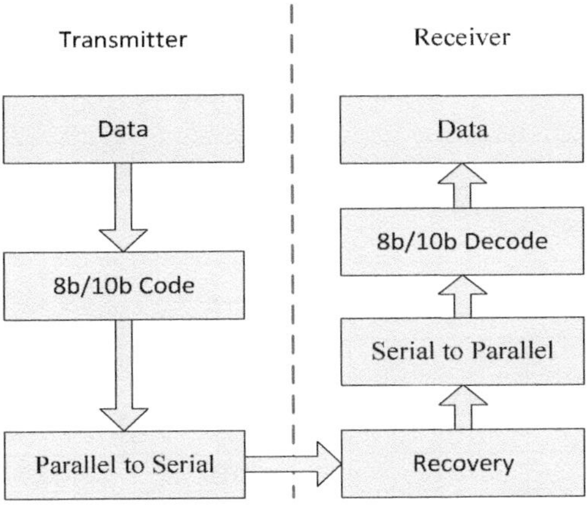

Fig. 1. Basic Block Diagram

The main parts of the scheme have three modules: the 8b/10b coding module at the transmitter, the clock recovery module and the 8b/10b decoding module at the receiver.

3 8B10B

8b/10b code is presented by IBM in the 80s of the 20th century and now has become the common coding technique used in the fields of the high-speed data transmission interface and bus (IEEE 802.3z, 36.2.4). The basic principle of the code is to encode 8 bits into 10 bits under a certain mapping. Then the number of the '0' of the bit stream can be as the same as that of the '1' of the bit stream. We call this "DC Balance". So the bit steam can be transmitted at high-speed asynchronously.

The data is divided into groups per 8bits. Each group is divided into two parts: 3bit and 5bit, which are mapped to 4bit and 6bit.(the 3b/4b and 5b/6b encoding). 8b/10b encoding defines not only the usual data coding, but also the "control code" to distinguish 12 special coding from the usual data, as shown in Figure 2:

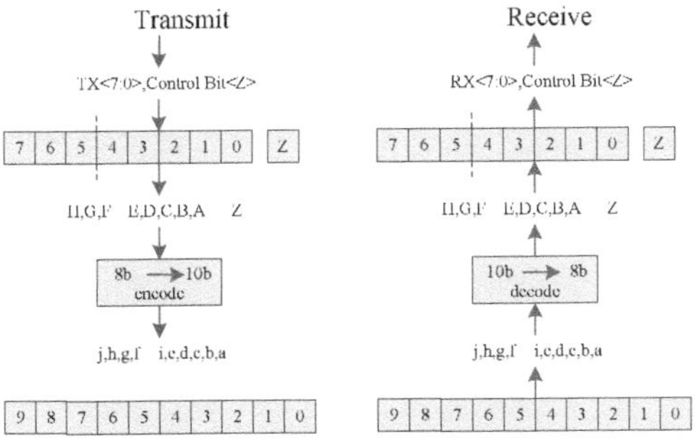

Fig. 2. 8b10b Coded Diagram

The 5b6b mapping rules are shown in Table 1:

Table 1. 5b6b mapping rules

Inputs		abcdei outputs		Rd'
Dx	EDCB A	rd+	rd-	
D0	00000	011000	100111	-rd
D1	00001	100010	011101	
D2	00010	010010	101101	
D3	00011	110001		rd
D4	00100	001010	110101	-rd
D5	00101	101001		rd
D6	00110	011001		
D7	00111	000111	111000	
D8	01000	000110	111001	-rd
D9	01001	100101		rd
D10	01010	010101		
D11	01011	110100		
D12	01100	001101		
D13	01101	101100		
D14	01110	011100		
D15	01111	101000	010111	-rd
D16	10000	100100	011011	
D17	10001	100011		rd
D18	10010	010011		
D19	10011	110010		
D20	10100	001011		
D21	10101	101010		
D22	10110	011010		
D23	10111	000101	111010	-rd
D24	11000	001100	110011	
D25	11001	100110		rd
D26	11010	010110		
D27	11011	001001	110110	-rd
D28	11100	001110		rd
D29	11101	010001	101110	-rd
D30	11110	100001	011110	
D31	11111	010100	101011	

The 3b4b mapping rules are shown in Table2:

Table 2. 3b4b mapping rules

Inputs		fghj outputs		Rd'
Dx.y	HGF	rd+	rd-	
Dx.0	000	0100	1011	-rd
Dx.1	001	1001		rd
Dx.2	010	0101		
Dx.3	011	0011	1100	
Dx.4	100	0010	1101	-rd
Dx.5	101	1010		rd
Dx.6	110	0110		
Dx.P7	111	0001	1110	-rd
Dx.A7	111	1000	0111	
NOTE: A7 replaces P7 if [(rd>0) and (e=i=0)] or [(rd<0) and (e=i=1)]				

According to the rules on the table, 8b are mapped to 10bit, which have only 3 possibilities: 5'0' with 5 '1', 6 '1' with 4 '0', or 6 '0' with 4 '1'. So the difference between the number of '0' and '1'in every 10bit data will be only - 2,0, +2. "Disparity" is defined to represent the difference, its range is: -2,0, +2.

In the process of high-speed serial communication, in order to avoid a continuous bit stream whose disparity take +2 (or -2) led to accumulation of uncertainty, parts of bit that have the opposite disparity are encoded by two ways. In the transmission process, the right mapping will be chosen according to the "Disparity" of the bits before, and thus offset the inequality. The method is called "Running Disparity" (RD). RD has two kinds of polarity: positive and negative, which is controlled by the "Disparity Control" (DC) module, as shown in Figure 3:

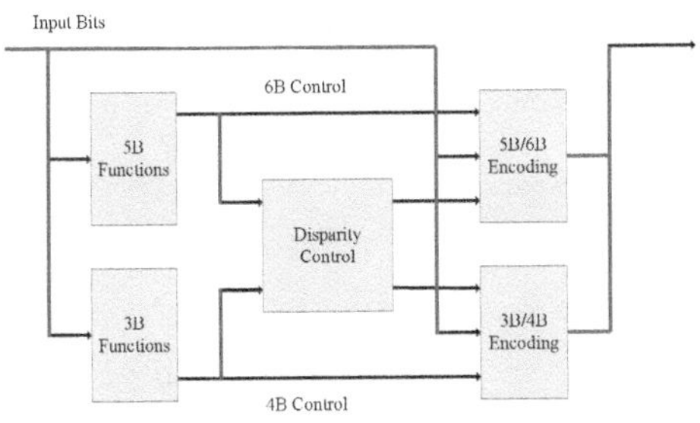

Fig. 3. 8b10b Flow Diagram

4 Data Recovery

In the receiver, the clock will be recovered by the "Recovery" module in order to transfer the incoming asynchronous bit streams into the synchronous bits, as shown in Figure 4:

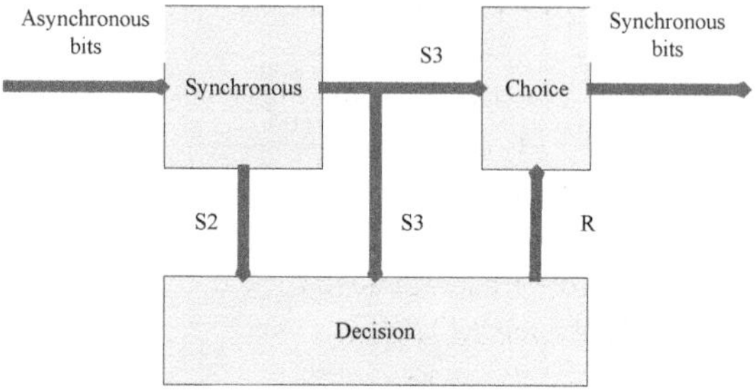

Fig. 4. Data Flow Diagram

A. Synchronous

First, there is no clock information received, a nominal clock that is called CLK(generated by local oscillator) is needed. It is either slightly faster or slightly slower than the incoming data streams. The CLK can generate another clock signal called CLK90 whose phase is slower 90 degrees than the CLK with the DLL component.

The next step, the input data will flow through an array of the D flip-flop, as shown in Figure 5:

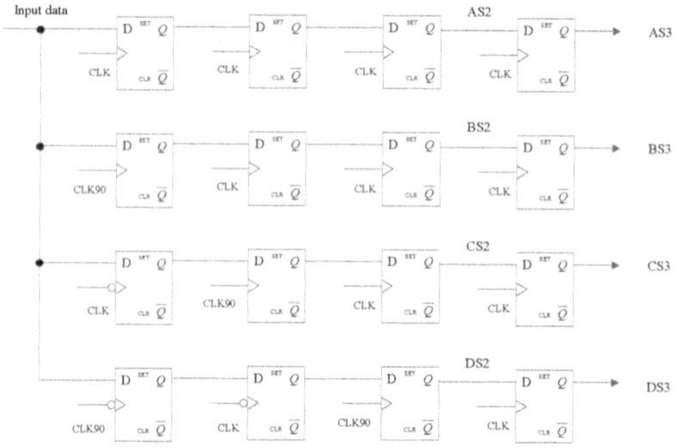

Fig. 5. D Flip-flop Array

As shown in Figure 5, the incoming data will flow through 4 flip-flops, two locked by CLK (one rising edge and one falling edge) and two locked by CLK90 (one rising edge and one falling edge).

The first flip-flops is clocked by the rising edge of CLK at time domain A; the second one is clocked by the rising edge of CLK90 at time domain B; the third one is clocked by the falling edge of CLK at time domain C; the last one is clocked by the falling edge of CLK90 at time domain D, as shown in Figure 6:

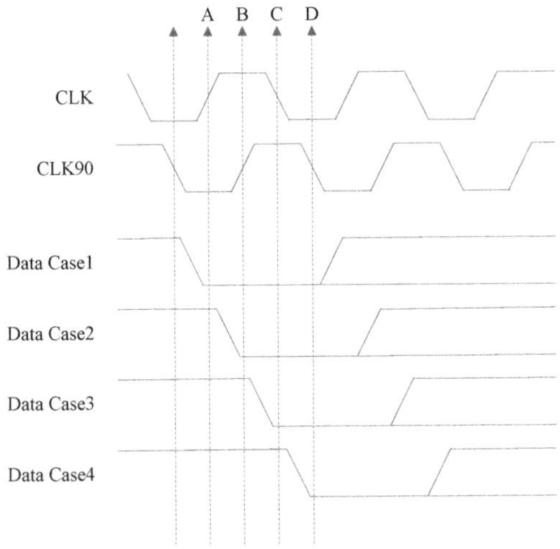

Fig. 6. Data Timing Diagram

B. Decision

The output of the third D flip-flop(AS2) and the last one (AS3) will be XOR. The result will flow through another D flip-flop to be synchronized, obtained the judgments signal AP, BP, CP, DP, AN, BN, CN, DN, then the position of the transition in the D flip-flop array will be determined, as shown in Figure 7:

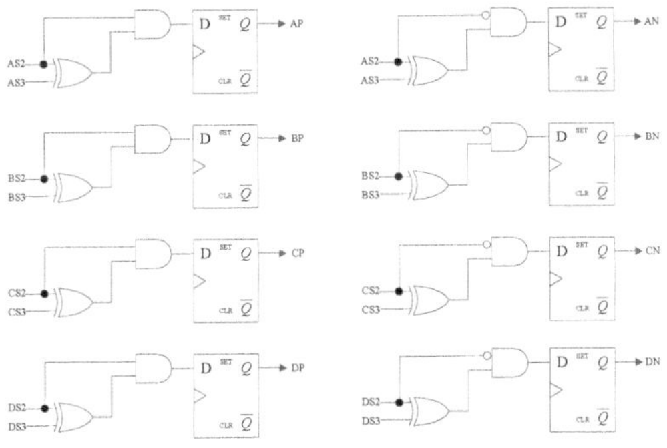

Fig. 7. Data Decision Diagram

a. AP=BP=CP=DP=1, or AN=BN=CN=1.Time domain A was the first to see the data.
b. AP=1 and BP=CP=DP=0, or AN=1 and BN=CN=DN=0.Time domain B was the first to see the data.
c. AP= BP=1 and CP=DP=0, or AN= BN=1 and CN=DN=0.Time domain C was the first to see the data.
d. AP= BP=CP=1 and DP=0, or AN= BN=CN=1 and DN=0.Time domain D was the first to see the data.

C. Choice

The data will be determined and sent to the output by multiple data selection device, as shown in Figure 8:

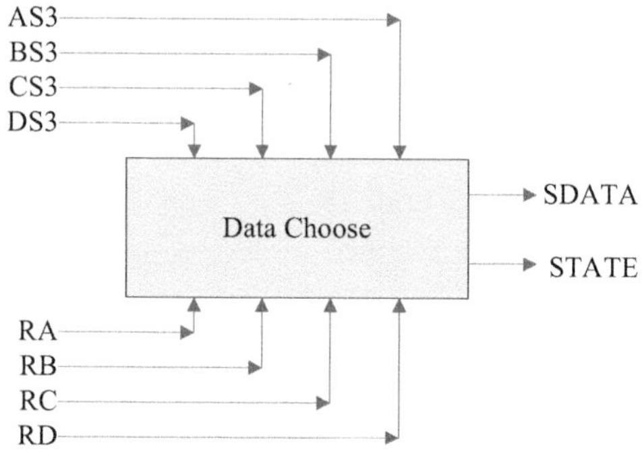

Fig. 8. Data Choose Diagram

As said previously, the local clock will be either a bit faster or slower than the incoming data stream, therefore, there will be 0, 1 or 2 signal transitions at the D flip-flop stage, and so 0, 1 or 2 bits of data will be received in one clock cycle. Therefore, a status signal (STATE) is needed to represent the following device.

The data received will be transmitted from serial data to parallel data (E.g.8 bits),the device will received 7 ,8 or 9 bits data in a period of time. For this a 9-bit shift register is needed.

Acknowledgment. This paper presents a new kind of high-speed asynchronous serial transmission scheme based on 8b10b and introduces the basic principle of two major parts: 8b10b and data recovery.

References

1. Culler, D.E., Singh, J.P., Gupta, A.: Parallel Computer Architecture: A Hardware/Software Approach, 2nd edn. China Machine Press (2002)
2. Noergaard, T.: Embedded Systems Architecture: A Comprehensive Guide for Engineers and Programmers. Posts & Telecom Press (2008)
3. Gajski, D.D., Vahid, F., Narayan, S., Gong, J.: Specification and Design of Embedded Systems. China Machine Press (2005)
4. Li, B.: Design for Reliability of Embedded Systems. Publishing House of Electronics Industry (2006)
5. Budruk, R., Anderson, D., Shanley, T.: PCI Express System Architecture. Publishing House of Electronics Industry (2005)
6. Sawyer, N.: Data Recovery. XAPP224 (v2.5), Xilinx (July 11, 2005)
7. Information on `http://www.ti.com.cn`

Research on Coordinated Evaluation of Regional Energy-Economy-Environment System*

Heshan Guan[1], Shuliang Zou[1], Xiaodong Zhou[1], and Zitong Huang[2]

[1] School of Economics and Management, University of South China, HengYang, China, 421001
[2] School of Politics and Public Administration, University of South China, HengYang, China, 421001
h.s.guan@tom.com

Abstract. The coordinated development of regional energy- economy-environment system is of vital important work for government management. This paper builds the coordinated evaluation way of regional energy-economy-environment system, which could depict the hierarchical structure composed of 19 indexes effectively, and then get some patterns of coordinated development for different districts in China. Group-ward clustering is used to group these patterns into some clusters. The experiment results show that our study can provide some useful decision gist for strategy making of government. The approach is to be a practical and effective method used to evaluate the coordinated degree of regional energy-economy-environment system.

Keywords: system, coordinated degree, clustering.

1 Introduction

Energy is becoming an important material foundation in the survival and development of human society. As we all know, with the rapid development of country's economy and rapid industrialization process in recent decades, the demand for energy increases year by year, but energy supply can't satisfy the demand for energy to some extent, which is a very serious problem for government managers. At the same time, environmental pollution has become a main obstacle of economic development and a threat to human living conditions. Therefore the coordinated development issue of energy, economy, and environment has become one of hot issues to governments, research institutions, experts and scholars.

Energy, economy and environment interact with one another and mutually promoted, coordinated evaluation of Energy-Economy-Environment (3E) system is a quantitative description of the relationship, which is very important for 3E system to ensure its healthy and sustainable development, government managers could usually make some useful decision based on the results of coordinated evaluation. Among the process of coordinated evaluation of 3E system, the coordinated degree model is its

* This work is partially supported by Philosophy and Social Sciences Foundation of Hunan Province (2010ZDA02) , Foundation of Hengyang science Department (2010KR05).

D. Zeng (Ed.): ICAIC 2011, Part II, CCIS 225, pp. 593–599, 2011.

key element. In general we could use some coordinated degree model consisted of some functions to measure the coordinated situation among the 3E system.

At present, there existing an amount of coordinated evaluations of system, For example, Q.C. Han built a coordinated model of water environment-economy-population composite system base on the multi-objects optimal thoughts and synergetic algorithm[1]; Z.S.Liu built a Evaluation model of urban-rural employment by upgraded grey thesis for appraisement, and then give out a qualitative analysis to Shaanxi province[2]; T.Zhao redefined the quantitative expression of system coordination and builds a set of coordinated evaluation models for 3E system[3]. The amount of existing evaluation researches on coordinated degree of system is large, mainly including distance type model of coordination degree, change-type model of coordination degree and integrated-type model of coordination degree[4]. Although these methods adopted by the experts and scholars have solved the evaluation problem of the coordinated degree of system to some extent, while the classification of coordinated development patterns is nearly little. Therefore, this paper firstly constructs the 3E coordinated degree evaluation index system, secondly build the coordinated degree function using hierarchical comprehensive evaluation method based on development rapid of systems, and finally use group-ward clustering to classify the patterns of coordinated development. This will not only assess the various layers of subsystems of complex system, but also carry out comprehensive assessment based on an assessment of subsystem, especially the coordinated development pattern has gotten more attention in this paper.

The remainder of this paper is organized as follows: In section 2 we discuss briefly the relevant knowledge about the normalization methods of data, and then built an coordinated degree mode based on development speed of systems, finally we use group-ward clustering to classify the patterns of coordinated development; we utilize a natural dataset to test the approach, analyze their results and give out some discussion in section 3; in section 4 we summarize this paper

2 Coordinated Development Patterns

A. Normalization of Original Data

This section mainly describes a strategy for standardization. Generally speaking the coordinated degree function is an ideal way for the synergy evaluation between indexes of system, and the standardization strategy described in this paper, which is to avoid dimensions' influence of their order of magnitude, is intended to be a basis for coordinated evaluation.

Let x_{ijt} be the jth index of ith sub-system, at some particular time t. We translate original data as formula (1)

$$x'_{ijt} = \frac{x_{ijt}}{\max_t \{x_{ijt}\}} \tag{1}$$

We will use the translated data for the following study.

B. Coordinated development degree

After we confirm the standardization way, we should pay more attention to build the coordinated evaluation mode consisted of some functions.

Firstly we try to build a coordinated development degree for each sub-system as formula (2)

$$x_{it} = \sum_{j=1}^{n_j} x'_{ijt}\omega_{ijt} \tag{2}$$

Where i=1,2,3, n_j represents the total number of indexes of the jth sub-system. The weight ω_{ijt} in multi-index system is just the parameter needed to decide previously. However there are already some ways that existed for weight choice, but these ways are subjective to some extent. Therefore we will briefly analyse data with equal weight in this paper. Furthermore we can also get the whole development degree of 3E system with equal weight as formula (3)

$$x_t = \frac{x_{1t} + x_{2t} + x_{3t}}{3} \tag{3}$$

Secondly, we will define the coordinated degree of sub-system and 3E system respectively based on development ratio[3]. The definitions of coordinated development degree for sub-system were given out as formula (4)

$$H_i = \begin{cases} \exp\left(\frac{dx_{it}}{dt} - \frac{dx_t}{dt}\right) \text{ if } \frac{dx_{it}}{dt} < \frac{dx_t}{dt} \\ 1 \text{ if } \frac{dx_{it}}{dt} < \frac{dx_t}{dt} \\ \exp\left(-\frac{dx_{it}}{dt} + \frac{dx_t}{dt}\right) \text{ if } \frac{dx_{it}}{dt} > \frac{dx_t}{dt} \end{cases} \tag{4}$$

Where i=1,2,3, $\frac{dx_{it}}{dt} = \frac{x_{i,t} - x_{i,t-1}}{x_{i,t-1}}$, $\frac{dx_t}{dt} = \frac{x_t - x_{t-1}}{x_{t-1}}$, and then we define the coordinated development degree for 3E system as formula(5)

$$H = \sqrt[3]{\prod_{i=1}^{3} H_i} \tag{5}$$

Now we could quantify the coordinated development degree of 3E system effectively with formula (5), which will be used for the experiments in section 3.

C. Group-ward Clustering

For some district, after calculate the coordinate degree year by year, we can get a corresponding time series. Of course it is easy to get a set of time series after calculating the coordinate degrees for all districts in China (there are 29 districts used for analyse in this paper), and each time series represents a kind of coordinated pattern respectively. Thus we try to classify the coordinated patterns using group-ward clustering, which can group the dataset of time series into "clusters", such that those within each cluster are more closely related to one another than objects assigned to different clusters.

J.H.Ward[5] proposed a clustering procedure seeking to form the partitions Pn, P n-1,........, P1 in a manner that minimizes the loss associated with each grouping, and to quantify that loss in a form that is readily interpretable, the procedure is called group-ward clustering in this paper. At each step in the analysis, the union of every possible cluster pair is considered and the two clusters whose fusion results in minimum increase in 'information loss' are combined. This paper will use group-ward clustering to group the coordinated patterns.

3 Experiments

We collect a dataset consisted of 30 districts in China between 2005 and 2008, excepting for XiZang district, because we can't get its relevant data of energy, and then use the above approach to analyze the data. The article firstly builds the coordinated degree evaluation indexes of 3E system composed of three modules, namely energy coordinated degree, economy coordinated degree and environment coordinated degree, Table 1 shows the evaluation modules and indexes of 3E system.

Table 1. Coordinated degree evaluation index system of regional 3e system

Module	Indexes
Energy coordinated degree	Total Electricity Consumption
	Energy Consummated by Per Unit of GDP
	Electricity Consummated by Per Unit of GDP
	Energy Consummated by Per Unit of Industrial Added Value
Economy coordinated degree	Total Output Value
	Residents' Consumption Level
	Total Capital
	Living Expenses
Environment coordinated degree	Total Cleaned Domestic Garbage
	Amount of Garbage Treatment Plants
	Capacity of Garbage Innocent Treatment
	Amount of Garbage Innocent Treatment
	Innocent Treatment Ratio of Domestic Garbage
	Amount of Waste Gas Treatment Plants
	Amount of Waste Water Treatment Plants
	Amount of Industrial Waste Water
	Amount of Domestic Waste Water
	Amount of Solid Waste
	Ratio of Nature Reserves Area to Total Area

According to the above regional 3E system coordinated degree evaluation index system and the previous coordinated development mode. We measure the coordinated degree of 30 provinces in China respectively, and omit the calculation process. Now, the final calculation results are given in table 2.

As we can see from table 2, although there are obvious difference between any two districts in terms of the values of coordinated development within the three years, but the extreme difference of coordinated degree of 3E system is 0.13(the biggest value is 0.97 and the smallest value is 0.84), which show that the coordinate development of different subsystem of 3E system is generally to be similar. However sometimes we would pay more attention to the coordinated development pattern of each district limited to a specified time period rather than the actual value of coordinated degree. In this paper we use SAS software[6] to classify the coordinated patterns into some clusters with group-ward clustering.

Table 2. Evaluation results and order of 3e system coordinated degree of 30 provinces in china between 2006 to 2008

Module Year/ ID districts		Economic Coordinated Degree			Energy Coordinated Degree			Envrionment Coordinated Degree			Coordinated Degree of 3E system		
		2006	2007	2008	2006	2007	2008	2006	2007	2008	2006	2007	2008
1	BeiJing	0.89	0.91	0.92	0.93	0.91	0.95	0.98	0.98	0.97	0.94	0.93	0.95
2	TianJin	0.89	0.91	0.82	0.95	0.93	0.88	0.98	0.99	0.95	0.94	0.94	0.88
3	HeBei	0.9	0.94	0.87	0.96	0.91	0.91	0.96	0.95	0.96	0.94	0.93	0.91
4	ShanXi	0.94	0.93	0.88	0.91	0.93	0.88	0.92	0.98	0.98	0.93	0.95	0.91
5	NeiMengGu	0.88	0.89	0.83	0.95	0.91	0.9	0.97	0.99	0.95	0.93	0.93	0.89
6	LiaoNing	0.92	0.92	0.78	0.95	0.9	0.91	1	0.95	0.89	0.96	0.92	0.86
7	JiLin	0.76	0.93	0.82	0.97	0.82	0.93	0.82	0.82	0.89	0.85	0.86	0.88
8	HeiLongJiang	0.88	0.88	0.84	0.99	0.91	0.94	0.92	0.99	0.89	0.93	0.93	0.89
9	ShangHai	0.95	0.91	0.91	0.92	0.89	0.95	0.95	0.95	0.96	0.94	0.91	0.94
10	JiangSu	0.91	0.9	0.86	0.94	0.93	0.92	0.99	0.99	0.95	0.95	0.94	0.91
11	ZheJiang	0.92	0.93	0.89	0.94	0.92	0.95	1	0.97	0.95	0.95	0.94	0.93
12	AnHui	0.94	0.97	0.85	0.92	0.85	0.93	0.93	0.83	0.93	0.93	0.88	0.9
13	FuJian	0.87	0.95	0.86	0.98	0.89	0.93	0.92	0.89	0.94	0.92	0.91	0.91
14	JiangXi	0.93	0.96	0.86	0.93	0.87	0.93	0.97	0.88	0.94	0.94	0.9	0.9
15	ShanDong	0.9	0.94	0.85	0.94	0.9	0.93	1	0.94	0.93	0.94	0.93	0.9
16	HeNan	0.85	0.97	0.86	0.98	0.88	0.94	0.91	0.88	0.93	0.91	0.91	0.91
17	HuBei	0.81	0.9	0.86	0.96	0.9	0.91	0.83	0.96	0.94	0.86	0.92	0.9
18	HuNan	0.9	0.95	0.82	0.95	0.88	0.9	0.98	0.9	0.93	0.94	0.91	0.88
19	GuangDong	0.93	0.96	0.87	0.94	0.88	0.94	0.99	0.88	0.94	0.95	0.91	0.92
20	GuangXi	0.88	0.93	0.87	0.95	0.87	0.92	0.96	0.9	0.96	0.93	0.9	0.91
21	HaiNan	0.85	0.95	0.84	0.99	0.9	0.95	0.88	0.92	0.9	0.9	0.92	0.89
22	ChongQing	0.92	0.9	0.74	0.97	0.9	0.86	0.97	0.95	0.92	0.95	0.92	0.84
23	SiChuan	0.89	0.92	0.86	0.94	0.88	0.92	0.98	0.93	0.94	0.94	0.91	0.91
24	GuiZhou	0.96	0.91	0.85	0.96	0.92	0.96	0.99	0.99	0.9	0.97	0.94	0.9
25	YunNan	0.88	0.98	0.95	0.99	0.87	0.99	0.88	0.82	0.96	0.92	0.89	0.97
26	ShanXi	0.91	0.86	0.84	0.9	0.89	0.88	0.93	0.99	0.98	0.92	0.91	0.9
27	GanSu	0.88	0.95	0.86	0.96	0.9	0.91	0.95	0.91	0.96	0.93	0.92	0.91
28	QingHai	0.93	0.86	0.88	0.99	0.99	0.91	0.96	0.89	0.98	0.96	0.91	0.92
29	NingXia	0.91	0.89	0.79	0.99	0.95	0.9	0.93	0.97	0.91	0.94	0.93	0.87

In Fig. 1 we can see the dendrogram of the hierarchical cluster tree of the coordinated patterns by using group-ward clustering. Usually the choice of the number of clusters in the data is sometimes subjective and depends on the researcher experience. However, we could choose the number of clusters with the values of six statistic parameters, or rather, the statistic parameters SPRSQ, ESQ, ERSQ, CCC, PSF and PST2 of group-ward clustering of SAS software. The group-ward clustering could split the series data into three clusters: C1={1,9,25,30}, C2={7,13,16,17,21,26} and C3={2,4,5,6,8,9,10,11,12,14,15,18,19,20,22,23,24,27,28,29}. Generally speaking, group C1 includes the decreasing firstly and then increasing sectors; group C2 includes the steady sectors; group C3 includes the decreasing sectors.

Aapproximately steady pattern is usually considered as the healthy and sustainable development of 3E system, that is to say the development pattern of group C2 is more reasonable than that of C1 and C3. Thus Government managers should pay more attention to the districts of Group C1 and Group C3, and take some useful practice to promote these districts to develop steadily, for example, fast increasing investment in energy supply, along with improving environment may be useful measure for government's decision.

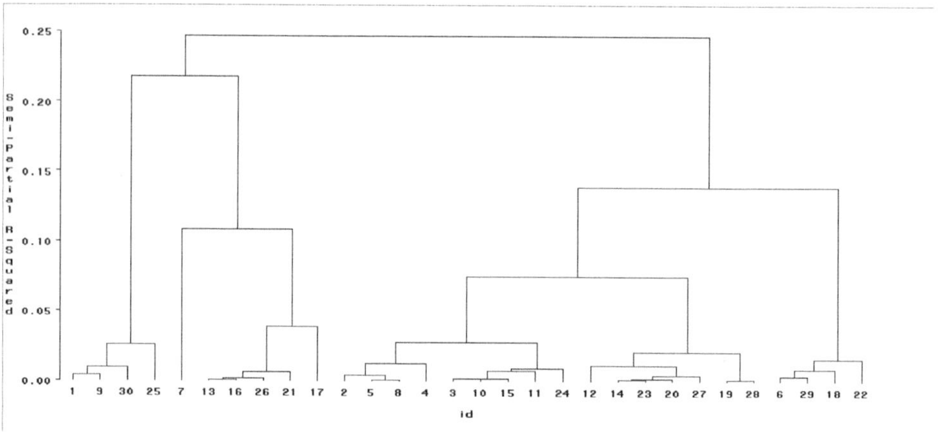

Fig. 1. The hierarchical cluster tree

4 Conclusions

The paper mainly builds the coordinated evaluation method for energy-economy-environment system and then use group-ward clustering to group the coordinated development patterns. Through the above empirical analysis it shows that the approach presented in this paper could provide a new thought to assess the 3E system. Evaluation results can reflect the situation of the regional energy-economy-environment system coordinated development and find out the characteristic of these coordinated patterns of 3E system development in China, which is very useful for government managers.

References

1. Han, Q.C., Xie, J.C., Wang, X.H., Si, Y.M., Pan, Y.: The Water Environment System Coordinated Degree Model and its Application on Xiao-Hua Region in the Yellow River. In: 2006 International Conference on Hybrid Information Technology (2006)
2. Liu, Z.S., Su, D.W., Pan, L.L.: Research on Coordinated Urban-rural Employment Evaluation Based on Index Grey Clustering. In: Proceedings of 2007 IEEE International Conference on Grey Systems and Intelligent Services (2007)

3. Zhao, T., Li, X.Y.: On the Coordinated Evaluation Model for Energy-Economy-Environment System. Jurnal of BeiJing Institute of Technology 10(2), 11–16 (2008) (in Chinese)
4. Wang, G.Z., Liu, J.: Research on evaluation of coordinated degree of regional energy-economy-environment system based on AHP-GRAM model. In: Proceedings of the Ninth International Conference on Machine Learning and Cybernetics (2010)
5. Ward, J.H.: Hierarchical Grouping to Optimize an Objective Function. Journal of the American Statistical Association 58, 236–244 (1963)
6. Gao, H.X.: SAS System SAS/STAT software guideline. Chinese Statistic Pressing (1997)

Design on Intelligent Controller of Magnetic Controlled Soft Starting for Reactive Compensation

Siyun Tan, Zhian Zhang, Jing Chen, Youxin Yuan, and Gang Chen

School of Automation
Wuhan University of Technology
Wuhan, Hubei, 430070, China
SiyunTan030@163.com

Abstract. Starting current is too large when the motor is started directly. The motor consumes large amounts of reactive power when it is connected to the electrical network. An intelligent controller of magnetic controlled soft starting is developed for reactive compensation. The controller used the 16-bit MSP430F1612 MCU as a control core. An algorithm with measuring phase angle calculated the power factor, active and reactive power, and could realize auto-tracking of reactive compensation and auto-switching of shunt capacitors. The paper put forward hardware structure and software design in detail. Experimental test demonstrates that the control system can dynamically adjust the output of reactive power compensation equipment, and effectively improve the system power factor, and reduce the motor starting current, so as to achieve the protection of the motor and equipment.

Keywords: soft starting, variable reactor, reactive power compensation, capacitors.

1 Introduction

With the continuous development of national economy, three phase asynchronous motors are widely used in metallurgy, petrochemical, power plants, building materials, coal and machinery processing and other industries. However, large-scale motor also caused a series of problems, such as starting impact, starting torque, low power factor and so on. When the motor is started directly with full voltage, the line current of the motor can be 5-7 times larger than that in the normal operation and the starting torque can be 2 times larger than that in the normal operation [1]. These will lead to a range of issues, which includes that the big starting current influences the normal operation of other equipments connected in the same power line, and even causes the electrical network to lose stability, and will be in the form of electromagnetic waves to interfere with the normal operation of electrical instruments [2]. The big starting torque can cause big mechanical impulse which will accelerate attrition on mechanical transmission components. The big starting current

D. Zeng (Ed.): ICAIC 2011, Part II, CCIS 225, pp. 600–607, 2011.

also can generate massive joule heat and damage the winding insulation, which reduces the use life of motor. Currently, most soft starters are suitable for soft starting of low-voltage motor. However, there are various methods of soft staring for high voltage motor, such as magnetic controlled soft starting [3-4], thyristor series soft starting [5-6], frequency converter soft starting [7] and so on. Thyristor series soft starting limits the scope of its application because it is difficult to withstand a high voltage for thyristor. Starting performance of the frequency converter soft starting device is well, but the characteristic of the device is high price and low reliability. Thyristor is used to control the voltage or current of low voltage winding by magnetic controlled soft starter, and then continuously change the voltage or current of high voltage winding. The voltage of motor is continuously changed to achieve the soft starting of motor.

Because of extensive use and long time running of three phase asynchronous motors, and they consume large amounts of reactive power in the running process. An additional reactive compensation cabinet is needed, which makes the device less concentrated, and too much space or excessive costs are likely to be increased. Therefore, a design on controller with integrated feature of reactive compensation and soft starting which is of high reliability, small size and low cost is put forward in this paper.

2 Hardware Design of Controller

Structure diagram of the controller of magnetic controlled soft starting of reactive compensation is shown in Fig. 1. Variable reactor is designed as linear transformer structure, which isolates high voltage and low voltage. The primary coil of variable reactor is connected with stator of asynchronous motor in series to form series circuit of the primary impedance. The secondary coil of variable reactor is connected with a pair of counter-parallel thyristors in series to form converting circuit of the secondary impedance. Before thyristor is conducted, the primary coil impedance of variable reactor is in the maximum value, so the voltage of motor is in the minimum value. When the thyristor is cut-off, the motor can be started for the designed variable reactor. When the motor is started, the firing angle of thyristor is controlled to continuously change the voltage of secondary coil in the variable reactor, then to change the voltage of primary coil in the variable reactor to achieve the purpose that continuously changing the terminal voltage of the motor, so that the motor can be started smoothly. When the soft staring is over, the switch K is closed and the motor run in rated speed. Through the sensor to detect the size of three-phase voltage, current and their phase, then control algorithm calculates the size of the power factor and the reactive power, and according to their size to reasonably control the switching of capacitors in the process of the starting and the running. Finally, the reactive power of the system is compensated.

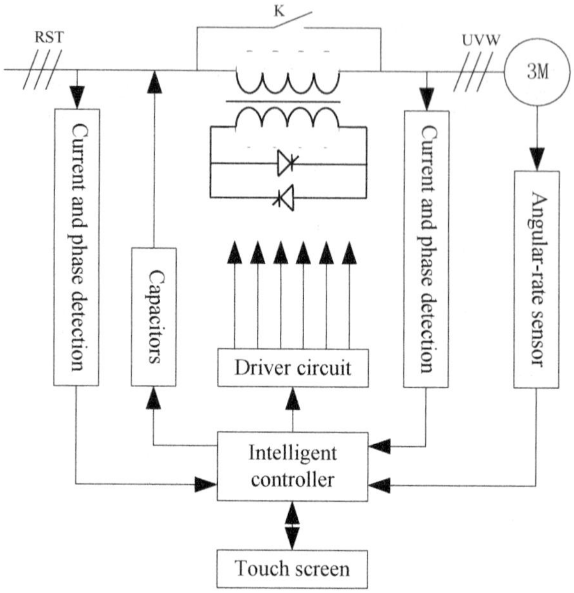

Fig. 1. Structure diagram of the controller of magnetic controlled soft starting for reactive compensation

A circuit board with large plate structure is used in the controller, which reduces its volume and improves its reliability. The hardware system block diagram is shown in Fig. 2.

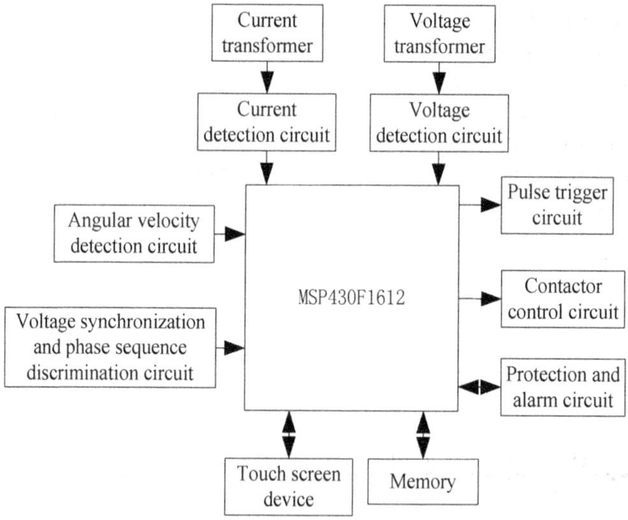

Fig. 2. Hardware system block diagram

The modular design is used in the hardware circuits, which can be divided into measurement module, memory module, communication module, pulse trigger module, protection alarm module, switching control module and power module. Power module provides power to the entire control system after transformer, rectifier and voltage regulator. Measurement module transforms the signals that are collected by the sensors into the others that are received by the MCU. Before the signals are transformed, they must be decoupled, filtered, phase adjusted and magnified. Main function of measurement module is collecting line current, line voltage, phase and angular velocity. Memory module is mainly used to record system data which include alarm actions data, per phase voltage, per phase current, reactive power, active power, power factor, switching time and switching times of capacitors and so on.

Protection alarm module is used to identify system faults (short circuit of motor, over current, phase loss, radiator overheating, etc) and protect system. Switching control module is used to control the switching of capacitors and the switching of main circuit (circuit breakers, contactors).

Communication module is used to communicate and transfer data with host computer by RS-485 and CAN communication bus. The soft-start/soft-stop parameters of motor and reactive compensation parameter can be installed by touch-screen device. It can also receive system data. The touch-screen device can display the information of system status and parameters so as to monitor the status of the motor.

Phase-shift trigger based on MCU is used in the digital trigger. The module includes A/D converter, voltage synchronization and phase sequence circuit, pulse forming, optical coupler isolation, pulse transformers etc. Opto isolation circuit is used for all digital inputs/outputs so as to improve system reliability.

Reactive compensation control unit is mainly composed of tuning reactors, contactors, compensation capacitors etc. The capacitor may be combined with the inductance of electrical network, which can constitute parallel resonant circuit for one frequency harmonic. The circuit can make the harmonic severely enlarged, and affect system stability. Or series resonant circuit can be constituted for one frequency harmonic. The circuit can make the harmonic current to be overloaded, and damage the capacitors. Therefore, the tuning reactors are needed to connect the capacitors in series, and restrain harmonic. They also can restrain impulse current which is generated while the capacitors switch is closed. They can protect the switch. The diagram of reactive compensation control unit is shown in Fig. 3.

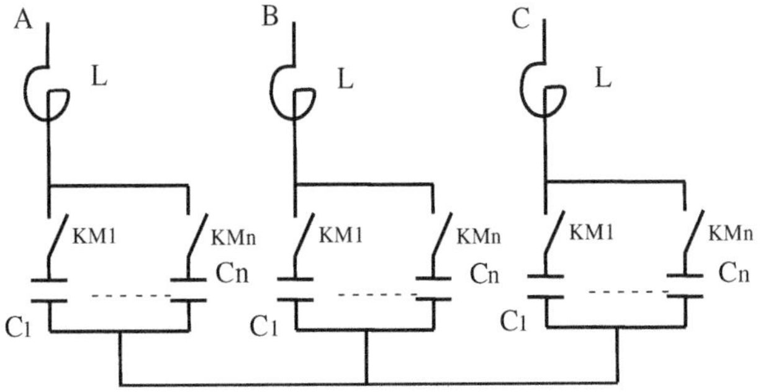

Fig. 3. Reactive compensation control unit

3 Software Design of the Controller

Controller software is designed as modular programming. Each functional module is designed as independent programming debugging block, which is not only good for developing the system function in the future, but also in favour of debugging and connecting the system, and is good for transplantation and modification of the module program. System software modules consist of measurement module, data storage module, switching control module, soft-start module, soft-stop module, fault detection module and communication module. The structure diagram of control system software is shown in Fig. 4.

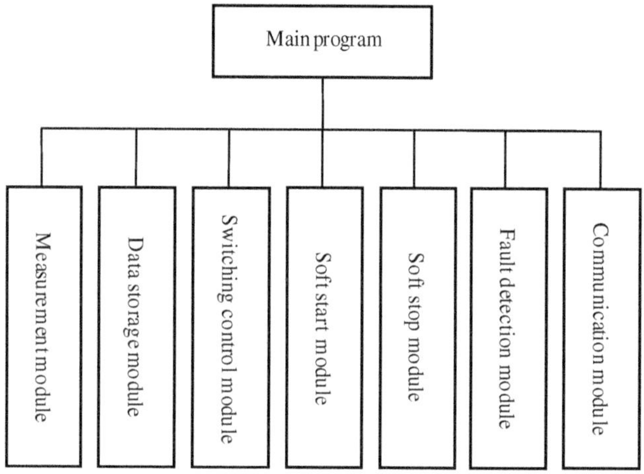

Fig. 4. The structure diagram of control system software

Main program controls the entire process of software system, and initializes each module. Measurement module completes collection and calculations of raw data which are needed by the controller. It calculates the line reactive power, power factor and other parameters for follow-up control according to the data that is collected from the line voltage and line current. Switching control module based on data which are obtained in real time controls the switch of capacitors. Data memory module is used to storage system data.

Soft-start module adjusts the firing angle of thyristor according to selecting starting mode and parameters, and continuously changes the voltage of low-voltage winding, and thus changes the voltage of high-voltage winding to achieve the purpose that continuously changing the terminal voltage of the motor, then achieves the soft staring of the motor. Soft-stop module adjusts the firing angle of thyristor according to selecting stopping mode and parameters, and continuously changes the voltage of the motor to achieve the soft stopping of motor. When the controller detects a system failure, alarm subroutine can control voice alarm. If a serious fault is emerged, the motor will be immediately stopped.

Communication module is used to control data transmission between the touch screen device and MCU. System parameters can be set through the touch screen. We can choose the model of soft starting of the motor (voltage slope starting, constant current starting, sudden jump pulse and voltage slope starting), compensation, the width of trigger pulse, min/max of the firing angle, start-up time, limited current of starting and so on. The process parameters of the system can be monitored by display screen when the motor is starting and running. These parameters include the start-up current of motor, angular speed, size of firing angle, reactive power, power factor, the switching status of capacitors, information of system alarm, the status information of controller hardware and so on. The flow chart of software main program is shown in Fig. 5.

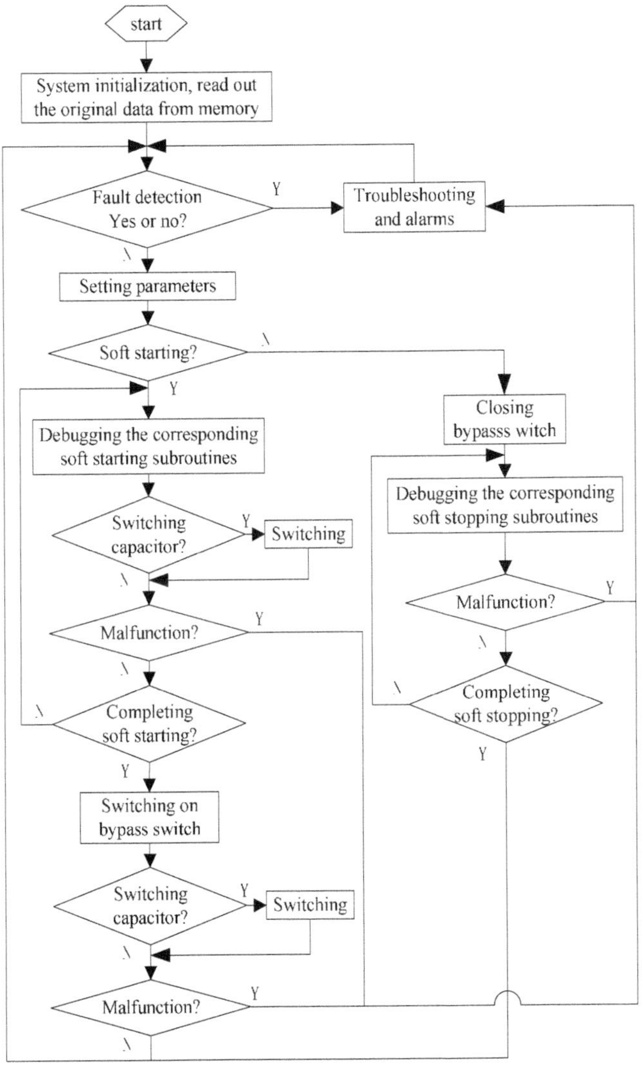

Fig. 5. The flow chart of software main program

4 Experimental Results

The controller, variable reactor and capacitors are used to design a control cabinet of magnetic controlled soft starting of reactive compensation in our laboratory. We also equip three-phase asynchronous motor with rated voltage 380V, rated current 23A, power 11KW as an experimental hardware environment. The control cabinet and the motor are connected together. After the controller powers up, the entire system parameters can be sited by man-machine interface. The motor can also be started by using the system default. The default mode of the slope soft starting is used to start the motor in the test, and the control effects of the controller are observed by starting the motor frequently. Experiments certify that the motor acceleration is slow and smooth. After the set time is reached, the speed of motor is in the maximum value. The soft starting is completed in this moment. The continuous changes of the motor soft starting current and voltage and the size of power factor are monitored in the display interface. The maximum starting current of the motor is limited to 3 times larger than rated current. The size of power factor is about 0.95. The controller of magnetic controlled soft starting for reactive compensation avoids that electrical network is impacted by the high current which is produced when the motor directly starts with full voltage. At the same time, it also reduces the impact injury of the motor when the motor is started. It compensates reactive power, and improves power factor, thus the power is saved.

5 Conclusion

The experiment shows that we can gain some obvious effects after using the controller in the three phase asynchronous motor, such as compensating reactive power, improving power factor, reducing the size of starting current and the damage of the motor and equipment. Apart from these, the production costs can be reduced and productivity can be improved. The application illustrates that the design of the controller of soft starting for reactive compensation is very reliable, and has high integration of the system hardware, compact structure, high performance price ratio and many other advantages. However, the harmonic is not completely eliminated by the control system. The presence of harmonics has a certain influence on the system. In the future, we will further study the method of harmonics suppression.

References

1. Yuan, y., Zhao, y., Ding, y., Zhang, y.: Study on AC motor soft starter based on fuzzy control. Electric Drive 36(5), 42–45 (2006)
2. Zhang, y.: Discussion the soft starting of asynchronous motor. China New Technologies and Products 18, 84 (2008)
3. Yuan, y., Wang, y., Pen, w., Yan, j.: The study of high-voltage soft starter based on variable reactor. Electric Power Automation Equipment 27(11), 38–41 (2007)

4. Gao, y.: Magnetic controlled soft starting of large-capacity motor. Journal of Wuhan University of Science and Technology(natural science edition) 28(1), 87–89 (2005)
5. Gan, s., Chu, j., Gu, w., Yu, y.: Based on SCR in series for AC motor soft start. Journal of Wuhan University of Technology (transportation science & engineering) 34(3), 520–523 (2010)
6. Wang, x., Xu, p., Duan, j., Liang, h.: Design electronic soft starter of thyristor based on MCU. Journal of Liaoning Technical University(natural science) 29(3), 447–450 (2010)
7. Ding, x., Jin, d.: The problems and solutions of variable frequency soft starting about AC motor. Power Electronic Technology 35(5), 1–3 (2001)

An Enhanced Study Instrument for Distance Education

Yancheng Zhu[1] and Meige Wang[2]

[1] College of Science, Wuhan University of Science and Technology, Wuhan, 430081, China
[2] School of Foreign Language, Hubei Normal University, Huang Shi, 435002, China
86400@home.news.cn

Abstract. The improving of new learning instrument for distance education in smart sensor systems, which for the first time will be based on an organic joining of all increased teaching approaches: e-studying and project-based learning is described in the paper. The project-based e-learning is based on novel, innovation research results obtained in the frame of FP6 program Marie Curie Chairs Excellence during the project Smart Sensors Systems Designs.

Keywords: education technique, distance learning, smart sensor system.

1 Introduction

The use of synchronous computer mediated communication in education brings some unique benefits. Small groups of students can be combined to make a viable class or expertise can be made available at different sites. When students are on campus they may be able to contact industry mentors and while on practicum keep in touch with teaching staff.

Travel expenses can be reduced as meetings or classes can take place virtually. There are also some unique problems. The logistics of arranging suitable meeting dates and times with appropriate partners can be daunting. When working in an international setting this is compounded by differences in time zones. Narrow bandwidth and lack of reliability can be major technical hurdle.Perhaps the major problem is that the interaction provided by using expensive technology may simply be the same occasional question. This paper looks at some of the strategies for increasing participation and interactionusing these technologies.

It is important to structure synchronous sessions in order to gain maximum participation. Components that can be adjusted include group size and type, duration, scheduling and outcome of task. In order to prevent anti-social outbursts it is he helpful to establish and publish rules for good'netiquette'. The moderator needs to model considerate behaviour as well as actively enforce the rules. To ease anxiety and enhance communication it is important to build a climate of trust and safety that encourages collaboration.

D. Zeng (Ed.): ICAIC 2011, Part II, CCIS 225, pp. 608–613, 2011.

2 Methodology and Pedagogical Approach

The e-learning is becoming one of the most popular and progressive forms of microelectronics education[2, 3]. However, often a prepared on-line course is integrated into a distance learning system; learners obtain an access, passed this course but teaching do not wishful result. Students or engineers have acquired the minimal volume of course material, have wasted time on training, but the strategic purposes are not reached.

What is the reason? The course has not been well methodically designed. The basic attention has been given to a technical realization and modern Internet and multimedia technologies, but not to a pedagogical approach. When beginning to create e-learning content, the pedagogical approaches need to be evaluated carefully. Simple pedagogical approaches make it easy to create content, but lack flexibility, richness and downstream functionality. On the other hand, complex pedagogical approaches can be difficult to set up and slow to develop, though they have the potential to provide more engaging learning experiences for learners.

Somewhere between these extremes is an ideal pedagogy that allows a particular educator to effectively create educational materials while simultaneously providing the most engaging educational experiences for students. First of all it is necessary to define precisely the objectives, which will be reached during the distance learning. Fussy objectives can lead to that accents in the course will be placed incorrectly and as a result, a decrease in a learning efficiency will be observed. Well understandable aim is a strong motivating factor, but it is also very important that learners should clear understand how they can demonstrate and use the obtained knowledge. Individual projects in the frame of PBL based on a distance access to appropriate remote laboratory will be the best solution.

3 Computer Modeling and Simulation

In order to help learners (attendees) to understand modern recently developed measuring methods an appropriate server side modeling software should be designed and run at the server in the frame of advanced developed pedagogical and methodological approach.

Presentation of the operation principle of modern measuring methods, integrated converters and measuring devices on its basis and their properties in the classical way is time consuming and poses difficulties for the learners because of the need to present considerable amounts of information in the different form. To illustrate the problems related to frequency (period)-to-digital converters functioning, many schemes, time courses of signals and characteristics describing relevant properties of the system are needed and have to be adapted for each new value of the parameters of the input signal or those of the path of the converter. A convenient solution is to be the application of integrated programming environments for simulation of the measuring systems.

In particular, the LabVIEW environment is to be commonly used at teaching process for realization of specific programs for presentation of various problems. The analysis of the converter operation and the effects of different parameters on its performance will be conducted in the interactive mode. An availability of special means and the built in opportunities, such as Internet server, Web Publishing Tool,LVRunTimeEng, LabVNC, etc., make the LabVIEW by the irreplaceable tool for use in the e-learning process (PBeL).

The LabVIEW environment for computer simulation of the operation principle and testing of modern frequency (period) to digital conversion (based on the method of the dependent count) similar to simulation of ADC principles described in [4] will be designed. Front panels should contain graphical objects, permitting interactive communication with the user. Animation effects allow illustration of the methods operation principle. The application permits performance of virtual experiments due to the possibility of control of converters parameters and the input signals parameters.

4 Developed Resources

Hardware and software resources for the proposed Project Based e-Learning (PBeL) conception based remote laboratory are shown in Figure 1. In addition to the computer modeling and simulation software, the online lecture course will be developed and stored at the server. As the environment could be used in both professional and educational audience, a sophisticated tutorial system must be developed. The online lecture course will cover the following main topics:

Smart and Quasi-Digital Sensors Sate of the Art;
Data Acquisition Methods for Sensor Systems;
Classical, Advanced and Self-Adapted Frequencyto-Digital Conversion Methods;
Digital Sensors and Program-Oriented Conversion Methods;
Smart Sensor Systems;
Sensor Buses, Protocols and Networks;

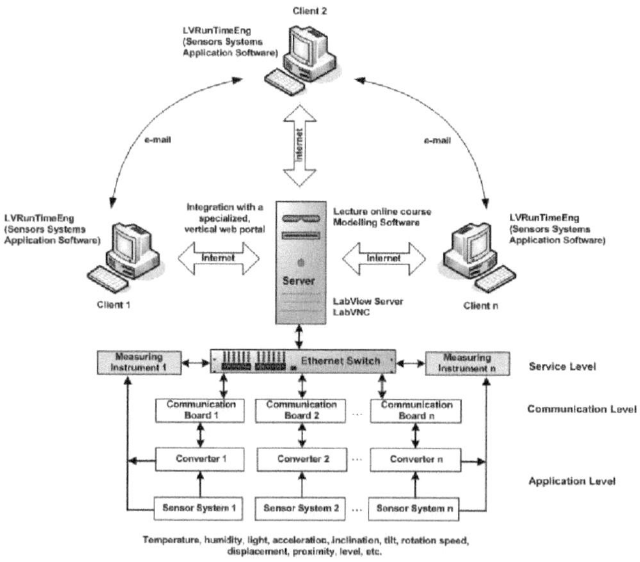

Fig. 1. Project based e-learing conception based remote laboratory

At the end of each course module, after passing the tutorials, a series of tests should be implemented, in order to check the level of knowledge.The LabView server (National Instruments) and LabVNC software will be also run at this server. The server realizes tasks of administration and user authentication; as well it allows learners to interact with the experimental equipments. Clients' computers will run different application software in web browsers

Remote laboratories are a rapidly growing innovation trend in distance engineering education.The mass proliferation of the Internet, together with the ever-increasing user-friendly computer controlled instrumentation, have favored the development of such remote laboratories with interactive experiments in the frame of PBeL conception. The remote laboratory's hardware should include standard equipments (Fig.1,)in grey color) such as a server, Ethernet switch or USB-to-LAN Hub, measuring instruments (universal frequency counter, functional generators, oscilloscope, etc.) with LAN communication ports (Service Level); and non standard equipments (in white color) such as universal communication boards with SPI or RS232 to Ethernet bridges (Communication Level) for communication and actuation purposes; and universal sensors and transducers converter and interface boards with connected various sensors systems (temperature, humidity, light, acceleration, inclination, tilt, rotation speed, displacement, proximity, level, etc.) at the Application Level. The possibility to control in a with the help of free LVRunTimeEng. remote way the measuring instruments at the Service

Level can be done also due to a free 'Agilent IO Control' application. This application, installed on the server allows sharing its instrumentation resources (GPIB, serial and USB ports, with the connected instruments) through the web with any computer using this same application configured as client. It is also expediently to use web cameras at sensors installations and actuators areas for additional visual control and verification of the correct operation of the lab equipment as a function of the control commands.

The universal sensors and transducers converter and interface board contains the integrated circuit of novel, programmable Universal Frequency to Digital Converter (UFDC-1), Figure 2 . It has 16 measuring modes of all frequency-time parameters of electric signals, one generating mode for calibration purpose and can work with all existing quasi-digital sensors and transducers. It means that many different laboratory activities and individual projects can be made on its basis. Only one difference between the hardware architecture of a hands-on laboratory and remote laboratory based on the proposal conception is that in the last one the Internet is the join between the end user and measuring devices. The rest of the architecture is the same. Using the remote laboratory the learners can work at home as if they were in the laboratory and working out an individual project according to the PBeL.

Fig. 2. The universal frequency converter

5 Summary

The use of proposed PBeL approach in different sensors related online courses such as ‘Smart Sensors Systems Design’, ‘Sensors and Signal Conditioning’, ‘Data Acquisition and Signal Processing for Smart Sensors’, ‘Smart Sensors and MEMS’ and others gives an unique opportunity to set up dozens remote laboratory activities based on a single universal sensors and transducers converter and interface board. Such approach will be also the best solution for different advanced online engineering courses and trainings in microelectronics. The most important impact of the application of the PBeL system will be on the living standard and general happiness of the people, who will be able to acquire a solid knowledge, high skill and strong courage to apply for and obtain good jobs in appropriate industry.

References

1. Yurish, S.Y.: Smart Sensors Systems: Project Research Results Dissemination, Exploiting and Using in Teaching Activities. In: Proceedings of Marie Curie Conference, European Science Open Forum (ESOF 2008), Barcelona, Spain, July 17-18, p. 297 (2008)
2. Calvo, B., Medrano, N., Sanz, M.T., Celma, S., Digón, D.: Microelectronics and Intelligent Instrumentation Master: Development of a Remote Laboratory for Distance Learning. In: Proceedings of 7th European Workshop on Microelectronics Education (EWME 2008), Hungary, Budapest, September 28-30 (2008)
3. García-Zubia, J., Hernández, U., Angulo, I., Orduña, P.: Realtime Remote Laboratory for Instrumentation Control. In: Proceedings of 7th European Workshop on Microelectronics Education (EWME 2008), Hungary, Budapest, September 28-30 (2008)

4. Odon, A., Krawiecki, Z.: Using LabView for Computer Modeling of Conversion Methods in Analog to Digital Converters. In: Proceedings of 12th IMEKO TC1 & TC7 Joint Symposium on Man Science & Measurement, Annecy, France, September 3- 5, pp. 395–400 (2008)
5. Yurish, S.Y.: Novel Universal Frequency-to-Digital Converters and Sensors Interfacing Integrated Circuits. Sensor Electronics and Microsystems Technologies 3, 80–90 (2008)
6. Sensors Web Portal, http://www.sensorsportal.com
7. http://www.ewh.ieee.org/tc/sensors/

A Heuristic Algorithm for Distributed Storage in P2P VOD Streaming System

Guangqing Deng[1], Ting Wei[1], Changjia Chen[1], Wei Zhu[2], Bin Wang[2], and Dengrong Wu[2]

[1] School of Electrical and Information Engineering, Beijing Jiaotong University, Beijing, China
[2] PPLive R&D, Shanghai Synacast Media Tech, Shanghai, China
GuangqingDeng033@163.com

Abstract. We do a network measurement towards PPLive VOD streaming system and find that server bandwidth saving rate (BSR) of a video is related to videos' popularity. Specifically, the video with low popularity suffers low BSR. In response to this phenomenon, we develop a heuristic replica replacement algorithm for distributed storage strategy in P2P VOD streaming system. We use daily weighted average upload flow of replicas in peers' as the criteria for deleting redundant replicas. We use dynamic probation of replicas to check the importance of replicas. Also, we design popularity coefficient and rate coefficient to regulate the replica distribution of unpopular videos and high coding rate videos in our distributed storage strategy, which improves the BSR of the P2P VOD streaming system.

Keywords: distributed storage, P2P, VOD, replacement.

1 Introduction

With the increase of network access bandwidth, more and more users watch online videos through P2P VOD streaming system. Recent years, great efforts [1, 4, 5] have been made on peer selection strategy, piece selection strategy and cache replacement algorithm in P2P streaming system. Huang et al. [4] compares the performance of Multiple Video Caching (MVC) and Single Video Caching (SVC) and choose LRU as the original disk replacement algorithm of PPLive. Tewari et al. [1][5] showed that having the number of replicas of each object proportional to the request rate for these objects has both per-node and network-wide advantages for P2P networks.

In order to reduce server bandwidth, every peer in VOD system will contribute some storage and bandwidth resource to help other peers. Because of the difference of videos' popularity and the asynchronism of peers' watching progress, the peer in a VOD system usually need to provide 1GB or 2 GB shared disk room. A peer's shared disk is used to store some video replicas and then upload multimedia content to those peers who watching the same video. There are two main methods to regulate replicas distribution over peers: disk replica replacement algorithm and pre-fetching [6]. Unlike P2P live streaming system, the concurrent users of a video in a VOD streaming system is usually not many except several most popular videos. So pre-fetching strategy is not wildly used and users' viewing behaviors mainly determine the replica distribution.

D. Zeng (Ed.): ICAIC 2011, Part II, CCIS 225, pp. 614–621, 2011.

In a large scale P2P VOD streaming system, such as PPLive VOD system [2], millions of peers' shared disk will form a huge distributed storage network. How to improve this distributed storage network's utilization is a challenging task not only for engineers but also for researchers. Because of the space limitation of shared disks, peers need to store those much-needed replicas to improve its upload bandwidth utilization which is the bottleneck of a P2P VOD system. If every peer does so, the content server's bandwidth will be reduced.

2 Network Measurement towards PPLive VOD System

Before designing a new replica replacement algorithm, we do a measurement towards PPLive VOD system which has more than 10 millions of users. The aim is described below. First, investigate the utilization of shared disk and upload bandwidth of each peer in PPLive VOD system. Second, analyze the advantages and disadvantages of current replica replacement algorithm in PPLive VOD system, which is beneficial to design a new algorithm.

A. PPLive Distributed Storage System

First, we measure the shared disk size and upload bandwidth of peers. As showed by figure 1, the size of PPLive VOD system is 2 GB. 55% of the users use up the shared disk; 45% of the users still have remaining shared disk. Obviously, 45% of users seldom watch videos since they installed the software. Also, this VOD system doesn't push multimedia content to peers. We will focus on active users because we can hardly change the content in the shared disk of inactive users. Besides shared disk, the distribution of peers' upload bandwidth is another factor affecting *BSR* [3]. As we know, the download of a peer in the P2P streaming system can be divided into two categories: download form peers and download from servers. Here, *BSR* is defined as the fraction of download from peers.

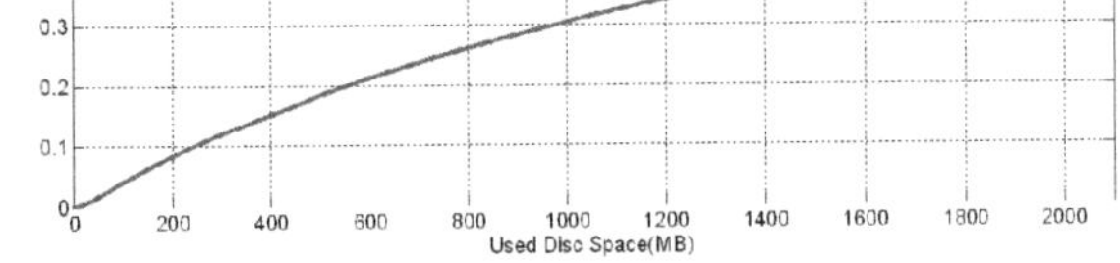

Fig. 1. Utilization of each peer's shared disk

It is difficult to measure peers' upload bandwidth because peers' number is too large. So we use the maximum upload speed of a peer to estimate its upload bandwidth. We also calculate peers' average upload speed and the result is showed by figure 2. The first and second peak of PDF curve about maximum upload speed corresponds to 1M and 2M ADSL users in mainland China. From the CDF figure, we can see that over 60% of peers have upload bandwidth more than 300 kbps but 90% of peers have average upload rate less than 200 kbps. It is clear that peers' upload bandwidth utilization is not high. There may be several reasons. First, PPLive VOD system implement a strategy to strictly limit peers' upload bandwidth utilization. So some users watching online videos also have enough available upload bandwidth to visit website or to play online games. Second, the replica replacement algorithm is inefficient. If a large number of obsolete replicas are stored in the distributed storage network, peers' upload bandwidth may always be idle. In this paper, we focus on the second reason. Also, we will evaluate the current replica replacement algorithm of PPLive VOD system in the next section.

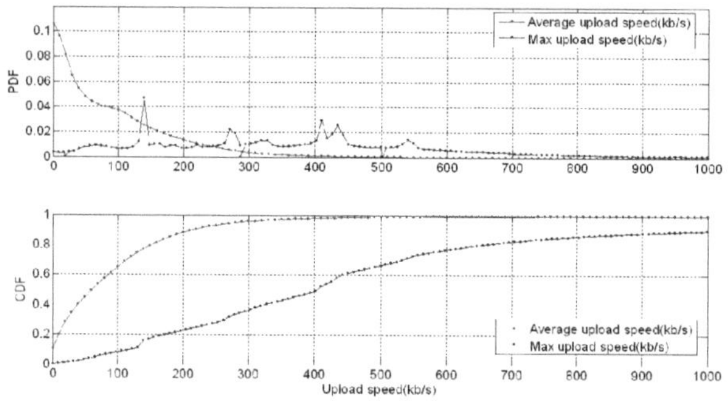

Fig. 2. The distribution of max and average upload speed of peers

B. Current Replica Replacement Algorithm

The size of videos in PPLive VOD streaming system is different. For example, the size of a blu-ray movie is over 1GB, but the size of a common video may be no more than 100 MB. In order to improve the efficiency of replica replacement algorithm, PPLive VOD system use chunk as the minimum unit for distributed storage system. Here, chunk size is about 20 MB and a video will be cut into chunks.

In figure 3, the horizontal axis relates to chunk sets in which every chunk has same viewing times; the vertical axis relates to the average *BSR* of the chunk set. The *BSR* of unpopular chunks is much lower than other chunks.

The *BSR* of videos with same popularity is also different. From figure 4, we can see that a chunk with daily hits 100 can get a high *BSR* if its replica number is larger

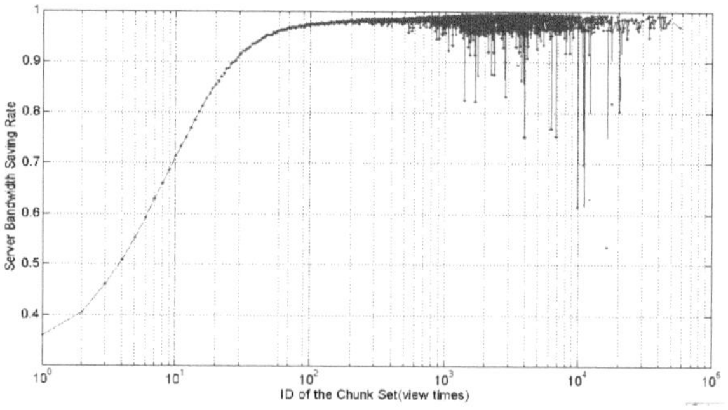

Fig. 3. The BSR of chunk sets

than 70. However, the *BSR* of a chunk can not be improved just only by increasing replica number blindly. For other chunks, the conclusion is also true. The *BSR* of a chunk with daily hits 500 can not be affected by increasing replica number if it is over 400. How to regulate the replica distribution of videos with different popularity and coding rate is the main issue we'll study in the next section.

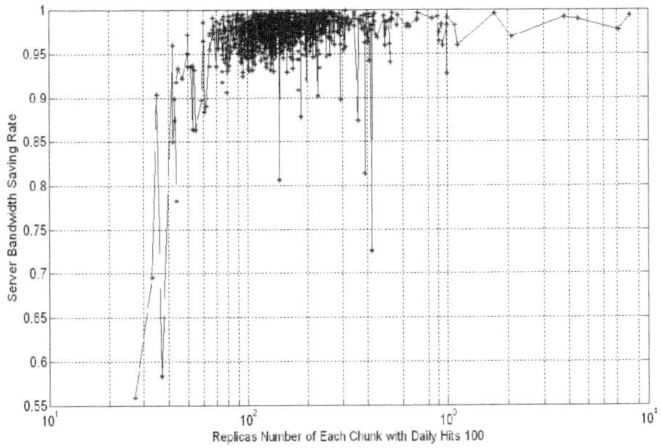

Fig. 4. The BSR of chunks with daily hits 100

3 A Heuristic Replica Replacement Algorithm

Through the large scale network measurement, we find several drawbacks of current PPLive VOD system. For example, the replica of unpopular videos is too few; the replica of popular videos is too many. [3] discusses the replica replacement algorithm ever used by PPLive VOD system. LRU algorithm also has the drawback as discussed above. Centralized replica algorithm can achieve ideal performance; however, the scalability of this kind of algorithm is poor. The users and online videos in PPLive VOD system are become more and more. In response to this phenomenon, we develop a new replica replacement algorithm.

A. Daily Weighted Average Upload Flow

First, we suppose the video which is watched currently is needed by P2P storage system and this video should be stored. If there is no storage space for new replicas, we have to delete the excess replica. In this paper, we choose the daily upload flow of a replica as the criteria for determining whether that replica needs to be deleted or not. The replica with large amounts of upload flow should be kept in shared disk because this replica is scarce. Scarce replica can increase peer's upload bandwidth utilization, which will improve the *BSR* of VOD system. We consider that the replica with little daily upload flow is useless for VOD streaming system. The reason is obvious. For a popular video, if the total number of its replicas is plethoric, the daily upload flow per peer will be little. For an unpopular video, because there are only a few people who need its replicas, the daily upload flow per peer will be little too. In both condition, we should delete the redundant replica with a little daily upload flow.

In fact, we consider the replica with lots of daily upload flow currently as the just needed replica in future. In other words, we speculate the popularity of a replica in future according to its current popularity. In order to improve the correctness of speculation, we will use *daily weighted average upload flow* (*DWAUF*) to estimate the scarcity of a replica. The larger the *DWAUF* is, the scarcer the replica is. The *DWAUF* V_{avg} is defined as below

$$V_{avg} = \sum_{i=1}^{I} V_i * (a_{I-i+1} / s_I) \tag{1}$$

Here, V_i is the quantity of upload flow of the replica i days before today. The range of I is from 1 to 7, which means that we at most use the upload flow in last 7 days to calculate *DWAUF*. Also,

$$a_i = 2^{(i-1)}, \quad s_i = 2^i - 1 \tag{2}$$

a_i is the general term of a geometric sequence with a_1 2 and common ratio 2. s_i is the sum of the first i terms of the geometric sequence. When the shared disk is full, the replica with least *DWAUF* will be removed first.

B. Popularity Coefficient

In order to improve the *BSR* of unpopular videos (especially those videos with daily hits less than 20 times), we need to save replicas of unpopular videos specially. Through network measurement, we find that a video's *BSR* can reach 90% if its replica number gets to 30. Based on this measurement result, we design a *popularity coefficient* R_p which just relates to the total replica number of a video in the P2P storage network.

$$R_p = \sqrt{30 / Func\_P(Cur\_Replica\_NO)} \tag{3}$$

$$Func\_P(x) = \begin{cases} x, & 1 \le x \le 30 \\ 30, & x > 30 \end{cases} \tag{4}$$

Here, $Cur\_Replica\_NO$ is the number of current online replica of a video. In PPLive VOD system, the tracker records the $Cur\_Replica\_NO$ and a peer can get it by sending a special request. Now, we will modify *DWAUF* V_{avg} by the *popularity coefficient* R_p,

$$V'_{avg} = V_{avg} * R_p \tag{5}$$

C. Rate Coefficient

In PPLive VOD streaming system, videos can be divided into three categories: common, high-definition and blu-ray. The typical coding rate of common, high-definition and blu-ray videos is 500 kbps, 800 kbps and 1300 kbps respectively. In

PPLive VOD system, the multimedia data is transferred through UDP connection. Suppose the average uploading speed of a UDP connection is fixed. The number of UDP connections needed by a blu-ray video is 2.6 times that needed by a common video. So the number of replicas needed by a blu-ray video is 2.6 times that needed by a common video too, since each peer will not store same replicas. The *rate coefficient* R_{rate} is defined as below,

$$R_{rate} = Func\_R(coding\_rate)/500 \tag{6}$$

$$Func\_R(x) = \begin{cases} 500, & x \le 500 \\ x, & 500 < x \le 1300 \\ 1300, & x > 1300 \end{cases} \tag{7}$$

$coding\_rate$ is the video's average coding rate whose unit is kbps. We use *rate coefficient* to modify $V_{avg}^{'}$ and the result is below,

$$V_{avg}^{''} = V_{avg}^{'} * R_{rate} = V_{avg} * R_p * R_{rate} \tag{8}$$

D. *Replica Probation*

When a peer downloads a new replica, we suppose this replica is needed by the distributed storage system and save it in the shared disk. Then we give a trial period (we call it *probation* for short) to this replica to check whether the replica is really redundant or not. During the probation, the replica will not be deleted in any condition. If the peer uploads large amounts of content to other peers in the probation, we consider that the replica is needed and continue to save it. Otherwise, we will delete the replica immediately and vacate the disk room to save other replicas.

We consider the probation of a replica is related to the peer's popularity. If the video is popular, the video's replica will be requested in a short while, so the probation of the replica will be short. Otherwise the replica's probation will be long. We set the longest probation 48 hours. The reason is that a replica should be deleted if it is not requested by other peers during 48 hours to avoid the waste of shared disk. Also, in order to avoid rash decision, we set the shortest probation 1 hour no matter how popular the replica is. We use the number of current online replica to estimate replicas' popularity and the *probation* T_p is defined as below,

$$T_p = 48 * \{30 / Func\_t(Cur\_Replica\_NO)\} \tag{9}$$

$$Func\_t(x) = \begin{cases} 30 & x \le 30 \\ x & 30 < x \le 1440 \\ 1440 & x > 1440 \end{cases} \tag{10}$$

E. *Distributed Replica Replacement Algorithm*

Based on *DWAUF*, *popularity coefficient*, *rate coefficient* and *replica probation* discussed above, the algorithm flow chart is showed as below.

A Heuristics Algorithm for Replica Replacement

A new replica need storing but the shared disk is full

1. get $Cur\_Replica\_NO$ for each replica

for (replica in peer's shared disk)

2. get the total uploading time T_{up} and replica probation T_p of the replica

if ($T_{up} < T_p$)

{continue;}

end if

4. calculate the weighted average daily upload flow V_{avg}

$$V_{avg} = \sum_{i=1}^{I} V_i * (a_{I-i+1} / s_I),$$

here, $a_i = 2^{(i-1)}, s_i = 2^i - 1$

5. calculate popularity coefficient R_p

$$R_p = \sqrt{30 / Func\_P(Cur\_Replica\_NO)}$$

$$Func\_P(x) = \begin{cases} x, & 1 \le x \le 30 \\ 30, & x > 30 \end{cases}$$

6. calculate rate coefficient R_{rate}

$$R_{rate} = Func\_R(coding\_rate) / 500$$

$$Func\_R(x) = \begin{cases} 500, & x \le 500 \\ x, & 500 < x \le 1300 \\ 1300, & x > 1300 \end{cases}$$

7. modify V_{avg} by R_p and R_{rate}

$$V''_{avg} = V_{avg} * R_p * R_{rate}$$

8. add [replica ID, V''_{avg}] to replacement list L

end for

9. delete the replica with smallest V''_{avg} in L

10. store the new replica and calculate the probation T_p of the new replica

$$T_p = 48 * \{30 / Func\_t(Cur\_Replica\_NO)\}$$

$$Func\_t(x) = \begin{cases} 30 & x \le 30 \\ x & 30 < x \le 1440 \\ 1440 & x > 1440 \end{cases}$$

4 Conclusion and Future Work

We do a large scale network measurement towards PPLive VOD system to get peers' characteristic, such as upload bandwidth and shared disk size. The conclusion is that the utilization of peers' upload bandwidth is low and more than half of users in

PPLive VOD system are active. The drawback of current replica replacement algorithm is that there are too many replicas of popular videos but too few replicas of unpopular videos. In response to this phenomenon, we develop a replica replacement algorithm to regulate the replica distribution in PPLive VOD system. In this distributed algorithm, peers needn't exchange any information between each other.

In our replica replacement algorithm, we use *daily weighted average upload flow* (*DWAUF*) to estimate the scarcity of replicas in PPLive VOD system. Just the replica with large amounts of upload flow will be stored in peers' shared disk. We design *popularity coefficient* and *rate coefficient* to regulate the number of replicas of unpopular videos and high coding rate videos.

Now, we are applying this new replica algorithm to PPLive VOD system. However, owing to time constraints, we have only applied *DWAUF* and *rate coefficient* to PPLive VOD system. We use the product of *DWAUF* and *rate coefficient* as the criteria for determining whether a replica needs to be deleted or not. With these improvements, the *BSR* of PPLive VOD system is over 90%. Also, the number of replicas of high coding rate videos is increased and the *BSR* of these videos is increased too. We continue applying *rate coefficient* and *probation* of replicas to PPLive VOD system. The results will be reported by our following papers.

Acknowledgment. This work was supported in part by China 973 2007CB307101, China NSFC 60672069 and 60772043.

References

1. Tewari, S., Kleinrock, L.: Proportional Replication in Peer-to-Peer Networks. In: Proc. of IEEE INFOCOM 1996 (1996)
2. PPLive, `http://www.pplive.com`
3. Li, H., Ke, X., PPLive R&D, Hu, P.: Towards Health of Replication in Large-Scale P2P-VoD Systems. In: Proc. of IPCCC (2009)
4. Huang, Y., Fu, T.Z.J., Chiu, D.M., Lui, J.C.S., Huang, C.: Challenges, Design and Analysis of a Large-scale P2P-VoD System. In: Proc. of SIGCOMM (2008)
5. Tewari, S., Kleinrock, L.: On Fairness, Optimal Download Performance and Proportional Replication in Peer-to-Peer Networks. In: Proc. of IFIP Networking (2005)
6. Cheng, B., Stein, L., Jin, H., Zhang, Z.: Towards Cinematic Internet Video-on-Demand. In: Proc. of EuroSys (2008)

On 3D Contour Matching Based on Geometry Features*

Caiying Luo[1] and Shucheng Zhou[2,**]

[1] Traffic College, Fujian Agriculture and Forestry University,
Fuzhou, Fujian, China
[2] Computer and Information College, Fujian Agriculture and Forestry University,
Fuzhou, Fujian, China
luocy53@yahoo.cn, zhoushuch@sohu.com

Abstract. In 3D objects recovering, geometry features of 3D contour curves were analyzed, method of 3D contour curves matching based geometry features was presented. Contour curves of 3D objects were extracted and represented by B-spline. According to geometry features of 3D contour, curvatures and torsions of point on curves were calculated after filtering and smoothing contour curves. Eigenvectors of 3D contours were constructed by curvatures and torsions. Features of corresponding points on two contour curves were compared to judge similarity of contour. Matching relation of contour was determined between tow contour curves based on similarity between two contours. Fragments mosaicing were realized for recovering broken objects based on contour curves matching. Experimental results show that that this method has high speed, accuracy and validity.

Keywords: 3D contour, Geometry feature, Eigenvector, Contour matching, Contour similarity.

1 Introduction

3D contour curve matching is an important researched problem in many applied disciplines, such as computer vision, pattern recognition and objects recovering. Exact and speed contour matching play an important role in object recognition and 3D reconstruction. Many researchers presented a lot of algorithms for 3D curves matching. Pajdla et al. [1] presented method of semi-differential invariants which distance between two points on curves and tangent vector are used to compare similarity of two points to completed 3D curve matching. Although high order derivatives are not calculated using this method, but geometry features can not be used adequately. Kishon et al. [2] used spline to completed curves matching. B-spline approximate contour curves firstly, then curves features are extracted, features are transform to 1D character string which are used to compared to completed curves

* This work is supported by the Natural Science Foundation of Fujian Province of China Grant #S0650005 to Caiying Luo and Science & Technology project of Educational Department of Fujian Province of China Grant #JA10119 to Shucheng Zhou.
** Corresponding author.

D. Zeng (Ed.): ICAIC 2011, Part II, CCIS 225, pp. 622–630, 2011.

matching. Because spline is not a invariant in describing contour curves when curves rotating and scaling, so applying area are limited. Zhang et al.[3] used Fourier transform to realize curves matching, although this method is small amount of calculation and Fourier descriptors can describe contour curves by certain precision, but Fourier descriptors can not describe local information of curves, local variety of curves can bring descriptors change abruptly, longest matching subsegment of curves are hard to be found. Chen et al. [4] used Fourier transform firstly and then wavelet transform, this method is large amount of calculation. Rodriguez et al. [5]projected 3D curves to 2D plane and used strings matching to measure the similarity between 3D curves. Calculation is complex by using this algorithm. Kong et al. [6] re-sampled the curves using polygonal approximation, got coarse alignment by using local features and then got fine alignment by using global features to complete curves matching. Several procedures were repeated by using this method, so matching speed is influenced. Copper et al. [7] presented algorithm based on Bayesian probability assessment to mosaic fragments. This method need more manual intervention, so it is low speed and is not enough precise. Sabalatnig et al. [8] classified attributes of fragments using a description language. According to results of comparing attributes of fragments to determine whether contours is matching between two fragments. Contours matching are completed based on that results are given and fragments are labeled using this method.

In 3D broken objects recovering, two adjacent fragments can be mosaiced according to relationship of matching contour curves. Geometry features of contour curves are rotation, scaling and translation invariant. The 3D curves matching algorithm based on geometry features of contour curves is presented for mosaicing fragments to realize broken objects recovering in this paper.

2 Preprocessing of Contour Curves

A. Representation of Contour

A discrete contour of fragment can be represented a curve which control points consist of points of contour edge. Contour curves are filtered and re-sampled equidistantly to get a series of same interval sampling points. Contour curve of a fragment is represented by a parameter curve.

Spline functions that have advantages of simple construction, easy to calculation are widely used in many disciplines. B spline curves have advantages of uniqueness, boundedness, continuity, local shape controllability and preservation of structures under affine transformation. Contour curves are represented by B spline in this paper.

Supposed that vector $\boldsymbol{P}$ is $n+1$ control points of characteristic polygon on contour curves. Vector $\boldsymbol{P}$ can be represented as

$$\boldsymbol{P} = \{\boldsymbol{P}_i \mid i = 0,1,\cdots,n\}$$

Given $m+1$ real valued t_i which are called knots, then the knot vector is represented as

$$\boldsymbol{t} = \{ t_0, t_1 \cdots, t_m \}$$

Vector $\boldsymbol{t}$ is a set of $m+1$ non-decreasing number sequence, namely $t_0 \le t_1 \le \ldots \le t_m$ and $\boldsymbol{t} \in [0,1]$, $i = 0,1,\cdots,m$.

B spline of degree k (order $k+1$) can be defined as

$$\boldsymbol{P}(t) = \sum_{i=0}^{n-1} \boldsymbol{P}_i N_{i,k}(t) \tag{1}$$

Functions $N_{i,k}(u)$ are the basis functions of B spline.

According to de Boor recursion formula [9], basis functions with parameter $\boldsymbol{t}$ ($t_i \in [0,1]$, $i = 0,1,\cdots,m$) can be defined as

$$B_{i,0}(t) = \begin{cases} 1, & t_i \le t \le t_{i+1} \\ 0, & \text{otherwise} \end{cases} \tag{2}$$

$$B_{i,k}(t) = \frac{t - t_i}{t_{i+k} - t_i} B_{i,k-1}(t) + \frac{t_{i+k+1} - t}{t_{i+k+1} - t_{i+1}} B_{i+1,k-1}(t) \tag{3}$$

Variant k is a positive integer, parameter t_i ($t_i \in [0,1]$, $i = 0,1,\cdots,m$) are values of knots on curve in formula 3 . We conclude that B spline of degree k are linear combination of two neighboring B spline of degree $k-1$.

Basis functions of cubic B spline are represented as

$$\begin{cases} B_{0,3}(t) = \frac{1}{6}(1-t)^3 \\ B_{1,3}(t) = \frac{1}{6}\left(3t^3 - 6t^2 + 4\right) \\ B_{2,3}(t) = \frac{1}{6}\left(-3t^3 + 3t^2 + 3^t + 1\right)^3 \\ B_{3,3}(t) = \frac{1}{6}t^3 \end{cases} \tag{4}$$

where the range of parameter t is $[0,1]$. B spline curves have well locality. B spline curves on i-th segment consist of $k+1$ control points ($\boldsymbol{P}_i, \boldsymbol{P}_{i+1}, \cdots, \boldsymbol{P}_{i+k}$). If this curve segment need to be modified, only need to modified $k+1$ control points relating to this curve.

B. Contour filtering

A contour curve can be represented with a discrete point set $\boldsymbol{P}$ after being sampled. Supposed that there have N sampling points on a contour curve represented by B spline, then a contour curve can be represented with a discrete points set $\boldsymbol{P}$ as

$$\boldsymbol{P} = \{\boldsymbol{P}_i \mid i = 0,1,\ldots,N-1\}, \boldsymbol{P}_i \in \boldsymbol{R}^3 \tag{5}$$

where $N = 2^j \mid_{j \in Z}$. Point $\boldsymbol{P}_i$ represents a point on a contour curve in equation 5. Its coordinate can be represented as (Px_i, Py_i, Pz_i). Point $\boldsymbol{P}_{N+i}$ is equal to point $\boldsymbol{P}_i$ because closed contour curve has property of periodicity. Extracting contour curves must be filtered for reducing effect of noise. Gaussian functions have characteristic of separability, which suit to remove most noise and remain detail information of contour curves. 1D zero mean Gaussian functions can be represented as

$$G(n) = \mathrm{e}^{-n^2/(2\sigma^2)}$$

where σ is distribution parameter of Gaussian functions. Filtering width is different along with distribution parameter σ changing. Convolution of three coordination components of a point on contour curve and Gaussian function is represented as

$$\begin{cases} P_x(n) = Px_i * G(i) = \sum_{i=0}^{N-1} Px_i \mathrm{e}^{-(n-i)^2/(2\sigma^2)} \\ P_y(n) = Py_i * G(i) = \sum_{i=0}^{N-1} Py_i \mathrm{e}^{-(n-i)^2/(2\sigma^2)} \\ P_z(n) = Pz_i * G(i) = \sum_{i=0}^{N-1} Pz_i \mathrm{e}^{-(n-i)^2/(2\sigma^2)} \end{cases} \tag{6}$$

According to equation 6, a point on contour curves is represented as

$$\boldsymbol{P}(n) = (P_x(n), P_y(n), P_z(n))$$

The bigger distribution parameter is, the more noise is removed, and smoothing effect is better, but detail information of contour is less. Contour curves can be described at different scale by adjusting distribution parameter to remain more detail information of contour curves and remove more noise, filtering effect is perfect.

3 Geometry Features Calculation

Parameters which have no relationship relative to coordination are needed in curves matching. Curvature and torsion, which are important geometry features of curve, are invariant relative to transformation, rotation and scaling. When a 3D curve is represented with a parameter equation, a point on curves is a function which is represented with arc-length parameter. Supposed arc-length is represented with t, vector function of curve is represented as

$$\mathbf{r}(t) = (x(t), y(t), z(t)) \tag{7}$$

In practical processing, contour curves are digitalized. Point set of 3D contour curve is gotten after filtering and resampling. Supposed that numbers of point are N which satisfy with $N = 2^n |_{n \in Z}$, contour curve is represented as

$$\mathbf{r} = (r(0), r(1), \cdots, r(N-1))$$

where $r(i)$ is represented i-th point on curve which coordination is represented (x_i, y_i, z_i). Closed curve can be regarded as being periodic, namely, $r(i+N) = r(i)$. In digitalized processing, curvature and torsion can be calculated with difference. Difference equations are represented as

$$\Delta x_i = x_i - x_{i-1},\ \Delta y_i = y_i - y_{i-1},\ \Delta z_i = z_i - z_{i-1}$$
$$\Delta t_i = \sqrt{(\Delta x_i)^2 + (\Delta y_i)^2 + (\Delta z_i)^2},\ \Delta \mathbf{r}_i = \mathbf{r}(i+1) - \mathbf{r}(i)$$

So arc-length of a point $r(i)$ can be represented as $t_i = \sum_{k=0}^{i} \Delta t_k$.

Point $r(i)$ is represented as

$$r(i) = (x(t_i), y(t_i), z(t_i))$$

If one tuple $(r(i), t_i)$ is gotten, curvature and torsion of the point can be calculated as.

$$k_i = \left|\Delta^2 \mathbf{r}(t_i)\right| = \sqrt{(\Delta^2 x(t_i))^2 + (\Delta^2 y(t_i))^2 + (\Delta^2 z(t_i))^2} \quad (8)$$

$$\tau_i = \frac{(\Delta \mathbf{r}(t_i), \Delta^2 \mathbf{r}(t_i), \Delta^3 \mathbf{r}(t_i))}{(\Delta^2 \mathbf{r}(t_i))^2} \quad (9)$$

where $\Delta \mathbf{r}(t_i)$ represent 1 order difference, sign (, , ,) represent mix product. Curvature and torsion of a point $r(i)$ construct a 2D vector which is represented as

$$\xi_i = [k_i, \tau_i]$$

2D vector ξ_i is consisted of curvature and torsion of a point on curves. Curvature and torsion are invariant that relative to rotation, transformation and scaling. Then vector ξ_i can be eigenvector in contour matching.

4 Algorithm of Contour Matching

Similarity measure is a key technique in contour matching. Contour curves are match whether or not that is determined by similarity between eigenvectors of two contour curves. Hausdorff distance is used to judge similarity between two contours in this paper.

Haudorff distance can measure matching degree between two point set. Supposed two subsegment of curve corresponding to two fragments S1 and S2 are **P** and **Q**, then **P** and **Q** are represented respectively as

$$\mathbf{P} = (\mathbf{p}_0, \mathbf{p}_2, \cdots, \mathbf{p}_n)$$
$$\mathbf{Q} = (\mathbf{q}_0, \mathbf{q}_{2,} \cdots, \mathbf{q}_n).$$

where $\mathbf{p}_i$ and $\mathbf{q}_i$ are simplified representations of eigenvectors which are gotten by calculating at one point on curves. Haudorff distance between two point sets **P** and **Q** is represented as

$$\mathrm{H}(\mathbf{P},\mathbf{Q}) = \max(\mathrm{h}(\mathbf{P},\mathbf{Q}), \mathrm{h}(\mathbf{Q},\mathbf{P})) = \max_{\mathbf{p}_i \in \mathbf{P}} \min_{\mathbf{q}_j \in \mathbf{Q}} \left(\left|\mathbf{p}_i - \mathbf{q}_j\right|\right) \quad (11)$$

where $\mathrm{h}(\mathbf{P},\mathbf{Q})$ is directional distance of **P** to **Q**, $\mathrm{h}(\mathbf{Q},\mathbf{P})$ is directional distance of **Q** to **P**. Calculating formulas of $\mathrm{h}(\mathbf{P},\mathbf{Q})$ and $\mathrm{h}(\mathbf{Q},\mathbf{P})$ are represented respectively as

$$\mathrm{h}(\mathbf{P},\mathbf{Q}) = \max_{p_i \in \mathbf{P}} \min_{q_j \in \mathbf{Q}} \left\|\mathbf{p}_i - \mathbf{q}_j\right\| \qquad i, j \in 1, 2, \cdots, n$$

$$\mathrm{h}(\mathbf{Q},\mathbf{P}) = \max_{\mathbf{q}_i} \min_{\mathbf{p}_j} \left\|\mathbf{q}_i - \mathbf{p}_j\right\| \qquad i, j \in 1, 2, \cdots, n$$

where sign $\|\cdot\|$ represent one distance norm. Euclidean distance was used in this paper. Euclidean distance of point $\mathbf{p}_i$ to point $\mathbf{q}_j$ is represented as

$$D_{i,j} = \left\| \mathbf{p}_i - \mathbf{q}_j \right\| = \left[\left(\mathbf{p}_i - \mathbf{q}_j \right)^T \left(\mathbf{p}_i - \mathbf{q}_j \right) \right]^{\frac{1}{2}}$$

Euclidean distance of point $\mathbf{q}_j$ to point $\mathbf{p}_i$ can be calculated according to same method.

Searching direction of two contour curves must be opposite, namely, one curve is searched clockwise direction, and another curve is searched counter-clockwise direction.

In broken objects recovering, only existing subsegments of curves are matching. So a contour curve is segmented into feature subsegments according curvature and torsion of points on contour curve. Because extracting contour curves should be influenced by calculating errors and losing or defect of fragments that would change shape of contour curve, so contour curves of adjacent fragments are not same exactly in practical application. Range of error, which is denoted as threshold ε, is permitted. If distance between two points which are on two contour curves respectively less than threshold ε, eigenvectors between two points are regarded to be equal.

Contour curve of broken object was pre-processed, resampled uniformly and re-parameterized. Curvatures and torsions of every point on contour curves are calculated to construct eigenvector of contour curves.

Supposed curve $\mathbf{P}$ is a curve segment belonging to fragment S1 after extracting to be regarded as a template curve, which is represented by B spline and is re-sampled uniformly. Curve $\mathbf{P}$ is re-parameterized. Curvatures and torsions of every point on contour curves are calculated to construct eigenvector of contour curves. There are $n+1$ sampling points on feature curve segment. Feature points set can be represented as

$$\mathbf{P} = \left(\mathbf{p}_0, \mathbf{p}_1, \cdots, \mathbf{p}_n \right)$$

Testing curve $\mathbf{Q}$ which belongs to fragment S2 is another feature curve segment which is represented by B spline and re-sampled uniformly. There are $n+1$ sampling points on curve $\mathbf{Q}$. Feature points set of curve $\mathbf{Q}$ is represented as

$$\mathbf{Q} = \left(\mathbf{q}_0, \mathbf{q}_1, \cdots, \mathbf{q}_n \right)$$

According to calculating method of feature points and distance between two points, matching algorithm is described as follows.

STEP 1: Hausdorff distance between curve segment $\mathbf{P}$ and $\mathbf{Q}$;

STEP 2: If $\mathrm{H}(\mathbf{P},\mathbf{Q}) < \varepsilon$, then $\mathbf{P}$ and $\mathbf{Q}$ are matching segment, positional information of matching segments are saved ; If two curve segments are not matching, go to STEP 4;

STEP 3: If all curve segments are not processed, next curve segment is gotten in turn to be new $\mathbf{P}$ and $\mathbf{Q}$, go to STEP 1;

STEP 4: If all segment of fragment S2 are not processed, next segment relative to current segment $\mathbf{Q}$ is regarded as new $\mathbf{Q}$, go to STEP 1;

STEP 5: If all segment fragment S1 are not processed, next segment relative to current segment **P** is regarded as new **P**, go to STEP1;

STEP 6: Matching calculating is end.

In broken object recovering, because more than one matching feature curve segment may be found in processing of curve matching calculation, two matching curves are tried to mosaice when two matching curve segments are found. If two matching feature curve segments belong to two adjacent fragments respectively, two fragments are mosaiced.

5 Analysis of Experimental Results

In experiment, broken objects are scanned by 3D scanner to get 3D scattered data points set, then contour curves of fragments are extracted. Fragments and contour curves of broken pottery are illustrated in fig. 1. Fragments of broken pottery are illustrated in fig. 1a, contour curves extracting from fragments that are projected on plane are illustrated in fig. 1b.

Matching of two contour curves and mosaicing two fragments are illustrated in fig. 2a.

Results of matching of three contour curves and mosaicing three fragments are illustrated in figure 2b.

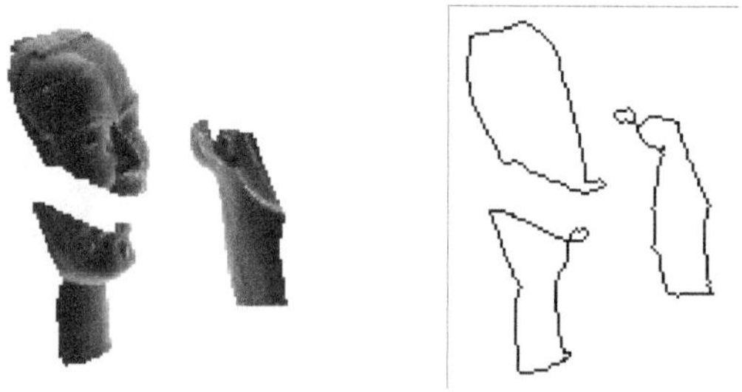

(a) Fragments (b) Contour curvesof fragment

Fig. 1. Fragments of broken object and contour curves

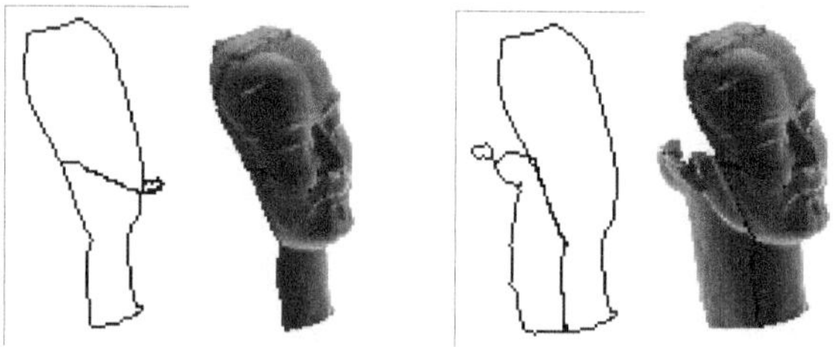

(a) Two contours matching (b) Three contours matching

Fig. 2. Contour matching and mosaicing of fragments

Experimental results show that the algorithm presented in this paper can realized various contour curves of fragments matching by comparing geometry features. Geometry features are invariant under rotating, translating and scaling. Geometry features construct eigenvector for contour matching. 3D broken objects can be reconstructed based on contour curves matching.

Eigenvector of point on curves only consist of two geometry features of contour curve. Parameters which used to match contour curve are less in processing of contour matching to reduce the computational complexity and amount of calculation. Matching segments can be found accurately, and results of mosaicing fragments are preferable. The algorithm presented in this paper has advantages of computational simplicity, high speed, robustness and flexibility.

6 Conclusion

Geometry features of contour curve are analyzed in researching restoration of fragmented objects in this paper. Technique of matching 3D curves and mosaicing fragmented objects based on geometry features is presented. Scattered point cloud data are obtained by scanning 3D fragments. B splines have advantage of uniqueness, boundedness, continuity, local shape controllability and preservation of structures under affine transformation, it is one of most effective method to represent curve. Contour curves are extracted and represented by B splines. Contour curves are filtered and re-sampled uniformly. Curvature and torsion, which are important geometry features of contour curve, are calculated to construct eigenvectors for matching 3D curves and mosaicing fragmented objects. Curvature and torsion of contour curve are rotation, translation and scaling invariant. Feature vectors consist of curvature and torsion. Similarity comparing of features is transformed to compare 1-D vector to reduce amount of calculation in process of contour matching. Matching relationship of contour curves is determined by similarity of contour curves. Adjacent fragments are mosaiced based on contour curves matching. Error influence of losing fragments and noise in process of collecting fragmented data are considered in matching and mosaicing process. Experimental results show that this algorithm has advantages of robust and fault tolerant. Contour curves matching and mosaicing adjacent fragments are accurately using this method.

References

1. Pajdla, T., Van Gool, L.: Matching of 3D curves using semi-differential invariants. In: Proceedings of the Fifth International Conference on Computer Vision (ICCV 1995), Boston, pp. 390–395 (1995)
2. Kishon, E., Wolfson, H., Hastie, T.: 3D curve matching using splines. Journal of Robotic Systems 8(6), 723–743 (1991)
3. Zhang, H., Eugene, F.: Shape Matching of 3-D Contours Using Normalized Fourier Descriptors. In: Proceedings of International Conference on Shape modeling and Application, Banff, pp. 261–270 (2002)
4. Chen, G., Tien, D.B.: Invariant Fourier wavelet descriptor for pattern recognition. Pattern Recognition 32(7), 1083–1088 (1999)

5. Rodriguez, W., Last, M., Kandel, A., et al.: 3-Dimensinal curve similarity using string matching. Robotics and Autonomous Systems 49, 165–172 (2004)
6. Kong, W., Kimia, B.B.: On solving 2D and 3D puzzles using curve matching. In: Proceedings of the IEEE Computer Society Conference on Computer Vision and Pattern Recognition, Hawaii, vol. 2(2), pp. 583–590 (2001)
7. Copper, D.B., Willis, A., Andrews, S., Baker, J., Cao, Y., Han, D., et al.: Bayesian pot-assembly from fragments as problems in perceptual-grouping and geometric-learning. In: Proceedings of the 16th International Conference on Pattern Recognition (ICPR 2002), Quebec, vol. 3, pp. 297–302 (2002)
8. Sablatnig, R., Menard, C., Kropatsch, W.: Classification of archaeological fragments using a description language. In: Proceedings of European Association for Signal Processing (Eusipco 1998), Island of Rhodas, Greece, vol. 2, pp. 1097–1100 (1998)
9. de Boor, C.: On Calculating with B splines. Journal of Approximation Theory 6(1), 50–62 (1972)
10. Zhu, X., et al.: Formative technique of free curves and surface. Science Press, Beijing (2000)

Center Point Detection of Fingerprint Based on Direction Mean Value Difference*

Shui-ping Wang[1,2], Xiao-fang Pei[3], Ling-zhi Liao[1], and Ying-nan Zhao[1]

[1] College of Computer Science & Technology ,
Nanjing University of Information Science & Technology Nanjing, China
[2] Department of Computer Science & Technology,
Nanjing University of Science & Technology, Nanjing, China
[3] College of Electronic & Information Engineering,
Nanjing University of Information Science & Technology Nanjing, China
ShuipingWang036@163.com

Abstract. Directional pattern is a representation of original fingerprint image. This paper proposes one algorithm based on direction mean value difference to detect fingerprint center point. To realize this algorithm mainly includes detecting point direction of each pixel and configuring block patterns with 8×8 and 16×16 ranges. Experimental results show the algorithm has good robustness and can easily wipe off the false center point detected from the fingerprints.

Keywords: fingerprints, directional field, core point.

1 Introduction

Fingerprint is the endpoint structure of the surface veins, which is invariable and unique. Automatic Fingerprint Identification System (AFIS) has become a research focus of national academic and industry area [1]. AFIS consists of fingerprint collection, classification, feature extraction and fingerprint matching. Fingerprint classification can be determined by the lines trend near its center. In the feature extraction module, the minutiae features are extracted in near center field. As a result, the accuracy and reliability of fingerprint center position detecting have great significance in AFIS.

2 Directional Pattern

Fingerprint pattern is a representation of the source image, which represents the fingerprint by the direction of curves. In the gray matrix, statistical volumes are calculated, such as gray-difference or gray-average, to identify the main direction of one small neighborhood range. There are two kinds of directional patterns. The first

* This work is partially supported by National Natural Science Foundation of China (60702076), and by Scientific Research Fund of Nuist(20070009, Y636).

D. Zeng (Ed.): ICAIC 2011, Part II, CCIS 225, pp. 631–637, 2011.

one is point pattern, which is the right direction of every pixel. The other one is block pattern, which consists of the general direction in every block.

A. Point Pattern

At every point (x, y) of fingerprint image, 16 directions, named from 0 to 15, are detected, shown as Fig.1. Before calculating the point direction of pixel (x, y), S_d-the gray scale along direction d is defined as shown in Equation 1.

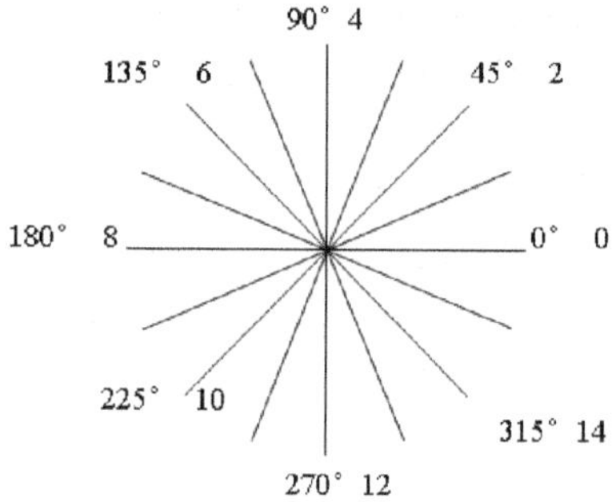

Fig. 1. The definition of point direction

$$S_d = \sum_{k=0}^{n-1} \left| \overline{f_d}(x, y) - f_d(x_k, y_k) \right| \quad d=0, 1, 2, \ldots, 15 \tag{1}$$

Where

$$\overline{f_d}(x, y) = \frac{1}{n} \sum_{k=0}^{n-1} f_d(x_k, y_k) \tag{2}$$

It is the n-point gray average in direction d. Eight neighbouring points are used, which are shown as Fig.2.

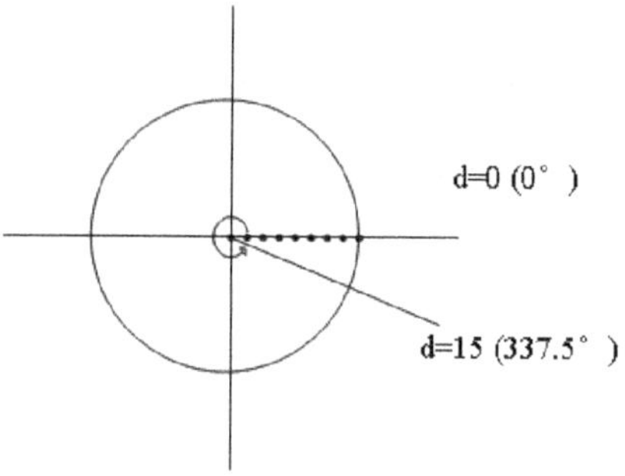

Fig. 2. The definition of adjacent points

The direction of point (x, y) is the d, whose S_d is the minimum one in 16 directions.

$$D(x, y) = d(\mathrm{Min}\{S_d\}) \quad d=0, 1, 2, \ldots, 15 \tag{3}$$

In the calculating procession of pattern S_d, the pixel in the other 15 directions, except 0 direction, can be determined by 0 degree coordinate rotation. For example, the k-point coordinates of each direction can be determined by Equation 4.

$$\begin{cases} x_{dk} = x_{0k}\cos(\alpha) - y_{0k}\sin(\alpha) \\ y_{dk} = x_{0k}\sin(\alpha) + y_{0k}\cos(\alpha) \end{cases} \quad (4)$$

Where α is the angle between 0 degree direction and d direction.

In this paper, a denoising operation is taken to smooth the result. In Fig.3, (a) is the original image, the point pattern picture of which is shown as (b).

(a) Original image (b) Point pattern picture

Fig. 3. Point pattern example

B. Block Pattern

The point pattern is divided into blocks of several size, such as 16×16 and 8×8. The direction histogram is detected, and the direction of the peak is the direction of the block. The original fingerprint picture, the 16×16 range block pattern and the 8×8 one are shown as Fig.4.

(a) Original image (b) 16×16 block (c) 8×8 block

Fig. 4. Block pattern example

3 Center Detection

Fingerprint singular points include the center (core) and the triangular points (delta). Accurate and reliable detection of singular points is of great significance to Fingerprint classification and fingerprint matching. Great deals of research and study have been done on the definition and detection of the fingerprint singular point. The mainstream of the singular point detection method is based on Poincare index. In this paper the other method based on the difference of the area average direction is proposed, which will be mentioned as follow.

A. Method Description

This method uses the direction angle changes to extract the core of fingerprint. $G(i, j)$ is the gradient vector field of point (i, j).

$$G(i,j)=\left[\frac{f(i+1,j)-f(i-1,j)}{2} \quad \frac{f(i,j+1)-f(i,j-1)}{2}\right]$$

$$(1<i<M,\quad 1<j<N) \tag{5}$$

Where M is the row number of fingerprint image, and N is the column number. $f(x, y)$ is the gray value of point (x, y). Then the direction θ of the field is defined as Equation 6.

$$\theta(i,j)=\frac{1}{2}\arg\left\{\sum_{m=i-\omega'}^{i+\omega'}\sum_{n=j-\omega'}^{j+\omega'}[G(m,n)]^2\right\}+\frac{\pi}{2} \tag{6}$$

Where $\omega' =$ (W-2) / 2, and W is the width of the field. If we use G_i to express $[f(i+1,j)-f(i-1,j)]/2$, and use G_j to express $[f(i,j+1)-f(i,j-1)]/2$, the Equation 6 can be described as follow.

$$\theta(i,j)=\frac{1}{2}\arg\left\{\sum_{m=i-\omega'}^{i+\omega'}\sum_{n=j-\omega'}^{j+\omega'}\left[G_i^2(m,n)+G_j^2(m,n)-2G_i(m,n)G_j(m,n)\right]\right\}+\frac{\pi}{2} \tag{7}$$

θ_1 of W1 × W1 block and θ_2 of W2 × W2 block are calculated. In the core area, there are relatively large angle changes, so large difference between θ_1 and θ_2 can be found.

As a result, if d is the Absolute value of difference between θ_1 and θ_2, and T is a threshold, the core point will be such areas which d is more than T.

B. Specific implementation

Based on the method mentioned in previous section, the specific processes of core area detecting can be mentioned as follow.

(1) Using the equations (1), (2) and (3), defined in section I, D(x, y) of every point in the fingerprint image are calculated.

(2) The obtained direction values are modulated by 8, and then the normalized direction values from 0 to 7 are got.

(3) To avoid false features, the image edge should be cut out.

(4) Considering 16×16 size block for regional, the regional average direction $\overline{D}_{16}(x, y)$ is obtained for each point of image.

$$\overline{D}_{16}(x, y) = \frac{1}{256} \sum_{m=x-7}^{x+8} \sum_{n=y-7}^{y+8} D'(m,n) \tag{8}$$

(5) Similar with step (4), the regional average direction $\overline{D}_8(x, y)$ in 8×8 size block for regional also is obtained.

$$\overline{D}_{8}(x, y) = \frac{1}{64} \sum_{m=x-3}^{x+4} \sum_{n=y-3}^{y+4} D'(m,n) \tag{9}$$

(6) The d value is defined as Equation 9. Pixels, except edge region, which the d value is more than a certain threshold, will be considered to be the core points.

$$d = |\overline{D}_{16} - \overline{D}_8| \tag{10}$$

The detection results based on the method of direction mean value difference are shown as Fig.5.

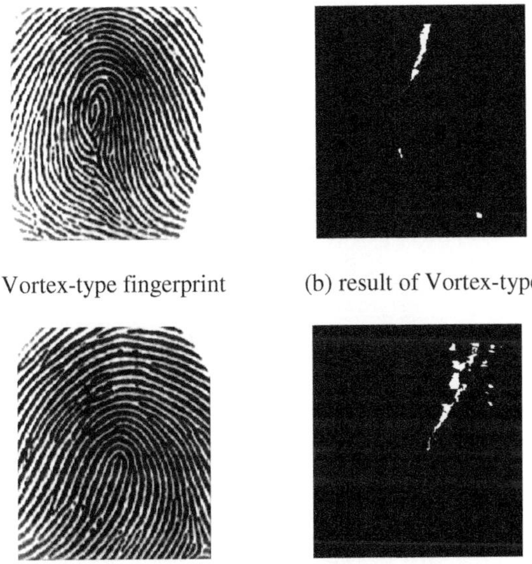

(a) A Vortex-type fingerprint (b) result of Vortex-type

(c) A Dustpan-type fingerprint (d) result of Dustpan-type

Fig. 5. Core located based on direction mean value difference

4 Experiment and Conclusion

The USB interfaced Polymath Fingerprint Scanner is used to acquire several fingerprint images, each of them are of 292×248 pixels. The experiments are done on Vortex-type, Dustpan-type and other different types of fingerprints. Picture (a) and (b) in Fig.5 are the detection results of a Vortex-type fingerprint, and the other two are the results of a Dustpan-type fingerprint. From them, we can conclude that, for the vast majority of fingerprint images, the post-processing will be relatively simple.

The results after post-processing are shown as Fig.6. Picture (a) is the result of Vortex-type image in Fig.6, and picture (b) is the result of Dustpan-type image, which is also shown in Fig.6.

(a) Result of Fig.5 (a) (b) result of Fig.5 (c)

Fig. 6. The core location after post-processing

In the algorithm tests, 400 fingerprint images are divided into 4 categories. The levels of them are very good, good, poor, and very poor. Several typical images of different levels are shown as Fig.7.

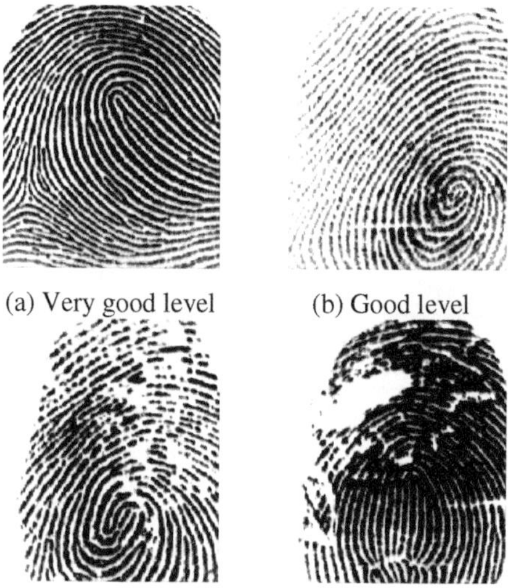

(a) Very good level (b) Good level

(c) Poor level (d) Very poor level

Fig. 7. Image examples of different lever

Table 1. Test Results of four different level images

Image Level	Number of Each Set	Results	
		Correct Number	Accuracy(%)
Very good	65	65	100.00%
Good	170	158	92.94%
Poor	144	113	78.47%
Very poor	21	10	47.62%
Overall	400	339	86.50%

The overall test result is shown in Table.1. There are different numbers of each set, and the total accuracy is about 86.50%. The test results of good quality fingerprint images are perfect, and the results of poor quality fingerprint images are not very bad. As a result, core point detection of fingerprint based on direction mean value difference, proposed in this paper, has good robustness.

Acknowledgment. We are grateful to Ling-zhi Liao, Ying-nan Zhao, and Rui Ma for their technical support. This study is supported in part by National Natural Science Foundation of China (60702076), and by School Foundation of Nanjing University of Information Science and Technology (20070009, Y636).

References

1. Yin, Y.-L., Ning, X.-B., Zhang, X.-M.: Development and application of automatic fingerprint identification technology. Journal of Nanjing University (Natural Sciences Edition) 38(1), 29–35 (2002)
2. Tang, T.-Z., Ning, X.-B., Ying, Y.-L.: A method for singularity detection in fingerprint images. Journal of Software 14(6), 1082–1088 (2003)
3. Jain, A.K., Prabhakar, S., Pankanti, S.: Matching and classification: A case study in fingerprint domain. Proceedings of the INSA-A(Indian National Science Academy) 67(2), 67–85 (2001)
4. Mei, Y., Sun, H.-J., De, D.-S.: Effective method for detection of fingerprints' singular points. Computer Engineering and Applications 44(28) (2008)
5. Wang, L., Dai, M.: Localization of singular points in fingerprint images based on the Gaussian-Hermit moments. Journal of Software 17(2), 242–249 (2006)
6. Duda, R.O., Hart, P.E., Stork, D.G.: Pattern Classification, 2nd edn. John Wiley & Sons, Chichester (2000)
7. Maltoni, D., Maio, D., Jain, A.K., Prabhaker, S.: Handbook of Fingerprint Recognition. Springer, New York (2003)
8. Zhang, W.-W., Wang, S., Wang, Y.-S.: Structure matching algorithm of fingerprint minutiae based on core point. Acta Automatica Sinica 29(6), 842–850 (2003)

A Bricklaying Best-Fit Heuristic Algorithm for the Orthogonal Rectangle Packing Problem*

Wenshui Lin[1], Jinping Xu[1], Jiandong Wang[1], and Xinyou Wu[2]

[1] School of Information Science and Technology, Xiamen University, Xiamen, Fujian 361005, China

[2] Zhangzhou Cambridge School, Zhangzhou, Fujian 363006, China

w.s_lin@yahoo.com.cn

Abstract. The present paper presents a novel heuristic algorithm, "BBF" algorithm, for the two-dimensional orthogonal rectangular packing problem. The BBF algorithm combines the main ideas of the Best-fit algorithm and the Bricklaying algorithm. Test data shows that the BBF algorithm compares with other heuristic algorithms in terms of both solution quality and efficiency.

Keywords: Rectangle Packing Problem, Heuristic Algorithm, Best-fit, Reference Brick.

1 Introduction

The orthogonal rectangle packing problem (also called two-dimensional orthogonal stock-cutting problem) is an important type of two-dimensional packing problems. It is to place rectangular items onto a rectangular sheet with infinite height in order to minimize the height of the nest, with restrict that all the items should have an edge orthogonal to an edge of the sheet, but rotations of 90 degrees of the items are allowed.

The orthogonal rectangle packing problem has numerous applications in the area of dynamic memory allocation, multiprocessor scheduling problems, and general layout problems (see [1-4]). Since A. E. Paull [5] proposed the newsprint layout problem, the orthogonal rectangle packing problem became a hot topic, and many algorithms were obtained [6-13]. As the orthogonal rectangle packing problem is NP hard, the algorithms can be broadly categorized into three methods: exact, heuristic, and meta-heuristic. Among the heuristic algorithms, the BL (Buttom-Left) algorithm [8] and BLF (Bottom-Left-Fill) algorithm [9] are most well-known, while the BF (Best-fit) algorithm [11] and the PH (Bricklaying) algorithm [12] seems to be the most outstanding. Our aim is to present a new heuristic algorithm that combines the main ideas of BF algorithm and PH algorithm.

* This work is partially supported by the Natural Science Foundation of Fujian Province of China Grant #2009J05015 to W. Lin and J. Wang.

D. Zeng (Ed.): ICAIC 2011, Part II, CCIS 225, pp. 638–643, 2011.

2 BF Algorithm and HP Algorithm

A. BF Algorithm

The main idea of the BF algorithm is, at every time step of the algorithm to find the lowest gap of the sheet and seek a "best-fitting" item to place within the gap. An remaining item is "best-fitting" for the gap if among all the remaining items it can occupy the width of the gap at the most, and its area is as large as possible. Once a best-fitting item is found, the item will be placed within the gap depending on a so-called niche-placement policy. If a gap is found for which no remaining rectangle can fit, then this is waste space, and the height of the gap will be raised up to the lowest neighbor.

B. HP Algorithm

The HP algorithm was motivated by the bricklaying method of construction workers during building a wall with bricks. The algorithm includes two stages, the preprocessing stage and packing stage. At the first stage the items are sorted into decreasing area order. At the packing stage, first place the first item in the list in the bottom-left of the sheet, and call the brick the "reference brick" and call the line along the upper edge of the item the "reference line". Then the sheet is separated into two parts, the lower space and upper space. Now place the former items in the list into the lower space from bottom to top and from left to right. When there is no item can be placed into the lower space, then place the first remaining item into the bottom-left of the upper space, and the item is referred to the new reference brick which gives rise to a higher reference line. Repeat the above procedure until all items are packed.

3 A New Heuristic Algorithm

A. An Overview

The main reason why the BF algorithm is regarded to be better than the BL and BLF algorithms is that, at every time step of packing it is not restricted to the first item encountered but to dynamically search the list for better candidate items for placement. However, in some time steps, the best-fitting items may occupy much height of the used sheet when their heights are much larger than their widths, which are expected to be rotated by 90 degrees before they are placed. On the other hand, the HP algorithm can overcomes the disadvantage of the BF algorithm by using the reference brick. Unfortunately, the HP algorithm does not adopt the best-fit strategy of the BF algorithm, i.e., at the time steps at the packing stage the HP algorithm does not dynamically select the best-fitting item for placement.

We present a new heuristic algorithm called BBF (Bricklaying Best-fit) algorithm. The BBF algorithm introduces the "reference brick" idea into the BF algorithm in order to overcome the disadvantage of the BF algorithm. Note that in our algorithm the best-fit strategy and the niche-placement policy are slightly different from those of the BF algorithm, and the placement policy of the reference bricks is different from that of the HP algorithm. In order to avoid confusion, we descript the procedure of our BBF algorithm below, and main teminologies that used by [11] and [12] are re-defined.

First we rotate each item so that its width is note less than its height, and sort the items into descending area order (resolving equal areas by decreasing widths). At the beginning of the packing process there will not be any items on the sheet. Place the first item (that has largest area) onto the bottom-left of the sheet. The placed brick is referred to a *reference brick*, and the line along the upper edge of the reference brick is called the *reference line*. A reference line is indexed by $y =$ its height from the bottom-left of the sheet. The space of the sheet below the reference line is called the *lower space*. Then find the *lowest gap* in the lower space. If there are more than one lowest gaps, we choose the widest one. If there are still more than one lowest gaps (with same width), then choose the most left one. Once a lowest gap is found, we examine the remaining items by their order in the item list to find the *best-fitting item*. The best-fitting item is the one that can be placed into the gap and after its placement its upper edge will not exceed the reference line, and with the largest area, and occupy the width of the lowest gap at the most. Note that, in the BF algorithm, the best-fit strategy priors to consider the width of the gap that the item occupies, instead of the area of the item. If a best-fitting item is found, then place the item within the lowest gap depending on the so-called *niche-placement policy*. That is, if after the placement of the item the height of the item's upper edge is more closed to that of the left neighbor's upper edge, then place the item next to the left neighbor, otherwise, place it next to the right neighbor. As the placement is restricted by the reference line, we call this a *reference packing* with the current reference line. If at some time step the reference packing is not available, then remove the current reference line and search the lowest gap for the placement of a new reference brick. In fact, the method to place of a new reference brick is a reference packing with reference line $y = +\infty$. If a gap for which no remaining rectangle can be reference packed within, even when the reference line is $y = +\infty$, then the gap is wasted space, and the relevant sheet array elements are raised up to the lower neighbour. We can now recheck for the lowest gap and continue with reference packing.

B. Implementation of BBF

Since floating-point data can be simply converted to integer format, we assume all the data (the widths and heights of the items, and the width of the sheet) are integers. The BBF algorithm includes two stages, the preprocessing stage and the packing stage.

Preprocessing Stage: The stage is almost the same with that of BF algorithm [11]. That is, we store the sheet as a linear array that has a number of elements equal to the width of the sheet. Each element of the array holds the total height of the packing at that x coordinate of the sheet as to reflect the skyline of the sheet. Therefore, the coordinate of the lowest space of the stock sheet can be found by locating the smallest-valued entry of the array. The width of the gap can be found by examining how many consecutive array items of equal value exist.

The item's data is defined as a list of rectangular items each denoted by a (width, height) pair. Note that different from the BF algorithm we sort the list of the items into decreasing area order (resolving equal area by decreasing width).

Packing Stage: First the reference line is $y = +\infty$, and the sheet is examined to find the lowest gap (initially at $x = 0$, $y = 0$, and lasting the entire sheet width). The item list is examined and the best-fitting item (at first time step at the packing stage it will be the first reference brick) returned. Reference pack the best-fitting item within the lowest gap. The item is assigned coordinates and removed from the item list. At the meantime, the relevant sheet array elements are incremented by the item height. If the placed item is a new reference brick, then the reference line is updated, in fact, just to $y =$ the maximum of the skyline. If a gap for which no remaining item can be reference packed within, then raise up the relevant sheet array elements to the lower neighbour, and update the reference line to be $y = +\infty$. The procedure continues until all the items are placed. The maximum of the final skyline returned as the solution of the problem.

Summary of Process: The whole process can be summarized by the following pseudo-code:

```
Obtain Sheet Width W
Obtain List of n items
Rotate each item so that Width ≥ Height
Sort Item List by Decreasing Area (resolving equal areas by decreasing widths)
Initialize Skyline Array of n Elements
Initialize Reference Line y = +∞
while Item Not Packed do
   Find Lowest Gap
   if(Find Best-Fitting Item = = True)
     Reference Pack Best-Fitting Item
     Raise Array to Appropriately Reflect Skyline
     Update Reference Line to y = Maximum of Skyline
   else
       if(Reference Line is Not y = +∞ )
          Update Reference Line to be y = +∞
       else
            Raise Gap to Lower Neighbor
       end if
   end if
end while
return Solution = maximum of Skyline
```

From the above pseudo-code, it is easy to see that, the time complexity is $O(n^2W)$. Futhermore, in order to improve the solution of the BBF algorithm we can adopt the same method with the HP algorithm [12] that change two certain items' order in the item list at the preprocessing stage. Thus the time complexity is $O(n^4W)$.

4 Test Data

In order to compare the relative performance of the pre-sented BBF algorithm to the BF and PH algorithms, we used several test problems from the literature. Perhaps the most extensive data sets given for this problem are found in Hopper and Turton [14], where 21 problem sets of rectangle data are presented in seven different-sized categories (each category has three problems of similar size and object dimension). Table 1 shows the test results. We also test the BBF algorithm with the ten instances with large item number n from literature [12]. The comparison of the BBF algorithm with the HP algorithm is shown in Table 2.

Table 1. Comparison of BBF and BF, HP with Data from [14]

Cat	Ins.	W	N	Opt	BF	PH	BBF
C1	C11	20	16	20	21	21	21
	C12		17		22	21	21
	C13		16		24	21	20
C2	C21	40	25	15	16	16	16
	C22		25		16	16	17
	C23		25		16	15	16
C3	C31	60	28	30	32	31	32
	C32		29		34	32	31
	C33		28		33	31	32
C4	C41	60	49	60	63	61	62
	C42		49		62	62	62
	C43		49		62	61	61
C5	C51	60	73	90	93	91	92
	C52		73		92	91	91
	C53		73		93	91	91
C6	C61	80	97	120	123	121	121
	C62		97		122	122	122
	C63		97		124	121	121
C7	C71	160	196	240	247	243	243
	C72		197		244	243	242
	C73		197		245	243	243

Table 2. Comparison of BBF and HP with Data from [12]

Ins.	W	N	Opt.	PH	BBF
RPP1	124	346	139	140	140
RPP2	148	406	131	132	132
RPP3	122	313	125	125	126
RPP4	143	403	139	140	140
RPP5	126	327	136	137	137
RPP6	139	418	146	147	147
RPP7	125	381	138	139	139
RPP8	140	390	126	127	127
RPP9	129	350	144	145	145
RPP10	123	375	141	142	142

References

1. Coffman Jr., E., Garey, M., Johnson, D.: An application of bin packing to multiprocessor scheduling. SIAM Journal on Computing 7(1), 1–17 (1978)
2. Coffman, E., Garey, M., Johason, D.: Approximation algorithms for bin packing problems – An update survey. In: Ausiello, G., Lucertini, M., Serafini, P. (eds.) Algorithm Design for Computer System Design, pp. 49–106. Springer, New York (1984)
3. Coffman Jr., E., Leighton, F.: A provably efficient algorithm for dynamic storage allocation. Journal of Computer and System Sciences 38(1), 2–35 (1989)
4. Dyckhoff, H.: A typology of cutting and packing problems. European Journal of Operational Research 44(2), 145–159 (1990)
5. Paull, A.: Linear programming: A key to optimum newsprint production. Pulp Paper Magazine of Canada 57, 85–90 (1956)
6. Christofides, N., Whitlock, C.: An algorithm for two-dimensional cutting problems. Operations Research 25(1), 30–44 (1977)
7. Beasley, J.: An exact two-dimensional non-guillotine cutting tree search procedure. Operations Research 33(1), 49–64 (1985)
8. Baker, B., Coffman, E., Rivest, R.L.: Orthogonal packing in two dimensions. SIAM Journal on Computing 9, 846–855 (1980)
9. Chazelle, B.: The bottom-left bin packing heuristic: An efficient implementation. IEEE Trans. Comput. 32(8), 697–707 (1983)
10. Liu, D., Teng, H.: An improved BL-algorithm for genetic algorithms of the orthogonal packing of rectangles. European Journal of Operational Reseach 112(2), 413–420 (1999)
11. Burke, E., Kendall, G., Whitwell, G.: A new placement heuristic for the orthogonal stock-cutting problem. Operatations Research 52(4), 655–671 (2004)
12. Zhang, D., Han, S., Jiang, Y.: A personification heuristic algorithm for the orthogonal stock-cutting problem. Chinese Journal of Computers 31(3), 509–513 (2008) (in Chinese)
13. Chen, D., Liu, J., Shang, M.: An intelligent enumerative algorithm for solving rectangle packing problem. Journal of Chongqing University of Post and Telecommunications (Natural Science) 20(4), 379–384 (2008) (in Chinese)
14. Hopper, E., Turton, B.: An empirical investigation of meta-heuristic and heuristic algorithms for a 2D packing problem. European Journal of Operational Research 128(1), 34–57 (2001)

Optimization and Control Researches into the Cooling System of Pneumatic Disc Brake*

Duan Zhengyong, Peng Yong, and Wu Heng

Xi'an Shiyou University
Shaanxi Xi'an, China
DuanZhengyong038@163.com

Abstract. This paper present some researches into the cooling and control system of pneumatic disc brake. Firstly, obtain that the friction torque has a great variation with the structure parameters of the pneumatic disc brake during running-in and non-running-in periods, then put forward the selection criterion of the friction torque when do strength check and thermal analysis for it. Secondly, based on the frictional mechanics, bring forward the heat transfer conditions when the brake works on the worst braking condition safely, furthermore, design some key parameters for this brake, to propose that to one type disc brake, its coefficient of rib must be more than 3.6. And finally, in terms of the design of the brake, determine that, for the cooling system, these parameters need to be monitored or controlled are the angular speed and torque of the shaft of the drawworks, the temperatures of the coolant at the inlet and the outlet, pressure of the compressed air which can be controlled manually, flows of the coolant and the cool water which controlled automatically.

Keywords: Pneumatic, Disc Brake, Cooling System, Optimization, Control.

1 Introduction

Recent years, hydraulic caliper disc brakes and pneumatic disc brakes, which are shown in Figures 1 and 2 respectively, are widespread used in oversea and inland oil fields, the former, driven by hydraulic pressure, usually is used as the main brake for rig, but the later, can be utilized as the main or an auxiliary brake.

Disk brakes have significant advantages in brake torque, control accuracy, reaction speed, safety and reliability, noises, maintenance such as replacement of vulnerable parts and labor intensity of operators[1-5], etc., so they are favored by oil fields operators. At present, these input products of pneumatic disc brake are dominant in domestic market. When design this type brake, there are four key factors: brake torque, heat dissipation of brake pairs, reaction speed, life of friction plate and its easy operation of replacement must be considered carefully. At the same time, this brake must equipped with huge forced cooling system and it must provides good cooling effect, otherwise, the temperatures of the friction pairs and the friction disc increase

* Supported Project: Specific Research Projects of Education Department of Shaanxi Province, Project NO: 08JK415.

D. Zeng (Ed.): ICAIC 2011, Part II, CCIS 225, pp. 644–652, 2011.

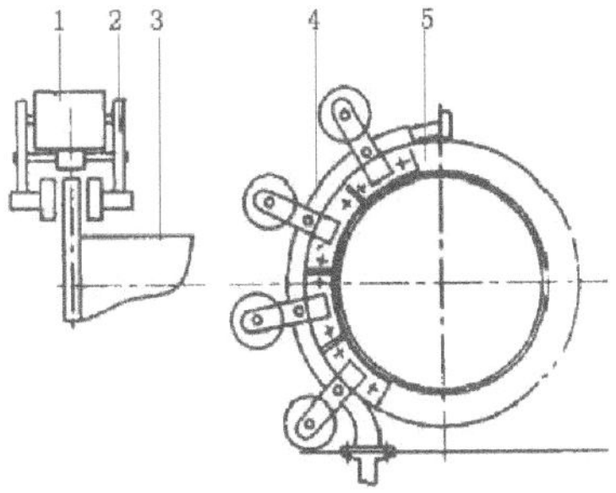

1-hydraulic cylinder, 2-caliper, 3-drum, 4-caliper carrier, 5-braking disc

Fig. 1. Schematic diagram of hydraulic disc brake

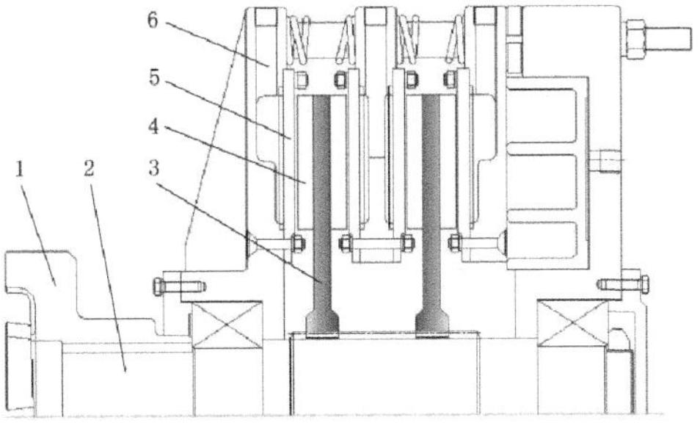

1-coupler, 2-shaft, 3-friction plate carrier, 4-friction plate, 5-friction heat transfer disc(disc for short form), 6-channel of coolant

Fig. 2. Schematic diagram of pneumatic disc brake

quickly, wear of the friction plate is serious, braking performance decrease sharply, which may result in tragic drilling accident, causing huge economic losses[6]. We can know that there are two factors affect the cooling effect of cooling system: internal and external. The internal factor is the friction disc's heat transfer power, which determined by the structure of the disc, and the external one is the parameters of the coolant and the heat exchanger. Therefore, in order to obtain a reliable cooling system, this paper has carried out some researches into the disc structure and the control system.

2 Thermal and Mechanics Analysis of the Brake

Because of the diversities of the well depths, maximum tensions in fast line and angular speeds of the drawworks drum shaft, the requirements of braking torque for the pneumatic disc brake are different. When single friction plate carrier(two friction pairs) can not provide the required braking torque, multiple carriers should be sued in the brake. For the safety of the braking system, in this situation, the action coefficient of each carrier must not be neglected, both in calculation of braking torque and the analysis of the cooling system. Following present the discussions of the heat transfer and mechanics of the cooling system.

2.1 Friction Braking Torque and Thermal Power

The friction braking torque and the friction thermal power both come from the resultant force of the pressure of pressed air and the spring's restoring force. By usage of the model of Figure 3 and Coulomb Friction Theory, these tow parameters can be computed.

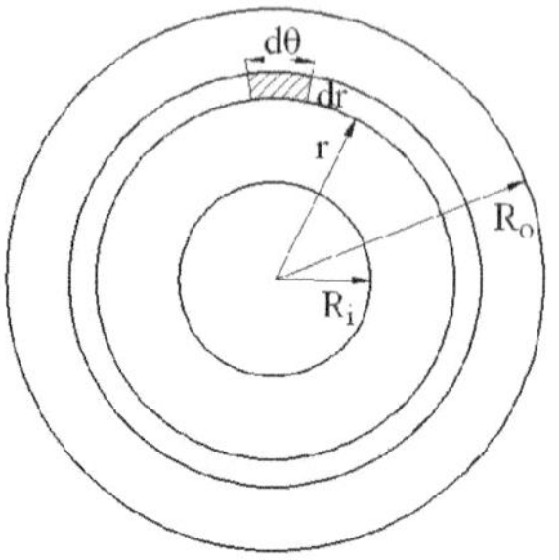

Fig. 3. Mechanics model of friction heat transfer disc

Where: R_o, R_i is the radii of the external and the internal respectively.

Provided that there is uniform pressure between the disc and the friction plate, the friction force generated on the infinitesimal area shown as the shaded zone during running-in period is:

$$df = \frac{\mu\left[\pi p(R_o^2 - R_i^2) - F_s\right]}{\pi(R_o^2 - R_i^2)} rdrd\theta \tag{1}$$

Where; μ is the coefficient of friction between the disc and the friction plate; p is the pressure of the pressed air. So, if there are x friction plate carriers, the total friction torque is:

$$T_1 = 2x\int_0^{2\pi}\int_{R_i}^{R_o} rdf = \frac{4\mu\varepsilon x\left[\pi p\left(R_o^2 - R_i^2\right) - F_s\right]}{3\left(R_o + R_i\right)}\left(R_o^2 + R_oR_i + R_i^2\right) \tag{2}$$

Where: ε is the action coefficient of each carrier.

After the running-in period, the torque can be computed by:

$$T_2 = \mu\varepsilon x\left[\pi p\left(R_o^2 - R_i^2\right) - F_s\right]\left(R_o + R_i\right) \tag{3}$$

So, the friction thermal power is:

During running-in period:

$$P_1 = \omega T_1 = \frac{2\pi\mu\varepsilon nx\left[\pi p\left(R_o^2 - R_i^2\right) - F_s\right]}{45\left(R_o + R_i\right)}\left(R_o^2 + R_oR_i + R_i^2\right) \tag{4}$$

After running-in period:

$$P_2 = \frac{\pi n}{30}T_2 = \frac{\pi n}{30}\mu\varepsilon x\left[\pi p\left(R_o^2 - R_i^2\right) - F_s\right]\left(R_o + R_i\right) \tag{5}$$

Where: n is the angular speed of the drawwworks' shaft.

Try

$$\eta = \frac{R_i}{R_o},$$

Theoretically,

$$0 \le \eta < 1,$$

obtained that

$$k = \frac{P_1}{P_2} = \frac{4}{3}\left[1 - \frac{\eta}{(1+\eta)^2}\right].$$

Obviously,

$$1 < k \le \frac{4}{3}.$$

The dimensionless curve of $k-\eta$ is shown as Figure 4.

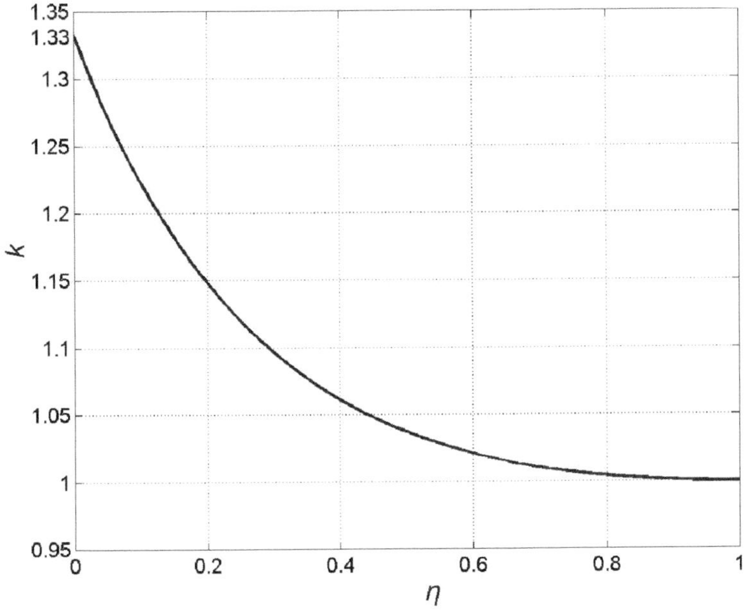

Fig. 4. Dimensionless curve of $k-\eta$

In practical brake design, η ranges 0.2-0.8, k varies in 1.01-1.15 correspondingly. So, when do check calculations of strength and thermal for brake, only using P_1 and P_2 respectively can promise the safety of the brake.

2.2 Technical Parameters of the Cooling System

Coefficient of heat transfer between the disc and the coolant is a very important technical parameter. It is related with the material and structure of the disc and the coefficient of heat transfer of the coolant. And it can be computed by:

$$k = \frac{1}{\delta\beta/\lambda + r + 1/h\eta}, \qquad (6)$$

Where: λ is the thermal conductivity of the disc material; h is the coefficient of heat transfer of coolant on the disc; δ is the thickness of the disc; r is the fouling factor; $\beta = A_2/A_1$, is called coefficient of rib, where A_1 is the area of the friction pair, A_2 is area of the contact surface between coolant and the ribs of the disc; η is the total efficiency of rib wall.

So, only the heat transfer power of the disc theoretically satisfies the following formula can it transfer the friction heat fully and timely, even neglecting the radiation heat and other heat dissipation forms.

$$P_c = k\beta A_1 \bullet \Delta t \geq P\zeta \tag{7}$$

Where: Δt is the temperature difference between both sides of the disc, $\Delta t = t_1 - t_2$, where t_2 is the temperature of the friction side of the disc(usually, when braking, it reaches to 300-1200°C[7-11]), and t_1 is the temperature of the another side; ζ is the efficiency of heat transfer; P is bigger one of P_1 and P_2.

Some things must point out that, μ is changed with t_2 , p and n , their relationships are: when t_2 increases, μ decreases, but t_2 reaches to a certain value, μ trends to a stable value; when p increases, μ decreases firstly, then trends to a constant, but finally would decrease; when n increases, μ increases or decreases, then increases and trends to a higher value compared with the initial one[9-11]. So, when the other conditions do not change, the more smaller t_2 is, that meaning the Δt is more smaller, the more μ is, equation (7) is more difficult to be satisfied; when t_2 and n do not change, the more smaller p is, the more smaller μ is, but p and μ vary in different magnitude, equation (7) is difficult to be satisfied still; at the same time, when t_2 and p do not change, n reaches to the maximum value, μ is still reaches to its maximum value, equation (7) is difficult to be satisfied. But, during pipe trips, when the load is much, the speed must be small, and vice versa.

From equations (4), (5) and (7) we know that if want to make the disc satisfy the requirement of heat transfer, we can use materials with good heat transfer performance, on one hand, by improving the β is another direct method.

At the same time, the flow q_1 of the coolant must can bring away these heat, i.e. q_1 must satisfies the following equation:

$$\frac{\Delta T C_o q_1}{60} \geq P\zeta \tag{8}$$

Where: C_0 is the specific heat of the coolant; ΔT is the temperature difference of the coolant at the outlet and the inlet.

2.3 Optimization of Key Parameters

Comparing with the pressed air pressure, the restoring force of the spring can be neglected. Furthermore, this means a safer design as to the thermal performance of the brake. Actually, after determinations of R_o and R_i, k depends on δ and β only. δ

can not be too smaller due to the strength problem. So, increasing β is the only way to improve k. But, too big β will results manufacture difficulties in the disc.

As far as one type of this brake concerned, the known parameters include: $n = 400$ r/min, $R_o = 0.38$ m, $R_i = 0.22$ m, $p = 0.6$ MPa, $\mu = 0.36$, $\Delta t = 800$ K, $\lambda = 398$ W/(mK), $h = 6000$ W/(m^{2}K), $r = 0.0002$ (m^{2}K/W), $\eta = \zeta = 0.9$, $\delta = 0.015$ m. In terms of available data, can determine Δt ranges in [200,900]°C usually, and considering the manufacturability of the disc, set β in [1,5]. In addition, definiteτ as the rate ratio of the heat transfer power of the disc and the friction heat power in the worst braking condition of the brake. So obtain the τ-β-Δt surface like Figure 5 during running-in period.

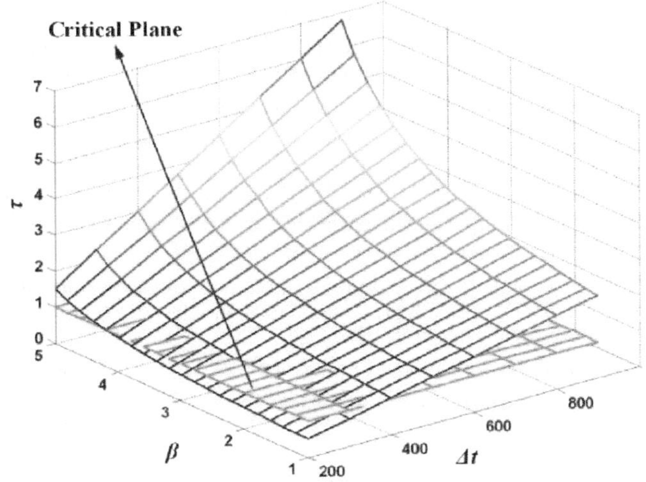

Fig. 5. τ-β-Δt

Shown as Figure 5, the surface of τ=1 is denoted as the critical plane. The figure indicates that most ofτ above the critical plane, that means the disc satisfies equation (7), further, the more the Δt is, the smaller the β is. So the key job is to guarantee the disc can satisfy equation (7) when Δt is small relatively. When Δt =250°C, β =3.6. After the running-in period, $p_2 < p_1$, the surface ofτ is higher than the one shown in Figure 5.

3 Control Issues of the Cooling System

The schematic diagram of the control system of the cooling system is shown as Figure 6.

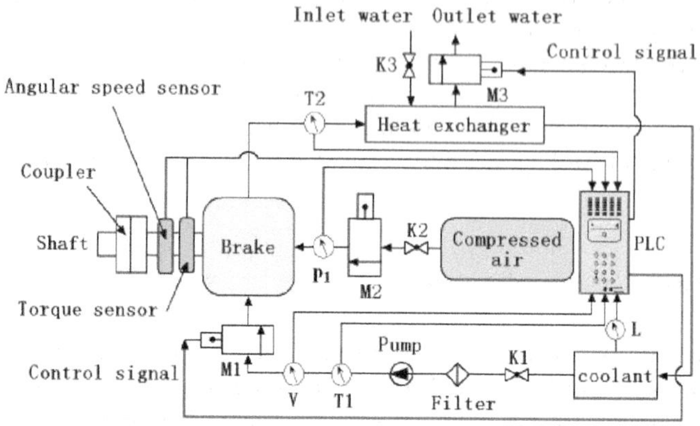

Fig. 6. Schematic diagram of the cooling system

Where: K1, K2, K3 are solenoid valves; M1 and M3 are automated valves of flow; M2 is manual valve of flow; T1 are T2 temperature sensors; V is the sensor of flow; p1 is pressure sensor and L is the sensor of liquid level.

According previous discussions and Figure 6, there are several key factors which affect te cooling effect:

- The friction heat power when braking *P*;
- The temperature difference of the coolant at the outlet and the inlet $\Delta T = T_2 - T_1$;
- The flow q_1 of the entry of the brake;
- The flow q_2 of the export of the heat exchanger and its performance.

During braking, the driller operates M2 to adjust the p1 to obtain different braking torque *T*, then slowing the angular speed of shaft of the drawworks. Actually, the friction heat power $P = T\pi n/30$. A good cooling effect, constraining ΔT in the acceptable range, need sufficient q_1 and q_2 . But after determinations of the structure of the brake, *P* depends on p1 and *n*, and *T* depends on p1, ΔT depends on *P*, q_1 and q_2 . So, when operating M2, the system must can monitor *n*, *T* and ΔT , use these signal to generate the control signals which are utilized to control q_1 and q_2 automatically. Here emphasize that the signal of the torque sensor indicates the practical friction torque generated from the friction pairs, which exhibits the performance of the brake directly, at the same time, it can give the driller immediate operation tips and can also be used as the warning signal when the brake is in braking failure. Therefore achieve the schematic diagram of the control system shown as Figure 7.

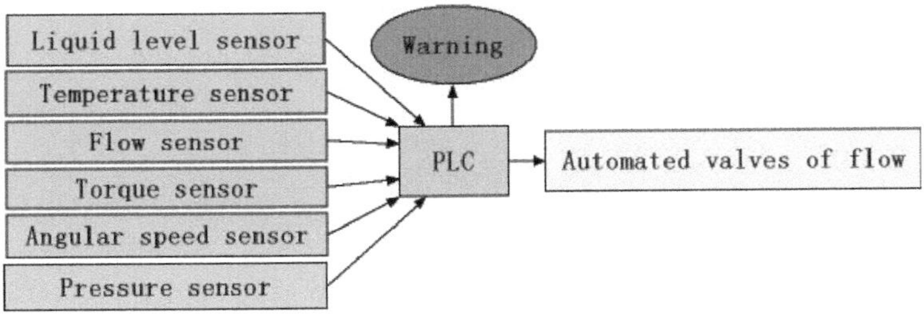

Fig. 7. Schematic diagram of control system

The control signal which is used to control q_1 can be obtained from equation (8) as:

$$q_1 = \frac{2\pi P n \zeta}{\Delta T C_o} \tag{9}$$

For the control signal of q_2, it can be determined after selection of the heat exchanger. These can avoid unnecessary large displacement to save energy under good cooling effect.

4 Conclusion

There are several key factors which must be put great attentions to when design or utilize a pneumatic disc brake, especially, the cooling system. The heat power which the disc can transfer and the coolant can bring away must be more than the friction heat power generated when the brake works on any situations. So the coefficient of rib of the disc should not be less than 3.6. At the same time, these parameters, such as the angular speed and torque of the shaft of the drawworks, the temperatures of the coolant at the inlet and the outlet, pressure of the compressed air, flows of the coolant and the cool water, are needed to monitored or controlled to insure the cooling effect of the system.

References

1. Cai, j., Zhang, W., Pu, X.: Elementary Introduction of Function and Application of Main and Auxiliary Brakes on Drawworks. Oil Filed Equipment 36(4), 25–27 (2007)
2. Gao, X., Ma, Q.: New Development on Technology of Disc Brake for Drilling Rig. Oil Filed Equipment 35(3), 92–94 (2006)
3. Suo, C.: The Development Status and Trends of the Brake System of Drilling Rig Drawworks. Oil Filed Equipment 34(1), 105–106 (2005)
4. Zhang, Y., Wang, A., Li, B., et al.: Discussion on Development and Application of Disc Brake. Oil Filed Equipment 30(6), 23–26 (2001)
5. Duan, Z., Zhang, H., Peng, Y., et al.: Development of Environment Pneumatic Disc brake for Rigs. Natural Gas Industry 28(9), 56–59 (2008)
6. Bai, W., Ji, Z., Yang, F., et al.: Forced Cooling System for brake Mechanism of Drawworks on Drilling Rigs. China Petroleum Machinery 31(7), 27–28, 55 (2003)

7. Zhang, W., Ma, g., Zhang, L., et al.: Application problem of disk brake on drilling rig drawworks and the solve. Oil Filed Equipment 35(1), 89–90 (2006)
8. Yim, Z., Han, Z.H., Chen, H., et al.: The Study on the Brake Friction Characteristic of Plasma Sp rayed Fe2Ni2Co2WC Coating. Tribology 16(2), 150–155 (1996)
9. Wang, X., Zhang, S., Fan, Q.: Experimental Study in Screening and Matching of Disc brake Pair Materials of Drilling Rig. Tribology 22(3), 197–201 (2002)
10. Zhai, Y., Zhang, S.: An Experimental Investigation of Friction and Wear Characteristics of Home-Made Drilling Rigs Brake Pairs. Journal of the University of Petroleum 2, 153–162 (1983)
11. Wang, X., Zhang, S., Wang, D.: Research on Friction and Wear Performance of Disc-brake Pair Materials for Drilling Rig. Lubrication Engineering 32(12), 5–8 (2007)

Application and Implementation of Campus Information Integrated Base on Web Service

Xiong Songquan

Ningbo Dahongying University
Ningbo, Zhejiang 315175, China
beginnow@home.news.cn

Abstract. In the current network, it is the key of the information integration research how to construct a new information service based on the old systems and databases. In the article the author put forward a solution to achieve campus information integration that how to use web service technology to connect the different networks and databases. The author analyzed the system requirements and the application environment, and compared the different technical solutions, and then selected a technical line which is based on non-real-time Web Service, and provided a system structure model, and implemented the key functions. By the test online, the web service realized the connections among the different networks and databases. The non-real-time method improved the system performance.

Keywords: Campus information integration, Web Service, College PE.

1 Overview

Following the teaching reforms development, more and more college PE courses being replaced by sports clubs; it is an increasingly urgent thing speeding up the information construction on college PE. On the one hand, college students on sports courses have the more individual demand in modern education. It is the trend that sports clubs are growing and traditional classes will be broken [1]. Comparing with the traditional teaching methods, the new teaching model increased the enormous difficulties and workload on teaching management and courses evaluation. If no information system implemented to help sports clubs, it would be difficult that the teaching reforms will be carried out; On the other hand, with growing in size of college, school students have risen sharply, but athletic facilities are lagging behind. The per capita share of athletic facilities declines, in order to reduce the investment costs and maximize the utilization of athletic facilities, it is necessary to implement information management based on campus information integration of the old system. In response to this situation, Ningbo Dahongying University has developed a set of programs called "College Student PE Management System (CSPEMS)", using the information technology to achieve integration of the campus card databases, clubs databases and teaching database. After system implemented,

D. Zeng (Ed.): ICAIC 2011, Part II, CCIS 225, pp. 653–659, 2011.

students could select their interested clubs in the ZFsoft teaching system, register their morning exercise activities and club activities by their campus card; teachers could get their weekly reports and term reports about students' register times, to converting the course grade to be imported the ZFsoft teaching system. Therefore, the CSPEMS greatly improves the efficiency of the original data collection, reducing the manual labor.

2 SYSTEM Requirement Analysis

There are two main types of college students sporting activities: Morning exercises and club activities. Morning exercises are that students must be the activities within the specified time, such as morning-running, shadowboxing, etc. Before students have sports, they must be swiped their campus card at the specified place as request; Club activities are the specific sports like as tennis ball, basketball, etc. Every student belongs to someone club, which is selected by himself with his hobby at the beginning of the semester. The clubs open on sometime every week on time. Before accessing or leaving a club, students must swipe their campus card through IC reader terminal. If the interval is less than the preset-time, leaving to swipe card will be decline.

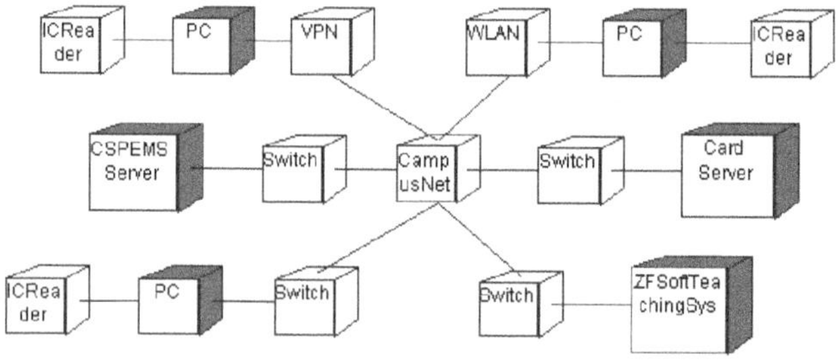

Fig. 1. The Campus Network of CSPEMS

System is running on the campus network environment, as shown in Figure 1. Considered the limits of being applied multi-campuses and wireless network, CSPEMS should be able to adapt to the entrance in VPN and WLAN, and the single response time must be less than 2 seconds for swiping card. All the clubs information, students information, teachers information and selection course information, come from ZFSoft Teaching System (Developed by ZFSoft CO., LPT), which uses the Oracle database system, and all the cards information come from Campus card system (Developed by Zhenyuan CO., LPT), which using SQL Server 2000 database. The CSPEMS should be to achieve data integration with these systems to process the data, and it is able to provide various reports to print for all the roles, has a configuration interface through web. In summary, the CSPEMS should include these functions: the interface program of ZFSoft teaching system, the interface program of Zhenyuan card system, the register program at client-side, the on-duty program for counting data at server side and the web sub-system for querying and configuration.

(1) The interface program of ZFSoft provides a interface to get the basic data from ZFSoft teaching system. However, selecting clubs is not the functions of CSPEMS, but ZFSoft.

(2) The main function of the interface program of Zhenyuan card system is providing the student NO corresponding to the card NO.

(3) The register program at client-side is running at PC connecting to the campus network. There are more than 30 computers as the client machines. Every computer has one or two IC Reader Terminal keeping connection. This program is responsible for sending the swiping card data to the CSPEMS to be stored.

(4) The on-duty program at the server side is responsible for processing the data. According to the rules, the program analyzes, clears up, transfers and sums to the data. It is be running at every midnight at the server side for counting daily and weekly and importing the data from ZFSoft system and Zhenyuan card system.

(5) The web sub-system provides the interface for querying and configuration.

3 System Design

A. Technology line

1) Communication mode

In the. NET distributed application systems, Web service,. NET Remoting, MSMQ, and Enterprise service are four usual technologies. Web service is a cross-platform technology, having nothing to do with programming language, operation system and various application systems. It's a perfect solution for high reusability of the distributed applications; . NET Remoting is a corresponding DCOM technology, which client-side program is able to access the server-side program's appdomain to get the server-side data. It is more safe and is implemented and configrationed easily, supporting for a lot of protocols; Enterprise service is a kinds of technology for managed code components and applications based on COM +, and integrates many .NET features, which makes it easier to access these services. Comparing with the structure, the client and the server are a loose coupling through Web Service communication. Each of the sides changing some function, the whole system's running is immunized, but the performance will decline by serialized and deserialized frequently. To the. NET Remoting, it often needs to open the specific ports to carry out binary data, which usually is forbidden by the network administrator, meanwhile it keeps a strong coupling relationship among the parts of the whole system. So .NET Remoting is more fit to apply at enterprise network environment. To Enterprise service, the main application is that it has high operational capability distributed transaction[2].

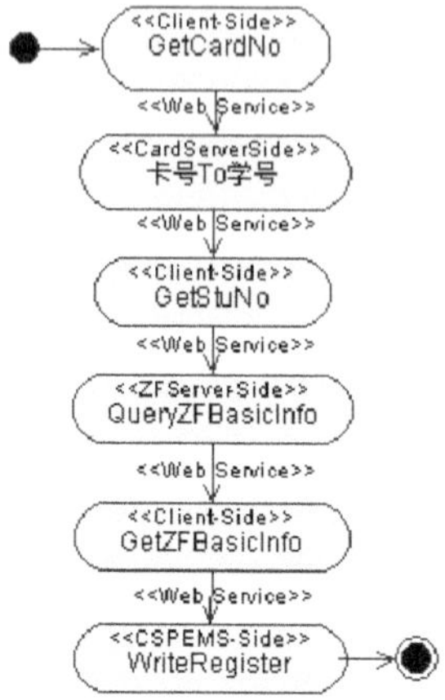

Fig. 2. The real-time swiping card models

To the CSPEMS, the ZFSoft teaching system and the Zhenyuan card system running independently just only provide the basic data. It is obviously appropriate that .NET Remoting technology is used in the CSPEMS, because the client can not be allowed to access the server application domain in this system. If doing this, it must be lead to high developing cost. What's more, every the swiping card just writes its own data to database, no involving in conflict of saving data to database at the same time. So it is no necessary to hold a high transaction level in the CSPEMS, but it is necessary to transfer the data between the different campuses. In summary, Web Service is the best choice.

2) Whether on real-time or not.

The process of swiping card may be designed on real-time, shown in Figure 2, the steps are these: ① The client program gets a card number by capturing the keyword events; ② The client program sends the card number to the interface program of Zhenyuan card system through the Web Service for converting to student ID; ③ The interface program of Zhenyuan card system returns the student ID through We Service; ④The client program sends the student ID to ZFSoft teaching system for requesting to query the basic teaching information through the Web Service;⑤The interface program returns the teaching data through Web Service; ⑥ After computing on the valid rules, the client program write the valid data to the CSPEMS server;⑦ The interface program returns the executed status of saving the data on the server side; ⑧ The client program displays the status message. There are sex times Web Service communication during one single swiping card, which requires four seconds. However, only permitted within two seconds, the design does not meet the requirements.

The most advantages of real-time systems are the accuracy of the online data. All the changes of the data in ZFSoft teaching system or Zhenyuan card system will be reflected to the CSPEMS in real. Considering non-real-time system, the process of swiping card is able to be designed, too. The basic data of ZFSoft teaching system and Zhenyuan card system are kept a copy on the CSPEMS server, which is refreshed at every midnight. By this way, the steps to communication is simplified greatly, the six times is reduced to two times using Web Service to communicate. But it will lead to some problems as follow: ① If someone student loses his campus card and regains a new card on the day, the CSPEMS will be not able to recognize the new card; ②If someone student's information has been changed, the changes will be unknown by the CSPEMS on the current day. Considered of the problems, the CSPEMS provides a new way to register by typing the student ID. After a number of tests, the system in non-real-time requires 0.9 seconds during one single swiping card.

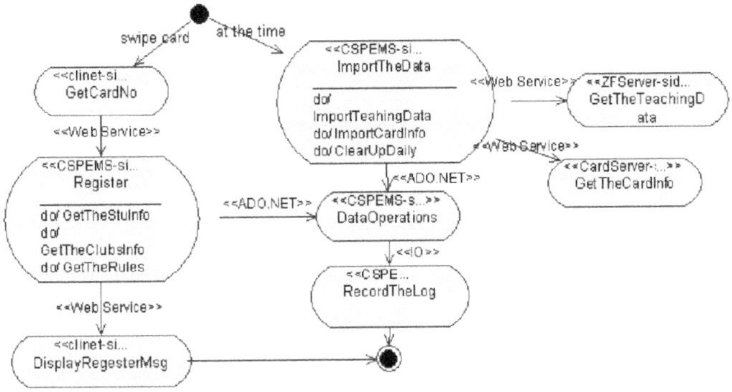

Fig. 3. Semi-real-time online credit card processes

B. System structure

In physics, this system consists of IC card reader terminal, pc client, the CSPEMS server, the ZFSoft teaching server and the Zhenyuan card server; In logic, is consists of ClubClient.exe, GtbLib.dll, GtbDal.dll, ZfOnduty.exe, CardDal.dll and ZfLib.dll. ClubClient.exe and GtbLib.dll are running in the client computers, and GtbDal.dll, ZfOnduty.exe, CardDal.dll, Zflib.dll and web sub-system are running in the CSPEMS server. The system architecture as shown in Figure 4:

(1) ClubClient.exe is the program of swiping card to register;
(2) GtbLib.dll is the interface program of the CSPEMS;
(3) GtbDal.dll is the data access program of the CSPEMS;
(4) ZfOnduty.exe is the on-duty program running in the CSPEMS server;
(5) CardDal.dll is the interface program of the Zhenyuan card system;
(6) ZfLib.dll is the interface program of the ZFSoft teaching system.

4 System Implementation

A. Applying SOAP header to achieve the security

For the Web service consumer programs, every Web Service method requests are related to security certification problems, in order to guarantee that only authorized users are allowed to call the Web Service method. In this system, a custom SOAP header class is declared deriving from the class of Web Service Authentication-Module, and it declarations user authentication information. Each calling the methods of Web Service method, the header objects information will be sent to the server with other soap message, and then is certified at the server side [3]. The achievement codes are as follows:

```
//Web service consumer-side call Web Service method
public GtbReader(string acc,string pwd)
{
_curAccount =
 new GtbLib.GtbWebService.AuthHeader();
_curAccount.UserName = acc;
_curAccount.Password = PubSystem.Safe.MD5(pwd);
_gtbService = new
GtbLib.GtbWebService.PtbDataService();
_gtbService.AuthHeaderValue = _curAccount;
}
//Web service consumer-side call Web Service method
public AuthHeader sHeader;
//Record credit card
[WebMethod]
[SoapHeader("sHeader")]
public string  RecordRefreshCard(......)
{
if (ValidAuthHeader(sHeader)) //If authorized
{ ......}
}
```

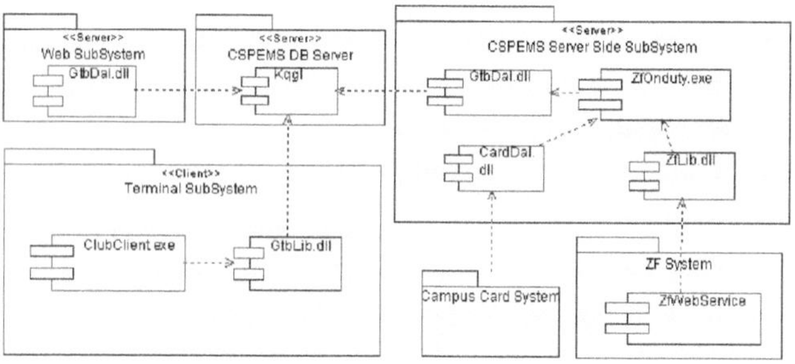

Fig. 4. The system architecture diagram of the CSPEMS

B. Process the swiping card input through global Hook

The IC card reader terminal is USB ID card reader, its type is 8H10D. The device with the PC is through the USB interface connected, and automatically converts the data inputting to keyboard events. On loading of the register program at client-side, the global keyboard hook is started to capture the keyboard events. The implementation codes are as follows:

```
if (!_isStarted && _hookType != 0)
{
//Create a hook callback delegate
_hookCallback =
new HookProc(HookCallbackProcedure);
//Creating Global Hook handle
_handleToHook =
SetWindowsHookEx(_hookType,_hookCallback,
Marshal.GetHINSTANCE(
Assembly.GetExecutingAssembly().GetModules()[0]),0);
}
```

5 Conclusions

This article gives a technology solution of campus information integration using Web Service and in non-real-time ways, which is implemented in the CSPEMS. It achieves good results that the utilization ratio of the athletic facilities has been risen and the intensity of the work of the original data collections has been reduced. The CSPEMS has a loose coupling structure to benefit to campus information integration. It has great application value.

References

1. Shu, N.: Multidimensional Thinking of University's Sports and Health Course Reform in the Transformation Period of the Society. Sichuan Sports Science 4, 140–142 (2005)
2. Ren, E.-m.: Analysis and Research of Distributed Communication Technology under.NET Framework. Information Technology & Informationization, 46–47 (May 2006)
3. Hu, F.-x., Zeng, y., Gao, M.: Web Services Application and Discussion. Computer Science 34(4), 75–77 (2007)

Resource Description and Discovery of Campus Grid

Zhi-yun Zheng, Jun-xia Zhao, Xing-jin Zhang, and Lun Li

School of Information Engineering
University of Zhengzhou
Zhengzhou 450001, Henan Province, China
Zhi.yzeng@ovi.com

Abstract. Campus grid construction of is an effective way to solve the island problem in the utilization of campus network resources. An appropriate and clear description is the prerequisite point. A set of categories describing RDF is proposed according to the characteristics of campus resources. In this paper, the description vocabulary set is first given based on a resource classification registration model. And then a query model is provided based on triples through the RDF triple model described in the analysis. Finally, the description, registration and retrieval of the resources are all realized by using the resource sharing prototype system of the campus grid. Experiments show that the described method is feasible and effective.

Keywords: Campus Grid, Resource Description, RDF, Vocabulary Set, Resource Discovery.

1 Introduction

The network resources play key roles in promoting scientific research with the popularity of the campus network, but also create new problems, such as the resource structures are unequal and the resource type is different in various sectors. Moreover, all departments have the resources polarization, making the campus network resources to a resource island, and then the full sharing of resources can not be done.

The grid integrates the whole internet into a super computer, realizing the comprehensive share of computing resources, storage resources, information resources, soft resources, communication resources, knowledge resources and specialist resource etc [1]. Campus grid is the application of grid technology on campus [2], and by integrating distributed and heterogeneous resources through grid technology and campus network, it constructs a unified platform to provide transparent service to teachers and students on campus, thus effectively solves the island problem of campus network resources.

The distributive and heterogeneous nature of the campus resources decides that the resource description be the key point in the construction of campus grid. An appropriate and clear description will surely improve the resource utilization. The commonly resource description languages are the Interface Definition Language(IDL)[3], the Job Description Language (JDL)[4], the Generic Grid Resource Description(GGRD), the Resource Specification Language(RSL), the Web Service Description Language(WSDL), the

D. Zeng (Ed.): ICAIC 2011, Part II, CCIS 225, pp. 660–668, 2011.

Resource Description Language (RDF), etc. The limitations with the above languages are all obvious. IDL has poor ability to extend. RSL and JDL, though strong in pertinence, are poor in universality [5]. GGRD stresses on resource description of and requirement, which can not be applied widely. WSDL can only describe coarse resources. However, RDF, with its open metadata description frame, can be used to describe resources of distributive nature [6]. Thus, RDF has the strong incomparable advantage in grid resource description. But RDF itself is only a resource description frame, which is domain-independent, and defines nothing as to what words can be used to describe such resources. Here in this paper, a method of describing resources by combining RDF and vocabulary classification is proposed, in an attempt to construct a resource vocabulary set. Finally, a classified registration model and a query model are given.

2 Description of Campus Network Resource

A. Resource of Campus Grid

The campus resources include two parts, the hard resources and the soft resources. Hard resources are physical entities that are tangible and visible, such as precision instruments for scientific research and experimentation in the colleges, data-storage equipment, high performance computers and personal computers. Soft resources are data, WebPages, scientific research records of colleges and electronic resources in the personal computer.

B. Description Method of Campus Grid Resource

RDF can be used to describe campus resources indistinguishably in a unified way. Meanwhile, a complete set of supporting metadata vocabularies is created on the campus soft and hard resources in order to facilitate data processing and to avoid conflicts. It is divided into three categories: public information, soft resources and hard resources.

Public information vocabulary defines common property of soft and hard resources which include URI, resources name, resource description, resource type, resource keys, the permission start time of access to the resource (acessStartTime), the permission end time of access to the resource (acessEndTime), the valid term of access to the resource (validTerms), campus domain name (domainName), resource publisher, resource reliability, resource registration date, download scores, etc.

Soft resource vocabulary set defines scientific research records of colleges, WebPages, courseware and all kinds of video and audio, which include resource format, resource size, the author of resource, resource language, resource type,permission right , the resource subject, etc..

Hard resource vocabulary set defines the various instruments, data-storage equipment, high-performance computer, personal computer, etc. Comparatively speaking, it is more complicated. For example, PC includes four kinds of resources which is CPU, memory, storage and main engine, and CPU vocabulary set also includes clock speed, multiple frequency, type, specifications, etc. While hard disk vocabulary set includes type, capacity, speed and other properties.

As is shown in table 1, soft resource vocabulary is taken for example.

Table 1. The vocabulary set of campus grid resources: soft resources

Resource Classification	Label	Comment
Soft resource	format	format of the resource
	size	size of the resource
	author	author of the resource
	language	language of the resource
	softType	type of the resource
	right	right of access to resources
	subject	subject of the resource

The followed is the grammar definition of related soft resource vocabulary. (Limited space, only part of the code is given)

```
<rdf:RDF>
   <rdf:Property
rdf:about="http://222.22.65.252/rdf/schema/softPredicate#format">
   <rdfs:label>format</rdfs:label>
   <rdfs:comment>The format of the resource   </rdfs:comment>
   <rdfs:range rdf:resouorce="&xsd;string"/>
   <rdfs:domain
rdf:resource="http://www.w3.org/2000/01/rdf-schema#Resource"/>
   </rdf:Property>
   ......
   </rdf:RDF>
```

RDF+vocabulary has the advantage of resource description as follows:

1) RDF description document, which can be used in a variety of platforms for data communication and exchange, is more formal than XML document and also it is more power in the documentation operations and resource description.

2) The vocabulary which is based on the features of campus resource is able to describe resource more exactly.

3) RDF supports user-defined vocabulary, and the RDF Schema provides powerful function for defining grammar and specification.

4) It is convenient to expand vocabulary and property, and the new addition predicate is irrespective of the existed document structure. What's more, there will be no redundant data added by describing property.

C. Framework of Campus Grid

According to the characteristic of campus resource, a three-layer network structure is designed which includes User Interface, Middle Layer and Resource Layer, as shown in Figure 1.

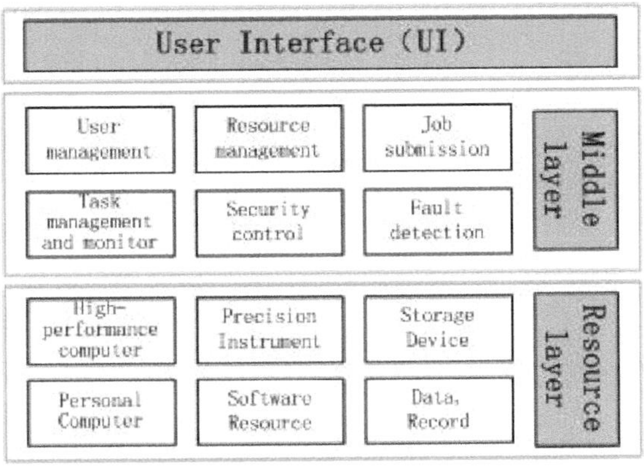

Fig. 1. The architecture diagram of campus grid

Particularly, the function of every layer is as follows:

User Interface: implementation of user requirement. With the interface supported by user layer grid users can submit and manage resources after logging in the grid platform or they can submit mission requirement to get enough resources.

Middle Layer: key part in the grid. It implements resource management, task scheduling, safety control and so on. It also shields isomerism of grid resource so that the platform could provide transparent and uniform service.

Resource Layer: basic layer of grid. It is made of all the resources and its function is to provide interface and method to control these resources.

3 Discovery Mechanism of Campus Grid Resource

A. Model of resource registration

In order to find resources and locate them quickly, a resource classification registration model is put forward which can be used to register resource by category, as shown in Figure 2.

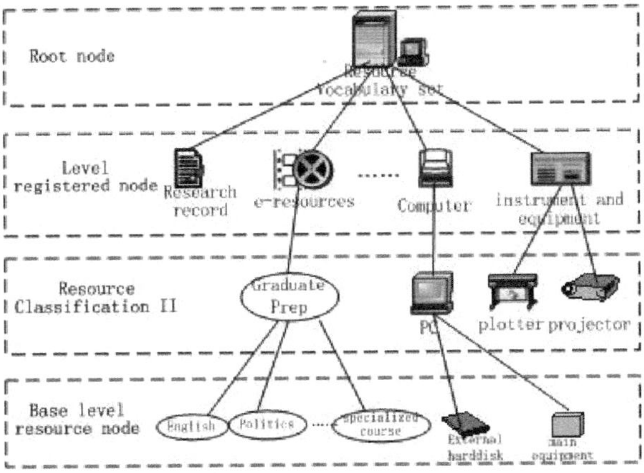

Fig. 2. The classification model of campus grid resources

Root Node: vocabulary registration center. The node includes vocabulary definition and resources Classificationon the submitted description.

Level registered Node: resource registration center node, classification information of the resource description stored in the node. According to description and classification of vocabulary, the node saves the description information into the appropriate node.

Resource Classification II Node: the further classification of resources can be extended several times, the more detail you give, the more precise resource you will find.

Base Level Resource Node: the description data of resource. Each node corresponds to a specific resource.

When resource information is registered, the root nodes will analysis the description of resources and save the description information into the appropriate storage node according to the related classification of resources. As the definition of vocabulary can describe the kernel information and the classification of resources very well, the resources that are submitted in the vocabulary registration center will be classified exactly. In the same way, if the definition of vocabulary is more precise and then the classification of resources is clearer, then the search result of resources will be more precise and more efficient.

B. Resource discovery

Resource discovery is the embodiment of user needs, and the accurate description of demand is prerequisite for effective resource discovery. It can be seen from the foregoing that the description of campus grid resources is based on the form of RDF+vocabulary. The vocabulary set which is obtained by the abstract characteristics of the resource itself would meet the formal description of resources, and RDF provides support for vocabulary and the search of the RDF document data characteristics. So the user needs can be described by using RDF triple model.

The format of RDF triple model is :< Subject, predicate, object> (abbreviated as <S, P, O>) [7], and requirement description can be expressed as <?S, ?P, ?O> ,and its meaning is to list all the grid resources. The "?" represents the variable prefix of query, and "S", "P", "O" is the result of query which returns the variable. If the return result is barring "?P", it means that the value of "?P" is not required in the query. Every item in the triple model has two functions:

1) Assign a value to it as a query parameter, and discover resources whose value satisfies the query.

2) Inquired as the characteristics of required resources and returned as the result of the query.

An instance that describes users' request for resources by RDF triples is given below.

Assume that job processing needs of resource users:

1) CPU, Clock Speed at 3GHz or more;

2) The resources belong to the aServer, and it is also said that the value of the domain is A1.

The relevant RDF description of demand can be expressed as the following shows:

<?res,mainFrenquency,”>3”>
<?res, domain,”A1”>

What it means to find clock speed at 3GHz or more and the CPU registered in the aServer.

The predicate is the most important part of the triple. The more extensive and complete the definition of the predicate is, the more precise the description of user needs will be.

4 The Realization of Campus Grid Resource Sharing Prototype System

To test the feasibility of the resource description based on RDF + vocabulary and resource classification register model system, a campus grid resource sharing prototype system is developed by using SQL Server 2000 database, and JBOSS as the server platform under the development environment including Eclipse,JDK6.0,Jena and ARQ distribution package.The prototype system is mainly achieved these functions, such as, description of the resource, resource registration, the generation of RDF document, queries of the resource, and so on.

A. Resource Registration

The interface of resource classification and registration is shown in Figure 3.The user can use it to register soft or hard resources.

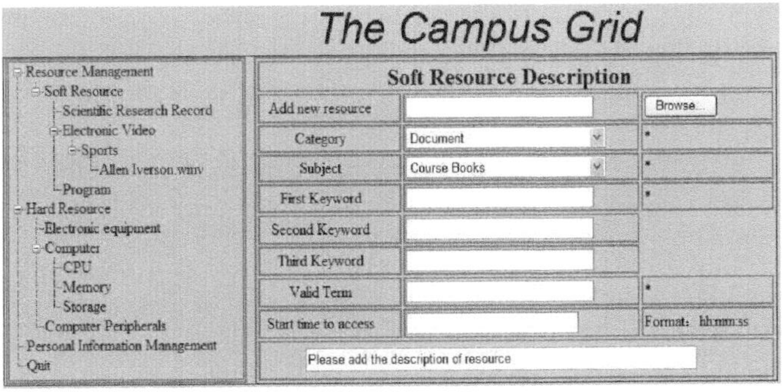

Fig. 3. The registration page of soft resources

After the server get the description information related to the resources, the information is translated into the RDF document handled by the core classes. For example, to register the soft resources – the video of "Grid Computing.mp3", its corresponding RDF document is as follows.

```
<rdf:RDF
  xmlns:rdf="http://www.w3.org/1999/02/22-rdf-syntax-ns#"
  xmlns:commonSoft="http://222.22.65.252/rdf/schema/softPredicate">
  <rdf:Description about="http://222.22.65.252/rdf/common/grid
computing.mp3">
 <commonSoft:name> Grid Computing Courseware </commonSoft:name>
 <commonSoft:description>It is the Video explanation of grid computing
courseware which is taught by Dr. Liu Peng ,Tsinghua University
</commonSoft:description>、 <commonSoft:type> soft resource
</commonSoft:type>
 <commonSoft:validTerm>5</commonSoft:validTerm>
 <commonSoft:format>.mp3</commonSoft:format>
 <commonSoft:size> 34988310</commonSoft:size>
 <commonSoft:softType> Electronic Video </commonSoft:softType>
 <commonSoft:subject> Courseware </commonSoft:subject>
 </rdf:Description>
</rdf:RDF>
```

When the resource had been described in the vocabulary, the descriptions will have the classified information of it, and then the registration stored it to the resource node "Electronic Video – Courseware" by matching its category.

B. Resource Query

Represented by RDF triple <?s?p,?o> to describe resource requirements, the queries of resources are mainly three types:

1) Querying by verb and object type, it means to query resource through characteristics of the resource and its property value.
2) Querying by verb or object type, choosing one of them, once two of them are selected, it would be the first query type.
3) Querying by object type.

The query algorithm is as follows:

{
fileAddress specifies the path to read the RDF file;
readRDF method read a RDF file to a model;
queryStr describes the demand for information;
output the resource which has been queried from ***model.***
}

queryStr model and relational database query are similar, such as,"?" represents for require or unknown, and followed by the parameter variable returned from the results, "where" statement is used to determine the subject, object or predicate of query parameters, "select" statement specifies the query results which are returned.

For example, when using object query type, if users only knows that the format of a resource is "chm", while cannot determine the characteristics label of the predicate "format", users can query by using object, the core code is as follows:

```
String fileAddress="D:\\jboss\\server\\default\\deploy\\RDF\\admin.rdf";
Model model=CommonRDF.readRDF(fileAddress);
String queryStr=      "PREFIX commonSoft:
<http://222.22.65.252/rdf/schema/softPredicate#/> "+
            "select ?s ?p where { ?s ?p \".chm\" .}";
Query query=QueryFactory.create(queryStr);
QueryExecution qe=QueryExecutionFactory.create(query,model);
try{
 ResultSet results=qe.execSelect();
 for(;results.hasNext();){
 QuerySolution soln=results.nextSolution();
 RDFNode x=soln.get("p");
 Resource r=soln.getResource("s");
 System.out.println("The resource URI to meet the requirements: "+r);
 System.out.println("The resource URI of the format predicate to meet the
requirements: "+x);
 }
}finally{
qe.close();
```

The output results are:

```
The resource URI to meet the requirements: http://222.22.65.252/grid/common/xml.chm
The resource URI of the format predicate to meet the requirements: http://222.22.65.252/schema/format
The resource URI to meet the requirements: http://222.22.65.252/grid/common/JDK_6.0_zh_CN.chm
The resource URI of the format predicate to meet the requirements: http://222.22.65.252/schema/format
```

5 Conclusion

This paper, by combining the grid technology and campus network, proposes to construct campus grid, in an attempt to realize information sharing, and thus to effectively solve the island problem of campus network resources. It carries on researches on description of campus grid and the discovery mechanism, and puts forward the RDF + vocabulary classification description method. Furthermore, it presents resource-classified registration model on vocabulary set and resource-query model based on triples. Finally, by developing campus grid prototype system, it implements the resource description, resource registration and resource query according to RDF + vocabulary set. Experiments show that the described method is feasible and effective. The future research will be oriented more on the enriching and extending of the resource-featured vocabulary set. The richer the vocabulary is, the more precise the resource description will be.

References

1. Ian, F., Carl, K., Jeffrey, M., et al.: The physiology of the grid: An open grid services architecture for distributed systems integration, University of Chicago, pp. 202–217 (2002)
2. Damandeep, K., Lokesh, S., Jyotsna, S.: Customized way of resource discovery in a campus grid. Journal of Advanced Networking and Applications 1(1), 51–56 (2009)
3. Nie, S.-q., Yu, S.-j., Le, J.-j., et al.: A kind of composite web service description language based on interface matching Computer. Applications and Software 26(1), 72–75 (2009)
4. Qi, C., Zhang, J., He, J.-h.: Implementation of Globus Resource Management Model Based on Service Level Agreement. Journal of Applied Sciences 26(5), 531–535 (2008)
5. Wu, H.-h., Qu, Y.-z.: Browsing RDF Data: State of Art Survey. Computer Science 36(2), 5–10 (2009)
6. Miguel, E., Isao, K., Said, M., et al.: Accessing RDF(S) data resources in service-based Grid infrastructures. Concurrency and Computation: Practice and Experience 21(8), 1029–1051 (2009)
7. Grigoris, A., Frank, v.: A Semantic Web Primer. The MIT Press, Cambridge (2008)

Mixed Programming between MATLAB and Other Programming Languages

Wang Jun[1,2], Jianyun Chen[2], and Zhuang Yan[2]

[1] Network Information Center
[2] Department of Computer Science and Technology
Nanjing University of Information Science & Technology
Nanjing, JiangSu Province, China
tangwang112@yeah.net

Abstract. The advantages and disadvantages about MATLAB are described in this paper. Several methods of mixed programming between MATLAB and other programming languages such as C#, FORTRAN, JAVA, DELPHI, VB and EXCEL are presented, including using COM Builder, MATLAB Compiler, MATLAB Engine, MATLAB Automation Server, Excel Link and other methods. Finally analysis and suggestions about the occasion adopting these methods are put forward.

Keywords: MATLAB, mixed programming, C#; COM Builder.

1 Introduction

MATLAB is Matrix Laboratory for short and it is a commercial mathematical software produced by The MathWorks which known as a major mathematical software the same as Mathematica and Maple. MATLAB can carry out matrix operations, drawing function and data, implementing algorithm, creating user interface, connecting to other procedures for programming languages and so on. MATLAB also has a very wide range of applications, mainly including signal and image processing, communications, control system design, test and measurement, financial modeling and analysis, and computational biology and many other applications.

- Friendly working platform and programming environment
- Simple and useful programming language
- Powerful capability of processing scientific computer data
- Excellent function of processing graphics
- Widely used toolkit modules
- Practical program interface and release platform
- Development of application software

Although the MATLAB language is very powerful, there are also some more difficult to overcome its own shortcomings: First, MATLAB language is an interpretive scripting language, so its program operating efficiency is low. Particularly,

D. Zeng (Ed.): ICAIC 2011, Part II, CCIS 225, pp. 669–676, 2011.

it is difficult to obtain desired effect that largely to use "for" or "while" statement. Second, M-file prepared by MATLAB is a text file that can be read easily, so that confidentiality is bad. Third, the program written by MATLAB only can be run on MATLAB platform, and it has not the ability to cross-platform, so that portable is poor. Finally, the capability of accessing to hardware is relatively poor, and the function of graphical user interface is not flexible enough.

2 The Purpose of Mixed Programming

Generally speaking, people always aim at achieving the desired by mixing programming between MATLAB and other programming languages probably confined to the following three kinds:

- Improving the program run-time speed
- Improving the program development speed
- Making the final program easy to release and deploy.

3 Mixed Programming between MATLAB and C#

A. Functions' redact

First, building a .M file (NetTest.m) as follow:

```
function NetTest(h,D)
for x=-5:0.2:5,
  if x>D
     y=h;
     hold on;
     plot(x,y)
  else if x<-D
     y=-h;
     hold on;
     plot(x,y)
  else
     y=h/(D*x);
     hold on;
     plot(x,y)
  end
end
```

B. Set MATLAB R2009a's compiling environment

MATLAB R2009a will compile M function into the external environment which in connection with the interface functions (such as C Shared library). Matlab R2009a has support external program compiler in the environment of windows : Lcc-win32 C

2.4.1, Microsoft Visual C++ 6.0, Visual C++ .NET 2003, Visual C++ 2005 SP1, Visual C++ 2008 Express, Visual C++ 2008 SP1 and so on. To set up the external compiler of Matlab. You can input command of mbuild –setup in the command window of Matlab, and then select compiler and complete setting according to instructions.

C. Using COM Builder to Complement Mixed Programming

MATLAB provides a COM compiler, so we can compile M file or MEX file into COM component, and generating RCW for COM component in .NET environment, and then to implement calling component in C #. Compiler-generated component object model is shown in Figure 1.

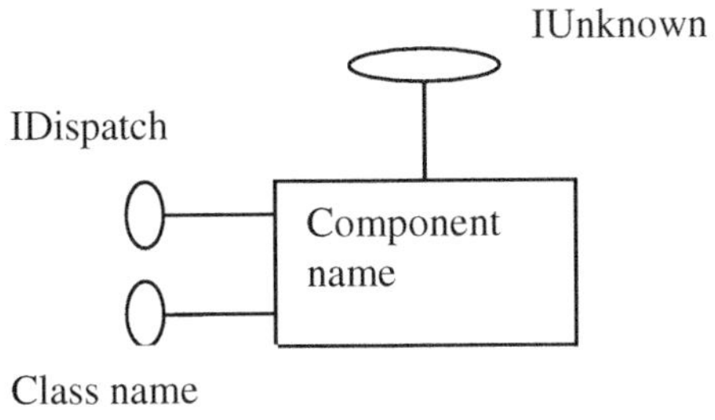

Fig. 1. Compiler-generated component object mode

It needs to create the instances of a class before calling the method of COM components in C#.NET. For example: Create the class MatlabNetclass's case: MatlabNetclass mt = new MatlabNetclass(). The parameter in C#.NET variable is not completely the same to MATLAB. It needs to turn the variable types in C#.NET into the corresponding variable types while invoke the methods of components.

D. Results

In Winform interface (*.Designer.cs) add three Lables, two textboxs and one Button inputs. And in *.cs document writing program as follow:

```
private void Btn_Click(object sender, EventArgs e)
{
MatlabNet.MatlabNetclass mt = new MatlabNetclass();
object m = this.Txt1.Text.Trim();
object n = this.Txt2.Text.Trim();
mt.nettest(m, n);
}
```

In the interface, want to input number m and n.Tow number cannot equal to zero, and m cannot equal to n. Shown in Figure 2. And result is shown in Figure 3.

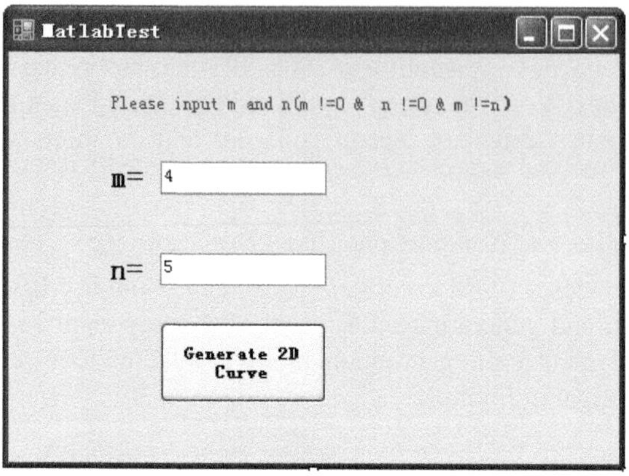

Fig. 2. Generate 2D Curve

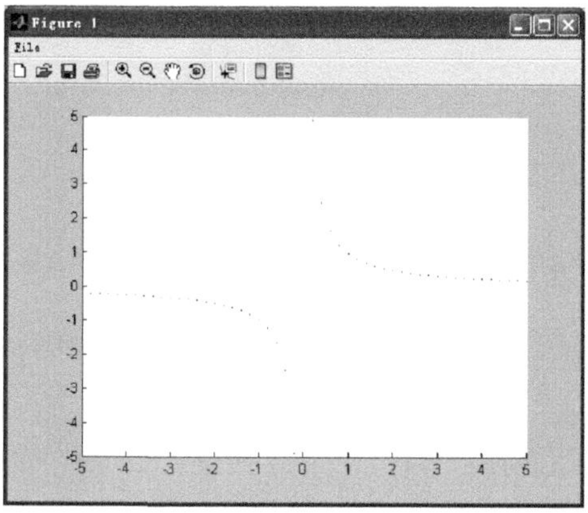

Fig. 3. Calls interface's generating

4 Mixed Programming between MATLAB and Fortran

A. Make use of Mex files

Mex is Matlab executable, in Windows environment It is called the DLL extension DLL, can be in the Mat The lab in direct call. Fortran Mex files usually consists of two The different parts: (1) completed subroutines, is calculated To the pure Fortran programs, usually for existing Fortran programs Code; (2) into the slot of program, which subroutines with Matlab environment The interface between, between to complete the call.

B. Make use of Matlab engine

It is simple and direct, Fortran programming of mixing process Calling sequence through Matlab, can function to engine. Matlab Command and control the running start, equivalent to Matlab engine Launched another Matlab process, this process runs in the background, User application through Matlab engine function library functions Complete with Matlab engine data exchange and ordered to deliver.

5 Mixed Programming between MATLAB and Java

C. Analysis Problems

A core problem of throughout mixing programming is that data whitch been transfered correctly in different programming software. As is known to all, JAVA is an object-oriented programming language. Each object is a known model and instantiation, so we must build a class, and put it into an object instance, and then use the I/O interface of JAVA. The object contains information stored in a file. And use JAVA interfacee of MATLAB predefined. To save the file after disposing the data. The file data information which stored in advance to write to the MATLAB environment. JAVA reads the file data is processed by MATLAB operations.

D. Set up Model

Mixing programming process is shown in Figure 4.

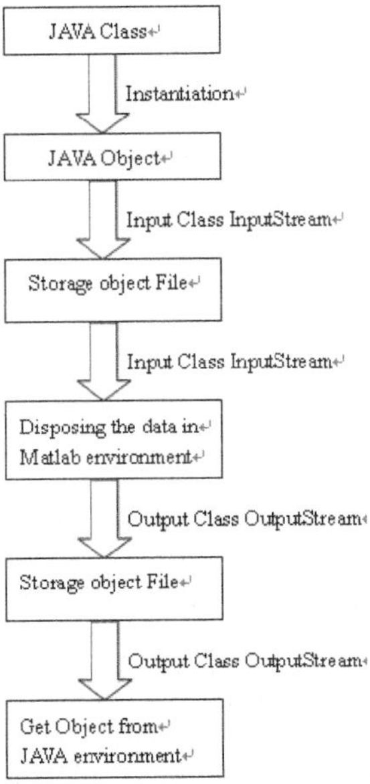

Fig. 4. Mixing programming process

6 Mixed Programming between MATLAB and DELPHI

Delphi calls Matlab has kinds of solution as follow:

① Matlab automation server. The server is the core automation based on Idispatch's interface. Matlab for automationserver, can be on any of the controller can be as automation applications on the Windows server platform, call needs to the Matlab environment.

② Matlab engine. Matlab does not provide interfaces between programs and Delphi, but we can use the interface function between Matlab and VC, make use of Matlab's engine to DLL for Delphi under VC.

③ Call formation of the DLL use Mideva. Mideva is MathTools which launched Matlab compiler development platform and provide for Matlab explanation and development environment. Its main characteristic is that turn M files into C code, and through the C++ compiled into an executable program or DLL.

④ Call the COM interface for programming. Matlab function automatically translates into a kind of COM components for other applications by using the tool that COM Builder in the Matlab. This mode can run from the Matlab environment.

7 Mixed Programming between MATLAB and VB

MATLAB provides programming interface for C/C + +, Fortran, etc, but does not provide a direct interface for VB. Currently, MATLAB with VB Mixed programming mainly uses the following methods:

By calling MATLAB Engine interface, VB using Windows OLE or ActiveX channel communication with MATLAB. Through the MATLAB statement of Object. Execute, we can directly use function library and graphics library provided by MATLAB in the program. This method can be faster to write out the procedure; but when you call the engine, it also starts a MATLAB process in the background which impacting for process speed and memory.

Using the way of embedding the VB matrix function library (Matrix VB) which is a third-party plug-in developed by MathTools. It embeds functions and grammars similar with MATLAB in VB, you can use these MATLAB functions and commands the same as VB. This method almost fully integrated into the VB language, which is relatively simple to use syntax and functions with higher programming efficient. Especially, graphic processing is very convenient to make up for the other methods in graphics processing weaknesses. However, it is still using the way of explaining the implementation, so that the implementation efficiency is low, and difficult to achieve real-time operation.

The interface between MATLAB and VC++ can transform MATLAB program file(.M)into dynamic-link library(.DLL), and VB can call the generated DLL function in the code to complement algorithm, for the convenience of separating from the MATLAB environment. As the part of algorithm is adopting C++ statement, run faster, can be applied to real-time computing. There are some certain requirements to the user for programming and set-up in this approach. The basic idea is shown in Figure 5. However, a DLL function can be call by VB must meet two conditions: first, the setting of the parameters convention called is _stdcall, but the default parameters convention in the editor of VC is _cdecl. This requires the corresponding conversion in the preparation of coding. Second, we must list export functions in the module(DEF)file. Carrying out the above operations can make DLL files normally called by VB, and then the mixed programming between VB and MATLAB is achieved.

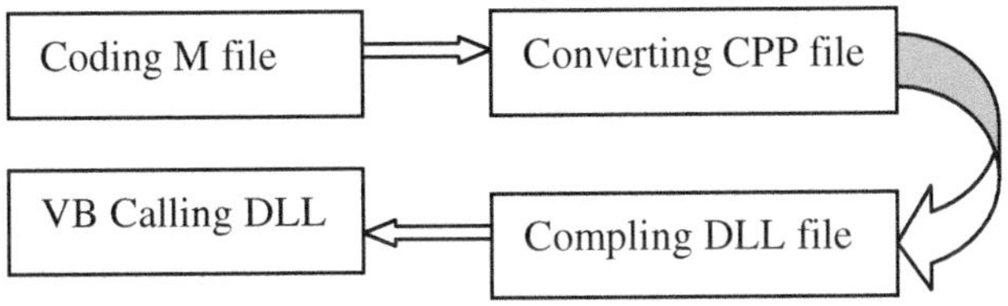

Fig. 5. The process diagram of basic idea for VB calling DLL

8 Mixed Programming between MATLAB and Excel

E. Make use of Excel Link

Excel Link is a plugin which realize the Excel link with Matlab in the Windows environment. By connecting Excel and Matlab, users can using Matlab's numerical calculation and graphical processing functions on the Excel worksheet space and the macro-program tools, and don't need from Excel environment. At the same time data that in your work environment exchange and update by Excel Link. ExcelLink working diagram is shown in Figure 6.

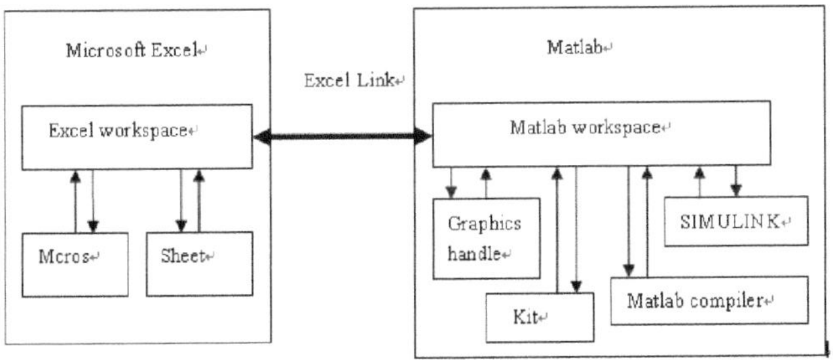

Fig. 6. ExcelLink working diagram

F. Excel Link's installation and Settings

First, intstall Excel software in the system. Then install the Matlab and Excel Link. By using Matlab installation disk to start the installation, choose custom installation, and selecte ExcelLink components.

Second, finish installing Excel Link, it needs to set in Excel for using. Start Excel, select the "add-ins" item under the "tools" of menu items.

9 Conclusion

In this paper, the mixed programming method of between MATLAB and C#, FORTRAN, JAVA, DELPHI, VB and EXCEL. The result shows that this method can make full use of existing resources of MATLAB, and play their own advantages to increase flexibility with a small amount of work to complete complex calculation, and then improve the program development and operational efficiency, thereby reduce program development cycle. The result shows that these methods can make full use of existing resources of MATLAB, and play their own advantages to increase flexibility with a small amount of work to complete complex calculation, and then improve the program development and operational efficiency, thereby reduce program development cycle. Finally, some suggestions about the application area are proposed based on different features of these methods.

Acknowledgment. This Project is supported by the Planning Project of The College of Software Engineering of Nanjing University of Information Science & Technology and the Program of Innovation Engieering on Information in Nanjing University.

References

1. Zhang, Y., Ma, Y.-j., Yuan, Q.-l.: Visual c # and Matlab programming and mixed, Northwest normal university (June 2008)
2. `http://www.docin.com/p-19795256.html` (2010.04.16)
3. Li, X., Tang, L.: Run the Other Executable Program in MATLAB. Control & Automation 22(9), 212–214 (2006)
4. Yao, G.-q., Chen, L.-p.: Integrated Programming between C# and Matlab Based on Component Object Model. Computer Engineering 34(14), 87–89 (2008)
5. Wang, Z., Crowcroft, J.: Quality of Service Routing for Supporting Multimedia Applications. IEEE Journal on Selected Areas in Communications 14(7), 1228–1234 (1996)
6. Turgay, K., Marwan, K.: Bandwidth-delay Constrained Path Selection Under Inaccurate State Information. IEEE/ACM Transaction on Networking 11(3), 384–398 (2003)
7. Moses, C., Joseph, N., Baruch, S.: Resource Optimization in QoS Multicast Routing of Real-time Multimedia. IEEE/ACM Transaction on Networking 12(2), 340–348 (2004)
8. Oh, J.S., Bae, S.I., Ahn, J.H., et al.: Routing Reinforcement for Efficient QoS Routing Based on Ant Algorithm. In: Proc. of International Conference on Information Networking, pp. 342–349. Springer, Berlin (2004)

Research on Airborne High Resolution SAR Image Classification*

Lianfei Duan[1], Luyi Yang[1], Jing Wang[2], and Zhanfeng An[2]

[1] The UAV T&R Section, Artillery Academy of PLA, Hefei, Anhui Province, China
[2] Library, Electronic Engineering Institute, Hefei, Anhui Province, China
lf_duan@yahoo.com.cn

Abstract. This paper studies the gray feature and texture feature including initial moment, energy based on gray level co-occurrence. An approach is proposed that feature is extracted and selected. Furthermore the BP neural network is applied to the image supervised classification. At least, the small areas are removed by morphological open operator. Considering the gray feature and texture feature of the SAR image , the method is more suitable for SAR image classification than the traditional method, which uses the texture feature only. The experimental results show the method can solve the airborne high resolution SAR image classification perfectly.

Keywords: Airborne SAR, Texture feature, Gray feature, Feature extracting, BP neural network.

1 Introduction

Because of the following characteristics: all weather condition and timeliness imaging, having little effect by the environment conditions, SAR has been widely used in disaster monitoring, environment surveying, military target surveillance, change detection, map surveying and mapping and other fields. It has drew extensive attention internationally. Because of the imaging principle's differences between SAR image and optical image, the traditional methods which used by optical image classification could not meet the need of SAR image classification well. How to use the SAR image's new information effectively to improve the accuracy of image classification is a hot subject which drew much attention all the time[1][2].

This paper studies and analyses the texture and gray feature, the main factors which affected the classification result, using the BP neural networks to realize the automatic supervised classification of Airborne High Resolution SAR Image.

* This work is partially supported by Grant #20070031 from key laboratory of Geo-Informatics of State Bureau of Surveying and Mapping.

D. Zeng (Ed.): ICAIC 2011, Part II, CCIS 225, pp. 677–683, 2011.

2 Feature Extraction and Feature Selection

A. Texture Feature

Texture is a structure made up of similar texture units or patterns, which refers to the gray, hue and colour between the neighbor pixels follow some kind of statistical regulation to form a certain spatial distribution. SAR image texture characteristics represent as shape, position, even degree, spatial relative relation between different objects and brightness contrast. It is un-separable between image textures and image gray, which representing as the change of brightness value[3][4][5].

Gray-level co-occurrence matrices(GLCM) of SAR image represents gray orientation of image, neighbour intervals, change extent and other integrated information. It is the foundation of image local pattern and array regulation analysis. Based on this, we can extract much more feature which described the texture of image. It is not study the single pixels, but describes the frequency of the relative pixels occurs. The gray-level co-occurrence matrices (GLCM) is defined as:

$$P(i,j)=[p(i,j\,|\,d,\theta)] \tag{1}$$

Where d is interval distance, θ is orientation. $P(i,j)$ is the probability of occurs the points which gray level is j apart d and orient θ from the point which gray level is i.

The Gray-level co-occurrence matrices (GLCM) could not distinguish texture features. For describing it , we need to extract the features from $p(i,j\,|\,d,\theta)$, which is describe the texture quantificationally. This paper choose the Energy, Inertia moment, Correlation, Entropy, local homogeneousness, variance and average value as the feature coefficient to describe texture.

Energy

$$E(d,\varphi)=\sum_i\sum_j[p(i,j\,|\,d,\varphi)]^2 \tag{2}$$

Inertia moment

$$I(d,\varphi)=\sum_k k^2[\sum_i\sum_j[p(i,j\,|\,d,\varphi)]^2] \tag{3}$$

$k=i-j$

Correlation

$$C(d,\varphi)=\frac{\sum_i\sum_j ijp(i,j\,|\,d,\varphi)-\mu_x\mu_y}{\sigma_x^2\sigma_y^2} \tag{4}$$

$$\mu_x=\sum_i i\sum_j p(i,j\,|\,d,\varphi) \qquad \mu_y=\sum_j j\sum_i p(i,j\,|\,d,\varphi)$$

$$\sigma_x^2=\sum_i(i-\mu_x)^2\sum_j p(i,j\,|\,d,\varphi) \qquad \sigma_y^2=\sum_j(j-\mu_y)^2\sum_i p(i,j\,|\,d,\varphi)$$

Entropy

$$H(d,\varphi) = -\sum_i \sum_j p(i,j \mid d,\varphi) \lg p(i,j \mid d,\varphi) \tag{5}$$

Local homogeneousness

$$L(d,\varphi) = \sum_i \sum_j \frac{1}{1+(i-j)^2} p(i,j \mid d,\varphi) \tag{6}$$

Texture mean

$$U(d,\varphi) = \sum_i \sum_j i \cdot p(i,j \mid d,\varphi) \tag{7}$$

Texture variance

$$V(d,\varphi) = \sum_i \sum_j (i - U(d,\varphi))^2 \cdot p(i,j \mid d,\varphi) \tag{8}$$

Energy is the even measurement of image gray. From the whole image, the texture is rougher, the same quality area is bigger and number of this homogeneous area is lager, the Energy is higher. The texture is thinner, the Energy is low. The image inertia moment could be understood as its definition. Correlation is the scale of the Gray-level co-occurrence matrices (GLCM) elements similarity on the row direction or column direction. When there is a linearity between the pixel pairs, the texture value is higher. Entropy describes the non-sequence of the image. When the object type number is high or the image texture is a clutter, the Entropy is higher. Local homogeneousness is estimate of image homogeneous area, the more number of the homogeneous areas, the lower of the local fixedness. Texture variance is the estimate of the image different areas.

In the experiment of the classification, combined considering the correlation and separability between texture features, we take the following five features: texture variance, entropy, local homogeneousness, Inertia moment, and gray mean.

B. Feature Selection

We set the type number is three, which is water body, residential area and background. To make the classification efficiently, we hope the shorter distance between the same types, the better and the further distance between the different types, the better[6]. So we chose the feature extraction algorithm based on the distance between same types or different types.

We hypothesize there are c types, set the D_w as the average square distance of all distances of same types. We has the following equation.

$$D_w = \sum_{i=1}^{c} (p_i \cdot d_i) \tag{9}$$

Where p_i is probability of No. i, d_i is average square distance of No. i. We set the D_b as average square distance of different types. So we have:

$$D_b = \sum_{i=1}^{c} p_i (m_i - m)^T (m_i - m) \tag{10}$$

Then, we can set up the equation to represent the minimal distance in the same kind and the maximal distance between the different kinds.

$$\varphi = D_b / D_w \tag{11}$$

During the feature extraction, we calculate the φ in equation (11) according to the known samples.

The statistics and analysis of image texture feature and gray feature are shown in the table 1 and table 2.

Table 1. Texture Feature Statistics and Analysis of All Types

	Water body	Residential area	Background
Energy	0.014891~ 0.043914	0.003692~ 0.00401	0.003839~ 0.004475
Entropy	1.947554~ 1.821326	2.439525~ 2.439525	2.377404~ 2.426895
Inertia moment	5.223776~ 17.05594	2205.622~ 5137.965	283.6084~ 493.7902
Variance	3.412795~ 8.268925	956.6534~ 2530.197	157.3962~ 272.136
Correlation	-0.00379~ 0.090195	-0.00008~ 0.00015	-0.00096~ 0.001692
Local homoge -neousness	0.234513~ 0.513261	0.026041~ 0.064967	0.048697~ 0.082376

Table 2. Gray Feature Statistics and Analysis of All Types

Types	Variance	Mean	Entropy
Water body	3.46~8.02	9.88~11.84	0.86~1.04
Residential area	908.65~2433.22	37.47~55.62	1.71~1.85
Background	161.24~280.58	48.47~58.69	1.59~1.71

According to the sample feature data and the method to choose feature, we choose the gray mean, texture variance, texture entropy, texture local homogeneousness and texture inertia moment as the classification features.

3 Back Propagation (BP) Neural Networks Classification

After feature extraction and feature selection, we use the back-propagation (BP) neural networks as the classifier to classify the airborne SAR image automatically. The typical back-propagation (BP) neural network normally contains neurons arranged in three types of layer: an input layer, a hidden layer(s), and an output layer. The adjacent neurons in the different layers are interconnected by the scalar weights, and there is no connection between neurons within layer[7][8]. In this paper, input layer is SAR image classification feature, output layer is classification result. Output layer result is zero represents water body, result is one represents residential area and result is two represents background. The choice of the neurons number of hidden layer is a difficult question. If the number is too less, the network is hard to process the result. If the number is too much, the learning process will spend too much time. Normally, we choose the two times of the input layer and add some redundancy. In this paper, the neurons number of the hidden layer is two times of feature number.

We choose the sigmoid transfer function as the initialized function of every neuron in the three layers back-propagation (BP) neural network:

$$v_i = \frac{1}{1+\exp(-u_i)} \tag{12}$$

Where u_i is the input signal of No. i neuron. v_i is the output signal.

In this paper, we design the number of the input layer is five, hidden layer is ten and output layer is two. Output result represents with binary number system to show the classification result. We change the binary system number into decimal system number.

From the classification error of checkout sample, we can draw the conclusion that the classification features of this paper, back-propagation (BP) neural network could realize the airborne high resolution SAR image classification automatically.

4 Experiment and Results

The experiment data of this paper is airborne SAR image, resolution is five meter. The experimental image area is the suburb of HeFei, which is shown as Figure 1. From the image, we can see that, in the residential area, the building array in the line and there is obviously regulation and orientation of the texture. In order to classify the water body, residential area and background automatically in the experiment, the pace of gray-level co-occurrence matrices (GLCM) is set as 13 pixels. In order to overcome the influence from the result caused by different orientation, gray-level co-occurrence matrices (GLCM) elements take the average value of $0°, 45°, 90°, 135°$.

We take texture variance, texture entropy, Local homogeneousness, inertia moment and gray mean, as the eigenvector to do the experiment. Three-Layers back-propagation (BP) neural network classification result is shown in Figure 2. Math morphological method is applied to slice parts processing. The result is shown in Figure 3.

Fig. 1. Airborne high resolution SAR image

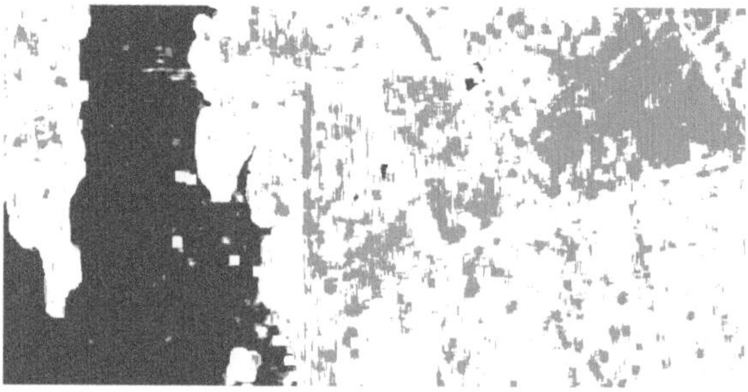

Fig. 2. Classification result of three-layers back-propagation (BP) neural network

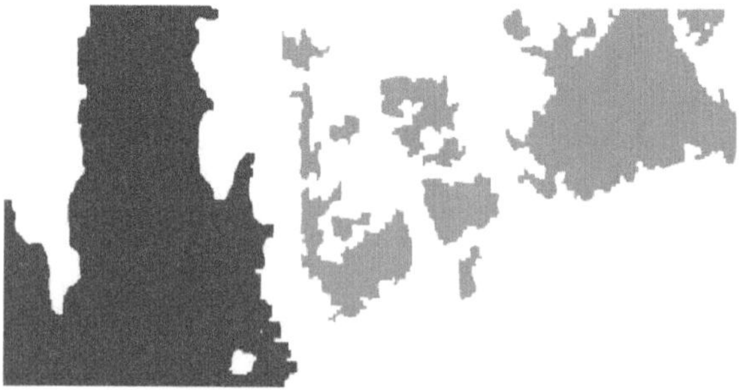

Fig. 3. Image after eliminating slice area

From the classification result, we can use the back-propagation (BP) neural network to extract the water body, residential area and background well, the misclassification is less.

5 Conclusions

Many methods are applied to classification of the SAR image. Considering the characteristics of airborne high resolution SAR image, we combine the gray statistical feature and the feature coefficient generated by gray-level co-occurrence matrices

(GLCM) to classify it. The two features are connected to each other and complemented to each other, improving the accuracy of classification. Among so many feature selection methods, the principle by calculating distance between kinds is a efficient one. It combines the classification features well. Multi-targets automatic classification problem can be well solved because of the characteristics of back-propagation (BP) neural networks: parallel structure, parallel processing, stability and self adaptability. From the result of this paper, if we take the hidden layer number two times of feature number, the image could be well classified. In the experiment, we can see that there are burr appeared in the fringe of the water body, residential area and background. Math morphologic operator could be used to filter the burr influence. However, much filter processing will reduce the accuracy. So, further studying should be made about the burr processing in the classification.

References

1. Sun, J.: Remote Sensing Principle and Application, pp. 204–210. WuHan University Press, Wuhan (2005)
2. Segl, K., Kaufman, H.: Detection of Small Objects From High-Resolution Panchromatic Satellite Imagery Based on Supervised Image Segmentation. IEEE Transaction on Geoscience and Remote Sensing 39(9), 89–112 (2001)
3. Guo, D.: Terrain Observation Theory and Application of Using Radar, pp. 214–217. Science Press, Beijing (2000)
4. Fan, W., Chao, W., Hong, Z.: Based-on texture features Residential area extraction from High resolution SAR image. Remote Sensing Technology and Application 20(1), 148–152 (2005)
5. Zeng, G.: Terrain classification based on multi-scale texture analysis using SAR image. The Master degree thesis of Wuhan University, pp.19–21 (2004)
6. Li, Y., Shi, Q., Zhang, Y., Zhao, R.: Study of SAR image automatic division methods. Journal of Electronic and information 28(5), 932–934 (2006)
7. Kohonen, T., Somervno, P.: How to Make Large Self-Organizing Maps for Nonvectorial Data. Neural Network 15, 945–952 (2002)
8. Cigizoglu, H.K.: Generalized Regression Neural Network in Modelling River Sediment Yield. Advances in Engineering Software 37, 63–68 (2006)

Utility-Based Network Adversary Model

Yunmeng Tan[1] and Shengbin Liao[2]

[1] Department of Electronics and Information Engineering,
Huazhong University of Science and Technology, China
[2] Engineering and Research Center of Information and Technology on Education,
Huazhong Normal University, China
YunmengTan@126.com

Abstract. The classic adversarial model is based on probabilistic analysis and queuing theory, which usually examine the stability of queuing networks and policies when the arrival process is adversarial. But widely used networks, such as Internet, are layered architecture, the adversarial model of queuing networks can not meet the needs of networks of layered architecture. This drives us to provide a new adversarial model, which deploys utility to measure the performance of networks and regards adversarial attacks as a type of denial of service attacks to study the stability and the performance.

Keywords: adversarial model, adversarial attacks, utility, network stability.

1 Introduction

With the rapid development of communication network technology and the wider use of Internet, information security has made a huge impact on our lives. The openness of Internet give rise to a target for a wide variety of attacks, which threaten the security of private information. In recent years, attackers have shown increasing sophistication in their ability to launch attacks that use a large number of hosts that are spread over a wide geographical area or multiple administrative domains [8]. For example, distributed denial-of-service attacks is a powerful attack method used by malicious network adversary which can inject much packets and lead to disrupt the service of legitimate users. Therefore, how to model the behavior of the network adversary is a key problem.

Due to the randomness of packets sent by network adversaries nodes selected by network adversaries to attack, classic method to model network adversary is queuing theory, i.e., adversarial queuing theory which gives the definition of an adversary model based on probability analysis and queuing theory. In this context, an adversary may inject some packets in any nodes in the networks, it may be able to compromise nodes' cryptographic secrets, degrade the packet delivery ratio, increase his control over traffic, and increase network delay. But adversarial queuing theory considers packet routing when packets are injected continuously into a network, its objective is to study stability and bounds on routing delay for various and scheduling policies.

In order to achieve its goals, an adversary is able to use some simple message manipulations: fabricated message injection, message deletion, message modification and re-ordering of message sequences [9]. Message injection can consume network

D. Zeng (Ed.): ICAIC 2011, Part II, CCIS 225, pp. 684–690, 2011.

bandwidth and lead that some services are not available. Therefore, this type of adversarial attacks can be seen as a type of denial of service attacks which can overload network system and hamper legitimate users from having access to network system by illegally injecting large amounts of data. In this paper, we focus on the adversary of denial of service attacks, i.e. their purpose is to flood the network with packets. We study routing and scheduling problems on Kelly type networks where the injection process is under the control of an adversary. We present a new model of adversaries based on utility which is borrowed from economics. The model of utility are often used to deal with the problem of resource allocation, in general, its goal is to maximize the total utility which measures the performance of networks, of course, the function of utility depends on different performance metrics such as sending rate, network bandwidth, and delay etc.

However, there is very little work which uses utility model to analyze the behavior of the network adversaries. General ideal is optimizing the performance of network with security requirements [10], i.e. this type of work combine the network utility maximization with security requirements. We will use random utility function to measure network nodes' utility. This is completely different form classic utility mode which is the function of definite variables. We will regard the process of packet injection of network adversary as stochastic process with very little assumption, and then, the nodes' utility is the function of random variables which denote sending rate, network delay etc. We also present the application of our utility adversarial model in network security. Our objective is optimizing the performance of network with security requirements using utility model presented in this paper, this is a new way of thinking for studying the problems of information security.

The rest of this paper is organized as follows. Section II first presents the classic adversarial model and utility model, and then, present our adversarial model. We introduce an application of utility adversarial model in Section III. Finally, Section IV concludes this paper.

2 Network Adversary Model

A. Classic adversarial model

Over the past few years, since the publication of the seminal paper [1] by Allan Borodin *et al.* in 2001, the framework of adversarial queuing theory has found many applications in communication networks. For more information, please refer to [1], [2], [3]. Roughly speaking, adversarial queuing model regards a routing network as a directed graph, and time evolution is in discrete model. At each time slot, the adversary inject some packets into a network from some nodes with predetermined routing paths. The biggest difference with classical multiclass queuing networks is that the injection of packets are controlled by network adversary in adversarial queuing network, but the arrival of packets is a stochastic process. In nature, multiclass queuing networks regard routing networks as queuing service system which provides the routing service for different routing paths and the service time distributions are all identical and independent of the routing paths as well as independent of the number of packets waiting for service at any time and any links of the network.

In adversarial network, during each time slot, a packet may be sent from its current node along with one of the outgoing paths. Moreover, only one packet may be scheduled along any link of the network in each time slot. When the packets arrives at its destination, it is absorbed by the system and assume the packets will not re-enter the network system. Any packets that wish to cross a link during a time step but are not sent wait in a queue for the link. In this context, the delay of a packet is the number of steps which the packet spends waiting in queues before it arrives its destination. A major issue arised in this class network is the stability and bounds on routing delay.

Usually, a classic adversarial model is defined as follows [1]:

Let w be an arbitrary positive integer, l any link in the network and τ any sequence of w consecutive time steps. And define $N(\tau,l)$ to be the number of paths injected by the adversary during time interval τ that cross link l of the network. For any $\rho>0$, define a (w,ρ) adversary that subject to the following load condition: for every sequence τ of w consecutive time steps and for every link l, $N(\tau,l)\leq\rho$. We say that such a (w,ρ) adversary injects packets at rate ρ with window size w. A rate ρ adversary is a (w,ρ) adversary for some w.

B. *Utility model*

The word utility comes from economics. Usually, the performance of networks are measured by sending rate, network delay, jitter of delay, bandwidth consumption and energy consumption etc. These metrics are often optimized individually, but with the increase of the complexity of network business, optimizing a metric may lead to the lack of performance of other metrics. So, in this context, the utility model is used to measure the total utility (performance) of networks . Usually, utility is a function of the above performance metrics, moreover, the function u(x) has the following characteristic[4]:

- U(x) is two times continuously differentiable.
- The derivatives u',u'' are bounded and $\lim_{x\to\infty} u'(x)=0$.
- U(0)=0, u'(0)=1.
- U'(x)>0 for any $x\in R$.
- $U''(x)<0 \text{ for any } x\in R$.

In recent years, Network Utility Maximization (NUM), a mathematical model of design and analysis of communication network based on optimization, has found many applications [5]. The basic model of NUM was firstly proposed by Kelly et al. in 1998 [6]. Low et al. present in [7] the following basic distributed optimization algorithm.

1) Model

Consider a communication network with J links, each with a fixed capacity of c_j bps. Let a route r be a non-empty subset of J, and R be the set of possible routes. Set $A_{jr} = 1$ if $j \in r$, so that the link j lies on route r, and set $Ajr = 0$ otherwise. This defines a 0-1 routing matrix $A = (A_{jr}, j \in J, r \in R)$.

Associate a route r with a user, and suppose that if a rate x_r is allocated to user r then this has utility $U_r(x_r)$ to the user. Assume that the utility $U_r(x_r)$ is increasing, strictly concave and continuously differentiable over the range $x_r > 0$.

Let $U = (U_r(x_r), r \in R)$ and $C = (c_j, j \in J)$. Under this model the network seeks a rate allocation $x = (x_r, r \in R)$ which solves the following optimization problem.

$$\begin{aligned} \max \quad & \sum_{r \in R} U_r(x_r) \\ \text{subject to} \quad & Ax \le C \\ & x \ge 0 \end{aligned} \tag{1}$$

2) Solving Optimization Problem using Lagrange Duality

The Lagrangian dual function for problem (1)

$$D(\mu) = \max_{x_r \ge 0} \left\{ \sum_{r \in R} U_r(x_r) - \mu^T(Ax - C) \right\} = \max_{x_r \ge 0} \left\{ \sum_{r \in R} U_r(x_r) - x_r \sum_{j \in J} A_{jr}\mu_j \right\} + \mu^T C$$

where $\mu = (\mu_j, j \in J)$ is a vector of lagrange multipliers. Thus, the dual problem for primal problem (1) is

$$\min_{\mu \ge 0} D(\mu) \tag{2}$$

In the dual formulation, Lagrange multiplier μ_j can be interpreted as congestion price on link j. A key observation is that sources can compute their optimal rate individually, based on the total congestion price , using the following source rate algorithm

$$x_r = \arg\max_{x_r \ge 0} \left\{ \sum_{r \in R} U_r(x_r) - x_r \sum_{j \in J} A_{jr}\mu_j \right\} \tag{3}$$

To solve the dual problem (2), one can use the following projected gradient method

$$\mu_j(t+1) = \left[\mu_j(t) - \alpha_j \left(c_j - \sum_{j \in J} A_{jr}x_r\right) \right]^+ . \tag{4}$$

where α_j is a positive scalar step size, and $[\alpha]^+$ denotes the projection of α onto the set R^+ of non-negative real numbers. according to the above formulation, the solution of the optimal source rates and congestion prices of links can be solved iteratively in equation (3) and (4), respectively. This suggests treating the network links and the sources as processors in a distributed computation system to solve the dual problem (2).

C. *Utility adversarial model*

The main objective of the classic adversarial model is study stability issues of network with different queueing policies. Stability is to deciding whether at any time the number of packets presented in network is bounded. According to the above definition of the classic adversarial model in II.A, the adversary must obey the following constraints [11], for every time interval I

$$N_e(I) \le r|I| + b \tag{5}$$

Where $N_e(I)$ denotes the number of packets injected by the network adversary that traverses the link e during the time I. Because different nodes may inject packets which have routing paths containing the link e, we introduce the following constraints:

$$\sum_{s(e)} N_e(I) \le c_l \tag{6}$$

The classic network adversarial model requires that the number of packets injected by an adversary is less than the capacity, from a perspective of network performance, this is equivalent to purse the maximization of the throughput. But optimizing the throughput of network may leads to the deterioration of other performance metrics such as network delay etc. Therefore, we present an network adversarial model based on network utility maximization as following:

$$\begin{aligned} &\max \sum_{s\in S} U_s(N_e(I)) \\ &subject\,to \\ &\sum_{S(e)} N_e(I) \le c_l \end{aligned} \tag{7}$$

The above model (7) is completely different from the model (1) presented in II.B, this is because the $N_e(I)$ in model is a random variable, the network adversary may inject packets in any way into the network. The model (7) is also different from the model (5) which only consider on class of packets, but mode (7) requires the total traffic constraints of the different data sources, the latter is more practical. Although the forms of the two models (1) and (7) are the same, in fact, the model (7) includes more richer content. Moreover, we will find a lot of the application of the model in stochastic network resource allocation and network information security. In the next section, we will present an practical application of the mode (7) in adversarial network attack.

3 Application of Utility Adversary Model

The most important tool for computer and network communication security is encryption and decryption, there are several scheme that can quantify the strength of an encryption method. Usually, the length of the key decide the level of the security, but in practical encryption application, the block length and key length are equal, hence, as specified in [12], we use the encryption block length to measure the level of the security, i.e. let a message consist of n frame with encrypted block length N_i bits for frame, so the security level is as following:

$$\bar{s} = \sum_{i=1}^{n} \log_2 N_i \tag{8}$$

Assume the transmission rate of the encryption block and the channel bit error probability are R_i and P_i, respectively, therefore, the throughput per block is $R_i(1-P_i)^{N_i}$, but because P_i is far less than 1, hence, we have

$$R_i(1-P_i)^{N_i} \approx R_i(1-N_iP_i) \tag{9}$$

So, we definite the utility of every data source as $U_s(R_i(1-P_i)^{N_i})$, and then, according to the model (7), we give our utility adversarial model as follows:

$$\begin{aligned} &\max \sum_{i=1}^{n} U_s(R_i(1-P_i)^{N_i}) \\ &subject\,to \\ &\sum_{i=1}^{n} \log_2 N_i = s_{level} \end{aligned} \tag{10}$$

Based on equation (9), the model (10) can be simplified as

$$\begin{aligned} &\max \sum_{i=1}^{n} U_s(R_i(1-N_iP_i)) \\ &subject\ to \\ &\sum_{i=1}^{n} \log_2 N_i = s_{level} \end{aligned} \tag{11}$$

Using similar algorithm presented in II.B, the Lagrange of model (11) is

$$L(N_i;\lambda) = \sum_{i=1}^{n} U_s(R_i(1-N_iP_i)) + \lambda(\sum_{i=1}^{n} \log_2 N_i - s_{level}) \tag{12}$$

Taking partial derivatives of equation (10) with respect to variable N_i and set them equal to zero, we have

$$\partial L(N_i;\lambda)/\partial N_i = U'_s(R_i(1-N_iP_i))(-R_iP_i) + (\lambda \log_e 2)/N_i = 0 \tag{13}$$

Utility function has many forms as specified in [13], here we define it as logarithmic function which has proportional fairness for different data flow. Substituting the utility function with logarithmic in equation (13), we obtain,

$$(-R_iP_i)/(1-N_iP_i) + (\lambda \log_e 2)/N_i = 0$$

So,

$$N_i^* = (\lambda^* \log_e 2)/(R_iP_i + \lambda^* P_i \log_e 2) \tag{14}$$

Where N_i^* and λ^* denote the optimal values of N_i and λ, respectively. The remaining problem is how to get the optimal value of λ. This can be finished by using the Lagrange optimization technique [14]. From equation (12), the dual function of problem (11) is given by

$$D(\lambda) = \max_{N_i} L(N_i;\lambda)$$

Hence, the dual problem of the primary problem (11) is given by

$$\begin{aligned} &\min D(\lambda) \\ &subject\ to\ \lambda \geq 0 \end{aligned} \tag{15}$$

From dual problem (15), we can obtain the optimal value λ^* of the dual variable λ.

4 Conclusion

In this paper, we consider the classic problem of modeling adversaries in all kinds of networks, but different from the classic model is that the performance of networks is measured by utility, moreover, we regard adversarial attacks as a type of denial of service attacks, define the security objectives of networks, and then present our adversarial modeling based on utility. Finally, we give a practical application of our model, the application obtains the optimal value of the encryption block length, this leads to the utility maximization which is the metric value of the network performance, therefore, this also validates the effectiveness of our adversarial modeling.

As the classical models are base on queuing theory, they do not consider the congestion control of the TCP layer and the MAC scheduling control, in the future, we will combine the adversarial model presented in this paper with the thinking of layered architecture, and study the stability of networks and delay of packets.

Acknowledgment. The work of this paper is supported by the National Natural Science Foundation of China (No.61072051), the grant from the National High Technology Research and Development Program of China(863 program) (No.2008AA01Z127), the Cultivation Fund of the Key Scientific and Technical Innovation Project Ministry of Education of ChinaNO.708065).

References

1. Borodin, A., Kleinberg, J., Raghavan, P., Sudan, M., Williamson, D.: Adversarial queuing theory. Journal of the ACM 48, 13–38 (2001)
2. Anshelevich, E., Kempe, D., Kleinberg, J.: Stability of load balancing alogrithms in dynamic adversarial systems. SIAM Journal of Comput. 37(5), 1656–1673 (2008)
3. Sethuraman, J., Teo, C.: Effective routing and scheduling in adversarial queueing networks. Algorithmica 43, 133–146 (2005)
4. Kallsen, J.: A utility maximization approach to hedging in incomplete markets. Mathematical Methods of Operations Research 50, 321–338 (1999)
5. Chiang, M., Low, S.H., Calderbank, A.R., Doyle, J.C.: Layering as optimization decomposition: A mathematical theory of network architectures. Proceedings of the IEEE 95(1), 255–312 (2007)
6. Kelly, F.P., Maulloo, A., Tan, D.: Rate control for communication networks: shadow prices, proportional fairness and stability. Journal of the Operational Research Society 49(3), 237–252 (1998)
7. Low, S.H., Lapsley, D.E.: Optimal flow control, I: Basic algorithm and convergence. IEEE/ACM Transactions on networking 7(6), 861–874 (1999)
8. Zhou, C., Leckie, C., Karunasekera, S.: A survey of coordinated attacks and collaborative. Computers & Security 29, 124–140 (2010)
9. Acs, G., Buttyan, L., Vajda, I.: Modelling adversaries and security objectives for routing protocols in wireless sensor networks. In: Proceedings of the Fourth ACM Workshop on Security of Ad Hoc and Sensor Networks, Alexandria, Virginia, USA, pp. 49–58 (2006)
10. Xie, T., Qin, X.: Performance evaluation of a newscheduling algorithm for distributed systems with security heterogeneity. Journal of Parallel and Distributed Computing 67(10), 1067–1081 (2007)
11. Alvarez, C., Blesa, M., Diaz, J., Serna, M.: Adversarial models for priority-based networks. Networks 49, 23–35 (2005)
12. Haleem, M., Mathur, C., Chandramouli, R., Subbalakshmi, K.: Opportunistic encryption: a trade-off between security and throughput in wireless networks. IEEE Transaction on Dependable and Secure Computing 4(4), 313–324 (2007)
13. Shenker, S.: Fundamental design issues for the future Internet. IEEE Journal of Select Area Communication 13(7), 1176–1188 (1995)
14. Nocedal, J., Wright, S.: Numerical optimization. Springer series in operations research. Springer, Heidelberg (1999)

A Survey on Mac Layer in Ad Hoc Wireless Networks

Liao Shengbin and Zhu Xiaoliang

National Engineering Center for e-learning
Huazhong Normal of University, Wuhan, China
passyou@home.news.cn

Abstract. The network architecture can be classified into two categories: a centralized network architecture and a decentralized network architecture. Ad hoc wireless networks have witnessed startling advances, which is a decentralized network architecture, due to the flexibility and freedom they offer. The survey is mainly concerned with mac layer in ad hoc wireless networks, the aim of which is to provide a comprehensive view of the role and details of the protocols that define and control access to the wireless channel, i.e., wireless media access control (MAC) protocols, and there are the inherent problems of the MAC protocols in shared wireless channel environments in ad hoc wireless networks and relevant solutions.

Keywords: ad hoc wireless network, MAC, backoff algorithm, cross-layer, markov model.

1 Introduction

A. MAC protocol overview

The MAC protocol determines when a node is allowed to transmit its packets, and typically controls all access to the physical layer. Since all the nodes use the same physical channel in ad hoc wireless networks, MAC plays an important role in coordinating channel access among the nodes so that information gets through from one node to another. The primary characteristic of an ad hoc wireless network is the absence of any predefined structure, nodes communicate directly with one another in a peer-to-peer fashion, so all network control, including channel access, must be distributed, which introduces several new challenges([1],[2],[6]).

Given its distributed nature, channel access in an ad hoc networks is achieved through the close cooperation between competing nodes. A good MAC protocol should provide an efficinet mechanism to share limited bandwidth resources, together with simplicity of operation, fairness in serving all stations, low delay and high throughput. Ad hoc MAC protocols includes two broad categories([4],[3]): contention-based MAC protocols and reservtion-based MAC protocols.

Contention-based MAC protocols are generally used in a distibuted network architecture, and are suitable for bursty data traffic under low network load because of their low delay characteristics, the simplicity of implementation of it is also a desirable feature in wireless ad hoc networks where no infrastructure access points

D. Zeng (Ed.): ICAIC 2011, Part II, CCIS 225, pp. 691–699, 2011.

exist([5],[6]). Our focus in the survey is on distributed contention-based MAC protocols, which mainly include the busy-tone like protocols which are characterized by an out of band feedback signal(busy-tone)[7] and the protocols based on carrier sensing and collision avoidance(CSMA/CA) principles([8],[9]). There are other protocols, such as MACA, MACAW, FAMA and DFWMAC, which incorporate the carrier sense multiple access(CSMA) technique as well as the RTS and CTS handshaking mechanism for collision avoidance(CA)([5],[9],[10]).

The most popular and important contention-based wireless MAC protocol, CSMA/CA, becomes the basis of the MAC protocol for the IEEE 802.11 standard([11],[12]), which is widely used and become the most mature technology for WLANs, in the following section we will introduce its basic operations.

B. *IEEE 802.11 medium access control*

The architecture of the MAC sublayer includes the distributed coordination function(DCF) and the point coordination function(PCF), DCF is the fundamental access mechanism in IEEE 802.11 MAC while PCF is used optionally[12]. We will only consider DCF.

The DCF allows automatic medium sharing between compatible PHYs through the use of CSMA/CA and a random backoff time following a busy medium condition. In addition, all directed traffic uses immediate positve acknowledgment(ACM frame) where retransmission is scheduled by the sender if no ACK is received. The random backoff algorithm is shown in Fig.1

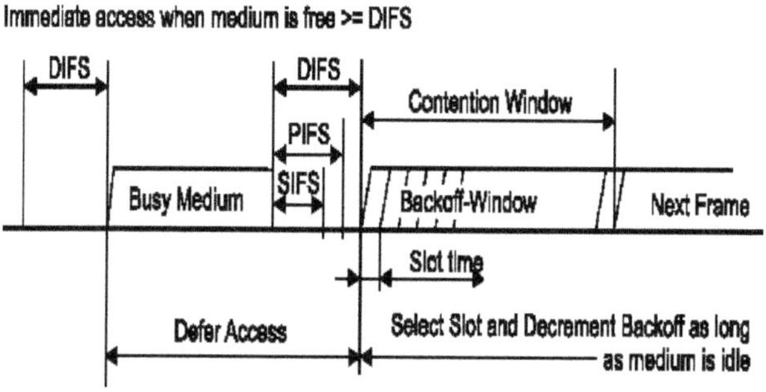

Fig. 1. Random backoff procedure

The DCF provides two access mechanisms. One is two-way handshake, i.e., DATA/ACK, and the other is four-way handshake, i.e., RTS/CTS/DATA/ACK. They are shown in Fig. 2 and Fig. 3 respectively.

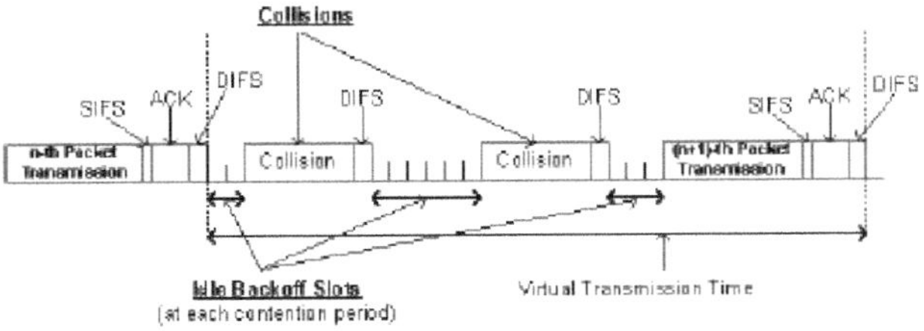

Fig. 2. Basic operations of CSMA/CA algorithm

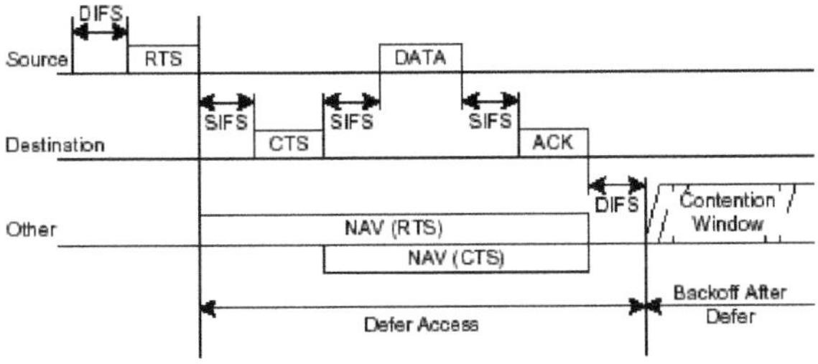

Fig. 3. RTS/CTS/DATA/ACK and NAV setting

2 The Key Mechanisms for MAC Design in Ad Hoc Wireless Networks and Problems

As mentioned earlier, the primary goal of MAC is to coordinate the channel access among multiple nodes to achieve high channel utilization, in other words, the coordination of channel access should minimize or eliminate the incidence of collisions, so how to effectively reduce collisions becomes a key issue for MAC desin in ad hoc wireless worknets.

Serverl mechanisms have been proposed to avoid collisions in medium access,namely carrier sense, handshake, and backoff mechanism.

C. Carrier sense, Handshake, and Backoff in the IEEE 802.11 DCF protocol

To reduce the collision possibility, it uses carrier sense functions and binary exponential backoff(BEB) mechanism[12]. In particular, two carrier sense functions, physical layer and virtual carrier-sense functions, are used to determine the state of the medium. The former is provided by the physical layer and the latter by the MAC layer, which if also referred to as the network allocation vector(NAV). The DCF provides two handshake mechanisms as showed in Fig. 2 and Fig. 3.

About backoff mechanisms, although BEB is widely used, there are a lot of paper discussing new backoff mechanisms such as[4], [9],[15],[16],[17]. This is a very important factor on the preformanc of the MAC protocol.

In the IEEE 802.11 it adopts the two-way DATA/ACK and four-way RTS/CTS/DATA/ACK handshake mechanisms, which are sender-initiated, but there are receiver-initiated[18] and hybrid channel access scheme[19].

D. Problems

- Hidden nodes and exposed nodes problems

  A hidden node is the one that can neither sense the transmission of a transmitter nor correctly receive the reservation packet from its corresponding receivr. An exposed node is the one that senses the transmission of a transmitter and can not interfere with the reception at the receiver. There are a extensive research about the topic([13], [14],[12]), which generally adopt control frame, such as RTS/CTS, to avoid the interaction of the hidden and exposed nodes.
- Energy efficiency and Transmission power control

  In wireless ad hoc networks the MAC protocols should reduce the collision as many as possible, and hence reduce the power consumption which is a critical issue due to a limited battery life. The MAC protocols with transmission power control can improve energy efficiency and channel reuse, and reduce channel interference. The two problems are ubiquitously researched in recent years, but I have no interested with it, if you do it, please see([20],[21],[22]).
- Fairness

  In wireless ad hoc networks the MAC protocols widely use the backoff mechanism to reduce collisions, and which may lead an unequal channel access opportunities between nodes with different backoff window. There are many fairness enhanced MAC control algorithm. But many researcher began to use a new method which is based on the flow contention graph([23],[24]) to solve the problems, about this modeling method I have clearly introduced and play it our inner server, which main thought is translating a contention relation into a optiamization problem.
- Rate adaptation

  Wireless channels are time-varying, depending on the spatial location of transmitting and receiving nodes, thus, rate adaptation is a powerful way to overcome channel variations, but IEEE 802.11 protocol only supports a single base rate[12]. According to the wireless channel condition the nodes dynamicaly adapt transmission rates before they transmit data packets is a trend in future. The fist commercial MAC that utilizes rate adaptation was the Auto Rate Fallbak(ARF) protocol[25], in which senders attempt to use higher transmission rates after transmission successes and revert to lower rates after failures. In [26], a full auto rate algorithm was proposed, which not only considered the rate of the data frame but also the rate of the control frame.In [27] and [28], to exploit the node deversity, present the opportunistic packet scheduling and auto rate protocol, the basic idea of which is to extend the functionality of the collision avoidance process to probe channel conditions of several receivers simultaneously.

3 Interaction with Other Layers

The interaction between the MAC layer and other layers in ad hoc networks are much more commplicate than in wired networks. First, according to the enhanced IEEE

802.11 protocols, the MAC algorithms must adapt to the different physical technology such as DSSS and FHSS which can provide different rate sets[12]. Sencond, network congestion and medium contentions are closely coupled since the channel is shared by all nodes in a region, the contention occurs not only in one pair of transmitting-receiving nodes, but also one to all nodes in the region. Thus, when desinging MAC protocols and algorithms, you should bear in mind the undesirable interactions beteween MAC and other layer algorithms, such as routing algorithms and tcp's end to end congestion control algorithms.

In term of the characteristic of wireless channels it is very important for improving the performanc of the mac layer protcols to joint different layer desin and analysis. In [29] and [30], unite tcp layer congestion control and MAC desin , and cross-layer optimization for ad hoc wireless networks, in [31] and[32] to improve tcp performance and link reliability also use cross-layer design methods. In [33] a routing algorithm was proposed by network layer and MAC layer protocols.

4 The Key Modeling Methods and Algorithms of the MAC Layer in Ad Hoc Wireless Networks

E. The Markov chain model based nodes state

As mentioned before, MAC layer mainly use carrier sense, handshake and backoff mechanism to avoid congestions, many algorithms were proposed to achieve these mechanism, but the following will introduce two kind of modeling methods and relevant algorithms, which is helpful for us to perceive and design the protocol.

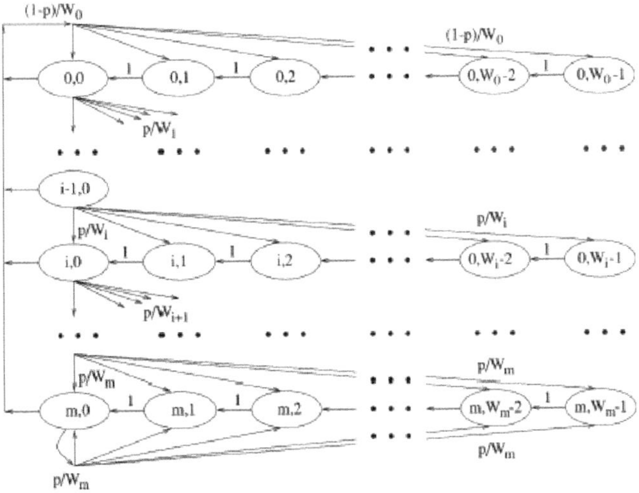

Fig. 4. Markov chain model

Let b(t) be the stochastic process representing the backoff time counter, and s(t) be the stochastic process representing the backoff stage (0,1,….m) of the station at time t in 802.11 MAC protocol. a disctete and integer time scale is adopted: t and t+1 correspond to the beginning of two consecutive slot times, and the backoff time counter of each station decrements at the beginning of each idle slot time. We assume that min contention window is W and each packet collides with constant and independent probability p. then we can model the bidimensional process { s(t), b(t) } with the discrete –time Markov chain to express the states of the station, which was depicted in Fig.4.

In this Markov chain, the only non null one-step transition probabilities are:

$$\begin{cases} P\{i,k \mid i,k+1\}=1 & k\in(0,W_i-2) \quad i\in(0,m) \\ P\{0,k \mid i,0\}=(1-p)/W_0 & k\in(0,W_0-1) \quad i\in(0,m) \\ P\{i,k \mid i-1,0\}=p/W_i & k\in(0,W_i-1) \quad i\in(1,m) \\ P\{m,k \mid m,0\}=p/W_m & k\in(0,W_m-1). \end{cases}$$

From above equation we can yields the probabilityτthat a station transmits in a randomly chosen slot time and packet tranmission probability p [34],then We can analyse the performance of the IEEE 802.11 and tune the parameters of the protocol.

Many papers([35],[36],[37],[15]) use above the modeling method to analyse the mac layer protocol and achieve some new conclusions and methods of the enhancement of the mac layer protocol.

Define abbreviations and acronyms the first time they are used in the text, even after they have been defined in the abstract. Abbreviations such as IEEE, SI, MKS, CGS, sc, dc, and rms do not have to be defined. Do not use abbreviations in the title or heads unless they are unavoidable.

F. The queue model based arrival patterns and packet size distributions

There are many papers they model each node as a discrete time queue,such as G/G/1[38], M/G/1[39], and derive the service time distribution. Their basic idea is to present analytic models for the queue characteristics in wireless network operating in the random access mode and analyzes and improve the performance of the MAC layer protocol.

If the probability of a collision be denoted by p and the min contention window is W in IEEE 802.11 MAC protocol, then, the probability that the contention window CW = W is given by[40].

$$\Pr\{CW=W\}=\begin{cases} p^{k-1}(1-p) & \text{for } W=2^{k-1}CW_{min} \\ p^m & \text{for } W=CW_{max} \end{cases}$$

The probability that back-off counter BC = i , is then given by[40]:

$$
\Pr\{BC=i\}=\begin{cases}\left[\sum_{k=0}^{m-1}\frac{p^k(1-p)}{2^kCW_{min}}+\frac{p^m}{CW_{max}}\right] & 1\leq i\leq CW_{min}\\ \left[\sum_{k=j}^{m-1}\frac{p^k(1-p)}{2^kCW_{min}}+\frac{p^m}{CW_{max}}\right] & 2^{j-1}CW_{min}+1\leq i\leq 2^jCW_{min}\\ \frac{p^m}{CW_{max}} & 2^{m-1}CW_{min}+1\leq i\leq CW_{max}\end{cases} \quad \ldots
$$

From above the equation we can also abtain the packet collision probaility p ([38]), but the method of which was different from the method we introduced in A.IV.

Acknowledgment. This work was supported by the NSF of China through the grant 61072051, supported by self-determined research funds of CCNU from the colleges' basic research and operation of MOE(N0. CCNU09A01018)", the Cultivation Fund of the Key Scientific and Technical Innovation Project, Ministry of Education of China (NO.708065).

References

1. Haas, Z.J., Deng, J., Liang, B., Papadimitratos, P., Sajama, S.: Wireless Ad Hoc Networks. In: Proakis, J.G. (ed.) Wiley Encyclopedia of Telecommunications. John Wiley & Sons, Chichester (2002)
2. Ivan, S.: Handbook of wireless networks and mobile computing. John Wiley & Sons, Chichester
3. Hongqiang, Z., Yuguang, F.: Medium control protocols in mobile ad hoc networks: challengs and solutons
4. Younggoo, K., Haniph, L.: A novel mac protocol with fast collision resolution for wireless LANs. IEEE, Los Alamitos (2003)
5. Bharghvan, V.: MACAW: A media access protocol for wireless LANs. In: SIGCOMM, London, England, pp. 212–225 (August 1994)
6. Gummalla, A.C.V., Limb, J.O.: Wireless medium access protocols. Submitted to IEEE Commn. Surveys
7. Tobagi, F.A., Kleinrock, L.: Packet switching in radio channels: Part II The Hidden Termianl problem in carrier sense multiple access and busy tone solution. IEEE Trans. Commun. COM-23, 1417–1433 (1975)
8. Kam, P.: MACA a new channel access method for packet radio. In: ARRL/CRRL, Amatuer Radio 9th Computer Networking Conference, September 22 (1990)
9. Gummalla, A., Limb, J.: Wireless collision detect(WCD): Multiple access with receiver initiated feedback and carrier detect signal. In: Proc. IEEE ICC 2000, New Orleans, LA, vol. L, pp. 397–401 (June 2000)
10. Fullmer, C., Garcia-Luna-Aceves, J.: Floor acquition multiple access for paket-ratio networks. In: Proc. SIGCOMM 1995, Cambridge, MA, pp. 262–273 (1995)
11. IEEE 802.11 Wireless LAN Medium Access Control(MAC) and Physical layer(PHY) specifications, IEEE (1997)

12. ISO/IEC, IEEE standard for wireless LAN medium access control(MAC) and physical layer(PHY) specifications, ISO/IEC 8802-11:1999(E) (1999)
13. Fullmer, C.L., Garcia-Luna_Aceves, J.J.: Solutions to Hidden terminal problems in wireless networks. In: Proc. ACM SIGCOMM 1997 (September 1997)
14. Tobagi, F.A., Kleinrock, L.: Packet switching in radio channels: Part II the hidden terminal problem in carrier sense multiple-access and the busy-tone solution. IEEE Trans. Commun. COM-23, 1417–1433 (1975)
15. Nah-Oak, S., Byung-jae, K., Miller, E.: Enhancement of IEEE 802.11 distributed coordination function with exponential increase exponential decrease backoff algorithm. IEEE, Los Alamitos (2003)
16. Younggoo, K., Yuguang, F.: Design of mac protocols with fast collision resolution for wireless local area networks. IEEE Transaction Wirreless Communications 3(3) (May 2004)
17. Federico, C., Enrico, G.: IEEE 802.11 protocol: Design and performance evaluation of an adaptive backoff mechanism. IEEE Journal Selected Areas in Communications 18(9) (September 2000)
18. Talucci, F., Gerla, M., Fratta, L.: MACA-BI a receiver oriented access protocol for wireless multihop networks. In: Proc. IEEE PIMRC 1997 (September 1997)
19. Wang, Y., Garica-luna-aceves, J.J.: A new hybrid channel access scheme for ad hoc networks. ACM WINET Journal, Special Issue on Ad Hoc Networks (July 2004)
20. Deng, J., Haas, Z.J.: Dual busy tone multihop access(DBTMA): a new medium access control for packet radio networks. In: Proc. IEEE ICUPC 1998 (October 1998)
21. You, T., Hassanein, H.: A new class of collision prevention MAC protocols for wireless ad hoc networks. In: Proc. IEEE ICC (May 2003)
22. Monks, J., Hwu, W.: A power controlled multiple access protocol for wireless packet networks. In: Proc. IEEE INFOCOM (April 2001)
23. Zuyang, F., Brahim, B.: Fair bandwidth sharing algorithms based on game theory frameworks for wireless ad hoc networks
24. Thyagarajan, N., Vaduvur, B.: Achieving MAC layer fairness in wireless packet networks. In: MOBICOM ACM (2000)
25. Kamerman, A., Monteban, L.: waveLan II: a high-performance wireless lan for the unlicensed band. Bell Labs Technical Journal (1997)
26. Li, Z., Nandi, S.: Full auto rate MAC protocol for wireless ad hoc networks. IEEE Proc. Commun. 152(3) (June 2005)
27. Wang, J., Zhai, H., Yuang, M.C.: Opportunistic media access control and rate adaptation for wireless ad hoc networks. In: Proc IEEE ICC 2004 (June 2004)
28. Sadeghi, B., Knightly, E.: Opportunistic media access for multirat ad hoc networks. In: Proc. ACM MOBICOM 2002 (2002)
29. Lijun, C., Low, S.H.: Joint congestion control and media access control desing for ad hoc wireless networks
30. Jiantao, W., Low, S.H., Doyle, J.C.: Cross-layer optimization in tcp/ip networks
31. Zhai, H., Fang, Y.: Tcp performance over mobile ad hoc networks. Canadian Journal of Electrical and Computer Engineer 29, 129–134 (2004)
32. Zhai, H., Chen, X.: Alleviating intra-flow and inter-flow contentions for reliable service in moble ad hoc networks. In: Proc. IEEE Milcom 2004 (November 2004)
33. Biswas, S., Morris, R.: Opportunistic routing in multi-hop wireless networks. In: Proc. The 2nd Workshop on Hot Topics in Networks. MIT, Cambridge (2003)
34. Giuseppe, B.: Performance analysis of the IEEE 802.11 distributed coordination function. IEEE Journal on Selected in Communication 18(3) (March 2000)

35. Hongqiang, Z., Yuguang, F.: Performance of wireless Lans based on IEEE 802.11 MAC protocols. IEEE, Los Alamitos (2003)
36. Chatzimisios, P., Boucouvalas, A.C., Vitsas, V.: IEEE 802.11 wireless Lans: performance analysis and protocol refinement. TURASIP Journal on Applied Singnal Processing 1, 67–78 (2005)
37. Younggoo, K., Haniph, L.: Improving transport layer performance by using a novel medium access control protocol with fast collision resolution in wireless Lans. In: MSWiM 2002 (September 2002)
38. Guoping, Z., Imrich, C.: A finite queeuing model for IEEE 802.11 mac protocol IEEE (2004)
39. Mustafa, O., Bruce, A.: An m/mmg1/1/k queuing model for IEEE 802.11 ad hoc networks. In: PE-WASUN 2004, October 7 (2004)
40. Omesh, T., Biplab, S.: Queueing analysis and delay mitiagation in IEEE 802.11 random access mac based wireless networks

DDGR: A Directed Diffusion Game Routing Protocol for Wireless Multimedia Sensor Networks*

Ke Zongwu, Guo Xiaoshan, Weng Yanqing, and Li Zhi

Department of Computer Science
Hubei Normal University
Huangshi 435002, China
KEZongwu@yeah.net

Abstract. The sensor node usually runs on battery, it may refuse to transmit data to avoid excessive using its energy, due to this potential selfish behavior, the research of routing protocols for wireless multimedia sensor networks can use game theories. In this paper, we provide a routing game model and a Directed Diffusion Game Routing protocol (DDGR). DDGR consists of four processes: interesting diffusion, QoS consultation, route reinforcement, and data transmission. The simulations indicate that DDGR can effectively improve the balance of energy consumption and prolong the networks lifetime.

Keywords: wireless multimedia sensor networks, game theory, routing.

1 Introduction

In the application of wireless multimedia sensor networks, it is usually transaction-oriented. The users need send data-query to the sensor nodes of specified areas, such as environmental monitoring, surveillance for rare animals and patient monitoring. Direct Diffusion routing [1, 2] is a typical routing protocol based on the data-query for wireless sensor networks. Different from the traditional algorithms, it's data-centric, and it's basic ideas is that the data is collected by sensor nodes will be defined as attribute-value pairs, the Sink node sends interests to request for returning the interesting data, then looking for an effective path from the source to the Sink node by diffusion dissemination algorithms. The purpose of DD routing is to establish a many-to-one path from one or many sources to Sink nodes. The routing process can be divided into three processes: interesting diffusion, data transmission and route reinforcement. It will startup the routing process through sink node broadcasting interests which different application needs.Getting a interests,a middle node establishing a gradient of data transmission path from a data source to the sink node, then forming multiple paths, finally depending on the optimizing criterion to generate an optimizing path by the strengthen mechanisms. In the process of establishing a

* The work was supported by Hubei provincial universities cooperation projects in P.R. China (Grant NO:CXY2009B031), and the plan for scientific and technological innovation team of excellent young and middle-aged in institute of high learning of Hubei Province in P.R. China (Grant No:T200806).

D. Zeng (Ed.): ICAIC 2011, Part II, CCIS 225, pp. 700–707, 2011.

rout, DD algorithm spreads interests by flooding, so it will use more energy and time, and it doesn't consider QoS demands. On the other hand, the multimedia sensor is limited resources, a high degree of redundancy, a floor of joint among different levels, and different QoS guarantee for different application requirements, these characteristics make the QoS routing more complicated [3]. Limited energy is one of the key features.The nodes'power is often provided by battery, and can not be replaced in wireless multimedia sensors networks. When a node's energy runs out, it will change network topology, and even cause network interruption., so there is a conflict between single node energy consumption and the entire networks' benefit. When a node's energy is lower, it may refuse to transmit. Under this potential "selfish" behavior, the routing problem of WMSNs is a natural fit for game theoretic. The goals of routing algorithm with game theory for WMSNs are to maximize benefit of networks and to reduce energy consumption.

This article will apply the basic ideas of game routing to the directed diffusion routing protocols, and a directed diffusion game routing protocol (DDGR) for multimedia sensors wireless networks is proposed. The simulation indicate that DDGR can effectively balance nodes energy consumption in the wireless multimedia sensor networks, and find a path to meet the QoS demands, and enhance the reliability of network by alternative paths.

2 Relational Research

Game theory in the the application of network routing technology has received extensive attention [4-10]. In [6] described the network energy and security about the game theory applications in sensor networks, and it presents the game theory has a wide range of applications in wireless sensor networking. In [7], Based on the game theory, a sensor-centric,reliable, and energy-constrained routing game model was present. In this model , sensors nodes are rational and intelligent agencies, they through co-operate find an optimized network structure to maximize the benefits, and minimize the theris cost during the routing . This routing is called a reliable query routing (RQR) game). Where,each node forwards data to an adjacent node, and each players 'pure strategy is a binary vector.In [8] discusses the transmiss data problem among the nodes in two different organizational sensor networks. Apparently, when transmitting other networks' data,sensor will cost its energy .so it will refuse and lead to be a low efficiency, this problem similars to the classic prisoner's dilemma. A "token" mechanism was provided ,and it encourages the cooperation among the sensor nodes in the different organizations to improve the network performance. In [9] , the path length and the routing energy consumption as two key aspects affect the network lifetime, and game theory was proposed to solve limited energy path and optimal problem .

All above game routing models are not providde for wireless multimedia sensor networks. With wireless multimedia sensor networks characters, we proposed a QoS game routing model. Based on it, we combine the game routing model with directed diffusion routing, and propose a Direct Diffusion Game Routing(DDGR) for wireless multimedia sensor networks.

3 BASIC IDEA of DDGR

DDGR consists of the following basic elements: Interests, Data Messages, Routing Game and Reinforcements. The interest message is the data inquiry information, which is produced by the sink node. This article takes the rare animals monitoring for example to show the structure of the interests:

type = golden monkeys
interval = 60 s
centre = [10, 10]
radius = 1 m
timestamp = 10:30:01 //hh:mm:ss

The duty is to search whether there is golden monkey at the speed of photographing once a minute in the region with point (10,10) as the center of circle, 1m as the radius .Interests will be broadcast to the sensor nodes of the specified area. Middle sensor nodes broadcast messages,and the method is described in Figure 1a. Many sensor network applications need know the source location, so we assume it will be known in this article. each node decide whether receive or forward interests depending on the distance ,which is whether less than the previous node to the target area. While each node is forwarding data, their own address and location information are added in to the interests. when interests reaches the target,it has a path from the source node to the Sink node.As is shown in Figure 1b, the distance to the target area from node 2 is less than from node 1, so node 1 chooses node 2 as the next hop.

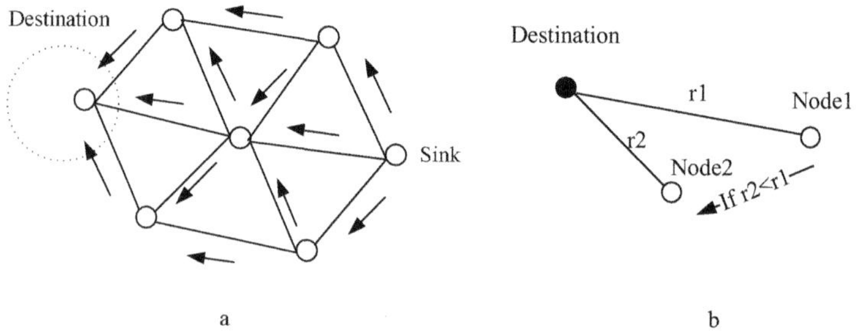

Fig. 1. Interests Diffusion in DDGR

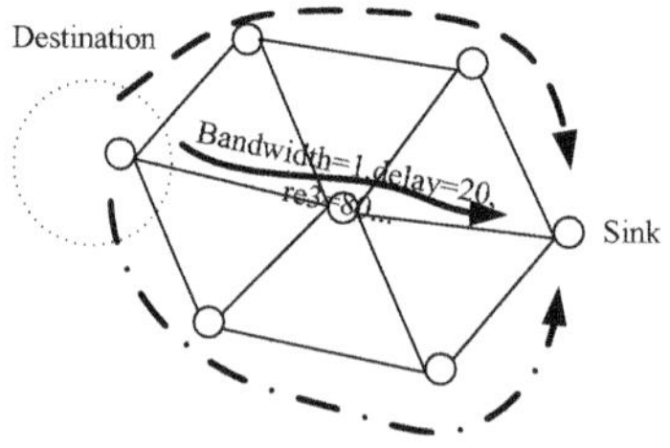

Fig. 2. QoS Negotiation in DDGR

The same interests will arrive at the destination node by different paths. When the first interestingmessage arrives, the destination node sets a specified time.If the interests arrives in the specified time ,the target will generate a QoS negotiation message ,which transmit to the sink node along the interests path.QoS negotiation message is primarily made up of QoS constraints, when it run through the middle nodes, the middle nodes will add with its residual energy, bandwidth, latency and other information and then forward. When the QoS negotiation information reaches the sink node ,it contains QoS constraints and related path information.sink node will set a specified time according to QoS constraints. If the QoS negotiation information arrives in the specified time, the path contains it will play game. This process is definded as QoS negotiation, and the process is shown in figure 2. as QoS delay constraint a standard to set the specified time and its value is defined as:

$$threshold = \alpha \times D_{\max}, 0 < \alpha \leq 1$$

Where, $D_{\max}$ is the maximum delay constraint. the threshold can prevent the path being selected in the routing if it does not satisfy the delay constraints, thereby reducing the additional burden and improving the performance.

The sink nodes deal with the path information,which contains the QoS negotiation information and receice in the specified time, removing the path does not meet the QoS requirements.Then,performing routing game operations, calculating the mixed Nash equilibrium solution about each path and selecting a forwarding highest probability path to transmit the enhanced information, which contains the path information and the path quality of service . When receiving the route reinforcement information.the target nodes start to use the enhanced path to transmit data,which contains complete path messages so that each node forward.During receiving data, if the sink node find the path failure or not meeting the QoS requirements, it will select a sub-optimal path to send the data from the spare path. The process of routing reinforcement is shown in figure 3:

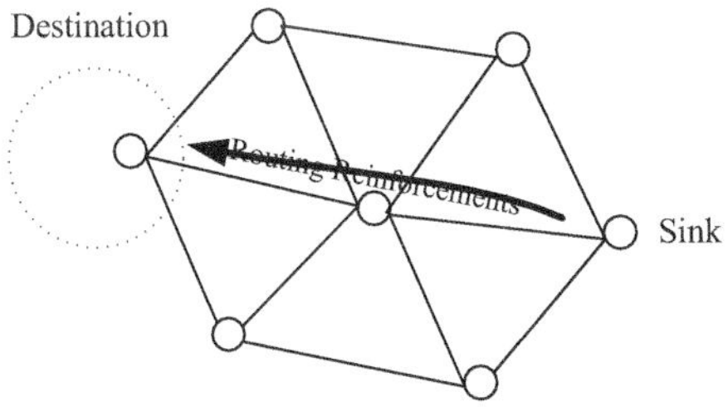

Fig. 3. Routing Reinforcements of DDGR

4 Game Model

Definition 1 Unicast QoS routing game of N players: $G = \{N, S, U\}$ is a unicast QoS routing game of N players, and the the game players are N paths, $N \geq 2$.The pure strategy set of each player is $S_i = \{0,1\}$, $S_i = 1$ indicates the path i transmits data, while $S_i = 0$ says it does not transmit. Each player's payoff function is defined as follow:

$$u(i) = \begin{cases} \gamma_i \times f(i) - c_i, & S_i = 1 \wedge S_{-i} = 0 \\ \gamma_j \times f(j), & \exists S_j = 1 \wedge S_{-j} = 0 \\ 0, & others \end{cases}$$

Where $\gamma_i = \frac{E_{i\min}}{E_0}$, c_i is the cost of path i. S_{-i} is the strategies of all paths except for the path i. $f(i)$ is when the path i is selected. The definition of $f(i)$ is as follow:

Definition 2 the network's overall revenue contribution of the path :

$$f(i) = \begin{cases} \varphi & ,b(i) \geq B_{\min} \wedge d(i) \leq D_{\max} \\ 0 & ,others \end{cases}$$

φ is a constant greater than zero, $b(i)$ is the minimum bandwidth of the path i, $d(i)$ is delay the path i, $B_{\min}$ is the bandwidth constraint, and $D_{\max}$ is the delay constraint. The definition shows that if a path meets the QoS guarantee to transmit, the network will get a benefit φ, and otherwise, the benefit is zero. So this path will not be selected in the game.

Due to the limited game players and limited strategies of each player, according to game theory, unicast routing game model must have the Nash equilibrium [11].

5 Simulation

Simulation is based on OMNET + + simulation platform, network-framework INETMANET based OMNET simulation is also used. INETMANET provides wireless network module , ADHOC simulation network module and the 802.15.4 simulation module. 802.15.4 is applied in the node's MAC layer, network layer takes a custom routing game module, application layer uses the DDGR, the QoS module nodes sever the QoS constraints, the energy module servrs the energy simulation of sensor nodes , which is provided by INETMANET. When receiving and sending messages, the node informs the energy module to modify the node energy. Simulation focuses on the energy balance, so we take the grid structure of wireless multimedia sensor network topology to be better for showing the distribution of energy. During the simulation,the sink node continuously generates data query requirements to its diagonal region, that is to say, the source and the destination node will not change.we assume that the initial energy of all nodes is set to the same value 100mAh. Without considing the node sleeping funtion in the simulation process, the transmitting power of the node is defined as 1.0mW. Figures 4 and 5 respectively show the distribution of each node's residual energy after runing the same period of time to the direct diffusion routing and DDGR.The pseudo-color pictures describes the hot spots, and three-dimensional grid graphs describes the energy distribution. In pseudo-color pictures,different colors represent the residual energy of different levels, x and y are position of the node. From figure 4 and figure 5,it can be seen that the direct diffusion routing don't consider the balance of energy, obviously, the hot zone shows that some nodes exit the excessive energy consumption, , while the node energy consumption is balance in DDGR routing, so the hot region is balance.

Figure 6 shows that the comparison of delay between two routing protocols, In the simulation, the sink node randomly generates a query ,and the node initial energy also randomly generates,the delay are used of the average delay by many experiments. In Figure 6, the delay characteristics are similar for two routing algorithms. Of course, considering the energy balance in DDGR, so the shortest delay doesn't think the only goal, and in the most cases, the delay is slightly larger than the direct routing algorithm.

Simulation shows that DDGR can effectively balance the node's energy and prolong the wireless multimedia sensor networks lifetime.

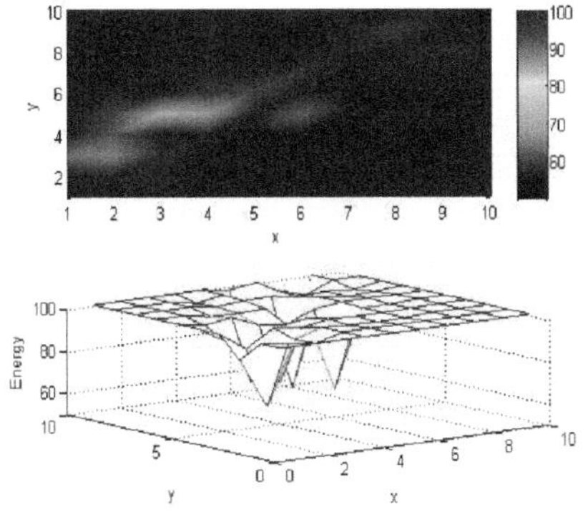

Fig. 4. The energy of each node under DD

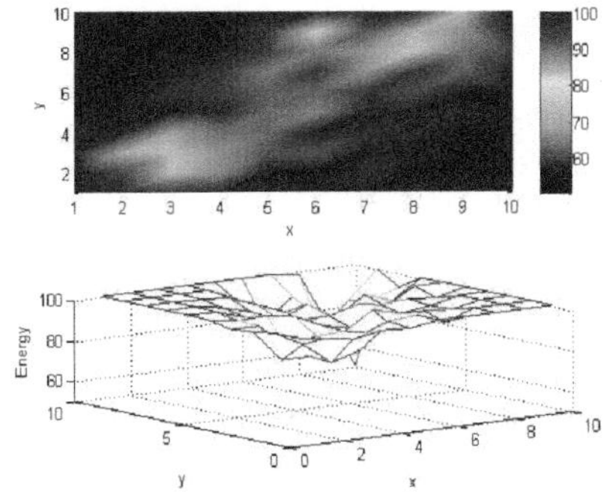

Fig. 5. The energy of each node under DDGR

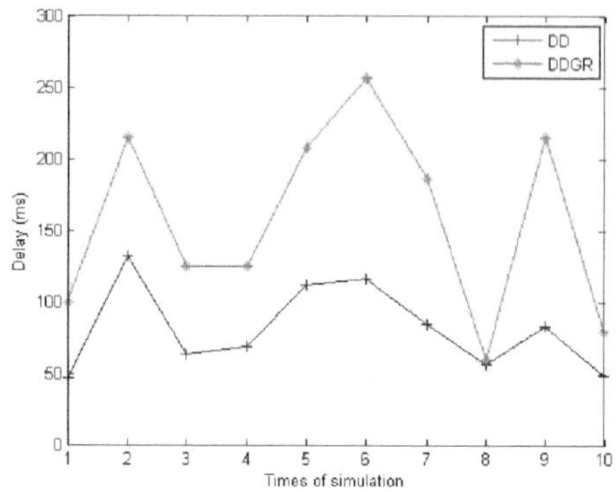

Fig. 6. The delays of routing generated by DD and DDGR

6 Conclusion

Routing game model has received a lot of scholars' attention.According to the characteristics of the networks ,this essay presents a routing game model for QoS routing of multimedia sensor network, and a directed diffusion routing protocol DDGR based on it. DDGR consists of four processes: interests' diffusion, QoS consultation, route reinforcement, and data transmission.Interest information diffusion effectively reduces the influence of storms by broadcast with the known location method. During the data transmission, the complete path information is added to the data so that nodes needn't store the routing table or look-up the table, as a result of reducing the demand for memory capacity. Routing reinforcement process employs the Nash equilibrium solution. The simulations indicate that DDGR can effectively improve the balance of energy consumption and prolong the networks lifetime.

Acknowledgment. The work was supported by Hubei provincial universities cooperation projects in P.R. China (Grant NO: CXY2009B031), and the plan for scientific and technological innovation team of excellent young and middle-aged in institute of high learning of Hubei Province in P.R. China (Grant No:T200806).

References

1. Intanagonwiwat, C., Govindan, R., Estrin, D., et al.: Directed diffusion for wireless sensor networking. IEEE-ACM Transactions on Networking 11(1), 2–16 (2003)
2. Intanagonwiwat, C., Govindan, R., Estrin, D.: Directed diffusion: a scalable and robust communication paradigm for sensor networks. In: Proceedings of the MobiCom 2000 Proceedings of the Sixth Annual International Conference on Mobile Computing and Networking, Boston. ACM, New York (2000)
3. Akyildiz, I.F., Melodia, T., Chowdhury, K.: A survey on wireless multimedia sensor networks. Computer Networks 51(4), 921–960 (2007)

4. Altman, E., Boulogne, T., El-Azouzi, R., et al.: A survey on networking games in telecommunications. Computers & Operations Research 33(2), 286–311 (2006)
5. Naserian, M., Tepe, K.: Game theoretic approach in routing protocol for wireless ad hoc networks. Ad Hoc Networks 7(3), 569–578 (2009)
6. Machado, R., Tekinay, S.: A survey of game-theoretic approaches in wireless sensor networks. Computer Networks 52(16), 3047–3061 (2008)
7. Kannan, R., Iyengar, S.: Game-theoretic models for reliable path-length and energy-constrained routing with data aggregation in wireless sensor networks. IEEE Journal on Selected Areas in Communications 22(6), 1141–1150 (2004)
8. Miller, D.A., Tilak, S., Fountain, T.: "Token" equilibria in sensor networks with multiple sponsors. In: Proceedings of the 2005 International Conference on Collaborative Computing: Networking, Applications and Worksharing and Workshops (IEEE Cat No 05EX1251C), San Jose, CA, USA, ACM, New York (2005)
9. Kannan, R., Ray, L., Kalidindi, R., Iyengar, S.S.: Max-min length-energy-constrained routing in wireless sensor networks. In: Karl, H., Wolisz, A., Willig, A. (eds.) EWSN 2004. LNCS, vol. 2920, pp. 234–249. Springer, Heidelberg (2004)
10. Wang, X., Wang, Q., Huang, M., Tian, Y.: QoS multicast routing mechanism based on the Fuzzy Integral and game theory. Journal of Software 19(7), 1743–1752 (2008)
11. Yang, R., Ter-Petrosyan, Li, S.: Dynamic cooperation - sophisticated game theory. China Press, Beijing (2007)

Research on Discretization Algorithm of Continuous Attribute Based on PCNN in a Bridge Erecting Machine Safety Monitoring System

Na Chen[1,2], Shaopu Yang[2], Cunzhi Pan[2], and Erfu Guo[2]

[1] School of Mechanical, Electronic and Control Engineering, Beijing Jiaotong University, Beijing, China
[2] Shijiazhuang Tiedao University, Shijiazhuang, China
NaChenA@yeah.net

Abstract. PCNN (Pulse Coupled Neural Network) model is suitable for implementing clustering algorithm because of its unique neighbor coupled feature. This paper presented a new discretization algorithm of continuous attribute based on simplified PCNN model. Measured signals of the bridge erecting machine safety monitoring system are computed on this algorithm. Then decision rules are gained by using Rough Sets model to analyze the discretized result, among which the ratio of deterministic rules is 100 percent. Obviously it improves the certainty of decision and analysis better.

Keywords: PCNN (Pulse Coupled Neural Network), discretization, a bridge erecting machine, safety monitoring, Rough Sets.

1 Introduction

The function of discretization of continuous attribute is to extend learning algorithms in discretized space to continuous feature space[2]. Through discretizion of feature space or dimension reduction, more simplified classifier can be obtained, which reduces data storage space, enhances learning speed and precision. It is a valid pretreatment method for many knowledge discovery systems such as Rough Sets theory[1]. Normal discretization algorithms are equal-width, equal- frequency, k mean value, etc. We apply some kinds of discretization algorithms to measured continuous signals of the bridge erecting machine safety monitoring system and combine feature extracted to divide the objects in the universe. As a result there are errors in the classification, which makes the presence of nondeterministic rules after decision and inference.

Therefore, in this paper we present a new discretization algorithm[3] of continuous attribute based on PCNN in order to decrease or remove the false division and the nondeterministic rules and to enhance the certainty in decision making. Simulation experiment is made by using measured signals of the bridge erecting machine safety monitoring system, and comparative analysis is also done among some different kinds of algorithms.

D. Zeng (Ed.): ICAIC 2011, Part II, CCIS 225, pp. 708–714, 2011.

2 Description of Discretization Problem

The process of discretization of continuous attribute can be described as following. By selecting proper division points the whole value domain of continuous attribute can be divided into several intervals, and each interval is given an integer to denote this interval's attribute value. At the same time, computing method for corresponding discretized interval of any point in the value domain is provided.

A decision information S=(U,A∪{d},V,f) consists of :U is a nonempty, finite set called universe; A is a nonempty, finite set of attributes; A is a finite set of condition attributes and{d} is a finite set of a decision attribute; V= V_a, in which V_a is called the domain of a; f $\bigcup_{a\in A}$ is an information function f: U×A→V.

For each a ∈ A, if $v_a = [l_a, r_a) \subset R$ is its value domain, in which R is a set of real numbers. A partition of v_a is defined as

$$p_a = \{[c_0^a, c_1^a], [c_1^a, c_2^a], \cdots, [c_{k_a}^a c_{k_a}^a]\} \text{ where}$$

$$l_a = c_0^a < c_1^a < c_2^a < \cdots < c_{k_a}^a < c_{k_{a+1}}^a = r_a; \qquad v_a = [c_0^a, c_1^a) \cup [c_1^a, c_2^a) \cup \cdots \cup [c_{k_a}^a, c_{k_{a+1}}^a).$$

The set of cuts of a is defined as $C = \{C_a \mid a \in A\}$. And a partition P of A is uniquely denoted as sets of sets C, shown as $P = \bigcup_{a\in A} p_a$.Obviously, P converts S into a discrete system S^P, and $S^P = (U, A^P \cup \{d\}, V, f)$ is the so-called discretization of S, where $A^P = \{a^p : a \in A\}$ and $a^P(x) = i \Leftrightarrow a(x) \in [c_i^a, c_{i+1}^a]$.

3 Clustering Algorithm Based on PCNN

PCNN is called as the 3rd generation artificial neural network[7], which is presented by Eckhorn according to synchronizing pulses of animals' cerebral vision cortex. PCNN has embodied its advantage and applied to fields such as image processing, object recognition, communication synchronization, pattern recognition and decision optimization. In this paper we use PCNN simplified model, in which the input of the model is one-dimensional attribute value and the classification result is described by the pulsed time.

A. PCNN simplified model

$$F_i[n] = S_i \tag{1}$$

$$L_i[n] = \sum_k w_{ik} Y_k[n-1] \tag{2}$$

$$U_i[n] = F_i[n](1 + \beta L_i[n]) \tag{3}$$

$$Y_i[n] = \begin{cases} 1, & U_i[n] > \theta_i[n] \\ 0, & U_i[n] \le \theta_i[n] \end{cases} \tag{4}$$

$$\theta_i[n] = \alpha\theta_i[n-1] \tag{5}$$

In the above equations, S_i is external input excitation; F_i is input item; L_i, U_i, Y and θ_i are respectively link input, interior activity item, pulse output and dynamic threshold; w denotes neighborhood neuron; β is link intensity; α is attenuation coefficient of threshold; n is iterative times.

B. Algorithm description

In the algorithm, threshold θ is initialized by the maximal attribute value, and each neuron is on the extinction state. With threshold decaying, the neuron of bigger attribute value is fired firstly when the result of (4) is 1. In next iteration, when n increases, (2) can generate link input L from fired neighbor neurons. And the generation of link input increases the interior activity U in (3). Then the neuron with increased interior activity maybe fired in (4). That is to say, neurons fired firstly increase their interior activity through their neighborhood link input. At the same time dynamic thresholds decrease with iteration going. Finally all the neurons are fired. Only but the time when each neuron is fired is different. And the order of firing time is the classificatory information we need.

4 Processing Measured Data on PCNN Model

In the bridge erecting machine wireless sensor safety monitoring system, the whole work process can be divided into 9 steps, and they are static preparation, front end lifting, first beam moving, rear end lifting, rear end lifting, second beam moving, beam placement, installing beam, lower leading beam moving and lengthways moving. According to theoretical and practical testing, the signals collected from the sensor point of inside of frange plate at the bending of rear landing leg are the most sensitive and valid[5]. Therefore, in this paper we verify discretization algorithm based on signals at this monitoring point.

A. Creating decision table

With collecting signals from the sensor position of inside of frange plate at the bending of rear landing leg and recording the whole work process of the bridge erecting machine, we extract features from measured signals in different work steps.

Table 1.

sampling	IF	CG	D_x	α_4	D
1	3.8230	0.1312	0.7402	0.8225	1
2	4.7386	0.1405	0.6686	2.7702	1
3	3.8326	0.1320	0.7362	0.8050	1
4	4.2204	0.1643	0.6210	0.6654	1
5	1.9428	0.0405	12.6971	2.8579e+003	2
6	1.9738	0.0468	0.6385	1.0178e+003	2
7	1.8814	0.0384	11.6192	3.0265e+003	2
8	2.0670	0.0486	2.2126	858.6449	2
9	4.7489	0.0870	16.7771	1.0695e+003	3
10	5.8826	0.0933	17.1638	651.7212	3
11	4.7502	0.0838	9.8428	485.4563	3

Table 1. *(continued)*

samplings		IF	CG	Dx	α_4	D
12	3.5733	0.0531	18.8545	670.8314	3
13	2.6048	0.0457	2.7753	545.2688	4
14	2.5867	0.0425	0.3228	413.8693	4
15	3.7712	0.0241	7.4257	3.4975e+003	4
16	5.2563	0.1600	1.2037	174.5679	4
17	1.4880	0.0553	0.5338	496.1297	5
18	2.8751	0.0501	0.5926	436.9572	5
19	1.6313	0.0340	6.5953	1.0539e+004	5
20	2.0872	0.0510	0.8104	893.7270	5
21	2.3612	0.0211	1.9801	6.5352e+003	6
22	4.3385	0.0493	6.3736	437.8209	6
23	1.9295	0.0173	0.4526	4.8963e+003	6
24	4.4534	0.0639	0.7362	536.5724	6
25	1.6866	0.0216	0.6209	4.3107e+003	7
26	6.2410	0.1731	0.3786	5.3436	7
27	2.4235	0.0245	1.1222	2.5721e+003	7
28	4.6895	0.0885	2.6811	43.3860	7
29	3.4916	0.0620	1.3613	2.1714e+003	8
30	5.1548	0.0777	0.6040	45.0176	8
31	2.7515	0.0298	1.0718	1.6041e+003	8
32	3.9822	0.0693	0.3746	48.7299	8
33	2.5219	0.0174	4.2128	1.5357e+004	9
34	5.8330	0.0927	0.6628	13.6522	9
35	2.5074	0.0172	4.0619	1.5419e+004	9
36	5.5663	0.0874	0.9764	28.3394	9

The feature IF denotes the complexity of FFT spectrum; CG is the central frequency of spectrum; D_x reflects square deviation of time series and displays unstable degree of time domain; α_4 denotes peak value of time series. We set feature set of {IF,CG, D_x, α_4} as condition attribute and set work step as decision attribute D. And information decision table[4] can be created and shown as table 1.

B. Discretization based on PCNN clustering algorithm

We treat many samplings of different work steps as nerve cells, make condition attribute set as neuron input and set up PCNN model. Applying pulsing time as classification standard, discretization results of each attribute can be gained, shown as table 2.

Table 2.

sampling	IF	CG	D_x	α_4	D
1	3	1	4	5	1
2	2	1	4	4	1
3	3	1	4	5	1
4	2	1	4	5	1
5	4	4	1	1	2
6	5	4	3	2	2
7	5	4	1	1	2
8	4	3	2	2	2
9	2	2	1	2	3
10	1	2	1	2	3
11	2	3	1	2	3
12	3	4	1	2	3
13	4	4	2	2	4
14	3	4	4	2	4
15	2	5	2	1	4
16	1	1	3	2	4
17	5	3	4	2	5
18	4	4	4	2	5
19	5	5	2	1	5
20	4	4	4	2	5
21	3	5	3	1	6
22	2	4	2	2	6
23	3	5	4	1	6
24	2	3	4	2	6
25	3	5	4	1	7
26	1	1	5	4	7
27	3	4	4	1	7
28	2	2	3	3	7
29	2	3	3	1	8
30	1	3	4	3	8
31	3	4	4	1	8
32	3	3	4	3	8
33	3	5	2	1	9
34	1	2	3	3	9
35	2	5	2	1	9
36	1	2	3	3	9

C. Setting Parameters

There are two parameters in our PCNN simplified model. They are respectively dynamic attenuation coefficient of threshold α and link intensity β. The dynamic attenuation coefficient of threshold α decides attenuation speed of threshold, and it also effects pulsing times of neurons. We can conclude from (3) that activity of neuron is not only decided by its input stimulation but also by the states of neurons in neighborhood. Therefore dynamic attenuation coefficient of threshold α has limit influence on pulsing time. In our model a same dynamic attenuation coefficient of threshold 0.45 is set according to different condition attribute. That is to say, the key of PCNN model considers neighborhood information and reflects increasing interior activity of neighboring neurons, which can make the time of neuron pulsed earlier and make neurons with similar value or close position being pulsed more possibly. And so link intension β is a very important parameter that affects pulsing time of all the neurons through link relationship among neighborhood neurons. Considering the four condition features from left to right in table 1, we set value 0.8, 0.7, 0.4, 0.1 to link intense parameter for each feature in turn.

D. Comparative analysis for several different algorithms

Decision table is gained after discretization based on PCNN clustering. Combining rough sets model we obtain decision rules by processing the discretized decision table[6]. According to attribute value of different feature the current work condition can be derived, which can direct controlling and monitoring better. 25 rules are gotten from our model, which are all deterministic and improve the certainty of decision. In this paper equal-frequency and FCM discretization algorithms are also applied to the same information decision system, in which rules are gained too. The number and certainty ratio of these rules are listed in table 3.

Table 3.

Discretization algorithm	**Number of rules**	**Number of deterministic rules**	**Number of indeterministic rules**
Equal-frequency	33	32	96.9%
Based on FCM	26	22	84.6%
Based on PCNN	25	25	100%

In the three algorithms above, each attribute is classified into 5 classes. Under the situation of similar knowledge granularity, the ratio of deterministic rules is 100 percent computed by discretization algorithm based on PCNN. It accounts for the validity of this algorithm in analyzing measured signals in the bridge erecting machine safety monitoring system.

5 Conclusion

PCNN is a kind of new neural network model, and its core is coupling mechanism in neighborhood. The order of pulsing time denotes the result of neuron clustering. In this paper we present a new discretization algorithm of continuous attribute based on PCNN clustering. Aiming at measured signals in the bridge erecting machine safety monitoring system, we have gained discretizied result. Combining rough sets model with discretized result, decision rules are obtained, in which the ratio of deterministic rules is 100 percent. Compared with other algorithms, the algorithm based on PCNN has improved the certainty of decision analysis. And it has a good guidance for controlling and monitoring in actual application system.

Although two parameters in our PCNN model are discussed, self-adaptive empirical parameters have not been realized in this paper, which should be improved later.

References

1. Pawlak, Z.: Rough set. International Journal of Computer and Information Science 11, 341–356 (1982)
2. Lenarcik, A., Piasta, Z.: Discretization of condition attributes space. In: Intelligent Decision Support, pp. 373–389. Kluwer, Dordrecht (1992)
3. Dougherty, J., Kohavi, R., Sahami, M.: Supervised and unsupervised discretiazation of continuous features. In: Proc. the 12th International Conferece, pp. 194–202 (1995)
4. Slowinski, R.: Intellligent Decision Support. In: Handbook of Applications and Advances of Rough Sets Theory. Kluwer Academic Publishing, Dordrecht (1992)
5. Chen, N., Yang, S., Pan, C., Tang, Y.: Research on Validity of Sampling Points in a Bridge Erecting Machine Safety Monitoring System using Variable Precision Rough Sets. In: 2010 3rd International Conference on Computer and Electrical Engineering (2010)
6. Ziarko, W., Slowinski, R.: Rough set approach to multiattribute decision analysis, Invited Review. European Journal of Operational Research 72, 443–459 (1994)
7. Johnson, J.L., Padgett, M.L.: PCNN Models and Applications. IEEE Trans. on Neural Networks 10(3), 480–498 (1999)

Research on Active Direction Control of Independent Wheel-Set of 100% LFLRV

Minghua Zhao, Jie Chen, and Zhigang Liu

Department of Electrical Engineering
Beijing Jiaotong University
Haidian District, Beijing. 100044, China
MinghuaZhao@ovi.com

Abstract. Dealing with the absence of active direction ability of independent wheel-set of 100%LFLRV, in this paper are presented the kinetic equation and the force condition of the wheel-set. Combined with the flux-oriented control strategy of traction motor, this paper analyzes the principles of direction and proposes a novel active direction control scheme with the detection of torque difference of the left and the right traction motors, embedding direction control into traction control. In the last part, a Matlab/Simulink model is built and is subjected to, for the test of the strategy proposed.

Keywords: 100% LFV, independent wheel-set, torque control, direction control approach.

1 Introduction

Along with the process of global industry and rapid growth of metropolis population, urban traffic has been worse day by day, therefore new urban transport systems are being taken into consideration by various countries. Among all of them, the world founds the application of light rail vehicle more and more widely, which means less pollution, lower power consumption, less investment and relatively larger passenger capacity[1]. To make better the inconvenience during getting-on and getting-off, and reduce the time delay, the design of Low-Floor Light Rail Vehicles(LFLRV) has been accepted more, with the distance between the floor and the ground varying from 180mm through 350mm[2], coming together with the advantages of more comforts, more convenience for the disabled, the elderly, the young, and the absence of common high platforms, hence more accessible and convenient.

For the lowering of inside floor to the utmost, a 70% and 100% low floor is often required, which means the reservation of the floor-passage space occupied by traction bogie, as is shown in Fig.1[3]. Therefore, the structure of independent wheel-sets is adopted, where two longitudinally-placed and two-end-outputting traction motor are mounted into one motive bogie, with one motor driving two wheels on the same side of vehicle. Such structure saves the space which was to be occupied by axles of the bogie and offers more accessibility during daily maintenance, repair, and disassembly.

D. Zeng (Ed.): ICAIC 2011, Part II, CCIS 225, pp. 715–726, 2011.

However, since independent wheel-sets comes without the coupling of motive axles, the absence of their active direction ability is inevitable. Specialists and engineers have come up with the research of different direction mechanisms[4-6]:

To design the tread profile, so that the direction ability is added by gravity restoring force; to develop the forced-direction bogie with latitudinal connecting ability; to adopt active suspension technology, making the bogie to be direction-able.

In this paper, the direction of 100% LFLRV is dealt with deeply, from the aspect of electrical control scheme and combined with the traction control procedure of vehicle.

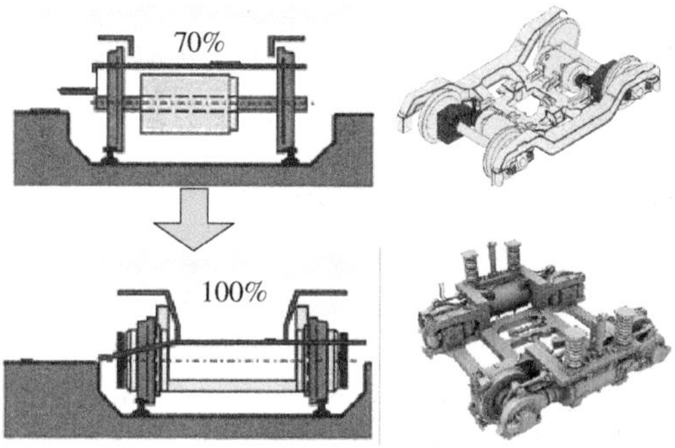

Fig. 1. The motive bogie of 70% and 100% LFLRV

2 The Kinetic Equation of Independent Wheel-Set

From the theory of wheel-rail contact[8], during the rolling contact of two elastic bodies, is the existence of creeping and creeping force at the contact patch, besides the common-known normal force and contact stress. By introducing x-y-z three-dimensional coordinate system, the creeping force on rail is with a positive sign, and rolling contact along rail with conventional rigid wheel-set is shown in Fig.2.

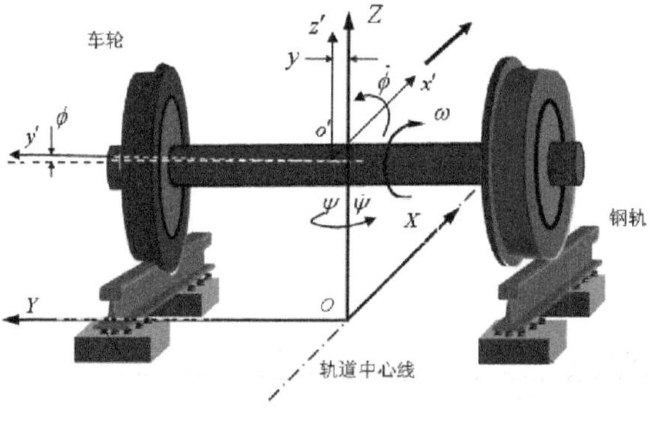

Fig. 2. Rolling contact along rail with conventional rigid wheel-set

In Fig.2, OXYZ is rail coordinate system, in which X is along the rolling direction of wheels, Y is the lateral direction of rail, and Z the vertical direction. The origin of o'x'y'z' coincides with the center of wheel-set, in which y' coincides with the axis of wheel-set, y is the lateral displacement between the center of wheel-set and rail, ϕ and $\dot{\phi}$ is the angle and angular velocity by which wheel-set rotates around x'. ψ and $\dot{\psi}$ is the angle and angular velocity by which wheel-set rotates around z', also known as wheel-set yaw angle. V is the rolling velocity of wheel-set.

During the lateral displacing of wheel-set, the difference between rolling radius of left and right wheel leads to opposite slight creeping of them along rail's longitudinal direction. Regardless of elastic deformation, the longitudinal, latitudinal creeping and relative rotating angle can be expressed with explicit expression including the parameters above.

Conventional wheel-set is entitled to be able to direct with wheel-rail creeping and creeping torque, which appears as central restoring along straight lane, and as curve directing along bend lane.

Without the mechanical coupling from axle, the left and right wheel rotates around their own center, so that an additional parameter should be added into its kinetic equation[8] . Finally, the kinetic equation for independent wheel-set is given as:

$$m\ddot{y} + \frac{2f_{22}}{V}\dot{y} - 2f_{22}\psi = 0 \tag{1}$$

$$I_{wz}\ddot{\psi} + \frac{2f_{11}L_g\lambda}{r_0}y + \frac{2f_{11}L_g^{\ 2}}{V}\dot{\psi} + \frac{2f_{11}L_g r_0}{V}\dot{\beta} = 0 \tag{2}$$

$$I_{wy}\ddot{\beta} + f_{11}\lambda y + \frac{f_{11}L_g r_0}{V}\dot{\psi} + \frac{f_{11}r_0^{\ 2}}{V}\dot{\beta} = 0 \tag{3}$$

Where $\dot{\beta}$ is the rotational speed difference between left and right wheel. Therefore, the rotational speed of left and right wheel are given as:

$$\omega_L = \frac{V}{r_0} + \dot{\beta}, \ \omega_R = \frac{V}{r_0} - \dot{\beta} \tag{4}$$

Then revising eq.(1) through (3) into state equation:

$$\dot{x} = A \cdot x \tag{5}$$

Where $x = \begin{bmatrix} \dot{y} & y & \dot{\psi} & \psi & \dot{\beta} & \beta \end{bmatrix}^T$, and if

$$a_{11} = -\frac{2f_{22}}{mV}, a_{14} = \frac{2f_{22}}{m}, a_{32} = -\frac{2f_{11}\lambda L_g}{I_{wz}r_0}, a_{33} = -\frac{2f_{11}L_g^{\ 2}}{I_{wz}V}, a_{35} = -\frac{2f_{11}r_0 L_g}{I_{wz}V},$$

$$a_{52} = -\frac{f_{11}\lambda}{I_{wy}}, a_{53} = -\frac{f_{11}r_0 L_g}{I_{wy}V},$$

$$a_{55} = -\frac{f_{11}r_0^2}{I_{wy}} \tag{6}$$

it gives

$$A = \begin{bmatrix} a_{11} & 0 & 0 & a_{14} & 0 & 0 \\ 1 & 0 & 0 & 0 & 0 & 0 \\ 0 & a_{32} & a_{33} & 0 & a_{35} & 0 \\ 0 & 0 & 1 & 0 & 0 & 0 \\ 0 & a_{52} & a_{53} & 0 & a_{55} & 0 \\ 0 & 0 & 0 & 0 & 1 & 0 \end{bmatrix} \tag{7}$$

From eq.(5), the characteristic polynomial of kinetic equation of independent wheel-set is:

$$\det[s1 - A] = 0 \tag{8}$$

Substituting eq.(6) and (7) into eq.(8),it gives:

$$\begin{aligned} &s^6 - (a_{11} + a_{33} + a_{55})s^5 + (a_{11}a_{33} + a_{33}a_{55} - a_{53}a_{35} \\ &+ a_{11}a_{55})s^4 + (a_{11}a_{35}a_{53} - a_{33}a_{11}a_{55})s^3 - a_{14}a_{32}s^2 + \\ &(a_{14}a_{32}a_{55} - a_{14}a_{35}a_{52})s = 0 \end{aligned} \tag{9}$$

For conventional wheel-set, the characteristic polynomial can be given as[9]:

$$s^4 - (a_{11} + a_{33})s^3 + a_{11}a_{33}s^2 - a_{14}a_{32} = 0 \tag{10}$$

The difference between eq.(9) and eq.(10) is that the characteristic root of eq.(10) contains a zero-root, which implies the existence of the system's uncertain movement. The movement can be towards center, when the rotational speed difference between the left and right wheel is zero, or deviating from center line, when the difference above is non-zero. All in all, eq.(10) explains the absence of direction ability of independent wheel-set, mathematically.

For independent wheel-set, the rotational speed of the left and right wheel can be given approximately as $\omega_L r_L = \omega_R r_R = V$, and with spin creeping neglected, the longitudinal and latitudinal creeping torque can be given as[9,10]:

$$\begin{aligned} &F_y = 2f_{22}(\frac{\dot{y}}{V} - \psi) \\ &M_z = 0 \end{aligned} \tag{11}$$

From eq.(11), for independent wheel-set, its latitudinal creeping is influenced mainly by ψ , and a non-zero ψ means its corresponding latitudinal creeping force, while longitudinal creeping force remains zero, as is the case with creeping torque.

3 The Analysis of Direction Based on Torque Control

It can be seen that the absence of direction ability of independent wheel-set is due to the absence of longitudinal creeping torque, from the kinetic analysis of conventional and independent wheel-sets. However, such disadvantage is highly likely to be eliminated with the torque control of the left and right traction motors, realizing approximately the same direction ability as conventional wheel-set.

Assume that the traction force exerted on left and right tread are FL and FR , respectively, it gives:

$$F_L = F + F_\beta \qquad F_R = F - F_\beta \tag{12}$$

And the longitudinal torque is:

$$M_z = (-F_L + F_R)L_g = -2F_\beta \cdot L_g \tag{13}$$

It should be noted that longitudinal direction torque exists only under latitudinal displacement. Since the relation between displacement y and longitudinal torque is linear, it can be given as:

$$M_z = K_T \cdot y \tag{14}$$

When passing along a circle with radius R, the outer wheel must travel a longer distance than the inner wheel. Due to the conical tread, the rolling radius of the outer and inner wheel can be different. The outer is larger when wheel-set displaces towards outer side, thus a pure rotation can be achieved, and a pure rotation curve is projected vertically from the curve along which the middle of wheel-set passes. The pure rotation curve is circular arc, which is parallel with the circular curve of the middle of rail, and the two share a common curvature center, however, the former is always outer than the latter, as is shown in Fig.3.

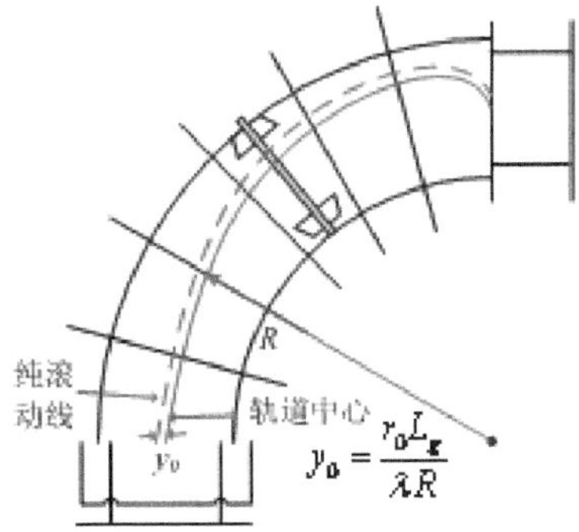

Fig. 3. Wheel-set passing along a curve lane

When passing along a circle with radius R, the outer wheel must travel a longer distance than the inner wheel. Due to the conical tread, the rolling radius of the outer and inner wheel can be different. The outer is larger when wheel-set displaces towards outer side, thus a pure rotation can be achieved, and a pure rotation curve is projected vertically from the curve along which the middle of wheel-set passes. The pure rotation curve is circular arc, which is parallel with the circular curve of the middle of rail, and the two share a common curvature center, however, the former is always outer than the latter, as is shown in Fig.3.

The distance between pure rotational curve and the middle of rail is influenced by the shape of tread, and the radius of the curve, which can be given as:

$$y_0 = r_0 L_g / \lambda R \tag{15}$$

The direction ability can be achieved by control the torque output of left and right traction motors. From eq. (10), (13) through (15), it gives:

$$F_\beta = f_{11} L_g / R \tag{16}$$

So direction ability can easily be achieved by the torques controlled by eq.(16).

The relationship between external force on wheel-set and its speed variation is expressed by the wheel-set's kinetic equation. There are three kinds of external force that is able to influence the wheel-set's kinetic stage: traction force, resisting force and braking force. Under the acceleration of dV/dt , the external force on wheel-set can be analyzed as:

1) F_P: Assume that the wheel-set is a particle, it gives:

$$F_P = mdV / dt \tag{17}$$

Where V is the speed of wheel-set, and m is its mass.

2) F_i : under dV/dt, the force needed to accelerate the traction wheel and its transmission mechanism:

$$F_i = \frac{\sum J_i \frac{d\omega_i}{dt}}{r_0} = \sum \frac{J_i}{r_0^2} \bullet \frac{dV}{dt} \tag{18}$$

Where r_0 is the radius of rotating circle, J_i is the inertia of wheel-set, and ω_i is angular velocity.

3) F_l : under dV/dt, the force required by angular acceleration:

$$F_l = \frac{\sum J_l \frac{d\omega_l}{dt}}{r_l} = \sum \frac{J_l}{r_l^2} \bullet \frac{dV}{dt} \tag{19}$$

Where r_l is the radius of rotating circle, J_i is the inertia of wheel-set, and ω_i is angular velocity.

4) F_D ': under dV/dt, the force required by angular acceleration of transmission mechanism of traction motors and of its axles:

$$F_D' = \frac{\sum J_D d\omega_D / dt}{r_D} \tag{20}$$

Where r_D is the radius of the motor's armature, J_D is the inertia of transmission mechanism of the armature and axle, and ω_i is angular velocity.

The former three kinds are forces on flange, and the last on armature. For the convenience of summing the forces up, F_D 'should be reduced to be on flange, hence it gives:

$$F_D = \mu_c (\frac{r_D}{r_0}) F_D' = \sum \frac{\mu_c J_D}{r_0} \frac{d\omega_D}{dt} \tag{21}$$

Where $\omega_D = \mu_c \omega_i = \mu_c V / r_0$, $\mu_c = r_2 / r_1$ is the transmission ratio of gearbox. Therefore, it comes:

$$F_D = \sum \frac{\mu_c^2 J_D}{r_0^2} \frac{dV}{dt} \tag{22}$$

And

$$\begin{aligned} F_s &= F_p + F_i + F_l + F_D \\ &= m\frac{dV}{dt} + (\sum \frac{J_i}{r_0^2} + \sum \frac{J_l}{r_l^2} + \sum \frac{\mu_c^2 J_D}{r_0^2}) \bullet \frac{dV}{dt} \\ &= m(1 + \sum \frac{J_i}{m r_0^2} + \sum \frac{J_l}{m r_l^2} + \sum \frac{\mu_c^2 J_D}{m r_0^2}) \bullet \frac{dV}{dt} \\ &= m(1+\gamma) \bullet \frac{dV}{dt} \end{aligned} \tag{23}$$

In eq.(23), F_s is the resultant force on wheel-set, γ is the ratio of the mass rotating vs. the mass all, normally set to 0.06.

Urban transport vehicles are driven by traction force originated from traction motor and through gear box, where the motor torque and traction force on tread can be given as:

$$\begin{aligned} T_e &= F \cdot r_0 / \mu_c \\ \omega_r &= \mu_c \cdot \omega = \mu_c V / r_0 \end{aligned} \tag{24}$$

Where ω_r is the angular speed of traction motor.

The resisting force during the rotation of wheel-set consists mainly of the friction between axle journal and bearing, between wheel and tread, between wheel and rail, the impact force and air resistance. The relation between load torque T_L and resistance F_z can be given as:

$$T_L = F_z \cdot r_0 / \mu_c \tag{25}$$

And the relation between the acceleration and its original force on wheel-set can be given as:

$$F - F_z = m(1+\gamma)dV / dt \tag{26}$$

The relation between torque and rotational speed can be derived with eq.(24) through (26), and given as:

$$T_L = T_e + A(s)\omega_r \tag{27}$$

where $A(s) = -\dfrac{m(1+\gamma)r_0^{\,2}}{\mu_c^{\,2}} s$

By substituting T_L, T_R and ω_{rL}, ω_{rR} into eq.(27), it gives

$$T_L - T_R = (T_{eL} - T_{eR}) + A(s) \cdot (\omega_{rL} - \omega_{rR}) \tag{28}$$

From eq.(28), it is obvious that the torque difference between the left and right traction motors can be derived by detecting the rotational speed difference between them, the torque difference are then compared with torque reference, and the output drives the wheel-set along a pure rolling curve.

4 The Scheme of Active Direction Control

From the analysis above, it proposes that by controlling the torque output of the left and right wheel in independent wheel-set, direction ability is realized. As is shown in Fig.4, the traction order defines the output traction force, which is then compared the actual traction force feedback from motors, and generate the required flux $\psi_{dr}^{\ e}$ and the torque component $i_{qs}^{\ e}$ of stator current. Direction control scheme detects the torque difference and is then compared with the direction force decided with actual track condition. After the comparison, the direction torque current $\Delta i_{qs}^{\ e}$ is generated, and it is then summed up with the left and right torque current, respectively.

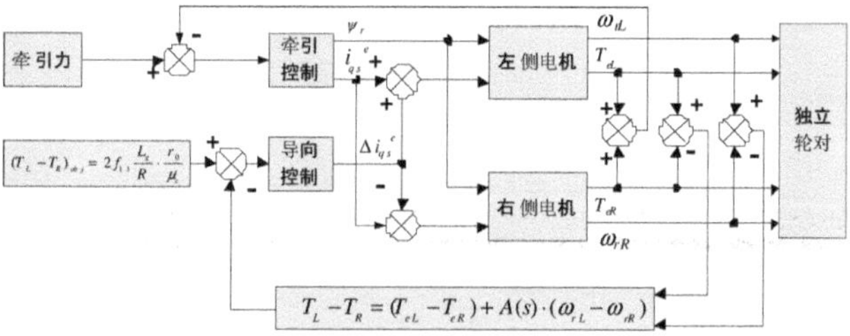

Fig. 4. Control diagram with direction control

By realizing the summed-up current, the independent wheel-set structure is entitled the ability of traction and direction ability at the same time.

Assume that the traction torque is T, and actual torque output from the left and right traction motors are given as:

$$T_L = T + \Delta T \,, T_R = T - \Delta T \tag{29}$$

With flux-oriented control principles, the traction torque of the left and right traction motors are:

$$\begin{aligned} T_{eL} &= \frac{3}{2} P \frac{L_m}{L_r} \psi_{dr}^{\ e} (i_{qs}^{\ e} + \Delta i_{qs}^{\ e}) \\ T_{eR} &= \frac{3}{2} P \frac{L_m}{L_r} \psi_{dr}^{\ e} (i_{qs}^{\ e} - \Delta i_{qs}^{\ e}) \end{aligned} \tag{30}$$

And the rotational speeds of them are:

$$\begin{aligned} \omega_{rL} &= \omega_r + \Delta\omega_r \\ \omega_{rR} &= \omega_r - \Delta\omega_r \end{aligned} \tag{31}$$

Combined with eq.(28), eq.(29) through (31), it gives:

$$\Delta T = A(s)\Delta\omega_r + B(s)\Delta i_{qs}^{\ e} \tag{32}$$

where

$$A(s) = -\frac{m(1+\gamma)r_0^2}{\mu_c^2} s \quad B(s) = \frac{3}{2} P \frac{L_m}{L_r} \psi_{dr}^{\ e},$$

Based on the analysis on independent wheel-set in literature[11], the transfer function of $\Delta\omega_r / \Delta T$ is

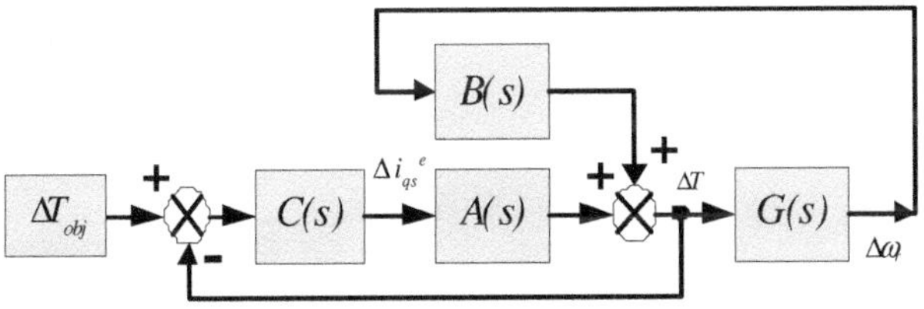

Fig. 5. The transfer function diagram with direction control

$\dfrac{\Delta\omega_r(s)}{\Delta T(s)} = G(s) = \dfrac{\mu_c}{I_{wy}s + f_{11}r_0^2 / V}$, and assuming the direction control function in Fig.4 as $C(s)$, Then the overall transfer function under the direction control above is derived as in Fig.5, which is key in the design of direction controller.

5 Simulation and Analysis

A simulation platform is built to testify the direction control scheme proposed, in which the parameters are set as is shown in tab.(1).

Table 1. Parameters used during the simulation

Para.	Val	Para	Val
λ	0.2	f11, f12	105 N
Iwv	75kg.m2	Lg	0.75 m
r0	0.30 m	R	200 m
μc	5.4	μg	0.97

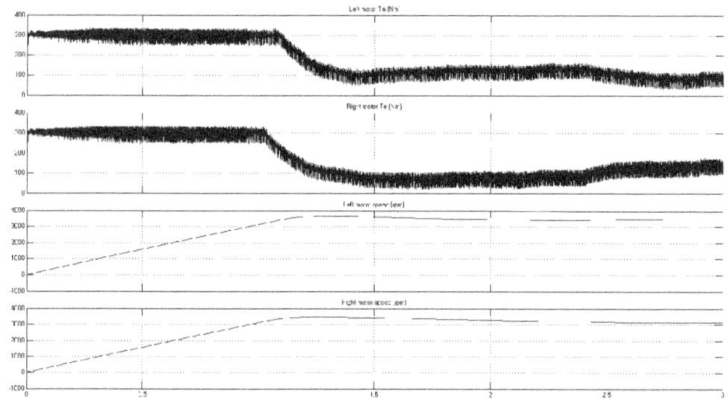

Fig. 6. Torque and velocity of the left and right motors

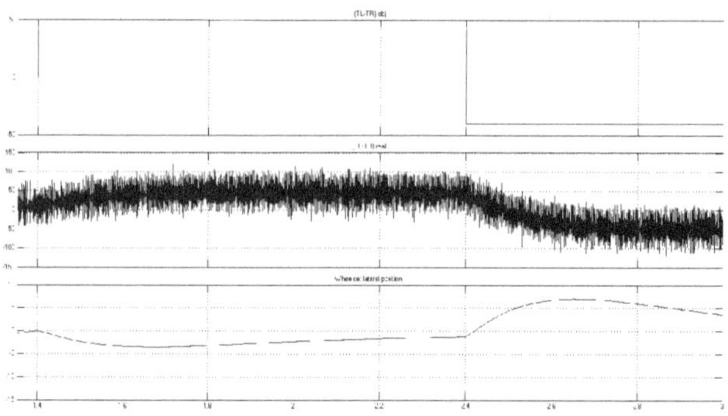

Fig. 7. Referenced and actual direction torque, lateral displacement

The simulated curves are shown in Fig.6. During the acceleration stage of 0s~1s, rotational speed of the wheel-set arrives at 70kmph, then torque output, which is approximately the same for the left and right motor is reduced for the vehicle to remain a constant velocity. At 1.4s, the vehicle enters a left-bend lane with the radius of 200m, when direction control comes into play, so that torque output becomes different for the left and right motor, with direction torque added. At 2.4s, the vehicle enters a left-bend lane with the radius of 240m, and torque output of the left motor starts decreasing, while that of the right motor increasing, driving the wheel-set leftwards. Under direction control, the torque output of the two motors supplies traction torque and direction torque at the same time.

In Fig.7 are shown the curves of direction torque reference, of actual direction torque, and of lateral displacement of wheel-set. From Fig.7 it can be seen that actual direction torque follows the reference exactly for both the left and the right wheels. For the realization of lateral displacement, rightwards direction is worse than leftwards, however, the difference is smaller with time passing by.

6 Conclusions

Being a key characteristics of 100% LFLRV, the adoption of independent wheel-set implies the absence of active direction ability due to the absence of conventional mechanical coupling of axles in the bogie. This means mechanical or electrical measures should be taken to realize direction ability. In this paper, the kinetic equation and the force condition of independent wheel-set are presented, which is also the case with direction control scheme based on torque-control mechanism. Here is proposed a novel active direction control scheme with the detection of torque difference of the left and the right traction motors, which enables the wheel-set to direct actively.

Acknowledgment. First of all, the author thanks my tutor Prof. Zhigang Liu who gave me all the guidance for this paper and my research in my years of PhD candidate studies. I'd also like to give my appreciation and thanks for my schoolmates and colleagues in my lab, especially for Lijun Diao, Lei Wang and others. At last, thank my family and friends who have been supported me all the way.

References

1. 张丽平，李芾，黄运华.国外城市轻轨发展及应用概况. 国外铁道车辆 5(3), 1–6 (2001)
2. 王新国, 张松江.低地板轻轨车辆的现状及发展. 电力机车与城轨车辆 4(26), 47–49 (2003)
3. 李芾, 张丽平，黄运华.城市轻轨车辆发展及其应用前景. 西南交通大学学报 2(37), 111–116 (2002)
4. Mei, T., Mei, T.x., Goodall, R.m.: Robust control for independently rotating wheelsets on a railway vehicle using practical sensors. IEEE Transactions on Control Systems Technology 9(4), 599–607 (2001)
5. Roger, G.: Advanced Control and Monitoring for Railway Vehicles for the suspensions and running gear. Excellence in railway systems engineering and integration, Derby (2005)
6. Selamat, H., Yusof, R., Goodall, R.m.: Self-tuning control for active steering of a railway vehicle with solid-axle wheelsets. Control Theory & Applications, IET 2(5), 374–383 (2008)

7. 陈泽深,王成国. 铁道车辆动力学与控制. 第1版. 北京:中国铁道出版社 (2004)
8. Bureika, G., Lingaitis, L.P., Mikaliūnas: Investigation of dynamic models of independently rotating wheels of wagons. Transport 19(1), 28–31 (2004)
9. 任利惠. 独立车轮导向技术研究[D]. 上海:同济大学 (2006)
10. Bosso, N., Soma, A., Gugliotta, A.: Introduction of a wheel-rail and wheel-roller contact model for independent wheels in a multibody code. In: ASME Railroad Conference, pp. 151–159 (2002)
11. Jérez, P., Busturia, J.M., Mei, T.X., et al.: Combined active steering and traction for mechatronic bogie vehicles with independently rotating wheels. Annual Reviews in Control 28(2), 207–221 (2004)

Research on Selection Arithmetic of Optimum Route in Road Military Logistics Based on Restricted Multi-objective Programming

Bu Chao

Department of Automobile Management
Automobile Management Institute
Bengbu, Anhui Province, China
BuChaoC@126.com

Abstract. According to the optimization of road military logistics transportation route,this paper set up a model of the traffic networks,using the method of OERI.It designed an arithmetic of the selection of randomly optimum route,based on OR.It also presented an experimental calculation and result analysis. This arithmetic has good reliability and practicability.

Keywords: military logistics, route, arithmetic.

1 Introduction

It is the key of accomplishing military logistics works that select the Optimum transportation route appropriately.At present,the research methods and programs in civilian logistics are in the highest flight in this field,unable to reflect the characteristic of military logistics. In road Military Logistics, the selection of Optimum Route is influenced by damage,speed, threat,traffic flow,etc.In the selection of Optimum transportation route in road Military Logistics, influence factor consists of safety,prompt and efficiency.

2 The Model of Selection of Optimum Route in Road Military Logistics

A. *Fuzzy Denotation Method of Time*

The time of completing a military transportation assignment is not in direct proportion to distance sometimes,especially when attacked or threatened by enemy.This time is influenced by road condition, damage,threat and traffic control stochastically, unable to be calculated exactly usually,but denoted by fuzzy words.

Reference[1] put forward a method called OERI(Overall Existence Ranking Index) for short.This paper brings forward a fuzzy denotation method of time,based on OERI.

Now,we express the time of an automobile force(unit) costs on a road in movement based on OERI.

D. Zeng (Ed.): ICAIC 2011, Part II, CCIS 225, pp. 727–732, 2011.

For example,(v_i,v_j) is a section of a road network.According to it,there is a triangular fuzzy number $t_{ij}=(t_1, t_2, t_3)$, t_1 and t_3 denote the left and right boundary, t_2 denotes the point whose membership degree is 1. $h_t(x)$ denotes the membership degree function of this fuzzy number.

$$h_t(x)=\begin{cases} 0,0\le x\le t_1 \\ \frac{x-t_1}{t_2-t_1},t_1<x\le t_2 \\ \frac{t_3-x}{t_3-t_2},t_2<x\le t_3 \\ 0,\ x>t_3 \end{cases} \quad (1)$$

Definition 1: $h_t^{-1}(x)$ is the antitonic function of $h_t(x)$, w is the enactive rank.Then the left and right antitonic function of can be denoted as follows:

$$h_{tL}^{-1}(w)=a_1+(b_1-a_1)w \quad (2)$$

$$h_{tR}^{-1}(w)=c_1-(c_1-b_1)w \quad (3)$$

Definition 2: F(t) is the OERI integral of t:

$$F(t)=\int_0^1[\lambda(w)L(w)h_{tL}^{-1}(w)+(1-\lambda(w))R(w)h_{tR}^{-1}(w)]dw \quad (4)$$

$$\lambda(w)\in(0,\ 1)$$

$\lambda(w)$ is the weight which distributed by decision-maker to rank w. L(w) and R(w) is the left and right weight which distributed by decision-maker to t. F(t) is the consecutive integral within [0, 1],which considers not only all ranks and their weight but left and right weight.Commanders can decide the value of $\lambda(w)$, L(w) and R(w) according to the situation.

B. The Model of Transportation Networks

We can predigest the road netwok in a certain area to a weighted directed graph *D(V,A)*:

$$D(V,A), V=\{v_i|1\le i\le n\}, A=\{a|a=(v_i,v_j),v_i,v_j\in V\}$$

Every arc in graph *D* has three parameters: p_{ij}, $F(t_{ij})$, c_{ij}.

1) p_{ij} denotes the probability of the automobile unit gets across a section of a road safely. p_{ij} is influenced by threat,conceal and camouflage condition,damage on a certain section of a road.It coule be forecast or analyzed based on the statistical information on this section of a road,but adjusted according to the situation in battlefield.

2)$F(t_{ij})$ is the OERI integral of time.

3) c_{ij} is the unit transportation cost of material on (v_i, v_j).

π_{ST} is the route which includes start "S" and terminal "T".

$$A(\pi_{ST})=\{a | a=(v_i,v_j),v_i,v_j\in\pi_{ST}\} \tag{5}$$

$$V(\pi_{ST})=\{v_j | v_j\in\pi_{ST}\} \tag{6}$$

The probability,time and unit transportation cost are denoted as follows:

*4)*Probability "$P(\pi_{ST})$":

$$P(\pi_{ST})=\prod_{(v_i,v_j)\in A(\pi_{ST})} p_{ij} \tag{7}$$

$$p'_{ij}=-\lg p_{ij} \tag{8}$$

$$P(\pi_{ST})=\sum_{(v_i,v_j)\in A(\pi_{ST})} p'_{ij} \tag{9}$$

*5)*Time which the automobile unit gets across π_{ST} "$F(\pi_{ST})$":

$$F(\pi_{ST})=\sum_{(v_i,v_j)\in A(\pi_{ST})} F(t_{ij}) \tag{10}$$

*6)*The unit transportation cost of material on π_{ST} "$c(\pi_{ST})$":

$$c(\pi_{ST})=\sum_{(v_i,v_j)\in A(\pi_{ST})} c_{ij} \tag{11}$$

We suppose $\pi^{\zeta}{}_{ST}$ to be a route in transportation networks D(V,A),and the transportation quantity on it to be x_{ζ}.

$$\min\{g_{ij} | (v_i,v_j)\in A(\pi_{ST})\}\geq x_{\zeta}\geq 0 \qquad \zeta=1,2,\cdots m \tag{12}$$

We suppose g_{ij} to be the capacity in (v_i, v_j).

$$\mathrm{Sgn}(x_{\zeta})=\begin{cases}0 & x_{\zeta}=0\\ 1 & x_{\zeta}>0\end{cases}$$

$$\mathrm{Sgn}(A(\pi^{\zeta}{}_{ST})\cap(v_i,v_j))=\begin{cases}0 & (v_i,v_j)\notin A(\pi^{\zeta}{}_{ST}\cap(v_i,v_j))\\ 1 & (v_i,v_j)\in A(\pi^{\zeta}{}_{ST}\cap(v_i,v_j))\end{cases} \tag{13}$$

If we plan to transport B material from S to T.We can decompose it to $\pi^1{}_{ST},\pi^2{}_{ST},\cdots,\pi^m{}_{ST}$.The model can be established as follows:

$$\mathrm{Lex}\min_{\zeta}[P(\pi^{\zeta}{}_{ST}),F(\pi^{\zeta}{}_{ST}),c(\pi^{\zeta}{}_{ST})]=\min_{\zeta}(q_1P(\pi^{\zeta}{}_{ST})+q_2F(\pi^{\zeta}{}_{ST})+q_3c(\pi^{\zeta}{}_{ST}))$$

$$s.t.\begin{cases}\sum_{(v_i,v_j)\in A(\pi_{ST})} t_{ij}\leq M\\ \sum(c(\pi^{\zeta}{}_{ST})x_{\zeta}\leq C\\ \sum_{\zeta=1}^{m} x_{\zeta}=A\\ 0\leq\sum(x_{\zeta}\mathrm{Sgn}(A(\pi^{\zeta}{}_{ST})\cap(v_i,v_j))\leq g_{ij}\end{cases} \tag{14}$$

q_1,q_2,q_3 are the decision weight of $P(\pi_{ST}), F(\pi_{ST}), c(\pi_{ST})$.They can be confirmed used Analytic Hierarchy Processand in practice.According to the objective function and constraint,it is a Multi-objective Programming.

3 Arithmetic

The course of arithmetic can be denoted as follows:

Step 1: $f^{(0)} = 0$.

If we get the least cost flow $f^{(K-1)}$ in K-l step and $B \le g(f^{(K-1)})$,then $f^{(K-1)}$ is the optimum. We can stop it.If $B > g(f^{(K-1)})$,we shall establish a weighted directed graph $D(f^{(K-1)})$,then go to Step 2.

Step 2:Search the multi-objective shortcut from v_s to v_T in $D(f^{(K-1)})$:

1) Every arc (v_i, v_j) has a weight coefficient w_{ij}.We change it into a vector $W_{ij}=(p_{ij},F(t_{ij}), c_{ij})$;

*2)*We can suppose that there is an arc between v_i and v_j.If $(v_i, v_j)\notin$ A in D,then add an arc (v_i, v_j), $W_{ij}=(+\infty, +\infty, +\infty)$;

3) Add a constraint to the circulation:

$$d^{(1)}(v_s, v_j) = \{W_{ij} | t_{ij} \le M, c_{ij} \le C\}$$

$$d^{(t)}(v_s, v_j) = \min_i \{d^{(t-1)}(v_s, v_i) + W_{ij} | t_i + t_{ij} \le M, c_i + c_{ij} \le C\} \tag{15}$$

4) $d^{(k)}(v_S, v_j)$ is the weight vector of shortcut if it was denoted as follows:

$$d^{(k)}(v_S, v_j) = d^{(k-1)}(v_S, v_j) \tag{16}$$

Step 3:If there is not shortcut and $B > g(f^{(K-1)})$,then $f^{(K-1)}$ is the minimum cost maximum flow of the transportation networks D.It has not optimum and goes to the end because the transportation networks can not satisfy the request of military logistics.

Step 4: If there is shortcut and $B > g(f^{(K-1)})$,then we can get the Augmenting Chain μ and go to Step 5.

Step 5:Adjust $f^{(K-1)}$ on the Augmenting Chain μ:

$$\theta = \min[\min_{\mu^+}(c_{ij} - f_{ij}^{(K-1)}), \min_{\mu^-}(f_{ij}^{(K-1)}), B - g(f^{(K-1)})] \tag{17}$$

Step 6: If $\sum_{(v_i,v_j)\in A(\pi_{ST})} t_{ij} \le M$ and $\sum_{\zeta=1}^{k}(\sum_{(v_i,v_j)\in \pi^k_{ST}} b^k_{ij} f^k_{ij}) \le C$, then we can get a new feasible flow $f^{(K)}$ and repeat the steps above-mentioned,or go to Step 2.

4 Sample

Fig.1 denotes the road military logistics networks in an area.An automobile unit plans to transport B material from V_1 to V_{10}.The time(see Table1), probability, unit transportation cost can be estimated according to war situation,road condition,statistic and experience,etc.

We can get the optimum result based on the arithmetic which this paper put forward,shown in Table 2. It could be concluded that the optimum route may change along with the value of q_1, q_2, q_3.Therefore decisionmakers have more freedom and self-determination.In war,the decisionmakers can modify the value of these parameters,according to the battle situation,mastery of the sky,exigence of logistics assignment,resource,etc.

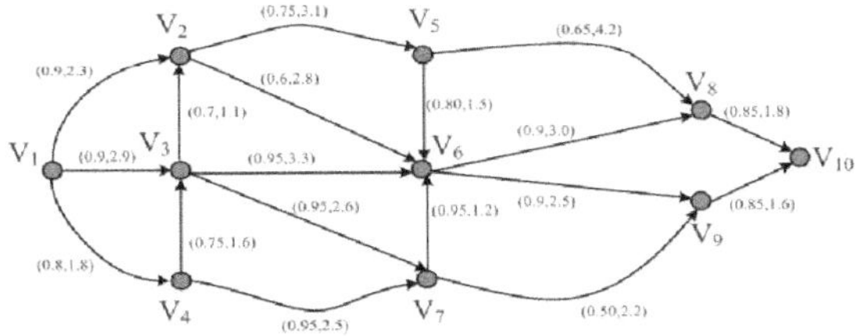

Fig. 1. Transportation networks

Table 1. Time which the automobile unit gets across

Section	Time(h)	Section	Tme(h)	Section	Time(h)
V_1~V_2	(4,6,8)	V_5~V_8	(4,10,16)	V_3~V_6	(4,6,8)
V_1~V_3	(1,4,7)	V_5~V_6	(3,5,7)	V_4~V_3	(1,3,5)
V_1~V_4	(1,3,5)	V_6~V_8	(6,8,10)	V_4~V_7	(3,6,9)
V_3~V_2	(2,4,6)	V_6~V_9	(3,4,5)	V_8~V_{10}	(2,4,6)
V_2~V_5	(5,7,9)	V_7~V_6	(1,2,3)	V_9~V_{10}	(2,3,4)
V_2~V_6	(6,8,10)	V_7~V_9	(7,9,11)		

Table 2. Optimum result

Parameter (q_1, q_2, q_3)	Optimum route	Probability	Time	Unit cost
(0.75,0.20,0.05)	V_1-V_3-V_6-V_8-V_{10}	0.654	(19,21,23)	11.0
(0.65,0.25,0.10)	V_1-V_3-V_6-V_8-V_{10}	0.654	(19,21,23)	11.0
(0.45,0.40,0.15)	V_1-V_4-V_7-V_6- V_9-V_{10}	0.552	(10,18,26)	9.6

5 Conclusion

This paper designs an arithmetic based on multi-objective programming to select the optimum route.It has many advantages,such as information integration,decision freedom.The decisionmakers in army can improve their preciseness,consult with this arithmetic.

References

1. Chang, P.T., Lee, E.S.: Fuzzy decision networks and deconvolution. Computers Math. Apphc. 37, 53–63 (1999)
2. Huang, X., Gao, J., Gao, Z.: New Methodology of Signed Digraph Modeling Based on Causal Ordering Analysis. Journal of Xi'an Jiaotong University 44(5), 25–30 (2010)

Author Index

GPSR Compliance

The European Union's (EU) General Product Safety Regulation (GPSR) is a set of rules that requires consumer products to be safe and our obligations to ensure this.

If you have any concerns about our products, you can contact us on ProductSafety@springernature.com

In case Publisher is established outside the EU, the EU authorized representative is:

Springer Nature Customer Service Center GmbH
Europaplatz 3
69115 Heidelberg, Germany

Batch number: 10838023

Printed by Printforce, the Netherlands